An Introduction to the Mechanics of Solids

Second Edition

Robert R. Archer

Nathan H. Cook

Stephen H. Crandall

Norman C. Dahl

Thomas J. Lardner

Frank A. McClintock

Ernest Rabinowicz

George S. Reichenbach

First Edition Edited by
Stephen H. Crandall and Norman C. Dahl

Second Edition Edited by
Thomas J. Lardner

Department of Mechanical Engineering
Massachusetts Institute of Technology

McGraw-Hill Book Company New York, St. Louis, San Francisco, Düsseldorf, Johannesberg, Kuala Lumpur, London, Mexico, Montreal, New Delhi, Panama, Rio de Janeiro, Singapore, Sydney, Toronto

This book was set in Times Roman, and was printed and bound by Kingsport Press, Inc. The designer was Barbara Ellwood; the drawings were done by John Cordes, J. & R. Technical Services, Inc. The editors were B. J. Clark and James W. Bradley. Peter D. Guilmette supervised production.

The photograph on the binding is of a 50-mm-square steel beam subjected to a bending test. Ingot developed by Otani Heavy Industrial Co., Ltd., Japan.

An Introduction to the Mechanics of Solids

Library of Congress Catalog Card Number 78-175183

07-013436-7

1 2 3 4 5 6 7 8 9 0 K P K P 7 9 8 7 6 5 4 3 2

Contents

Preface to the Second Edition

The reader is advised to read the preface to the first edition. The aim and emphasis of the book have not changed: the principles underlying the mechanics of rigid and deformable solids in equilibrium have not changed.

We have resisted the temptation to increase by a great amount the material covered, or to emphasize formalism and rigor in place of the emphasis on constructing idealized models to represent actual physical situations. We believe that the reader must appreciate that engineering is the finding of solutions, i.e., the determination of answers to physical problems. The second edition has maintained the spirit and tradition of the first in this regard. We hope, too, that the book has maintained the tradition of engineering thinking, a tradition which M. A. Biot[1] refers to as the ". . . tradition of clarity, simplicity, intuitive understanding, unpretentious depth, and a shunning of the irrelevant."

Changes have been made; these changes, however, are more in the spirit of reform than of revolution. New material dealing with energy, hydrostatics, postbuckling behavior, and indicial notation has been introduced. There is also

[1] M. A. Biot, Science and the Engineer, *Appl. Mech. Rev.*, vol. 16, no. 2, pp. 89–90, February 1963.

a discussion of the role of computers in structural analysis. In this regard we have tried to emphasize that the computer can be used as a tool in the *solution* of problems. The *physical understanding* and *formulation* of a problem, however, are the most important parts of the solution, and the basic principles still reside in the three steps of Eq. (2.1). Many sections have been revised and a number of chapters reorganized to improve previous expositions.

A number of new problems have been added, and an effort has been made to show the variety of situations to which the principles contained in this book may be applied, from biology to the design of nuclear-reactor containment vessels.

We wish to thank the many readers who have submitted lists of misprints and comments and our many colleagues who have found the book useful during the last twelve years. Professor W. M. Murray is owed thanks for his contribution to Sec. 4.14.

THOMAS J. LARDNER

Preface to the First Edition

This book is concerned with the mechanics of rigid and deformable solids in equilibrium. It has been prepared by members of the Mechanical Engineering Department at the Massachusetts Institute of Technology for use as a text in the first course in applied mechanics.

The central aim has been to treat this subject as an *engineering science*. To this end we have clearly identified three fundamental physical considerations which govern the mechanics of solids in equilibrium, and we have explicitly related all discussion and theoretical development to these three basic considerations. We have focused on these fundamentals in an effort to bring unity to an elementary presentation of our subject.

A further aspect upon which we have put considerable emphasis is the process of constructing idealized models to represent actual physical situations. This is one of the central problems of engineering, and throughout the book we have attempted to give it attention commensurate with its importance.

We have assumed that the reader has already studied mechanics as part of a program in physics and that he is familiar with the differential and integral calculus. We further assume that the reader is acquainted with vector notation and with the algebraic operations of addition and multiplication of vectors.

The first chapter is devoted to a discussion of the fundamental principles of mechanics and to an exposition of the requirements of equilibrium. In the second chapter the basic principles are stated explicitly in Eq. (2.1) in the form of three steps and are illustrated by application to lumped parameter models and one-dimensional continua. The next three chapters are devoted to extending the depth of meaning contained in the basic principles. An important facet of this development is the extension of the fundamental concepts to three-dimensional continuous media. In the final four chapters simple but important problems involving these concepts are solved. There are problems for the reader at the end of each chapter. Some of these include extensions of the text material. Answers to approximately one-third of the problems are given at the rear of the book.

In endeavoring to emphasize the basic principles, we have, of necessity, had to omit many interesting applications. We have not attempted to provide a compendium of useful results, but rather we have selected a limited number of particular applications and have examined these with more than usual care. It is our opinion that a course based on this text will provide an appropriate introduction to the more advanced disciplines of elasticity and plasticity. With equal conviction we believe that a course based on this text will provide a firm foundation for subsequent design courses in this field.

Many people have participated directly and indirectly in the preparation of this book. In addition to the authors, many present and former members of our staff have contributed ideas concerning methods of presentation and problems from examinations. We wish to acknowledge, in particular, the cooperation of R. J. Fitzgerald in working out problem solutions and the help of Miss Pauline Harris in typing the manuscript.

There was a preliminary edition in 1957 (with a supplement in 1958); it enabled us to experiment with presenting this material in semipermanent book form. We wish to thank those members of the M.I.T. classes of 1960 and 1961 who used the preliminary editions and who by their comments and criticisms helped to make this book better than it otherwise would have been.

STEPHEN H. CRANDALL
NORMAN C. DAHL

1

Fundamental Principles of Mechanics

1.1 INTRODUCTION

Mechanics is the science of *forces* and *motions*. It involves a relatively small number of basic concepts such as force, mass, length, and time. From a few experimentally based postulates and assumptions regarding the connections between these concepts, logical deduction leads to quite detailed predictions of the consequences. Mechanics is one of the oldest physical sciences, dating back to the time of Archimedes (287–212 B.C.). A delightful account of the use of mechanics by Archimedes in the defense of Syracuse against the Romans may be found in *Plutarch's Lives*.[1] As a science, mechanics has intrigued almost all of the great scientists, e.g., Stevin, Galileo, Newton, d'Alembert, Lagrange, Laplace, Euler, Einstein, to name only a few whose names are familiar. It continues to be a fascinating subject by continually expanding its areas of application. The reader interested in the history of mechanics will easily find a number of interesting books on this topic.

[1] Plutarch, "The Lives of the Noble Grecians and Romans," Marcellus, pp. 376–380, Modern Library, Inc., New York.

Applied mechanics is the science of applying the principles of mechanics to systems of practical interest in order (1) to understand their behavior and (2) to develop rational rules for their design. This book is an introduction to *applied* mechanics of solids. The logical structure of the principles of mechanics will be briefly developed as needed, but the main emphasis here is on the rational applications of these principles. It is assumed that the reader is familiar with the broad outline of newtonian mechanics from his studies in physics.

1.2 GENERALIZED PROCEDURE

The general method of attack in solving problems in applied mechanics is similar to that in any scientific investigation. The steps may be outlined as follows:

1. Select system of interest.
2. Postulate characteristics of system. This usually involves idealization and simplification of the real situation.
3. Apply principles of mechanics to the idealized model. Deduce the consequences.
4. Compare these predictions with the behavior of the actual system. This usually involves recourse to tests and experiments.
5. If satisfactory agreement is not achieved, the foregoing steps must be reconsidered. Very often progress is made by altering the assumptions regarding characteristics of the system, i.e., by constructing a different idealized model of the system.

This generalized approach applies to the problems treated in this book and equally well to problems on the frontiers of research. The design engineer who must deal with mechanics follows a similar sequence but with a somewhat different motive in that it is his job to accomplish a certain desired function. He must first create a possible design, either by invention or by adaptation of prior designs, before he can analyze its behavior as in steps 1, 2, and 3. If this behavior is not compatible with the desired function, he must modify or redesign the system and repeat the analysis until an acceptable result is obtained. The criteria of acceptability include not only satisfactory technical operation but such factors as economy, minimum weight, or ease of fabrication. Acceptability may also require consideration of pollution and/or ecological factors.

Since this is an introductory text, we have devoted most of our space to the first three of the above steps. We have, however, made occasional reference to the other steps. Examples of cases where there is not at present satisfactory agreement between theory and experiment have been given to illustrate the tentative nature of scientific reasoning and also to acquaint the reader with the fact that, despite its fundamental importance in the scientific development of the last 300 years, mechanics is still a vigorous, growing field with many frontiers being actively extended.

Let us consider further the first two steps in the above outline: the selection of a system and the idealization of its characteristics. In research investigations these are usually the most difficult steps. The trick is to set up a model which is simple enough to analyze and yet still exhibits the phenomena under consideration. The more we learn, the more detailed become our models of reality.

For example, the reader will observe in the following pages the increasing sophistication with which we select and isolate systems for intensive analysis. In the simplest situations an entire structure can be treated as a whole. We later find it necessary to consider subassemblies (e.g., a single member or joint, or half of an original structure) as isolated systems. Later we obtain more detailed information by selecting infinitesimal elements inside structural members as systems for analysis.

The reader will also observe the increasing sophistication we employ in idealizing the characteristics of a system. For example, a large part of our work deals with ordinary engineering structural members: rods, beams, shafts, etc. These members are ordinarily relatively rigid, so we begin by using the idealized concept of a perfectly *rigid* body. A certain plateau of understanding is reached on this basis. Then to answer further questions it is necessary to consider the deformation of a member under load. By assuming the deformations to be *elastic* we reach another plateau of understanding. Then further enlightenment comes when we include assumptions regarding *plastic* behavior.

1.3 THE FUNDAMENTAL PRINCIPLES OF MECHANICS

Having selected a system and set up a conceptual model of its behavior, we next ask, What are the principles of mechanics and how are they applied? In broad outline they are very simple. Mechanics deals with forces and motions. We must therefore study the forces, and we must study the motions. Finally, we connect the forces with the motions by using hypotheses concerning the dependence of motion on force.

One of the most important of our basic concepts is *force*. In the next section we begin a review of the properties of force.

The study of *motion* involves geometry and, in general, time. It is possible to distinguish two different types of movement which are important in the mechanics of solids. The first type involves gross overall changes in position with time, while the second type involves local distortions of shape. For example, an automotive-engine connecting rod has a complicated overall motion in which one end moves up and down while the other end travels in a circle. Simultaneous with this overall motion is a very small change in the shape of the rod; the rod alternately elongates and shortens as it first pulls the piston and then is pushed by the piston. This second type of movement, involving change in shape, we call *deformation*. In this book we shall consider situations in which there is deformation, but we shall not usually be concerned with gross overall motions. Detailed examination of overall motion may be found in texts on *dynamics*, *kinetics*, and *kinematics*.

Both types of movement are influenced by forces. The hypotheses connecting force and motion that we employ are those of *newtonian* mechanics. While this theory must be extended to cope with very large velocities, there is ample experimental evidence for the validity of the newtonian postulates in the realm of ordinary engineering where all velocities are small compared with the speed of light. A basic tenet of newtonian mechanics is the proportionality of force and acceleration for a particle. Actually, in this book we deal only with a degenerate case of this: the case of no acceleration which occurs when there is no unbalanced force.

The hypotheses relating force and deformation within solids are considerably more varied. Several aspects of this question will be surveyed in Chap. 5 and exploited in the following chapters.

Application of these hypotheses to a particular system permits us to predict the motion and deformation if the forces are known or, conversely, to determine the forces if the motions and deformations are known. At the design stage of a structure or machine this gives us information on which to base a judgment as to whether the system will perform in a safe and efficient manner.

Every mechanical analysis involves the three steps described above, which we list again for emphasis:

1. Study of forces
2. Study of motion and deformation
3. Application of laws relating the forces to the motion and deformation

In most situations all three of the above steps require careful analysis. In special cases one or more of the steps become trivial. For example, when we assume a member to be perfectly rigid, we automatically rule out considerations concerning the deformation of the member. If in addition the member is constrained to remain at rest, no considerations regarding the motion are necessary.

The problems treated in this book generally will not involve overall motion. As a consequence, the basic steps in the analysis may be simplified as follows:

1. Study of forces
2. Study of deformations
3. Application of laws relating the forces to the deformations

In considering the forces we will have to take into account the requirement that there should be a state of balance. In considering the deformations we will have to take into account the requirement that the deformations of the individual parts of a structure should be consistent with the overall deformation. In relating the forces to the deformation we will have to take into account the special properties of the particular materials involved. These three fundamental steps underlie the development of the remainder of this book. They form the central core.

1.4 THE CONCEPT OF FORCE

It is assumed that the reader already has an intuitive notion of force, and what a force can do. The development of the idea of force in mechanics has provided us with an effective means for describing a very complex physical interaction between "bodies" in terms of a simple, convenient concept.

Force is a directed interaction; i.e., it is a *vector* interaction. (The reader may find it convenient to recall some of the properties of a vector by considering Probs. 1.1 to 1.5 at the end of the chapter.) For example, in Fig. 1.1*a* there is an attraction between the airplane and the center of the earth indicated by the pair of vectors $\mathbf{F}_1$ and $\mathbf{F}_2$. In Fig. 1.1*b* there is a force between the spring and the weight indicated by the pair of vectors $\mathbf{F}_1$ and $\mathbf{F}_2$. Newton, in his third law, postulated equal and opposite effectiveness of force on the two interacting systems; that is, $\mathbf{F}_1$ and $\mathbf{F}_2$ are equal and opposite vectors along the same line of action in Fig. 1.1.

It has become customary to apply the term *force* indiscriminately to either the pair $\mathbf{F}_1$, $\mathbf{F}_2$ or to the single vectors separately. When we analyze an isolated system such as the airplane in Fig. 1.1*a*, we represent the interaction with the earth by the vector $\mathbf{F}_1$, calling it the force exerted by the earth upon the airplane. Similarly, in isolating the spring in Fig. 1.1*b* the interaction with the weight would be represented by $\mathbf{F}_1$, the force which the weight exerts on the spring.

Force interactions may occur when there is direct contact between systems, as illustrated by Fig. 1.1*b*. Force interactions may also occur between systems which are physically separated, as in Fig. 1.1*a*. *Electric*, *magnetic*, and *gravitational* forces are of this type. The force of the earth on an object at or near the surface is called the *weight* of the object.

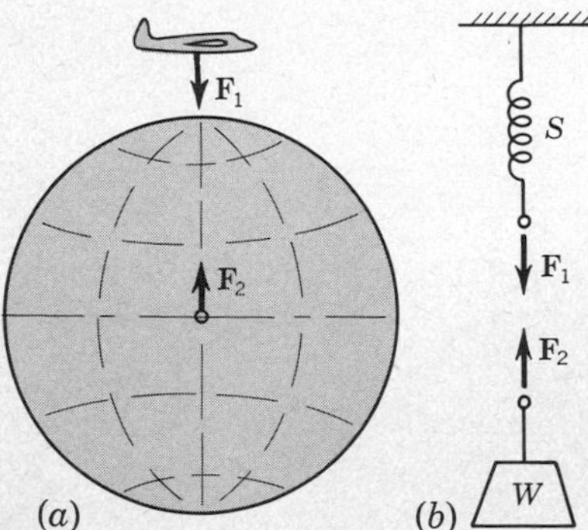

Fig. 1.1 Force interactions (*a*) at a distance and (*b*) by direct connection.

Force interactions have two principal effects: they tend to alter the motion of the systems involved, and they tend to deform or distort the shape of the systems involved. In Fig. 1.2*a* the attraction of the earth has a tendency to alter the motion of the airplane from a level flight to a vertical dive. The application of a force to the deformable spring in Fig. 1.2*b* tends to stretch it. Either of these effects can be used as the basis for a *quantitative* measure of the magnitude of force. The definitions of most *units* of force are based on the alteration of motions of standard systems. For example, the *pound force* is defined as that force which gives an acceleration of 32.1740 ft/sec^2 to a mass which is 1/2.2046 part of a certain piece of platinum (in possession of the International Committee for Weights and Measures) known as the *standard kilogram.*

The system of units used frequently by engineers in the United States is based on a gravitational system in which length, force, and time—*foot, pound,* and *second*—are considered the fundamental quantities. The engineer prefers to use force as a fundamental quantity because he feels an intuitive grasp of force. The unit of mass in this system has the dimensions of lb-sec^2/ft, which is occasionally called a *slug*.

The International System of Units, officially abbreviated SI, which is slowly being adopted, is a modernized version of the metric system. It was established by international agreement to provide a logical and interconnected framework for all measurements in science, industry, and commerce. In this system the basic quantities are length, mass, and time—*meter, kilogram,* and *second.* The unit of *force* in this system is the *newton.* Thus a newton is the force which gives an acceleration of 1 m/sec^2 to a mass of one kilogram. One newton equals approximately two-tenths of a pound of force.

It is likely that both these systems of units will continue to be used over the next few years and one will need to be able to switch from one system to the other.

Table 1.1 contains a listing of the common systems of units and some conversion factors between systems. It should be mentioned that one's so-called

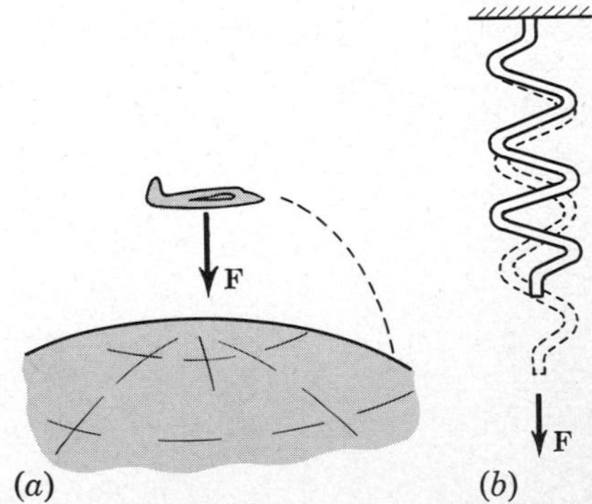

Fig. 1.2 Forces tend to alter motion or distort shape.

intuitive grasp of the order of magnitude of physical quantities depends on one's system of units. Let us now leave the discussion of units and return to our discussion of force.

A very important property of force is that the superposition of forces satisfies the laws of *vector addition*. This is a fundamental postulate based on experimental observation. Thus, if force is defined in terms of the rate of change of momentum of a standard body, it is found that when two bodies interact with the standard body, the rate of change of momentum is the vector sum of the individual rates of change of momentum resulting when each body separately interacts with the standard body.

This implies that if two forces $\mathbf{F}_1$ and $\mathbf{F}_2$ have the same point of application P in Fig. 1.3*a*, then we may replace them by their vector sum $\mathbf{F}_1 + \mathbf{F}_2$ with no observable effect on the system. It also means that any force $\mathbf{F}$ in Fig. 1.3*b* can be replaced by its components along any three mutually perpendicular axes through the point of application P. It is assumed that the reader is familiar with the algebra of vectors and their representation in terms of unit vectors $\mathbf{i}$, $\mathbf{j}$, and $\mathbf{k}$ along coordinate axes.

Summarizing the above discussion, we can say that

1. Force is a vector interaction which can be characterized by a pair of equal and opposite vectors having the same line of action.
2. The magnitude of a force can be established in terms of a standardized experiment.

Table 1.1 System of units and some conversion factors

Units

	United States	*CGS*	*SI*
Length:	inch	centimeter	meter (m)
Force:	pound force	dyne	newton (N)
Time:	second	second	second (s)
Mass:	pound mass, slug	gram	kilogram (kg)

Conversion factors

Length:	1 in. =	0.0254	meter (m)
Pressure:	1 psi =	6.895×10^3	newton/meter² (N/m²)
	1 psf =	47.88	newton/meter² (N/m²)
Density:	$1\ \dfrac{\text{lbf-sec}^2}{\text{in.}^4} =$	10.69×10^6	kilogram/meter³ (kg/m³)
Force:	1 lbf =	4.448	newton (N)
Mass:	1 lbm =	0.4536	kilogram (kg)

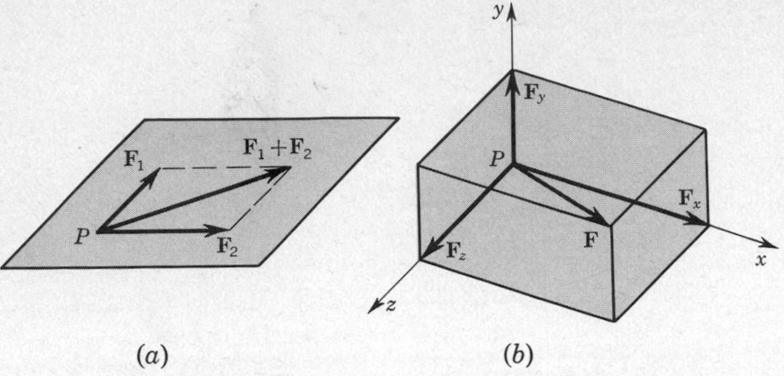

Fig. 1.3 Vector properties of force.

3. When two or more forces act simultaneously, at one point, the effect is the same as if a single force equal to the vector sum of the individual forces were acting.

If we isolate a system S, as shown in Fig. 1.4, the interactions with external systems can be indicated by vectors $\mathbf{F}_1, \mathbf{F}_2, \ldots$, which show the forces *exerted by* the external agencies that interact with S. This set of forces is often referred to simply as the external forces acting on the system S. Each individual force is characterized by a magnitude and a vector direction. Furthermore, in dealing with an extended system it is necessary to specify the *point of application* of each force. Throughout this book we shall be continually focusing our attention on systems of forces like that shown in Fig. 1.4 in which we *isolate* the system from its environment, and replace the effect of the environment with a system of *external forces*. In the mechanics of solids, the system which we isolate is a specific physical part or a group of parts. In the mechanics of *fluids*, it is often more useful to isolate a particular *control volume* in space rather than isolating particular particles which are flowing through a volume.

At this point it is appropriate to consider the "scale" or absolute size of the system under study, and how scale affects the nature of force interactions.

In Fig. 1.1*b* we showed force interaction by direct connection between a spring and a weight. On a finer (atomic) scale, we would see that there is no "direct connection," but rather, interaction between the electron fields of neigh-

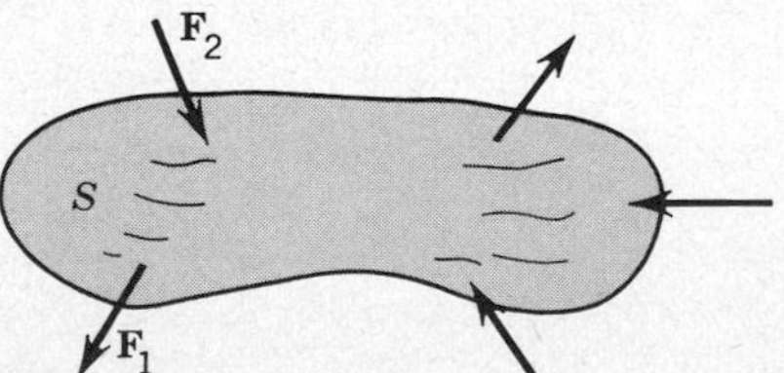

Fig. 1.4 Isolated system with external forces.

boring atoms, more like the "at a distance" interaction of Fig. 1.1*a*. On this fine scale, we would see that the contact force is the vector sum of all the atomic interaction forces, each of which varies continually due to atomic thermal motion.

In Fig. 1.3, we considered forces acting "at a point." While mathematically acceptable, this concept is physically only approximate in that all interactions involve a finite area. The point approximation grows poorer as the size of the interaction area approaches the size of the system under study.

In later sections we shall consider the matter within a solid to be continuously distributed. This is an approximation which is valid only when the solid is large compared to atomic dimensions. By the same reasoning, although we see our universe as composed of discrete stars, etc., studies on a cosmic scale may well consider the matter of the universe to be uniformly distributed.

Whenever we model a physical system for study, we must select a "scale" which is fine enough to provide critical results, yet as coarse as possible to minimize analytical effort.

1.5 THE MOMENT OF A FORCE

In Fig. 1.5 let $\mathbf{F}$ be a force vector applied at P and let O be a fixed point in space. The *moment* or *torque* of $\mathbf{F}$ about the point O is defined as the vector *cross product* $\mathbf{r} \times \mathbf{F}$, where $\mathbf{r}$ is the displacement vector from O to P.

The moment itself is a vector quantity. Its direction is perpendicular to the plane determined by OP and $\mathbf{F}$. The sense is fixed by the *right-hand rule:* When the fingers of the right hand curl in the direction that $\mathbf{F}$ tends to turn about O, the right thumb points in the direction of the moment vector. An alternate method of fixing the sense is to imagine a right-handed screw at O pointing perpendicular to the plane AOB. The direction in which this screw advances when turned by $\mathbf{F}$ is the direction of the moment $\mathbf{r} \times \mathbf{F}$.

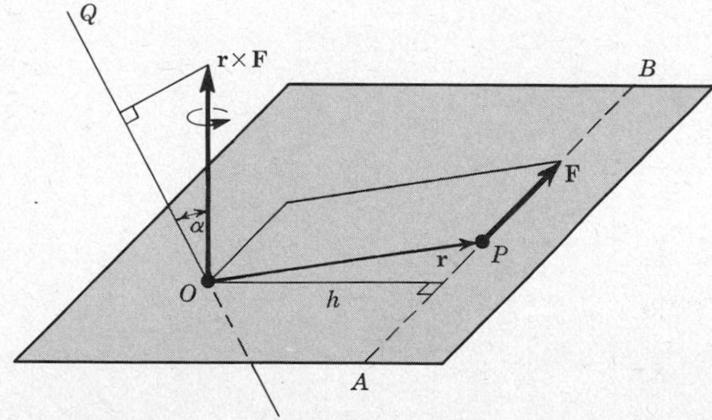

Fig. 1.5 The moment of a force $\mathbf{F}$ about a *point* O is $\mathbf{r} \times \mathbf{F}$.

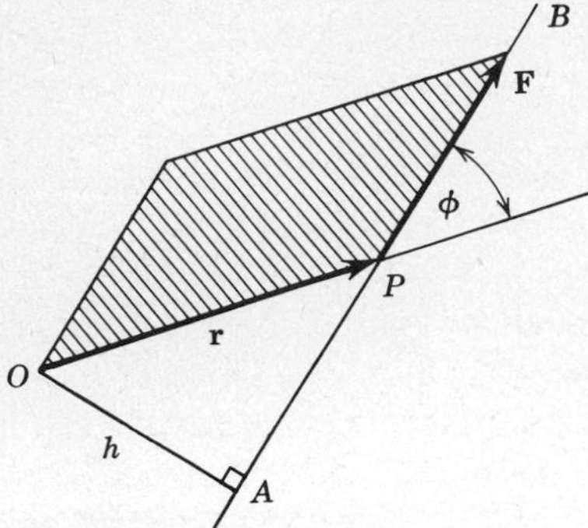

Fig. 1.6 Magnitude of cross product $\mathbf{r} \times \mathbf{F}$ is the area of parallelogram.

Recall from calculus that the *magnitude* of the cross product $\mathbf{r} \times \mathbf{F}$ is given by

$$Fr \sin \phi$$

where F and r are the magnitudes of the vectors $\mathbf{F}$ and $\mathbf{r}$ and ϕ is the angle between $\mathbf{r}$ and $\mathbf{F}$ shown in Fig. 1.6. The magnitude of the moment is therefore the area of the parallelogram having $\mathbf{r}$ and $\mathbf{F}$ as sides. Note that the magnitude is independent of the position of P along AB; that is, the moment of a force about a given point is invariant under the operation of sliding the force along its line of action. In simplest form the magnitude of the moment is $h|\mathbf{F}|$, where h is the length of the perpendicular dropped from O to AB and $|\mathbf{F}|$ is the magnitude of the force vector $\mathbf{F}$. Commonly used *units* for moments are the *foot-pound* and the *inch-pound*.

If we consider an idealized two-dimensional structure shown in Fig. 1.7, the moment of the force $\mathbf{F}$ about the point O is

$$\mathbf{M} = \mathbf{r} \times \mathbf{F} = \mathbf{k}h|\mathbf{F}|$$

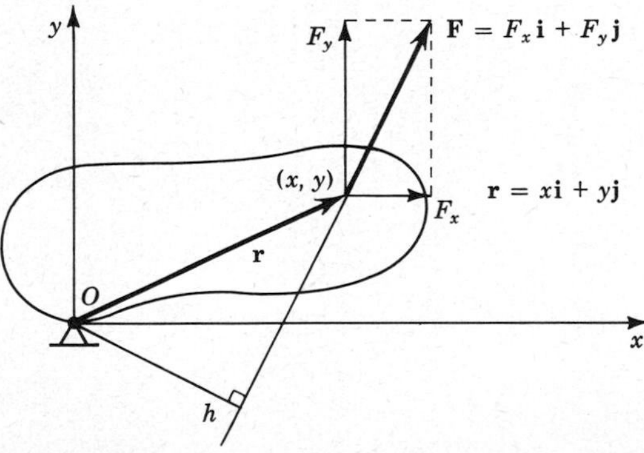

Fig. 1.7 Moment about O.

where $\mathbf{k}$ is the unit vector in the z direction perpendicular to the plane of x and y. Alternatively, if we write out the vectors $\mathbf{r}$ and $\mathbf{F}$ in component form, we have

$$\begin{aligned}\mathbf{M} &= (x\mathbf{i} + y\mathbf{j}) \times (F_x\mathbf{i} + F_y\mathbf{j}) \\ &= \mathbf{k}(xF_y - yF_x)\end{aligned}$$

We see that the magnitude of the moment is given by the algebraic sum of the magnitudes of the moments of the components about O. Very often it is convenient, especially in two-dimensional problems, to work with the moments of components.

It should be emphasized that what we have just defined is the moment of a force *about a point*. The direction of the axis of the moment is perpendicular to the plane containing the force and the point. If another line OQ in Fig. 1.5 passes through O, the component of $\mathbf{r} \times \mathbf{F}$ along OQ is called the moment of $\mathbf{F}$ *about the line or axis OQ*. The magnitude of this component along the line OQ is the projection of the vector $\mathbf{M}$ along OQ. This is given by the *dot product* of $\mathbf{M}$ and a unit vector in the direction of OQ. The magnitude of this component is $|\mathbf{r} \times \mathbf{F}| \cos \alpha$ or $h|\mathbf{F}| \cos \alpha$.

Example 1.1 As an example of the determination of the moment about a line, let us consider Fig. 1.8 and determine the moment M about the shaft axis OO due to the force $\mathbf{P}$ applied to the crank handle as shown.

To find the moment about OO we need first the moment about the point A and then its component in the direction of OO. For the set of

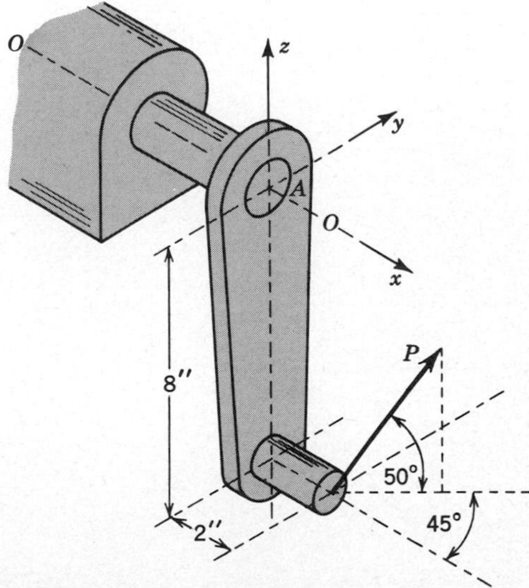

Fig. 1.8 Example 1.1.

coordinate axes shown this component is in the x direction. The component of the moment about the line OO is therefore

$$\begin{aligned} M &= \mathbf{i} \cdot [\mathbf{r} \times \mathbf{F}] \\ &= \mathbf{i} \cdot [(2\mathbf{i} - 8\mathbf{k}) \times P\,(\cos 50^\circ \cos 45^\circ\,\mathbf{i} + \cos 50^\circ \sin 45^\circ\,\mathbf{j} + \sin 50^\circ\,\mathbf{k})] \\ &= 8P \cos 50^\circ \sin 45^\circ \end{aligned} \tag{a}$$

The result (a) is equal to the "lever arm" between OO and the point of application of $\mathbf{P}$ multiplied by the component of $\mathbf{P}$ normal to the plane passing through OO and the point of application of $\mathbf{P}$.

When several forces $\mathbf{F}_1, \mathbf{F}_2, \ldots, \mathbf{F}_n$ act, their total moment or torque about a fixed point O is defined as the sum

$$\mathbf{r}_1 \times \mathbf{F}_1 + \mathbf{r}_2 \times \mathbf{F}_2 + \cdots + \mathbf{r}_n \times \mathbf{F}_n = \sum_j \mathbf{r}_j \times \mathbf{F}_j \tag{1.1}$$

where the $\mathbf{r}_j$ are displacement vectors from O to points on the lines of action of the $\mathbf{F}_j$. A particularly interesting case occurs when there are two equal and parallel forces $\mathbf{F}_1$ and $\mathbf{F}_2$ which have opposite sense, as shown in Fig. 1.9. Such a configuration of forces is called a *couple*. Let us determine the sum of the moments of $\mathbf{F}_1$ and $\mathbf{F}_2$ about O. The operation is indicated schematically in Fig. 1.9. Denoting the total moment by $\mathbf{M}$, we have

$$\begin{aligned} \mathbf{M} &= \mathbf{r}_1 \times \mathbf{F}_1 + \mathbf{r}_2 \times \mathbf{F}_2 \\ &= (\mathbf{r}_2 + \mathbf{a}) \times \mathbf{F}_1 + \mathbf{r}_2 \times \mathbf{F}_2 \\ &= \mathbf{r}_2 \times (\mathbf{F}_1 + \mathbf{F}_2) + \mathbf{a} \times \mathbf{F}_1 \end{aligned} \tag{1.2}$$

where $\mathbf{r}_1$ and $\mathbf{r}_2$ are vectors to arbitrary points on the lines of action of $\mathbf{F}_1$ and $\mathbf{F}_2$.

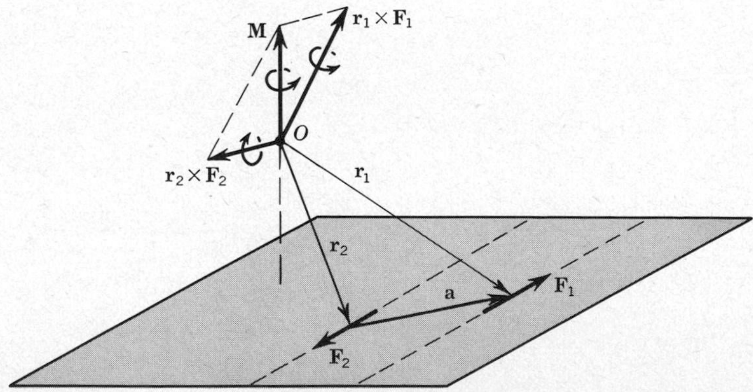

Fig. 1.9 The moment of a couple about the point O.

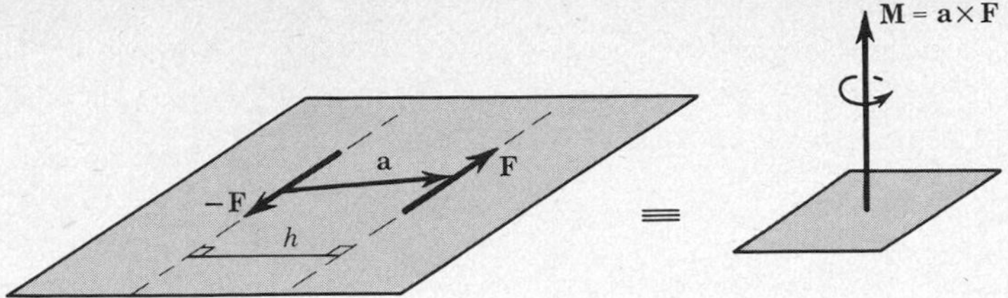

Fig. 1.10 A couple is represented by a moment vector.

Now $\mathbf{F}_1$ and $\mathbf{F}_2$ are of equal magnitude and opposite sense and therefore cancel when added at the same point. The result of (1.2) is then simply

$$\mathbf{M} = \mathbf{a} \times \mathbf{F}_1 \tag{1.3}$$

where **a** is a displacement vector going from an arbitrary point on $\mathbf{F}_2$ to an arbitrary point on $\mathbf{F}_1$. The important thing about this result is that it is *independent of the location of O:* The moment of a couple is the same about all points in space. A *couple* may be characterized by a moment vector without specification of the moment center O as indicated in Fig. 1.10. The magnitude of the moment is most simply computed as $h|\mathbf{F}|$, where h is the perpendicular distance between the vectors **F** and $-\mathbf{F}$. It is often convenient to distinguish between vectors representing the moments of couples and vectors representing forces by using some notational device. We shall use the encircling arrow shown in Fig. 1.10 to indicate the moment of a couple. When sketching a plane figure acted on by a couple whose axis is perpendicular to that plane, the notation of Fig. 1.11 is commonly used.

1.6 CONDITIONS FOR EQUILIBRIUM

According to Newton's law of motion, a particle has no acceleration if the resultant force acting on it is zero. We say that such a particle is in *equilibrium.* Although zero acceleration implies only *constant* velocity, the case that we deal with most

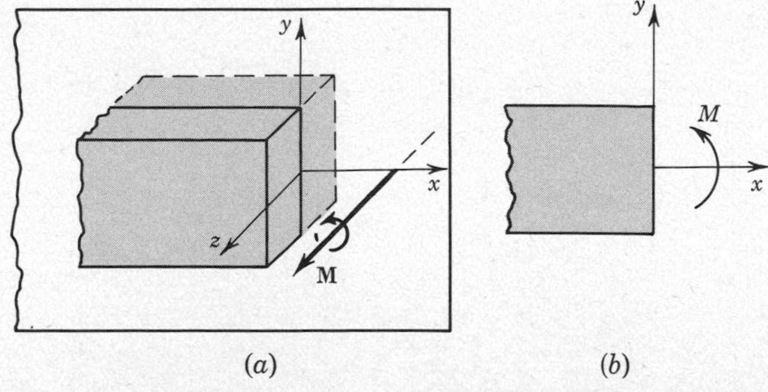

Fig. 1.11 Representation of a couple in a plane sketch.

frequently is that of zero velocity. The study of forces in systems at rest is called *statics*. If several forces $\mathbf{F}_1, \mathbf{F}_2, \ldots, \mathbf{F}_n$ act on a particle, the necessary and sufficient condition for the particle to be in equilibrium is

$$\mathbf{F}_1 + \mathbf{F}_2 + \cdots + \mathbf{F}_n = \sum_j \mathbf{F}_j = 0 \tag{1.4}$$

Under these circumstances we say that the forces are *balanced* or are in *equilibrium*.

One of the striking features of newtonian mechanics is that the postulates are made in terms of the simplest bodies, namely, *particles*, and then logical deduction is used to extend the theory to collections of particles and to solids and fluids. As an example of this extension process we next outline how the concept of equilibrium is extended from a single particle to a general collection of particles.

Consider an isolated system of particles as indicated in Fig. 1.12. We say that such a system is in equilibrium if *every* one of its constituent particles is in equilibrium. Now the forces acting on each particle are of two kinds, *external* and *internal*. The internal forces represent interactions with other particles in the system. Because of our fundamental postulate about the nature of force interactions, we can represent these internal interactions by equal and opposite vectors having the same line of action.

If each particle in Fig. 1.12 is in equilibrium, the resultant force on it is zero. Now let us consider *all* the forces in Fig. 1.12 as a single set of vectors. The vector sum of all the forces is clearly zero since the vector sum of each cluster around a particle must separately be zero. In the process of adding all vectors, however, we find that the internal forces occur in self-canceling pairs, and thus we are left with the result that *if a set of particles is in equilibrium the vector sum of the external forces must be zero;* i.e.,

$$\mathbf{F}_1 + \mathbf{F}_2 + \cdots + \mathbf{F}_n = \sum_j \mathbf{F}_j = 0 \tag{1.5}$$

Let us further consider the total moment of *all* the forces in Fig. 1.12 about an arbitrary point O. The total moment must be zero since the vector sum of forces acting on each particle is separately zero. In the process of forming the total moment of all the vectors, however, we find that the internal forces occur in self-canceling pairs having the same line of action and hence give no contribution to the total moment. We are left with the result that *if a set of particles is in equilibrium the total moment of all the external forces about an arbitrary point O must be zero;* i.e.,

$$\mathbf{r}_1 \times \mathbf{F}_1 + \mathbf{r}_2 \times \mathbf{F}_2 + \cdots + \mathbf{r}_n \times \mathbf{F}_n = \sum_j \mathbf{r}_j \times \mathbf{F}_j = 0 \tag{1.6}$$

where $\mathbf{r}_j$ stands for a position vector extending from O to an arbitrary point on the line of action of the external force $\mathbf{F}_j$.

The conditions (1.5) and (1.6) are *necessary* conditions for equilibrium; i.e., if the system is in equilibrium, *then* (1.5) and (1.6) must be satisfied. This is the way in which we shall employ these conditions in this book. We shall know that

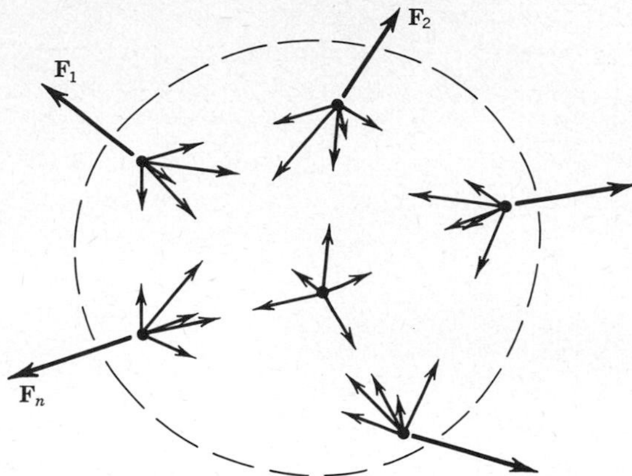

Fig. 1.12 An isolated system of particles showing external and internal forces.

our system is in equilibrium (usually from the fact that the system is at rest), and we shall use (1.5) and (1.6) to obtain information about the forces.

It is interesting, however, to consider the converse problem. Suppose we know that the external forces acting on a system of particles satisfy both (1.5) and (1.6). Can we then conclude that every one of the constituent particles is in equilibrium? The answer is, in general, no.

For example, in Fig. 1.13 a system of two particles is shown acted upon by an equilibrium set of external forces $\mathbf{F}$ and $-\mathbf{F}$. The internal forces $\mathbf{F}_i$ and $-\mathbf{F}_i$ are also an equilibrium set, but the particles will be in equilibrium only when $\mathbf{F} = \mathbf{F}_i$. If, instead of two particles, we consider a rubber band as our system in Fig. 1.13, we can easily perform the indicated experiment. If equal and opposite forces are applied to the ends of an unstretched rubber band, it does not remain in equilibrium. The ends of the band begin to accelerate away from one another, and the band begins to stretch.

This example suggests that if the system of particles was perfectly rigid so that no pair of particles could separate, the internal forces might automatically

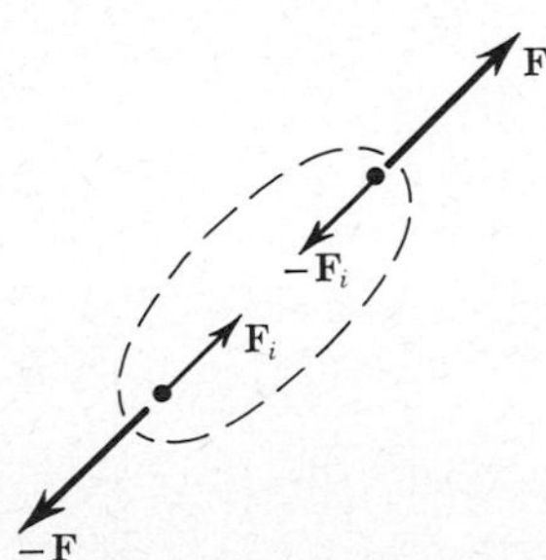

Fig. 1.13 Illustration of a system which is *not* in equilibrium even though the external forces balance.

adjust themselves so as to provide internal equilibrium whenever the external forces make up an equilibrium set. This can in fact be proved.[1] A rigorous proof requires a careful analysis of the possible motions of a rigid body. We shall not go into the details here but shall simply state the final result: *The necessary and sufficient conditions for a perfectly rigid body to be in equilibrium are that the vector sum of all the external forces should be zero and that the sum of the moments of all the external forces about an arbitrary point together with any external applied moments should be zero.*

A necessary and sufficient condition for the equilibrium of a *deformable* system is that the sets of external forces which act on the system *and on every possible subsystem isolated out of the original system* should all be sets of forces which satisfy both (1.5) and (1.6).

It is important to emphasize that our two previous statements of equilibrium for perfectly rigid bodies and deformable systems are the essence of the theory of equilibrium. We will be using the concepts embodied in these statements continually throughout this book. As we mentioned in the introduction, our emphasis will be on the rational *applications* of the concepts. We will first treat systems of particles or engineering structural members which are relatively rigid so that if our system is in equilibrium, Eqs. (1.5) and (1.6) are valid. Later, in discussing deformable systems we will find that the equations of equilibrium for infinitesimal subsystems will be differential equations. Of course, on a sufficiently fine scale, the microscopic particles which constitute a system are generally not in equilibrium, even though the assembly of particles is in a state of macroscopic equilibrium. This is the case in any "static" piece of metal, liquid, gas, etc. The study of effects produced by the nonequilibrium particles is found in texts on *statistical mechanics*.

The two vector equations (1.5) and (1.6) are equivalent to six scalar equations so that in general we can solve for six scalar unknowns in each set of external forces. There are several simple special cases which deserve explicit mention.

Two-force member In Fig. 1.14 a system is in equilibrium under the action of only two external forces applied at A and B. The two forces cannot have random orientation, as shown in Fig. 1.14*a*, but must be directed along AB. This is proved by using (1.6) and taking moments about A and B. In order for the moment about A to vanish, the line of action of $\mathbf{F}_B$ must pass through A. Similarly, the line of action of $\mathbf{F}_A$ must pass through B. We must also have $\mathbf{F}_A = -\mathbf{F}_B$ in order for (1.5) to be satisfied.

Three-force member In Fig. 1.15 a system is in equilibrium under the action of only three external forces applied at A, B, and C. The three forces cannot have

[1] See, for example, J. L. Synge and B. A. Griffith, "Principles of Mechanics," 3d ed., p. 60, McGraw-Hill Book Company, New York, 1959.

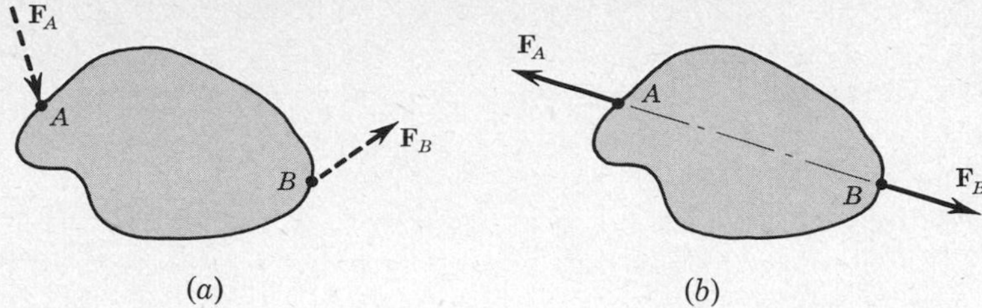

Fig. 1.14 The forces $\mathbf{F}_A$ and $\mathbf{F}_B$ must be equal and opposite and directed along AB if the system is in equilibrium.

random orientation, as shown in Fig. 1.15*a*. They must all lie in the plane ABC if the total moment about each of the points A, B, and C is to vanish. Furthermore, they must all intersect in a common point O, otherwise the total moment about the intersection of any two of the lines of action could not vanish. This result that the three forces must intersect at a common point is a useful one to keep in mind. An interesting exercise in vector analysis is to prove the above statements. A limiting case occurs when point O moves off at great distance from A, B, and C, in which case the forces $\mathbf{F}_A$, $\mathbf{F}_B$, and $\mathbf{F}_C$ become *parallel* coplanar forces.

General coplanar force system In Fig. 1.16 the external forces acting on a system in equilibrium all lie in the plane of the sketch. In this case three of the six general scalar equations of equilibrium are immediately satisfied: there are no force components perpendicular to the plane, and if moments are taken about a point O lying in the plane, the only moment components will be perpendicular to the plane. This leaves only *three independent* scalar conditions of equilibrium for

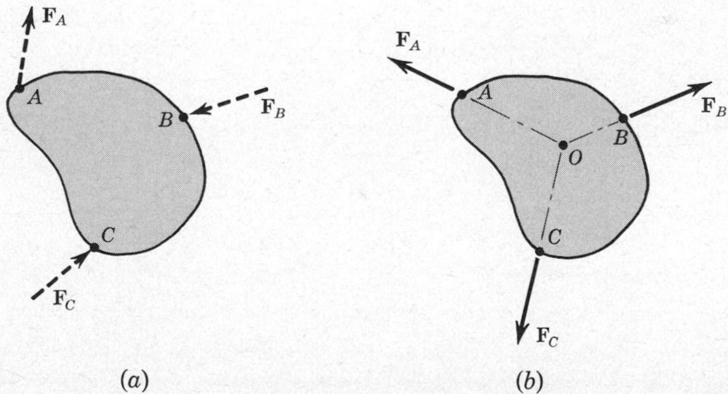

Fig. 1.15 The forces $\mathbf{F}_A$, $\mathbf{F}_B$, and $\mathbf{F}_C$ must be coplanar and intersect at a common point O if the system is in equilibrium.

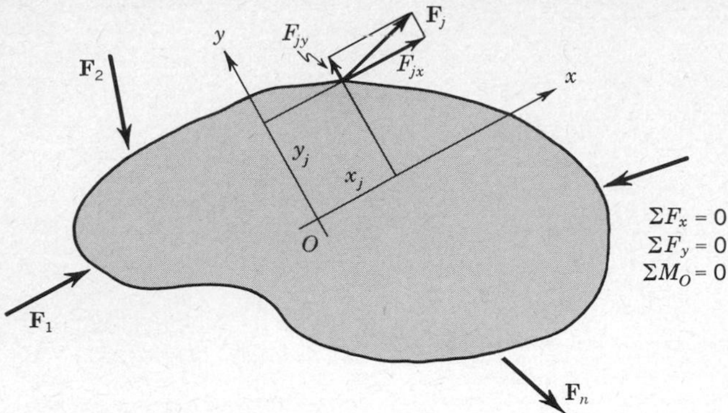

Fig. 1.16 Equilibrium conditions for a coplanar force system.

two-dimensional problems. Taking an arbitrary point O in the plane and an arbitrary orientation of the xy axes in the plane, the condition for the vector sum of the external forces to vanish is simply

$$\sum_j \mathbf{F}_j = \sum_j (F_{jx}\mathbf{i} + F_{jy}\mathbf{j}) = 0$$

or each component of the resultant force vector must vanish:

$$\sum_j F_{jx} = 0 \qquad \sum_j F_{jy} = 0 \tag{1.7}$$

The condition that the total moment about O should vanish may be written

$$\begin{aligned}\sum_j \mathbf{r}_j \times \mathbf{F}_j &= \sum_j (x_j\mathbf{i} + y_j\mathbf{j}) \times (F_{jx}\mathbf{i} + F_{jy}\mathbf{j}) \\ &= \mathbf{k} \sum_j (x_jF_{jy} - y_jF_{jx}) = 0\end{aligned} \tag{1.8}$$

where x_j and y_j are the coordinates of a point on the line of action of $\mathbf{F}_j$, and F_{jx} and F_{jy} are the x and y components of $\mathbf{F}_j$. See Prob. 1.7 for alternate formulations of the conditions for equilibrium for coplanar forces.

1.7 ENGINEERING APPLICATIONS

Many practical engineering problems involve structures or machines in equilibrium. Certain forces, usually loads, are specified, and it is necessary to determine the reactions which come into play to balance the loads.

The general method of analysis that is followed throughout this book involves the preliminary steps:

1. Selection of system
2. Idealization of system characteristics

These are followed by an analysis based on the principles of mechanics, which includes the following steps:

1. Study of forces and equilibrium requirements
2. Study of deformation and conditions of geometric fit
3. Applications of force-deformation relations

In some systems it is possible to determine all the forces involved by using only the equilibrium requirements without regard to the deformations. Such systems are called *statically determinate.* In this chapter we restrict ourselves to statically determinate systems. Our method of analysis then involves selection of appropriate systems, idealization of their characteristics, study of the forces, and the use of the equilibrium conditions to solve for the unknown forces in terms of the known forces.

The conditions of equilibrium (1.5) and (1.6) provide us with relations that must be satisfied by the external forces acting on an *isolated* system in equilibrium. The difficulty in applying these to practical cases usually centers around the process of isolation itself. This is the key step upon which everything else depends. There is the difficulty of deciding what system or subsystem to isolate, and there is the difficulty of ensuring that a true isolation has been accomplished and that all external forces have been accounted for.

In simple cases it is obvious what system should be isolated; usually a single isolation suffices to solve the problem. In complex analyses many different isolations may be required, and an intricate pattern of partial results may have to be assembled before the problem can be completely solved.

The best way to perform an isolation is to draw a reasonably careful sketch of the periphery of the isolated subsystem and then to show all external forces acting. A systematic way of doing this is to recall that forces either (1) act from a distance or (2) act through direct contact and to account first for any possible forces, such as gravity, which can act from afar. Then go carefully around the entire periphery, indicating all forces which make direct contact with the system. The sketch of the isolated system and all the external forces acting on it is often called a *free-body diagram.* The reader is strongly urged to adopt the habit of attempting to draw clear and complete free-body diagrams for *every* mechanics problem which he undertakes to solve. We use the word "attempt" because we recognize that this is indeed the most difficult and most important step!

In constructing a free-body diagram for part of an engineering system, it is often useful to make simplifying assumptions or *idealizations* concerning the nature of the forces which act. For example, if a relatively light column carries a large load, we can obtain a useful engineering estimate of the forces in the column by neglecting its own weight. In this case the idealization is convenient but not

Table 1.2 Force-transmitting properties of some idealized mechanical elements

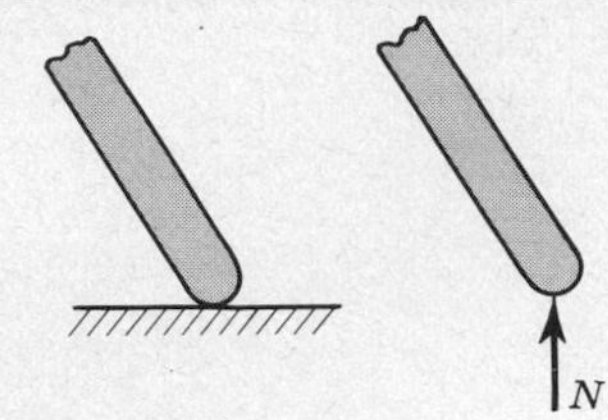

(*a*) A *frictionless* surface can exert only a normal contact force N.

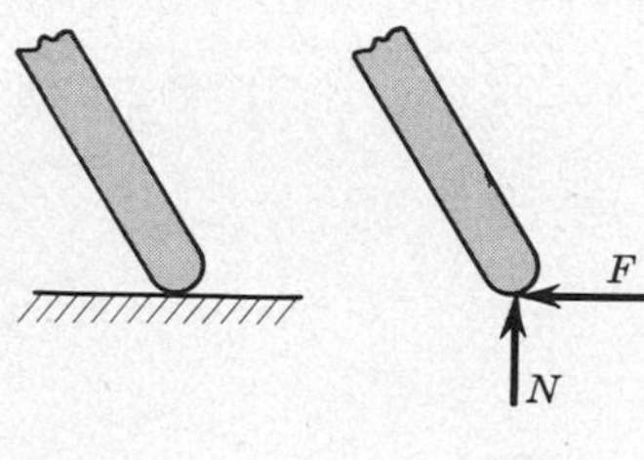

(*b*) When there is *friction* the surface can exert a tangential force F as well as a normal force N. The force F assumes *any* value necessary to prevent motion up to a maximum value $F_s = f_s N$, where f_s is the *coefficient of friction*.

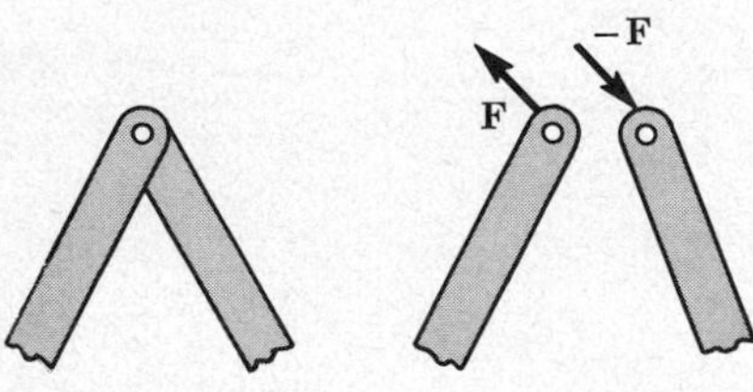

(*c*) A *frictionless pinned joint* transmits a force **F** which passes through the pin. No torque about the pin is transmitted.

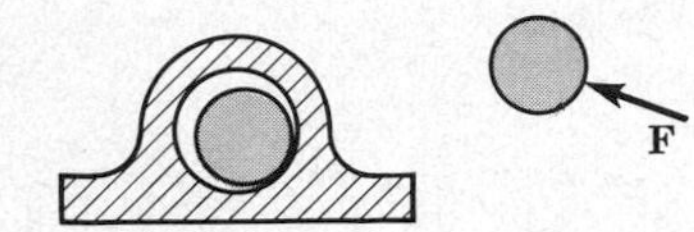

(*d*) A *frictionless bearing* exerts a force **F** on the shaft, which passes through the center of the shaft. No torque about the shaft is transmitted.

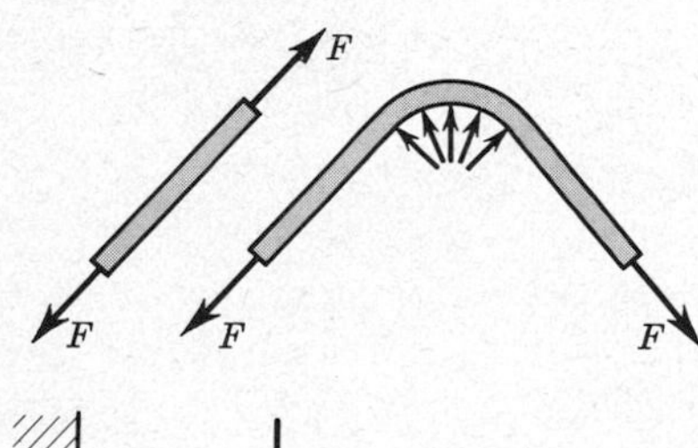

(*e*) A weightless *flexible string* or *cable* transmits force along its length. Each element is subjected to equal and opposite tensile forces F along the string. Compressive forces cannot be sustained. If the string passes over a frictionless peg or pulley, the direction of the force in the string is altered but its magnitude remains constant.

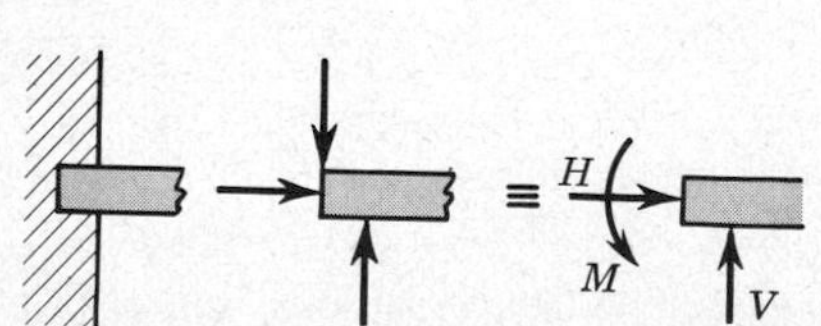

(*f*) An ideal *clamped* support provides complete restraint against longitudinal or transverse motion and against rotation. It can supply force reactions H and V and a moment reaction M.

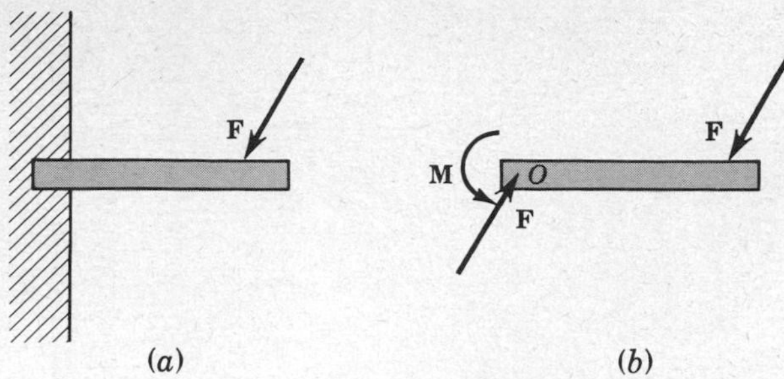

Fig. 1.17 The forces at an ideal clamped support are equivalent to a force and moment.

absolutely necessary, because we can, if required, include the weight in our analysis. In other cases, our ignorance of the actual forces is such that we cannot obtain quantitative estimates without making idealizing assumptions.

Common idealizations include the perfectly rigid body and the inextensible but perfectly flexible string or cable. In Table 1.2 the force-transmitting properties of several mechanical elements are shown.

We will discuss the case of friction which is shown in case (*b*) of Table 1.2 in the next section.

In case (*f*) of Table 1.2 we have shown an ideal clamped support which might, for example, occur at the end of a cantilever beam shown in Fig. 1.17.

If we draw a free-body diagram of the beam as shown in Fig. 1.17*b*, the effect of the wall support on the beam is idealized as a net force acting at the beam end passing through a point O. From the force equilibrium requirement this force is equal to $\mathbf{F}$; further, for moment equilibrium there must be a moment acting at the support. Figure 1.17*b* is equivalent to case (*f*) in Table 1.2. As can be seen, we have considerably idealized the actual support conditions at the wall as far as the details of the interaction between the wall and the beam are concerned. However, for many purposes this simplification is sufficiently correct.

1.8 FRICTION

One of the important forces in mechanics is that due to *friction*. Friction forces are set up whenever a tangential force is applied to a body pressed normally against the surface of another. Thus, in Fig. 1.18*a*, if a normal force P presses body A against the surface of B, and a tangential force T is also applied to body A, then a friction force F will be generated at the interface tending to prevent movement under the action of T. This is indicated in the free-body diagrams in Fig. 1.18*b* and *c*.

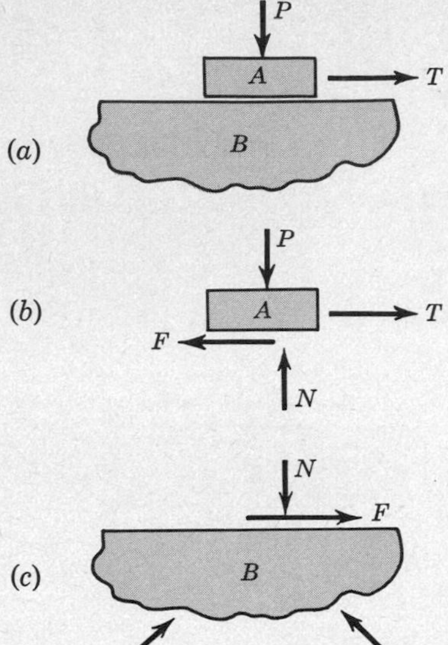

Fig. 1.18 (*a*) Body A pressed against B; (*b*) free-body diagram of body A; (*c*) free-body diagram of body B.

The friction force arises from the interaction of the surface layers of bodies A and B. This interaction will, in general, be made up of a number of processes, including, in particular, the adhesion of surface atoms. A detailed description of friction phenomena is very complicated, and attempts to obtain a complete understanding of friction is a very active area of research in physics and applied mechanics.[1] The outline given below is only an approximate description of the behavior of the total friction force between two surfaces.

The main properties of the friction force F acting on A in Fig. 1.18*b* are:

1. If there is *no relative motion* between A and B, then the friction force F is exactly equal and opposite to the applied tangential force T. This condition can be maintained for any magnitude of T between zero and a certain *limiting* value F_s, called the *static friction force*. If T is greater than F_s, sliding will occur.
2. If body A slides on body B, then the friction force F will have a direction opposite to the velocity of A relative to B, and its magnitude will be F_k, called the *kinetic friction force*.

It has been found that for a given pair of surfaces the forces F_s and F_k are proportional to the normal force N. We can thus introduce two constants of

[1] See, for example, "Friction, Selected Reprints," American Institute of Physics, New York, 1964, and J. J. O'Connor and J. Boyd (eds.), "Standard Handbook of Lubrication Engineering," chaps. 1, 2, McGraw-Hill Book Company, New York, 1968.

proportionality f_s and f_k, which are called the *static and kinetic coefficients of friction*, according to the equations

$$\begin{aligned} F_s &= f_s N \\ F_k &= f_k N \end{aligned} \tag{1.9}$$

These coefficients are intrinsic properties of the interface between the materials A and B, being determined by the materials A and B and by the state of lubrication or contamination at the interface. Further, it has been found that:

1. Both coefficients of friction are nearly independent of the area of the interface. In particular, if body A in Fig. 1.18 were tipped up so that only an edge or a corner was in contact with B, we should still find approximately the same coefficients of friction. Note that under these circumstances the tangential and normal directions are determined only by the surface of B.
2. Both coefficients are nearly independent of the roughnesses of the two surfaces, although this is a conclusion which many people find hard to accept.
3. The static coefficient f_s is nearly independent of the time of contact of the surfaces at rest. Similarly, the kinetic coefficient f_k is nearly independent of the relative velocity of the two surfaces. Figure 1.19 shows a schematic representation of typical static-friction–time and kinetic-friction–velocity plots.

The effect of lubrication and sliding velocity on the friction coefficient for steel on steel surfaces is shown in Fig. 1.20. The top curve is for unlubricated surfaces and the bottom curve is for surfaces well lubricated by a fatty soap. The curves in between represent steel surfaces which are imperfectly lubricated. In all cases, changing the sliding velocity by a factor of 10 changes the friction by no more than about 10 percent.

Figure 1.20 also shows that in the case of unlubricated or poorly lubricated surfaces, the friction goes down as the sliding speed goes up (a negative characteristic). This can lead to frictional oscillations, often called stick-slip, and this

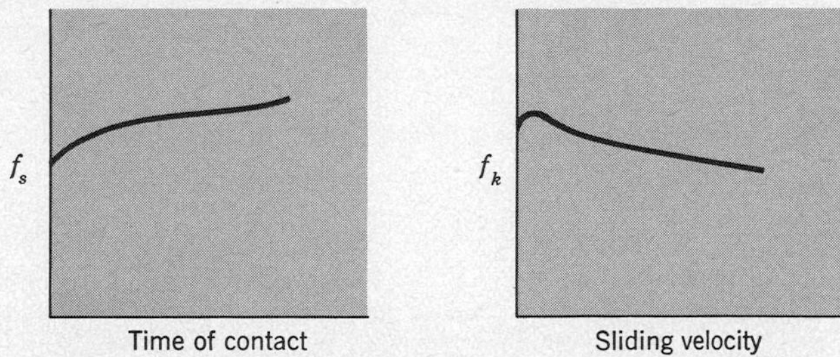

Fig. 1.19 Schematic representation of the variation of friction coefficients.

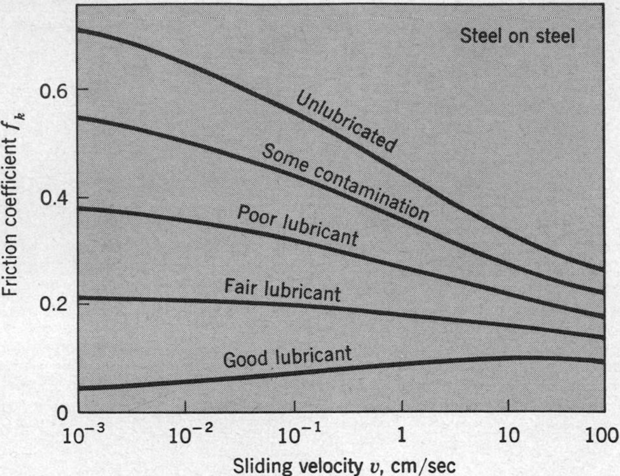

Fig. 1.20 Variation of kinetic friction coefficient with sliding velocity.

phenomenon is responsible for many of the noises of our environment, including the creaking of doors, the squeaking of brakes, and the music of violins.

Since the difference between static and kinetic friction values is not great, and the effects of the time of stick and sliding velocity are relatively small, it has proved possible to give friction-coefficient values which are applicable to almost all sliding conditions. A schematic representation of typical friction coefficients is given in Fig. 1.21 for nonmetal on nonmetal or nonmetal on metal, such as leather on wood, or nylon on steel. The extent of the shading shows the probable range of values.

It can be seen that for any state of lubrication, there is a range of about a factor of 2 between the maximum and minimum friction values that might be encountered. In most mechanics calculations, this uncertainty in friction is the

Table 1.3 Coefficients of friction

Materials	*Surface conditions*	f_s	f_k
Metals on metals (e.g., steel on steel, copper on aluminum)	Carefully cleaned	0.4–1.0	0.3–1.0
	Unlubricated, but not cleaned	0.2–0.4	0.15–0.3
	Well lubricated	0.05–0.12	0.05–0.12
Nonmetals on nonmetals (e.g., leather on wood, rubber on concrete)	Unlubricated	0.4–0.9	0.3–0.8
	Well lubricated	0.1–0.2	0.1–0.15
Metals on nonmetals	Unlubricated	0.4–0.6	0.3–0.5
	Well lubricated	0.05–0.12	0.05–0.12

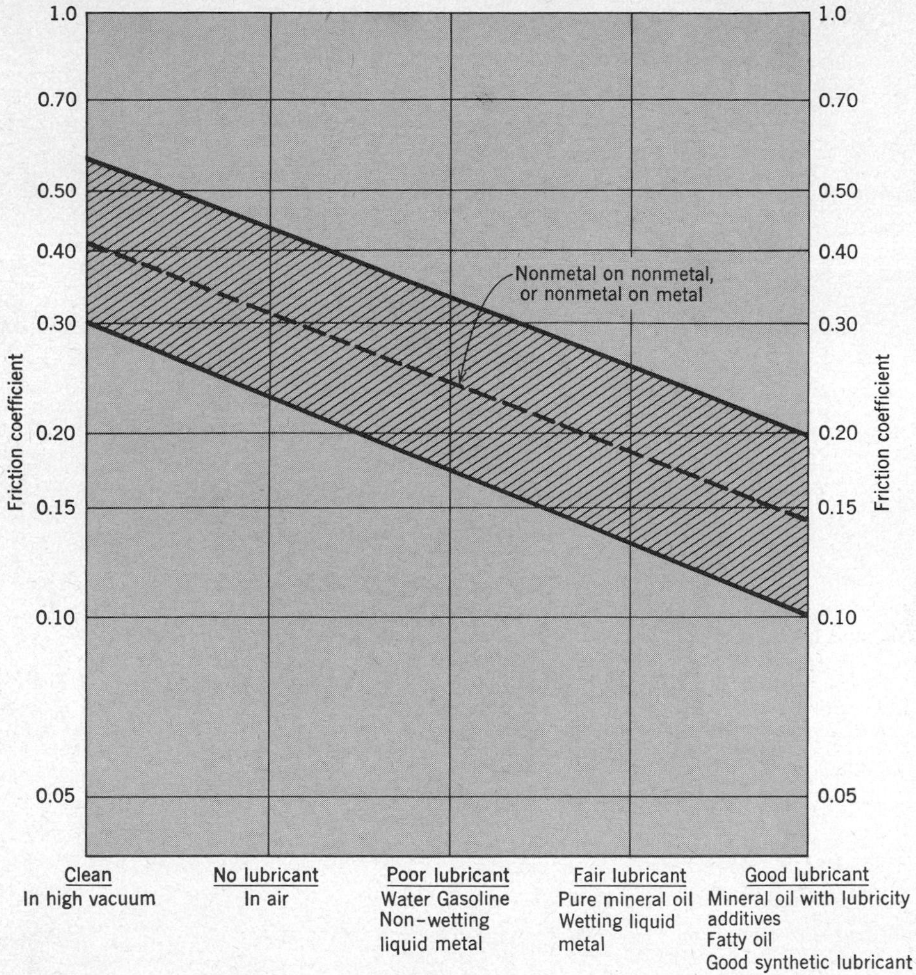

Fig. 1.21 General-purpose friction chart.

factor which limits the overall accuracy of the calculation, since other parameters are generally known within a few percent.

Similar curves can be drawn for similar and dissimilar metals in contact.[1] Some further typical values of friction coefficients are shown in Table 1.3; in practical applications care must be taken in determining or estimating the friction coefficient.

[1] E. Rabinowicz, Surface Energy Approach to Friction and Wear, *Prod. Eng.*, March 15, 1965, p. 95.

1.9 EXAMPLES

To illustrate the concepts of analysis discussed above, we now consider several examples. In these examples we shall focus attention on the problem of selecting a system and idealizing its characteristics, as well as on the method of analyzing the idealized system.

Example 1.2 Let us first consider a highly idealized problem shown in Fig. 1.22*a* to illustrate the construction of free-body diagrams and the concept of impending motion for frictional forces. We are asked to find the range of values of W which will hold the block of weight $W_0 = 100$ lb in equilibrium on the inclined plane if the coefficient of static friction is $f_s = 0.5$. We will assume that the cable is weightless and that the pulley is frictionless.

In Fig. 1.22*b* we show a free-body diagram of the block with motion impending down the plane so that the frictional force F is shown opposite to the direction of motion (see Fig. 1.18*b*). In Fig. 1.22*c* and *d* are shown the free-body diagrams of the weight W and the flexible weightless cable.

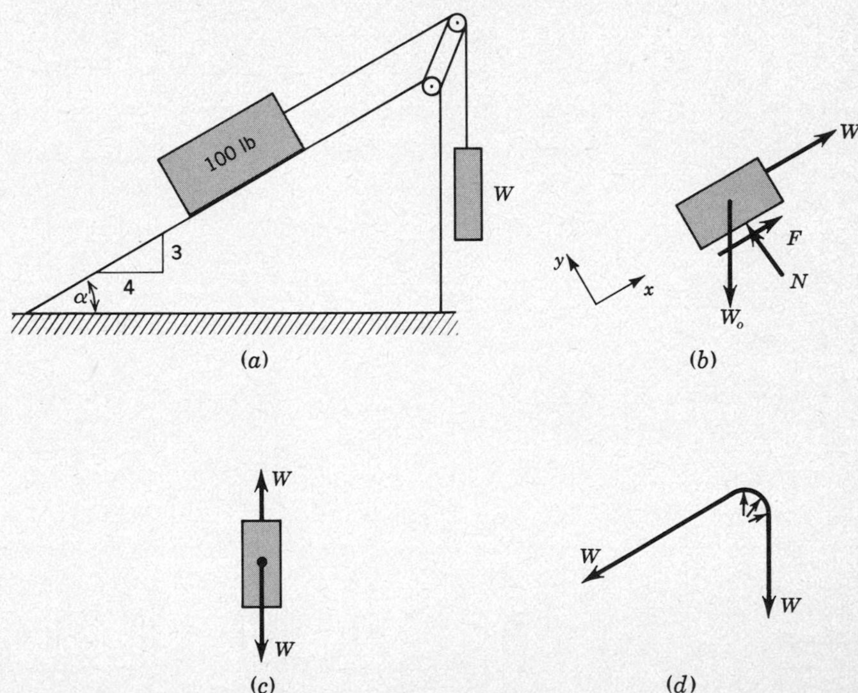

Fig. 1.22 Example 1.2.

If we use the force-equilibrium equations (1.7) and the orientation of the coordinate axes as shown in Fig. 1.22*b*, we have

$$\Sigma F_x = 0 \qquad W + F - W_0 \sin \alpha = 0 \tag{a}$$

$$\Sigma F_y = 0 \qquad N - W_0 \cos \alpha = 0 \tag{b}$$

For motion about to start down the plane we have

$$F = f_s N \tag{c}$$

If we now solve for the unknown value of W from (*a*), (*b*), and (*c*), we find

$$\frac{W}{W_0} = \sin \alpha - f_s \cos \alpha \tag{d}$$

From (*d*), therefore, we find, upon substituting for the numerical values,

$$W = 100[\tfrac{3}{5} - (\tfrac{1}{2})(\tfrac{4}{5})] = 20 \text{ lb} \tag{e}$$

For values of W less than 20 lb, the weight W_0 will no longer be in equilibrium and w ill slide down the inclined plane.

For values of W greater than 20 lb, the weight W_0 will be in equilibrium until the value of W is reached such that the weight is about to move up the plane. For this situation, the free-body diagram of Fig. 1.22*b* again holds except that now the frictional force F is pointed in the opposite direction. The weight W for motion of the block up the plane is then given by (*d*) with the sign of f_s reversed

$$\frac{W}{W_0} = \sin \alpha + f_s \cos \alpha \tag{f}$$

Evaluation of W from (*f*) gives

$$W = 100 \text{ lb}$$

For values of W greater than 100 lb, the block will no longer be in equilibrium and will move up the inclined plane.

Therefore the range of values of W for equilibrium of the block is

$$20 \text{ lb} \leqq W \leqq 100 \text{ lb}$$

Example 1.3 The simple triangular frame shown in Fig. 1.23*a* is used to support a small chain hoist. We are asked to predict the forces acting on the wall at B and C when the chain hoist is supporting its rated capacity of 5,000 lb. The rod BD is pinned at its ends. The member CD is pinned at D and secured with four bolts at C.

In Fig. 1.23*b* a first attack is made by drawing an isolated free-body diagram of the frame. *Note carefully* that the system we have isolated is *not*, by itself, *rigid*; i.e., by itself it will collapse. This isolation is perfectly proper because, in fact, the external forces ($\mathbf{F}_B$, $\mathbf{F}_C$, and $\mathbf{M}_C$) are just sufficient to prevent collapse. We have at this stage neglected the weight of the frame and of the chain hoist. With this idealization the only force and moment interactions occur at *B*, *C*, and *D*. At *D* we show the vertical load of 5,000 lb. At *B* there is a pinned joint. The force interaction there could be a force vector in all three dimensions plus an equally general moment vector. In this case, however, since the frame lies in a single plane and the load also is in this plane, it is reasonable to expect that all forces of any consequence will also lie in this plane. We have accordingly shown $\mathbf{F}_B$ as

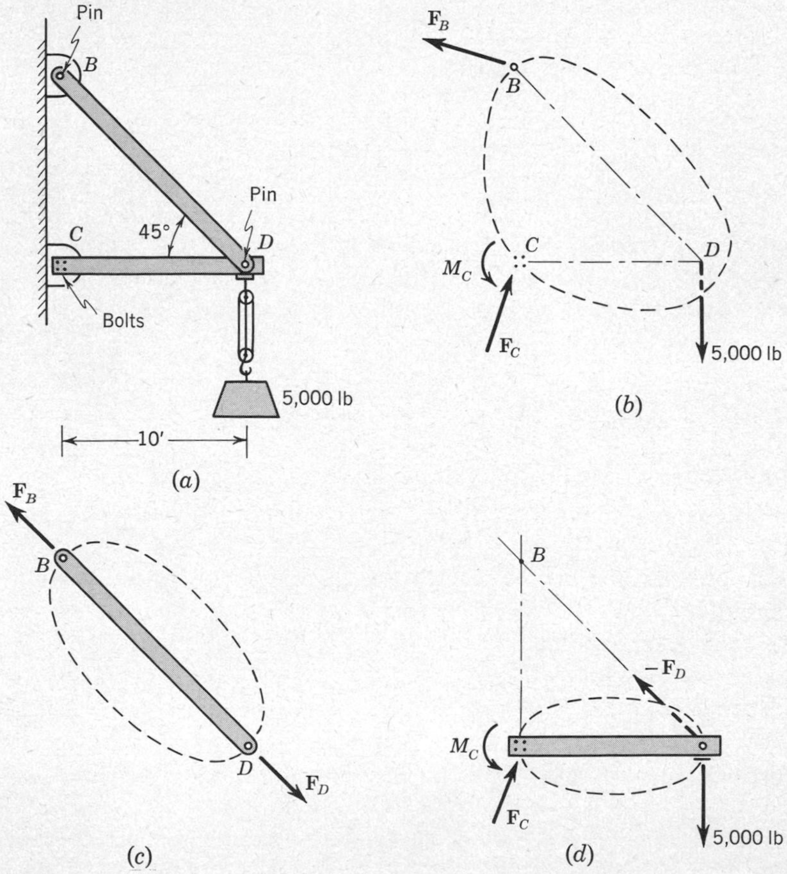

Fig. 1.23 Example 1.3.

lying in the plane of the frame. The orientation within the plane is unknown. A couple with moment vector perpendicular to this plane could be transmitted if there were friction forces around the pin. We have, however, made the idealization that this moment can be neglected on the basis of the following consideration: if there are frictional forces acting at the periphery of the pin, they will produce a frictional moment equal to the frictional force (fN) times the pin radius. When a force and a moment act at the same point in this manner, their effect is equal to that due to a single force, displaced sideways (Prob. 1.10). For this system the necessary sideways displacement is simply the friction coefficient times the pin radius. Therefore, for a typical friction coefficient of $\frac{1}{3}$, the *greatest* effect of friction at pin B would be to displace $\mathbf{F}_B$ sideways $\frac{1}{3}$ of the pin radius. In this example, such a small displacement appears insignificant to the general geometry, and is neglected.

Coming next to point C, where bar CD is joined to the wall by four bolts, a similar argument leads us to the conclusion that a force $\mathbf{F}_C$ and a moment component M_C in the plane of the frame could be acting on the frame. Here we are not willing to neglect the moment since a bolted joint could possibly transmit moments of considerable size in comparison with a pinned joint.

Our free-body diagram in Fig. 1.23*b* then contains two unknown force vectors (each with two components) and one unknown moment component, due to forces all lying in one plane. According to (1.7) and (1.8), three independent equilibrium conditions are available for a coplanar system. Since we have five unknown components, we *cannot* obtain a complete solution from Fig. 1.23*b* alone.

In an attempt to get additional relations we must isolate subsystems. We show in Fig. 1.23*c* a free-body diagram of bar BD. Since both ends are pinned and we are neglecting the weight of the bar itself, we can say that BD is a two-force member, and hence, as shown in Fig. 1.14, the forces $\mathbf{F}_B$ and $\mathbf{F}_D$ must be equal and opposite vectors along BD.

Next, in Fig. 1.23*d* we show a free-body diagram of bar CD. At D we now can show the orientation of the interaction with BD since it must be equal and opposite to the force $\mathbf{F}_D$ in Fig. 1.23*c*. The direction of $\mathbf{F}_C$ still remains unknown. Counting unknowns in Fig. 1.23*d*, we have the two components of $\mathbf{F}_C$ and the magnitudes of M_C and $\mathbf{F}_D$, or a total of *four* scalar unknowns. Again we cannot obtain a complete solution from the *three* independent conditions of equilibrium. This time we are up against a stone wall. Having isolated each bar separately as well as the combination of both bars together, we have exhausted all possibilities. We must conclude that the conditions of equilibrium alone are insufficient to analyze our model. This is, in fact, the case. The frame model of Fig. 1.23*b* is *statically indeterminate*.

We then have two courses open to us (besides giving up in despair).

We can consider a more highly idealized model which is statically determinate, or we can develop a theory for handling statically indeterminate structures. In this book we shall actually do both. In the following paragraphs we shall discuss a simplified model. In subsequent chapters we shall develop a theory which will permit us to return to this problem again in Chap. 8 and to estimate the errors committed in employing the simpler model.

The most ambiguous part of the model of Fig. 1.23 was the moment M_C at the bolted joint. This moment may be quite small if the bolts are loosely fitted and are not tightened up. This consideration leads us to adopt the simplified model of Fig. 1.24*a* where we have idealized the bolted joint into a pinned joint. In the free-body diagram of Fig. 1.24*b* there will be forces $\mathbf{F}_B$ and $\mathbf{F}_C$ at the wall-support points, but no moments. The directions

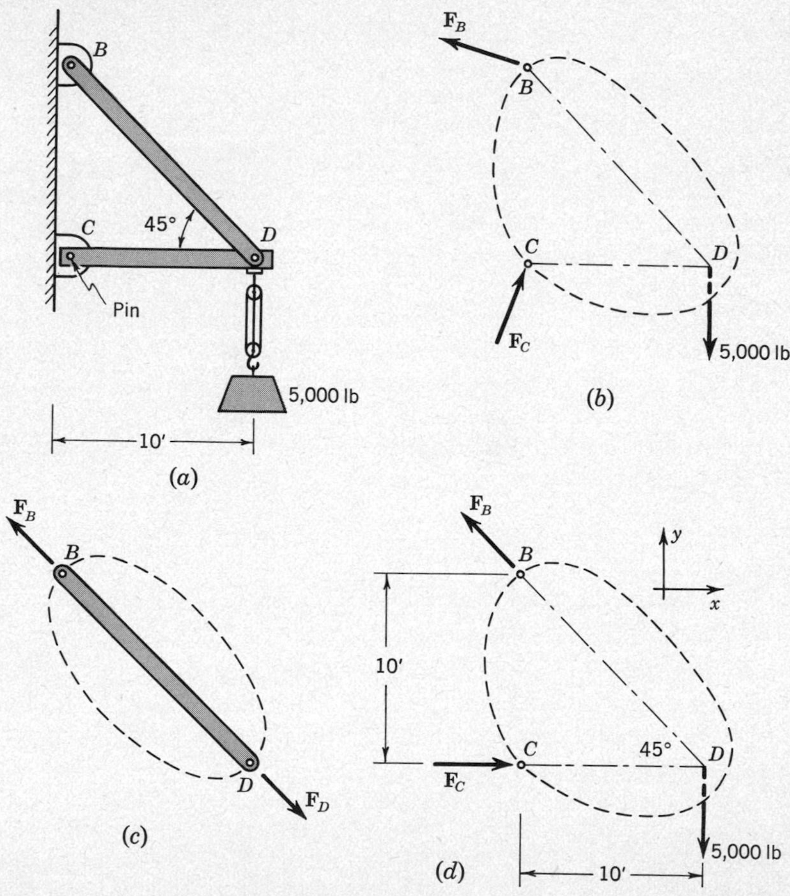

Fig. 1.24 Idealized model of system of Fig. 1.23.

of $\mathbf{F}_B$ and $\mathbf{F}_C$ are unknown in Fig. 1.24*b*. Taking advantage of our previous experience with Fig. 1.23, we show a free-body diagram of bar *BD* in Fig. 1.24*c*. Since this is a two-force member, $\mathbf{F}_B$ must be along the line *BD*. Returning to the entire frame in Fig. 1.24*d* with this information, we conclude that since there are only three forces acting on the isolated free body and since $\mathbf{F}_B$ and the load intersect at *D*, then $\mathbf{F}_C$ must also be collinear with *D* as shown. We could draw the same conclusion by noting that bar *CD* is a three-force member. It now remains only to find the magnitudes of $\mathbf{F}_B$ and $\mathbf{F}_C$. These can be determined in several ways by applying the equilibrium requirements of (1.7) and (1.8). For example, if we require that the total moment about *B* should be zero, we have

$$\Sigma \mathbf{M}_B = -10\mathbf{j} \times \mathbf{F}_C + 10\mathbf{i} \times (-5{,}000\mathbf{j}) = 0 \tag{a}$$

where **i** and **j** are unit vectors in the *x* and *y* directions. Expressing $\mathbf{F}_C$ as $F_C\mathbf{i}$, where F_C is the scalar magnitude of $\mathbf{F}_C$, we easily find from (*a*)

$$F_C = 5{,}000 \text{ lb} \tag{b}$$

Summing vertical force components yields

$$\Sigma F_y = 0 = F_B \sin 45^\circ - 5{,}000 \tag{c}$$

or

$$F_B = 7{,}070 \text{ lb} \tag{d}$$

Thus we have determined the forces acting *on* the isolated frame in Fig. 1.24*b*. The forces acting on the wall supports *from* the frame are equal and oppositely directed.

Our analysis has been based on several assumptions and idealizations. We have neglected the weight of the frame and the hoist. We have neglected frictional moments at pinned joints, and we have made the additional idealization, in going from Fig. 1.23*a* to Fig. 1.24*a*, that the bolted joint could be treated as a pinned joint. It will take us until Chap. 8 before we can fully assess the significance of these simplifying assumptions. There we shall see that the results obtained above do actually constitute a very useful engineering approximation.

Example 1.4 A pinned truss is shown in equilibrium in Fig. 1.25. It is a plane structure consisting of relatively rigid links connected by pinned joints. It carries loads at *E* and *F* as shown; it is pinned to a rigid foundation at *A* and is supported on a roller support at *B*. The primary problem is to determine the forces at *A* and *B* due to the loads at *E* and *F*. A secondary problem is to determine the forces in the individual links of the truss.

To obtain the reactions at *A* and *B* we isolate the entire truss in the free-body diagram of Fig. 1.26. We have made the idealization that the

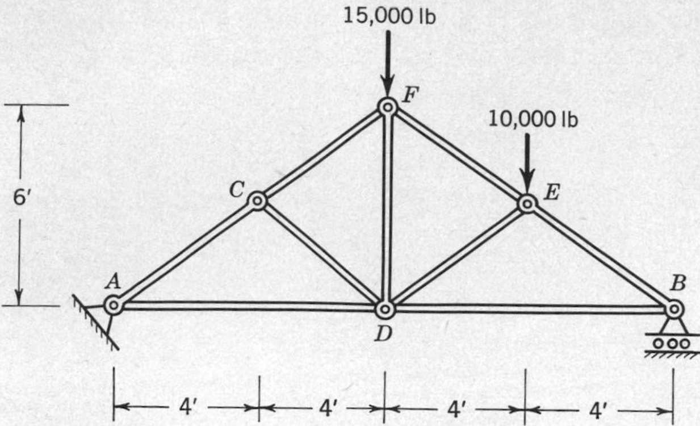

Fig. 1.25 Example 1.4.

weight of the truss can be neglected. Tracing the periphery of the isolated system, we have included the loads at E and F. At B we have idealized the roller support by showing a *vertical* reaction $\mathbf{F}_B$. Our rationalization is that if the support has been designed to permit horizontal motion it should not provide *much* horizontal resisting force. At the pinned joint of A we have shown a reaction $\mathbf{F}_A$ which passes through the pin. Again we have made an idealization by neglecting the possibility of a frictional moment around the pin. Our rationalization is that even if there is some friction the smallnessof the pin implies that the effect of the friction moment about the center of the pin will remain *small*.

Since the truss is a planar system in equilibrium, the external forces shown in Fig. 1.26 must satisfy (1.7) and (1.8). We note that $\mathbf{F}_A$ (magnitude and direction unknown) and $\mathbf{F}_B$ (magnitude unknown) represent three

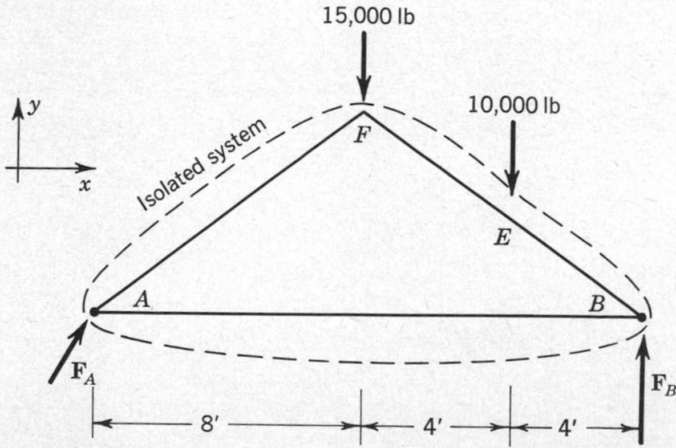

Fig. 1.26 Free-body diagram of truss of Fig. 1.25.

unknown scalar quantities, and thus the three independent conditions for equilibrium of coplanar forces are sufficient to determine $\mathbf{F}_A$ and $\mathbf{F}_B$. Taking A as our moment center, we have, according to (1.8),

$$\Sigma M_A = 16F_B - 12(10{,}000) - 8(15{,}000) = 0$$

$$F_B = 15{,}000 \text{ lb} \qquad (a)$$

Letting $\mathbf{F}_A = \mathbf{i}A_x + \mathbf{j}A_y$, where $\mathbf{i}$ and $\mathbf{j}$ are unit vectors in the x and y directions, we next apply (1.7) to get

$$\Sigma F_x = A_x = 0$$

$$\Sigma F_y = A_y + 15{,}000 - 15{,}000 - 10{,}000 = 0 \qquad (b)$$

$$A_y = 10{,}000 \text{ lb}$$

Thus the reactions at A and B are both vertically upward, with magnitudes of 10,000 and 15,000 lb, respectively.

This solution for the reactions makes no use of the particular design of the truss within the isolated system of Fig. 1.26. All that is required is that the truss be in equilibrium. The designer of the truss, however, is interested in how the loads are transmitted by the various members so that he can be sure that each member is strong enough. To obtain this kind of information we must consider free-body diagrams of subassemblies within the truss. As an illustration we show in Fig. 1.27 how the forces in members AC and AD can be determined. In Fig. 1.27a and b free-body diagrams of the bars AC and AD show that (if the weights of the bars and the frictional moments around the pin joints are neglected) they are two-force members, and hence

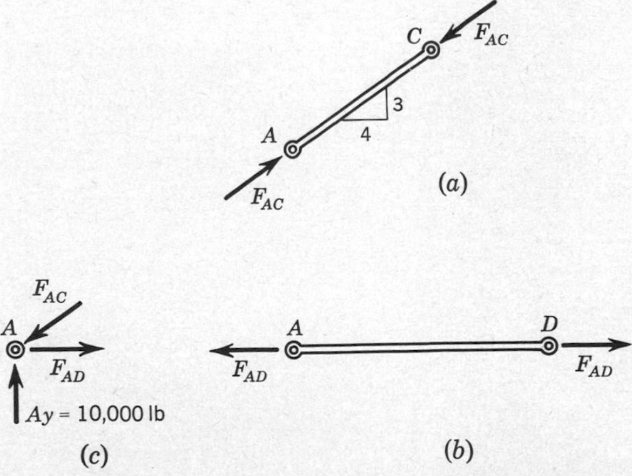

Fig. 1.27 Isolations of (a) bar AC, (b) bar AD, and (c) pin A.

that F_{AC} and F_{AD} must be directed along the links. In Fig. 1.27*c* a free-body diagram of the joint at *A* shows these same forces acting on the pin. Note that the force on the pin by the bar is equal and opposite to the force on the bar by the pin according to Newton's third law. Since the pin is in equilibrium, we have, according to (1.7),

$$\begin{aligned} F_{AD} - \tfrac{4}{5}F_{AC} &= 0 \\ 10{,}000 - \tfrac{3}{5}F_{AC} &= 0 \end{aligned} \qquad (c)$$

from which we find $F_{AC} = 16{,}670$ lb and $F_{AD} = 13{,}330$ lb. The force F_{AC} tends to shorten the bar *AC* and is called a *compressive* force; the force F_{AD} tends to extend the bar *AD* and is called a *tensile* force.

Example 1.5 Figure 1.28 shows a 3,000-lb load held in equilibrium by a 6-ft derrick boom *ABC* supported by the guy wires *BD* and *BE* and a ball-and-socket joint at *C*. The points *C*, *D*, and *E* all lie in the *xy* plane as shown. It is desired to determine the reactions at *C*, *D*, and *E* due to the 3,000-lb load at *A*.

The sketch of Fig. 1.28 can be used to represent a free-body diagram of the boom and guy wires if we indicate the external forces acting at *A*, *C*, *D*, and *E*. We idealize the situation by neglecting the weight of the boom and

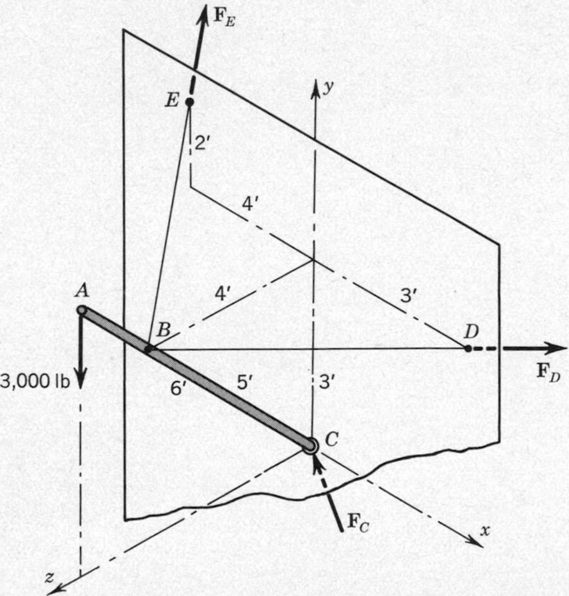

Fig. 1.28 Example 1.5. Derrick boom *ABC* is supported by ball-and-socket joint at *C* and guy wires at *D* and *E*.

guy wires. At C we show a force $\mathbf{F}_C$ acting on the ball joint with unknown orientation; we neglect the possibility of a frictional moment. At D and E we make use of the property of an ideally flexible cable given in Table 1.2 to show $\mathbf{F}_D$ and $\mathbf{F}_E$ acting on the wires *along the directions BD and BE*. Thus, if F_D and F_E are the magnitudes of these forces and **i**, **j**, and **k** represent unit vectors in the x, y, and z directions, we may write

$$\begin{aligned} \mathbf{F}_D &= F_D(\tfrac{3}{5}\mathbf{i} - \tfrac{4}{5}\mathbf{k}) \\ \mathbf{F}_E &= F_E(-\tfrac{4}{6}\mathbf{i} + \tfrac{2}{6}\mathbf{j} - \tfrac{4}{6}\mathbf{k}) \end{aligned} \qquad (a)$$

Since the derrick is in equilibrium, the forces in Fig. 1.28 must satisfy (1.5) and (1.6). A convenient method of application is to take C as a moment center for (1.6). This has the advantage of eliminating the three components of $\mathbf{F}_C$.

$$\Sigma\mathbf{M}_C = \mathbf{CA} \times (-3{,}000\mathbf{j}) + \mathbf{CB} \times \mathbf{F}_D + \mathbf{CB} \times \mathbf{F}_E = 0 \qquad (b)$$

Using the determinant representation illustrated in Prob. 1.4 at the end of this chapter, the vector cross products in (*b*) can be expanded as follows:

$$\begin{vmatrix} \mathbf{i} & \mathbf{j} & \mathbf{k} \\ 0 & 6\left(\frac{3}{5}\right) & 6\left(\frac{4}{5}\right) \\ 0 & -3{,}000 & 0 \end{vmatrix} + F_D \begin{vmatrix} \mathbf{i} & \mathbf{j} & \mathbf{k} \\ 0 & 3 & 4 \\ \frac{3}{5} & 0 & -\frac{4}{5} \end{vmatrix} + F_E \begin{vmatrix} \mathbf{i} & \mathbf{j} & \mathbf{k} \\ 0 & 3 & 4 \\ -\frac{2}{3} & \frac{1}{3} & -\frac{2}{3} \end{vmatrix} = 0$$

$$(14{,}400 - \tfrac{12}{5}F_D - \tfrac{10}{3}F_E)\,\mathbf{i} + (\tfrac{12}{5}F_D - \tfrac{8}{3}F_E)\,\mathbf{j} + (-\tfrac{9}{5}F_D + 2F_E)\,\mathbf{k} = 0 \qquad (c)$$

Setting the components of **i**, **j**, and **k** separately equal to zero yields simple simultaneous equations[1] for F_D and F_E with the solution

$$\begin{aligned} F_D &= 2{,}670 \text{ lb} \\ F_E &= 2{,}400 \text{ lb} \end{aligned} \qquad (d)$$

[1] This example is unusual in that we have *three* simultaneous equations with only *two* unknowns. The solution (*d*) satisfies all three equations. The physical reason for this apparent paradox lies in the fact that the particular loading we have considered in Fig. 1.28 involves a set of forces, all of which pass through *ABC*, and therefore whose moment vectors about C can only be perpendicular to *ABC*. This leaves only two independent scalar conditions of moment balance. The three conditions of (*c*) are actually equivalent to two since balance of moment components along *ABC* is automatically ensured. In fact, the structure of Fig. 1.28 is incapable of supporting a moment around *ABC*; the boom would spin in its socket and the guy wires would wind around at *B*. Twist could be prevented by replacing one of the guy wires with a rigid bar welded to the boom at *B*.

In (*b*) we have taken the guy-wire tensions as acting at point *B*. We would obtain the same final result (*d*) if we let the forces act on the total system at *D* and *E* as shown in Fig. 1.28. Equation (*b*) would then become

$$\Sigma \mathbf{M}_C = \mathbf{CA} \times (-3{,}000\mathbf{j}) + \mathbf{CD} \times \mathbf{F}_D + \mathbf{CE} \times \mathbf{F}_E$$

The force $\mathbf{F}_C$ may now be obtained by applying (1.5)

$$\Sigma \mathbf{F} = \mathbf{F}_C + \mathbf{F}_D + \mathbf{F}_E - 3{,}000\mathbf{j} = 0$$

$$\mathbf{F}_C = -2{,}400(\tfrac{3}{5}\mathbf{i} - \tfrac{4}{5}\mathbf{k}) - 2{,}670(-\tfrac{2}{3}\mathbf{i} + \tfrac{1}{3}\mathbf{j} - \tfrac{2}{3}\mathbf{k}) + 3{,}000\mathbf{j} \qquad (e)$$
$$= 2{,}200\mathbf{j} + 3{,}730\mathbf{k} \text{ lb}$$

The complete solution for the reactions thus consists of the guy-wire tensions (*d*) and the ball-and-socket force (*e*).

While the above solutions employing vector-analysis methods of solution are applicable to all problems, and essential to some, it is frequently easier, and much more instructive to solve the equations of equilibrium in the component form; i.e., the sum of force components in each of three orthogonal directions must be zero, and the sum of moment components about the axes must be zero. Often, careful selection of axes for writing force and moment equations can simplify the solution. In Fig. 1.28, we see that we can solve for F_D and F_E by writing moment equations about the x and y axes.

$$\Sigma M_x = (6 \times \tfrac{4}{5})(3{,}000) - 3(\tfrac{4}{5}F_D) - 5(\tfrac{4}{6}F_E) = 0$$

$$\Sigma M_y = 3(\tfrac{4}{5}F_D) - 4(\tfrac{4}{6}F_E) = 0 \qquad (f)$$

$$\therefore\ F_E = 2{,}400 \text{ lb} \qquad F_D = 2{,}670 \text{ lb}$$

The x, y, and z force components of F_C can now be found by equating the sum of forces in those directions to zero.

$$\Sigma F_x = F_{C_x} + \tfrac{3}{5}F_D - \tfrac{4}{6}F_E = 0 \qquad F_{C_x} = 0$$

$$\Sigma F_y = F_{C_y} + \tfrac{2}{6}F_E - 3{,}000 = 0 \qquad F_{C_y} = 2{,}200 \text{ lb} \qquad (g)$$

$$\Sigma F_z = F_{C_z} - \tfrac{4}{5}F_D - \tfrac{4}{6}F_E = 0 \qquad F_{C_z} = 3{,}730 \text{ lb}$$

Example 1.6 A screw jack, which is frequently used to raise or lower weights, is shown schematically in Fig. 1.29*a*. The screw is characterized by a thread pitch p and diameter d (Fig. 1.29*b*). We wish to determine the operating characteristics in the presence of a coefficient of friction f between the screw threads and the jack body. In particular we wish to determine the relationship between the moment necessary to raise and lower the weight W

and the frictional and geometrical characteristics of the jack. In Fig. 1.29*c* the screw has been isolated, and highly idealized, showing the distributed thread loads as acting at one point for convenience of analysis. We see that each portion of the screw must slide up an incline at a *helix angle* α, where

$$\tan \alpha = \frac{p}{\pi d}$$

In Fig. 1.29*d* we have introduced the concept of a *friction angle* β, where $\tan \beta = f$. The resultant R of the normal component N and the frictional component fN acts at an angle β to the normal to the screw thread. Thus, R acts at $\alpha \pm \beta$ to the vertical, depending on whether the jack is being raised, as

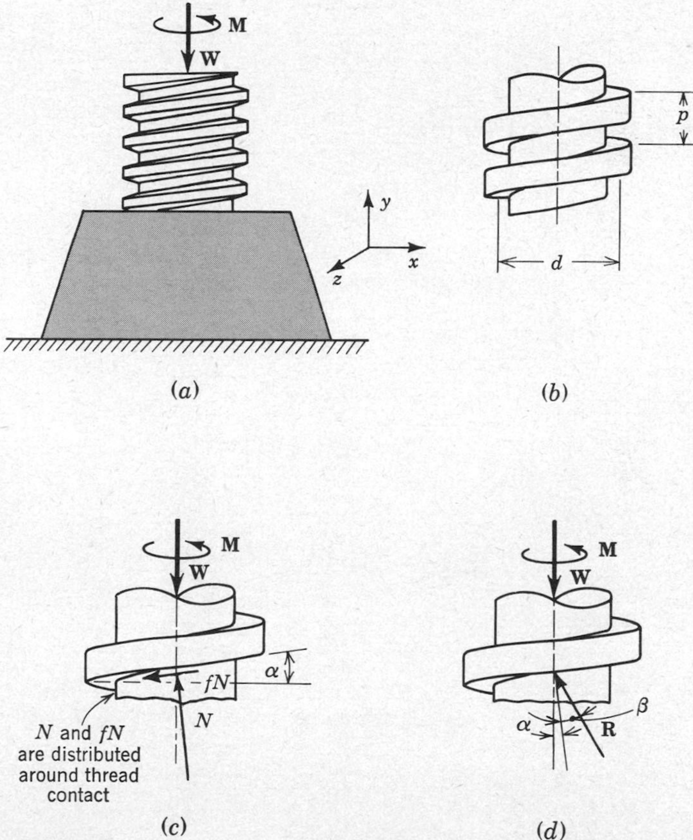

Fig. 1.29 Example 1.6.

shown in Fig. 1.29*d*, or is being lowered. If we now sum forces along the y axis and moments about the y axis, we have

$$\Sigma F_y = R\cos(\alpha \pm \beta) - W = 0$$

$$\Sigma M_y = M - \frac{d}{2} R \sin(\alpha \pm \beta) = 0 \qquad (a)$$

Thus

$$M = \frac{Wd}{2}\tan(\alpha \pm \beta) \qquad (b)$$

Equation (*b*) with the plus sign gives the moment necessary to move the screw upward. The moment necessary to lower or unwind the screw is given by (*b*) when the minus sign is taken

$$M = \frac{Wd}{2}\tan(\alpha - \beta) \qquad (c)$$

When $\beta = \alpha$ in (*c*), the moment M vanishes for equilibrium, and the screw will support the weight W without unwinding. If $\beta > \alpha$ in (*c*), a negative M is required to lower the weight. A jack that has $\beta \geq \alpha$ is said to be *self-locking*, a desirable property for a jack to have.

For a system of this type we can define an efficiency η as the ratio of work input to useful work output (the difference being due to wasted frictional heating). Here, the work input per revolution is $2\pi M$, while the useful work of raising the weight is pW; thus

$$\eta = \frac{pW}{2\pi M} = \frac{\tan\alpha}{\tan(\alpha + \beta)} \qquad (d)$$

We see that the efficiency for small α and β is approximately

$$\eta \approx \frac{\alpha}{\alpha + \beta}$$

And thus for a self-locking device, $\beta \geqq \alpha$, the efficiency cannot surpass 50 percent.

Example 1.7 In Fig. 1.30*a* a light stepladder is shown resting on the floor. We wish to estimate the force in link AB when a 210-lb man stands on top of the ladder.

We begin this example by selecting the ladder and idealizing its characteristics in Fig. 1.30*b*. Many idealizations have been made. In view of the lightness of the ladder compared with the man, the ladder has been represented by *weightless* bars in the plane of the sketch. The joints at A, B, and C, which in most ladders are relatively free, have been idealized as *frictionless* pinned joints. The interactions between the legs and the floor, which in practice can have a large variation in friction, have been taken as *frictionless*, and therefore the reactions N_1 and N_2 have been shown normal to the floor. The effect of the man's weight is indicated by the 210-lb vertical force at C. These idealizations permit us to obtain a good first approximation of the actual case where the ladder is light but not weightless, where the floor is not completely frictionless, where the pinned joints are reasonably free but not completely frictionless, etc. To illustrate the effect of neglecting friction between the floor and the ladder, we shall later repeat the analysis for the case where there is a coefficient of friction of 0.20.

First, considering the frictionless case of Fig. 1.30, we study the forces and equilibrium requirements. Since the desired result is the force in AB,

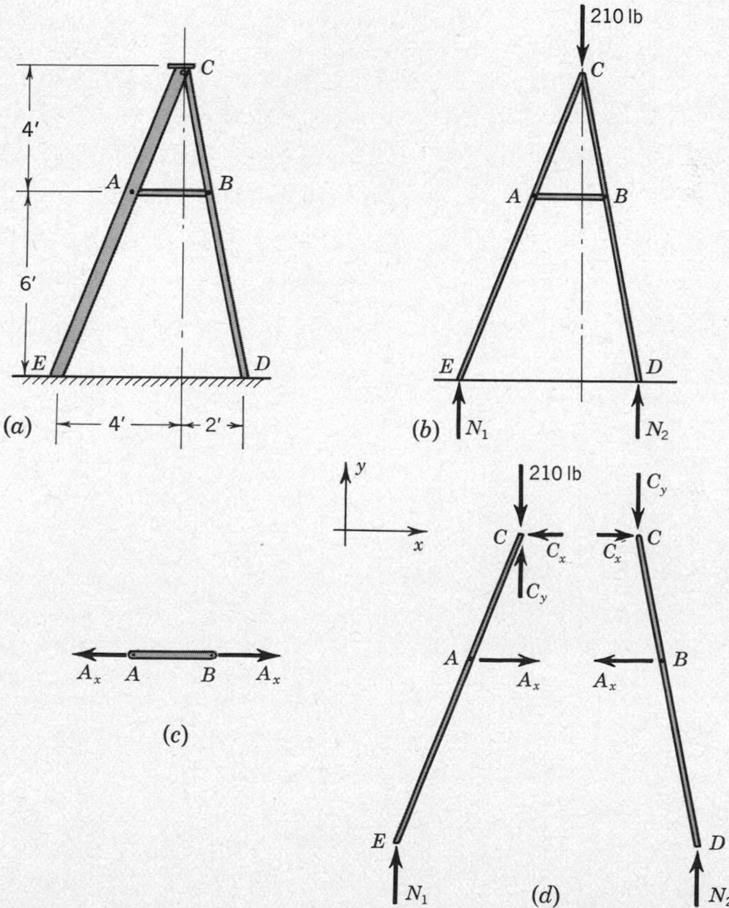

Fig. 1.30 Example 1.7.

we begin by isolating the link AB in the free-body diagram of Fig. 1.30*c*. Because AB is a two-force member, the forces at A and B must be equal and opposite and directed along AB. We cannot, however, find the numerical value of A_x by considering only the free body of Fig. 1.30*c*.

We next try isolating other parts of the system. In Fig. 1.30*d* each leg of the ladder is isolated. The pin force at C is represented by the unknown components C_x and C_y. Note that the unknown forces A_x, C_x, and C_y are shown consistently according to Newton's third law; e.g., the action of the leg CE on the link AB is A_x to the left in Fig. 1.30*c*, while the action of the link AB on the leg CE is A_x to the right in Fig. 1.30*d*. We do not know the true sense of A_x (it may be negative), but we *are* consistent.

The isolated free bodies in Fig. 1.30*d* are planar systems, and thus there are three independent equilibrium conditions which may be applied to each. There are, however, *four* unknown magnitudes in each free body, so that we cannot obtain A_x by considering only one of the legs as a free body. We *can* get enough equations by writing the equilibrium equations for *both* legs. An alternate procedure is to make use of Fig. 1.30*b* in which the entire structure is isolated. The external forces, N_1, N_2, and the 210-lb load must satisfy the equilibrium conditions (1.7) and (1.8). For the moment about D to vanish we must have

$$\begin{aligned} \Sigma M_D &= 6N_1 - 2(210) = 0 \\ N_1 &= 70 \text{ lb} \end{aligned} \tag{a}$$

Returning to leg CE in Fig. 1.30*d* with $N_1 = 70$ lb, we determine A_x most simply by asking for the moment about C to vanish.

$$\begin{aligned} \Sigma M_C &= 4N_1 - 4A_x = 0 \\ A_x &= N_1 = 70 \text{ lb} \end{aligned} \tag{b}$$

Thus the tension in the link is 70 lb.

Let us reexamine this example for the case where there is friction between the ladder and the floor. In Fig. 1.31 we show isolated free bodies of the entire ladder and of the left leg with friction forces F_1 and F_2 acting. By applying (1.7) and (1.8) to Fig. 1.31*a*, we learn that $F_1 = F_2$, that Eq. (*a*) still holds, and that $N_2 = 140$ lb. The equilibrium requirements do not, however, fix the magnitude or sense of the friction force. The law of static friction tells us the upper limit of the magnitude. If the coefficient of friction is $f_s = 0.20$, we must have

$$\begin{aligned} |F_1| &\leqq 0.20(70) = 14 \text{ lb} \\ |F_2| &\leqq 0.20(140) = 28 \text{ lb} \end{aligned} \tag{c}$$

and since $F_1 = F_2$, this becomes

$$|F_1| = |F_2| \leqq 14 \text{ lb} \tag{d}$$

The law of static friction does not tell us in which sense the friction acts
Actually, the force F_1 in Fig. 1.31 can have any value satisfying

$$-14 \text{ lb} \leqq F_1 \leqq 14 \text{ lb} \tag{e}$$

depending upon the previous history of the ladder. For example, suppose that with the man on the ladder the link AB had been disconnected and the legs allowed to spread a little, and then external agents had slowly forced the legs back together until AB could just be reconnected. Under these circumstances $F_1 = 14$ lb. On the other hand, if AB had been disconnected and the legs forced inward and then slowly pulled out again until AB could just be reconnected, we would have $F_1 = -14$ lb.

Without more specific information all we can do is solve for A_x in terms of F_1 and obtain the range of values for A_x which corresponds with the range (*e*). Taking moments about C in Fig. 1.31*b* yields

$$\begin{aligned} \Sigma M_C &= 4A_x - 70 - 10F_1 = 0 \\ A_x &= 70 + 2.5F_1 \end{aligned} \tag{f}$$

When F_1 lies in the range (*e*), the force A_x in the link AB lies in the range

$$35 \text{ lb} \leqq A_x \leqq 105 \text{ lb} \tag{g}$$

Thus the presence of friction can, at worst, give rise to a 50 percent increase in A_x as compared with the frictionless case.

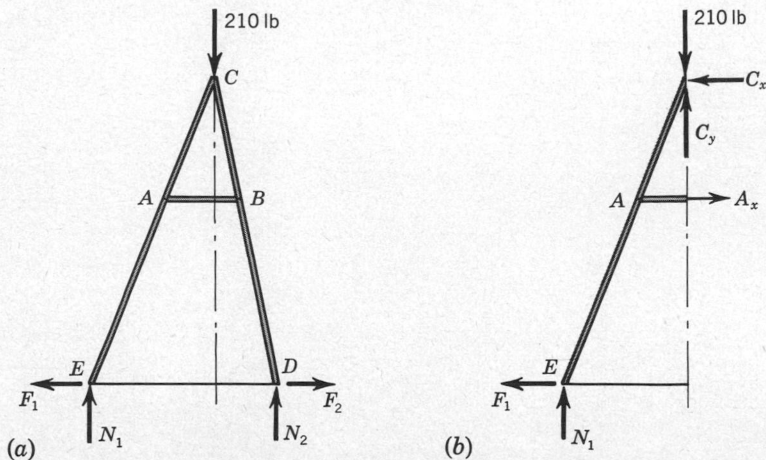

Fig. 1.31 Introduction of friction between the ladder and floor.

1.10 HOOKE'S JOINT

In this section we are going to discuss the problem of a universal joint. Figure 1.32 shows a Hooke's joint (universal joint) which is sometimes used to transmit torque between two shafts that meet at an angle. The shafts A and B lie in the xz plane and make the angle θ as shown. It is required to estimate the torque M_B to *balance* a given torque M_A for the configuration shown under the assumption that friction between the moving parts can be neglected. We shall first show how an over-idealized model coupled with an incomplete analysis can lead to—not surprisingly—an incorrect conclusion. That is, we will first solve the problem incorrectly.

Incorrect solution In Fig. 1.33*a* a free-body diagram of both shafts and the connecting cross is shown. If we neglect the weights of the parts, the external interactions are the torques $\mathbf{M}_A$ and $\mathbf{M}_B$ and the bearing reactions at A and B. Let us first make the idealization that these bearing reactions may be taken, as in Table 1.2*d*, as single forces perpendicular to the surface of the shaft. These forces would then have to lie in the shaded planes shown in Fig. 1.33*a*. We have shown these reactions split into horizontal and vertical components: H_A and V_A at A, and H_B and V_B at B.

There are five unknown scalar components in Fig. 1.33*a*: the four forces and the magnitude M_B of the moment $\mathbf{M}_B$. We can solve for these by applying the conditions of equilibrium (1.5) and (1.6). It is not difficult to show that the conditions of force balance (1.5) lead to $H_A = H_B = 0$ and $V_B = -V_A$. Using Fig. 1.33*b*, we see that the pair of bearing forces separated by the distance d is equivalent to a couple of magnitude $V_B d$. In Fig. 1.33*c* this couple is combined vectorially with the known $\mathbf{M}_A$ and the given direction of $\mathbf{M}_B$ to form a closed triangle representing the condition (1.6) for balance of moments. From this

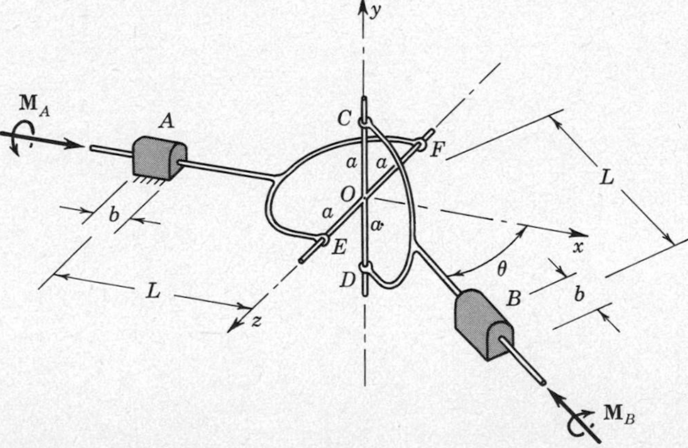

Fig. 1.32 Hooke's joint mechanism.

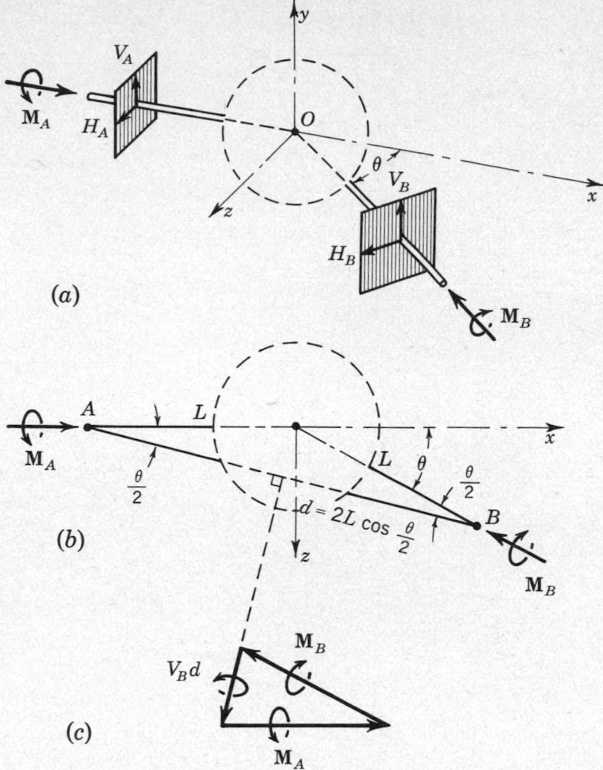

Fig. 1.33 Isolation of entire mechanism.

triangle it is clear that

$$M_B = M_A \tag{a}$$

Alternatively, if we proceed directly with moment equilibrium about the point O, we have

$$M_A\mathbf{i} - M_B(\mathbf{i}\cos\theta + \mathbf{k}\sin\theta) - L\mathbf{i} \times V_A\mathbf{j} - L(\mathbf{i}\cos\theta + \mathbf{k}\sin\theta) \times V_A\mathbf{j} = 0$$

Upon working out the cross products, we find

$$V_A L\,(1 + \cos\theta) = -M_B \sin\theta$$

$$M_A + V_A L\sin\theta = M_B\cos\theta$$

from which we may find M_B and V_A in terms of M_A. After a little algebra we find the result (a) as above and

$$V_A L = \frac{-M_A\sin\theta}{1 + \cos\theta}$$

Remember, this is an *incorrect* solution! But where did the analysis go wrong?

In the foregoing analysis we used only the free body of both shafts and the cross taken together. Suppose we were to continue our analysis further by considering the shaft B separately as in Fig. 1.34. We show the torque $\mathbf{M}_B$ and the force V_B already discussed. In addition we show the forces at C and D. Here

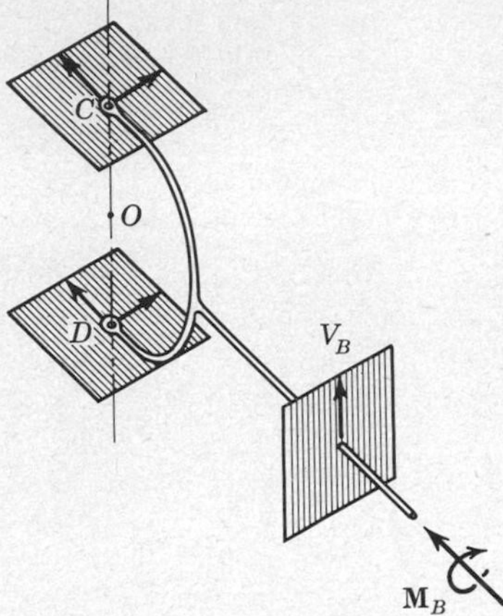

Fig. 1.34 Isolated free-body diagram of the shaft B.

again, relying on the idealization of Table 1.2*d*, the interactions must be single-contact forces perpendicular to the vertical arm of the cross, i.e., they must lie in the horizontal shaded planes at C and D in Fig. 1.34. Now, suddenly, we are struck by a contradiction. The system of Fig. 1.34 cannot possibly be in equilibrium because there is nothing to balance V_B in the vertical direction. Thus, although Fig. 1.33 is in equilibrium, the individual elements are *not*. Our idealizations of the nature of the bearing interactions have led us to a contradiction; that is, we made a mistake. An agonizing reappraisal is necessary.

Correct solution Possibly our first suggestion would be to permit vertical force components at C and D (these could occur if there were shoulders on the vertical crossarm). By including vertical forces at C and D it would be possible to have equilibrium in Fig. 1.34. However, if we continued on to the cross, we would find that the cross could not possibly be in equilibrium under these circumstances.

We then return to the bearing at B. After some consideration we postulate the behavior sketched in Fig. 1.35. The shaft tends to tip or cock in the bearing, resulting in a pair of contact forces $\mathbf{F}_1$ and $\mathbf{F}_2$. The cocking can take place in any plane through the axis of the shaft.

Introducing this new model of the bearing behavior at B (and A) requires a fresh start on the analysis. This time we consider isolated free bodies of the connecting cross and of each shaft separately in Fig. 1.36. To provide for the double-contact type of reaction shown in Fig. 1.35, we have shown the four com-

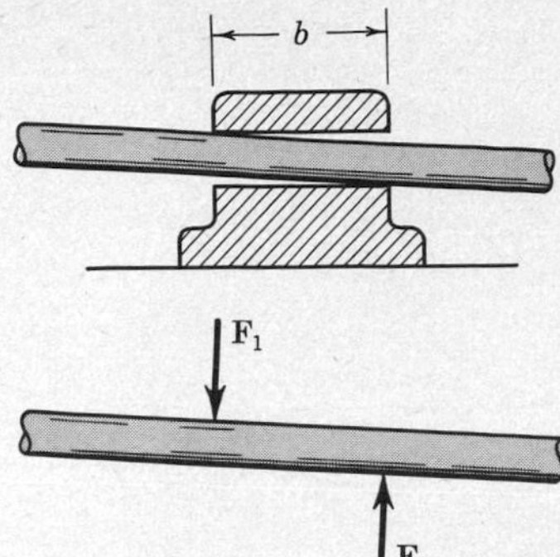

Fig. 1.35 When a shaft cocks slightly in a long bearing, there is double contact with two reactions F_1 and F_2.

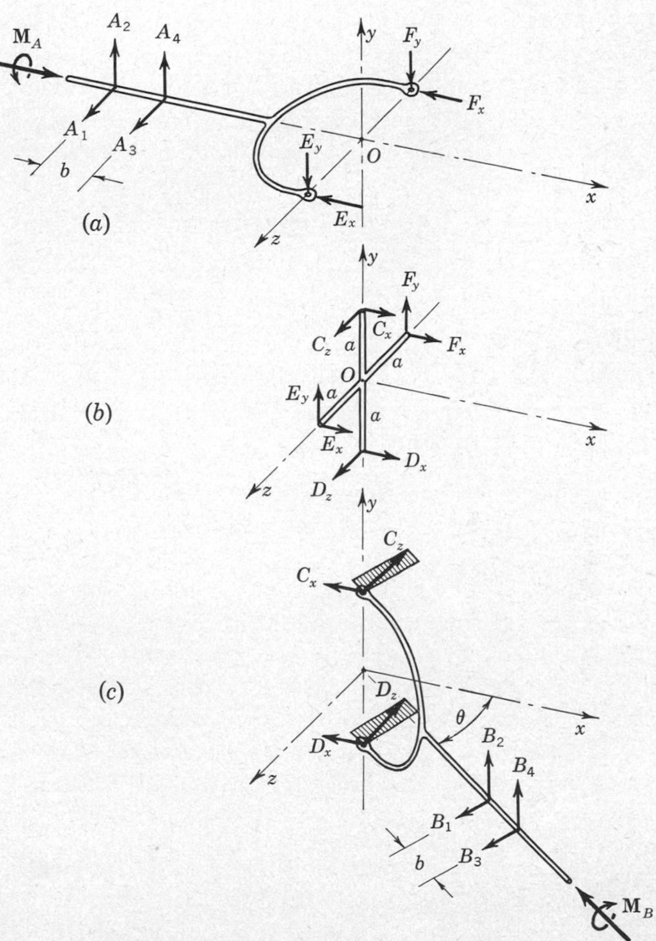

Fig. 1.36 Free-body diagrams of the three parts of the Hooke's joint.

ponents A_1, A_2, A_3, and A_4 at bearing A and the four components B_1, B_2, B_3, and B_4 at bearing B. Since at this stage we do not know in which direction the shafts will cock, we have assigned arbitrary directions to these components.

The interactions at the cross have, however, been taken, as before, to be single-contact forces perpendicular to the crossarms. This is based on the assumption that there are no shoulders on the crossarms. Furthermore, if the arm CD tends to cock in its bearing at C, we assume that the bearing at D provides sufficient restraint so that the double contact of Fig. 1.35 cannot occur. Note that the interactions at C, D, E, and F are shown consistently in the three free-body diagrams of Fig. 1.36. For example, the force component C_x acting on the cross in Fig. 1.36*b* is equal and opposite to the force component C_x acting on the shaft B in Fig. 1.36*c*.

There are a total of 16 unknown force components and one unknown moment component in the three free-body diagrams of Fig. 1.36. The conditions for equilibrium (1.5) and (1.6) are equivalent to six scalar equations for each free body, or a total of 18 equations. Systematic application of the conditions for equilibrium will provide a complete solution for all the unknown force and moment components.

For example, the statements of force and moment balance for the cross (Fig. 1.36*b*) are

$$\begin{aligned}\Sigma\mathbf{F} &= (C_x + D_x + E_x + F_x)\mathbf{i} + (E_y + F_y)\mathbf{j} + (C_z + D_z)\mathbf{k} = 0 \\ \Sigma\mathbf{M}_0 &= (aC_z - aD_z - aE_y + aF_y)\mathbf{i} + (aE_x - aF_x)\mathbf{j} \\ &\qquad + (-aC_x + aD_x)\mathbf{k} = 0\end{aligned} \tag{b}$$

Setting each component separately equal to zero yields six equations for eight unknowns. The best we can do is to express all eight unknowns in terms of two. It is simple to obtain the relations

$$\begin{aligned}C_x &= D_x = -E_x = -F_x \\ C_z &= -D_z = E_y = -F_y\end{aligned} \tag{c}$$

from (*b*). These may be considered as giving all the cross interactions in terms of F_x and F_y.

We can evaluate F_x and F_y by considering the free-body diagram of the shaft A in Fig. 1.36*a* and making use of results (*c*). For example, equilibrium of forces parallel to x yields $F_x = 0$, and equilibrium of moments about the x axis yields $2aF_y = M_A$. Inserting in (*c*) gives

$$\begin{aligned}C_x &= D_x = -E_x = -F_x = 0 \\ C_z &= -D_z = E_y = -F_y = -\frac{M_A}{2a}\end{aligned} \tag{d}$$

We are now ready to go to Fig. 1.36*c* and evaluate M_B. If, however, for completeness we were to consider the other equilibrium conditions in Fig. 1.36*a*, we would find that $A_1 = A_2 = A_3 = A_4 = 0$.

Using the values (*d*) in Fig. 1.35*c*, we evaluate M_B from the condition of moment equilibrium around the shaft axis.

$$M_B - aD_z \cos\theta + aC_z \cos\theta = 0$$
$$M_B = M_A \cos\theta \qquad (e)$$

For completeness the other equilibrium conditions can be used to obtain $B_1 = B_3 = 0$ and $B_4 = -B_2 = (M_A/b) \sin\theta$. To aid visualization the complete solution is shown in Fig. 1.37. Compare this figure with Fig. 1.33 and note how

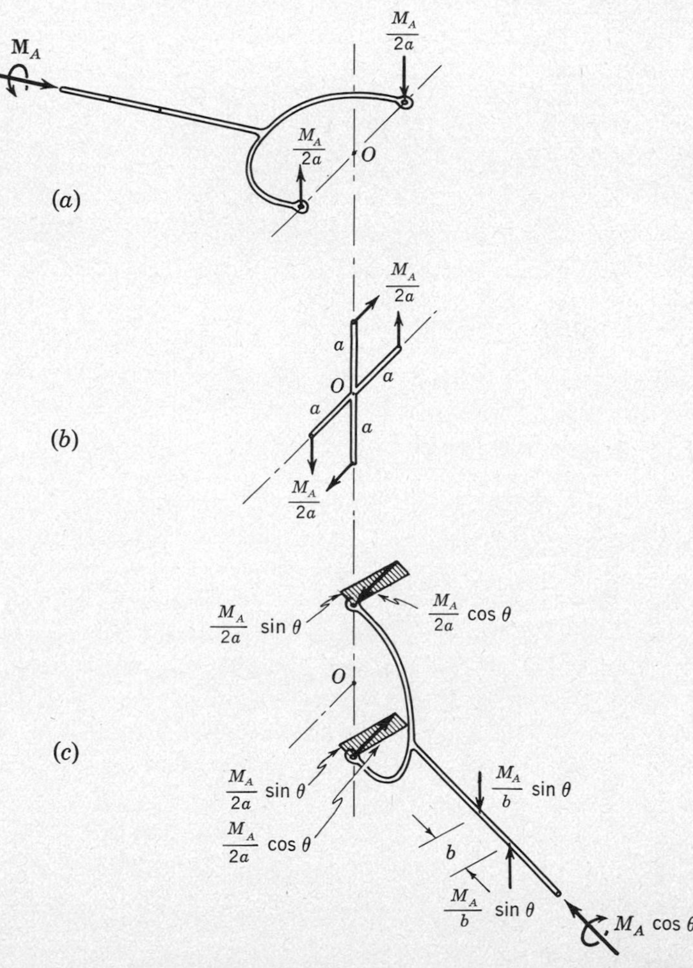

Fig. 1.37 Equilibrium solution of the free-body diagrams of Fig. 1.36.

the difference in type of bearing reaction has led to the different solutions of (*a*) and (*e*) above. In both solutions there is equilibrium of the entire assembly. In the first solution our assumption about the nature of the interactions in the cross and at the bearings made it impossible for the individual elements to be in equilibrium. In the second solution equilibrium of all parts was achieved by altering the assumptions regarding the bearing interaction. This second solution does give an accurate representation of the behavior of the Hooke's joint.

It is important to emphasize that our solution (*e*) is for the configuration shown in Fig. 1.32. If we twist the shaft A through some angle such that the crosspiece CD is no longer vertical, then a new analysis is necessary to obtain the relation between M_A and M_B for equilibrium. An exact solution for arbitrary angle of orientation ϕ of shaft A measured from EF in the direction of twist of M_A (Fig. 1.32) can be found.[1] The result is

$$M_B = \frac{\sin^2 \phi + \cos^2 \theta \cos^2 \phi}{\cos \theta} M_A \qquad (f)$$

When $\phi = 0$, the result (*f*) reduces to (*e*).

It is of interest to sketch the variation of (M_B/M_A) with ϕ in order to see the relation between input and output moments. However, we will leave this to the interested reader. It may also be of interest to build a model of a Hooke's joint to show the configuration and to run some simple quantitative experiments to verify (*f*).

1.11 FINAL REMARKS

Now that we have worked a number of examples, it is of value to review the procedures we discussed in Sec. 1.2 for the solution of problems. The general method of attack in solving problems of applied mechanics is similar to that in any branch of applied science. Most problems are complicated enough so that they can be approached in a variety of ways. No universal method of attack can be given, but it is possible to give a general outline of the steps. Such an outline should be used consciously whenever there is difficulty in obtaining a solution. Revising the outline in view of problems you have tried or solved will not only be of aid to further work but will be of philosophical and psychological interest. A tentative outline follows:

1. Define your objective either in terms of variables or relations; i.e., what is the problem?

[1] J. L. Synge and B. A. Griffith, "Principles of Mechanics," 3d ed., p. 267, McGraw-Hill Book Company, New York, 1959.

2. Select, define, and sketch the system of interest so that it involves as many of the desired variables and given data as possible, and as few other variables or quantities. It is a good practice to draw diagrams to scale so that geometric relations become more evident.
3. Postulate the characteristics of the system. This usually involves idealization and simplification of the real situation and specification and/or elimination of some variables.
4. Apply the principles of mechanics to the idealized model. Compare the number of unknowns with the number of *independent* equations to determine whether a solution can be obtained. If not, and you are sure that a complete set of physical principles and equations has been applied, return to step 2 and select additional or alternative systems to consider.
5. Reduce the desired result preferably to symbolic form. Check the equation for its behavior as the values of the variables are changed. Do they go to the right limiting cases? Are the equations dimensionally consistent?
6. Substitute numerical values. Compare the result with what you would expect from intuition. You should form the habit of scrutinizing the answer to every engineering calculation to see if the result is in accord with your intuition. Does it make sense? Is it the right order of magnitude? This is difficult in the beginning when one's intuition is undeveloped, but it is just this process which develops the intuition and forms the foundation for future judgments. A good engineer gains something from every computation he makes. For instance, would you expect a bar of certain dimensions to support a mosquito, a mouse, a man, a car, or a truck? Reexamine the idealizations and the assumptions in the light of the results; for instance, is it still reasonable to neglect friction in the light of the normal forces which are now known?
7. Compare the predictions with the behavior in the actual system by test and experiment.
8. If at any point satisfactory results are not achieved, reconsider the previous steps. A frequent difficulty is a failure to select an appropriate system or systems and to define the actions on it by its surroundings. Alternatively, it may be necessary to alter the assumptions regarding the characteristics of the system, i.e., to construct a different idealized model of the system.

PROBLEMS

1.1. The angles between the vector $\mathbf{F} = F_x\mathbf{i} + F_y\mathbf{j} + F_z\mathbf{k}$ and the coordinate axes are θ_x, θ_y, and θ_z. The cosines of these angles are known as *direction cosines*. Evaluate the direction cosines in terms of the components of **F**. Show that

$$\cos^2\theta_x + \cos^2\theta_y + \cos^2\theta_z = 1$$

1.2. A vector $\mathbf{F} = F_x\mathbf{i} + F_y\mathbf{j} = F_a\mathbf{a} + F_b\mathbf{b}$, where **i**, **j** and **a**, **b** are pairs of perpendicular unit vectors in a plane.

Show that

$$F_a = F_x \cos\theta + F_y \sin\theta$$
$$F_b = -F_x \sin\theta + F_y \cos\theta$$

and

$$F_x = F_a \cos\theta - F_b \sin\theta$$
$$F_y = F_a \sin\theta + F_b \cos\theta$$

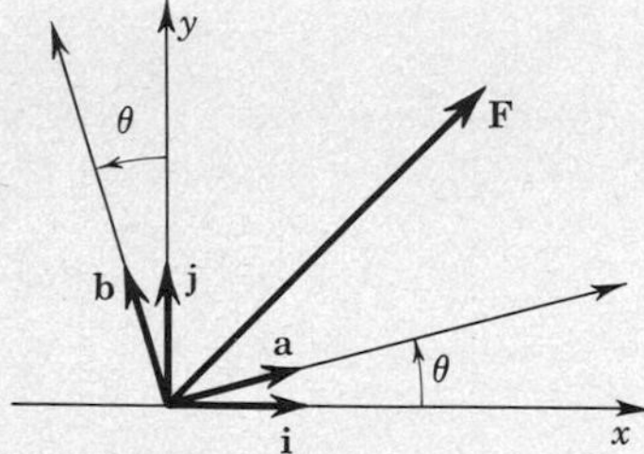

Prob. 1.2

These equations may also be written in the matrix form

$$\begin{bmatrix} F_a \\ F_b \end{bmatrix} = \begin{bmatrix} \cos\theta & \sin\theta \\ -\sin\theta & \cos\theta \end{bmatrix} \begin{bmatrix} F_x \\ F_y \end{bmatrix} \qquad \begin{bmatrix} F_x \\ F_y \end{bmatrix} = \begin{bmatrix} \cos\theta & -\sin\theta \\ \sin\theta & \cos\theta \end{bmatrix} \begin{bmatrix} F_a \\ F_b \end{bmatrix}$$

or

$$\mathbf{F}' = \mathbf{TF} \qquad \mathbf{F} = \mathbf{T}^{-1}\mathbf{F}' \qquad \mathbf{T}^{tr} = \mathbf{T}^{-1}$$

1.3. According to the distributive law for vector cross products

$$\mathbf{r} \times \mathbf{F}_1 + \mathbf{r} \times \mathbf{F}_2 = \mathbf{r} \times (\mathbf{F}_1 + \mathbf{F}_2)$$

This states that the sum of the moments of two concurrent forces about a point is equal to the moment of the vector sum of the forces about the same point. Verify this by simple geometry in the special case where $\mathbf{F}_1$ and $\mathbf{F}_2$ lie in the xy plane and intersect at the point P, and **r** is the displacement vector OP.

1.4. Let **F** be a force vector which passes through the point $P(x,y,z)$ and which has components F_x, F_y, and F_z. Show that the moment of **F** about the origin of coordinates O can be represented by the determinant

$$\mathbf{M}_o = \begin{vmatrix} \mathbf{i} & \mathbf{j} & \mathbf{k} \\ x & y & z \\ F_x & F_y & F_z \end{vmatrix}$$

1.5. Let **F** be an arbitrary vector through the origin and let **a**, **b**, **c** be three arbitrary noncoplanar *unit* vectors passing through the origin. It is desired to decompose **F** into vectors parallel to **a**, **b**, and **c**; that is, it is desired to find magnitudes L_a, L_b, and L_c such that

$$\mathbf{F} = L_a\mathbf{a} + L_b\mathbf{b} + L_c\mathbf{c}$$

Sketch this and show that it involves a parallelepiped whose edges are parallel to **a**, **b**, and **c** with **F** as the diagonal. Use the properties of the scalar triple product to show that

$$L_a = \frac{\mathbf{F} \times \mathbf{b} \cdot \mathbf{c}}{\mathbf{a} \times \mathbf{b} \cdot \mathbf{c}}$$

with similar expressions for L_b and L_c.

1.6. Find the force and moment which must be applied at O to hold the light bar shown in equilibrium.

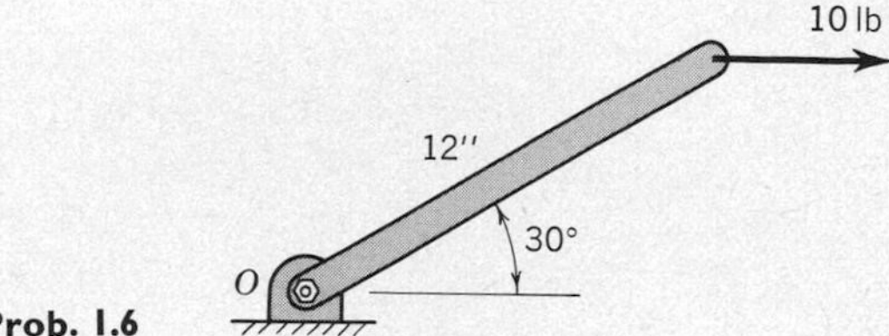

Prob. 1.6

1.7. For a set of coplanar forces show that any of the following alternative conditions for equilibrium imply Eqs. (1.7) and (1.8).

(*a*) The sum of all forces in any *two* nonparallel directions in the plane is zero, and the resultant moment about any point in the plane is zero.

(*b*) The total moments of all forces about *two* points P and Q in the plane are zero, and the sum of the force components parallel to PQ is zero.

(*c*) The total moments of all forces about *three* noncollinear points O, P, and Q are zero.

Note that the conclusion here for coplanar forces is that there exist three *independent* equations from equilibrium considerations. If three equations are *not* sufficient for the determination of all the unknowns, the system is *statically indeterminate*.

1.8. Find the reactive forces and the moment at the wall for the cantilever beam supported as shown in the figure.

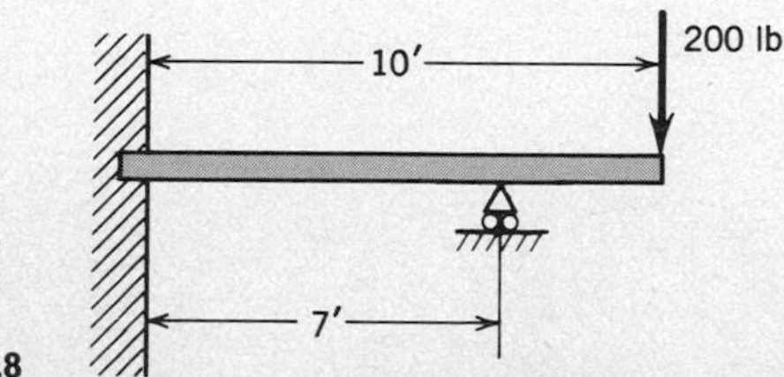

Prob. 1.8

1.9. Two equal cylinders, each weighing 200 lb, are placed in a box as shown. Neglecting friction between the cylinders and the box, estimate the reactions at A, B, and C.

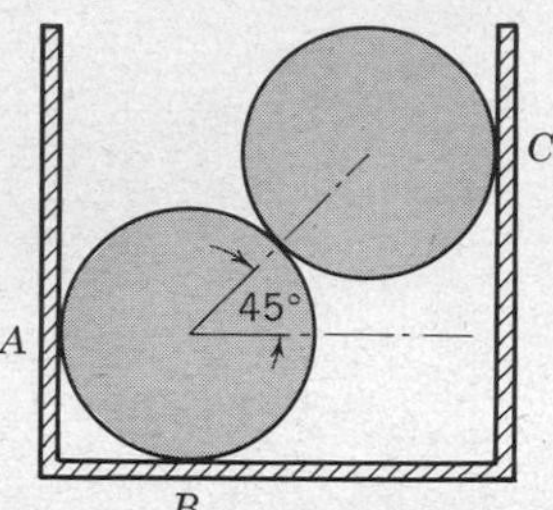

Prob. 1.9

1.10. Frequently, a force **F** and a moment **M** act at the same point as shown in the figure for a coplanar system. Show that a coplanar force and moment may be replaced with an equal force displaced sideways a specified distance $a = |\mathbf{M}|/|\mathbf{F}|$. The steps are shown in the figure.

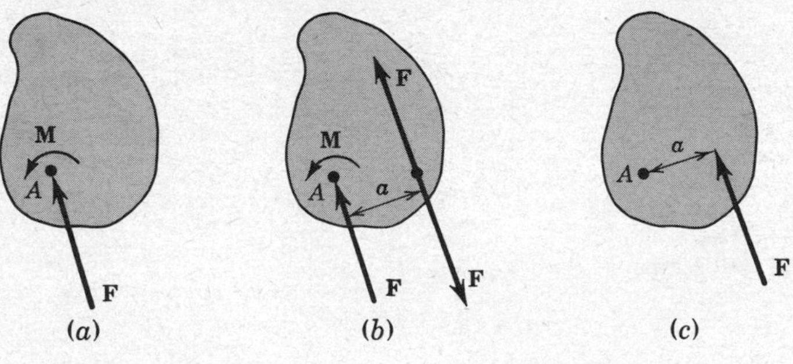

Prob. 1.10

1.11. Find the force carried in each bar of the hinged equilateral triangle when loaded as shown.

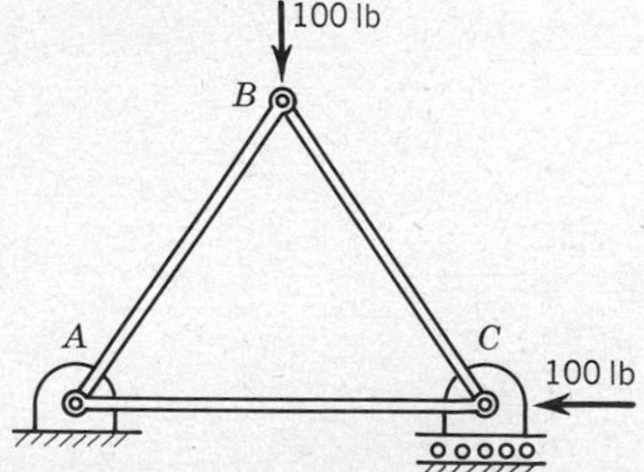

Prob. 1.11

1.12. Estimate the force in link AB when the weight of the boat supported by the davit is 1,500 lb.

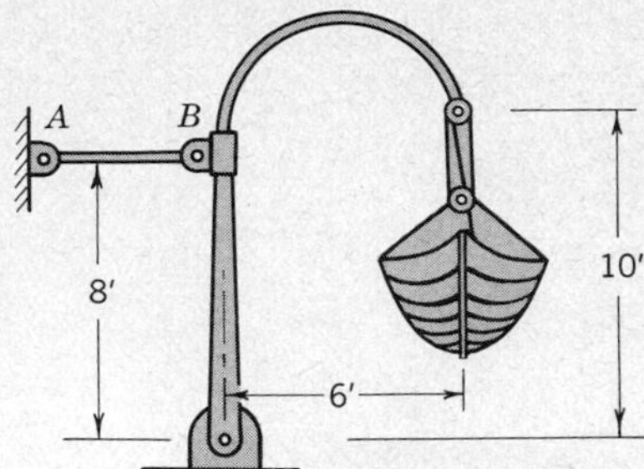

Prob. 1.12

1.13. Compare the forces F required to just start the 200-lb lawn roller over a 3-in. step when (*a*) the roller is pushed and (*b*) the roller is pulled.

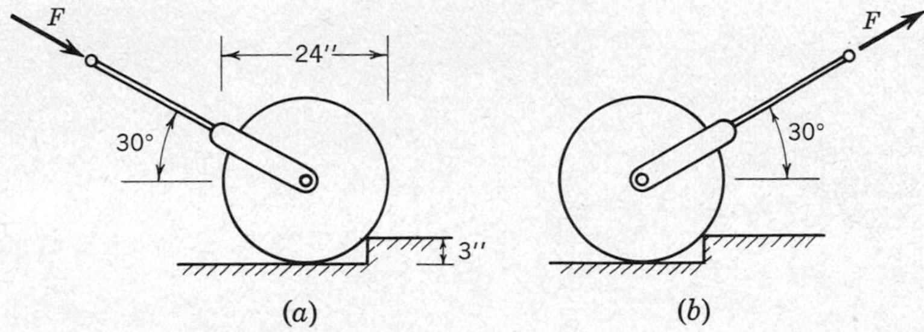

Prob. 1.13

1.14. The bracket ABC is free to swing out horizontally on the vertical rod. Estimate the forces transmitted to the vertical rod at A and B when a 200-lb load is supported at C. Show magnitudes and directions on a clear sketch.

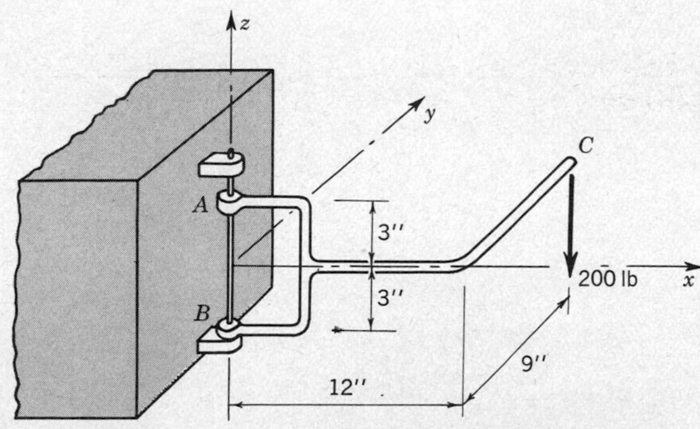

Prob. 1.14

1.15. A 25-lb force is required to operate the foot pedal as shown. Determine the force in the connecting link and the force exerted by the lever on the bearing at O. Neglect the weight of the lever.

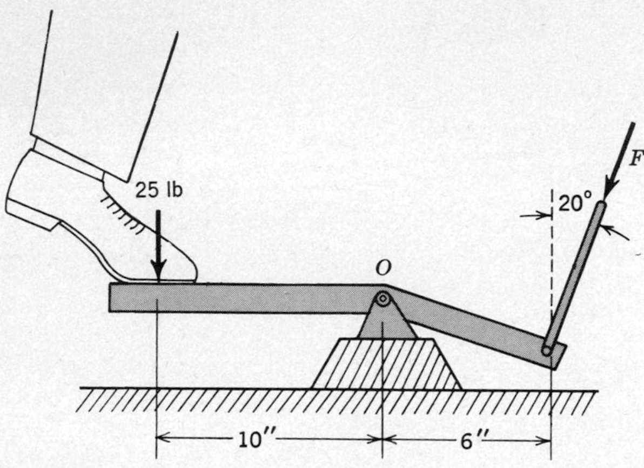

Prob. 1.15

1.16. A spot weld which holds the bracket to the plate at point A as shown in the figure can withstand a maximum twist in the plane of the plate of 80 ft-lb. Determine the maximum load W.

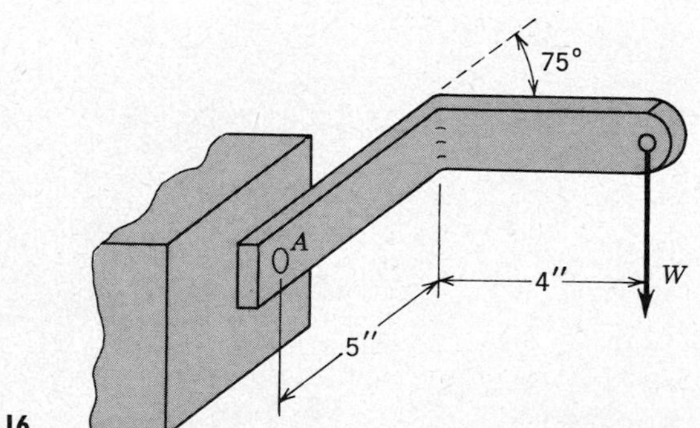

Prob. 1.16

1.17. Find the forces in the remaining bars of Example 1.4.

1.18. During a hurricane, some of the wires attached to a power pole are broken so that the loading of the pole is as shown in the sketch. There are two wires still attached to the crossarm, exerting loads of 400 lb and 500 lb parallel to the x axis. There is a transformer weighing 1,000 lb whose center of gravity lies in the yz plane a distance of 20 ft above the ground and 2 ft from the center of the pole. Neglect the weight of the pole. The pole is buried in the ground for a depth of 8 ft. Find the forces and moments which act *on* the buried section GA at the ground level G.

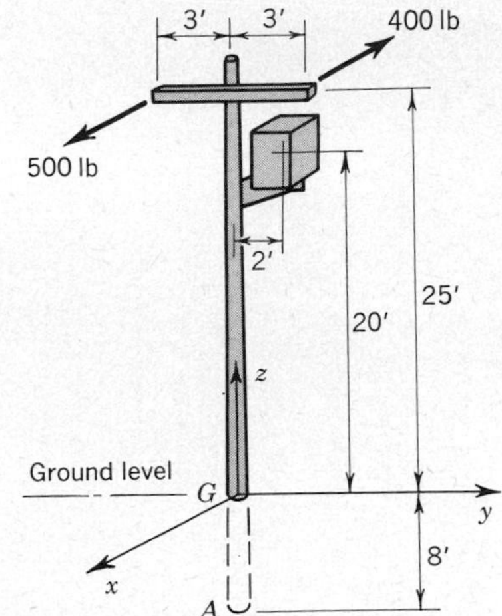

Prob. 1.18

1.19. An airplane engine pod is suspended from the wing by the strut AG shown. The propeller turns clockwise when viewed from behind. The weight of the engine is 2,500 lb and may be assumed to act at G. Find the force and moment exerted *by the strut onto the wing* at A when the engine is delivering 4,000 lb thrust and 15,000 ft-lb of torque.

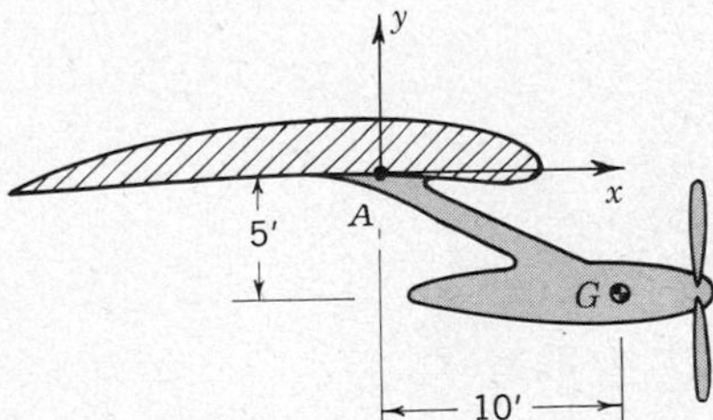

Prob. 1.19

1.20. The top of a tin can is removed, and the empty can is inverted over a pair of billiard balls on a table as shown in the sketch. For certain combinations of sizes and weights the configuration shown is stable. For other combinations the can tips over when released. It is proposed to set up a demonstration for a temperance lecture by using in sequence a frozen-orange-juice can and a beer can with the same pair of billiard balls. Investigate whether tipping will occur for the following sizes and weights.

	Orange juice	*Beer*
Diameter of ball	1¾″	1¾″
Weight of ball	7 oz	7 oz
Diameter of can	2″	2½″
Weight of empty can with lid removed	2 oz	3½ oz

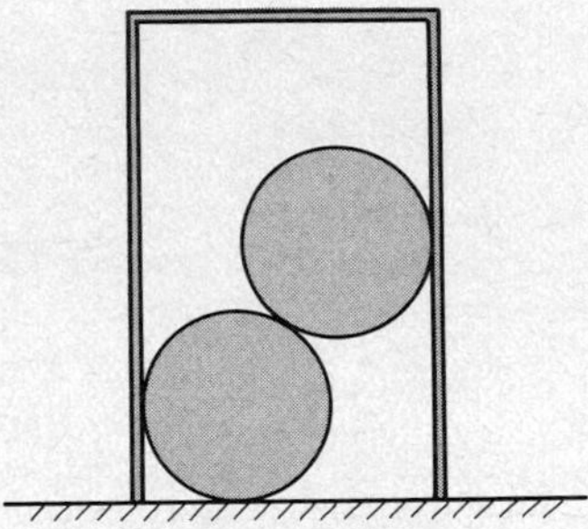

Prob. 1.20

1.21. A rigid rod with negligible weight and small transverse dimensions carries a load W whose position is adjustable. The rod rests on a small roller at A and bears against the vertical wall at B. Determine the distance x for any given value of θ such that the rod will be in equilibrium. Assume that friction is negligible.

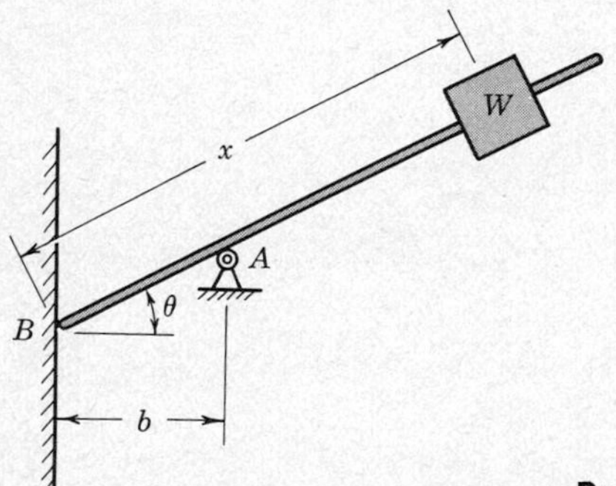

Prob. 1.21

1.22. A light frame is hinged at A and B and held up by a temporary prop at C. Find the reactions at A, B, and C when a 2,000-lb load is supported at D.

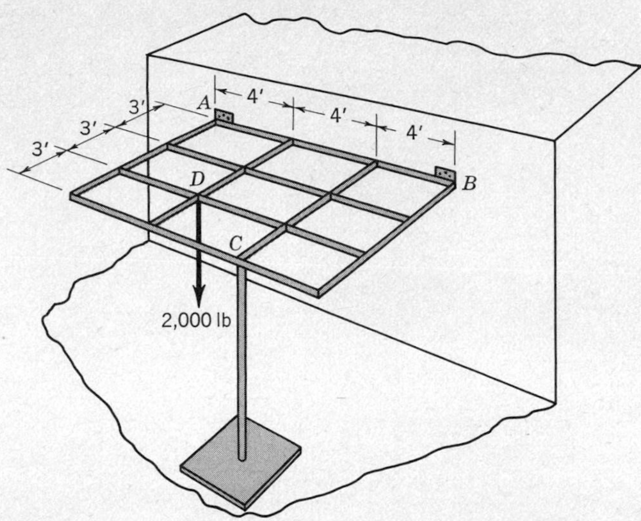

Prob. 1.22

1.23. It is desired to lift the wheelbarrow shown with one hand at the handle A by applying at A a vertical force **F** and a twisting moment **M** about the axis of the handle. Estimate the magnitudes of **F** and **M**.

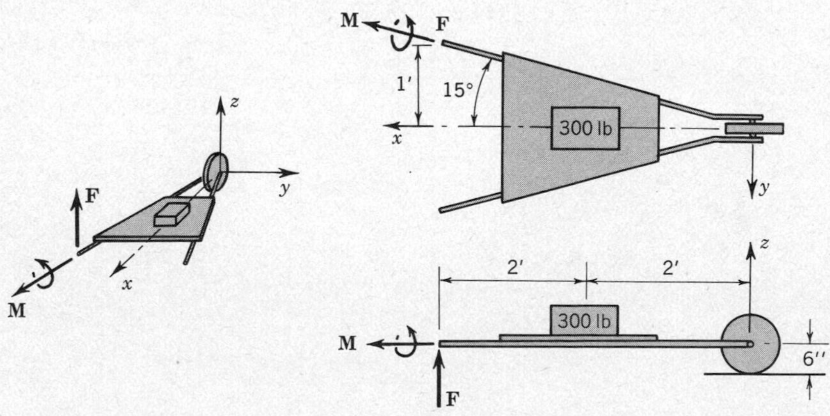

Prob. 1.23

1.24. In building construction it is common to build a floor or a roof on temporary supports which permit "leveling up" before setting the permanent columns in place. The sketch below shows one of the ways in which this "leveling up" is performed. The temporary column C supports a weight of 2,000 lb. Driving in the wedge at B lifts one end of the rigid bar AB and hence lifts C by half as much.

(*a*) Estimate the minimum force between hammer and wedge which will cause the wedge to move farther in, assuming all coefficients of friction are 0.3.

(*b*) What would happen if the coefficients of friction were too small? What is the value of the coefficient of friction which marks the border line between desirable and undesirable performance?

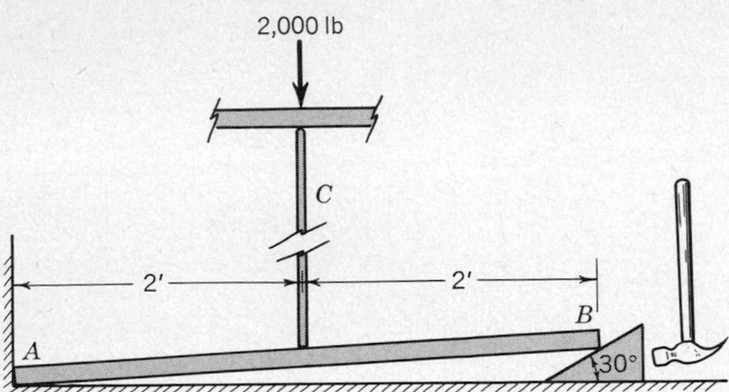

Prob. 1.24

1.25. A freely pivoted light rod of length l is pressed against a rotating wheel by a force P applied to its middle. The friction coefficient between the rod and wheel materials is f. Compute, for both directions of rotation, the friction force F as a function of the variables l, P, and f, and any others which are relevant. One of these two situations is sometimes referred to as a *friction lock*. Which one, and why?

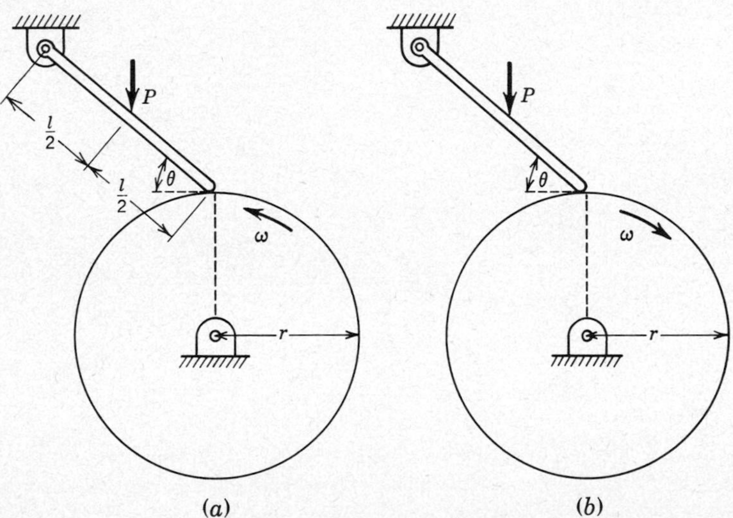

Prob. 1.25

1.26. The drawing shows a section through the latch of a screen door.

(*a*) Find the force P required to just start the latch sliding under the following assumptions:

1. The coefficient of friction between all surfaces shown in the sketch is 0.3.
2. The hinges of the door are well oiled, and their distance from the latch is large compared with the dimensions in the sketch.
3. The spring of the latch exerts a force of 1 lb, due to its compression, along its center line.

(*b*) What would P be if there were no friction?

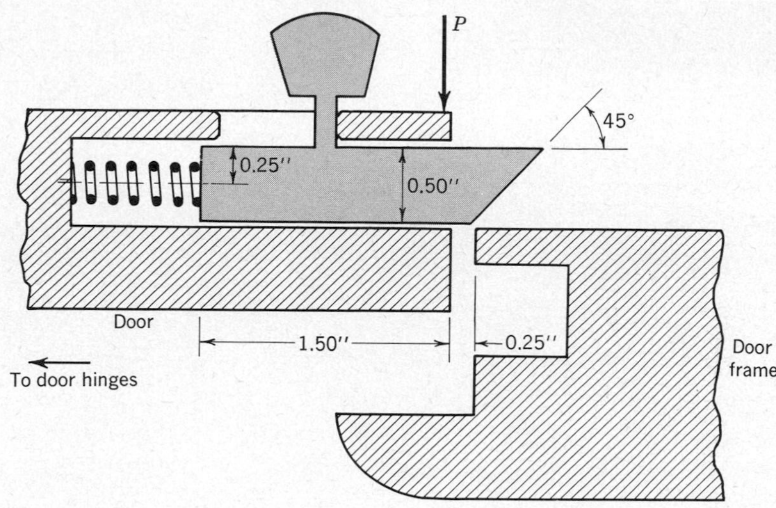

Prob. 1.26

1.27. A four-engine jet transport, which weighs 230,000 lb fully loaded, has its center of gravity at the location shown in the sketch. Before taking off for Europe the pilot must test the engines by operating them, one at a time, at a thrust of about 8,000 lb. As he checks the left outboard engine, the other three engines idle at negligible thrust. The rear-wheel brakes are locked during the test, but the nose wheel has no brakes. In addition the nose wheel is mounted on a caster, so it cannot resist a sidewise force.

(*a*) What forces does the ground exert on the landing wheels during the test?

(*b*) What must the coefficient of friction between ground and wheels be to prevent the rear wheels from slipping?

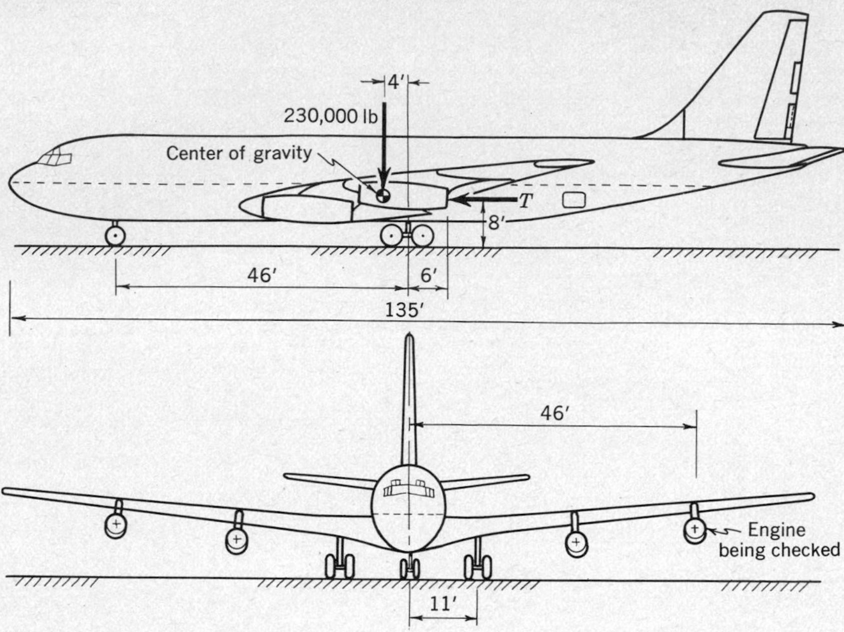

Prob. 1.27

1.28. A 50:1 worm-gear reducer is bolted down at A and B. An input torque M_i of 100 in.-lb turns the worm at a steady rate in the direction shown. The output shaft rotates as shown *against* a resisting torque M_o. Neglecting friction in the gears, estimate the forces acting on the reducer housing exerted by the bolts at A and B when the above torques are acting.

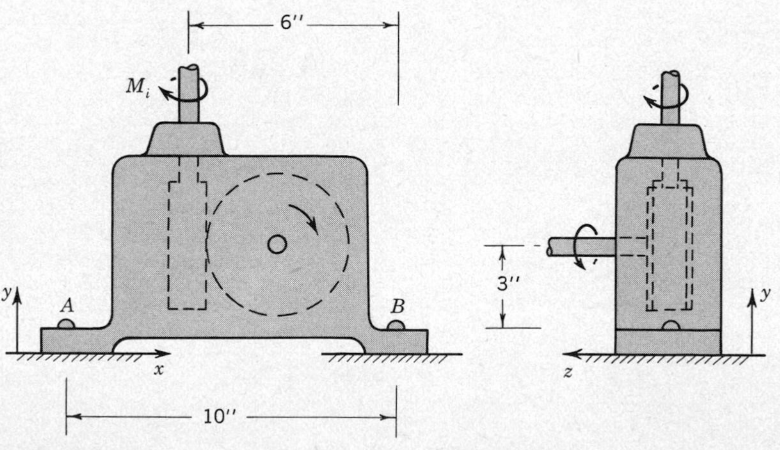

Prob. 1.28

1.29. An electric motor is mounted in a three-point support as shown. The motor weighs 20 lb, which may be assumed to act at the center of the motor. Before starting, the belt tensions are 28 lb each. When running, the motor is delivering a torque of 2 ft-lb. What are the reactions at the supports A, B, and C when the motor is running?

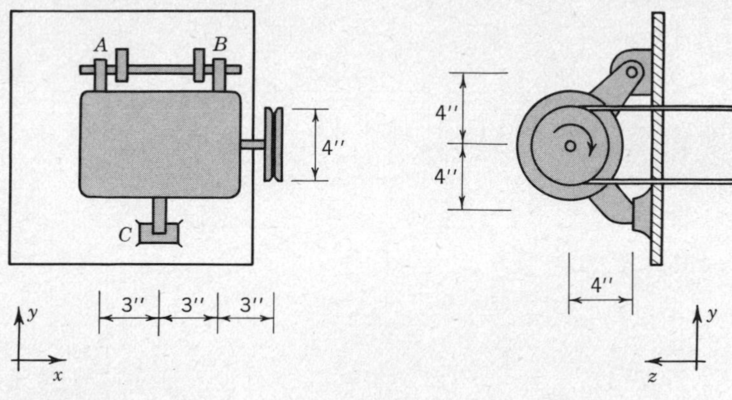

Prob. 1.29

1.30. The crane shown is supported by cables BD and BE. Determine the cable tensions.

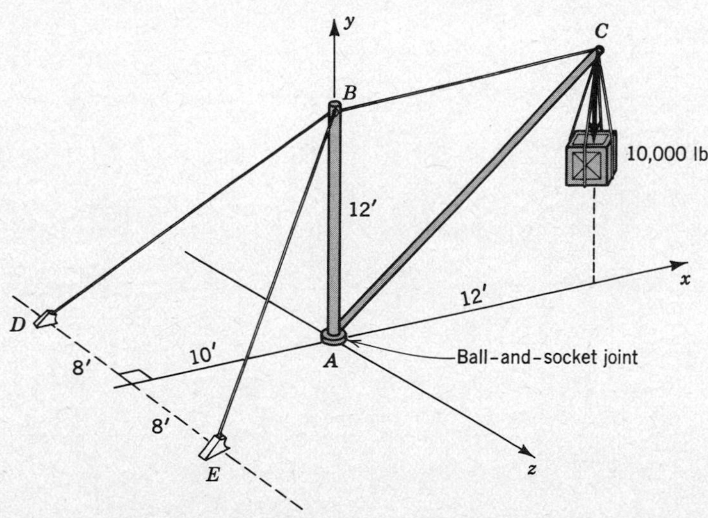

Prob. 1.30

1.31. Determine the forces in the six members of the truss shown.

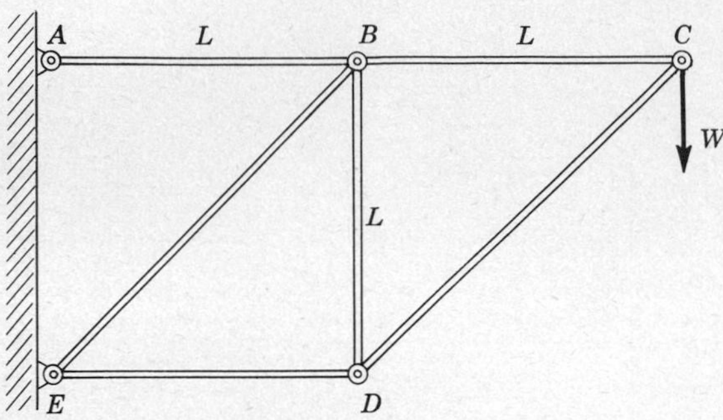

Prob. 1.31

1.32. In this problem you are to determine the forces on the tip of the needle in a record player. Consider the case shown in the figure where the needle is tracking an 8-in.-diameter groove. The groove and needle geometry are as shown. The arm holding the needle is first statically balanced so that its center of gravity lies exactly at the pivot point. A 2-g weight is then placed directly above the needle tip to supply the required tracking force. The arm pivot is a frictionless ball joint. The coefficient of friction between the needle and the groove is 0.2.

Determine the forces acting on the needle tip and show these clearly on a sketch.

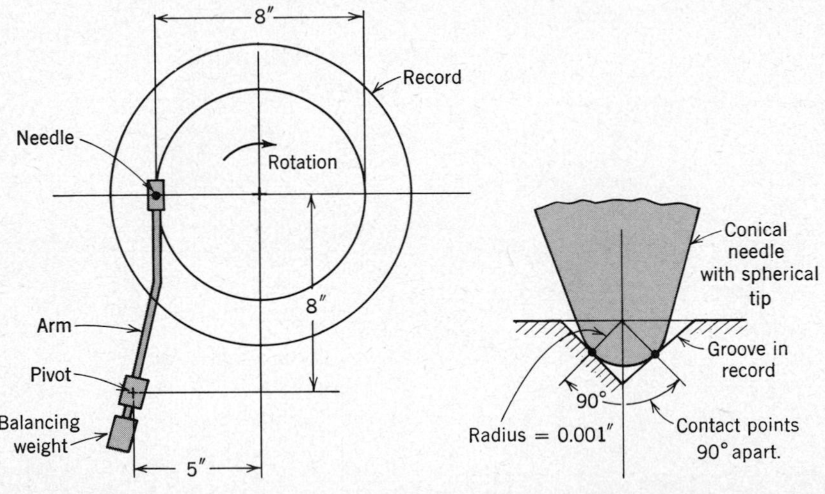

Prob. 1.32

1.33. Adjustable supports that can be slid up and down vertical posts are very useful in many applications. Such a support is shown, with pertinent dimensions. If the coefficient of friction between post and support is 0.30, and if a load 50 times the weight of the hanger is to be placed on the hanger, what is the minimum value of x for no slipping of the hanger?

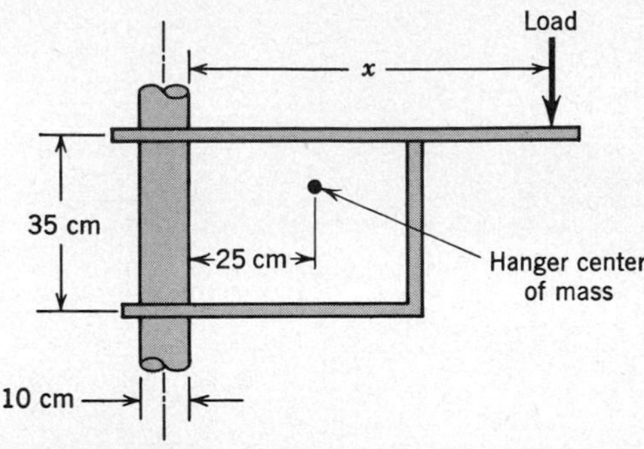

Prob. 1.33

1.34. The clean-air car shown has the following characteristics:

Wheelbase $L = 100$ in.
Weight $W = 2{,}500$ lb
Weight distribution (on level ground), 60 percent on rear wheels
Horsepower h.p. = 100, rear-wheel drive
Height of center of gravity $h = 2$ ft
Wheel diameter $d = 2\frac{1}{2}$ ft

If the coefficient of friction between tires and road is $f = 0.7$, what is the maximum hill angle θ that can be climbed?

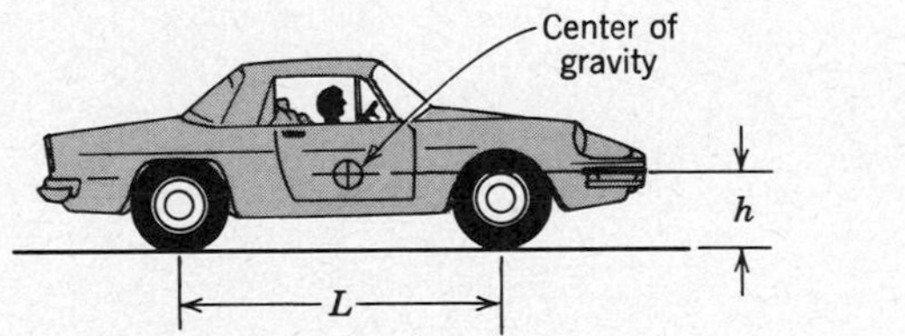

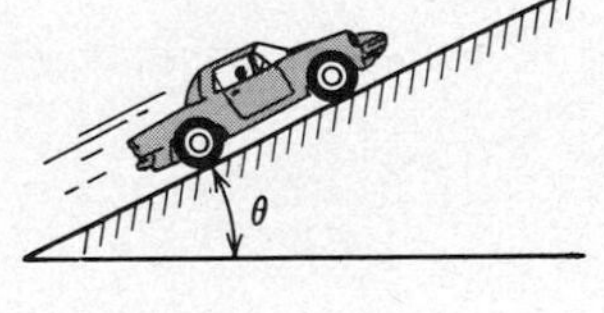

Prob. 1.34

1.35. Obstetric forceps are medical instruments designed for the extraction under certain conditions of a child from the mother during delivery. The instruments vary considerably in size and shape but basically it is a forcep or lever with a blade designed to allow for a firm hold upon the fetal head. Additional force is exerted on the fetal head during delivery by the walls of the birth canal. Forceps in crude versions existed in the 12th century and the forerunner of the modern forceps was devised in the latter part of the 16th century.

Two designs of forceps are shown in the figure. If the applied traction (pulling) force F_T is 30 lb for each design and the clamping force in the crossed-lever design is 5 lb, determine the forces applied by the forcep blades on the fetal head. Assume that the birth canal exerts a constant force on the fetal head. Which design is better?

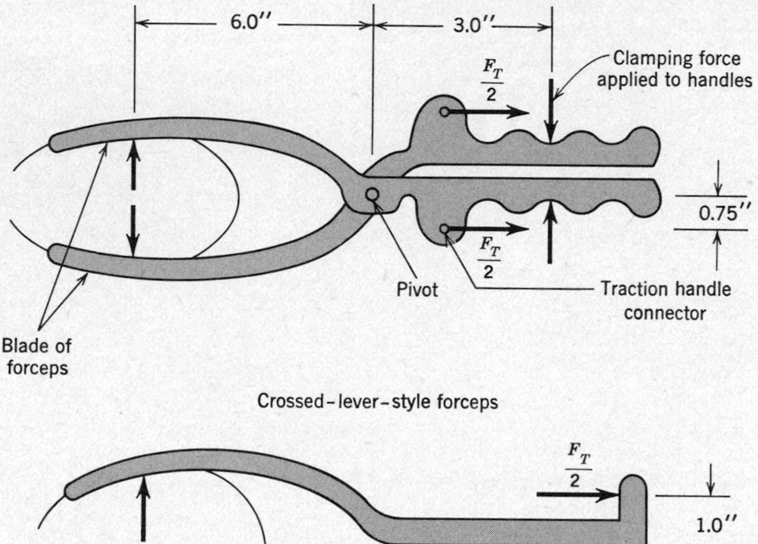

Prob. 1.35

1.36. Assume that frictionless, smooth, identical logs are piled in a box truck (sides perpendicular to the bottom). The truck is forced off the highway and comes to rest on an even keel lengthwise but with the bed at an angle θ with the horizontal as shown. As the truck is unloaded, what is the *least* angle for θ so that when the fourth log is removed the three remaining logs stay where they are?

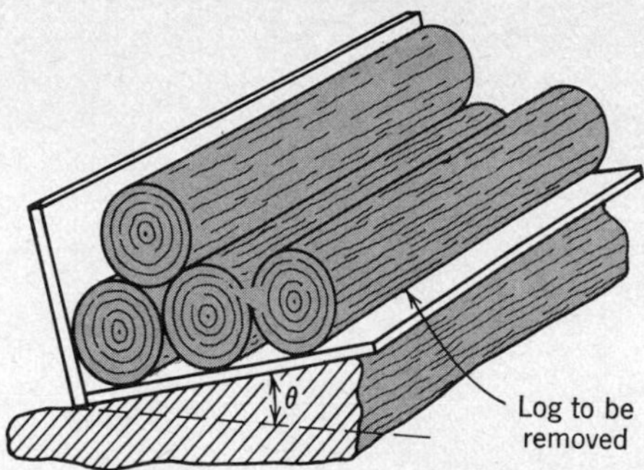

Prob. 1.36

1.37. A circular cylinder A rests on top of two half-circular cylinders B and C, all having the same radius r. The weight of A is W and that of B and C is $\frac{1}{2}$ W each. Assume that the coefficient of friction between the flat surfaces of the half-cylinders and the horizontal table top is f. Determine the maximum distance d between the centers of the half-cylinders to maintain equilibrium.

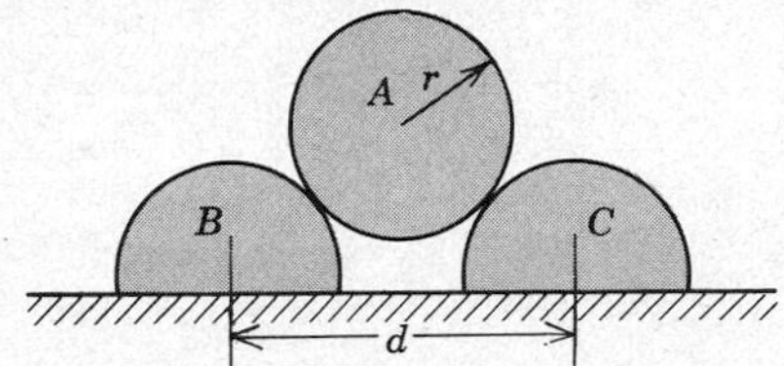

Prob. 1.37

1.38. Determine the force exerted on each side of a bicycle chain link by the bolt cutters shown in the figure if the handles are subjected to a force of 80 lb.

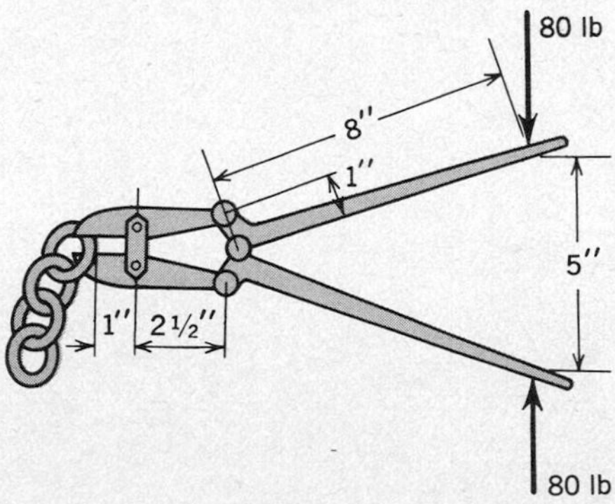

Prob. 1.38

1.39. A block of weight W rests on an inclined plane which makes an angle $\theta = \tan^{-1} \frac{3}{4}$ as shown. A force P, parallel to the x axis, is applied to the block and gradually increased from zero; when P reaches the value $0.4W$ the block begins to slide. What is the coefficient of friction between the block and the inclined plane?

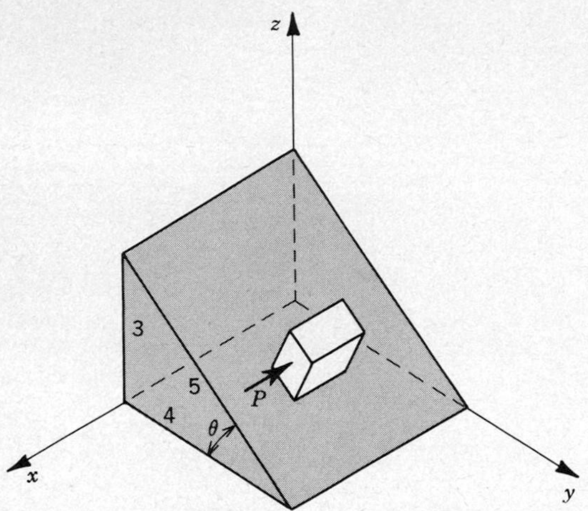

Prob. 1.39

1.40. A lightweight portable crane for mountain bridge construction is needed. Experience with other cranes has indicated that the simple design and erection shown in the figure is convenient

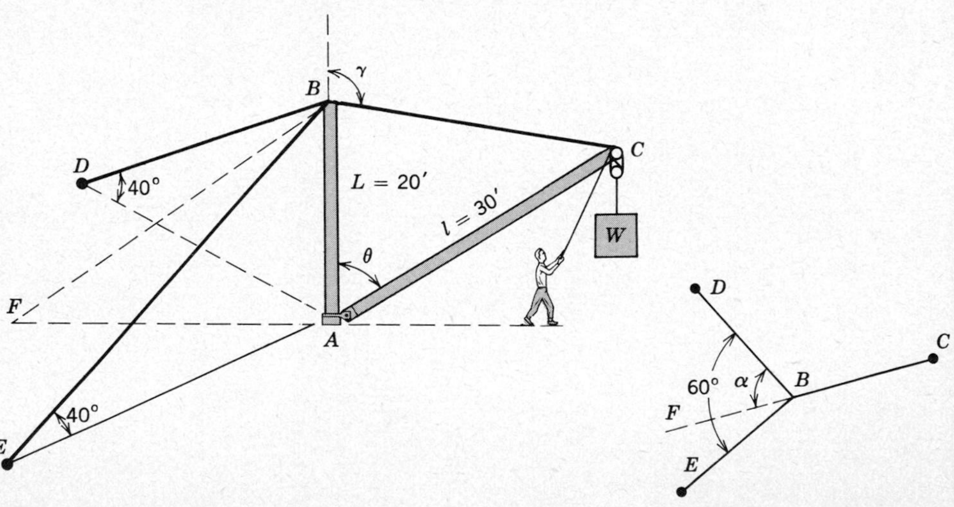

Prob. 1.40

for the field. In particular, the required application of the crane is in loading and unloading bridge parts weighing 1,000 lb. To save weight, it was decided to design the crane system such that the cables BD, BE, and BC would carry (when $W=1{,}000$ lb) no more than 1,200 lb by adjusting the angle γ and limiting the angle of swing of the boom AC about the mast AB. The length of cable BC is adjustable. What is needed is a simple sheet, possibly with a graph, showing the operator in the field what angle γ for a given W will give him a maximum swing α or what the limits on α are for a given γ. Prepare such an information sheet.

The vertical mast AB of the crane is rigidly supported by two guy wires BD and BE. The line of intersection of the vertical plane ABC with the horizontal plane ADE is defined by α.

1.41. The mast AD is acted on by a 4,000-lb force and supported by cables CE and DF as shown. Find the reaction force exerted on the mast by the frictionless ball-and-socket joint at A and also the tensions in the cables CE and DF.

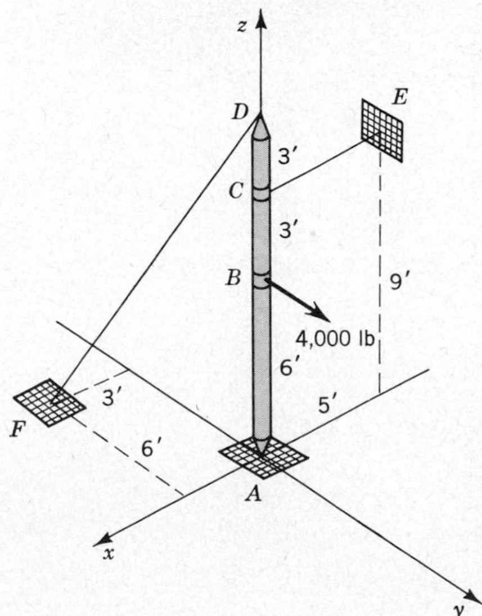

Prob. 1.41

1.42. A man holds a 20-lb weight in his hand. The forearm and hand weigh 4 lb, and the elbow is flexed to a right angle as shown. Compute first the force required in the flexor muscle and the force in the humerus against the ulna to support the load neglecting friction.

If the coefficient of friction in the elbow joint is 0.015, what change in the previously determined muscle force is required (*a*) to raise the load, and (*b*) to just support it. The radius of curvature of the joint is 0.75 in.

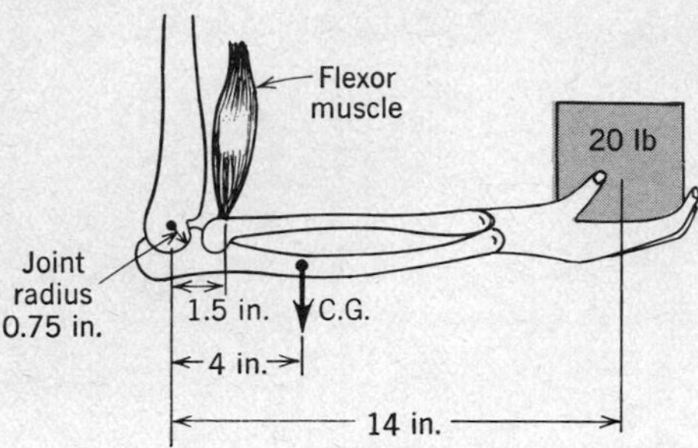

Prob. 1.42

1.43. A folding camp stool rests upon a horizontal floor (neglect friction) and is loaded as shown in the figure. Determine the magnitude of the shear force on the pin A and the position of the load on the bar BC to make this force a maximum.

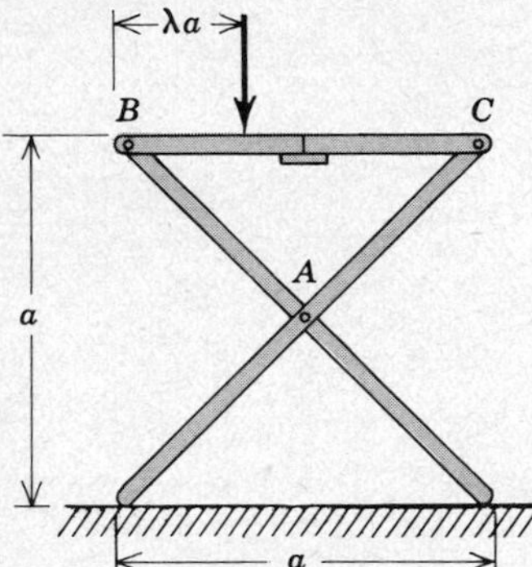

Prob. 1.43

1.44. A window air-conditioning unit is supported by a round rod as shown. For what angle θ will the required *cost* of the rod be a minimum?

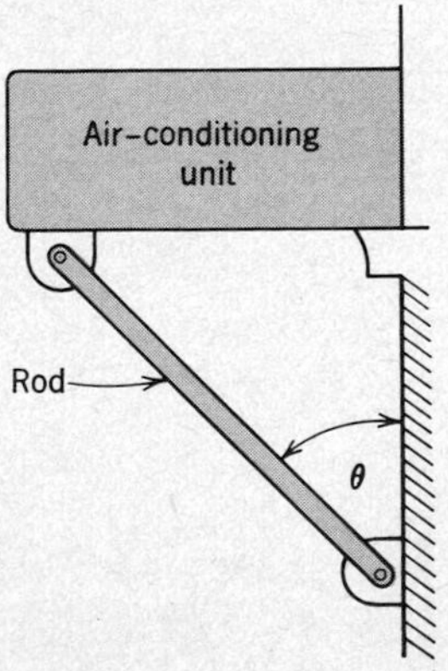

Prob. 1.44

1.45. A longshoreman can barely start pushing a trunk up a 30° concrete ramp. He can barely hold it from sliding back when the slope is 60°. What is the coefficient of static friction between the trunk and the concrete?

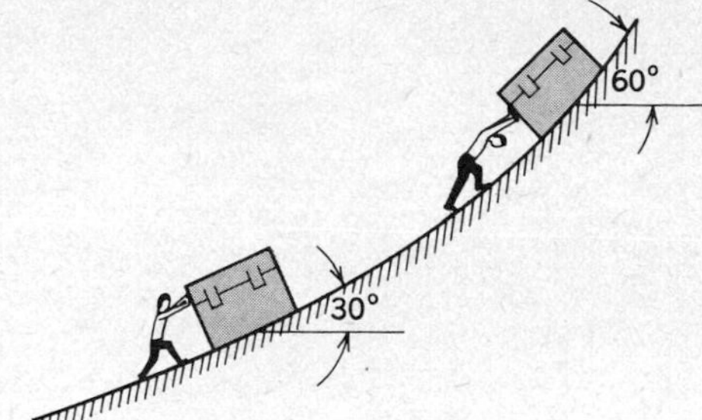

Prob. 1.45

1.46. Tensile tests are being run on a specimen of high-strength carbon-filament material with the configuration shown. The pins are *fixed in the grips* but loose in the heads. The coefficient

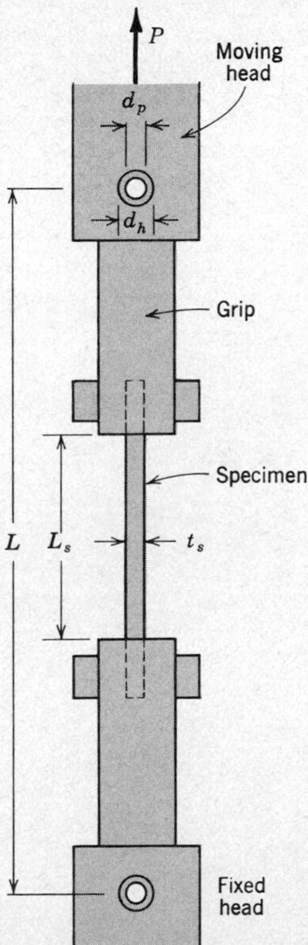

Prob. 1.46

of friction is 0.15 and the materials may be assumed rigid. As the heads are moved apart by the motion of the testing machine, how far might the line of application of the load be displaced from the center of the specimen? Assume the worst possible location of the pins in the holes. Evaluate for the following values of the dimensions:

$d_p = 0.250$ in.
$d_h = 0.255$ in.
$t_s = 0.08$ in.
$L = 24$ in.
$L_s = 4$ in.
$P = 5{,}000$ lb

1.47. Figure *a* shows the external forces assumed to be acting on a version of an SST just prior to touchdown. Figure *b* shows the pertinent dimensions. The following information is given:

(*a*) The "Canard" (forward) control surface is set at its zero-lift angle of attack, and the drag force on it is $\mathbf{D}_c = 100\mathbf{i}$ lb, which is applied at AC_c (Fig. *b*).

(*b*) The aircraft weight is $\mathbf{W} = -400{,}000\mathbf{j}$ lb, acting at the CG.

(*c*) The lift and drag forces on the wing, $\mathbf{L}_w = L_w\mathbf{j}$ and $\mathbf{D}_w = D_w\mathbf{i}$, act at AC_w. The aerodynamic moment about AC_w can be neglected.

(*d*) The lift and drag forces on the tail, $\mathbf{L}_t = L_t\mathbf{j}$, $\mathbf{D}_t = D_t\mathbf{i}$, act at AC_t. Further, the lift-to-drag ratio for the tail is given as $(L/D)_t = 1.2$. The aerodynamic moment about AC_t can also be neglected.

(*e*) The thrust $\mathbf{T}$ has a magnitude of 160,000 lb and acts along the *thrust axis*, a line parallel to and 20 ft below the aircraft center line (Fig. *b*).

(*f*) The aircraft is assumed to be in static equilibrium. Determine the numerical values of the tail lift L_t and the lift-to-drag ratio for the wing.

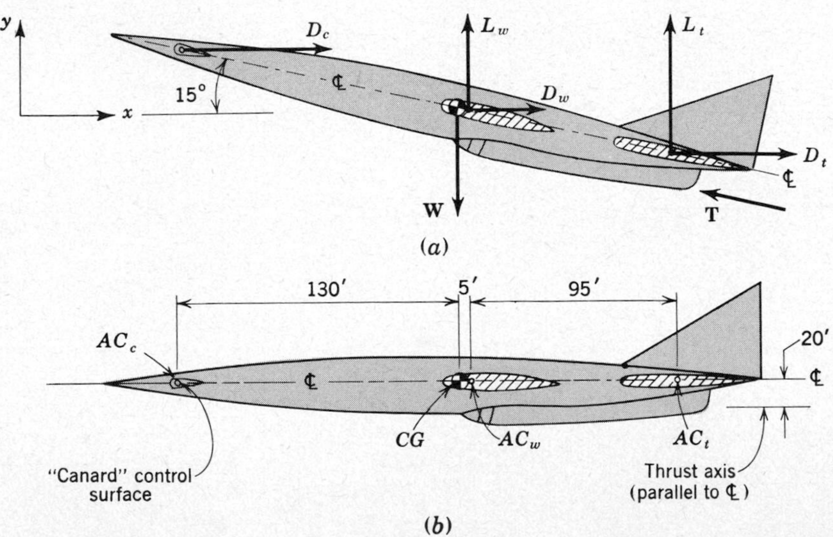

Prob. 1.47

1.48. In building an orbiting space laboratory it will be necessary to drill holes in the flat steel wall of a space vehicle. The astronaut doing the drilling is unable to apply any appreciable amount either of force or of torque during the drilling so that it is necessary to mount the drill in a holder with three legs terminating in magnets, which grip the wall of the space vehicle. If the drilling torque is 100 in.-lb, and the normal force is 12 lb, compute the minimum allowable holding force at each leg if the friction coefficient between the legs and space vehicle is 0.4.

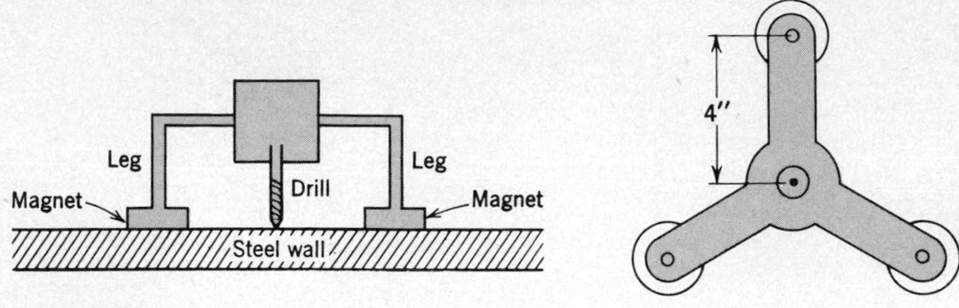

Prob. 1.48

2
Introduction to Mechanics of Deformable Bodies

2.1 ANALYSIS OF DEFORMABLE BODIES

As pointed out in Chap. 1, the attack on a problem in applied mechanics begins with the selection of the system to be analyzed, a starting point common with scientific investigations in all fields. This process of selection usually is a dual one of *identification* and *simplification*—identification of a system we recognize as representing a reasonably complete description of the interrelationships of the actual physical situation, and simplification of this system until we have a model which we are capable of analyzing. In this book we shall restrict ourselves to situations in which the acceleration is zero and where the movements of the system are restricted to deformations. Consequently, when we have selected our model we shall analyze this model, as outlined in Sec. 1.7, by the following three steps:

1. Study of forces and equilibrium requirements

2. Study of deformation and conditions of geometric fit **(2.1)**

3. Application of force-deformation relations

We dignify these three steps with an equation number because they are fundamental to all work in the mechanics of deformable bodies; we shall make frequent reference to them.

In Chap. 1 the problems were limited to cases which involved only the first of the above three steps. In this chapter we shall consider cases for which the analysis will contain all three of the above steps.

We can illustrate the use of Eq. (2.1) in the analysis of deformable systems by considering the following three examples.

Example 2.1 A machine part carrying a load F terminates in a piston which fits into a cavity, as shown in Fig. 2.1. Within the cavity are two springs arranged coaxial with each other. Each spring has the characteristic that the force required to deflect it is proportional to the amount of deflection. Such a spring is said to have a *linear* force-deflection relation, and the amount of force required to produce a unit deflection is called the *spring constant* of the spring. We use the symbols k_A and k_B to denote the spring constants of the two springs in the cavity. When the springs are unloaded, each has the same length L. We wish to know how much of the load F is carried by the spring with constant k_A.

The first step in the analysis is to select our model. This is a relatively simple step in this situation; the model consists simply of the piston and the two springs, as shown in the sketches in Fig. 2.1b. We have assumed that the springs have been made with flat ends such that the compressing force, which is distributed around the periphery of the spring, can be considered to

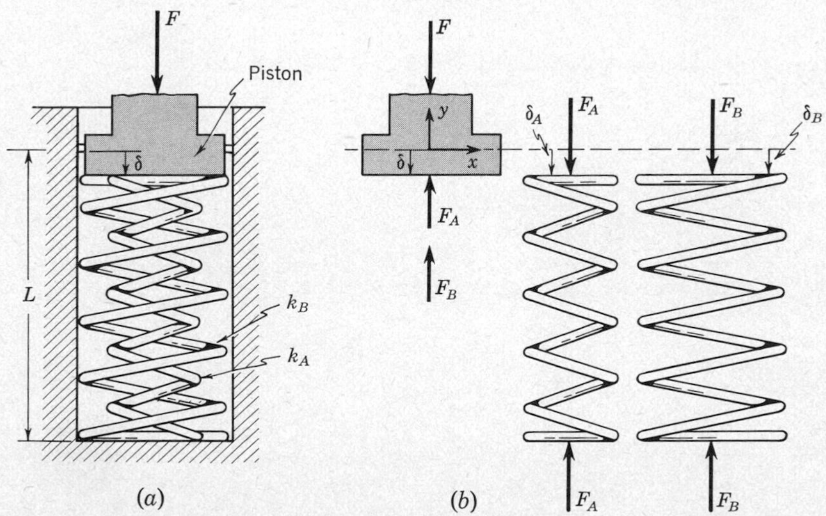

Fig. 2.1 Example 2.1.

act along the spring axis. We have also assumed that gravity effects can be ignored without substantially changing the problem. We now apply the steps of (2.1) to this model.

STUDY OF FORCES AND EQUILIBRIUM REQUIREMENTS

To ensure that a body is in equilibrium we *isolate* that body from its surrounding environment and replace the environment with the forces which are the sole effect of the environment on the body. As noted in Chap. 1, the sketch which shows the body and all the forces acting on it is called the free-body sketch; in Fig. 2.1*b* are free-body sketches of the piston and of the two springs. Equilibrium of a given free body requires that Eqs. (1.5) and (1.6) be satisfied. Thus, for each of the three free bodies in Fig. 2.1*b*, we must have

$$\Sigma \mathbf{F} = 0$$

$$\Sigma \mathbf{M} = 0$$

Considering first the piston, we see that the forces acting on it all have the same line of action and thus $\Sigma \mathbf{M} = 0$. Since we have only forces in the y direction, the equation $\Sigma \mathbf{F} = 0$ is satisfied when

$$\Sigma F_y = F_A + F_B - F = 0 \qquad (a)$$

By taking equal and opposite forces at the ends of each spring in Fig. 2.1*b*, we have satisfied the conditions $\Sigma \mathbf{M} = 0$ and $\Sigma \mathbf{F} = 0$ for the springs.

STUDY OF DEFORMATION AND CONDITIONS OF GEOMETRIC FIT

Our purpose in studying the nature of the deformation and the conditions of geometric fit is to determine what constraints or restrictions these factors impose upon the deformations. Put another way, we want to find what are the requirements for *geometric compatibility* with the restraints. It will be necessary to express these requirements in a quantitative or analytical manner. In the problem at hand the action of the piston is to cause both springs to move the *same* amount as the piston, and thus the requirement for geometric compatibility is simply

$$\delta_A = \delta_B = \delta \qquad (b)$$

APPLICATION OF FORCE-DEFORMATION RELATIONS

In order to deal precisely with the manner in which the deformation of a physical body is related to the forces acting on it, we must express this relation quantitatively, either by equations or by graphs. For this problem the force-deflection relation is a simple one: The force in each spring is linearly proportional to the deflection of the spring, and the constant of proportionality is the spring constant. Thus,

$$\begin{aligned} F_A &= k_A \delta_A \\ F_B &= k_B \delta_B \end{aligned} \qquad (c)$$

We note that the spring constant has the units of force per unit length, e.g., lb/in., or newton/m. With Eqs. (*a*), (*b*), and (*c*) we have in quantitative form all the information we can write down about the force balance, the geometric fit, and the force-deflection characteristics of the system shown in Fig. 2.1*a*. We have at this stage completed the *physical* part of the analysis. We now manipulate these equations *mathematically*, eliminating the deflections, to obtain the desired result.

$$\frac{F_A}{F} = \frac{k_A}{k_A + k_B} \tag{d}$$

The total deflection of the piston is therefore

$$\delta = \frac{F_A}{k_A} = \frac{F}{k_A + k_B}$$

Example 2.2 A very light and stiff wood plank of length $2L$ is attached to two similar springs of spring constant k, as shown in Fig. 2.2*a*. The springs are of length h when the plank is resting on them. Suppose that a man steps up on the middle of the plank and begins to walk slowly toward one end. We should like to know how far he can walk before one end of the plank touches the ground; that is, we want to know the distance b in Fig. 2.2*b*, when the right end E of the plank is just in contact with the ground. Note that the springs can exert tension as well as compression.

Again we begin by selecting a model to represent the system. Our model is shown in Fig. 2.2*c* and *d*. In this model we show no force at the right end E because we are interested in the limiting case where the plank just comes in contact with the ground. We represent the man by his weight W located at the distance b from the center. Because the plank is described as being light, we neglect its weight; it follows as a consequence of this assumption that the springs in Fig. 2.2*a* are exerting no force on the plank, and thus h is the free length of each spring. Finally, we have assumed that the stiffness of the plank is such that we can consider it to remain absolutely straight so the deflections of the springs are as illustrated in Fig. 2.2*d*. We now analyze the model using the three steps of (2.1).

FORCE EQUILIBRIUM

Applying the conditions

$$\Sigma \mathbf{F} = 0$$

$$\Sigma \mathbf{M} = 0$$

to the free body of the plank in Fig. 2.2*c*, we find that the first of these is satisfied when

$$\Sigma F_y = F_C + F_D - W = 0 \tag{a}$$

and the second is satisfied when

$$\Sigma M_C = 2aF_D - (a + b)W = 0 \tag{b}$$

The springs in Fig. 2.2*d* satisfy the equilibrium requirements of two-force members.

STUDY OF GEOMETRY OF DEFORMATION AND REQUIREMENTS OF GEOMETRIC COMPATIBILITY

When the plank remains straight, we see from the similar triangles in Fig. 2.2*d* that the lengths of the springs have the following ratio:

$$\frac{h_C}{h_D} = \frac{L + a}{L - a} \tag{c}$$

Also, the deflections of the springs are

$$\begin{aligned} \delta_C &= h - h_C \\ \delta_D &= h - h_D \end{aligned} \tag{d}$$

RELATIONS BETWEEN FORCES AND DEFLECTIONS

Here, since both springs have the same spring constant, the force deflection relations are

$$\begin{aligned} F_C &= k\delta_C \\ F_D &= k\delta_D \end{aligned} \tag{e}$$

Equations (*a*), (*b*), (*c*), (*d*), and (*e*) give us seven independent relations for the seven unknowns F_C, F_D, h_C, h_D, δ_C, δ_D, and b. Solving these equations, we find that the value of b is given by

$$b = \frac{a^2}{L}\left(\frac{2kh}{W} - 1\right) \tag{f}$$

It also is of interest to calculate the spring deflections in terms of the value of b determined by Eq. (*f*). These become

$$\begin{aligned} \delta_C &= \frac{W}{2k}\left(1 - \frac{b}{a}\right) \\ \delta_D &= \frac{W}{2k}\left(1 + \frac{b}{a}\right) \end{aligned} \tag{g}$$

We see that δ_D is always positive in the sense defined in Fig. 2.2*d*. δ_C is positive so long as $b < a$. When $b = a$, the man is directly over the spring D, and, as would be expected, all the load is taken by the spring D, and the deflection and force in the spring C are zero. When $b > a$, then δ_C is negative (i.e., the spring extends). In Fig. 2.2*c* we assumed that $b > a$ and that the spring C is compressed. If after making these assumptions in a particular case the result from (f) was that $b > a$, then we would find from our algebra that both δ_C and F_C were negative; we would interpret these negative results to mean that the actual δ_C and F_C were in directions opposite to those defined as positive in Fig. 2.2*d*.

A number of additional remarks can be made concerning (f). If we rewrite (f) in a slightly different form

$$\frac{b}{L} = \left(\frac{a}{L}\right)^2 \left(\frac{h}{W/2k} - 1\right) \tag{h}$$

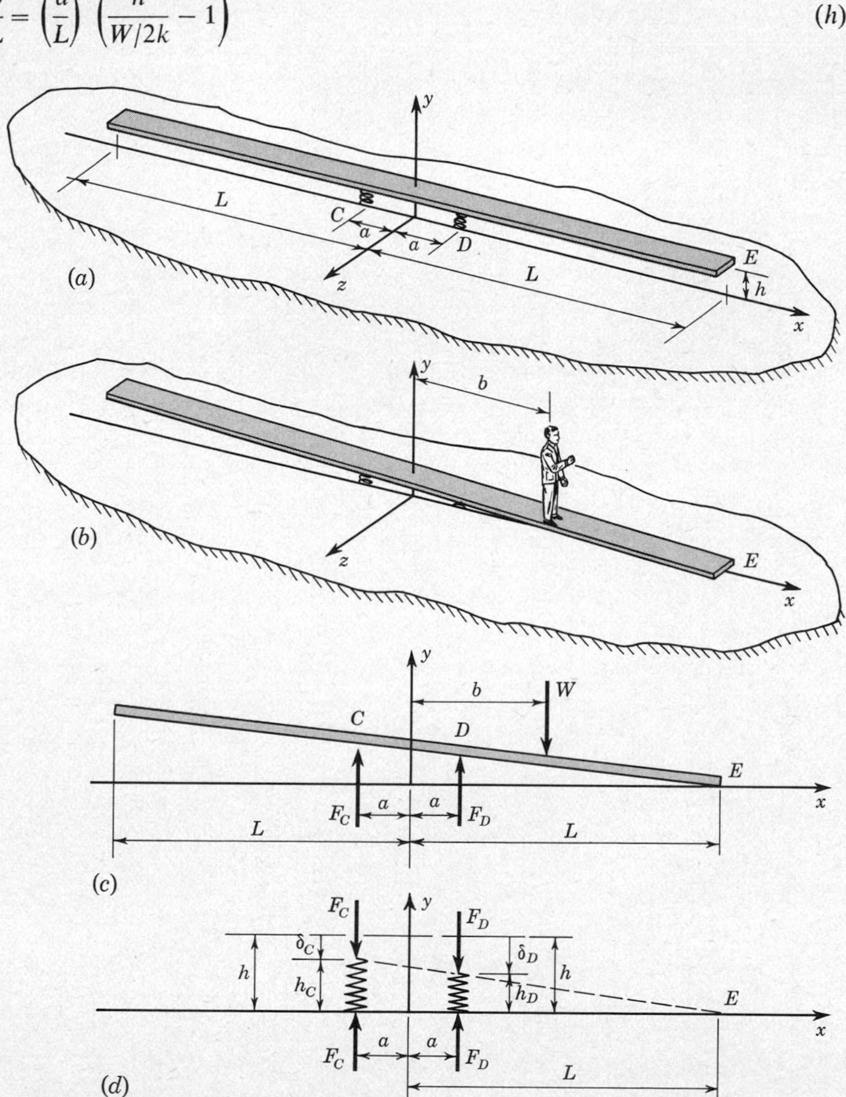

Fig. 2.2 Example 2.2.

we see the basic *nondimensional* ratios of the problem. It is always good practice in any engineering problem to consider the physical meaning of the nondimensional quantities which appear in the problem. The ratio $(W/2k)/h$ is the ratio of the deflection of the springs to the original length of the springs when $b = 0$, that is, when the man is standing midway between the springs. If this ratio is small, then b/L might exceed unity and our analysis does not apply. If this ratio is near 1, i.e., if the springs are soft, then b/L is near zero.

Another question of interest is the effect of different spring constants on the value of b. We expect that the result for b/L will depend on the ratios a/L, $W/2k_Ch$, and on the ratio of the spring constants, for example, $\gamma = k_C/k_D$. The result for b/L is (Prob. 2.40)

$$\frac{b}{L} = \frac{(a/L)^2}{1 - \tfrac{1}{2}(1-\gamma)(1 + a/L)}\left[\frac{2k_Ch}{W} - 1 + \frac{1}{2}(1-\gamma)\left(1 + \frac{L}{a}\right)\right] \qquad (i)$$

Equation (i) is a surprisingly complicated function of the ratio of the spring constants. Note that (i) reduces to (h) when $\gamma = 1$.

Example 2.3 A light rigid bar ABC is supported by three springs, as shown in Fig. 2.3*a*. Before the load P is applied, the bar is horizontal. The distance from the center spring to the point of application of P is λa, where λ is a dimensionless parameter which can vary between $\lambda = -1$ and $\lambda = 1$. The problem is to determine the deflections in the three springs as functions of the load position parameter λ. We shall obtain a general solution for arbitrary values of the spring constants and display the results for the particular set $k_A = 1/2k$, $k_B = k$, and $k_C = 3/2k$.

The system is modeled by a rigid weightless bar and three linear-elastic springs. We analyze this model by following the steps of (2.1).

STUDY OF FORCES AND EQUILIBRIUM REQUIREMENTS

Free-body diagrams for the springs and for the bar are shown in Fig. 2.3*b* and Fig. 2.3*c*. We note that there are three unknown parallel forces acting on the bar in Fig. 2.3*c* and only two independent equilibrium requirements; i.e., the problem is statically indeterminate. The best we can do is to use the equilibrium conditions to express two of the forces in terms of the third. If we take the middle force F_B as our primary unknown, we can conveniently obtain F_A and F_C in terms of F_B by requiring balance of moments about C and A, respectively,

$$\begin{aligned} \Sigma M_C &= 0 \qquad 2aF_A = (1-\lambda)aP - aF_B \\ \Sigma M_A &= 0 \qquad 2aF_C = (1+\lambda)aP - aF_B \end{aligned} \qquad (a)$$

STUDY OF GEOMETRIC COMPATIBILITY REQUIREMENTS

In order for the springs to remain connected to the bar, it is necessary for the spring deflections in Fig. 2.3*b* to be the same as the corresponding bar

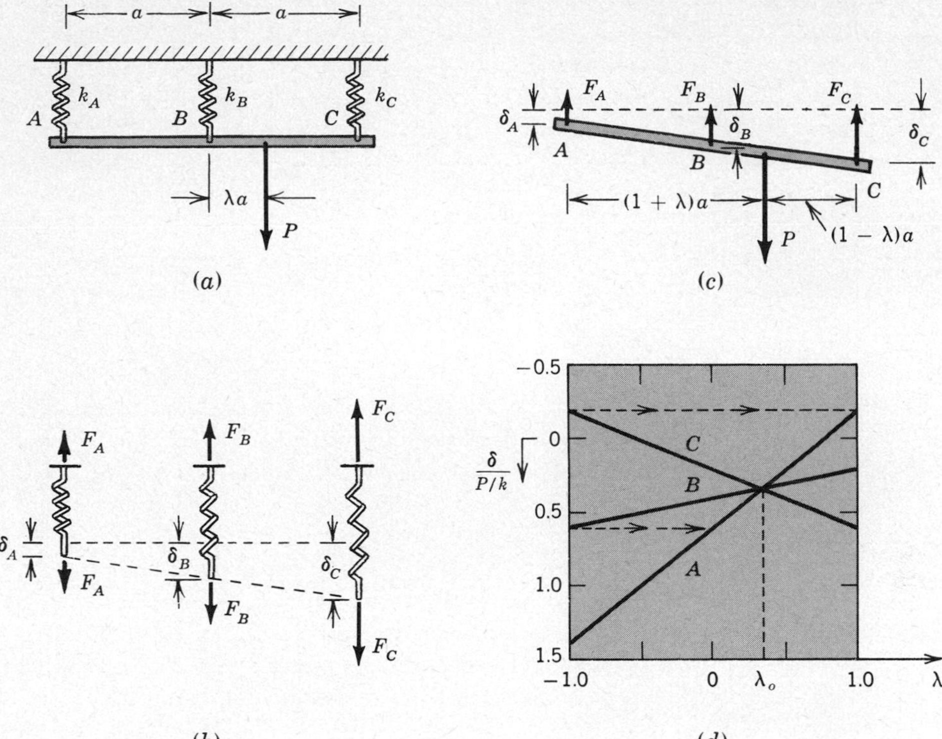

Fig. 2.3 Example 2.3.

deflections in Fig. 2.3*c*. Since the bar *ABC* is rigid, it is necessary that the points *A*, *B*, and *C* in Fig. 2.3*c* remain collinear. Because two points determine a straight line, one of the three deflections can be expressed in terms of the other two. Thus, if the end deflections δ_A and δ_C are given, the midpoint deflection must be

$$\delta_B = 1/2(\delta_A + \delta_C) \tag{b}$$

RELATIONS BETWEEN FORCES AND DEFORMATIONS

For the linear springs in Fig. 2.3*b* we have

$$\delta_A = \frac{F_A}{k_A} \qquad \delta_B = \frac{F_B}{k_B} \qquad \delta_C = \frac{F_C}{k_C} \tag{c}$$

Equations (*a*), (*b*), and (*c*) are six independent relations among the six unknowns: the three forces and the three deflections. The equations are conveniently solved by substituting (*a*) into (*c*) to obtain all the deflections in terms of F_B and then inserting these deflections into (*b*) to obtain a single equation for F_B. Once F_B is known, F_A and F_C are given by (*a*). Finally, when the forces are known, the deflections are given by (*c*). For general values of the spring constants we find

$$\begin{aligned} \delta_A &= P\,\frac{2k_C - \lambda(k_B + 2k_C)}{k_A k_B + 4k_A k_C + k_B k_C} \\ \delta_B &= P\,\frac{k_A + k_C + \lambda(k_A - k_C)}{k_A k_B + 4k_A k_C + k_B k_C} \\ \delta_C &= P\,\frac{2k_A + \lambda(k_B + 2k_A)}{k_A k_B + 4k_A k_C + k_B k_C} \end{aligned} \tag{d}$$

Note that each deflection is a *linear* function of the load position parameter λ. Figure 2.3*d* shows how these deflections vary for the particular case $k_A = \frac{1}{2}k$, $k_B = k$, and $k_C = \frac{3}{2}k$. Note that for each position of the load, the middle deflection at B is always midway between the end deflections A and C. Note also that when the load is near one end, the deflection at the other end is negative. In order for our solution to be valid in these ranges it is necessary for the springs to be capable of working in compression.

It is interesting to observe that when the load is at the position indicated by λ_O in Fig. 2.3*d*, all three spring deflections are equal. This means that the bar deflects without tipping when the load is applied at this position. For any other value of λ the bar tips.

As derived, the line marked A in Fig. 2.3*d* represents the deflection $\delta_A(\lambda)$ as the load position parameter λ is varied. We can, however, give an alternate interpretation to this line. It also represents the actual position of the bar when the load is applied at point A. To see this we note that when the load is at point A the deflections at the three spring locations are given by the ordinates of the three lines at $\lambda = -1$. If we transfer the ordinate $\delta_B(-1)$ to the location of point B (that is, $\lambda = 0$) and transfer the ordinate $\delta_C(-1)$ to the location of point C (that is, $\lambda = 1$), we note that the transferred ordinates (see horizontal dashed transfer lines) lie precisely on the line marked A. The lines marked B and C in Fig. 2.3*d* also have this same reciprocal property. They represent either the deflection at the labeled position when the load is varied, or they represent the position of the bar when the load is applied at the labeled position. The position marked λ_O then has an additional significance. It is the one point on the bar whose deflection is independent of the load location. It acts as a pivot point for the bar as the load position is varied.

For arbitrary values of the spring constants in Fig. 2.3*a*, the qualitative behavior is similar to that described for the particular set of spring constants used to construct Fig. 2.3*d*. It can be shown that, in general, the pivot location is given by

$$\lambda_O = \frac{k_C - k_A}{k_A + k_B + k_C} \tag{e}$$

and the invariant deflection at this location is

$$\delta(\lambda_O) = \frac{P}{k_A + k_B + k_C} \tag{f}$$

In these three examples we have found it necessary to consider all three steps of (2.1) in order to obtain the desired information about the system behavior. In Example 2.2, for a *given* value of b we could have found the values of F_C and F_D by considering only step 1. As noted in Chap. 1, such systems in which we can determine the forces without reference to the deformations or the force-deformation relations are called *statically determinate* systems. In other situations, such as in Examples 2.1 and 2.3, we find that the forces cannot be determined without considering the deformations, and we call these *statically indeterminate* systems; in such cases it always will be necessary to consider all three steps, even though we are interested only in the forces.

Although the three steps of (2.1) underlie the analysis of *all* problems in the mechanics of deformable bodies, the knowledge of their existence does not ensure that a solution will be found to any *given* problem. In any given situation the possibility of finding a solution depends upon the relative complexity of the physical situation, the difficulty in defining a sufficiently accurate and yet sufficiently simple model, and the problems involved in formulating the three analytical steps when studying the model.

2.2 UNIAXIAL LOADING AND DEFORMATION

The basic type of deformation which will be considered in most of the problem situations in this chapter is shown in Fig. 2.4*a*. A bar is loaded by two forces, and we are interested in the relative motion of the points of application of the two forces. Consider the deformation of three rods of identical material, but having different lengths and cross-sectional areas as shown. Assume that for each bar the load is gradually increased from zero, and at several values of the load a measurement is made of the elongation δ. If the maximum elongation is very small (say, not greater than 0.1 percent of the original length[1]), then for most materials the results of the three tests will be represented by a plot like Fig. 2.4*b* or like Fig. 2.4*c*. The relative positions of the three curves in each plot are what we would expect from the experience we have had with easily deformable bodies such as rubber bands.

[1] In Chap. 5 there is a discussion of the load-elongation curves for larger elongations.

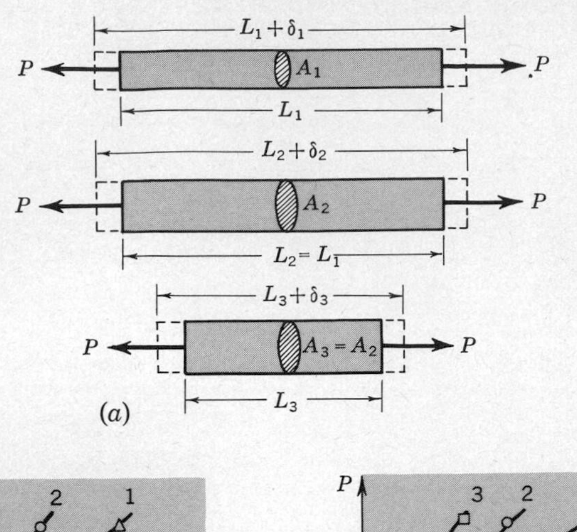

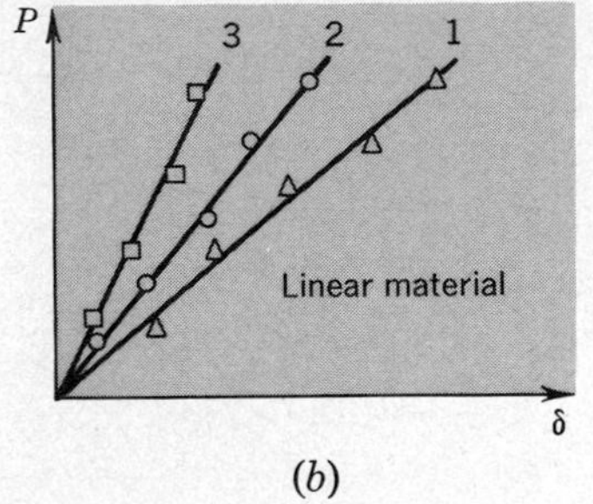

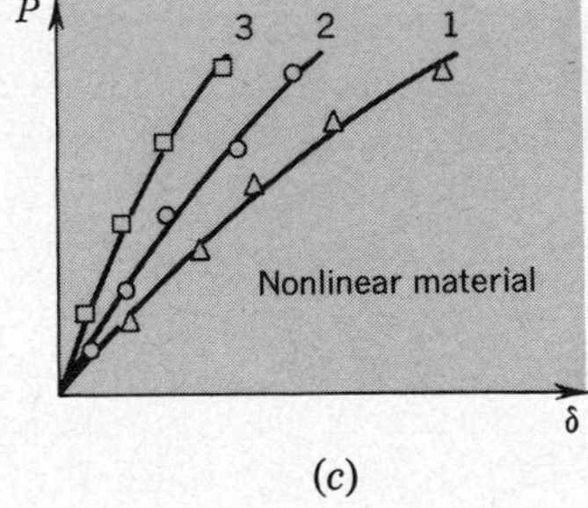

Fig. 2.4 Uniaxial loading.

If the experimental data are replotted with load over area as ordinate and elongation over original length as abscissa, the test results for the three bars can be represented by a single curve, as shown in Fig. 2.5*a* or *b*. The fact that this replotting brings all the data from different test specimens into common agreement greatly simplifies the problem of determining the load-deformation behavior of materials. Thus, to obtain the uniaxial load-elongation characteristics of a particular material, we can test a single specimen and present the results as a plot of P/A against δ/L, as illustrated in Fig. 2.5.

If the uniaxial load-elongation relation of the material is linear, then this relation can be expressed by giving the slope of the straight line in Fig. 2.5*a*. This slope is called the *modulus of elasticity* and usually is denoted by the symbol E. In terms of the coordinates of Fig. 2.5*a*, E is defined by

$$E = \frac{P/A}{\delta/L}$$

Since P/A has the dimensions of force per unit area and δ/L is dimensionless, the dimensions of E are those of force per unit area (in Chap. 4 we will define a terminology in which P/A is the average *stress* across the area A, while δ/L is the average *strain* along the length L). Using the units of pound for force and inch for length,

Table 2.1

Material	E, psi[1]
Tungsten carbide	$60\text{–}100 \times 10^6$
Tungsten	58×10^6
Molybdenum	40×10^6
Aluminum oxide	47×10^6
Steel and iron	$28\text{–}30 \times 10^6$
Brass	15×10^6
Aluminum	10×10^6
Glass	10×10^6
Cast iron	$10\text{–}20 \times 10^6$
Wood	$1\text{–}2 \times 10^6$
Nylon, epoxy, etc.	$4\text{–}8 \times 10^4$
Collagen	$2\text{–}15 \times 10^3$
Soft rubber	$2\text{–}8 \times 10^2$
Smooth muscle	2–150
Elastin	50–100

[1] Conversion factor:

$1 \text{ psi} = 6.9 \times 10^3 \text{ N/m}^2$

E has the dimensions lb/in.^{2} (usually abbreviated as psi). Typical values of E for a few materials are given in Table 2.1.

If we rewrite the definition for E, we obtain an expression for δ

$$\delta = \frac{PL}{AE} \tag{2.2}$$

Equation (2.2) is a simple form of Hooke's law,[1] so named after Robert Hooke who was the first to record that many materials have a linear relation between load and

[1] Hooke's law for loading in more than one direction will be discussed in Chap. 5.

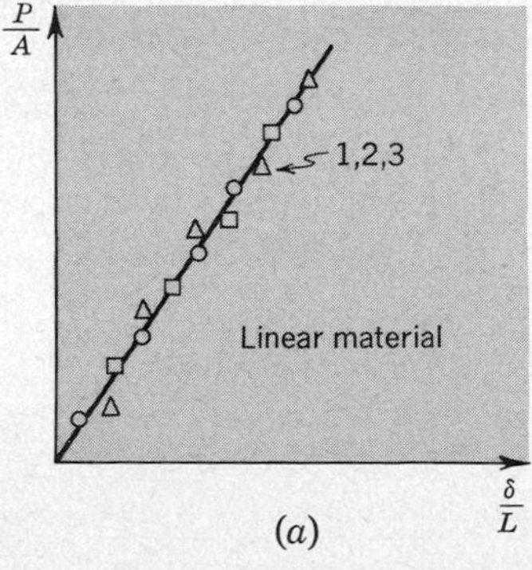

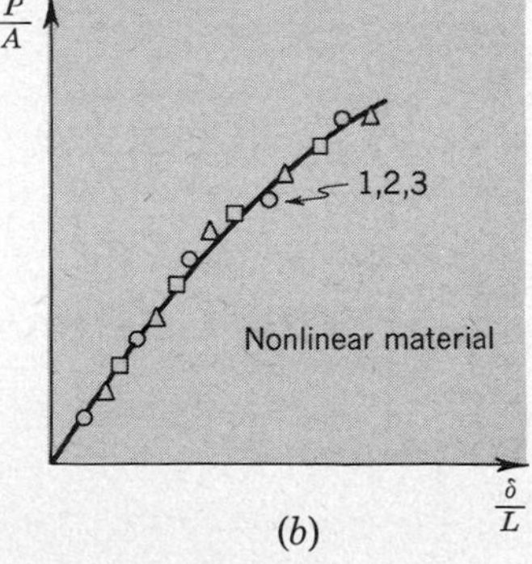

Fig. 2.5 Uniaxial-loading data of Fig. 2.4b and c plotted as P/A versus δ/L.

deformation. (He also was the inventor of the Hooke's joint discussed in Sec. 1.10.) It should be noted that, when the load-deflection curve is linear, a solid bar subjected to end loads acts in the same manner as the coiled springs with which we are familiar. If we had a "spring" made of a bar of steel 1 in.^{2} in area and 1 in. long, the spring constant would be, from Eq. (2.2),

$$k = \frac{P}{\delta} = \frac{AE}{L} = \frac{1(30 \times 10^6)}{1} = 30 \times 10^6 \text{ lb/in.}$$

If the material is nonlinear, then it will not be possible to represent the uniaxial load-elongation relation by a single constant; in fact, it will be necessary to specify this relation by an actual curve such as that illustrated in Fig. 2.5*b*. Analytical work with a nonlinear load-elongation curve is generally more complicated than that with a linear material whose load-elongation relation can be expressed by the simple analytical expression of (2.2). For this reason, when materials are only slightly nonlinear, it has become common practice to approximate the data by a straight line with a slope which will fit the nonlinear behavior as well as possible. Cast iron, copper, and zinc are examples of slightly nonlinear materials for which one finds tabulated values of the modulus of elasticity. In the following sections, in addition to the linear problems, we shall consider a problem in which we use the actual load-elongation curve of a nonlinear material. We believe that the use of nonlinear load-elongation curves gives some insights which would not be obtained if only linear materials were dealt with.

For most materials, experiments with small deformations show that the shortening of a rod due to a compressive force is equal to the extension due to a tensile force of the same magnitude. We shall assume, therefore, that Eq. (2.2) and curves such as Fig. 2.5 represent behavior in compression as well as in tension.

2.3 STATICALLY DETERMINATE SITUATIONS

The most straightforward way to become familiar with the ideas implied in (2.1) is to use them in the solution of several problems. In this section we shall consider statically determinate problems, i.e., ones in which the forces can be obtained without reference to the geometry of deformation.

Example 2.4 Figure 2.6 shows a triangular frame supporting a load of 5,000 lb. This is the same frame that was considered in Example 1.3; in the present instance the type and size of members and the nature of the connections are specified in greater detail. Our aim is to estimate the displacement at the point D due to the 5,000-lb load carried by the chain hoist.

As a model for the behavior of this system we shall take the same one that was finally chosen in Example 1.3. Essentially, this model (Fig. 2.7*a*) is one in which the bolted connection at C is treated as a frictionless pinned joint. Following (2.1), we analyze this model.

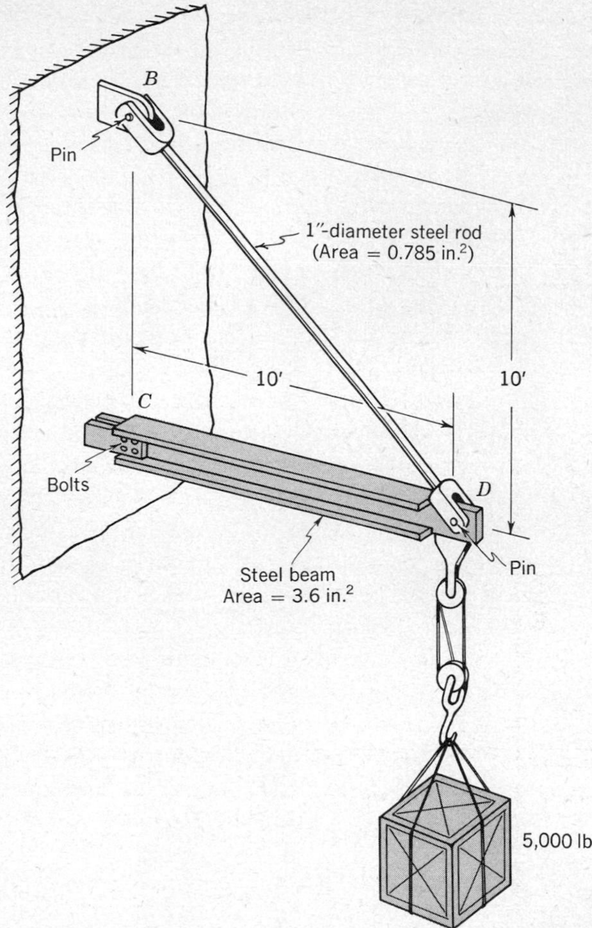

Fig. 2.6 Example 2.4.

FORCE EQUILIBRIUM

The forces in the members were determined in Example 1.3. These are shown in the free-body sketch of Fig. 2.7*b*.

FORCE-DEFORMATION RELATIONS

Equation (2.2) gives the deformations of the members *BD* and *CD* due to the forces acting on the ends of the members. These deformations are

$$\delta_{BD} = \left(\frac{FL}{AE}\right)_{BD} = \frac{7{,}070(14.14)(12)}{0.785(30 \times 10^6)} = 0.0510 \text{ in.} \qquad \text{extension}$$

$$\delta_{CD} = \left(\frac{FL}{AE}\right)_{CD} = \frac{5{,}000(10)(12)}{3.6(30 \times 10^6)} = 0.0056 \text{ in.} \qquad \text{compression} \tag{a}$$

GEOMETRIC COMPATIBILITY

Geometric compatibility of the deformations requires that the bars BD and CD move in such a way that, while the bars change lengths by the amounts calculated above, they remain straight and also remain fastened together at D. The mechanism by which this can be accomplished is illustrated in Fig. 2.7*c*. Assume, for the moment, that the bars are uncoupled at D and allowed to change lengths by δ_{BD} and δ_{CD} so that the bars now are of lengths BD_1 and CD_2, respectively. We see from the sketch that we can bring the ends D_1 and D_2 into coincidence without further change in the lengths of the bars by rotating the bars BD_1 and CD_2 about B and C as centers. Thus, due to the action of the 5,000-lb load, the point D moves to D_3, and the deformed shape of the structure is as shown in the dotted position in Fig. 2.7*c*.

Locating the point D_3 at the intersection of the two arcs in Fig. 2.7*c* is a rather lengthy calculation. Fortunately, since the deformations of the bars are only very small fractions of the lengths (these deformations are exaggerated greatly in Fig. 2.7), we can, with great accuracy, replace the arcs by the tangents to the arcs at D_1 and D_2 and obtain the intersection D_4 as an approximation to the location of D_3. Because of its simplicity and accuracy, this approximation is used in practically all engineering calculations of deflections of structures.

Employing this approximation of replacing the arcs with tangents, we illustrate in Fig. 2.7*d* the calculation of the horizontal and vertical displacements by which the point D moves to D_4. We begin by laying off δ_{BD} and δ_{CD} to locate the points D_1 and D_2. At D_1 and D_2 we erect the perpendiculars (tangents) D_1G and D_2F; these perpendiculars intersect in the desired point D_4. From the geometry of Fig. 2.7*d* we then can write

$$\begin{aligned} \delta_H &= \delta_{CD} = 0.0056 \text{ in.} \\ \delta_V &= D_2F + FD_4 \\ &= DG + FG \\ &= \sqrt{2}\,\delta_{BD} + \delta_{CD} = 0.0778 \text{ in.} \end{aligned} \tag{b}$$

We thus have accomplished our objective of making an estimate of the displacement of the point D of the frame in Fig. 2.6. As stated in Chap. 1, we shall return to this problem again in Chap. 8, using another model of the actual structure; at that time we shall find that the displacements calculated above represent a very good engineering estimate.

In analyzing a deformable structure according to the steps (2.1), the equilibrium requirements of the first step should be satisfied in the deformed equilibrium configuration. In most engineering applications the deformations are so small that it is sufficiently accurate to apply the equilibrium requirements to the *undeformed* configuration. As an example of this approximation the forces in Fig. 2.7*b* were obtained by applying the equilibrium requirements to the undeformed frame. If the equilibrium requirements are applied to the deformed shape of Fig. 2.7*d*, the forces in the bars are not significantly different (the tension in BD is decreased by 2.1 lb, and the compression in CD is decreased by 0.24 lb). In Chap. 9 we shall consider a class of problems where it *is* essential to apply the equilibrium requirements in deformed configurations.

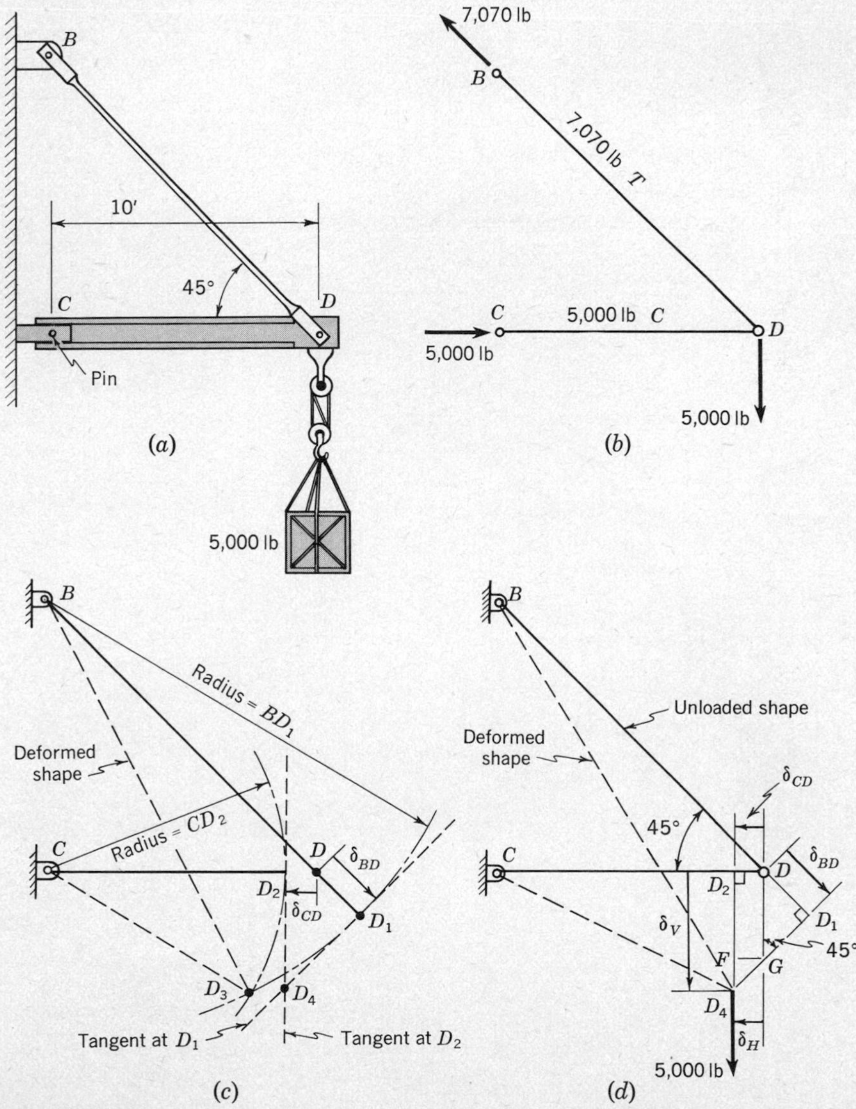

Fig. 2.7 Example 2.4.

Example 2.5 Figure 2.8*a* shows the truss of Example 1.4 with exactly the same loads. The truss material is aluminum; all the outer members of the truss have a cross-sectional area of 4 in.^{2}, and each of the three inner members has an area of 2 in.^{2} We wish to determine how much the length of each member changes due to the loads shown in Fig. 2.8*a*.

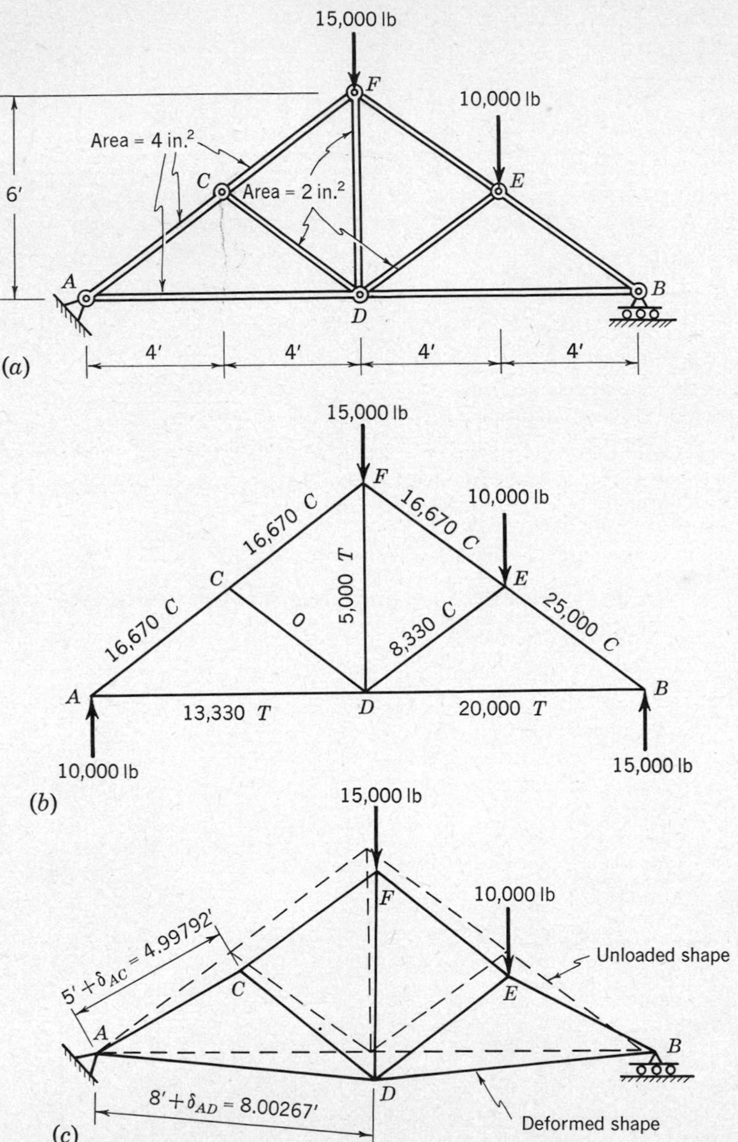

Fig. 2.8 Example 2.5.

We adopt as a model the same one chosen in Example 1.4; namely, one where the roller support at B can exert only a vertical reaction, and one where the force in each member is directed along the member. Applying the steps (2.1) to our model, we find the following.

FORCE EQUILIBRIUM

By continuing in a similar manner the analysis of bar forces started in Example 1.4, the force in each member of the truss may be obtained. The results of such an analysis are shown in Fig. 2.8*b*, where the symbols T and C are used to indicate whether the force is tensile or compressive.

GEOMETRIC COMPATIBILITY OF DEFORMATION

The members of the truss make up a series of triangles, and the truss is the sum of these triangles. If the pins are frictionless and allow free rotation, then the three members which form any one triangle also will form a triangle if the three members change their lengths. Furthermore, adjacent triangles can distort independently of each other in the present example, since the roller support at B is free to move horizontally to accommodate the distortion of the triangles which make up the truss. From this we conclude that each member of the truss is free to lengthen or shorten without any restraint being imposed by the other members of the truss. The overall behavior of the truss is indicated in Fig. 2.8*c*, where the deflections have been exaggerated greatly for purposes of illustration.

RELATIONS BETWEEN FORCES AND DEFORMATIONS

The deformation of each member is obtained from Eq. (2.2). For example, the deformations of the three members which make up the triangle ACD are

$$\delta_{AC} = \left(\frac{FL}{AE}\right)_{AC} = \frac{16{,}670(5)}{4(10 \times 10^6)} = 0.00208 \text{ ft} \qquad \text{compression}$$

$$\delta_{AD} = \left(\frac{FL}{AE}\right)_{AD} = \frac{13{,}330(8)}{4(10 \times 10^6)} = 0.00267 \text{ ft} \qquad \text{extension}$$

$$\delta_{CD} = \left(\frac{FL}{AE}\right)_{CD} = 0$$

The changes in length of the other members of the truss can be obtained from similar calculations.

The determination of the displacement of any point, say D, in the truss of Fig. 2.8*c* is a fairly cumbersome calculation. However, the principles involved in such a calculation are quite simple, as was illustrated in Example 2.4. Computer programs have been devised to facilitate the calculation of displacements. This will be illustrated in Sec. 2.5. In Sec. 2.6 we shall develop an *energy method* which provides a convenient means of calculating deflections of linearly elastic structures.

Example 2.6 The stiff horizontal beam *AB* in Fig. 2.9*a* is supported by two soft copper rods *AC* and *BD* of the same cross-sectional area but of different lengths. The load-deformation diagram for the copper is shown in Fig. 2.9*b*. A vertical load of 34,000 lb is to be suspended from a roller which rides on the horizontal beam. We do not want the roller to move after the load is put on, so we wish to find out where to locate the roller so that the beam will still be horizontal in the deflected position. Also, we should like to know if the location would be the same if the load is increased from 34,000 lb to 68,000 lb.

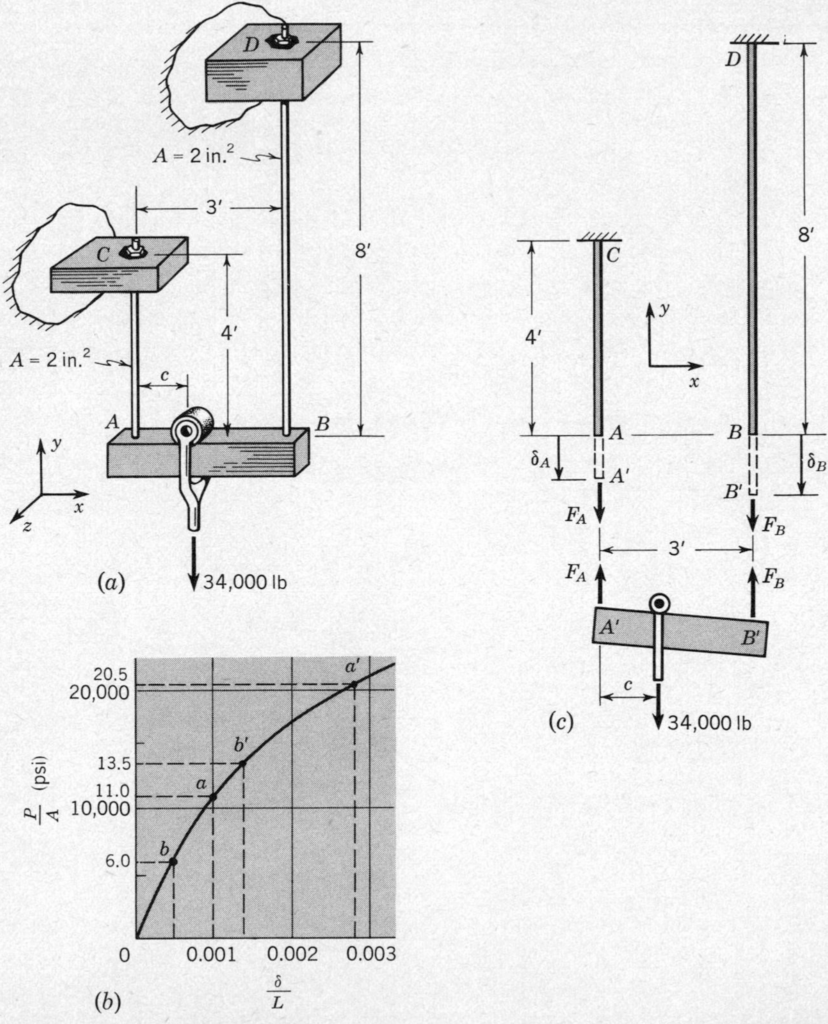

Fig. 2.9 Example 2.6.

Before proceeding with the solution to the problem, we should like to point out that this situation is fundamentally the same as that treated in Example 2.2: Find the location of a load which will produce a specified deflection of a stiff beam supported on two "springs." The present situation is computationally more complicated than Example 2.2 because of the nonlinear nature of the "springs."

We select the model in Fig. 2.9*c* as representing the behavior of the system. In this model we have assumed that the points A and B deflect vertically to A' and B', and that the beam is stiff enough to be considered rigid. Also, we have assumed that there are no horizontal forces or couples acting between the beam and the bars. Applying to this model the three steps of (2.1), we obtain the following.

FORCE EQUILIBRIUM

The equilibrium requirements of $\Sigma\mathbf{F} = 0$ and $\Sigma\mathbf{M} = 0$ will be satisfied for the free body of the beam in Fig. 2.9*c* when

$$\begin{aligned} \Sigma F_y &= F_A + F_B - 34{,}000 = 0 \\ \Sigma M_{A'} &= 3F_B - c(34{,}000) = 0 \end{aligned} \qquad (a)$$

GEOMETRIC COMPATIBILITY

Since the beam moves down without rotating, we must have

$$\delta_A = \delta_B \qquad (b)$$

Using this equality, we see from the lengths of the bars in Fig. 2.9*a* that

$$\frac{\delta_A}{L_A} = \frac{\delta_B}{L_A} = \frac{\delta_B}{48} = 2\frac{\delta_B}{96} = 2\frac{\delta_B}{L_B} \qquad (c)$$

RELATION BETWEEN FORCE AND DEFORMATION

The diagram in Fig. 2.9*b* gives the relation between force and elongation. For example, if we enter the diagram with $F_B/A_B =$ 10,000 psi, we find $\delta_B/L_B = 0.0009$ (*d*)

The relations (*a*), (*c*), and (*d*) represent the formulation of the three steps of (2.1); that is, they represent our analysis of the *physics* of the problem. We now must combine (*a*), (*c*), and (*d*) mathematically to find the correct location of the roller. Dividing the first of Eqs. (*a*) by A_A, we have

$$\frac{F_A}{A_A} + \frac{F_B}{A_A} = \frac{34{,}000}{A_A}$$

Substituting $A_A = A_B = 2 \text{ in.}^2$, we obtain the following relation.

$$\frac{F_A}{A_A} + \frac{F_B}{A_B} = 17{,}000 \text{ psi} \qquad (e)$$

We now select an arbitrary value of δ_B/L_B. Then, using (*c*) to obtain δ_A/L_A, we enter the diagram in Fig. 2.9*b* and obtain F_B/A_B and F_A/A_A. We then check to see if these values satisfy (*e*). If (*e*) is not satisfied, we make a new guess for δ_B/L_B and obtain new values for F_A/A_A and F_B/A_B. Proceeding in this way, we find the points *a* and *b* in Fig. 2.9*b*. From these points,

$$\frac{F_A}{A_A} = 11{,}000 \text{ psi} \qquad F_A = 22{,}000 \text{ lb}$$

$$\frac{F_B}{A_B} = 6{,}000 \text{ psi} \qquad F_B = 12{,}000 \text{ lb} \tag{f}$$

$$\frac{\delta_A}{L_A} = 0.001 \text{ in./in.} \qquad \delta_A = \delta_B = 0.048 \text{ in.}$$

Substituting this value for F_B in the second of Eqs. (*a*), we obtain the required location of the roller.

$$c = 1.06 \text{ ft} \tag{g}$$

If we repeated the analysis for a load of 68,000 lb, the solution would be represented by the points a' and b' in Fig. 2.9*b*, with the results

$$\begin{aligned} F_A &= 41{,}000 \text{ lb} \\ F_B &= 27{,}000 \text{ lb} \\ \delta_A &= \delta_B = 0.13 \text{ in.} \\ c &= 1.19 \text{ ft} \end{aligned} \tag{h}$$

It is not surprising that this value of *c* differs from the previous result, since a nonlinear material does not produce equal increments of elongation for equal increments of load. For example, in Fig. 2.9*b* a value of P/A of 10,000 psi produces a δ/L of 0.0009, whereas the next increment of 10,000 psi produces an increment in δ/L of 0.0017.

Example 2.7 A thin ring of internal radius r, thickness t, and width b is subjected to a uniform pressure p (psi) over the entire internal surface, as shown in Fig. 2.10*a*. A view looking down the axis of the ring is sketched in Fig. 2.10*b*. We should like to determine the forces in the ring. We also should like to determine the deformation of the ring due to the internal pressure.

The model which we assume to represent the behavior of the hoop is shown in Fig. 2.10*c*, which is a free body obtained by cutting the hoop on a diameter. We assume that at the cut sections 1 and 2 there are acting tangential and radial forces F_T and F_R which resist the internal pressure. Note that if, at the section 1, we arbitrarily assign to F_T and F_R the directions shown in Fig. 2.10*c*, then the directions shown at section 2 in the same sketch follow automatically from the symmetry of the hoop and its loading. We now apply (2.1) to our model.

FORCE EQUILIBRIUM

Before examining in detail the free body of Fig. 2.10*c*, it will be instructive to examine the free body of the other half of the hoop, shown in Fig. 2.10*d*. The directions of the forces F_T and F_R in Fig. 2.10*d* follow directly from those in Fig. 2.10*c* according to Newton's third law. We observe that the forces F_T act in similar manner on the two halves of the hoop, but the forces F_R act inward on the upper half and outward on the lower half. This action of the forces F_R violates the symmetry which we expect to find in the two halves of the hoop. We can resolve this paradox in only one way; we must conclude that the radial forces F_R are zero, and that on any radial cut made across the hoop there is acting only a tangential force F_T.

Returning now to the free body of Fig. 2.10*c*, we see that moment equilibrium is satisfied about the hoop center. Also, force equilibrium in the

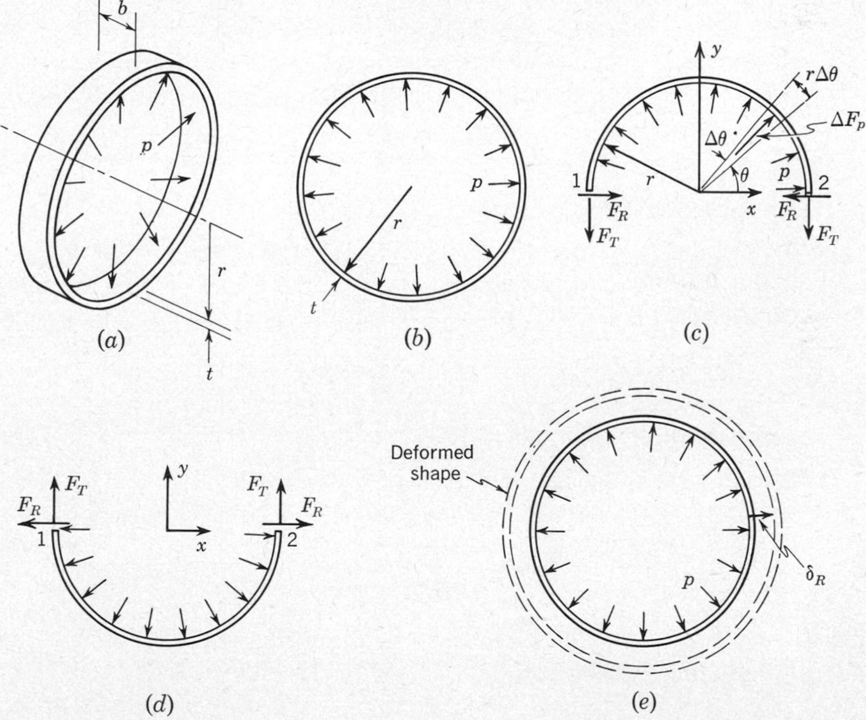

Fig. 2.10 Example 2.7.

x direction is satisfied as a result of the symmetry, and thus to ensure equilibrium we need only require force balance in the y direction. Considering an arc length $r\,\Delta\theta$ on the inner surface of the ring, there will be acting on this arc a radial force

$$\Delta F_p = p[b(r\,\Delta\theta)] \tag{a}$$

The component in the y direction of this radial force is

$$\Delta F_y = \Delta F_p \sin\theta = p[b(r\,\Delta\theta)]\sin\theta \tag{b}$$

In the limit as $\Delta\theta \to 0$ the sum of the forces ΔF_y acting on the free body of Fig. 2.10c becomes the integral in the following force-balance equation

$$\Sigma F_y = \int_{\theta=0}^{\theta=\pi} pbr \sin\theta\, d\theta - 2F_T = 0 \tag{c}$$

Integrating (c) we find

$$F_T = prb \tag{d}$$

It is of interest to note that the quantity $[(r\,\Delta\theta)\sin\theta]$ in (b) is the projection on the x axis of the arc length $r\,\Delta\theta$. Thus the force ΔF_y can be seen to be numerically equal to the pressure p acting over the projected area which is seen when we look in the y direction. Since this is true for each element of arc length, it must also be true for the entire half hoop for which the projected area is $2rb$, and our equilibrium equation for the half hoop can be written directly as

$$\Sigma F_y = p(2rb) - 2F_T = 0 \tag{e}$$

FORCE-DEFORMATION RELATION

The hoop may be thought of as a flat plate of thickness t, width b, and length $2\pi(r + t/2)$ subjected to a tensile force F_T given by (d). Using this model, we can calculate from (2.2) the increase in the circumference of the hoop, δ_T.

$$\delta_T = \frac{F_T[2\pi(r + t/2)]}{(bt)E} = \frac{2\pi p r^2}{tE}\left(1 + \frac{t}{2r}\right) \tag{f}$$

GEOMETRIC COMPATIBILITY

Since the circumference of a circle of radius r is equal to $2\pi r$, an increase in circumference of δ_T must be accompanied by a radial expansion δ_R, as shown in Fig. 2.10e, where

$$\delta_R = \frac{\delta_T}{2\pi} \tag{g}$$

Substituting (f), we find

$$\delta_R = \frac{pr^2}{tE}\left(1 + \frac{t}{2r}\right) \tag{h}$$

If we have a *thin* hoop we can neglect $t/2r$ compared to unity, and thus arrive at the following result which is used in engineering calculations involving thin hoops.

$$\delta_R = \frac{pr^2}{tE} \tag{i}$$

The approximations inherent in the "thin-ring" treatment above are typical of many "engineering approximations." In order to realistically employ approximate methods, we must learn to judge a good approximation from a poor one. We develop this capability by comparing the approximate result with either an exact result or an experimental result. For rings, where exact analytical solutions are available for "thick rings," we find that the approximate solutions are good when $t/r < 0.1$ (see Sec. 5.7).

Example 2.8 This problem situation is illustrated in Fig. 2.11*a* and *b*. In a test on an engine, a braking force is supplied through a lever arm *EF* to a steel brake band *CBAD* which is in contact with half the circumference of a 24-in.-diameter flywheel. The brake band is $\frac{1}{16}$ in. thick and 2 in. wide and is lined with a relatively soft material which has a kinetic coefficient of friction of $f = 0.4$ with respect to the rotating flywheel. The operator wishes to predict how much elongation there will be in the section *AB* of the brake band when the braking force is such that there is a tension of 9,000 lb in the section *BC* of the band.

We begin to formulate our model by drawing a free-body sketch of the section *AB* of the brake band, shown in Fig. 2.11*c*. There will be forces acting at all points of contact between the band and the drum. These forces are shown as components normal and tangential to the surface of contact, where we assume the tangential component is caused by friction between the flywheel and the lining and thus is shown acting in the same direction as the motion of the flywheel past the band. Figure 2.11*d* shows a free-body sketch of an element of the band. In this sketch it is assumed that the force in the band is a tangential force T which varies along the circumference, changing by an amount ΔT over the length $R\,\Delta\theta$. The total radial component of force on the element is given the symbol ΔN, and the tangential component caused by the friction then is $f\,\Delta N$. We shall assume that the force T is carried entirely by the steel band and not at all by the lining, which was described as being relatively soft. We now have specified a model which is explicit in its provision for carrying of the loads and thus should be a suitable basis for calculating the desired elongation.

It is important to emphasize that because the tension varies along the brake band we will derive a *differential equation* for its variation along the band. We do this by considering the differential element shown in Fig. 2.11*d* and letting the size of the element shrink to zero. By letting the element size become vanishingly small, we are finding the conditions which must be satisfied at a *point* on the band.

Proceeding in the same manner as in the previous examples, we analyze this model with the use of Eq. (2.1).

FORCE EQUILIBRIUM

All the forces on the small element in Fig. 2.11*d* may be considered concurrent, and thus equilibrium is satisfied when $\Sigma\mathbf{F} = 0$. All forces are parallel to the $r\theta$ plane, and $\Sigma\mathbf{F} = 0$ can be satisfied conveniently by requiring that the sum of the force components in the r and θ directions be zero.

$$\begin{aligned}\Sigma F_r &= \Delta N - T \sin\frac{\Delta\theta}{2} - (T + \Delta T)\sin\frac{\Delta\theta}{2} = 0 \\ \Sigma F_\theta &= (T + \Delta T)\cos\frac{\Delta\theta}{2} - T\cos\frac{\Delta\theta}{2} - f\,\Delta N = 0\end{aligned} \tag{a}$$

Considering the free body in Fig. 2.11*d*, we note that the angle $\Delta\theta$ is small (in the limit, zero). For small angles it is frequently convenient to make the following approximations.

$$\sin\theta \approx \theta$$

$$\tan\theta \approx \theta$$

$$\cos\theta \approx 1$$

These approximations are accurate up to surprisingly large values of θ as

Table 2.2

θ		$\sin\theta$	$\tan\theta$	$\cos\theta$
Degrees	*Radians*			
0	0	0	0	1
5	0.0873	0.0872	0.0875	0.9962
10	0.1745	0.1736	0.1763	0.9848
15	0.2618	0.2588	0.2679	0.9659

shown in Table 2.2. Using these approximations, Eqs. (*a*) become

$$\begin{aligned}\Delta N - T\frac{\Delta\theta}{2} - (T + \Delta T)\frac{\Delta\theta}{2} &= 0 \\ (T + \Delta T) - T - f\,\Delta N &= 0\end{aligned} \tag{b}$$

Neglecting ΔT compared to $2T$ in the first of (*b*), and eliminating ΔN between the two equations, we obtain

$$\frac{\Delta T}{\Delta\theta} = fT \tag{c}$$

In the limit as $\Delta\theta \to 0$ this becomes the derivative

$$\frac{dT}{d\theta} = fT \tag{d}$$

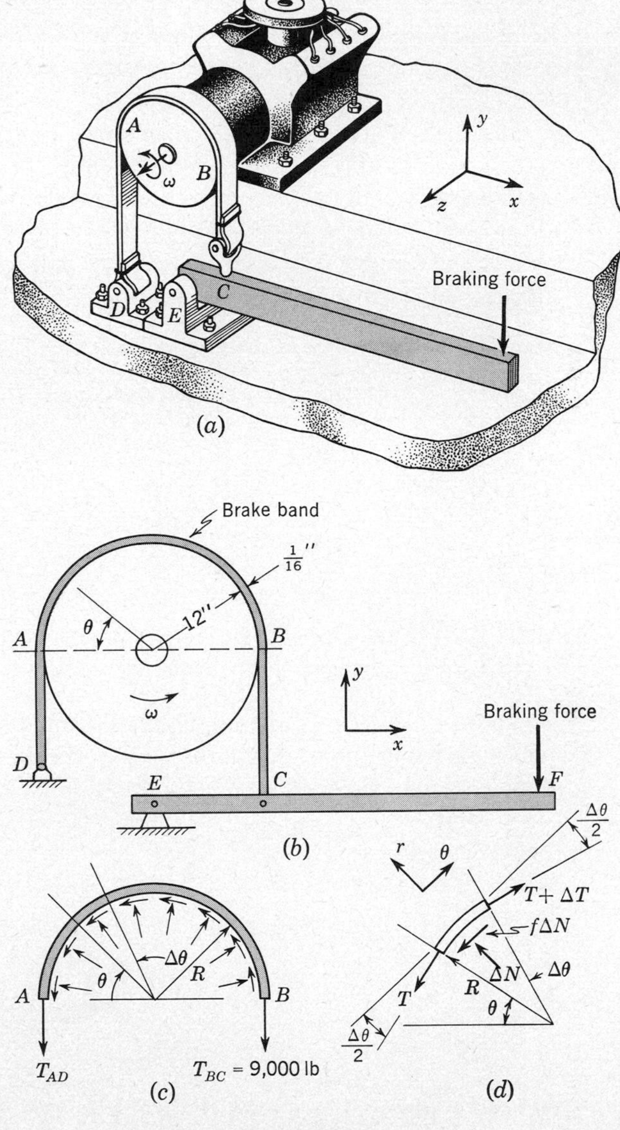

Fig. 2.11 Example 2.8.

Integrating (*d*) and satisfying the boundary condition that $T = T_{AD}$ at $\theta = 0$, we obtain

$$T = T_{AD}e^{f\theta} \tag{e}$$

where e is the base of natural logarithms. We now can calculate T_{AD} from the condition that $T = T_{BC} = 9{,}000$ lb at $\theta = \pi$. Then,

$$T = 2560e^{0.4\theta} \text{ lb} \tag{f}$$

By application of equilibrium alone, we have found the force in the brake band and thus the problem is statically determinate. Before going on to consider the elongation, we should like to emphasize the surprising nature of the variation in tension in the brake band. The tension varies *exponentially* with angular location, the tension at B being 3.5 times that at A. If the band extended all around the circumference of the flywheel, the ratio of the tensions at the ends would be 12.3! Many machines employ this frictional behavior to advantage, much as the sailor uses it to halt the motion of a large ship by taking a few turns of a rope about a piling.

RELATION BETWEEN FORCE AND DEFORMATION

Applying (2.2) to the small element of length $R\,\Delta\theta$, we find the elongation $\Delta\delta$ to be

$$\Delta\delta = \frac{TR\,\Delta\theta}{AE} \tag{g}$$

We see that the elongation varies with position along the band. To calculate total deflection, we need to consider the integral of the incremental variations along the band.

GEOMETRIC COMPATIBILITY

The total elongation of the brake band from A to B, δ_{AB}, is the sum of the tangential elongations $\Delta\delta$ of the small elements of length $R\,\Delta\theta$ shown in Fig. 2.11*d*. In the limit as $\Delta\theta \to 0$ this sum becomes the following integral:

$$\delta_{AB} = \int_{\theta=0}^{\theta=\pi} d\delta = \int_{\theta=0}^{\theta=\pi} \frac{TR\,d\theta}{AE} \tag{h}$$

Substituting (*e*) and integrating, we obtain an estimate of the elongation of the section AB.

$$\begin{aligned}\delta_{AB} &= \frac{T_{AD}R}{AE}\int_0^{\pi} e^{f\theta}\,d\theta = \frac{T_{AD}R}{AEf}(e^{f\pi} - 1) \\ &= \frac{2560(12)(e^{0.4\pi} - 1)}{0.125(30 \times 10^6)(0.4)} = 0.051 \text{ in.}\end{aligned} \tag{i}$$

2.4 STATICALLY INDETERMINATE SITUATIONS

We shall now consider two examples in which we must examine the deformation of the system in order to determine the manner in which the forces are distributed within the system.

Example 2.9 Figure 2.12*a* shows the pendulum of a clock which has a 3-lb weight suspended by three rods of 30-in. length. Two of the rods are made of brass and the third of steel. We wish to know how much of the 3-lb suspended weight is carried by each rod.

Our model of the system is shown in Fig. 2.12*b*. We assume that the support at the top and the weight at the bottom are stiff and act as rigid members. Because of the symmetry of the rod arrangement and the loading,

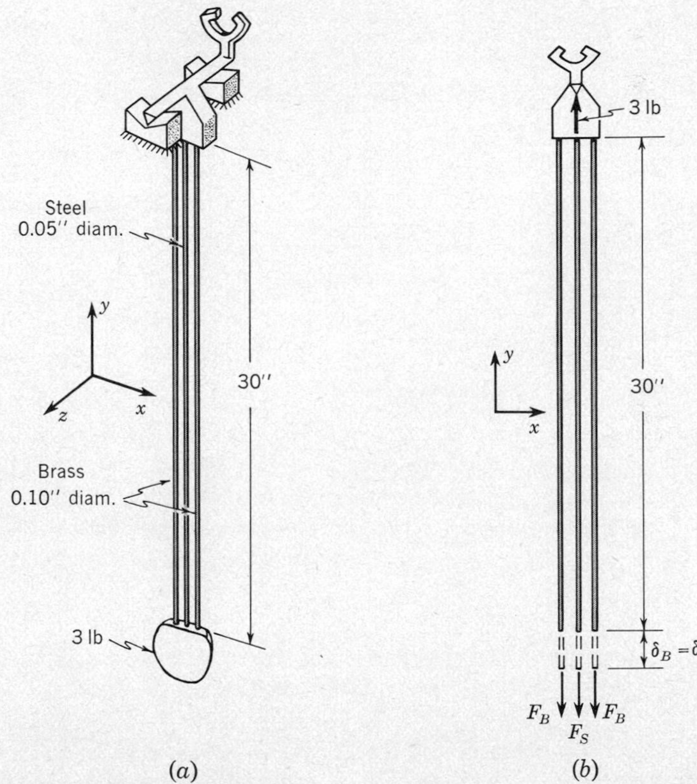

Fig. 2.12 Example 2.9.

each brass rod will carry the same load and all three rods will elongate the same amount. Applying the steps of (2.1) to our model, we obtain the following results:

FORCE EQUILIBRIUM

$\Sigma\mathbf{M} = 0$ is satisfied because of the symmetry of the force system. $\Sigma\mathbf{F} = 0$ is satisfied by

$$\Sigma F_y = 3 - F_S - 2F_B = 0 \tag{a}$$

GEOMETRIC COMPATIBILITY

The rods extend equal amounts, and so

$$\delta_S = \delta_B \tag{b}$$

RELATION BETWEEN FORCES AND DEFORMATIONS

Using Eq. (2.2), we obtain

$$\delta_S = \frac{F_S L_S}{A_S E_S} \qquad \delta_B = \frac{F_B L_B}{A_B E_B} \tag{c}$$

Combining (*b*) and (*c*),

$$\begin{aligned} F_S &= \frac{A_S}{A_B}\frac{E_S}{E_B}\frac{L_B}{L_S}F_B \\ &= \frac{(0.05)^2}{(0.10)^2}\frac{(30 \times 10^6)}{(15 \times 10^6)}\frac{30}{30}F_B = 0.50F_B \end{aligned} \tag{d}$$

Combining (*a*) and (*d*), we find

$$F_S = 0.60 \text{ lb}$$

$$F_B = 1.20 \text{ lb}$$

Example 2.10 Figure 2.13*a* shows an instrument suspension consisting of two aluminum bars and one steel rod mounted in a stiff frame, together with a spring EA which is inclined at 45° to BA. In assembly the nut on the steel rod at D is tightened so there is no slack in the line BAD, and then the spring EA is installed with sufficient extension to produce a force of 10 lb. We wish to find the deflection of the joint A (relative to the frame) caused by the spring loading.

A simple model of the system is shown in Fig. 2.13*b*. We assume that the frame is essentially rigid compared to the aluminum bars and the steel rod; thus the points B, C, and D can be considered as fixed, and the deflection may be measured relative to these fixed points. Also, we shall consider the steel rod to be pinned at point D; this is a good approximation since AD is such a slender member (its length is 53 times its diameter) that any rotational restraint offered by the nut and washer will be negligible. Finally, we assume

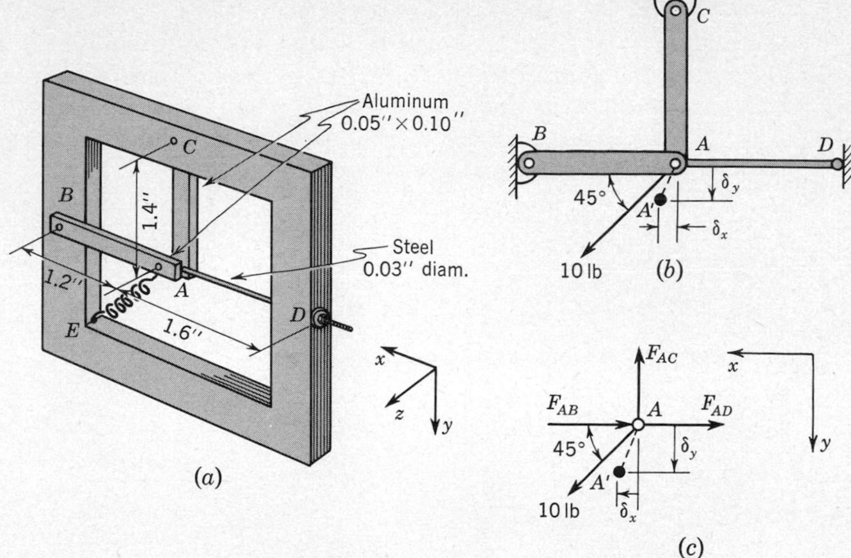

Fig. 2.13 Example 2.10.

that the action of the 10-lb spring force will be to move the point A to a new position A', as shown in Fig. 2.13*b*. Applying (2.1) to our model, we find the following:

FORCE EQUILIBRIUM

We assume that the forces in the members AC and AD are tensile and the force in AB is compressive, as illustrated in the free body of joint A in Fig. 2.13*c*. (It should be noted that, if we obtain a negative value for one of these forces, it will mean only that we assumed the wrong direction for the force; thus, in complicated situations one should not waste at lot of time deciding which "sense" to give to an unknown reaction, since the algebraic sign of the result will tell whether the original assumption was coreect.) For the free body in Fig. 2.13*c* the requirement of $\Sigma \mathbf{M} = 0$ is satisfied by the concurrence of all the forces, and $\Sigma \mathbf{F} = 0$ can be satisfied by

$$\begin{aligned} \Sigma F_x &= \frac{10}{\sqrt{2}} - F_{AD} - F_{AB} = 0 \\ \Sigma F_y &= \frac{10}{\sqrt{2}} - F_{AC} = 0 \end{aligned} \tag{a}$$

The statically indeterminate nature of the situation is indicated by the fact that we cannot determine the values of F_{AB} and F_{AD} from these equilibrium equations alone.

GEOMETRIC COMPATIBILITY

The deflections will be very small compared with the lengths of the bars. Thus, using the approximation developed in Example 2.4, we may assume that a movement of the end of a bar in a direction perpendicular to the axis of the bar can be accomplished without any change in the length of the bar. Then the extensions and contractions of the bars are

$$\begin{aligned} \delta_{AC} &= \delta_y \qquad \text{extension} \\ \delta_{AD} &= \delta_x \qquad \text{extension} \\ \delta_{AB} &= \delta_x \qquad \text{compression} \end{aligned} \tag{b}$$

RELATION BETWEEN FORCES AND DEFORMATIONS

$$\begin{aligned} \delta_{AC} &= \left(\frac{FL}{AE}\right)_{AC} = \frac{7.07(1.4)}{0.005(10 \times 10^6)} = 0.00020 \text{ in.} \\ \delta_{AD} &= \left(\frac{FL}{AE}\right)_{AD} = \frac{F_{AD}1.6}{0.00071(30 \times 10^6)} \\ \delta_{AB} &= \left(\frac{FL}{AE}\right)_{AB} = \frac{F_{AB}1.2}{0.005(10 \times 10^6)} \end{aligned} \tag{c}$$

Solving (*a*), (*b*), and (*c*) simultaneously, we obtain

$$\begin{aligned} F_{AD} &= 1.72 \text{ lb} \qquad \text{tension} \\ F_{AB} &= 5.35 \text{ lb} \qquad \text{compression} \\ \delta_y &= 0.00020 \text{ in.} \\ \delta_x &= 0.00013 \text{ in.} \end{aligned} \tag{d}$$

2.5 COMPUTER ANALYSIS OF TRUSSES

In Example 2.5 we considered the calculation of the deflection of the pin joints of a plane truss. We did not carry the calculations very far, except to note that the procedure, while straightforward, is a fairly "cumbersome" calculation. Procedures of this type are ideal for computer applications. There are many computer programs available for the analysis of structures. In this section we will consider a specific program and use it for the analysis of plane truss problems. It is not our intention to present the method used in the program except to say that most computer programs use a matrix formulation[1] which encompass the

[1] There are many books available on matrix methods for structural analysis; see, for example, H. Martin, "Introduction to Matrix Methods of Structural Analysis," McGraw-Hill Book Company, New York, 1966. E. C. Pestel and F. A. Leckie, "Matrix Methods in Elastomechanics," McGraw-Hill Book Company, New York, 1963. O. Zienkiewicz and Y. K. Cheung, "The Finite Element Method in Structural and Continuum Mechanics," McGraw-Hill Book Company, New York, 1967.

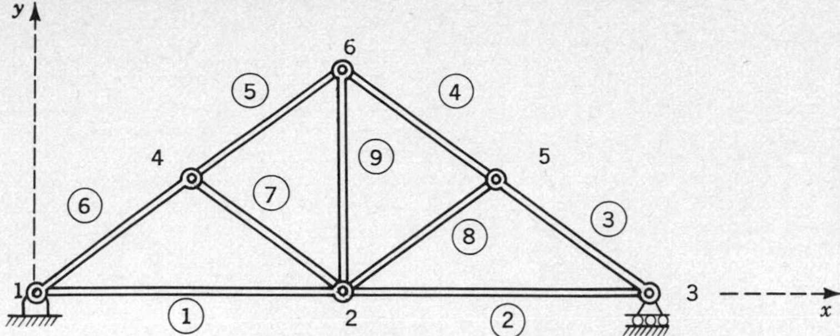

Fig. 2.14 Member and joint numbering for truss of Example 2.5.

three steps of (2.1). Our discussion is to provide a general view of the *use* of a specific program to obtain forces and deflections in trusses.

The program we will briefly discuss is the IBM STRESS program developed in the Civil Engineering Department at The Massachusetts Institute of Technology. Further discussion may be found in the user's manual.[1]

In general, the STRESS program is a convenient program for the calculation of forces and moments in two- and three-dimensional structures consisting of prismatic bars with either pinned or rigid joints, and subjected to concentrated or distributed loads, support motions, or temperature effects. We are interested in this section in the case of pinned joints corresponding to truss structures. In the STRESS version for the IBM 1130, the program is limited to a maximum number of 125 joints and a maximum number of 250 members. For large structures the use of the 1130 may be inconvenient and time-consuming, and a larger machine is suggested. The complete limitations on the program may be found in the user's manual.

Let us return once again to the truss problem discussed in Example 2.5 (Fig. 2.8) and analyze it by using STRESS. It is clear that we must input to the computer precisely those quantities which are required to solve for the bar forces and deformations, namely:

1. Structure geometry, i.e., coordinates of joints, supports, and information as to which members connect which joints
2. Loadings
3. Member sizes and elastic characteristics

In Fig. 2.14 we have established a *global* coordinate system, and have numbered each joint (1 → 6) and each member (① → ⑨). Table 2.3 shows the input

[1] Structural Engineering System Solver (STRESS) for the IBM 1130, Model 2B (1130-EC-03X), Version 2, User's Manual, H20-0340-2, August, 1969, IBM, New York.

Table 2.3 Input data for Example 2.5

```
STRUCTURE 1130 STRESS SAMPLE ON EXAMPLE 2.5
* COMMENT CARD.NOT PART OF PROGRAM.FOR EXAMPLE DISCUSSION
TYPE PLANE TRUSS
NUMBER OF JOINTS 6
NUMBER OF MEMBERS 9
NUMBER OF SUPPORTS 2
NUMBER OF LOADINGS 1
JOINT COORDINATES
* COORDINATES GLOBAL IN INCHES
1    0.0     0.0  S
2   96.0     0.0
3  192.0     0.0  S
4   48.0    36.0
5  144.0    36.0
6   96.0    72.0
MEMBER INCIDENCES
* ESTABLISHES POSITIVE DIRECTION LOCAL COORDS
1  1  2
2  2  3
3  3  5
4  5  6
5  6  4
6  4  1
7  4  2
8  5  2
9  6  2
JOINT RELEASES
* INDICATES IF SUPPORT JOINT CAN MOVE
3 FORCE X
* JOINT 3 CAN MOVE IN X DIRECTION
MEMBER PROPERTIES PRISMATIC
1 THRU 6  AX  4.0
7         AX  2.0
8         AX  2.0
9         AX  2.0
CONSTANTS E 10000.0 ALL
* MODULUS IN KIPS PER SQ.IN
LOADING 1 VERTICAL CONCENTRATED LOADS
* JOINT LOADS IN KIPS
JOINT LOADS
5 FORCE  Y -10.0
6 FORCE  Y -15.0
TABULATE ALL
* OUTPUT MEMBER FORCES AND JOINT DISPLACEMENTS
SOLVE
PROBLEM CORRECTLY SPECIFIED, EXECUTION TO PROCEED.
```

data for this example. While the input statements are almost self-explanatory, we shall briefly discuss them. Each line corresponds to an input-data card:

1. STRUCTURE statement describes the problem. The word STRUCTURE is required as the first word of the first card of the input after the input-control cards. Identification of the problem then follows on this card.
2. TYPE of structure. The TYPE statement describes the type of structure, which in our case is a *plane truss*.
3. General description of structure and number of loadings. Here we need four cards as indicated. The NUMBER OF LOADINGS card indicates how

many different *load systems* will be put on the structure, not how many individual external loads.

4. JOINT COORDINATES. Each joint is numbered as shown in Fig. 2.14 and its *xy* coordinates in *inches* with respect to the global coordinate axes is given. The letter *S* next to joints 1 and 3 indicates a support joint.
5. MEMBER INCIDENCES. Each member is connected to two joints which are indicated on a card for each member.
6. JOINT RELEASES. This statement specifies which support joints, if any, are free to move. Joint 3 is free to move in the *x* direction because of the roller, and we specify with a separate card under this heading that joint 3, a support joint, has no *X* force acting.
7. MEMBER PROPERTIES PRISMATIC. These set of cards indicate that each bar is prismatic. For each member the cross-sectional area *AX* in square inches must be given. In the present problem, members 1 through 6 have cross-sectional area 4.0 in.^{2}, and members 7 through 9 have 2.0 in.^{2}.
8. CONSTANTS. For a truss we must specify the elastic modulus *E* of each member in kips/in.^{2} (1 kip = 1,000 lb, a quaint civil engineering unit). Here all bars have the same modulus; we specify this by the statement ALL.
9. LOADING. This card indicates the name of the loading. We have only one loading which consists of vertical concentrated loads. The next card indicates that we have joint loads, and the next two cards indicate the joints at which the loads act, the magnitude of the loads in kips, and the direction of the loads with respect to the global coordinate axes.
10. TABULATE ALL. This card indicates the type of output required.
11. SOLVE. This is the final card of the input deck which is then followed by control cards.

If the problem is specified correctly, the last statement of Table 2.3, PROBLEM CORRECTLY SPECIFIED, EXECUTION TO PROCEED, is typed out on the 1130 console after the input data are processed.

A check on the input geometry can be made with STRESS in that it is possible to sketch the truss on a plotter connected to the computer to confirm the correctness of the input. For more complicated problems than the present one, this is a convenient and useful check.

The *output* format for the STRESS program is almost self-explanatory and is shown in Table 2.4. All loads are in kips, and deflections are in inches. The member loads are given for each loading; in this case we have only one loading.

In order to determine whether a bar is in tension or compression we must associate with each member a *local* coordinate system which is determined by the order of the specification of the joint incidences for each member in the input statements. For example, for member 8, the positive direction of the local axis is from 5 to 2 (check input table), Fig. 2.15.

Table 2.4 Output for Example 2.5

```
LOADING 1 VERTICAL CONCENTRATED LOADS
 ==========================================:


          MEMBER FORCES

 MEMBER JOINT    AXIAL FORCE
    1     1        -13.333
    1     2         13.333
    2     2        -19.999
    2     3         19.999
    3     3         24.999
    3     5        -24.999
    4     5         16.666
    4     6        -16.666
    5     6         16.666
    5     4        -16.666
    6     4         16.666
    6     1        -16.666
    7     4          0.000
    7     2          0.000
    8     5          8.333
    8     2         -8.333
    9     6         -4.999
    9     2          4.999


        APPLIED JOINT LOADS,   FREE JOINTS

JOINT    FORCE X      FORCE Y
   2      -0.000        0.000
   4      -0.000       -0.000
   5      -0.000      -10.000
   6      -0.000      -14.999


      REACTIONS,APPLIED LOADS SUPPORT JOINTS

JOINT    FORCE X      FORCE Y
   1       0.000       10.000
   3       0.000       14.999


          FREE JOINT DISPLACEMENTS

JOINT  X-DISPLACEMENT  Y-DISPLACEMENT
   2       0.0319          -0.1650
   4       0.0622          -0.1247
   5       0.0019          -0.1666
   6       0.0478          -0.1470


          SUPPORT JOINT DISPLACEMENTS

JOINT  X-DISPLACEMENT  Y-DISPLACEMENT
   1       0.0000          0.0000
   3       0.0799          0.0000
```

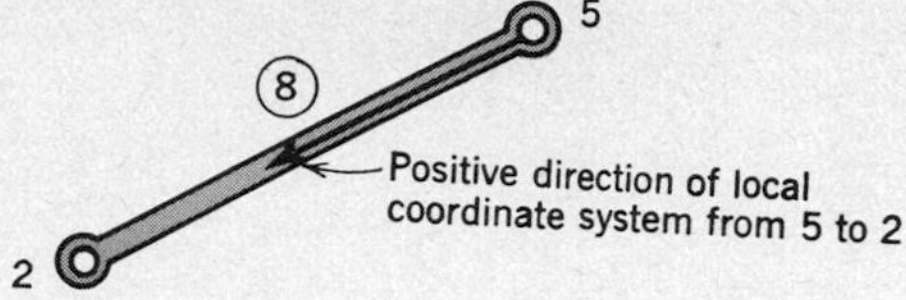

Fig. 2.15 Local coordinate system for member 8 of Fig. 2.14.

The output statement of the member forces for member 8 is

MEMBER	JOINT	AXIAL FORCE
8	5	8.333
8	2	−8.333

At joint 5, a positive force acting on the bar is a compressive force and a negative force at joint 2 again indicates compression.

After the listing of member forces in the output statement we find "Applied Joint Loads, Free Joints." These results check on the values of the applied forces at the free joints by using the final answers from each member force. This calculation is a convenient check on the accuracy of the program.

Next we find the reactive forces at the supports in terms of components in the global coordinate system.

Finally, the joint displacements in inches of each joint are given.

It should be clear from our efforts in Sec. 2.3 that the problem just solved was *statically determinate*. That is, we could solve for all forces without considering the deformations. If, however, both truss supports were rigid pin joints, such that point *B* could not move, the problem would become indeterminate (Fig. 2.16). In general, using the methods of Sec. 2.3, we would have a very tiresome time attempting to find the forces in all members and the deflections of all joints. Using the

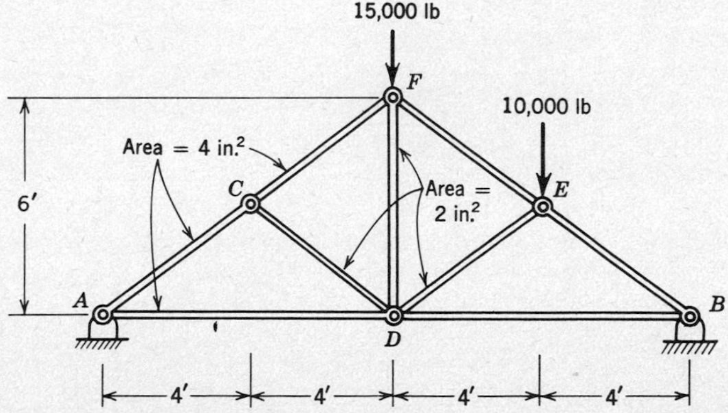

Fig. 2.16 Statically indeterminate version of truss in Example 2.5.

STRESS program, however, all we have to do is to rerun the program with point B fixed. This is accomplished simply by omitting the JOINT RELEASE CARDS. The output data are shown in Table 2.5.

The use of STRESS (or any similar computer program[1] to which the reader may have access) in these two examples indicates the convenience of the use of standard structural-mechanics computer programs. An engineer must often decide whether to use a hand-computation procedure or to turn to a computer. The choice depends on the availability and expense of the computer as compared with the value of the engineer's time. In preliminary design stages, an engineer often gains additional insights by performing rough calculations himself. When the overall design layout has been settled, detailed calculations for ranges of parameter values are often turned over to the computer.

2.6 ELASTIC ENERGY; CASTIGLIANO'S THEOREM

In this section we give a brief introduction to the concept of energy and develop a powerful tool for calculating deflections of elastic systems.

The *work* done by a force vector **F** when its point of application moves through a displacement vector **ds** is the scalar, or dot, product

$$\mathbf{F} \cdot \mathbf{ds} = F \cos \theta \, ds$$

where θ is the angle between **F** and **ds**. In general, **F** will vary as the point of application follows a certain path so that the total work done by **F** is given by an integral

$$\int \mathbf{F} \cdot \mathbf{ds}$$

When work is done by an external force on certain systems, their internal geometric states are altered in such a way that they have the potential to give back equal amounts of work whenever they are returned to their original configurations. Such systems are called *conservative*, and the work done on them is said to be stored in the form of *potential energy*. For example, the work done in lifting a weight is said to be stored as gravitational potential energy. The work done in deforming an elastic spring is said to be stored as elastic potential energy. By contrast, the work done in sliding a block against friction is not recoverable; i.e., friction is a nonconservative mechanism.

Consider the elastic, but not necessarily linear, spring in Fig. 2.17. Let the spring undergo a gradual elongation process during which the external force F remains in equilibrium with the internal tension. The *potential energy* U associated

[1] See, for example, F. W. Beaufait, W. H. Rowan, Jr., P. G. Hoadley, R. M. Hackett, "Computer Methods of Structural Analysis," Prentice-Hall, Inc., Englewood Cliffs, N.J., 1970.

Table 2.5 Output for statically indeterminate version of Example 2.5

```
LOADING 1 VERTICAL CONCENTRATED LOADS
 ===========================================


          MEMBER FORCES

 MEMBER JOINT      AXIAL FORCE
    1     1            3.333
    1     2           -3.333
    2     2           -3.333
    2     3            3.333
    3     3           25.000
    3     5          -25.000
    4     5           16.666
    4     6          -16.666
    5     6           16.666
    5     4          -16.666
    6     4           16.666
    6     1          -16.666
    7     4            0.000
    7     2           -0.000
    8     5            8.333
    8     2           -8.333
    9     6           -4.999
    9     2            4.999


        APPLIED JOINT LOADS,    FREE JOINTS

JOINT    FORCE X       FORCE Y
  2        0.000         0.000
  4        0.000         0.000
  5       -0.000       -10.000
  6       -0.000       -15.000


      REACTIONS,APPLIED LOADS SUPPORT JOINTS

JOINT    FORCE X       FORCE Y
  1       16.666        10.000
  3      -16.666        14.999


          FREE JOINT DISPLACEMENTS

JOINT  X-DISPLACEMENT  Y-DISPLACEMENT
  2      -0.0080          -0.1117
  4       0.0222          -0.0713
  5      -0.0380          -0.1132
  6       0.0078          -0.0937
```

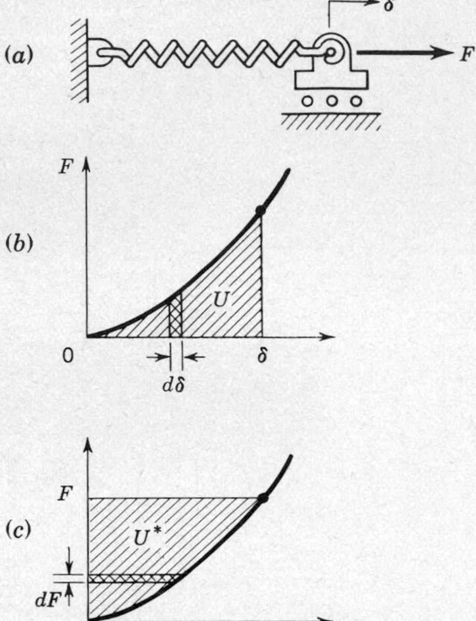

Fig. 2.17 Nonlinear spring (*a*) has deflection δ due to force F. Potential energy U is area *under* force-deflection curve (*b*). Complementary energy U^* is area *above* curve (*c*).

with an elongation δ is defined to be the work done by F in this process

$$\int \mathbf{F} \cdot \mathbf{ds} = \int_0^\delta F\, d\delta = U \tag{2.3}$$

In Fig. 2.17*b*, this energy appears as the shaded area under the force-deflection curve. Note that U is a function of the elongation δ. If this spring should happen to be part of a larger elastic system, it will always contribute the energy (2.3) to the total stored energy of the system whenever its individual elongation is δ.

The magnitude of the energy that a given spring or member can store is sometimes an important consideration in mechanical design. Parts which are subjected to impact loads are often chosen on the basis of their capacity to absorb energy.

Next, consider the general elastic system of Fig. 2.18*a*, which can be loaded by an arbitrary number of loads. At a typical loading point A_i the load is $\mathbf{P}_i$, and the equilibrium displacement due to all the loads is $\mathbf{s}_i$. If, during the loading process, the displacements $\mathbf{s}_i$ are permitted to grow slowly through a sequence of equilibrium configurations, the total work done by all the external loads will equal[1] the total potential energy U stored by all the internal elastic members

[1] See Prob. 2.53.

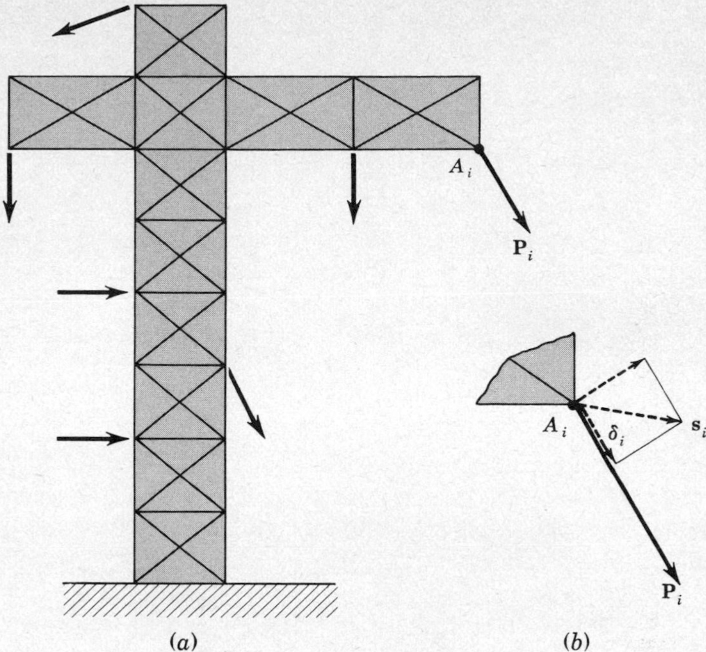

Fig. 2.18 General elastic structure (*a*) subjected to loads $\mathbf{P}_i$ applied at points A_i; (*b*) enlarged view showing displacements $\mathbf{s}_i$ at points A_i.

$$\sum_i \int_0^{\mathbf{s}_i} \mathbf{P}_i \cdot d\mathbf{s}_i = U \tag{2.4}$$

We next introduce the concepts of complementary work and complementary energy to obtain a result parallel to (2.4) but with the roles of force and deflection interchanged.

When the point of application of a variable force $\mathbf{F}$ undergoes a displacement $\mathbf{s}$, the *complementary work* is

$$\int \mathbf{s} \cdot d\mathbf{F}$$

When complementary work is done on certain systems, their internal force states are altered in such a way that they are capable of giving up equal amounts of complementary work when they are returned to their original force states. Under these circumstances the complementary work done on such a system is said to be stored as *complementary energy*. The class of systems which store complementary energy is not as wide as that which stores potential energy, but it does include all elastic systems for which the equilibrium requirements can be applied in the undeformed configuration.[1]

[1] See, for example, T. M. Charlton, "Energy Methods in Applied Statics," p. 63, Blackie and Son, Ltd., London, 1959.

Let us reconsider the gradual loading process for the nonlinear spring of Fig. 2.17*a*. The *complementary energy* U^* associated with a force F is defined to be the complementary work done by F

$$\int \mathbf{s} \cdot \mathbf{dF} = \int_0^F \delta \, dF = U^* \tag{2.5}$$

In Fig. 2.17*c* this appears as the shaded area above the force-deflection curve. Note that U^* is a function of the force F. If this spring should happen to be part of a larger elastic system, it will always contribute the complementary energy (2.5) to the total system complementary energy whenever the force in it has the value F.

Now we return to the general elastic structure of Fig. 2.18*a*. To facilitate the calculation of complementary work we show, in Fig. 2.18*b*, the displacement $\mathbf{s}_i$ decomposed into components parallel and perpendicular to $\mathbf{P}_i$. The parallel, or *in-line*, component is called δ_i. Now if the loads in Fig. 2.18*a* are gradually increased from zero so that the system passes through a succession of equilibrium states, the total complementary work done by all the external loads will equal[1] the total complementary energy U^* stored by all the internal elastic members.

$$\sum_i \int_0^{\mathbf{P}_i} \mathbf{s}_i \cdot \mathbf{dP}_i = \sum_i \int_0^{\mathbf{P}_i} \delta_i \, dP_i = U^* \tag{2.6}$$

The energy functions U and U^* have many uses in mechanics. They are used to construct variational principles[2] which provide alternatives to the direct application of the steps (2.1). These principles in turn form the starting point for a variety[3] of approximate solution techniques. We shall not, however, discuss these matters further in this book. Our reason for introducing the energy functions here is that we wish to prove a theorem which provides a simple but powerful tool for the calculation of deflections in elastic systems.

The theorem follows almost immediately from (2.6). Suppose that the system in Fig. 2.18*a* is in its equilibrium position with the complementary energy (2.6). We now consider a small increment ΔP_i to the load P_i while all the other loads remain fixed. The internal forces will change slightly to maintain force equilibrium, and the increment in complementary work will equal the increment in complementary energy ΔU^*. For small ΔP_i we have, approximately

$$\delta_i \, \Delta P_i = \Delta U^*$$

[1] See Prob. 2.54.

[2] See, for example, chap. 1 of S. H. Crandall, D. C. Karnopp, E. F. Kurtz, Jr., and D. C. Pridmore-Brown, "Dynamics of Mechanical and Electromechanical Systems," McGraw-Hill Book Company, New York, 1968.

[3] See, for example, H. L. Langhaar, "Energy Methods in Applied Mechanics," John Wiley & Sons, Inc., New York, 1962, and S. H. Crandall, "Engineering Analysis," McGraw-Hill Book Company, New York, 1956.

or

$$\frac{\Delta U^*}{\Delta P_i} = \delta_i$$

In the limit as $\Delta P_i \to 0$ this approaches a derivative which we indicate as a partial derivative since all the other loads were held fixed

$$\frac{\partial U^*}{\partial P_i} = \delta_i \tag{2.7}$$

This result is a form of *Castigliano's theorem* (extended to nonlinear systems). It states that if the total complementary energy U^* of a loaded elastic system is expressed in terms of the loads, the in-line deflection at any particular loading point is obtained by differentiating U^* with respect to the load at that point. The theorem can be extended to include moment loads M_i as well as force loads P_i. In the case of a moment load, the in-line displacement is the angle of rotation ϕ_i about the axis of the moment vector M_i, and in place of (2.7) we have

$$\frac{\partial U^*}{\partial M_i} = \phi_i \tag{2.8}$$

Although the theorem just proved applies to nonlinear elastic systems, we shall in this book use it only in connection with *linear* systems. In linear systems there is an essential simplification. In Fig. 2.17 we see that, in general, $U^* \neq U$, but when the force-deformation relation is linear the two shaded regions become triangles of *equal* area; that is, $U^* = U$. This means that for linear systems it is not essential to make a distinction between complementary energy and potential energy.

For example, consider a linear spring with the force-deflection law

$$F = k\delta \tag{2.9}$$

where k is the spring constant. According to (2.3), $U = \frac{1}{2}k\delta^2$, and according to (2.5), $U^* = F^2/2k$, but because of (2.9) these are equal in magnitude. As a consequence we shall henceforth discontinue making a distinction between potential and complementary energies. We shall use the nonspecific appellation *elastic energy* for any expression which has the same magnitude as $U = U^*$. Thus for the linear spring we say that the elastic energy is

$$U = \tfrac{1}{2}k\delta^2 = \tfrac{1}{2}F\delta = \frac{F^2}{2k} \tag{2.10}$$

Similarly, for the linear uniaxial member illustrated in Figs. 2.4 and 2.5 the elastic energy is

$$U = \frac{EA}{2L}\delta^2 = \tfrac{1}{2}P\delta = \frac{P^2L}{2EA} \tag{2.11}$$

In subsequent chapters we shall derive analogous expressions for elastic energy in members with more complex loading patterns.

To apply Castigliano's theorem to a linear-elastic system it is necessary to express the total elastic energy of the system in terms of the loads. This requires using the equilibrium requirements to express the internal member forces in terms of the applied loads. Then from formulas such as (2.10) and (2.11) the energy of each internal member is obtained. The total elastic energy U results from adding the energies of all the internal members. Finally, the in-line deflection δ_i at any loading point A_i is obtained by differentiation with respect to the load P_i

$$\delta_i = \frac{\partial U}{\partial P_i} \tag{2.12}$$

Example 2.11 Consider the system of two springs shown in Fig. 2.19. We shall use Castigliano's theorem to obtain the deflections δ_1 and δ_2 which are due to the external loads P_1 and P_2. To satisfy the equilibrium requirements the internal spring forces must be

$$\begin{aligned} F_1 &= P_1 + P_2 \\ F_2 &= P_2 \end{aligned} \tag{a}$$

The total elastic energy, using (2.10), is

$$U = U_1 + U_2 = \frac{(P_1 + P_2)^2}{2k_1} + \frac{P_2{}^2}{2k_2} \tag{b}$$

The deflections then follow from (2.12)

$$\begin{aligned} \delta_1 &= \frac{\partial U}{\partial P_1} = \frac{P_1 + P_2}{k_1} \\ \delta_2 &= \frac{\partial U}{\partial P_2} = \frac{P_1 + P_2}{k_1} + \frac{P_2}{k_2} \end{aligned} \tag{c}$$

For this case it is easy to verify that this solution satisfies all the requirements of the steps (2.1).

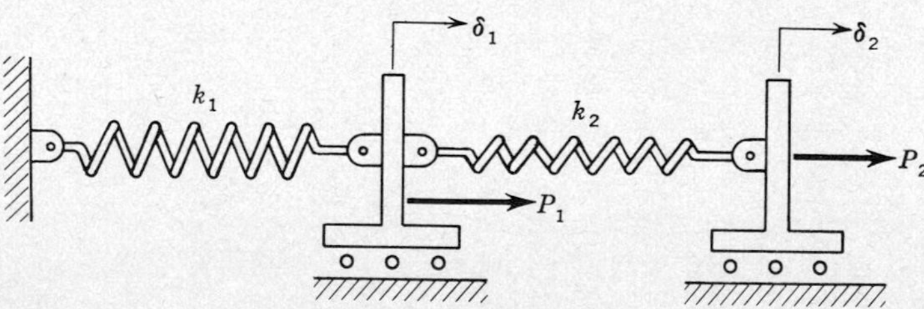

Fig. 2.19 Example 2.11.

Example 2.12 Let us consider again Example 2.4 (also Example 1.3), and determine the deflections using Castigliano's theorem. In Fig. 2.20 the isolated system from Example 2.4 is shown together with the applied loads. Because we will treat the members of the frame as springs, their "constants" are also given.

The total stored energy is the sum of the energy stored in the two members. We should note that the two energies *add* together even though one member is in tension and one in compression. We use the equilibrium requirements to express the member forces F_1 and F_2 in terms of the load P so that the total energy is

$$U = U_1 + U_2 = \frac{P_1{}^2}{2k_1} + \frac{P_2{}^2}{2k_2} = \frac{2P^2}{2k_1} + \frac{P^2}{2k_2} \tag{a}$$

We can calculate directly the deflection of point D in the direction of P (positive downward) from Eq. (2.12).

$$\delta_P = \frac{\partial U}{\partial P} = \frac{\partial}{\partial P}\left(\frac{P^2}{k_1} + \frac{P^2}{2k_2}\right) = 2P\left(\frac{1}{k_1} + \frac{1}{2k_2}\right) \tag{b}$$

$$\delta_P = 2 \times 5{,}000[7.20 + 0.556] \times 10^{-6} = 0.0776 \text{ in.}$$

In order to calculate the horizontal deflection at point D using Castigliano's theorem, there must be a horizontal force at D—but, alas, the horizontal force at D is zero. We can satisfy both requirements by applying a fictitious horizontal force Q and, after we have determined the horizontal displacement $\partial U/\partial Q$ in terms of P and Q, then setting $Q = 0$. Figure 2.21 shows the frame isolated with both P and Q applied. The forces in the individual bars which are required to satisfy the equilibrium requirements are indicated. The total energy in terms of the loads P and Q is

$$U = \frac{2P^2}{2k_1} + \frac{1}{2k_2}(P - Q)^2 \tag{c}$$

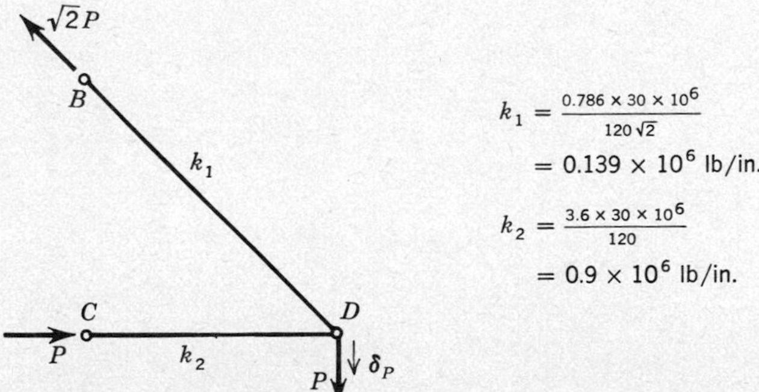

Fig. 2.20 Example 2.12.

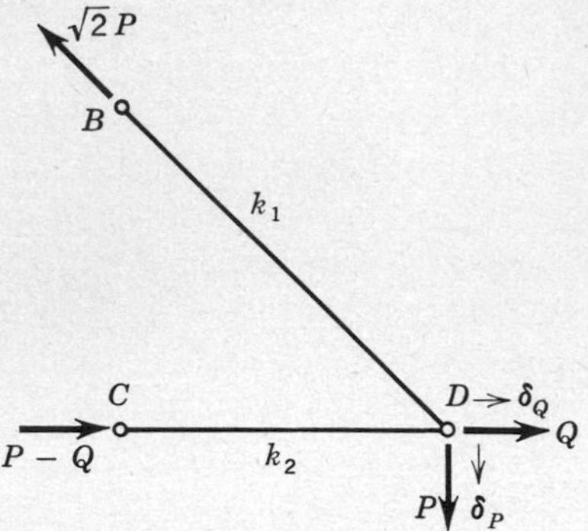

Fig. 2.21 Structure of Fig. 2.20 with fictitious load Q at D.

and

$$\delta_Q = \frac{\partial U}{\partial Q} = 0 - \frac{P - Q}{k_2} \tag{d}$$

but $Q = 0$, therefore

$$\delta_Q = \frac{-P}{k_2} = -0.00556 \text{ in.}$$

Note the ease by which the deflections can be obtained by the use of the energy method when compared to the method used in Example 2.4.

Example 2.13 As a final example in this chapter, let us use Castigliano's theorem to determine deflections in the truss problem that we considered in Example 2.5 and in the computer-solution example of Sec. 2.5.

The method of solution here is exactly the same as in the previous example; i.e., we calculate the total energy in terms of the real and fictitious loads, and determine deflections by taking appropriate derivatives. However, because a larger number of members is involved, it is worthwhile to set up a system that permits the necessary manipulations with the least effort.

If a truss is made of n axially loaded members, the energy stored in the ith member, according to (2.11), is

$$U_i = \frac{F_i^2 L_i}{2A_i E_i} \tag{a}$$

and the total energy in the system of n members is

$$U = \sum_{i=1}^{n} U_i \tag{b}$$

The deflection at *any* external (applied or fictitious) load P, in the *direction of* P, is simply

$$\delta_P = \frac{\partial U}{\partial P} = \frac{\partial}{\partial P}\sum_{i=1}^{n}\frac{F_i^2 L_i}{2A_i E_i} = \sum_{i=1}^{n}\frac{F_i L_i}{A_i E_i}\frac{\partial F_i}{\partial P} \tag{c}$$

The quantity $\partial F_i/\partial P$, which represents the rate of change of the force in the ith member with load P, can be thought of as the load in the ith member due to a *unit* load at P. Why? For convenience, we can rewrite (c) in terms of three quantities which we shall tabulate separately

$$\delta_P = \sum_{i=1}^{n} F_i \frac{L_i}{A_i E_i}\frac{\partial F_i}{\partial P} \tag{d}$$

We will again number the members as shown in Fig. 2.22. In Fig. 2.22 we show the truss properly supported by R_1, R_2, and R_3 and loaded by *fictitious* forces P and Q at the two points where we wish to determine deflection information. In Example 2.5 we solved for the forces F_i due to the actual applied loads. We can now set up a system for evaluating (d). In order to evaluate, using (d), the deflection at the joint at which the fictitious load P is applied, it appears that we need to find the forces F_i in each member as a function of the actual applied loads and in terms of P. However, once the member forces are found, we set $P = 0$ in (d). Therefore, we can use immediately the member forces F_i from the actual loads and the forces for a unit load at P to evaluate $\partial F_i/\partial P$.

In Table 2.6 we have tabulated the individual quantities in (d) as well as their products. In the first column the F_i are the *actual* forces in each member; that is, P does not appear here because it is actually zero. Ignore for now the appearance of Q in the first two rows. This will be used later to solve the statically indeterminate case where Q is not zero. The second column is self-explanatory. The third and fourth columns show the load in each of the

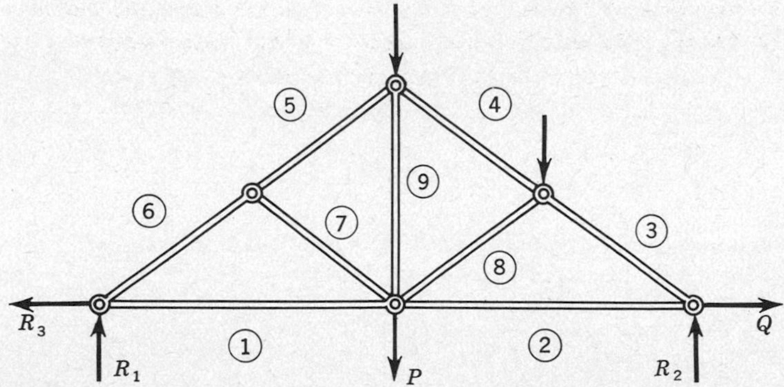

Fig. 2.22 Example 2.13.

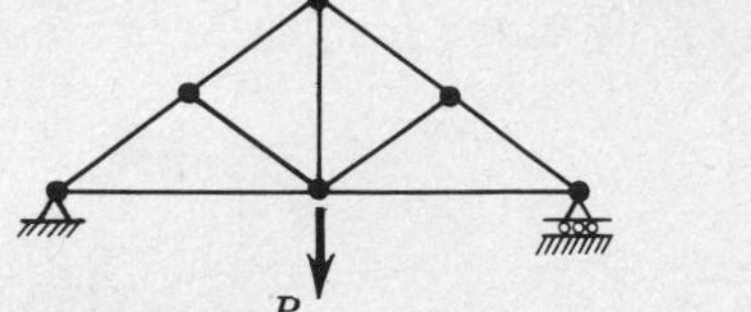

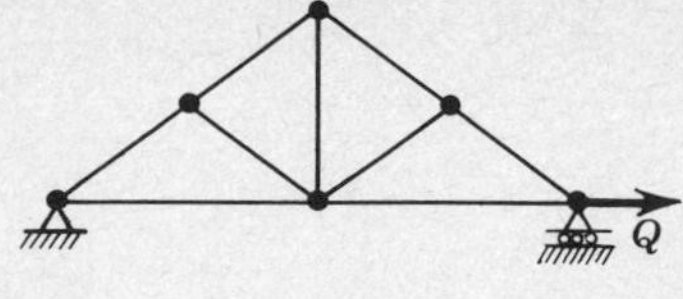

Fig. 2.23 Unit loads on truss of Example 2.13.

i members due only to a *unit* load P and for a *unit* load Q (Fig. 2.23). In the last columns we tabulate the appropriate products and their sums to get displacements. The power of this method should now be apparent.

Table 2.6 Truss solution by energy methods

i	F_i 10^3 lb	$(L/AE)^*$ in./lb	$\frac{\partial F_i}{\partial P}$	$\frac{\partial F_i}{\partial Q}$	$\left(\frac{FL}{AE}\frac{\partial F}{\partial P}\right)_i^\dagger$	$\left(\frac{FL}{AE}\frac{\partial F}{\partial Q}\right)_i^\dagger$
1	$+13.33+Q$	2.4×10^{-6}	$+2/3$	$+1$	21.36×10^{-3}	32.0×10^{-3}
2	$+20.0+Q$	2.4×10^{-6}	$+2/3$	$+1$	31.95×10^{-3}	48.0×10^{-3}
3	-25.0	1.5×10^{-6}	$-5/6$	0	31.26×10^{-3}	
4	-16.67	1.5×10^{-6}	$-5/6$	0	20.85×10^{-3}	
5	-16.67	1.5×10^{-6}	$-5/6$	0	20.85×10^{-3}	
6	-16.67	1.5×10^{-6}	$-5/6$	0	20.85×10^{-3}	
7	0	1.5×10^{-6}	0	0	0	
8	-8.33	3.0×10^{-6}	0	0	0	
9	$+5.0$	3.6×10^{-6}	$+1$	0	18.00×10^{-3}	
					$\Sigma=0.1651$ in. $=\delta_y$	$\Sigma=0.080$ in. $=\delta_x$

* Calculated for $E=10\times10^6$ lb/in.²
† $Q=0$.

If we wish, we can also solve the statically indeterminate case of the truss which we considered in our discussion of the computer solution in Sec. 2.5 (Fig. 2.16). We simply require that $\partial U/\partial Q=0$ as there is no horizontal motion at the point at which Q acts. Thus from Eq. (*d*) and Table 2.6

$$\Sigma F_i \frac{L_i}{A_i E_i}\frac{\partial F_i}{\partial Q} = 0 = [13.33\times10^3 + Q + 20\times10^3 + Q][2.4\times10^{-6}]$$

or

$$Q = -16.67\times10^3 \text{ lb}$$

With very little effort we have determined the indeterminate support force.

If now we wish to solve for the deflection at P, we must reevaluate the products in rows 1 and 2 of Table 2.6 with Q at its actual value as determined above. These new values are

i	$\left(\frac{FL}{AE}\frac{\partial F}{\partial P}\right)_{i,\,Q\neq 0}$
1	$21.36 \times 10^{-3} + 1.6Q \times 10^{-6}$
2	$31.95 \times 10^{-3} + 1.6Q \times 10^{-6}$

The values for members 3 through 9 do not change since they carry no Q. Therefore

$$\delta_P = \frac{\partial U}{\partial P} = 0.1651 + 3.2Q \times 10^{-6}$$

but

$$Q = -16.67 \times 10^3$$

Thus

$$\delta_P = 0.1651 - 0.0534 = 0.1117 \text{ in.}$$

Note that this agrees with value given by the computer output in Table 2.5.

2.7 SUMMARY

We have considered in this chapter several examples of quite different situations involving deformable bodies, and we have found that all of them could be handled within the same framework of model formulation followed by application of the steps (2.1). By now it must be apparent that the reduction of a given physical system to an idealized model is an essentially *creative* step and that there are no simple rules for mastering this art. As we gain factual knowledge and develop resourcefulness by applying this knowledge to a wide variety of situations, we become more skillful, both in creating simple but effective models and in analyzing them.

It should also be clear that although the steps (2.1) are sufficient to set up problem solutions, the mathematical evaluation of the solutions may be tedious. We saw that the computer is of tremendous help in solving complex problems involving many loads or many members. We also saw that for certain problems of moderate complexity, an energy method can greatly facilitate the solution.

In the following chapters we shall consider more complex situations of loading, structural shape, and material behavior. As we proceed to more sophisticated problems, we must be careful not to confuse the complexity of the physical *facts* in any given situation with the physical *principles* involved. In all cases the principles are simply those contained in the three steps of (2.1).

PROBLEMS

2.1. A wood diving board is hinged at one end and supported 5 ft from this end by a spring with a constant of 200 lb/in. How much will the spring deflect if a girl weighing 120 lb stands at the end of the board? Will the spring deflection be altered if the board is made very rigid?

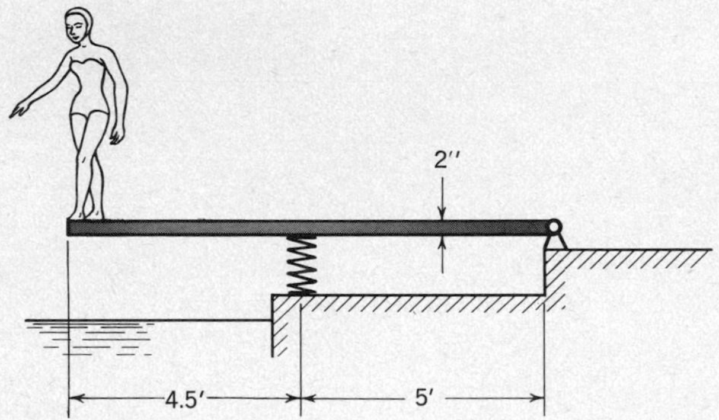

Prob. 2.1

2.2. A safety valve for a pressure system has a discharge hole of 2-in. diameter. The spring has a free length of 10 in. and a spring constant of 670 lb/in. At what pressure will the valve open?

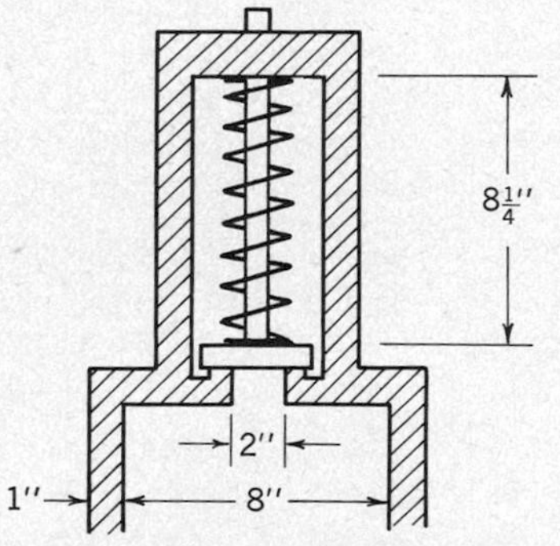

Prob. 2.2

2.3. An 8-ft-diameter sound baffle weighing 240 lb is to be hung from a ceiling with three springs which are to be mounted on radii making angles of 120° with each other, as shown in the sketch. Three springs, each 10 in. long, are delivered to the job. Springs a and b have a spring constant of 80 lb/in. and spring c one of 90 lb/in. If the springs a and b are mounted 3 ft from the center, how far from the center should the spring c be mounted if the sound baffle is to hang level?

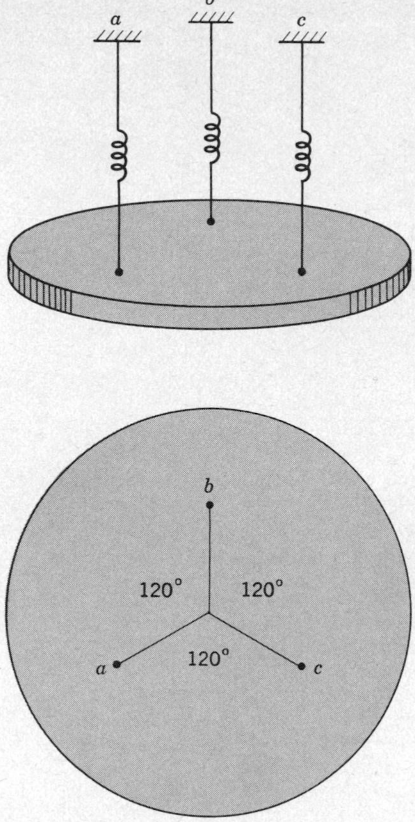

Prob. 2.3

2.4. An operator of a punch press operates part of the press by pushing a foot lever. The lever has a spring to return it to position after each push. The operator has complained that she gets tired pushing the lever. Can you suggest a change in the spring which may make the operator's job easier?

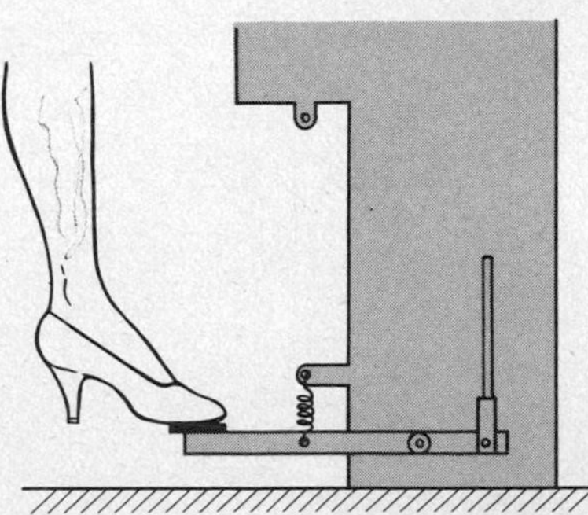

Prob. 2.4

2.5. High-speed rail-transportation design requires a knowledge of deflection characteristics of the various components of the roadway. If the member AB of a truss section as shown is assumed rigid, estimate the angle the member AB makes with the horizontal when a load acts at the position indicated. Bars AC, BC, and BD are steel with the cross sections indicated.

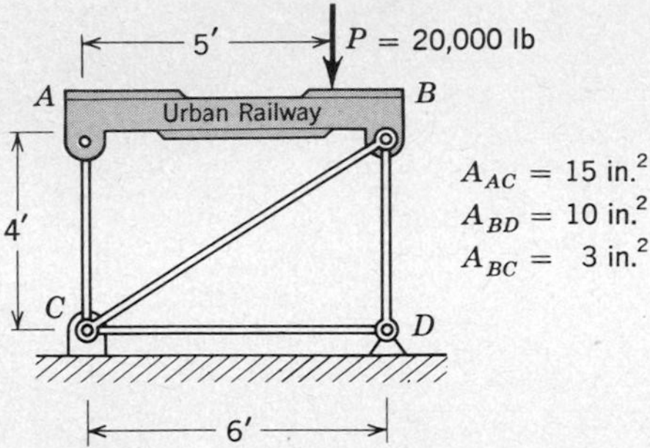

Prob. 2.5

2.6. The stiff member AB is horizontal before the load of 5,000 lb is applied at A. The three steel bars ED, BD, and BC are fastened with pins at their ends. Find (*a*) the force in the bar BD, and (*b*) the horizontal and vertical movement of the point A.

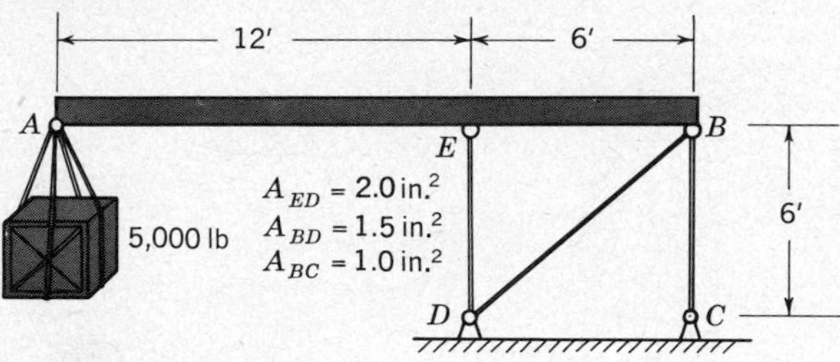

Prob. 2.6

2.7. A small railroad bridge is constructed of steel members, all of which have a cross-sectional area of 5 in.² A train stops on the bridge, and the loads applied to the truss on one side of the bridge are as shown in the sketch. Estimate how much the point R moves horizontally because of this loading.

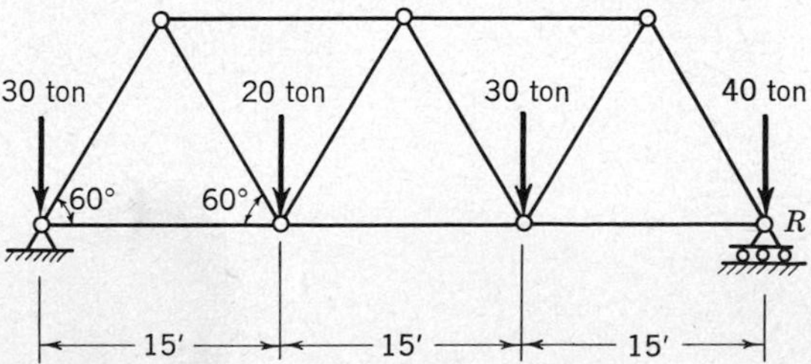

Prob. 2.7

2.8. Estimate how much the point B of the truss in Example 2.5 moves horizontally.

2.9. Determine the elongation of member BC due to the force of $P = 100{,}000$ lb. BC is steel and is 2.5 in.^{2} in cross-sectional area.

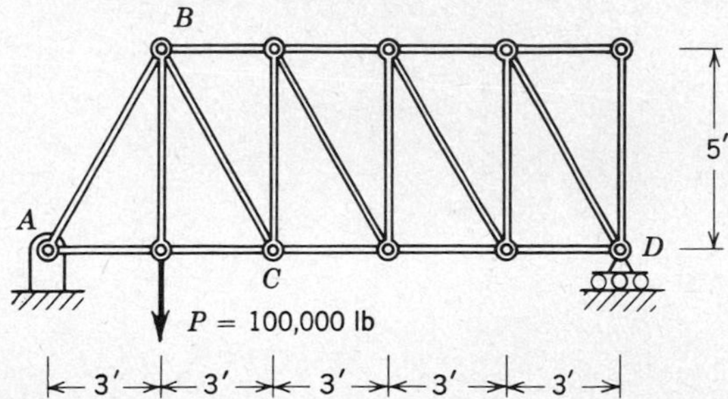

Prob. 2.9

2.10. In the pin-jointed cantilever truss shown, all the members have a cross-sectional area A and elastic modulus E. Find:

(*a*) the forces in the rods due to the load W, distinguishing between tension $(+, T)$ and compression $(-, C)$.

(*b*) the vertical deflection of the loaded joint.

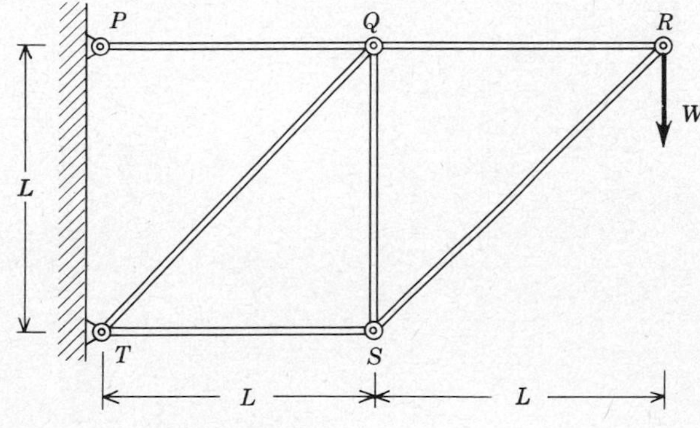

Prob. 2.10

2.11. A square reinforced-concrete pier 1×1 ft in cross section and 4 ft high is loaded as shown in the figure. The concrete is strengthened by the addition of eight vertical 1×1 in. square steel reinforcing bars placed symmetrically about the vertical axis of the pier. Find the stress (force/unit area) in the steel and concrete and the deflection. For concrete, take $E = 2.5 \times 10^6$ psi.

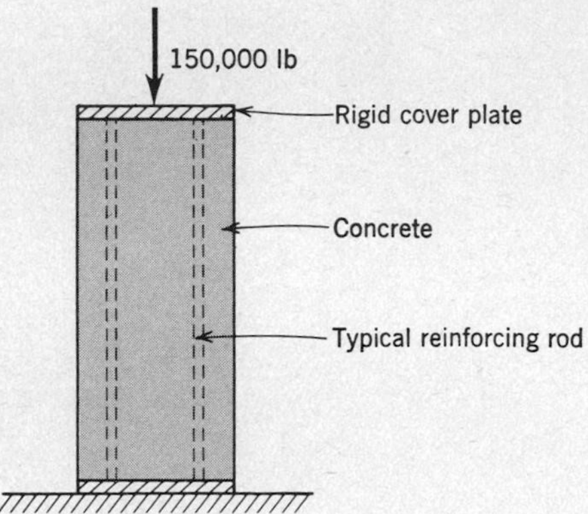

Prob. 2.11

2.12. Consider the pin-connected framework loaded as shown in the figure. Find the axial force in each bar. The two outer bars are identical with cross-sectional area A_0; the inner bar has a cross-sectional area A. All bars have the same modulus of elasticity E.

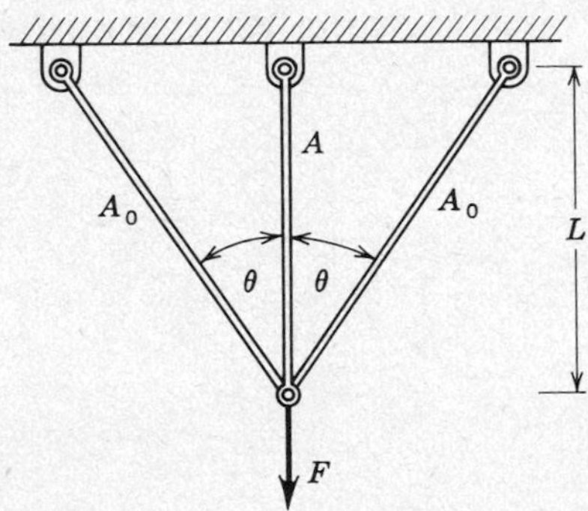

Prob. 2.12

2.13. In Example 2.3 (see Fig. 2.3), for general values of the spring constants show that the deflection at point A when the load is at $x = \lambda a$ as given by Eq. (*d*) is also equal to the deflection at the location $x = \lambda a$ when the load is at point A.

2.14. In a particular machine it is necessary to have a very stiff spring with a "kink" in the load-deflection curve. The suggested design consists of a 6-in.-diameter brass cylinder with a 0.25-in. wall thickness and a 10-in.-diameter aluminum cylinder with 0.25-in. wall thickness, the aluminum cylinder being made 0.003 in. shorter than the brass cylinder.

Sketch accurately the graph of the load-deflection relation for this spring.

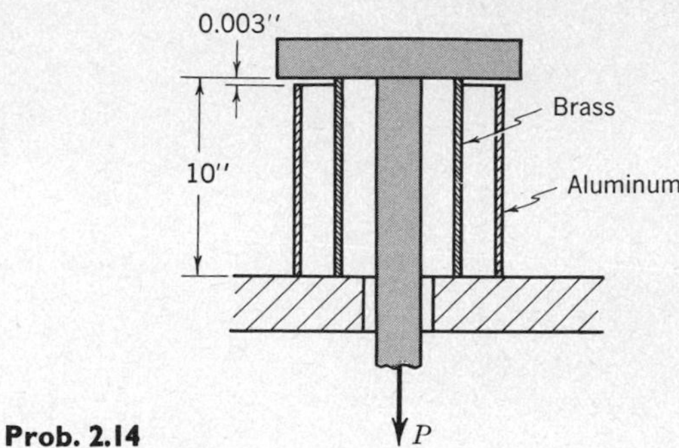

Prob. 2.14

2.15. Some miners are trapped 6,000 ft below the surface. They make their way to the bottom of an abandoned shaft. At the surface is a hoist with 5,980 ft of 1-in.-diameter standard plow-steel hoisting rope. A 1-ft length of this rope weighs 1.60 lb and has a spring constant (including the effect of untwisting) of about 10^6 lb/in. If you think the miners can be hoisted to the surface, explain quantitatively how this can be done.

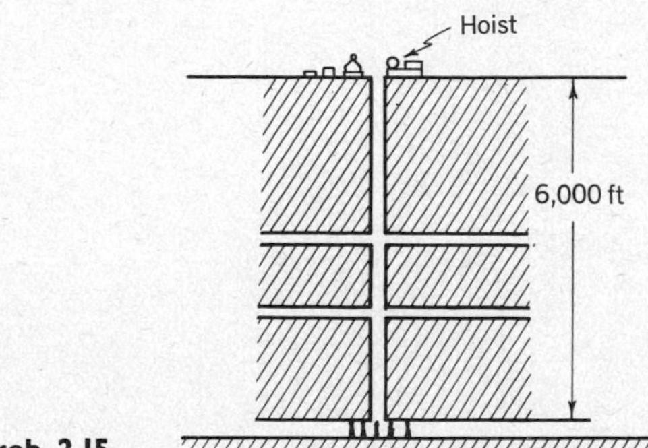

Prob. 2.15

2.16. The rigging on the mainmast of a sloop consists of two ⅜-in.-diameter stainless-steel wire ropes on either side of the mast. The cables go from the top of the mast to a spreader and then down to the deck. The wood mast has a cross-sectional area of 20 in.^{2} A 1-ft length of the wire rope has a spring constant (including the effect of untwisting) of 140,000 lb/in. In mounting the rigging the four wire ropes are brought up snug and then the turnbuckles (which have 20 threads per inch) are turned 15 more turns. Estimate the compressive force in the mast after the turnbuckles have been tightened.

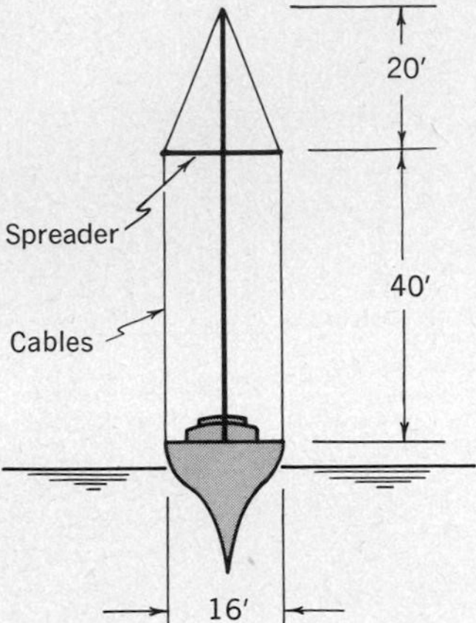

Prob. 2.16

2.17. A very stiff horizontal member is supported by two vertical steel rods of different cross-sectional area and length. If a vertical load of 30,000 lb is applied to the horizontal beam at point *B*, estimate the vertical deflection of the point *B*.

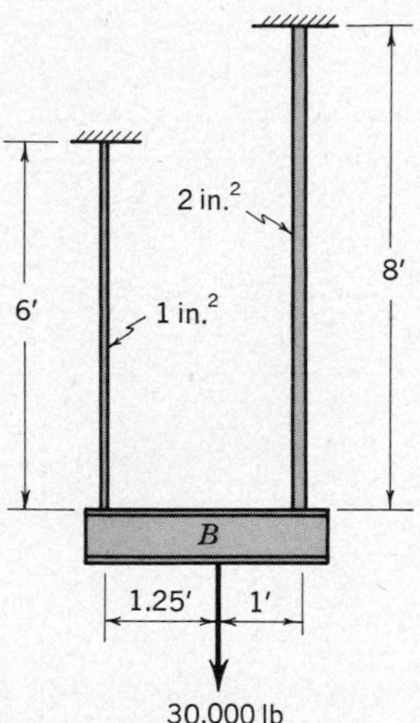

Prob. 2.17

2.18. Two linear springs of different spring constant are connected in series as shown. Calculate the overall spring constant of the assembly.

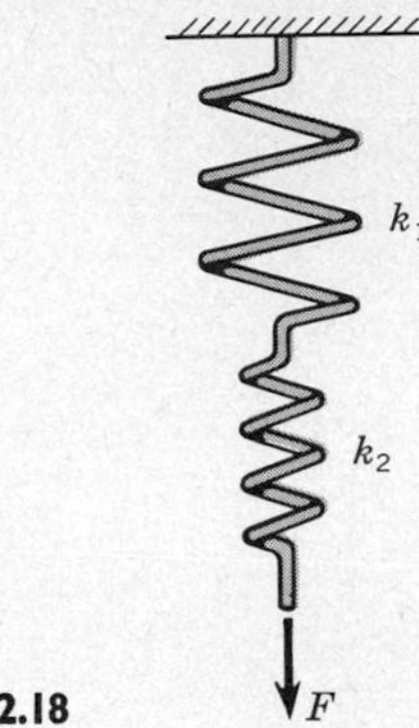

Prob. 2.18

2.19. A stiff beam is hinged at one end and supported by two springs of spring constant k. Where should a force P be applied so that the spring constant of the system (P divided by the deflection under P) is $^{20}/_{9}k$?

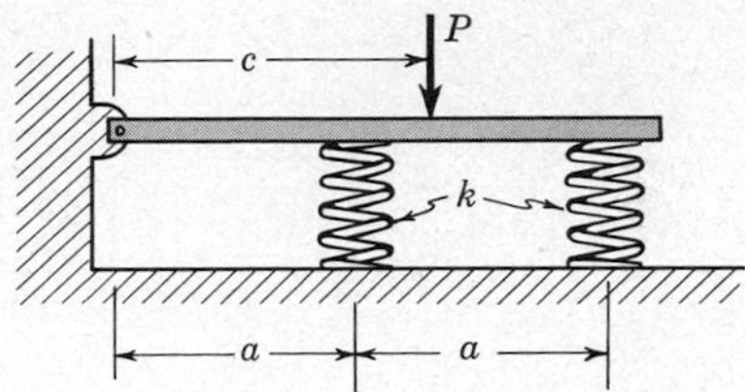

Prob. 2.19

2.20. A stiff horizontal bar AB is supported by three springs with different spring constants, arranged as shown. Where should a force P be applied so as to keep the bar AB horizontal? With P in this position how much does the bar move down because of P?

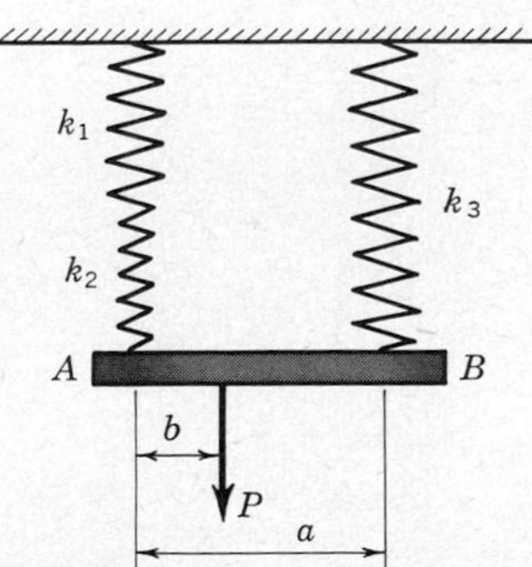

Prob. 2.20

2.21. An inventor devises a springboard playground toy for children which consists of a tough, lightweight plastic board attached to a heavy steel frame by three springs. The free lengths of all three springs are 12 in. and the spring constants are $k_1 = k_3 = 100$ lb/in.; $k_2 = 120$ lb/in.

In order to test the strength of the connections, the inventor, who weighs 240 lb, climbs onto the board and stands at several places. As he stands at the location shown, what forces are transmitted to the frame connections? What angle does the board make with the horizontal?

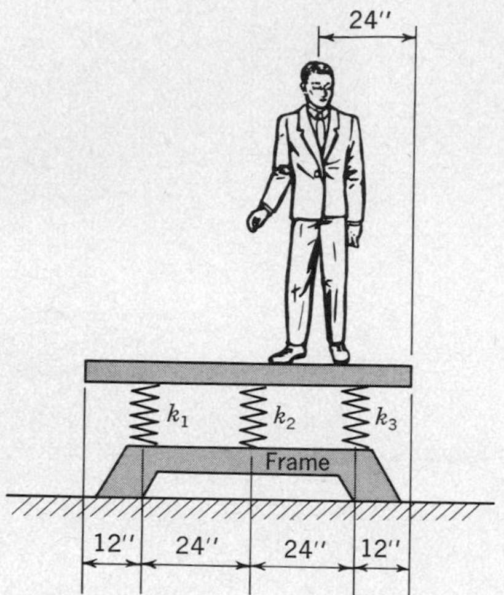

Prob. 2.21

2.22. A 90-ft flagpole is made of 6-in.-diameter steel pipe. It is attached to its foundation by a ball-and-socket joint and is supported in the upright position by four ¼-in.-diameter high-strength steel wires, as shown in the sketch. When there is no wind, the tension in the wires is negligible. In a hurricane the wind blows hard from the south, and its effect can be represented by a horizontal force of 900 lb at the mid-height of the pole.

Estimate how far the top of the pole moves from its original position, which was vertically above the base.

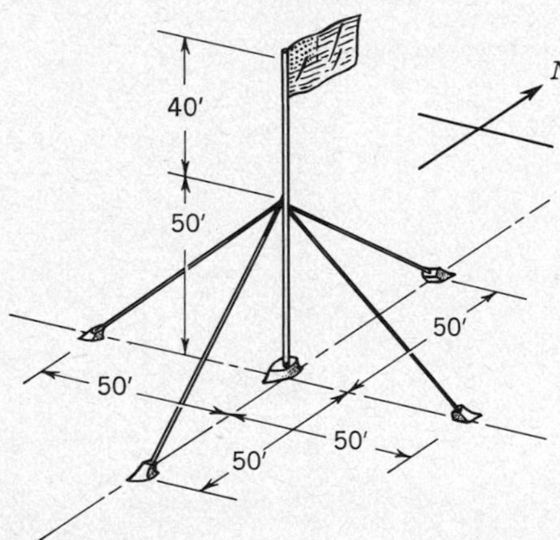

Prob. 2.22

2.23. In the structure shown in (*a*) the member *AB* is very rigid in comparison to *BC*. It is desired to estimate the vertical deflection at *B* when a load of 100 lb is supported at *B*. It is known that when the 100-lb load is supported entirely by *BC*, as in (*b*), the deflection at *B* is 0.9 in.

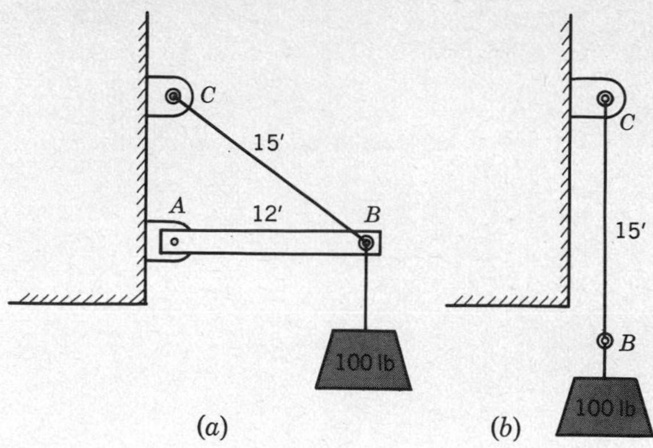

Prob. 2.23

2.24. A bolt is threaded through a tubular sleeve, and the nut is turned up just tight by hand as shown. Using wrenches, the nut is then turned further, the bolt being put in tension and the sleeve in compression. If the bolt has 16 threads per inch, and the nut is given an extra quarter turn (90°) by the wrenches, estimate the tensile force in the bolt if both the bolt and sleeve are of steel and the cross-sectional areas are

Bolt area = 1.00 in.^{2}
Sleeve area = 0.60 in.^{2}

6″

Prob. 2.24

2.25. A rigid beam AC is supported at its left end A by a pin. At its right end C it is supported by another rigid bar CF, which is in turn supported by an aluminum rod at D and a steel rod at E. Before any loads are applied the rigid bars both are level. A known load P is applied at point F and an unknown load Q at point B.

Find Q in terms of P if the rigid bar CF is to be level after the two loads are applied.

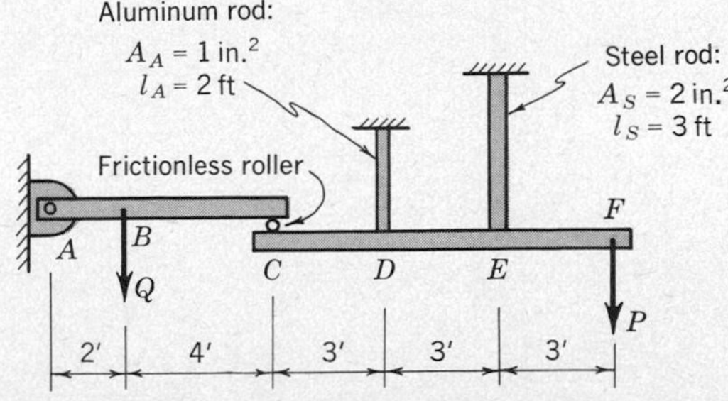

Prob. 2.25

2.26. A water pipe is made of longitudinal wooden staves held together with circumferential steel bars of 1-in. diameter, as shown in the sketch. The pressure of the water in the pipe is 100 psi. Because of the danger of leakage between the staves, the diameter D of the rod centerline cannot be allowed to increase more than 0.03 in. due to the water pressure.

Estimate the maximum allowable longitudinal spacing s between the circumferential rods.

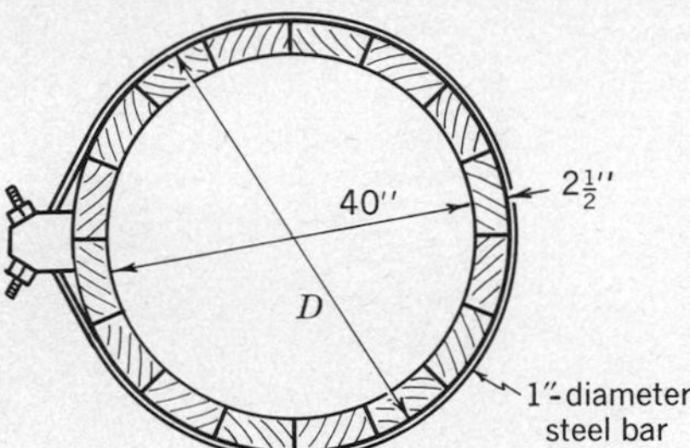

Prob. 2.26

2.27. A lightweight rope of area A and modulus of elasticity E is hung over a stationary shaft. A weight W is attached to the longer end, and, at the same time, the rope is forced against the shaft with a horizontal force P just sufficient to prevent the weight from dropping. Find the value of P if the static coefficient of friction between the rope and the shaft is f.

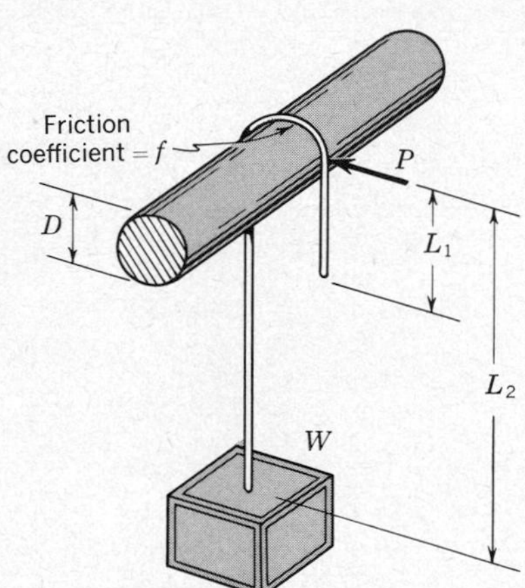

Prob. 2.27

2.28. A brake is designed as shown. A $1 \times \frac{1}{16}$ in. steel band restrains the wheel from turning when a 2,000 in.-lb torque is applied. The friction coefficient is 0.4. Find the tensions T_1 and T_2 that just keep the wheel from rotating.

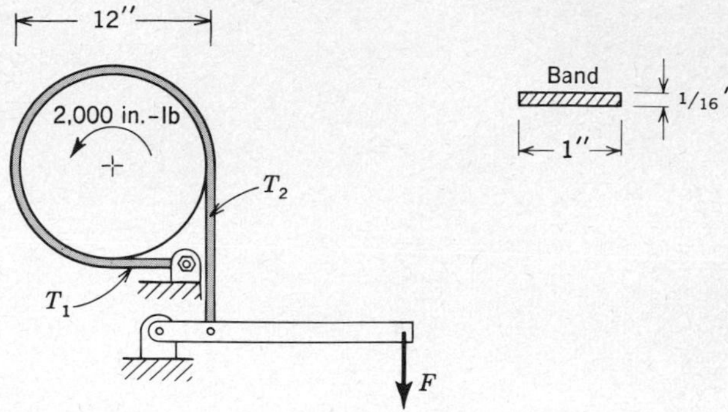

Prob. 2.28

2.29. A hawser from a ship is wrapped four times around a rotating capstan as shown in the figure. The dockworker pulls with a force of 40 lb. What is the maximum force the man can exert on the boat if the coefficient of friction between the capstan and hawser is 0.3?

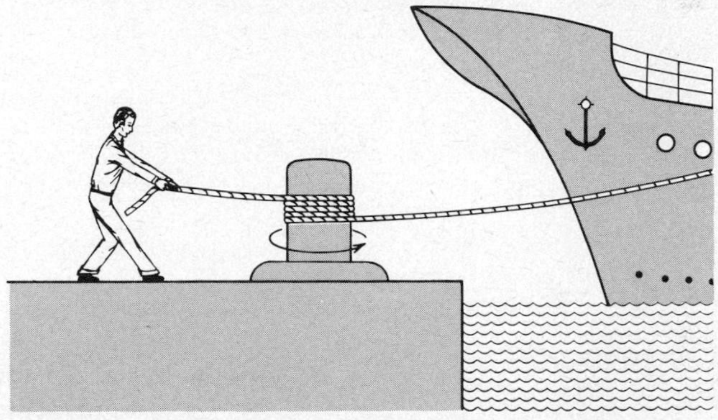

Prob. 2.29

2.30. Calculate the frictional resistance to rotation of a dry thrust bearing maintaining a load F as shown in the figure.

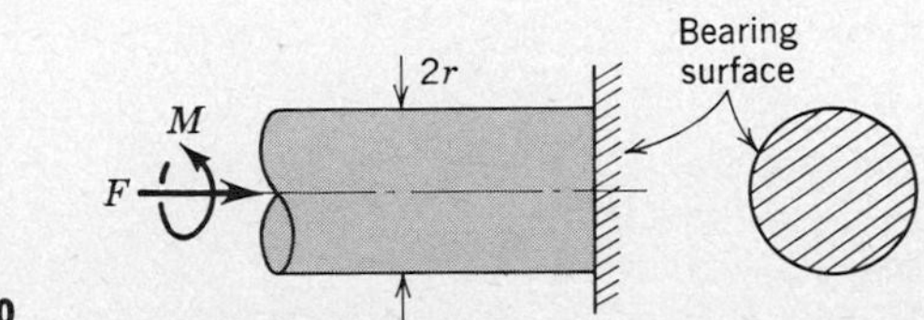

Prob. 2.30

2.31. A composite hoop consists of a brass hoop of 10-in. internal radius and 0.1-in. thickness, and a steel hoop of 10.1-in. internal radius and 0.2-in. radial thickness. Both hoops are 2-in. thick normal to the plane of the hoop. If a radial pressure of 200 psi is put in the brass hoop, estimate the tangential forces in the brass and steel hoops.

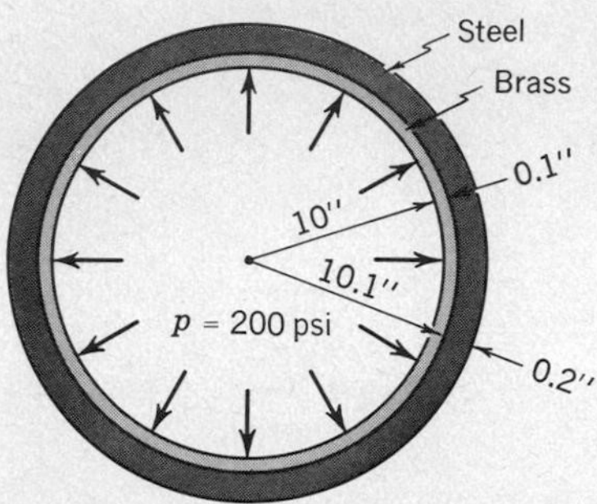

Prob. 2.31

2.32. What will be the radial expansion of the brass cylinder in Prob. 2.31 when the pressure is 200 psi?

2.33. Show that in Example 2.8 the distribution of radial force per unit length along the circumference is

$$\frac{dN}{R\,d\theta} = \frac{T}{R} = \frac{T_{AD}e^{f\theta}}{R}$$

2.34. A group of students, 22 in all, have a tug of war using a manila rope of ½-in. diameter. They start with a rope initially 50 ft long and dispose themselves as shown in the sketch. A team will win when they pull their end of the rope over the edge of the field. When the rope is stretched, they want a clearance of 4 ft at each end between the end of the rope and the edge of the field. They estimate that each man can pull about 100 lb. A 1-ft length of the rope has a spring constant of 29,400 lb/in., including the effect of the untwisting of the rope.

How long should they make the field?

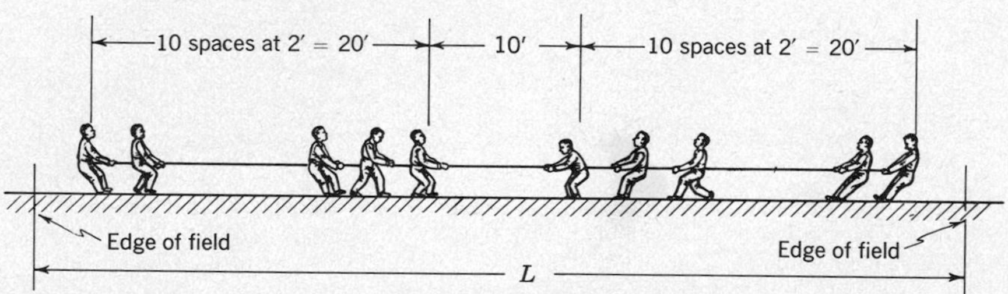

Prob. 2.34

2.35. Illustrated is a schematic diagram of a cable-control system for the rudder of a subsonic jet trainer aircraft. The rudder lever arm is connected to the pilot's foot control by $\frac{3}{16}$-in.-diameter extra-flexible stainless-steel cable, a 1-in. length of which has a spring constant of 330,000 lb/in. (including the effect of untwisting). The cables have an initial tension of 300 lb.

Cable length from rudder lever arm to the pilot's foot control is about 20 ft. The pilot can push on his foot control with a force of about 150 lb. In a static test of the rudder control system a force was exerted on the rudder and gradually increased until the pilot could no longer hold his foot control stationary. Through what angle had the rudder rotated when the force reached the level which just caused the pilot's foot control to move? Would this angle have been different if there had been no initial tension in the cables?

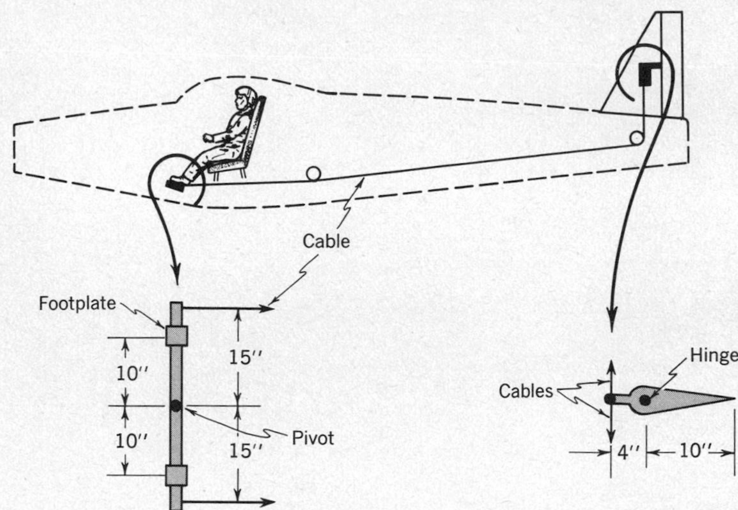

Prob. 2.35

2.36. A packing case weighing 10 tons is unloaded on a ramp making an angle of 30° with the horizontal. The static coefficient of friction between box and ramp is 0.2. To keep the box from sliding down the ramp a 2 × 2 in. piece of wood is placed between the case and the ramp as shown. How much does the piece of wood shorten because of its being used for this purpose?

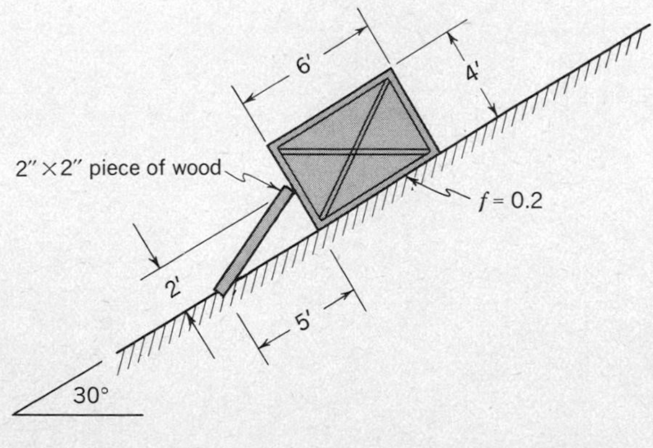

Prob. 2.36

2.37. A 1-in.-outside-diameter brass tube is to be compressed 0.03 in. by means of a steel screw clamp, each screw of which has 20 threads per inch and an effective cross-sectional area of 0.1 in.^{2} It is known that it will take 1,200 lb to compress the brass tube 0.03 in. The tube is put into the clamp with the jaws parallel and just touching the tube. How many turns must be given the screw C to compress the tube 0.03 in.?

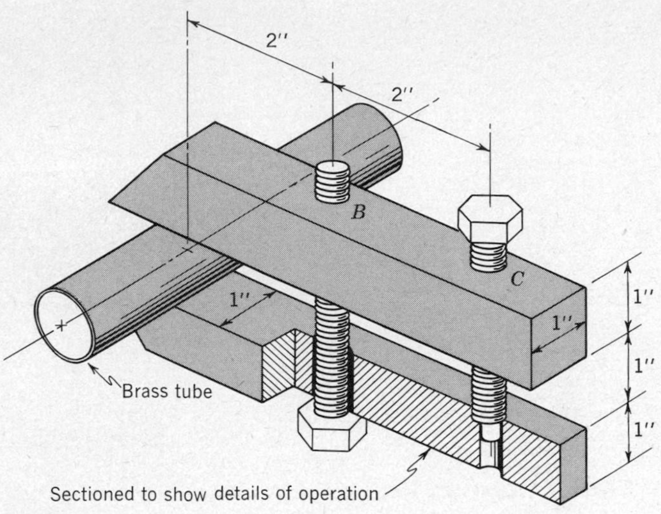

Prob. 2.37

2.38. When designing electrical equipment which involves high-amperage currents, it is necessary to consider the magnetic forces on the conductors. For instance, in a synchrotron the copper coils alternately expand and contract due to magnetic forces. Consider a case in which the copper coil is placed in a steel ring, as shown in the sketch. Estimate the tangential force in the copper coil when the magnetic force reaches a value of 5,000 lb per foot of circumference, directed radially outward. (Take the modulus of elasticity of copper to be 17×10^6 psi.)

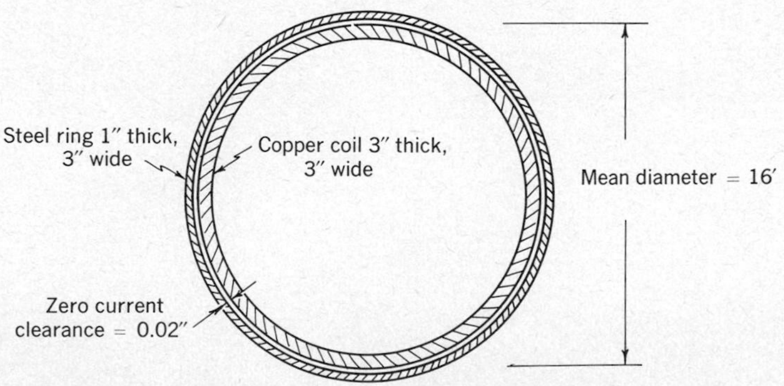

Prob. 2.38

2.39. To obtain a particular nonlinear spring behavior, an instrument designer uses a linear material B and a nonlinear material C in the design shown in the sketch. A rod of material B is used in tension between the yoke and the upper support, and a rod of material C is used in compression between the yoke and the bottom support. The assembly is manufactured with great precision so that there is no slack in the system when $P = 0$. If a load of 325 lb is carried by the yoke, how much does the yoke move? Also, how much of the load is carried by the rod of material B?

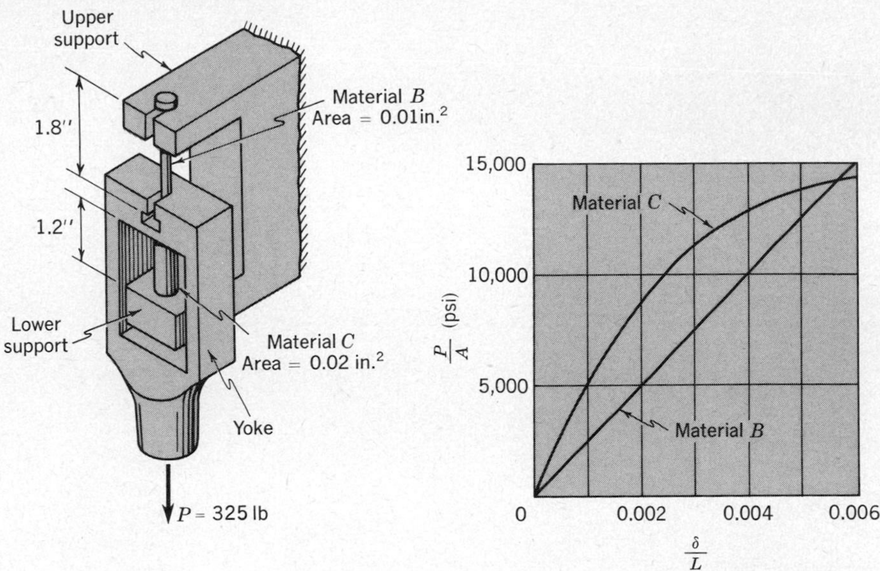

Prob. 2.39

2.40. Consider Example 2.2 in which the spring constants are now k_C and k_D. Verify that Eq. (i) is the solution.

2.41. For the system of Example 2.5 calculate the vertical displacement of point F using the methods of Example 2.4 and using the energy method of Example 2.11.

2.42. A statically indeterminate 45° truss is made as shown. Construction is such that the members fit together with negligible interference when P is zero. All struts are of the same material and same cross section. Using energy methods, calculate the load in member CD.

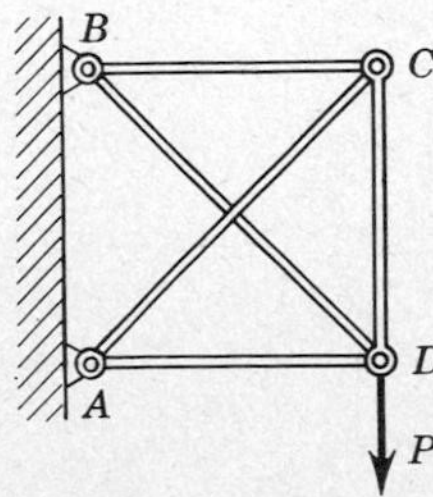

Prob. 2.42

2.43. Find the horizontal and vertical movement at A due to a 1,000-lb vertical load at that point. Members AB and AC are steel tubes with an area of 2.1 in.^{2} each. Member AD is an aluminum rod with an area of 1 in.^{2} The members are fastened to a rigid base with bolted connections, and are pinned together at A. Explicitly state the assumptions used in developing your model.

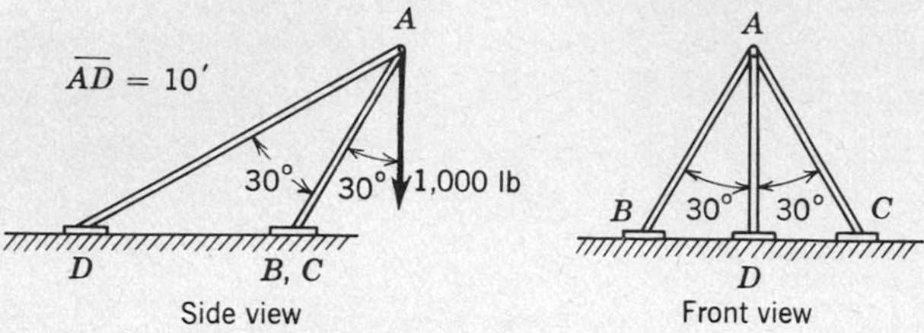

Prob. 2.43

2.44. A very rigid beam ABC is supported at point A by an elastic hinge and at point B by two springs connected in series. A force P can be applied to the beam at a variable distance x from point A. The beam can be assumed to be weightless. Determine:

(*a*) The force acting on the beam at support B in terms of the force P, the spring constants k_1, k_2, and k_3, and the variable x

(*b*) The angular displacement of the beam about the hinge at A (θ_A) in terms of P, k_1, k_2, k_3, and the variable x

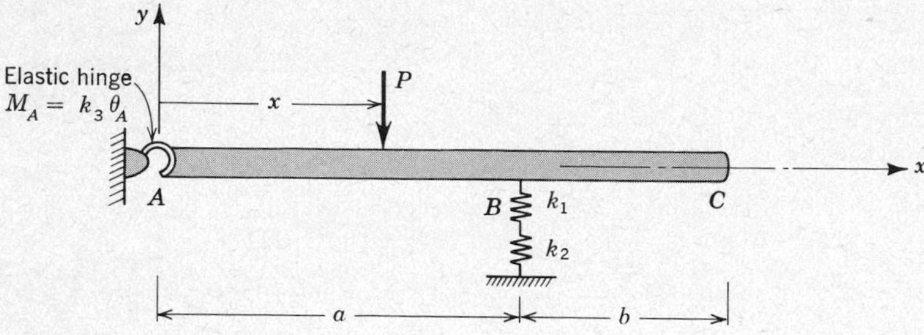

Prob. 2.44

2.45. Calculate the energy stored in a uniform bar under a constant load F_0 with and without the inclusion of the weight of the bar. The weight per unit length of the bar is assumed to be w.

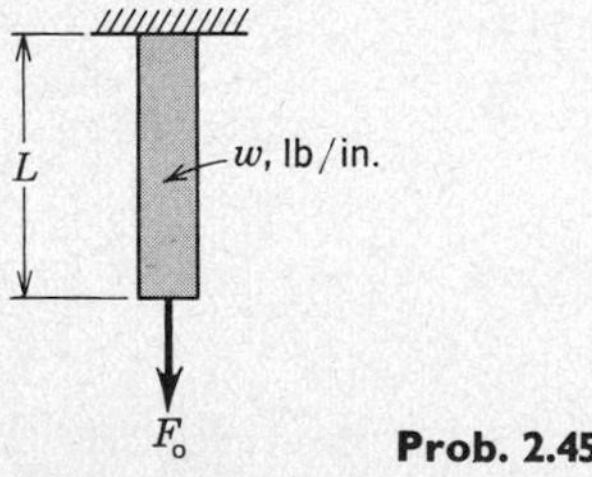

Prob. 2.45

2.46. The optimal design of trusses for minimum weight very often is required. In optimal design for minimum weight many factors must be considered, among them the stresses in bars in tension, forces in bars in compression which might cause large sideward deflections (buckling), and the availability of structural members for the design. The cantilever truss shown in Fig. (*a*) supported at A and B is to carry the loads shown at joints C and D. It is required to redesign, if possible, the truss for minimum weight such that the maximum stress in any tension member does not exceed 20,000 psi and the maximum force P_c in any compression member is less than

$$P_c \leqq \frac{10^8 I}{L^2} \text{ (lb)}; I \text{ (in.}^4\text{)}; L \text{ (in.)}$$

where I is the cross-sectional moment of inertia of the member and L is the length of the member.

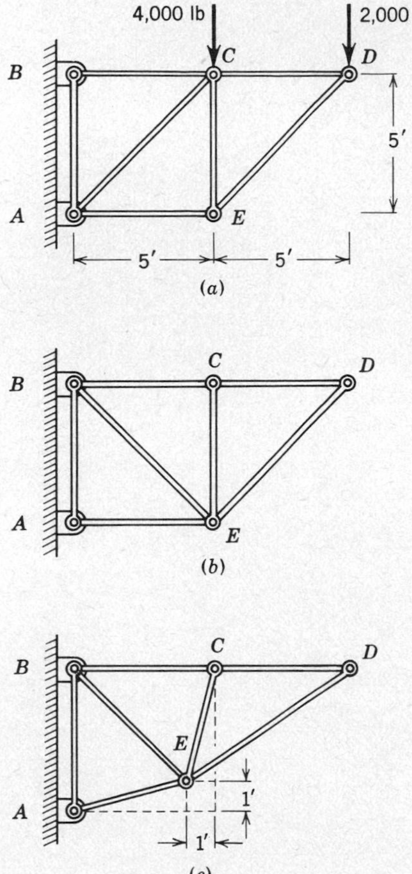

Prob. 2.46

A complete optimum solution for this truss structure providing the optimum properties of the members is not a realistic task because there are only two aluminum structural members available for construction. They have the properties:

(1) $A = 2.17$ in.^{2}; $I = 0.59$ in.^{4}; weight/ft $= 7.5$ lb
(2) $A = 2.76$ in.^{2}; $I = 0.91$ in.^{4}; weight/ft $= 9.50$ lb

Two possible redesigns which hold the support and loaded joints fixed are shown in (*b*) and (*c*). Find a design of the truss with minimum weight consistent with the above constraints.

2.47. A prestressed concrete reactor vessel for a high-temperature gas-cooled nuclear reactor is shown. The vessel is prestressed, with wire "tendons" axially, circumferentially (hoop), and across the heads. For this problem, consider only the circumferential or hoop loadings. There are 310 steel hoop tendons as shown. They are distributed around in the wall sections. Each tendon has 170 wires $\frac{1}{4}$ in. in diameter.

Estimate the prestress tensile force which may be applied to each hoop tendon prior to pressurization. Estimate how large a pressure can be applied before the prestressed concrete goes into tension. Use a "thin-ring" approximation. The properties are

Concrete in compression $E = 1.5 \times 10^6$ psi, maximum compressive stress $\sigma_f = 6{,}000$ psi
The steel tendon wire $E = 30 \times 10^6$ psi, maximum allowable stress 150,000 psi

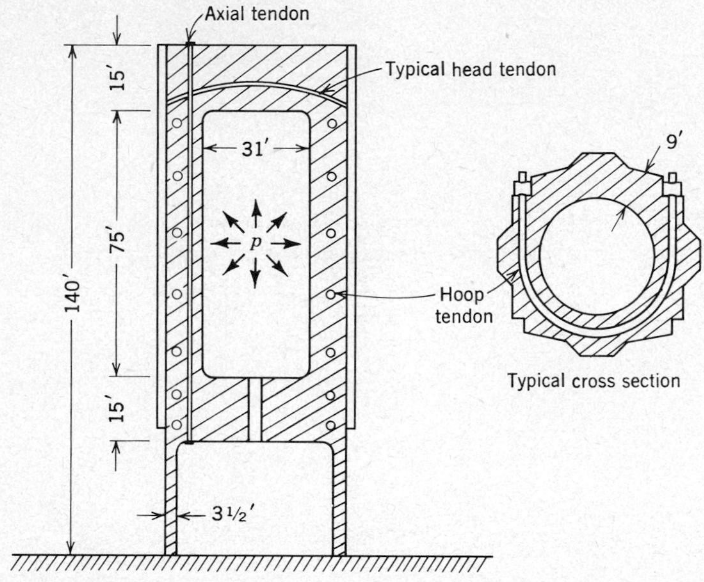

Prob. 2.47

2.48. When long pipes are incorporated in a hydraulic or fluid system, one of the factors that determines the speed at which the system can respond is the speed of propagation of pressure waves in the pipes. This speed is essentially the same as the speed of sound and is given by

$$c = \text{speed} = \frac{1}{(\rho k)^{1/2}}$$

where

ρ = mass density of the fluid
k = compressibility = $(\Delta V/V)/\Delta P$

where ΔV is the change in the volume V due to a pressure ΔP. Thus materials with low k are favored for hydraulic systems where speed is important.

All this is just background. Now here is the problem:

One of the factors that places a lower limit on k is the change in volume of the pipe itself. Consider a steel pipe of diameter D and wall thickness t.

If a pressure ΔP is applied to the fluid in the pipe, the volume inside the pipe will increase by an amount ΔV due to the expansion of the pipe. Calculate $k = (\Delta V/V)/\Delta P$ for the pipe in terms of D, L, t, E, and V. The pipe is very long compared to its diameter. State clearly any assumption you make.

For $D = 0.5$ in., $t = 0.05$ in., and $L = 20$ ft, what is the lower limit to k?

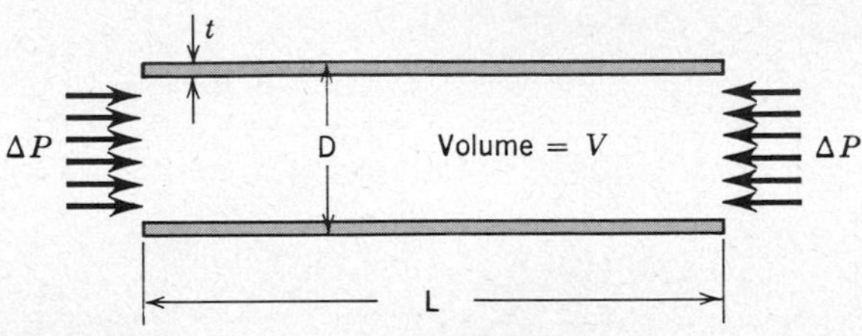

Prob. 2.48

2.49. During normal child delivery the fetal head is pushed through the birth canal by the increased pressure in the uterus during contraction. However, in certain cases (lower uterine spasm) no descent by the fetal head is noticed. If the amniotic-fluid pressure in the uterus during a contraction is approximately 60 mm Hg, find the driving force on the fetal head when there is no motion of the head.

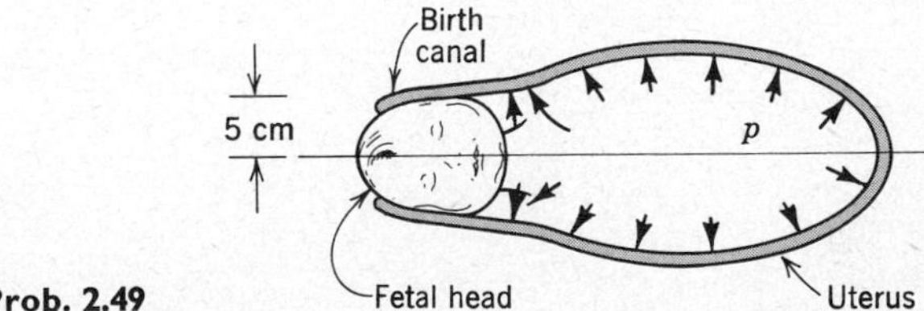

Prob. 2.49

2.50. In order to increase its pressure capability, a copper tube is tightly wrapped with a single layer of square stainless-steel wire. During wrapping, the wire tension is maintained at 100 lb.

(*a*) How much does the tube radius decrease during wrapping? (Neglect any axial force in the tube.)

(*b*) Determine the internal pressure which will just cause the stress to reach the maximum allowable value in either the tube or the wire. (Neglect any axial force in the tube.)

Tube: Copper
Inner radius 0.9 in.
Thickness 0.1 in.
$E = 20 \times 10^6$ psi
σ max $= 30{,}000$ psi

Wire: Stainless steel
Square wire, 0.05×0.05 in.
$E = 30 \times 10^6$ psi
σ max $= 150{,}000$ psi

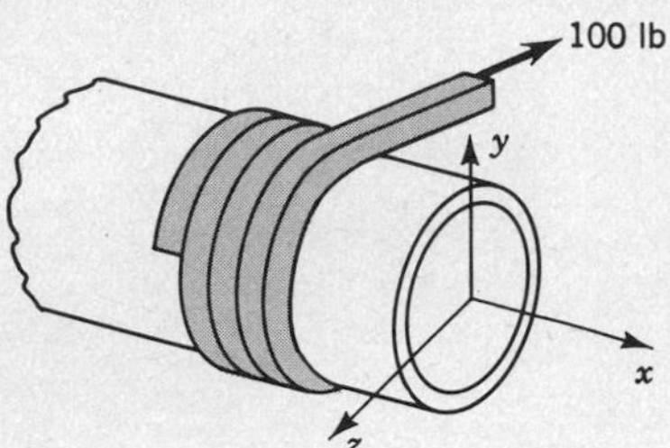

Prob. 2.50

2.51. An inventor has filed a patent for a weighing device with adjustable sensitivity. He claims that by adjusting the pre-tension in the springs A and B by screwing up on the nut C the sensitivity (inches of pointer movement per ton of load) of the device can be altered. What is the sensitivity when $L = 24$ in.? Is the device more or less sensitive when $L = 26$ in.? How much?

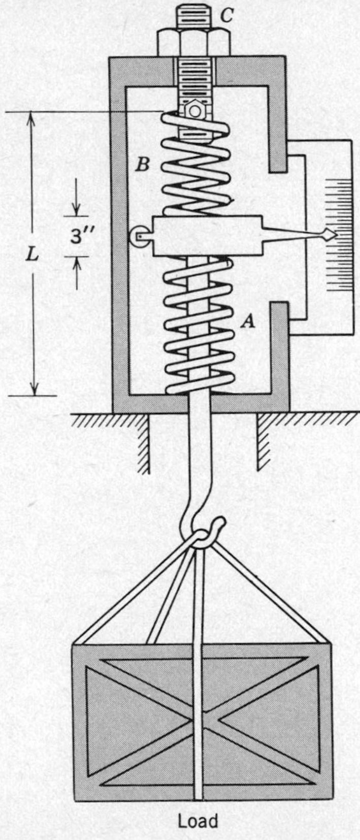

Spring	Free length	Spring constant
A	10″	8,000 lb/in.
B	8″	5,000 lb/in.

Prob. 2.51

2.52. A steel cable hangs under its own weight. The diameter of the cable is not constant but varies in a manner that makes the tensile stress at all points along the cable the same. Derive the differential equation that describes the variation of cable diameter with position along the cable. Solve the equation and find the expression for d_1 in terms of d_2, L, σ_0, and γ (the weight per unit volume of the cable).

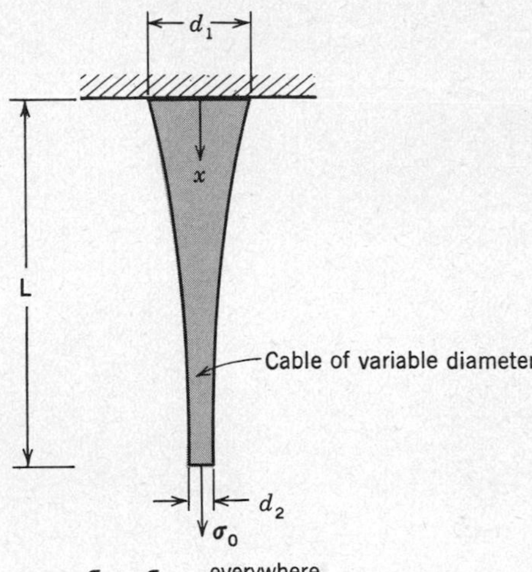

Prob. 2.52

2.53. The system shown consists of three nonlinear-elastic springs interconnected as shown. Show that for this system the statement (2.4) has the form

$$\int_0^\delta P\,d\delta = \sum_{i=1}^{3} \int_0^{e_i} F_i\,de_i$$

where e_i is the elongation and F_i is the force in the ith element. Prove the truth of this statement directly by using the requirements of *equilibrium* and *geometric compatibility* to relate the e_i and F_i to δ and P.

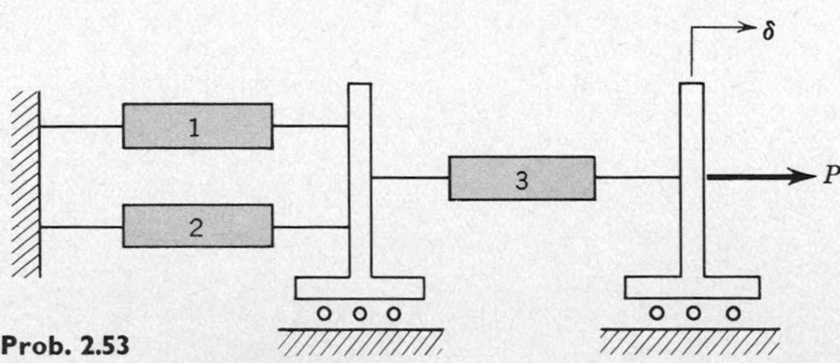

Prob. 2.53

2.54. Show that for the system of Prob. 2.53 the statement (2.6) has the form

$$\int_0^P \delta \, dP = \sum_{i=1}^{3} \int_0^{F_i} e_i \, dF_i$$

Prove the truth of this statement directly by using the requirements of *equilibrium* and *geometric compatibility* to relate the e_i and F_i to δ and P. Note that in Probs. 2.53 and 2.54 it is unnecessary to know the force-deflection laws for the springs.

2.55. Show for the system of Fig. 2.18 that if the component of $\mathbf{P}_i$ which is in-line with $\mathbf{s}_i$ is called f_i then

$$f_i = \frac{\partial U}{\partial s_i}$$

where U is the total elastic potential energy of the system expressed in terms of the displacements of the loading points. This theorem, which is complementary to (2.7), is called *Castigliano's first theorem*, whereas (2.12) is called *Castigliano's second theorem*.

3
Forces and Moments Transmitted by Slender Members

3.1 INTRODUCTION

In the previous chapter we considered the fundamental basis for the study of mechanics of solids, summarized in the three steps of (2.1). While it is true that considerations of forces, deformations, and force-deformation relations are all that are required in the analysis of deformable solids, it is also true that there is a depth of sophistication in each of these considerations which we have not yet plumbed. In this and the following two chapters we shall reexamine the significance of the separate steps of (2.1) in order to lay a more secure foundation for our subsequent study of complete problems which again require the simultaneous consideration of all three steps.

In this chapter we shall be concerned only with step 1, the study of forces and the equilibrium requirements, as applied to slender members. In Chap. 4 we shall extend this study to solids of arbitrary configuration and also investigate the geometry of deformation of solids. In Chap. 5 we shall give a more extended description of the force-deformation relationships of solids. We shall then be ready to study, in the final four chapters, complete problems of fundamental engineering importance.

If we look critically at any engineering structure, be it a bridge, an automobile, or a house, we shall note that a large portion of the load-carrying members can be classified as *slender members*. By a slender member we mean any part whose length is much greater (say at least five times greater) than either of its cross-sectional dimensions. This classification includes such things as beams, columns, shafts, rods, stringers, struts, and links. Even if a long, thin rod is formed into a hoop or a coil spring whose diameter is large compared with the thickness of the rod, it still retains its identity as a slender member.

A slender member can be pulled, bent, and twisted. We have already considered tensile and compressive loadings along the axis of a member. We now turn to a study of forces and moments which tend to twist or bend the member.

3.2 GENERAL METHOD

A general method for determining the forces and moments acting across any section of a slender member which is in equilibrium is to imagine a *hypothetical cut* or *section* across the member at the point of interest. If we then consider either part of the member as an isolated free body, the force and moment required at the section to keep that part of the member in equilibrium can be obtained by applying the conditions for equilibrium. In general, there will be both a force and a moment acting across the section.

For convenience, we usually resolve the force and the moment into components normal and parallel to the axis of the member (Fig. 3.1). In Fig. 3.1 the x axis has been oriented so as to coincide with the longitudinal axis of the member. The y and z axes lie in the plane of the cross section; the choice of their particular orientation within the cross section usually is governed by the shape of the section or by the direction of the transverse loading being carried by the member.

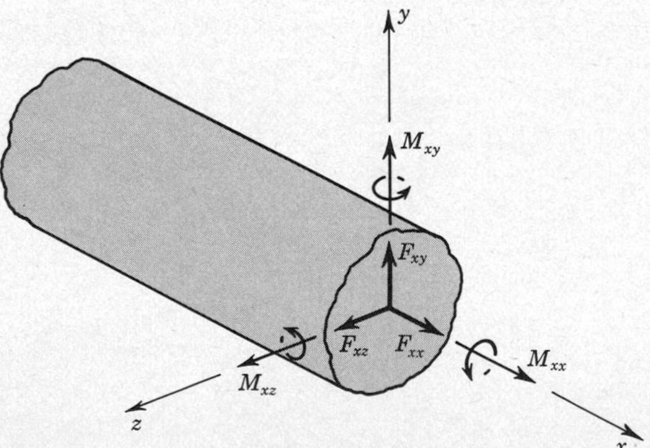

Fig. 3.1 Forces and moments acting on a section of a member.

The notation F_{xx}, ..., etc., of the components in Fig. 3.1 is used to indicate both the orientation of the cross section and the direction of the particular force or moment component. The first subscript indicates the direction of the outwardly directed normal vector to the face of the cross section. The cross-sectional *face* will be called *positive* when the outward normal points in a positive coordinate direction and *negative* when its outwardly directed normal vector points in the negative coordinate direction; thus, the cross-sectional face in Fig. 3.1 is a positive face. The second subscript indicates the coordinate direction of the force or moment component. Thus, F_{xy} is the force acting on the x section in the y direction, and M_{xz} is the moment component in the z direction. These different components have different effects on the member, and hence they have been given special names, as follows:

F_{xx} *Axial force.* This component tends to elongate the member and is often given the symbol F or F_x. We discussed such forces in Chaps. 1 and 2.

F_{xy}, F_{xz} *Shear force.* These components tend to shear one part of the member relative to the adjacent part and are often given the symbols V, or V_y and V_z.

M_{xx} *Twisting moment.* This component is responsible for the twisting of the member about its axis and is often given the symbol M_t or M_{tx}.

M_{xy}, M_{xz} *Bending moments.* These components cause the member to bend and are often given the symbols M_b, or M_{by} and M_{bz}.

In Chaps. 6, 7, and 8 we shall investigate the relation of these force and moment components to the stresses and deformations in the member. Our primary concern now is with the calculation of the magnitude of these components.

To ensure consistency and reproducibility of analyses it will be convenient to define a sign convention for the axial force, shear force, twisting moment, and bending moment. We shall define these to be positive when the force or moment component acts on a *positive face in a positive coordinate direction*; the force and moment components shown in Fig. 3.1 all are positive according to this convention. Note that Newton's third law concerning action and reaction implies that a *positive* component also results when a *negatively directed component* acts on a *negative* face. For example, in Fig. 3.2 let S and S' be sections of a slender member obtained by making a hypothetical cut and separating the surfaces. By Newton's third law the axial forces F_x are equal and opposite. On the section S, F_x is a positive axial force since a force component in the positive x direction acts on a cross section whose outward normal is in the positive x direction. On the section S', F_x is also a positive axial force since a force component in the negative x direction acts on a cross section whose outward normal is in the negative x direction.

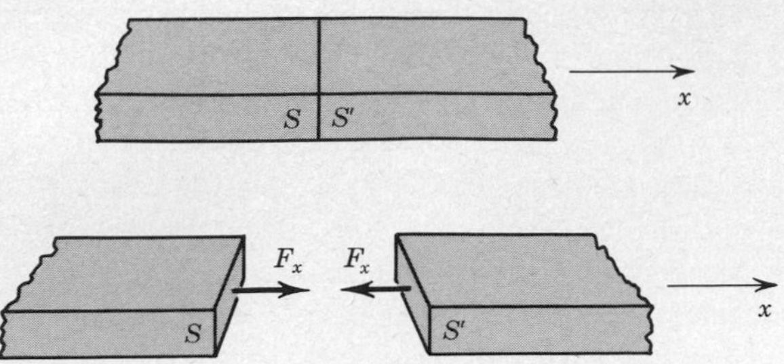

Fig. 3.2 Positive axial force F_x is a tensile force.

Figure 3.1 illustrates the general case where six components of force and moment act across a section. In many instances the problem is considerably simpler in that all forces act in one plane. If the plane of loading is the xy plane, only three components occur: the axial force F_{xx} (F), the shear force F_{xy} (V), and the bending moment M_{xz} (M_b), Fig. 3.3.

As we mentioned above, the determination of stresses and deformations in a slender member requires a knowledge of the forces and moments. The steps involved in solving for the forces and moments in a slender member may be organized as follows:

1. Idealize the actual problem, i.e., create a model of the system, and isolate the main structure, showing all forces acting on the structure.
2. Using the equations of equilibrium ($\Sigma \mathbf{F} = 0$ and $\Sigma \mathbf{M} = 0$), calculate any unknown external or support forces.
3. Cut the member at a section of interest, isolate one of the segments, and repeat step 2 on that segment.

Example 3.1 As an example, let us consider a beam supporting a weight near the center and resting on two other beams, as shown in Fig. 3.4*a*. It is desired to find the forces and moments acting at section *C*.

In this particular problem we have to exercise judgment as to the nature of the support forces. If the beam is not completely rigid, it will tend to

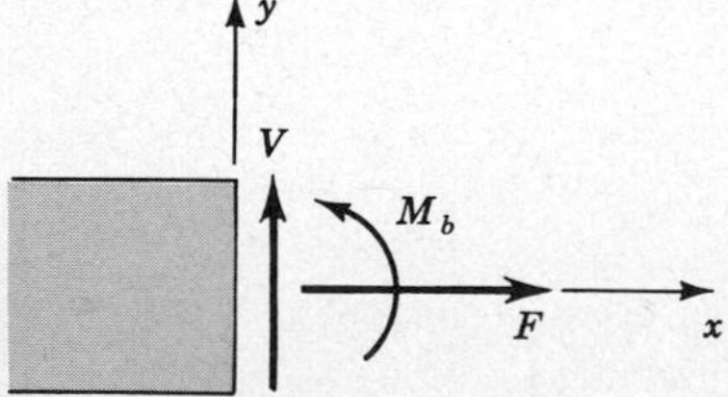

Fig. 3.3 Force and moment components in two dimensions.

bend slightly, as in Fig. 3.4*b*. The reactions between the beam and the supports will then be forces passing through the inner corners of the supports as shown. These reactions will in general have a normal component perpendicular to the beam and a frictional component tangent to the beam. We can say very little about the magnitude and sense of the friction forces. Like the stepladder of Example 1.7, the sense of the friction forces depends on the previous history of the system. We *can* say that the friction forces are limited by the static coefficient of friction. When the coefficient of friction is small, we can be satisfied that the friction forces will be small compared with the normal forces. On the basis of these considerations we idealize the system in Fig. 3.4*c*, where we have shown *vertical* reactions at A and B. There is now no ambiguity; we will get definite answers for this idealization. These results would provide a quantitative framework upon which to base estimates of the indeterminate longitudinal forces in the manner of Example 1.7. In Fig. 3.4*c* we have also neglected the weight of the beam itself on the basis that it is small in comparison with the load W. The following discussion applies to the idealized model of Fig. 3.4*c*.

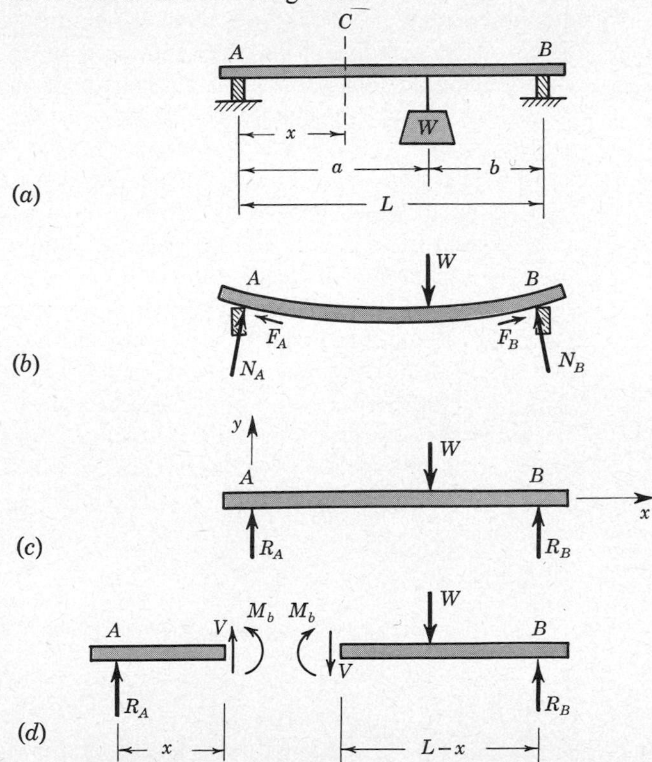

Fig. 3.4 Example 3.1. Calculation of shear force and bending moment at a section of a beam.

In order to calculate the support forces at A and B we apply the requirements for equilibrium to the isolated body AB in Fig. 3.4*c*. Since all the forces are parallel to the y axis, the requirement $\Sigma \mathbf{F} = 0$ is met by

$$\Sigma F_y = R_A + R_B - W = 0 \tag{a}$$

and the requirement $\Sigma \mathbf{M} = 0$ is met by

$$\Sigma M_A = R_B L - Wa = 0 \tag{b}$$

Although it is not difficult to solve (*a*) and (*b*) simultaneously for R_A and R_B, we may note that it is often possible to avoid simultaneous equations by using alternative forms of the equilibrium requirements.[1] For example, here we could use, in addition to (*b*), the requirement

$$\Sigma M_B = Wb - R_A L = 0 \tag{c}$$

which gives directly $R_A = Wb/L$, $R_B = Wa/L$. We exclude an unknown force when we balance moments about a point on the line of action of that force. If the lines of action of two unknown forces intersect, a moment equation written about that point will not contain either force. By judicious choice of the equilibrium requirements used in any particular problem we can often simplify the subsequent mathematical manipulations.

Thus, we have calculated the support forces; we know all the forces *external* to our system. We now desire to know the *internal* forces at C. To do this, we make the forces at C *external* by cutting the beam at C and isolating either part. For completeness in this example we show both parts isolated in Fig. 3.4*d*. In general, we should have forces in the x, y, and z directions and moments about the three axes, as shown in Fig. 3.1. Since our model is two dimensional with no horizontal forces, we need only V and M_b. Note that V and M_b are oppositely directed in the two free-body diagrams in accordance with Newton's third law but bear the same labels, since in the left-hand free body the face of the cut has an outward normal directed in the positive x direction while in the right-hand free body the face of the cut has an outward normal directed in the negative x direction. We now apply the equilibrium requirements to either part (the left part is easier) and obtain

$$\begin{aligned} V &= -R_A = -\frac{Wb}{L} \\ M_b &= R_A x = \frac{Wb}{L}x \end{aligned} \tag{d}$$

for the shear force and bending moment at x.

[1] See Prob. 1.7 for alternative conditions of equilibrium.

For some purposes it is adequate to know the internal forces and moments at a single section. More often it is necessary to know how these internal forces and moments *vary* along the length of the member. For example, a designer usually needs an estimate of the magnitude and location of the *maximum* force and moment in order to design a member of suitable cross section and material. When we consider the deflection of a shaft or a beam in Chaps. 6 and 8, we shall need complete descriptions of the twisting and bending moments as functions along the length.

A graph which shows shear force plotted against distance along a beam is called a *shear-force diagram.* A similar graph showing bending moment as a function of distance is called a *bending-moment diagram.* Axial-force diagrams and twisting-moment diagrams are also employed in discussing slender members.

Shear-force and bending-moment diagrams can be constructed by extending the technique described above for obtaining the shear force and bending moment at a single location. It is only necessary to consider the location of the cut as the independent variable and plot the shear force and bending moment obtained as functions of this variable. To illustrate this we return to the idealized model of Fig. 3.4*c*.

Example 3.2 It is desired to obtain the shear-force and bending-moment diagrams for the idealized beam of Fig. 3.4*c* which is redrawn in Fig. 3.5*a*. In Example 3.1 we obtained the values

$$
\begin{aligned}
V &= -\frac{Wb}{L} \\
M_b &= \frac{Wb}{L}x
\end{aligned}
\qquad (a)
$$

as the shear force and bending moment at the section C, a distance x from the left end. These results are valid for any value of x between $x = 0$ and $x = a$. We can thus consider Eqs. (*a*) to define the shear-force and bending-moment diagrams in the range $0 < x < a$. These equations are sketched in the left-hand portions of Fig. 3.5*b* and *c*.

To complete the diagrams, we consider in Fig. 3.6 a free-body diagram of the right element of the beam when the cut has been made at a distance x where now $a < x < L$. The equilibrium conditions for this element yield

$$
\begin{aligned}
V &= R_B = \frac{Wa}{L} \\
M_b &= R_B(L - x) = \frac{Wa}{L}(L - x)
\end{aligned}
\qquad (b)
$$

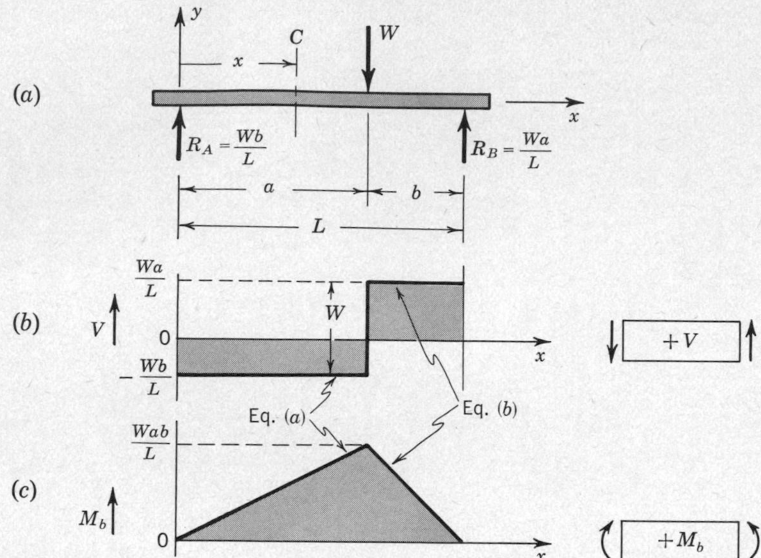

Fig. 3.5 Example 3.2. Shear-force and bending-moment diagrams for beam of Fig. 3.4*c*.

These equations are represented in the right-hand portion of the shear-force and bending-moment diagrams of Fig. 3.5. The sign convention of positive force or moment when a positive component acts on a positive face or when a negative component acts on a negative face is shown for this system in the small sketches to the right of the diagrams.

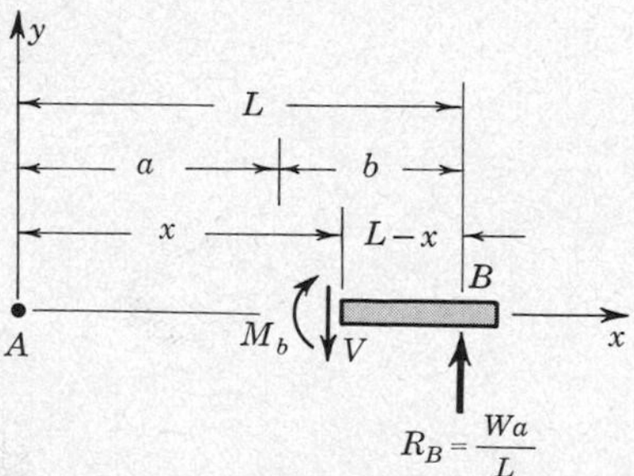

Fig. 3.6 Example 3.2. Free-body diagram for computing V and M_b when $a < x < L$.

3.3 DISTRIBUTED LOADS

In the previous section it was assumed that the load acting on the slender member and the support forces were concentrated or "point" forces. Another idealization which is commonly employed is the concept of a *continuously distributed* loading.

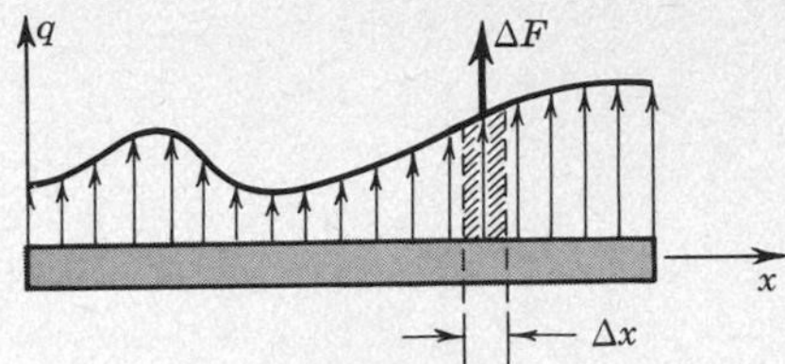

Fig. 3.7 Distributed load of intensity q.

In Fig. 3.7 a beam is subjected to a distributed loading of parallel forces. Such forces might arise from fluid or gas pressures, or from magnetic or gravitational attractions. Let the total force on a length Δx be denoted by ΔF; then the *intensity of loading* q is defined as the limit

$$q = \lim_{\Delta x \to 0} \frac{\Delta F}{\Delta x} \tag{3.1}$$

and has the dimensions of force per unit length. The intensity of loading will, in general, vary with position. In engineering work the most common distributions are the *uniform distribution* where $q(x)$ is constant and the *linearly varying* distribution where $q(x)$ has the form $Ax + B$.

To avoid error in dealing with distributed loading, it is convenient to adopt a sign convention for loading. We shall consider a loading to be positive when it acts in the direction of a positive coordinate axis. Positive loading forces F_y and F_z are shown in Fig. 3.8.

Example 3.3 Consider the cantilever beam AB, built in at the right end, shown in Fig. 3.9*a*. Bricks having a total weight W have been piled up in triangular fashion. It is desired to obtain shear-force and bending-moment diagrams.

In Fig. 3.9*b* the loading has been idealized as a continuous linearly varying distribution of intensity $q = -w = -w_o x/L$. The maximum intensity w_o is at present unknown, but we shall relate it to the total weight W. The interaction with the wall is represented by the reaction R_B and the clamping moment M_B. Figure 3.9*b* is thus a free-body diagram of the entire system.

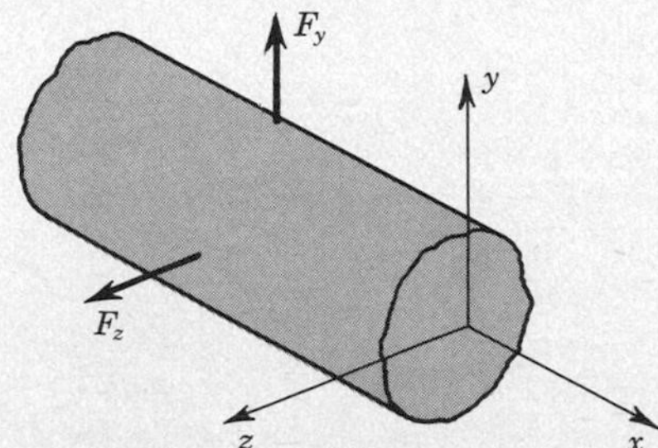

Fig. 3.8 Sign convention for positive loading forces on a member.

To obtain the support reactions we apply the conditions for equilibrium to Fig. 3.9*b*. The reaction R_B can be obtained from the requirement of vertical-force balance. In Fig. 3.9*c* we show a small element of length Δx on which the total load is $w\,\Delta x$. In the limit as $\Delta x \to 0$ the sum of these forces acting on the beam can be represented as an integral over the length of the beam. The statement of vertical equilibrium is then

$$\Sigma F_y = R_B - \int_0^L w\,dx = 0 \tag{a}$$

Substituting $w_o x/L$ for w and integrating gives

$$R_B = \int_0^L \frac{w_o}{L}\,x\,dx = \frac{w_o L}{2} \tag{b}$$

We may note parenthetically here that the integral in (*a*) and (*b*) also represents the total load W, so that we have in addition found the connection between w_o and W.

$$w_o = 2\,\frac{W}{L} \tag{c}$$

To obtain M_B we use the requirement of moment equilibrium about point B. Since R_B passes through B, it does not contribute to the moment around B. To obtain the moment of the distributed load, we again use integration. The small element in Fig. 3.9*c* has a load of $w\,\Delta x$ with a lever arm $(L - x)$ about point B. In the limit the total (counterclockwise) moment about B is then

$$\int_0^L w(L - x)\,dx + M_B \tag{d}$$

which must *vanish* for equilibrium. Substituting for w and integrating yields

$$\begin{aligned} -M_B &= \int_0^L \frac{w_o}{L}\,x(L - x)\,dx \\ &= \frac{w_o}{L}\left(\frac{L^3}{2} - \frac{L^3}{3}\right) \\ &= 2\,\frac{W}{L^2}\,\frac{L^3}{6} = \frac{WL}{3} \end{aligned} \tag{e}$$

The support reactions are thus given by (*b*) and (*e*).

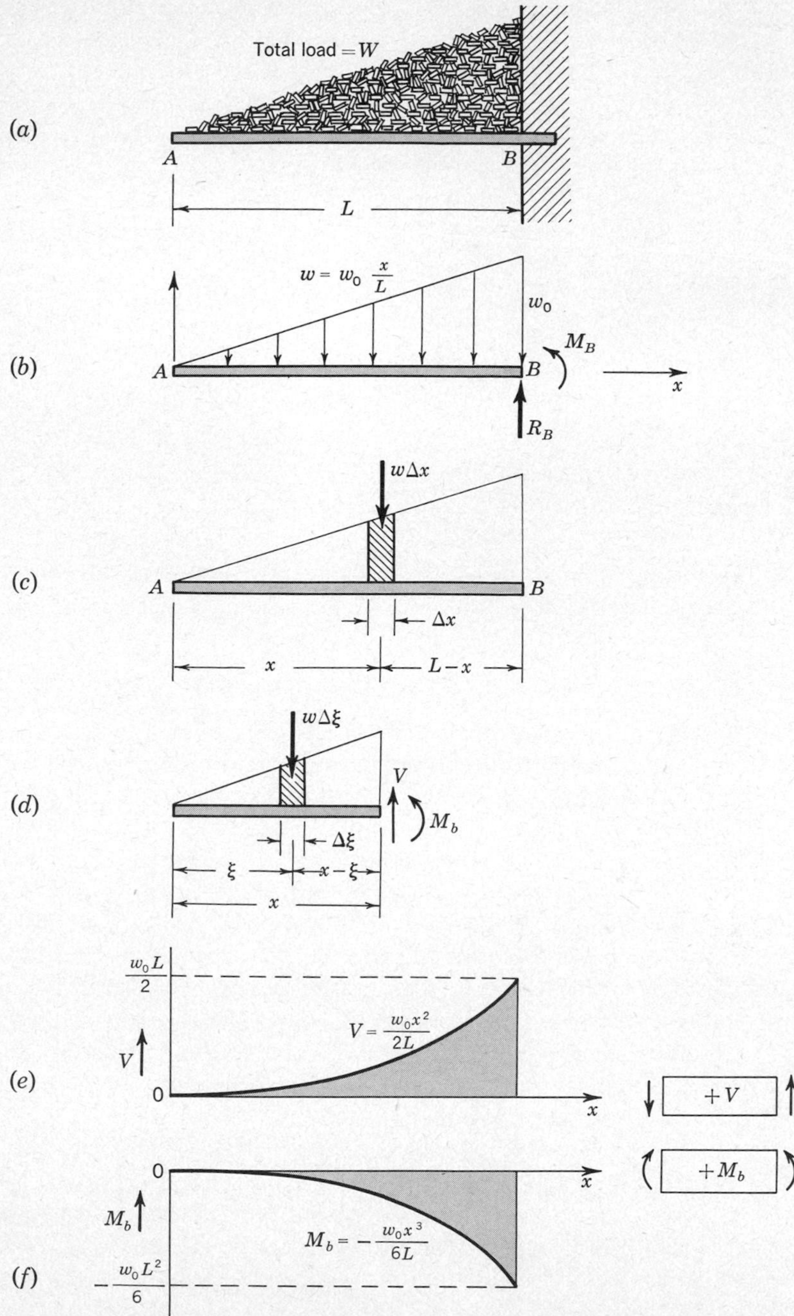

Fig. 3.9 Example 3.3. Distributed load handled by integration.

To obtain the internal force and moment at an arbitrary section x, we consider the free body in Fig. 3.9d. We can evaluate the shear force V and the bending moment M_b by applying the requirements of equilibrium. The contributions from the distributed load are again obtained by using integration. The variable ξ is introduced as a dummy variable in the integration to avoid confusion with x which is here the length of the segment.

$$\Sigma F_y = V - \int_0^x w\, d\xi = 0$$

$$\Sigma M_x = M_b + \int_0^x w(x - \xi)\, d\xi = 0 \qquad (f)$$

Inserting $w_o\xi/L$ for w and integrating leads easily to

$$V = w_o \frac{x^2}{2L}$$

$$M_b = -w_o \frac{x^3}{6L} \qquad (g)$$

which are valid for any value of x between 0 and L. These equations are sketched in the shear-force and bending-moment diagrams in Fig. 3.9e and f.

3.4 RESULTANTS OF DISTRIBUTED LOADS

Two systems of forces are said to be *statically equivalent* if it takes the *same* set of additional forces to reduce each system to equilibrium. A single force which is statically equivalent to a distribution of forces is called the *resultant* of the distributed force system.

In solving problems where the loading is distributed, it is often more convenient to work with the *resultant* of the distributed load on the member rather than to work with the actual distribution. This is permissible only when we are evaluating *external* reactions on the member; it is not allowable when calculating internal forces and moments. This restriction can be illustrated by considering two similar beams, loaded as in Fig. 3.10. It is clear that the loading on beam (b) is the resultant of the loading on beam (a) and that the external support forces are the same in both cases. The internal forces and moments and the deformations of the beams are, however, quite different.

Consider a one-dimensional loading of parallel forces of intensity $q(x)$ in Fig. 3.11. To determine the magnitude of its resultant R and its location $\bar{x}$, we write the equations of equilibrium twice, once using the actual load $q(x)$ and again using the resultant R at $\bar{x}$. The two sets of equations must give identical reaction forces if R is to be the resultant of the distributed load.

$$\Sigma F_y = \int_0^L q\, dx - R_A - R_B = 0$$

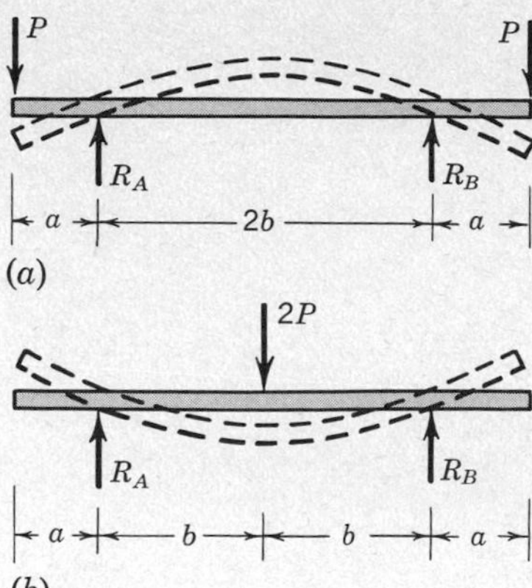

Fig. 3.10 A given loading (*a*), when replaced by its resultant (*b*), produces the same support reactions but not the same internal forces and moments nor the same deflections.

and

$$\Sigma F_y = R - R_A - R_B = 0$$

$$\Sigma M_A = \int_0^L x(q\,dx) - R_B L = 0$$

and

$$\Sigma M_A = R\bar{x} - R_B L = 0$$

Thus, the conditions on R and $\bar{x}$ are

$$R = \int_0^L q\,dx \qquad \text{and} \qquad \bar{x} = \frac{\int_0^L xq\,dx}{R} \tag{3.2}$$

These results have a simple interpretation if we consider the curve of loading intensity $q(x)$ plotted against x to make up a *loading diagram*. The first of (3.2) then states that the resultant is equal to the *total area of the loading diagram*,

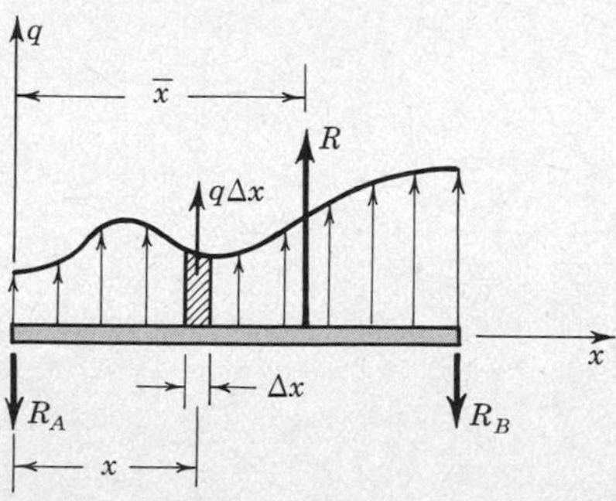

Fig. 3.11 The resultant R of the distributed loading $q(x)$.

while the second states that the line of action of the resultant passes *through the centroid of the loading diagram.*

The reader is reminded that the centroid of an area in the xy plane has the coordinates

$$\bar{x} = \frac{\int x\,dA}{\int dA} \qquad \bar{y} = \frac{\int y\,dA}{\int dA} \tag{3.3}$$

where the integrals extend over the area in question, and that the centroid of a volume has the coordinates

$$\bar{x} = \frac{\int x\,dV}{\int dV} \qquad \bar{y} = \frac{\int y\,dV}{\int dV} \qquad \bar{z} = \frac{\int z\,dV}{\int dV} \tag{3.4}$$

where the integrals extend over the volume in question.

The above treatment for a loading which varies in one direction can be extended to two and three dimensions. If a distributed load of parallel forces acts on a given area A in the xy plane with an intensity p (force per unit *area*), the resultant R is a single force parallel to the given distribution with magnitude

$$R = \int p\,dA \tag{3.5}$$

The line of action of the resultant pierces the xy plane in the point with coordinates

$$\bar{x} = \frac{\int xp\,dA}{R} \qquad \bar{y} = \frac{\int yp\,dA}{R} \tag{3.6}$$

In each case the integral is taken over the given area A.

Similarly, if a distributed load of parallel forces acts throughout a given volume V with an intensity γ (force per unit *volume*), the resultant R is a single force parallel to the given distribution with magnitude

$$R = \int \gamma\,dV \tag{3.7}$$

The line of action of the resultant passes through the point with coordinates

$$\bar{x} = \frac{\int x\gamma\,dV}{R} \qquad \bar{y} = \frac{\int y\gamma\,dV}{R} \qquad \bar{z} = \frac{\int z\gamma\,dV}{R} \tag{3.8}$$

Here each integral extends over the given volume V. The most important application of this last case occurs when the distributed loading is due to *gravity* acting on the material within V. In this case the resultant (3.7) is called the *weight*, and the point determined by (3.8) is called the *center of gravity*. If the weight *density* γ is constant throughout the volume, then the center of gravity (3.8) coincides with the centroid of volume (3.4).

To show the application of resultants to the analysis of beams with distributed loading, we reconsider the system of Example 3.3.

Example 3.4 Figure 3.12*a*, which is the same as Fig. 3.9*b*, shows the free-body diagram of the cantilever beam AB with a linearly varying distributed load. In Fig. 3.12*b* the distributed load has been replaced by a single resultant R at the location $\bar{x}$. Since the loading diagram is a triangle, its area is half the product of base times altitude, and its centroid is two-thirds the distance from vertex to midpoint of opposite side. Thus, without further calculation, we have

$$R = \frac{w_o L}{2} \qquad \bar{x} = \frac{2L}{3} \tag{a}$$

The *external* supports R_B and M_B are now easily obtained by applying the conditions of equilibrium to Fig. 3.12*b*.

$$\Sigma F_y = R_B - R = 0 \qquad \text{or} \qquad R_B = \frac{w_o L}{2} \tag{b}$$

$$\Sigma M_B = R(L - \bar{x}) + M_B = 0 \qquad \text{or} \qquad M_B = -\frac{w_o L^2}{6} \tag{c}$$

These values are identical with those obtained in Example 3.3.

It is *not* permissible to use the above resultant R to calculate shear force and bending moments *within* the beams. We can, however, section the beam at an arbitrary point x, as in Fig. 3.13*a*, and then the shear force and bending moment at the section become *external* forces for the isolated beam element of Fig. 3.13*b*. We may replace the distributed force acting on the

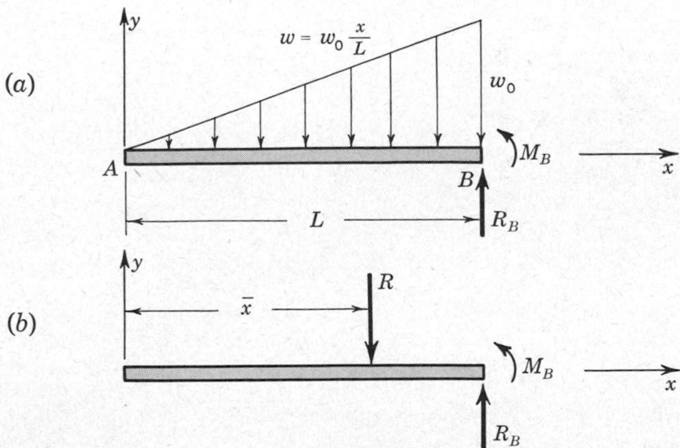

Fig. 3.12 Example 3.4. A distributed loading is replaced by its resultant.

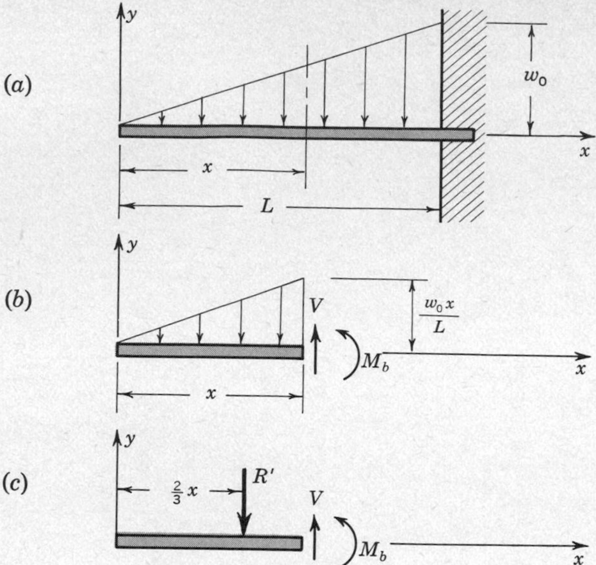

Fig. 3.13 Example 3.4. Distributed loading on a segment of a beam is replaced by its resultant.

portion of the beam, shown in Fig. 3.13*b*, by its resultant R'. Now, using the same technique as in the first part of this example, we obtain from the equilibrium conditions applied to Fig. 3.13*c*

$$V = R' = \frac{w_o x}{L}\frac{x}{2} = \frac{w_o x^2}{2L}$$

$$M_b = -R'\frac{x}{3} = -\frac{w_o x^3}{6L} \qquad (d)$$

These values are equivalent to those obtained in Example 3.3, which were used in sketching the shear-force and bending-moment diagrams of Fig. 3.9.

3.5 DIFFERENTIAL EQUILIBRIUM RELATIONSHIPS

We now turn to an alternative procedure for obtaining internal forces and moments along a slender element. Instead of cutting a beam in two and applying the equilibrium conditions to one of the segments, we consider a very small element of the beam as a free body. The conditions of equilibrium combined with a limiting process will lead us to *differential equations* connecting the load, the shear force, and the bending moment. Integration of these relationships for particular cases

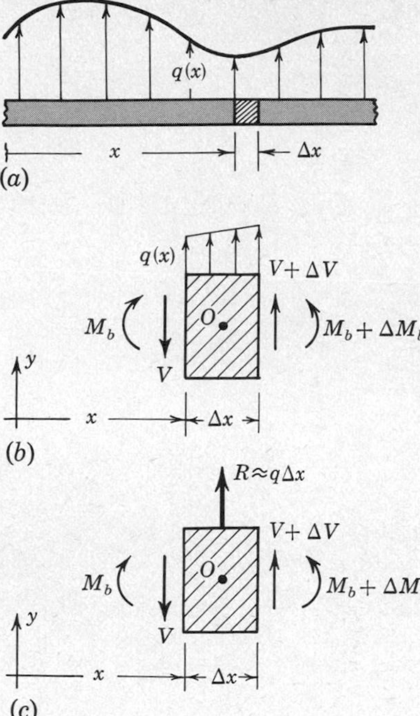

Fig. 3.14 Free-body diagram of small element isolated from a beam under distributed loading.

furnishes us with an alternative method for evaluating shear forces and bending moments.

Figure 3.14 shows a beam element of length Δx. The external actions on this element are the distributed load of intensity q acting over the length Δx and the shear forces and bending moments on the two faces as shown in Fig. 3.14*b*. In Fig. 3.14*c* we have replaced the distributed loading by its resultant R. Strictly, we should compute R and its location according to (3.2). It is, however, clear that if the variation of $q(x)$ is smooth and if Δx is very small then R is very nearly given by $q(x)\,\Delta x$, and the line of action of R will very nearly pass through the midpoint O of the element. In the interests of simplicity and clarity (and at the expense of mathematical rigor) we shall assume in writing the equilibrium conditions that Δx is already so small that we can safely take R to have the magnitude $q\,\Delta x$ and to pass through O. The conditions of equilibrium applied to Fig. 3.14*c* are then

$$\begin{aligned} \Sigma F_y &= V + \Delta V + q\,\Delta x - V = 0 \\ \Sigma M_o &= M_b + \Delta M_b + (V + \Delta V)\frac{\Delta x}{2} + V\frac{\Delta x}{2} - M_b = 0 \end{aligned} \tag{3.9}$$

Before completing the limiting process, we rearrange (3.9) as follows:

$$\begin{aligned} &\frac{\Delta V}{\Delta x} + q(x) = 0 \\ &\frac{\Delta M_b}{\Delta x} + V = -\frac{\Delta V}{2} \end{aligned} \tag{3.10}$$

Now as Δx approaches zero, so also do ΔV and ΔM_b. The ratios in (3.10) tend to differential quotients or derivatives. Thus the limiting forms of (3.10) when Δx goes to zero are

$$\frac{dV}{dx} + q = 0 \tag{3.11}$$

$$\frac{dM_b}{dx} + V = 0 \tag{3.12}$$

These are the basic differential equations relating the load intensity $q(x)$ with the shear force $V(x)$ and bending moment $M_b(x)$ in a beam. Equations (3.11) and (3.12) can be integrated from a section at $x = x_1$, where the shear force and bending moment take the values $V(x_1)$ and $M_b(x_1)$, to a section $x = x_2$, where the corresponding values are $V(x_2)$ and $M_b(x_2)$,

$$V(x_2) - V(x_1) + \int_{x_1}^{x_2} q\, dx = 0 \tag{3.13}$$

$$M_b(x_2) - M_b(x_1) + \int_{x_1}^{x_2} V\, dx = 0 \tag{3.14}$$

To illustrate the application of these equations, we consider the following example.

Example 3.5 In Fig. 3.15*a* a beam carrying a uniformly distributed load of intensity $q = -w_o$ is supported by a pinned joint at A and a roller support at B. We shall obtain shear-force and bending-moment diagrams by integration of the differential relationships (3.11) and (3.12).

The free-body diagram of the whole beam in Fig. 3.15*b* shows the reaction R_B directed vertically because of the roller joint. Then, since the loading is also vertical, the remaining reaction R_A at the pin can only be vertical if the beam is to be in equilibrium. In Fig. 3.15*c* we have sketched the loading diagram $q = -w_o$. We now write (3.11) for this case and integrate

$$\frac{dV}{dx} - w_o = 0$$

$$V - w_o x = C_1 \tag{a}$$

We could evaluate the constant of integration C_1 if we knew the shear force at any particular value of x. We do know that at the end $x = 0$, $V = -R_A$, and at the end $x = L$, $V = R_B$, but, since we have not yet evaluated these reactions, let us retain C_1 as an unknown constant. We next write (3.12) for this case, using for V its value as given by (*a*).

$$\frac{dM_b}{dx} + w_o x + C_1 = 0 \tag{b}$$

Integrating (*b*) introduces another constant of integration C_2.

$$M_b + \tfrac{1}{2} w_o x^2 + C_1 x = C_2 \tag{c}$$

We have two boundary conditions available to fix C_1 and C_2. There is no moment restraint at either end of the beam, hence

$$\begin{aligned} M_b &= 0 \qquad \text{at } x = 0 \\ M_b &= 0 \qquad \text{at } x = L \end{aligned} \tag{d}$$

Inserting these boundary conditions in (*c*) yields $C_1 = -\tfrac{1}{2} w_o L$ and $C_2 = 0$.

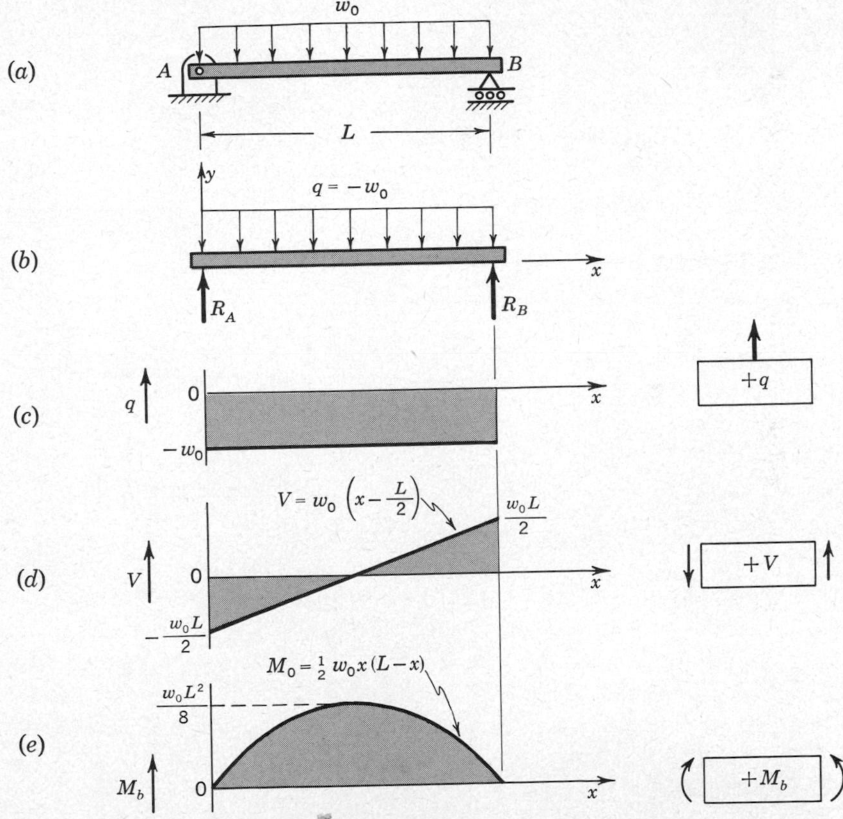

Fig. 3.15 Example 3.5.

The shear force and bending moment are then given by (*a*) and (*c*) as

$$V = w_o\left(x - \frac{L}{2}\right)$$
$$M_b = \tfrac{1}{2}w_o x(L - x) \qquad (e)$$

These relations are sketched in the shear-force and bending-moment diagrams in Fig. 3.15*d* and *e*. Notice that in this procedure we did not need to calculate the reactions. Now that we have the solution (*e*), we can obtain the reactions because, *except for sign*, the reactions are just the values of V at $x = 0$ and $x = L$. It follows from Fig. 3.15*d* that

$$R_A = R_B = \frac{w_o L}{2} \qquad (f)$$

In this particular problem the result (*f*) is almost self-evident from symmetry, and we might have used this to evaluate C_1 in (*a*).

The results of this particular problem are of considerable practical importance because the uniformly distributed load is so common. For example, if the only load on a beam is its own weight, we have a uniformly distributed load whenever the beam is of uniform cross section. In designing roof beams and floor joists in buildings, it is customary to design on a basis of uniformly distributed loading.

There are many alternative ways of using the differential relationships (3.11) and (3.12) to assist in obtaining shear-force and bending-moment diagrams. Notice in Fig. 3.15 that the slope of the bending-moment curve is the negative ordinate of the shear-force curve. In particular, the bending moment is maximum when the shear force is zero. The slope of the shear-force curve is the negative ordinate of the loading diagram. It is often possible to use these relations in a qualitative way to predict the general shape of the diagrams and then to calculate a few key points (such as ends) in order to tie down the curves in a quantitative manner.

Example 3.6 Consider the beam shown in Fig. 3.16*a* with simple transverse supports at A and B and loaded with a uniformly distributed load $q = -w_o$ over a portion of the length. It is desired to obtain the shear-force and bending-moment diagrams. In contrast with the previous example, it is not possible[1] to write a single differential equation for V and M which will be valid over the complete length of the beam. Instead let subscripts 1 and 2 indicate values of variables in the loaded and unloaded segments of the beam.

[1] At least not without inventing a special notation, as will be done in the next section.

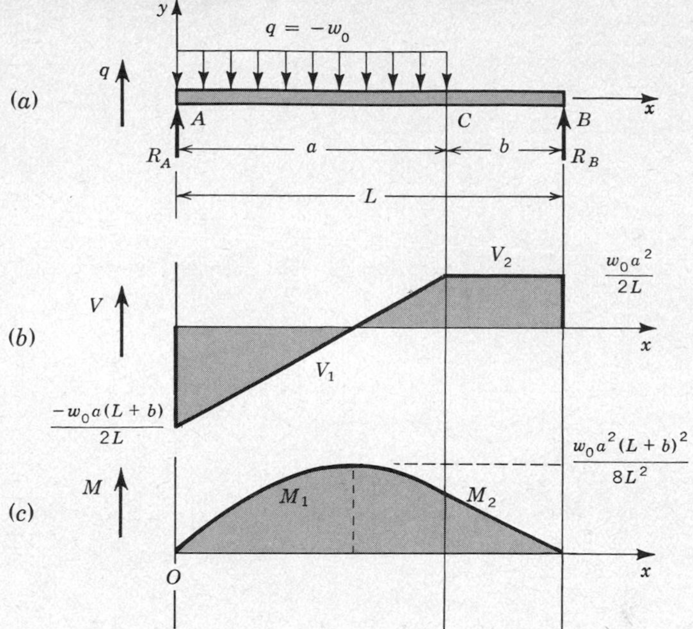

Fig. 3.16 Example 3.6.

Using (3.11) in each segment and integrating gives

$$\frac{dV_1}{dx} - w_o = 0 \qquad \frac{dV_2}{dx} = 0$$

$$V_1 - w_o x = C_1 \qquad V_2 = C_2 \tag{a}$$

Next we write (3.12) in each segment, using the V's from (*a*)

$$\frac{dM_{b1}}{dx} + w_o x + C_1 = 0 \qquad \frac{dM_{b2}}{dx} + C_2 = 0 \tag{b}$$

and integrating again gives

$$M_{b1} + \tfrac{1}{2} w_o x^2 + C_1 x = C_3 \qquad M_{b2} + C_2 x = C_4 \tag{c}$$

We have, as in the previous example, the boundary conditions at each end

$$M_{b1} = 0 \text{ at } x = 0 \qquad M_{b2} = 0 \text{ at } x = L \tag{d}$$

However, we need two additional conditions in order to determine the remaining two arbitrary constants. These follow from equilibrium requirements at the *junction* of the two segments

$$\begin{aligned} V_1 &= V_2 && \text{at } x = a \\ M_{b1} &= M_{b2} && \text{at } x = a \end{aligned} \tag{e}$$

Inserting these boundary conditions into (a) and (c) leads to

$$C_3 = 0 \qquad C_4 = LC_2 = \tfrac{1}{2}w_o a^2$$

$$C_1 = C_2 - w_o a \qquad C_2 = \tfrac{1}{2}\frac{w_o a^2}{L} \tag{f}$$

$$C_1 = \tfrac{1}{2}w_o a\left(\frac{a}{L} - 2\right) = -\tfrac{1}{2}w_o a\,\frac{(L+b)}{L}$$

The shear-force and bending-moment diagrams can be constructed from

$$\begin{aligned} V_1 &= w_o x - \frac{w_o a}{2L}(L+b) && 0 \leqq x \leqq a \\ V_2 &= \tfrac{1}{2}\frac{w_o a^2}{L} && a \leqq x \leqq L \end{aligned} \tag{g}$$

and

$$\begin{aligned} M_{b1} &= +\frac{w_o a}{2L}(L+b)x - \tfrac{1}{2}w_o x^2 && 0 \leqq x \leqq a \\ M_{b2} &= \tfrac{1}{2}w_o a^2 - \tfrac{1}{2}\frac{w_o a^2 x}{L} && a \leqq x \leqq L \end{aligned} \tag{h}$$

as shown in Fig. 3.16b and c.

Clearly, if the loading requires separate representations for a number of segments each with its own differential equation form, it becomes very awkward to carry along the additional arbitrary constants which are later eliminated by matching the V's and M's at the junctions of the segments. In the next section a notation is introduced which greatly facilitates the handling of multisegment problems.

3.6 SINGULARITY FUNCTIONS

In the foregoing section we demonstrated how the relatively routine procedure of integration could be used to obtain shear-force and bending-moment diagrams for beams with distributed loads. Where there are concentrated-force and concentrated-moment loadings or where the distributed load suddenly changes its magnitude, we have seen that the procedure just outlined becomes fairly cumbersome unless a special mathematical apparatus is available to handle discontinuous loadings. In this section we introduce a family of singularity functions specifically designed for this purpose.

Figure 3.17 shows five members of the family.[1]

[1] This use of a family of singularity functions is common in engineering and physics. Our notation is patterned after that used by W. H. Macauley, Note on the Deflection of Beams, *Mes. Math.*, vol. 48, pp. 129–130, 1919. A review of the use of such functions for beam problems is found in W. D. Pilkey, Clebsch's Method for Beam Deflections, *J. Eng. Educ.*, vol. 54, pp. 170–174, 1964.

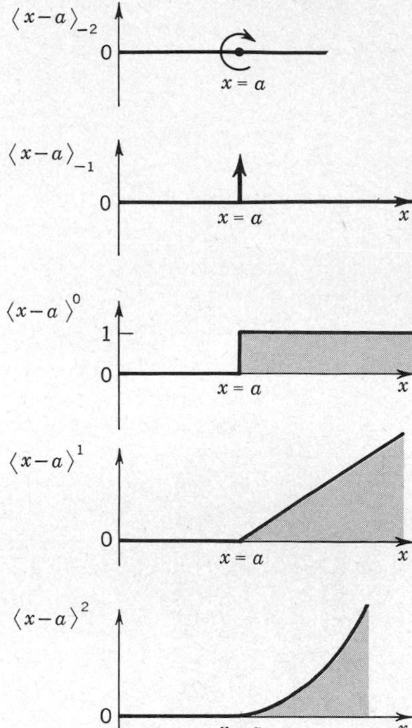

Fig. 3.17 Family of singularity functions.

$$f_n(x) = \langle x - a \rangle^n \tag{3.15}$$

When $n \geqq 0$ the notation (3.15) has the following significance: If the expression in the *pointed* brackets is negative (i.e., if $x < a$), the value of $f_n(x)$ is zero; if the expression in the *pointed* brackets is positive (i.e., if $x > a$), then the value of $f_n(x)$ is just $(x - a)^n$. Thus the pointed brackets are like ordinary brackets except for the one idiosyncrasy that they are blind to negative quantities. The function $\langle x - a \rangle^0$ is called the *unit step* starting at $x = a$, and the function $\langle x - a \rangle^1$ is called the *unit ramp* starting at $x = a$. The integration law for these functions is

$$\int_{-\infty}^{x} \langle x - a \rangle^n \, dx = \frac{\langle x - a \rangle^{n+1}}{n + 1} \qquad n \geqq 0 \tag{3.16}$$

The first two members of the family shown in Fig. 3.17 are exceptional. To emphasize this, the exponent is written below the bracket instead of above. These functions are zero everywhere except at $x = a$ where they are infinite. They are, however, infinite[1] in such a way that

[1] For a more rigorous treatment see Probs. 3.44, 3.45, and 3.46 at the end of this chapter.

$$\int_{-\infty}^{x} \langle x - a \rangle_{-2}\, dx = \langle x - a \rangle_{-1}$$

$$\int_{-\infty}^{x} \langle x - a \rangle_{-1}\, dx = \langle x - a \rangle^{0} \tag{3.17}$$

The function $\langle x - a \rangle_{-1}$ is called the *unit concentrated load* or the *unit impulse* function. In physics texts it is known as the *Dirac delta* function. The function $\langle x - a \rangle_{-2}$ is called the *unit concentrated moment* or the *unit doublet* function.

The integration laws (3.16) and (3.17) permit us to obtain shear forces and bending moments by integration from any loading distribution which we are able to represent in terms of the family (3.15). Figure 3.18 illustrates some examples of load-intensity distributions and how they are represented by singularity functions. Most practical cases of beam loading can be built up by superposition of the cases shown in Fig. 3.18. The following examples illustrate the process.

Example 3.7 We consider the problem studied in Example 3.6 again, but we shall utilize the singularity functions.

In Fig. 3.19*b* the load intensity q is given in a form which permits easy translation into the singularity functions. The load $q = -w_o$, which stops

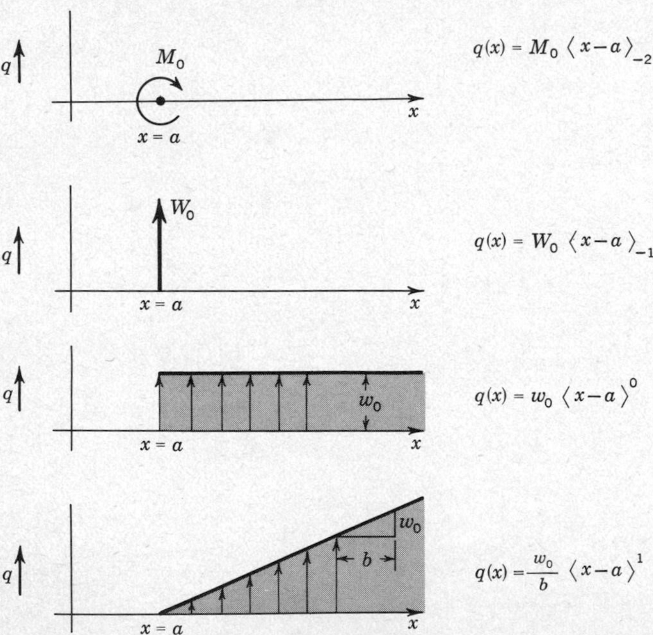

Fig. 3.18 Examples of loading intensities represented by singularity functions.

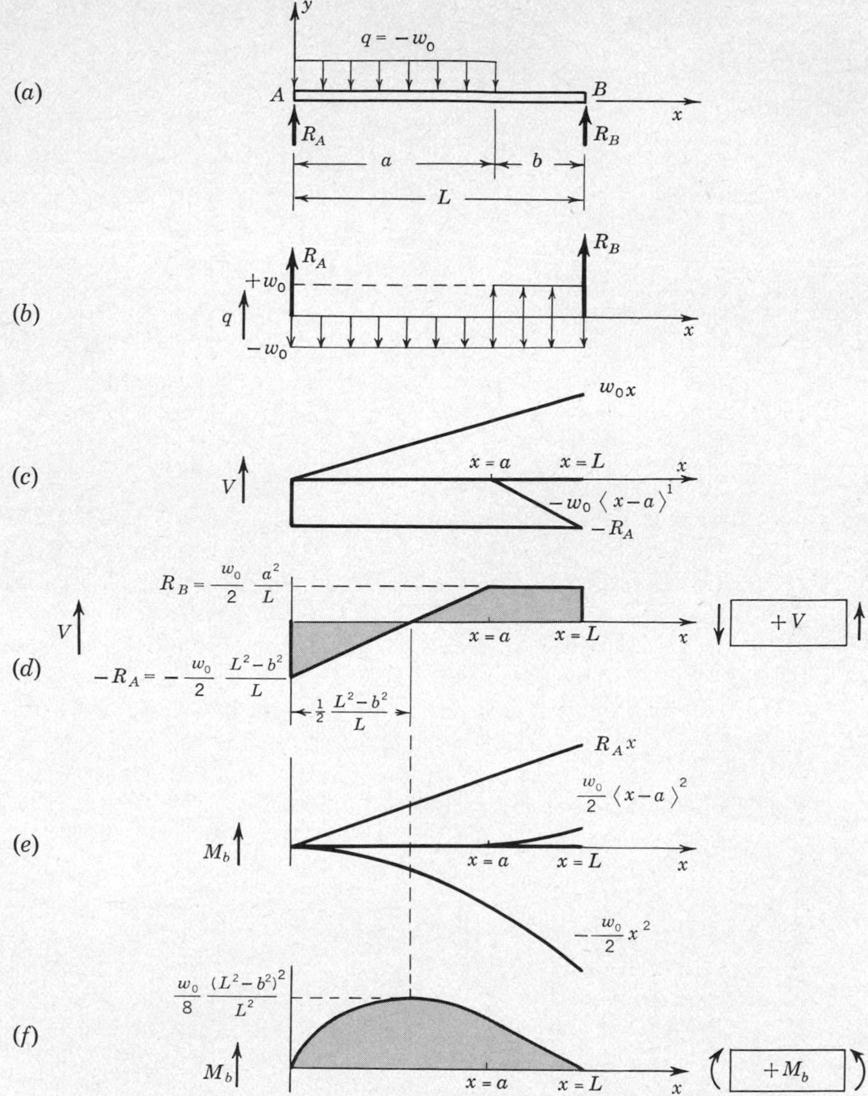

Fig. 3.19 Example 3.7.

at $x = a$ in Fig. 3.19*a*, is shown continuing on to *B*, but at $x = a$ the supplementary load $q = w_o$ is started. This cancels the other load, leaving no net distributed load after $x = a$. Now, with the aid of Fig. 3.18, we write the load-intensity function

$$q(x) = -w_o + w_o\langle x - a\rangle^0 \qquad (a)$$

which is valid for $0 < x < L$. Substituting (*a*) into (3.11) and integrating we find

$$V(x) = w_o x - w_o \langle x - a \rangle^1 + C_1 \tag{b}$$

In particular, $V(0) = C_1$, but this is just $-R_A$, which is easily found from a moment-balance equation about point B to give

$$-R_A = -\frac{w_o a}{L}\left(b + \frac{a}{2}\right) \tag{c}$$

Integrating (*b*) in accordance with (3.12) with C_1 known gives

$$M_b = -\frac{w_o x^2}{2} + \frac{w_o}{2}\langle x - a \rangle^2 + \frac{w_o a}{L}\left(b + \frac{a}{2}\right)x + C_2 \tag{d}$$

where $C_2 = 0$ since $M_b(0) = 0$.

The shear-force and bending-moment diagrams are easily constructed using (*b*) and (*d*) with C_1 and C_2 known. To help the reader interpret (*b*), each term has been sketched separately in Fig. 3.19*c* before showing the resultant shear-force diagram in Fig. 3.19*d*. In similar fashion each term of (*d*) has been sketched separately in Fig. 3.19*e* before showing the resultant bending-moment diagram in Fig. 3.19*f*.

An important practical result in a problem like this is the magnitude and location of the maximum bending moment. The location is given by the point where the shear force passes through zero in Fig. 3.19*d*, and the magnitude is obtained by evaluating (*d*) at this location.

There are many alternative techniques for solving problems like this one. The method we have shown involved the separate evaluation of a support reaction which was used in evaluating an arbitrary constant of integration. An alternate procedure involves the introduction of the support reactions into the loading term as unknowns and their determination from the two boundary conditions on the moments at the ends of the beam.

Let us work through this example again, this time including the reactive forces in the loading term $q(x)$. With the aid of Fig. 3.18, we write the load-intensity function

$$q(x) = R_A\langle x \rangle_{-1} - w_o\langle x \rangle^0 + w_o\langle x - a \rangle^0 + R_B\langle x - L \rangle_{-1} \tag{e}$$

This representation is valid for all x, since it gives $q = 0$ when x is outside of the segment between A and B. Now, according to (3.13) and the integration laws (3.16) and (3.17), we have, since $V = 0$ at $x = -\infty$,

$$-V(x) = \int_{-\infty}^{x} q\,dx = R_A\langle x \rangle^0 - w_o\langle x \rangle^1 + w_o\langle x - a \rangle^1 + R_B\langle x - L \rangle^0 \tag{f}$$

A second integration using (3.14), and the fact that $M_b = 0$ at $x = -\infty$, yields

$$M_b(x) = -\int_{-\infty}^{x} V\,dx = R_A\langle x\rangle^1 - \frac{w_o}{2}\langle x\rangle^2 + \frac{w_o}{2}\langle x-a\rangle^2 + R_B\langle x-L\rangle^1 \tag{g}$$

If R_A and R_B were known, (f) and (g) would furnish the complete solution for the shear force and bending moment. It is not difficult to obtain the reactions from a separate calculation, and in some cases this may prove to be the simplest procedure. We can, however, use our results to determine the reactions by making use of our observation that there should be no internal forces and moments outside of the segment AB. If we take x just slightly larger than $x = L$, the shear force (f) should vanish, that is,

$$\begin{aligned} R_A - w_o L + w_o(L-a) + R_B &= 0 \\ R_A + R_B &= w_o a \end{aligned} \tag{h}$$

and the bending moment (g) should also vanish, that is,

$$\begin{aligned} R_A L - \frac{w_o}{2}L^2 + \frac{w_o}{2}(L-a)^2 &= 0 \\ R_A &= \frac{w_o}{2}\frac{L^2 - b^2}{L} \end{aligned} \tag{i}$$

Equations (h) and (i) furnish two relations for determining the two reactions R_A and R_B. Note that these relations are, in fact, the conditions for equilibrium of the entire beam. Vertical-force balance is indicated by (h), and balance of moments about point B is indicated by (i). What this means is simply that the satisfaction of the equilibrium requirements for *every* differential element of the beam implies satisfaction of the equilibrium requirements of the entire beam.

Example 3.8 In Fig. 3.20*a* the frame BAC is built-in at B and subjected to a load P at C. It is desired to obtain shear-force and bending-moment diagrams for the segment AB.

In Fig. 3.20*b* a free-body analysis of the segment AC has been made. In order to maintain equilibrium, there must be a force P and a moment $PL/2$ transmitted across the cut at A. Figure 3.20*c* is a free-body diagram of the segment AB. At A the force and moment shown were obtained from Newton's third law and the results of Fig. 3.20*b*. At B we have shown the force reaction R_B and the moment reaction M_B from the built-in support.

Now the force and moment at A are actually transmitted across the dimension b, and the force and moment at B are actually transmitted along

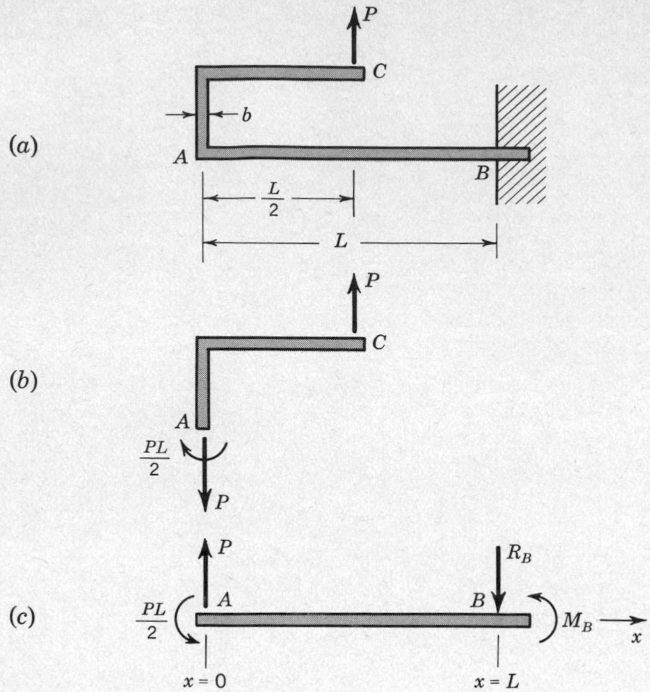

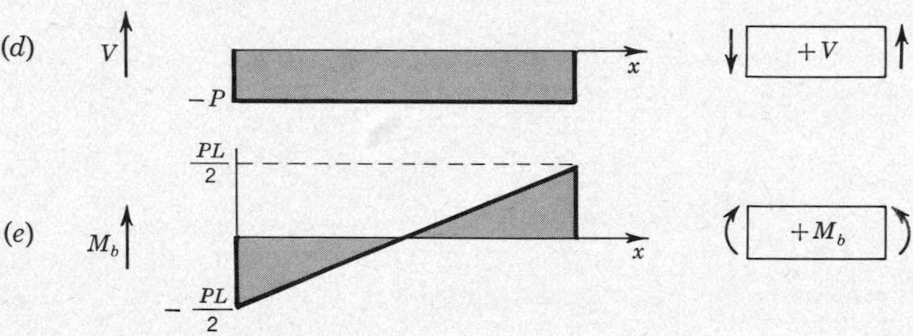

Fig. 3.20 Example 3.8.

some small distance into the wall. We shall, however, idealize these and take them to be concentrated forces and moments at the points A and B. Since there is no loading between A and B, $q(x) = 0$ and from (3.11)

$$V(x) = C_1 \qquad (a)$$

where $C_1 = -P$ because of the assumed concentrated force at A. Integrating again using (3.12) we find

$$M_b(x) = Px + C_2 \qquad (b)$$

and setting $M_b(0) = -PL/2$ gives $C_2 = -PL/2$. With C_1 and C_2 known, we can sketch the shear-force and bending-moment diagrams as shown in Fig. 3.20*d* and *e*.

The preceding examples have shown the advantage of the singularity-function method. However, the reader may be left with the question of when to include the reactive forces into the loading-intensity function, or, if the reactive forces are to be included in $q(x)$, should they be evaluated first from overall equilibrium requirements or should they be evaluated as in the second part of the solution to Example 3.7 after integration of (3.11) and (3.12). There are no definite answers to these questions; all that can be said is that the best way to proceed depends upon the problem. However, in general, the algebraic work is simplified if all the reactions are determined first from overall equilibrium (assuming that this can be done). Inclusion of these now-known reactive forces into the loading function will still require a decision. It is to be emphasized again, however, that whatever route is followed, all constants of integration must be evaluated carefully from the support conditions. Let us consider another example in which it is necessary to include the reactive forces into the loading term.

Example 3.9 The loading on a beam is assumed to have the shape shown in Fig. 3.21*a*. It is required to find the location of the supports *A* and *B* such that the bending moment at the midpoint is zero.

As our first step the reactive forces at *A* and *B* will be found. From symmetry

$$R_A = R_B = \tfrac{1}{2}R$$

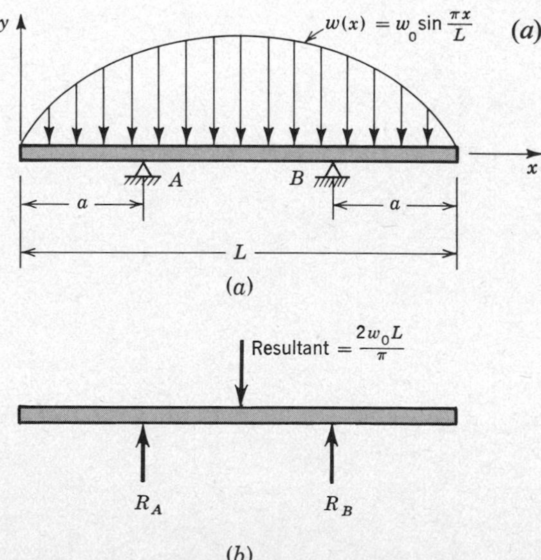

Fig. 3.21 Example 3.9.

where R is the resultant of the load-distribution curve (Fig. 3.21*b*)

$$R = \int_0^L w(x)\,dx = w_o \int_0^L \sin\frac{\pi x}{L}\,dx = \frac{2w_o L}{\pi} \tag{b}$$

The load-intensity function can now be written in the form

$$q(x) = -w_o \sin\frac{\pi x}{L} + \frac{w_o L}{\pi}\langle x - a\rangle_{-1} + \frac{w_o L}{\pi}\langle x - (L-a)\rangle_{-1} \tag{c}$$

Upon integration of (*c*) using (3.11), we find

$$V = -\frac{w_o L}{\pi}\cos\frac{\pi x}{L} - \frac{w_o L}{\pi}\langle x - a\rangle^0 - \frac{w_o L}{\pi}\langle x - (L-a)\rangle^0 + C_1 \tag{d}$$

where C_1 is the constant of integration. But $V = 0$ at $x = 0$, so that

$$C_1 = \frac{w_o L}{\pi} \tag{e}$$

Upon integration of (*d*) using (*e*) and (3.12), we find

$$M_b = -\frac{w_o L}{\pi}\left(x - \frac{L}{\pi}\sin\frac{\pi x}{L}\right) + \frac{w_o L}{\pi}\langle x - a\rangle^1 + \frac{w_o L}{\pi}\langle x - (L-a)\rangle^1 + C_2 \tag{f}$$

The constant $C_2 = 0$ since M_b at $x = 0$ is zero. Therefore M_b will vanish at $x = L/2$ if

$$a = \frac{L}{\pi} \tag{g}$$

3.7 FLUID FORCES

In many applications structural components are subjected to forces due to fluids in contact with the structure.

Fluids at rest, for example, water in a water-filled balloon, offer no appreciable resistance to changes of shape if the forces causing the change of shape are applied slowly enough. This suggests that in a fluid at rest there are no frictional forces between the particles of fluid. As a consequence of this observation we can say that the force per unit area, or *hydrostatic pressure* p, at a point in a fluid at rest is normal to any surface passing through that point. That is, we can say that in a liquid at rest the pressure at a point is the same in all directions. Further, the pressure on a surface acts in the opposite direction to the outward-pointing normal to the surface (see Fig. 3.22*a*). We are familiar with these results from our knowledge of pressure in a gas.

A simple equilibrium consideration for a fluid under the action of gravity will show that the pressure is a linear function of distance from the free surface. In Fig. 3.22*b* a small cylindrical element of fluid is shown in equilibrium under fluid pressures and the weight of the fluid element $\gamma\,\Delta A\,\Delta z$, where γ is the weight density of the fluid, ΔA is the cross-sectional area of the cylindrical element, and Δz is the vertical thickness of the element.

Force equilibrium in the horizontal plane is satisfied identically from symmetry, and equilibrium in the z direction requires

$$p\,\Delta A + \gamma\,\Delta A\,\Delta z - (p + \Delta p)\,\Delta A = 0$$

or in the limit

$$\frac{dp}{dz} = \gamma$$

If the reference pressure at $z = 0$ is taken as p_o, for example, atmospheric pressure at the surface $z = 0$, then

$$p = \gamma z + p_o$$

We will now consider an example employing the results that fluid pressure acts normal to a surface and is a linear function of depth.

Example 3.10 Fig. 3.23 shows a 5-ft-square gate which is retaining the water at half the length of the gate as shown. If it is assumed that the total pressure load on the gate is transmitted to the supports at A, B, D, and E by means of symmetrically located simply supported beams AB and DE, find the maximum bending moment in the beams. The bottom edge DA of the gate is 2 ft below the water line, and $\gamma = 62.4$ lb/ft^3.

The fluid pressure acts normal to the gate, is uniform on lines parallel to DA, and varies linearly from zero pressure (above atmospheric) at the middle of the gate to a maximum pressure p_A along the bottom edge where

$$p_A = \gamma z_A = (62.4)(2) = 124.8 \text{ lb/ft}^2 \qquad (a)$$

To obtain the loading on the beams we assume that the pressure loading on the shaded strip in Fig. 3.23b is carried equally by the two beams. If the

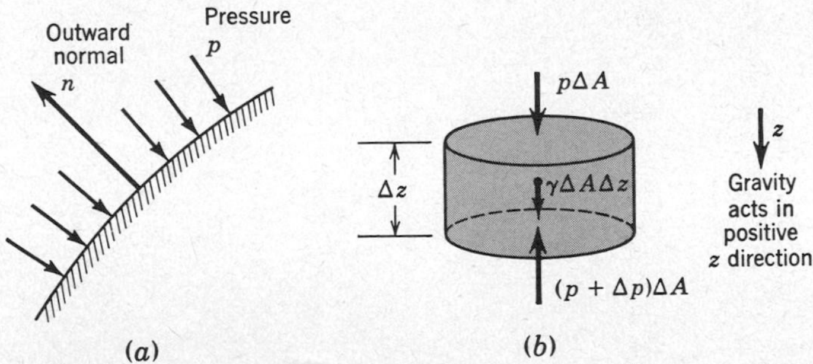

Fig. 3.22 (a) Fluid pressure acts normal to the surface; (b) element of fluid in a gravity field.

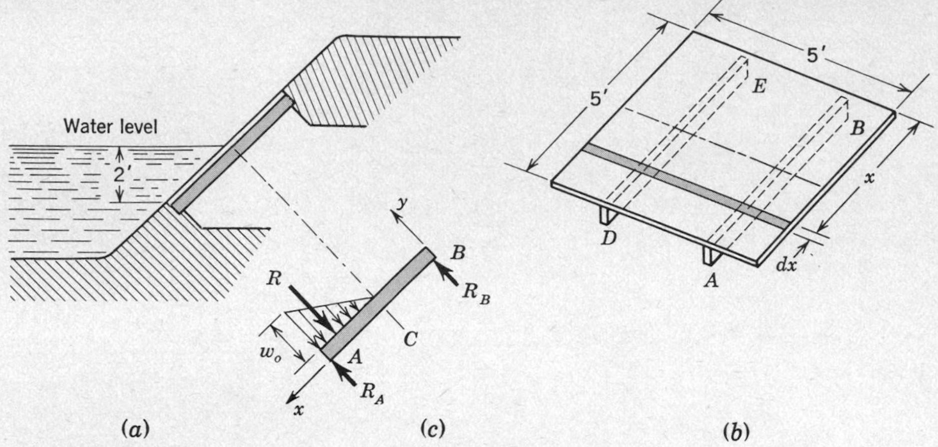

Fig. 3.23 Example 3.10

pressure at this location is $p(x)$, the load per unit length $w(x)$ carried by one of the beams is

$$w(x) = \tfrac{5}{2}p(x) \tag{b}$$

This implies that $w(x)$ varies along the length of a beam in the same fashion as $p(x)$, as indicated in Fig. 3.23*c* for the beam *AB*. This loading can be represented by the expression

$$q(x) = -\tfrac{2}{5}w_o\langle x - \tfrac{5}{2}\rangle^1 \tag{c}$$

where w_o is obtained by evaluating (*b*) at the bottom of the gate

$$w_o = \tfrac{5}{2}p_A = 312 \text{ lb/ft} \tag{d}$$

The resultant R of this loading is $(\tfrac{1}{2}w_o)\tfrac{5}{2} = \tfrac{5}{4}w_o$, and its line of action passes through the centroid of the loading diagram which is $\tfrac{1}{3}(\tfrac{5}{2}) = \tfrac{5}{6}$ ft from A.

Integrating (*c*) using (3.11) gives

$$-V(x) = -\frac{w_o}{5}\langle x - \tfrac{5}{2}\rangle^2 + C_1 \tag{e}$$

Now

$$V(0) = -R_B = -C_1 \tag{f}$$

and R_B is computed by taking moments about A and using the resultant, R, to get

$$\begin{aligned} 5R_B &= R(\tfrac{5}{6}) \\ R_B &= \tfrac{5}{24}w_o \end{aligned} \tag{g}$$

Integrating (e) using (3.12) leads to

$$M_b(x) = -\frac{w_o}{15}\langle x - \tfrac{5}{2}\rangle^3 + \tfrac{5}{24}w_o x + C_2 \qquad (h)$$

and $C_2 = 0$ in order that the moment vanish at $x = 0$. Now the maximum bending moment is located in this case between A and C at the point where $V = 0$. Solving (e) for x_o such that $V(x_o) = 0$ gives $x_o = \tfrac{5}{2}(1 + \sqrt{\tfrac{1}{6}})$. Substitution into ($h$) gives for the maximum bending moment

$$M_b(x_o) = 207 \text{ ft-lb} \qquad (i)$$

3.8 THREE-DIMENSIONAL PROBLEMS

The foregoing treatment of straight slender members subjected to forces lying in a single plane passing through the member may be extended to handle arbitrary three-dimensional loadings of slender members. In the general case, as shown in Fig. 3.1, there will be a vector force and vector moment acting at any section of the member. The force and moment can be obtained by applying the equilibrium requirements to either segment of the cut member or by applying the equilibrium requirements to a differential element and integrating. The three-dimensional aspects of the problem can be handled by using vector notation or by reducing the problem to three two-dimensional problems by resolving all forces and moments into three components. This will be illustrated in the following examples.

Example 3.11 As an example of a curved slender member we show (Fig. 3.24a) a piece of refinery piping AB anchored at B and bent into a quadrant of a circle of center O and radius a. It is desired to obtain force and moment diagrams for this segment of pipe when a transverse load P is acting as shown.

The same basic procedure is used as before. A cut is made at C, and the segment AC is isolated as a free body in Fig. 3.24b. To help describe the forces and moments, we have introduced the rectangular coordinate system (x,r,s) at the face of the cut. Here $s = a\theta$ is arc length along the pipe, and r is radial distance from O. The force P at the free end A can be held in equilibrium by the force V_x and the torques M_{ss} and M_{sr} acting on the section C providing

$$\Sigma\mathbf{F} = P\mathbf{i} + V_x\mathbf{i} = 0 \qquad (a)$$

$$\Sigma\mathbf{M}_C = M_{ss}\mathbf{u}_s + M_{sr}\mathbf{u}_r + \mathbf{r}_{CA} \times (P\mathbf{i}) = 0 \qquad (b)$$

where $\mathbf{i}$, $\mathbf{u}_r$, and $\mathbf{u}_s$ are unit vectors in the x, r, and s directions and $\mathbf{r}_{CA}$ is the vector distance from C to A. In order to express $\mathbf{r}_{CA}$ in terms of the unit vectors, we may note from Fig. 3.24a that

$$\mathbf{r}_{CA} = \mathbf{r}_{OA} - \mathbf{r}_{OC} = a\mathbf{j} - a\mathbf{u}_r \qquad (c)$$

where $\mathbf{j}$ is the unit vector in the y direction, and that, from Fig. 3.24c,

$$\mathbf{j} = \mathbf{u}_r \cos\theta - \mathbf{u}_s \sin\theta \qquad (d)$$

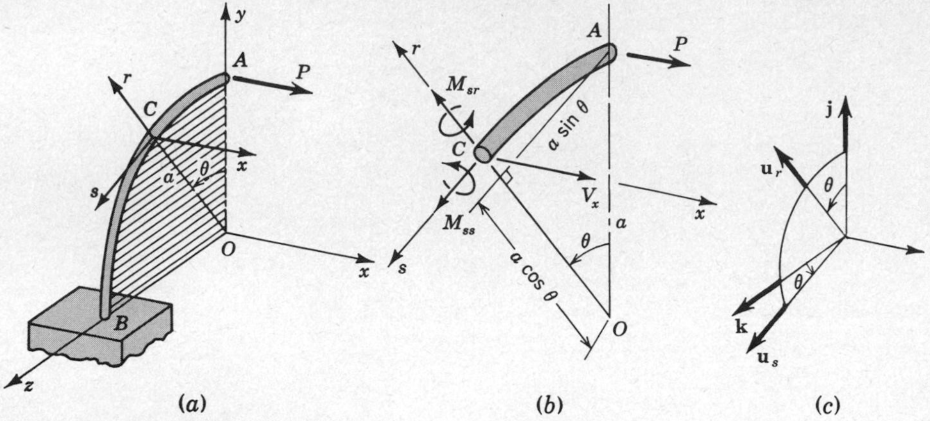

Fig. 3.24 Example 3.11.

Therefore

$$\mathbf{r}_{CA} = -a(1 - \cos\theta)\mathbf{u}_r - a\sin\theta\mathbf{u}_s \tag{e}$$

Equation (*e*) may also be obtained directly from Fig. 3.24*b*. Upon substitution of (*e*) into (*b*) and evaluation of the cross products, we find

$$\Sigma\mathbf{M}_C = M_{ss}\mathbf{u}_s + M_{sr}\mathbf{u}_r + P[a(1 - \cos\theta)\mathbf{u}_s - a\sin\theta\mathbf{u}_r] = 0 \tag{f}$$

Setting the coefficient of each unit vector separately equal to zero in (*a*) and (*f*) yields

$$\begin{aligned} V_x &= -P \\ M_{sr} &= Pa\sin\theta \\ M_{ss} &= -Pa(1 - \cos\theta) \end{aligned} \tag{g}$$

These are the shear force, bending moment, and twisting moment acting at the general angle θ. Figure 3.25 shows their variation with arc length $s = a\theta$ along with sketches of the sign conventions employed.

Example 3.12 Consider the offset bell-crank mechanism in Fig. 3.26*a*. A shaft supported in journal bearings at *A* and *D* is loaded as shown and has offset links attached at *B* and *C*. The problem is to obtain diagrams showing the variation of shear force, bending moment, and twisting moment in the shaft *AD*.

We first idealize the problem in Fig. 3.26*b* by assuming that the bearings exert *forces* perpendicular to the shaft but do not exert any *moments*. The magnitude of the reactions at *A* and *D* can be determined by applying the equilibrium requirements to the set of forces shown in Fig. 3.26*b*. If these forces are in equilibrium, their resultant moment about any point is zero. It is most convenient to take moments about points *A* and *D* because we thereby exclude two unknown forces from each equation and thus obviate the need for solving simultaneous equations. Using **i**, **j**, and **k** to represent unit vectors in the x, y, and z directions, we have

$$\Sigma \mathbf{M}_A = (6\mathbf{i} + 6\mathbf{k}) \times (-P\mathbf{j}) + (14\mathbf{i} + 9\mathbf{j}) \times (-200\mathbf{k}) + (20\mathbf{i}) \times (D_y\mathbf{j} + D_z\mathbf{k}) = 0$$

with a similar equation $\Sigma \mathbf{M}_D$. Working out the vector products we obtain

$$\begin{aligned} \Sigma \mathbf{M}_A &= 6P\mathbf{i} - 6P\mathbf{k} - 9(200)\mathbf{i} + 14(200)\mathbf{j} + 20D_y\mathbf{k} - 20D_z\mathbf{j} = 0 \\ \Sigma \mathbf{M}_D &= 6P\mathbf{i} + 14P\mathbf{k} + 20A_z\mathbf{j} - 20A_y\mathbf{k} - 6(200)\mathbf{j} - 9(200)\mathbf{i} = 0 \end{aligned} \tag{a}$$

Collecting terms in these relations leads to

$$\begin{aligned} \Sigma \mathbf{M}_A &= \mathbf{i}[6P - 9(200)] + \mathbf{j}[14(200) - 20D_z] + \mathbf{k}(20D_y - 6P) = 0 \\ \Sigma \mathbf{M}_D &= \mathbf{i}[6P - 9(200)] + \mathbf{j}[20A_z - 6(200)] + \mathbf{k}(14P - 20A_y) = 0 \end{aligned} \tag{b}$$

and since all three components of a vector must vanish if the vector is to vanish, we find

$$\begin{aligned} P &= 300 \text{ lb} \\ A_y &= 210 \text{ lb} \\ A_z &= 60 \text{ lb} \\ D_y &= 90 \text{ lb} \\ D_z &= 140 \text{ lb} \end{aligned} \tag{c}$$

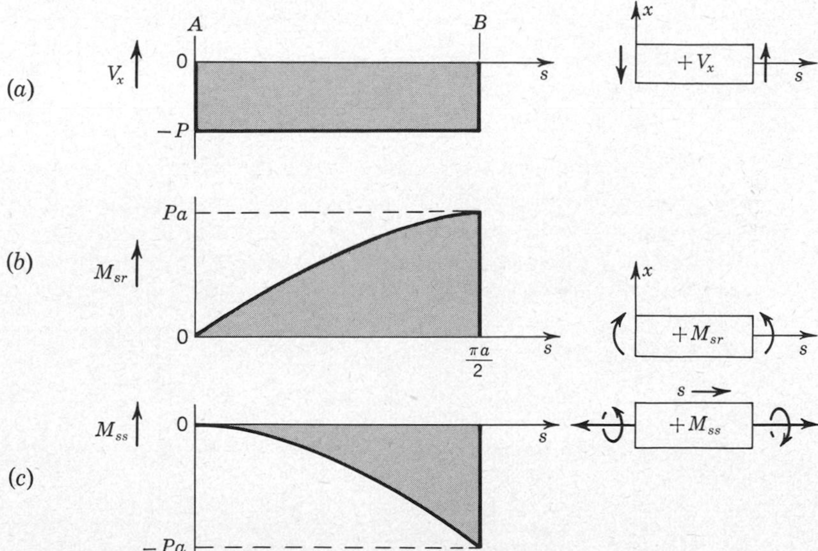

Fig. 3.25 Shear-force, bending-moment, and twisting-moment diagrams for Example 3.11.

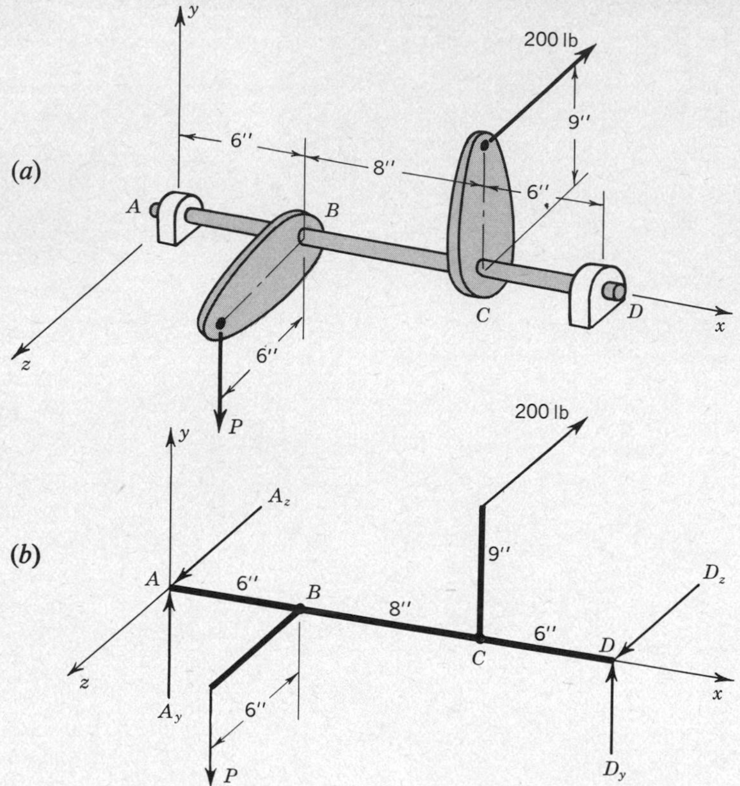

Fig. 3.26 Example 3.12. (*a*) A three-dimensional crank mechanism; (*b*) an idealized model of the mechanism.

Having obtained the reactions, we can now determine the forces and moments at any section of the shaft. For example, in Fig. 3.27 we have sketched a free body of one segment of the shaft after a cut has been made at some point O which lies between B and C. At the cut are shown three unknown force components F_x, V_y, and V_z and three unknown moment components M_t, M_{by}, and M_{bz}. These can be evaluated by requiring that the system of Fig. 3.27 be in equilibrium.

$$\Sigma \mathbf{F} = \mathbf{i}F_x + \mathbf{j}(210 - 300 + V_y) + \mathbf{k}(60 + V_z) = 0$$

$$\begin{aligned}\Sigma \mathbf{M}_o &= \mathbf{i}M_t + \mathbf{j}M_{by} + \mathbf{k}M_{bz} + 60x\mathbf{j} - 210x\mathbf{k} + 6(300)\mathbf{i} + (x - 6)(300)\mathbf{k} = 0 \\ &= \mathbf{i}[M_t + 6(300)] + \mathbf{j}(M_{by} + 60x) + \mathbf{k}[M_{bz} - 210x + (x - 6)(300)] = 0\end{aligned} \qquad (d)$$

Setting the coefficients of **i**, **j**, and **k** separately equal to zero provides us with the following forces and moments at any section for which 6 in. $< x <$ 14 in.

$$F_x = 0 \qquad M_t = -1{,}800 \text{ in.-lb}$$
$$V_y = 90 \text{ lb} \qquad M_{by} = -60x \text{ in.-lb} \qquad (e)$$
$$V_z = -60 \text{ lb} \qquad M_{bz} = 1800 - 90x \text{ in.-lb}$$

If we repeat this analysis for the ranges $0 < x < 6$ in. and 14 in. $< x <$ 20 in., we then have all the information required to construct the force and moment diagrams illustrated in Fig. 3.28.

An alternate procedure for arriving at these same diagrams is to return to the idealized three-dimensional system of Fig. 3.26 and consider the three two-dimensional projections shown in Fig. 3.29. The forces shown in each projection must constitute an equilibrium set. Furthermore, by using Fig. 3.29*a* alone, it is possible to determine P and the value of the twisting moment between points B and C; by using Fig. 3.29*b* alone, it is possible to determine the reactions A_z and D_z and the values of V_z and M_{by} at any section; and by using Fig. 3.29*c* along with the value of P previously obtained, it is possible to determine the reactions A_y and D_y and the values of V_y and M_{bz} at any section. The reader should pause long enough here to verify that this process is actually equivalent to that previously outlined. The stage at which a three-dimensional vector system is resolved into components is a matter of personal preference.

The shear forces and bending moments which are given in terms of y and z components in Fig. 3.28 can be vectorially combined and represented in a three-dimensional manner, as sketched in Fig. 3.30*a*. This information may also be presented as in Fig. 3.30*b* where we plot the magnitude of the resultant shear force and bending moment together with the angles which define the planes in which these quantities act. It is of interest to note that in the central portion BC the plane of the resultant shear force is not perpendicular to the resultant bending-moment vector.

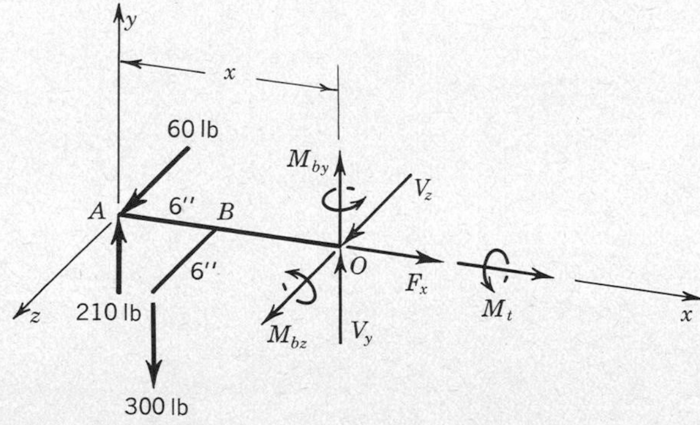

Fig. 3.27 Example 3.12. Forces and moments at a general point O of the idealized model of Fig. 3.26*b*.

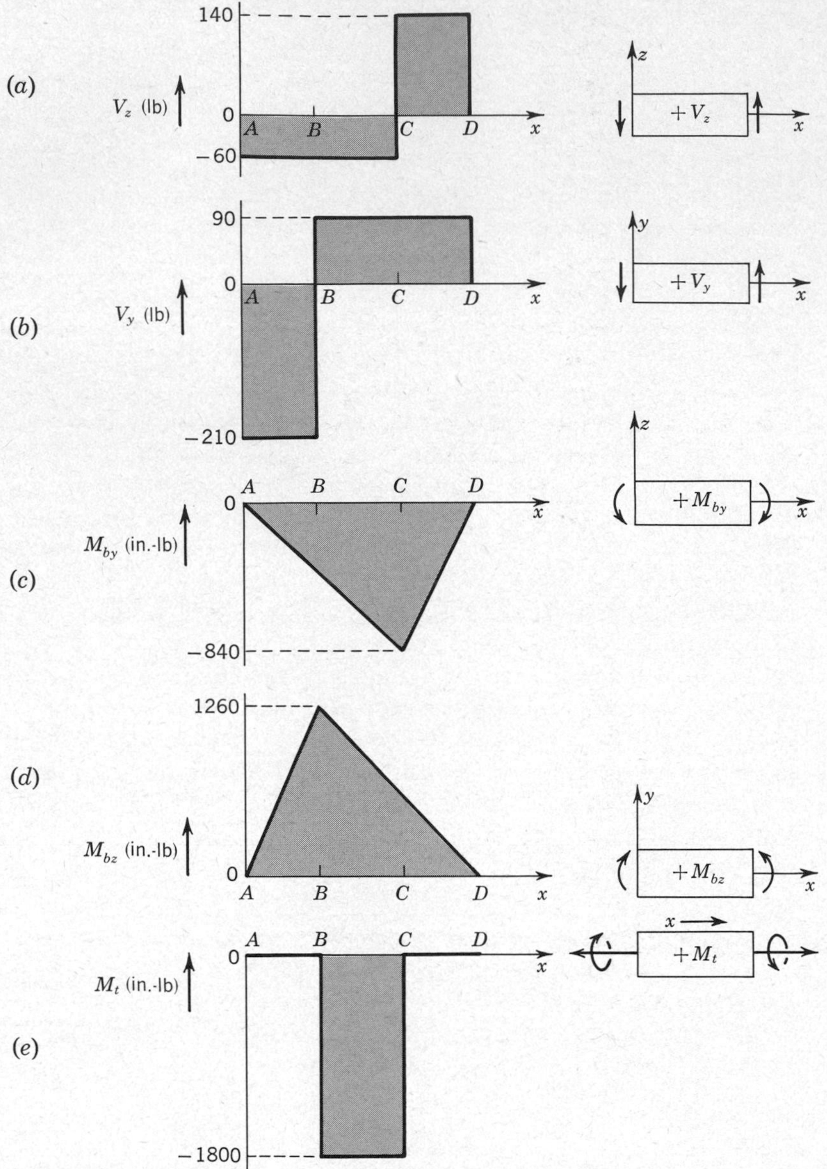

Fig. 3.28 Example 3.12. Force and moment diagrams for the shaft of Fig. 3.26.

The results of this example suggest that *the maximum bending moment along a straight member subjected to point loads* (no distributed loads) *must always occur at a loading point* rather than at a point between loads. This is easy to see when all the forces lie in one plane, for then the moment diagram consists entirely of straight-line segments. It is not immediately obvious that this should still be true in the general three-dimensional case where the resultant moment diagram may have curved segments, as in Fig. 3.30*b*. It can be shown,[1] however, that where the resultant moment diagram is not made up of straight lines it is made up of curves which are *concave* outward, and hence that a maximum never occurs between loads. It is possible for a minimum to occur between loading points.

The examples in this chapter have illustrated two general approaches to obtaining internal forces and moments in slender members. In all cases we have capitalized on the fact that when a system is in equilibrium then any isolated subsystem of the original system is also in equilibrium. In one procedure the subsystem was one of the two segments resulting from an imaginary cut or section of a slender member. In the other procedure the subsystem was an infinitesimal element. This latter procedure required integration in order to obtain the forces and moments.

In all the examples treated in this chapter it was possible to obtain complete solutions using only the conditions of equilibrium; i.e., these situations were all *statically determinate*. In later chapters we shall meet *statically indeterminate*

[1] See Prob. 3.43.

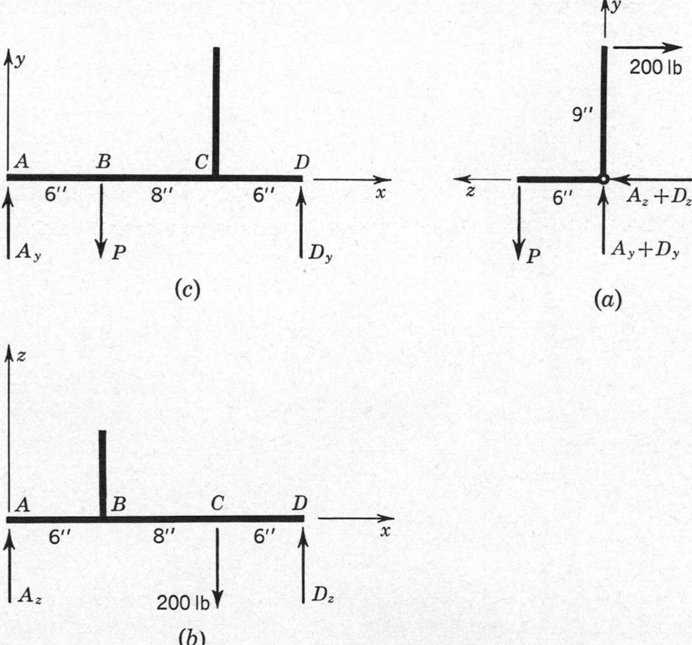

Fig. 3.29 Example 3.12. Projections of the forces in Fig. 3.26*b* on the three coordinate planes.

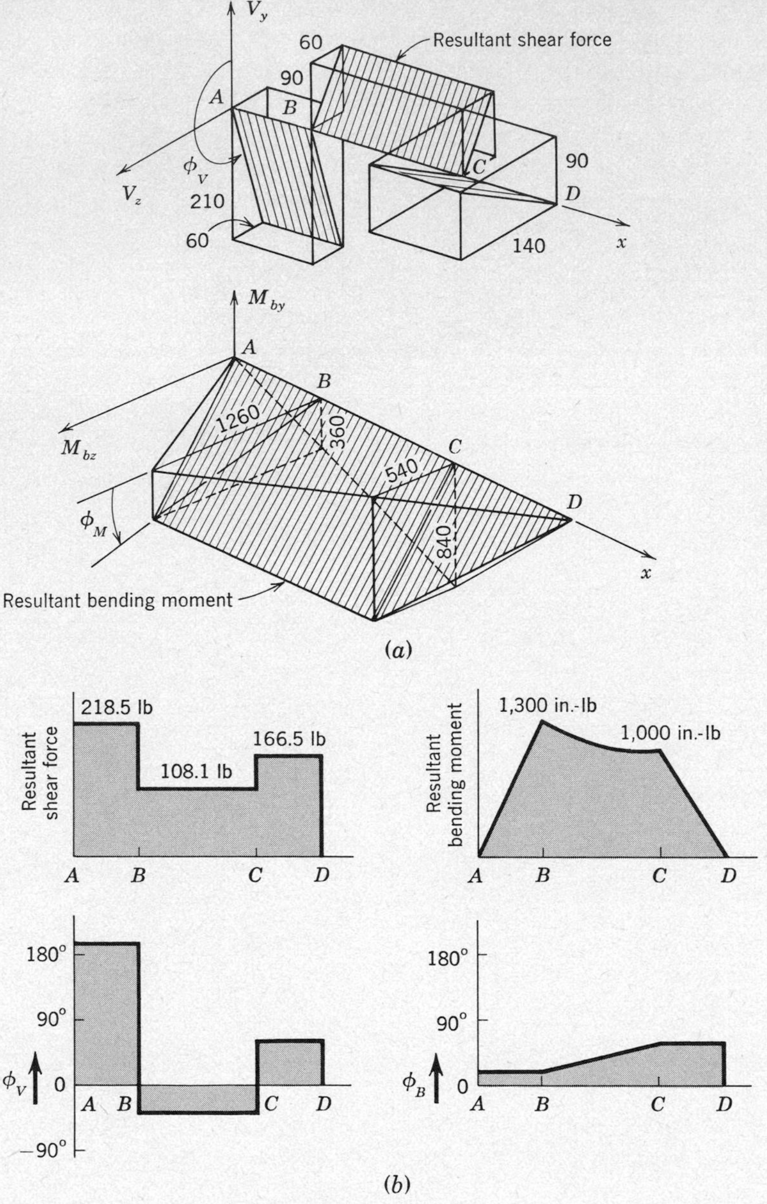

Fig. 3.30 Example 3.12. Resultant shear-force and bending-moment diagrams for the crank mechanism of Fig. 3.26.

slender members. In these cases we must use the equilibrium conditions, just as we have in this chapter, but we also must include steps 2 and 3 of (2.1). Solutions will be obtained by simultaneously satisfying the requirements of *all three* steps.

PROBLEMS

3.1–3.8. In each case, sketch shear-force and bending-moment diagrams. Indicate sign convention employed and label important values.

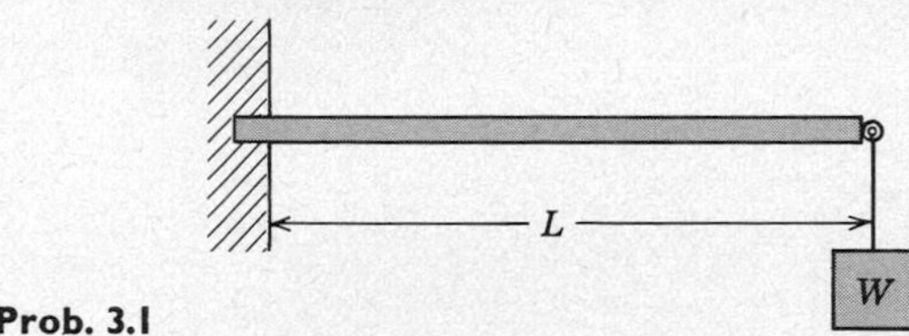

Prob. 3.1

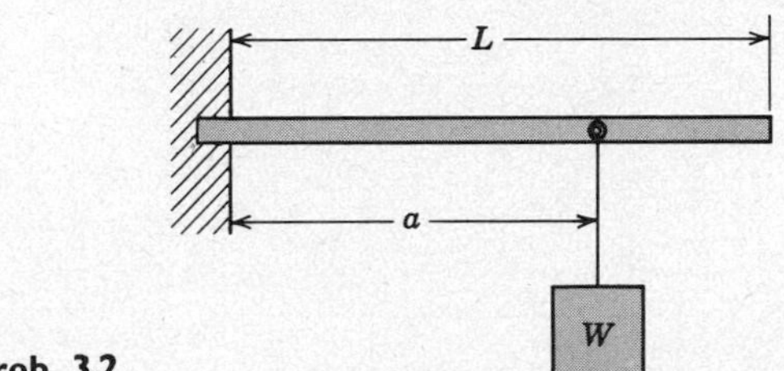

Prob. 3.2

Prob. 3.3

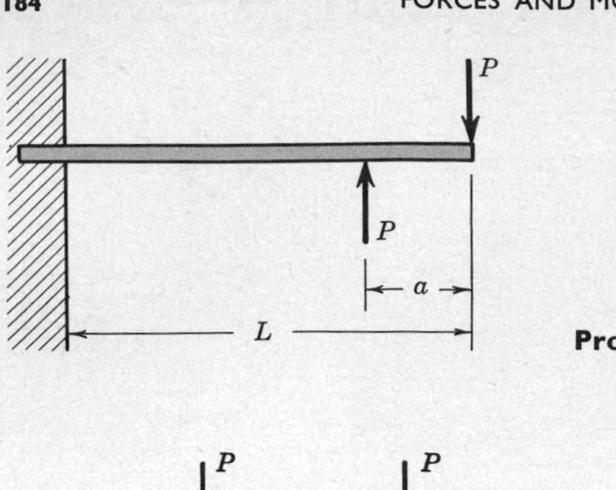

Prob. 3.4

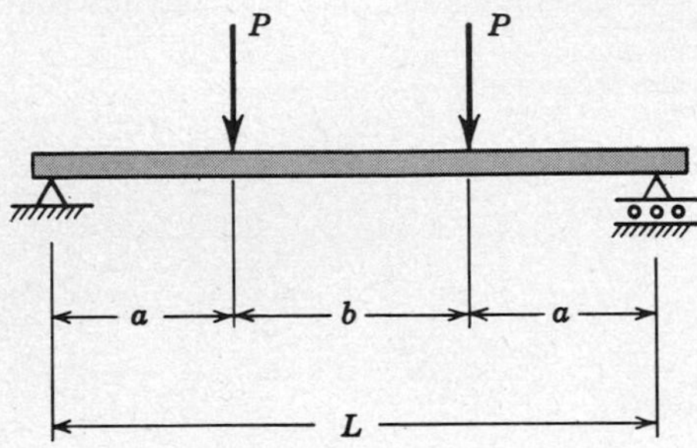

Prob. 3.5

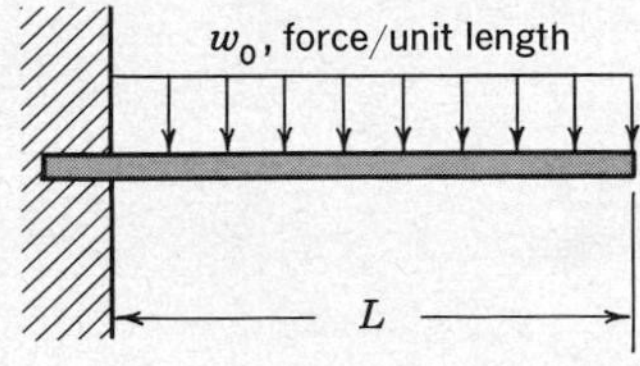

Prob. 3.6

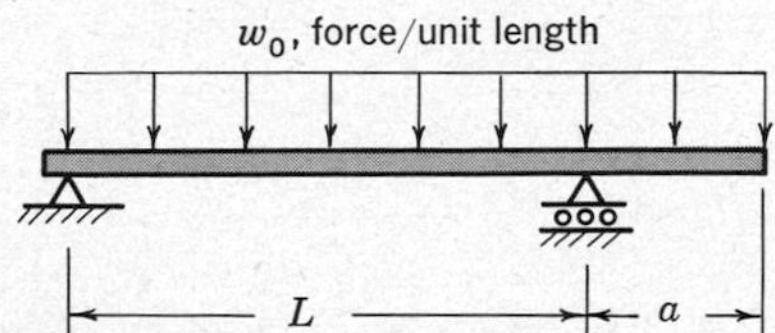

Prob. 3.7

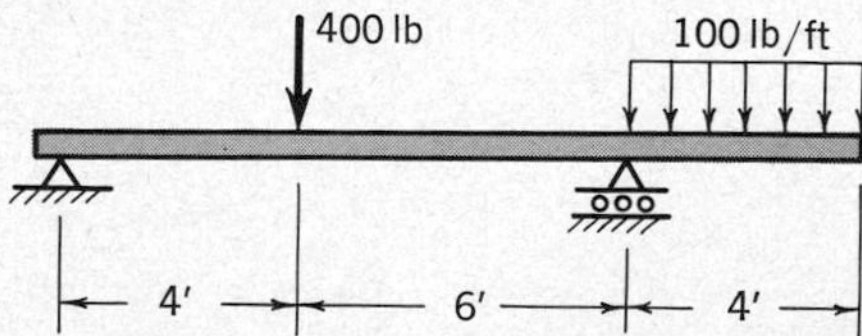

Prob. 3.8

3.9. Determine the axial force, the shear force, and the bending moment acting at any section θ in the circular arc AB.

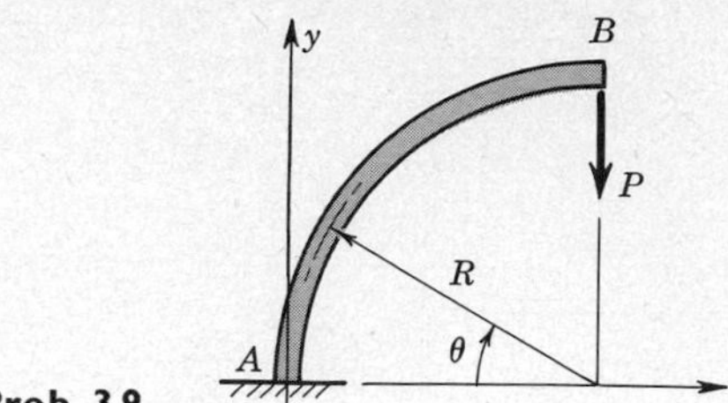

Prob. 3.9

3.10. Find the reactions and expressions for the shear force and bending moment as functions of distance along the beam.

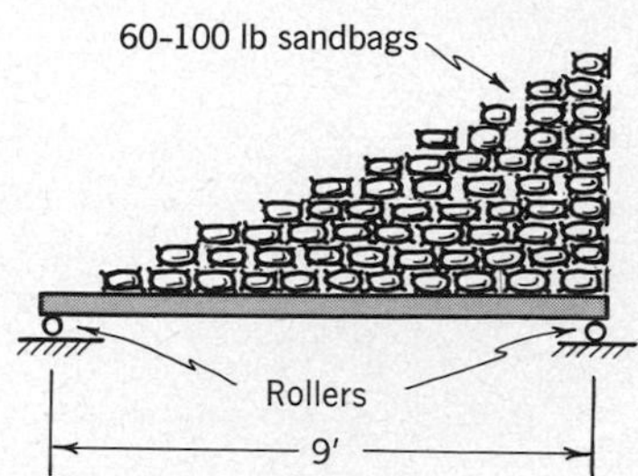

Prob. 3.10

3.11. Sketch the diagram of bending moment as a function of θ for the semicircular member.

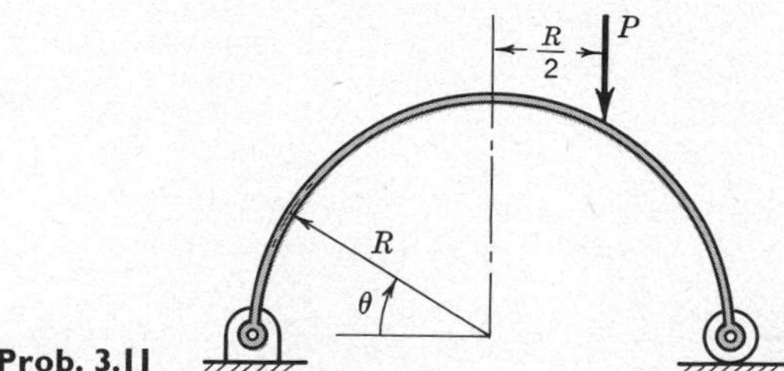

Prob. 3.11

3.12. Sketch shear-force and bending-moment diagrams for the cantilever beam which carries a concentrated force P and a distributed load of intensity w_o force per unit length.

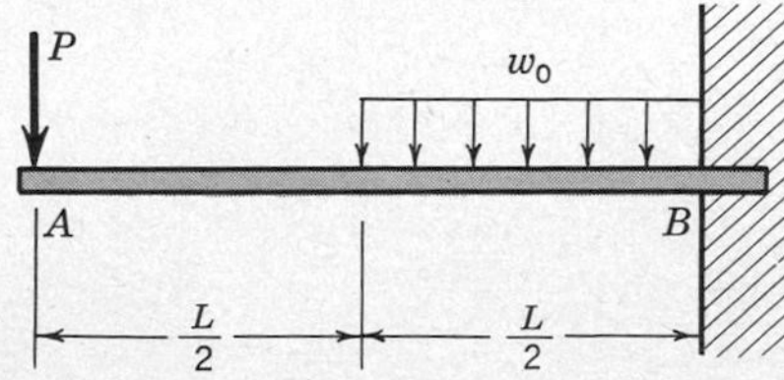

Prob. 3.12

3.13. Draw sketches showing the internal forces and moments acting at sections 1, 2, and 3 in the member shown.

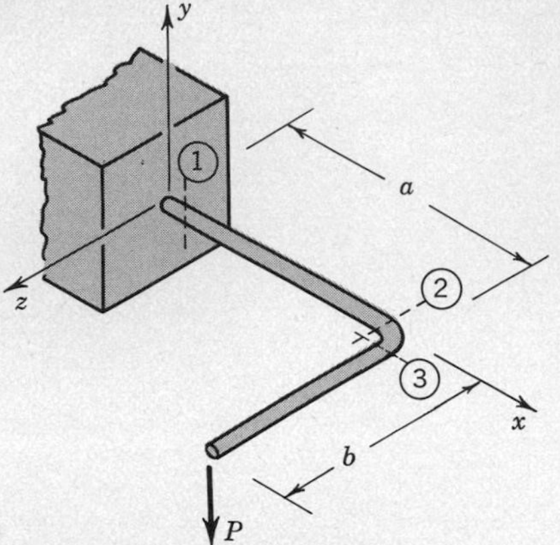

Prob. 3.13

3.14. Calculate the internal forces and moments acting at sections 1 and 2 in the structure shown.

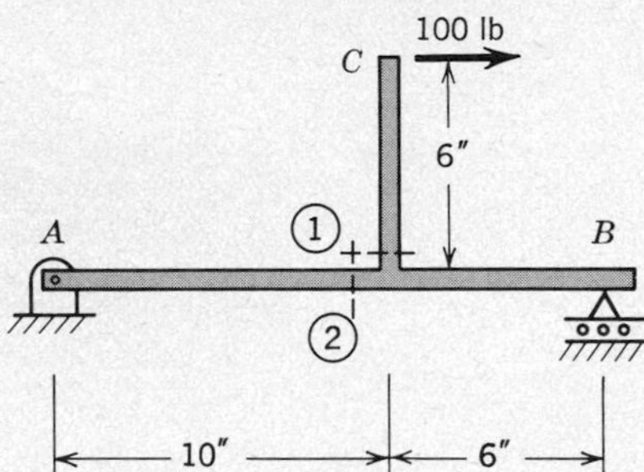

Prob. 3.14

3.15. Calculate the internal forces and moments acting at sections 1 and 2 in the pinned framework shown.

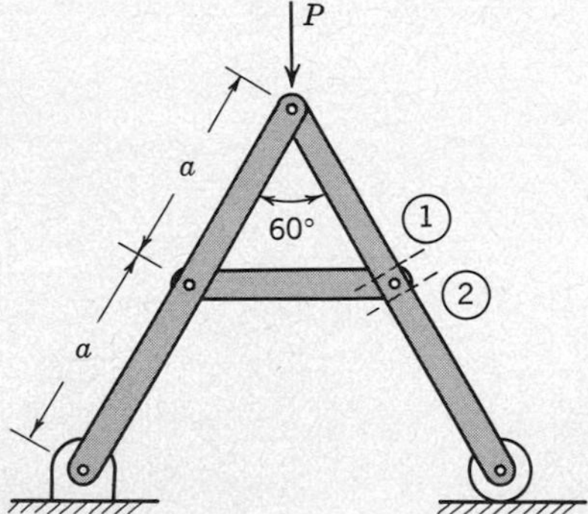

Prob. 3.15

3.16. The sketch shows a possible set of muscle forces acting on the femur of a man who is running upstairs. Find the unknown reactions R_A and R_D in terms of P and show how the transverse force varies along the femoral shaft. Show how the bending moment varies along the shaft, and comment on the compensating effect of the muscles attached at B and C in terms of reducing the bending moments in the shaft.

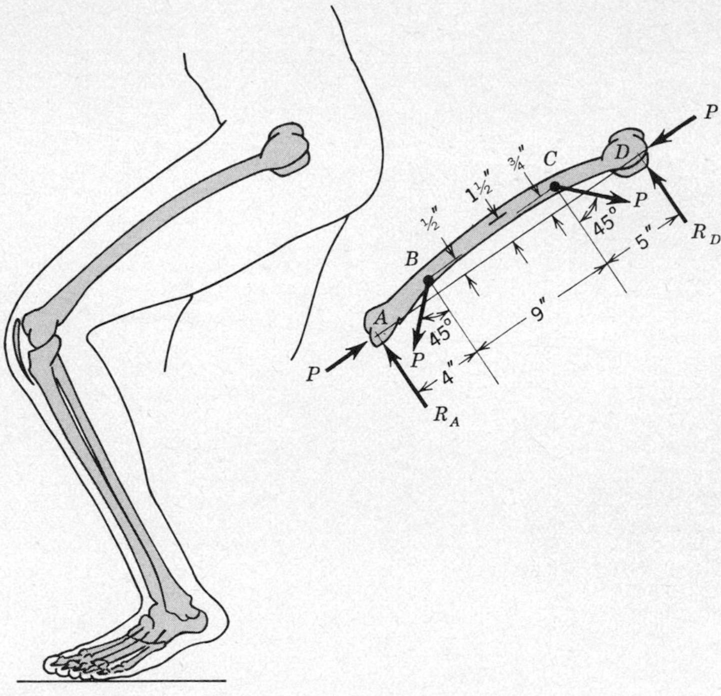

Prob. 3.16

3.17. A carpenter with a power saw has a 20-ft plank of uniform weight per unit length w_o and two sawhorses. He wishes to cut a 6-ft length from the plank, but in order to minimize splitting of the ends he wants to cut it at a point where the bending moment in the plank is zero. If he places one sawhorse at one end of the plank, where should he put the other so that the bending moment will be zero 6 ft from the other end of the plank?

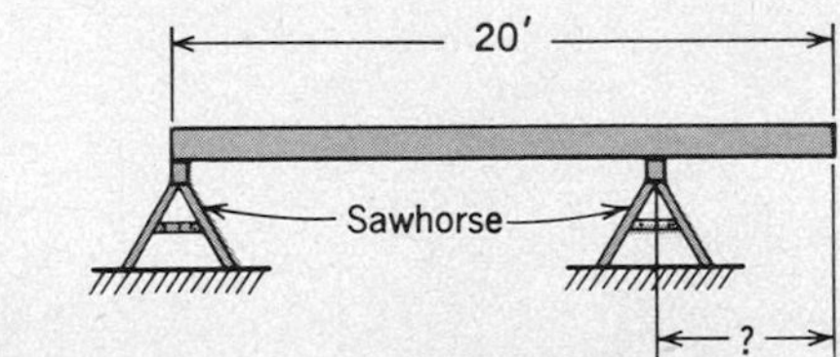

Prob. 3.17

3.18. A section of a scaffold consists of a plank laid across two supports and extending a distance a on either side of the supports. A mason working at the center of the plank thinks that he should stack his supply of bricks on the ends of the plank in order to minimize the bending moment in the plank. Is he correct? If equal numbers of bricks are stacked at each end of the plank, for what weight of bricks, W_B, is the maximum bending moment in the plank a minimum? The man weighs W_M.

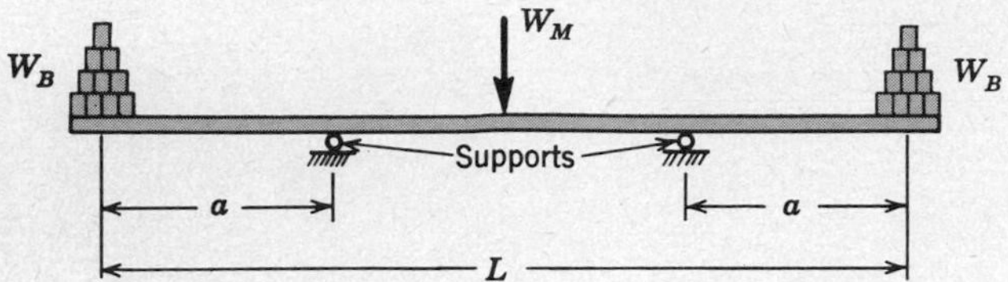

Prob. 3.18

3.19. The rocket shown experiences a wind gust during its vertical ascent which results in the loading shown. Rotation of the system may be prevented if the resultant moment about the center of mass of the system vanishes. This is to be achieved by varying the orientation of the thrust vector T with respect to the vertical axis.

(*a*) What relationship must exist among T, α, p_o, and L in order that this requirement be satisfied?

(*b*) Determine the internal shear force and bending moment at $L/4$ and $3L/4$ in terms of p_o.

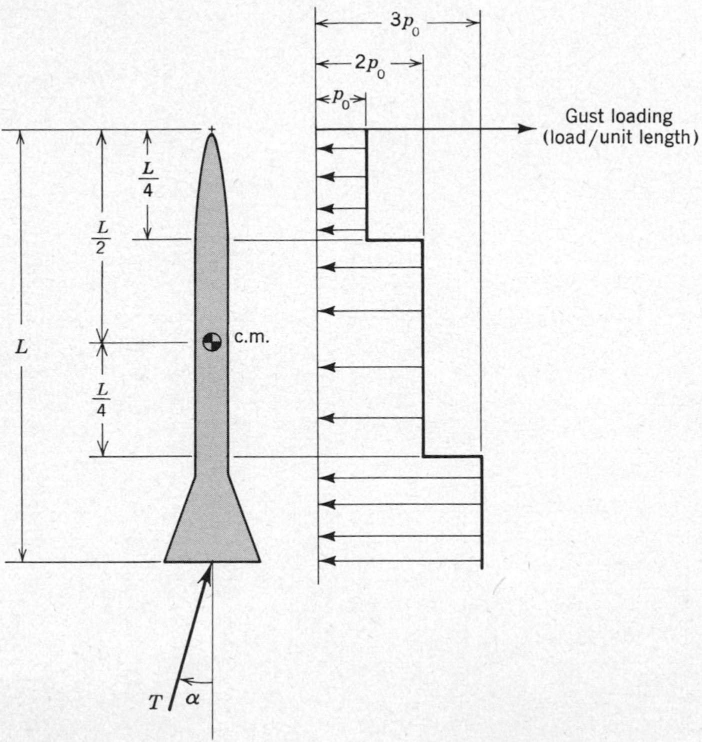

Prob. 3.19

3.20. A portion of an airplane landing gear is shown, with a coplanar force system present. Frictionless pin joints with their axes perpendicular to the plane of the paper are located at A, B, and C.

(*a*) Find reaction forces at A, B, and C.

(*b*) Find all forces (i.e., all axial loads, shears, and moments) along each member AD, DE, EG, DC, and CB; show in the sketch for each member the positive sense of each force and moment.

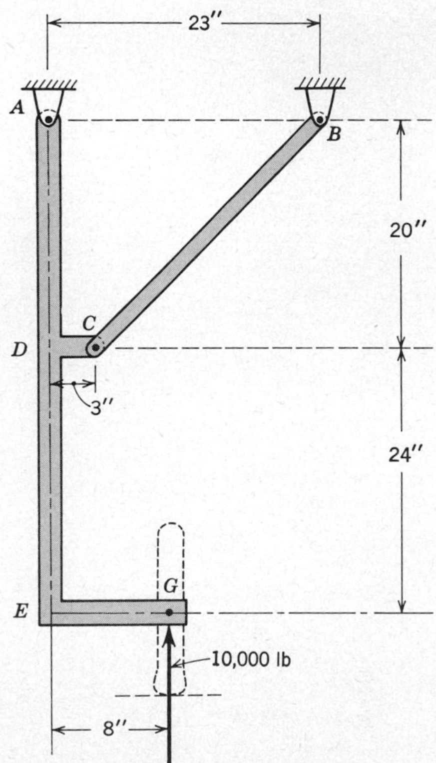

Prob. 3.20

3.21. (*a*) Cilia are motile hairlike appendages on the free surfaces of certain cells. They are present in the trachea and in the reproductive tracts of humans as well as in lower animals. Their motion can be considered as made up of an effective stroke which can be thought of as a pendular motion with constant angular velocity through an angle of approximately 140° and a return or recovery stroke as shown in the figure. For the configuration shown in the figure in which the cilium is *arrested* by a force $F = 2.2 \times 10^{-4}$ dyne, calculate the moment at the cell boundary.

(*b*) If a cilium moving in a viscous fluid rotates through its effective stroke, estimate the driving moment at the cell boundary. The viscous force on an *element of length* of the cilium may be taken to be proportional to the length of the element, the angular velocity, the viscosity, and to a function which depends upon the position along the cilium.

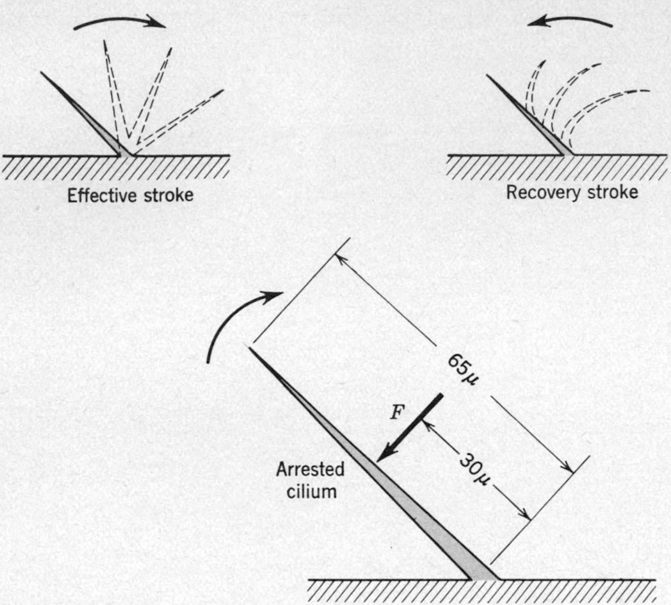

Prob. 3.21

3.22. Construct shear-force and bending-moment diagrams for the case where the concentrated moment is $M_o = PL/4$.

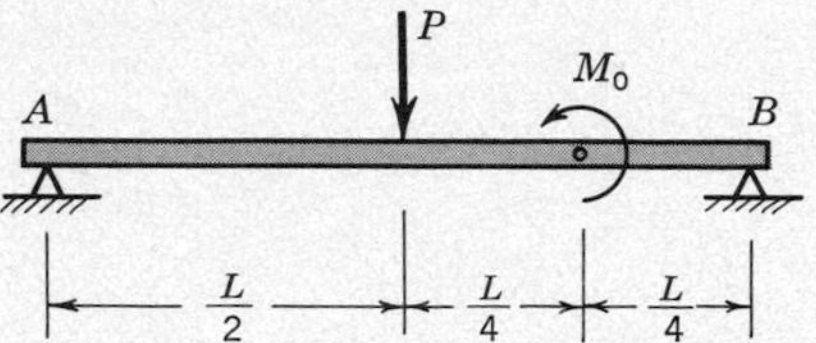

Prob. 3.22

3.23. (*a*)–(*h*). Solve Probs. 3.1–3.8 again, but this time obtain the shear-force and bending-moment diagrams by using the singularity-function notation.

3.24. A diving board is supported by a hinged joint at the left end and a simple support near the center. How should the distance a be varied in order that the maximum bending moment should be the same for divers of all weights?

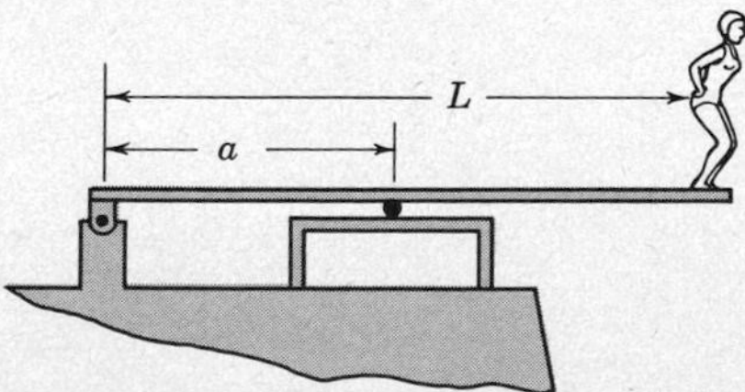

Prob. 3.24

3.25. A bookshelf is made by placing a wooden plank on two brick supports. Where should the bricks be placed so as to make the maximum bending moment as small as possible?

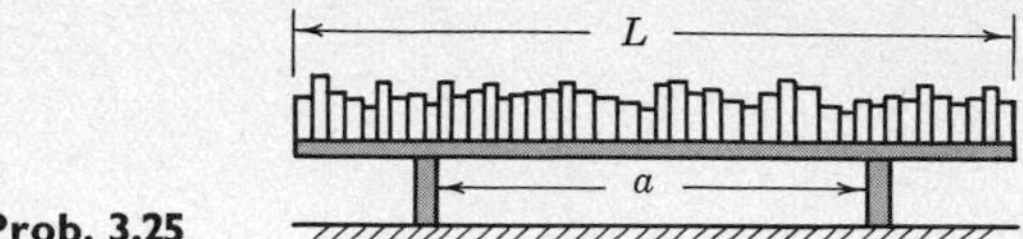

Prob. 3.25

3.26. A pivoted flagpole is to be erected by using a gin pole and winch as shown. Where should the rope be attached to the flagpole so that during erection the maximum bending moment is as small as possible?

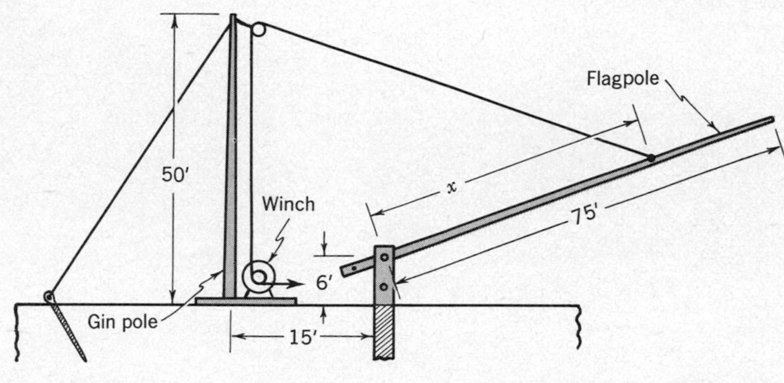

Prob. 3.26

3.27. A metal rod is bent into a circular arc of radius R with an arc of 270°. One end of the rod is fixed so that the arc lies in a horizontal plane. Give the magnitude and location of the maximum bending and twisting moments when a weight W is hung first at A, then B, then C.

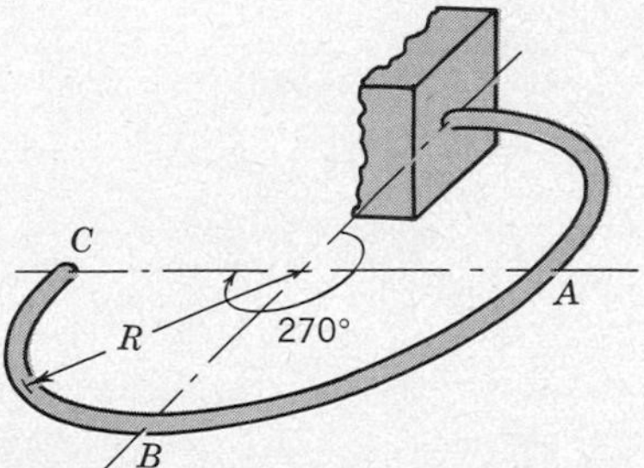

Prob. 3.27

3.28. A radio antenna protrudes 18 in. above the fuselage of an airplane. A guy wire is attached to the end of the antenna to strengthen it against drag forces. Assuming that the drag force is uniformly distributed and has a total resultant D, find the force the guy wire should exert in order to minimize the bending moment in the antenna.

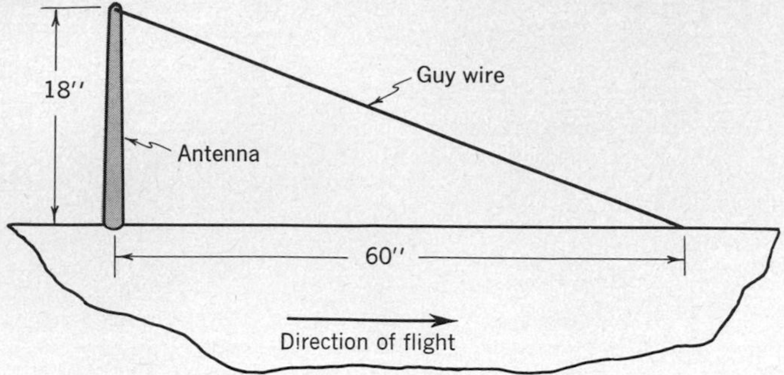

Prob. 3.28

3.29. When the load P is applied, the coil spring shown has a pitch equal to the coil radius. Find the maximum bending and twisting moments in the coil.

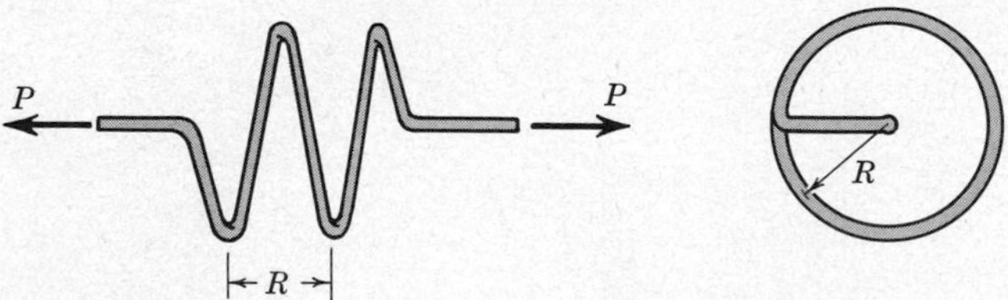

Prob. 3.29

3.30. A number of small, two-wheeled boat trailers have a suspension system similar to that shown. Estimate the maximum twisting and bending moments in the 18-in. bar and the 24-in. bar when the wheels are carrying loads of 500 lb each. A wheel with a 5.00–8 tire has an outside diameter of 5 in. + 8 in. + 5 in. = 18 in.

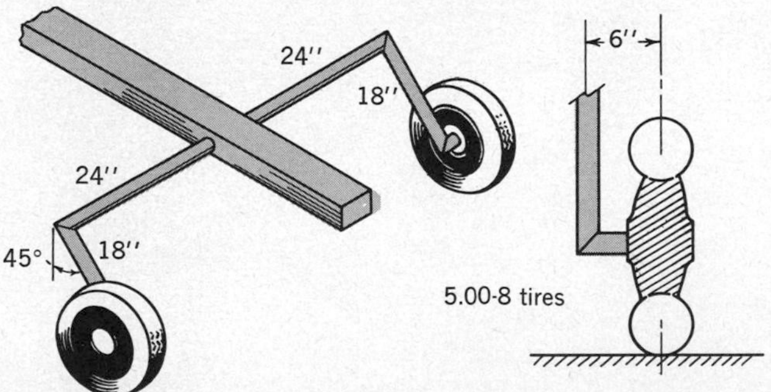

Prob. 3.30

3.31. A *flexible* cable is supported at two points A and B a distance $2L$ apart. The cable is loaded with weights such that in each horizontal unit distance there is a total weight of w_o. The cable is observed to sag a distance c below the supports. Consider a free-body sketch of a segment of cable from $x = 0$ to $x = x$. Show that the horizontal component H of the cable tension $T(x)$ is independent of x. Show that the slope of the cable at x is

$$\frac{dy}{dx} = \frac{w_o x}{H}$$

Integrate this equation to obtain the equation of the cable. Evaluate H in terms of w_o and the dimensions of the cable.

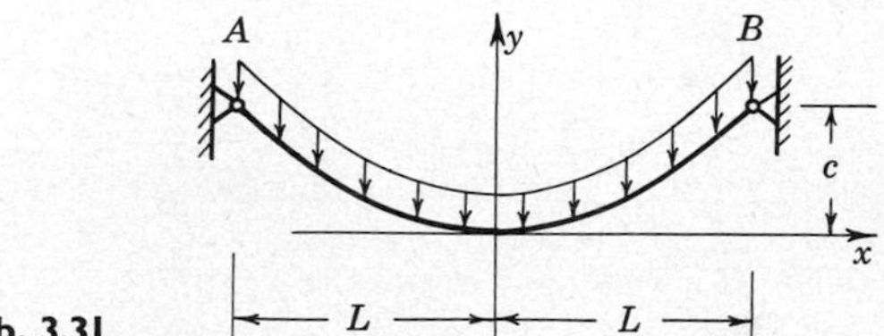

Prob. 3.31

3.32. A curved shaft capable of transmitting bending and twisting moments is supported in bearings at A and B which are only capable of exerting single force reactions perpendicular to the shaft. Twisting moments M_A and M_B are applied to the ends of the shaft. Find the force reactions at A and B and the magnitude of M_B required for equilibrium, all in terms of the magnitude of M_A. Sketch twisting- and bending-moment diagrams.

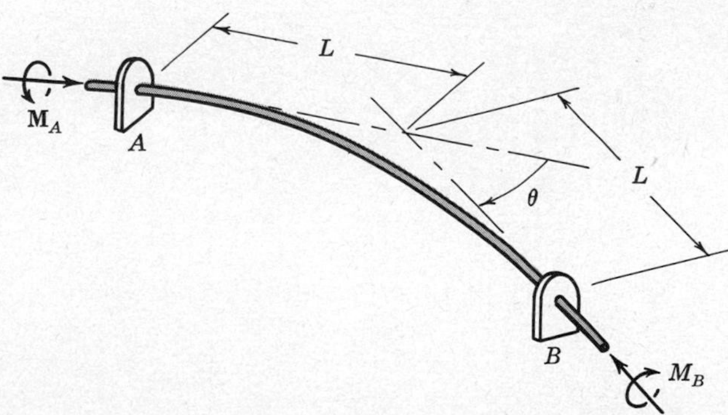

Prob. 3.32

3.33. The differential equations (3.11) and (3.12) were derived for a slender member extending in the x direction with loading in the y direction. To emphasize this, we can write (3.11) and (3.12) as follows:

$$\text{(a)} \qquad \begin{aligned} \frac{dV_y}{dx} + q_y &= 0 \\ \frac{dM_{bz}}{dx} + V_y &= 0 \end{aligned}$$

Show that for a slender member extending in the y direction with loading in the x direction the corresponding equations are

$$\text{(b)} \qquad \begin{aligned} \frac{dV_x}{dy} + q_x &= 0 \\ \frac{dM_{bz}}{dy} - V_x &= 0 \end{aligned}$$

There are six different combinations of slender members extending in one coordinate direction with transverse loading in another coordinate direction. Verify that for three of these the differential equations corresponding to (3.11) and (3.12) have the sign pattern of (*a*) and that for the other three the sign pattern is that of (*b*).

3.34. A wooden dam is made of planks fastened to uprights which are driven into the river bed. The uprights are a distance 8 ft apart, and the water is 6 ft deep. Draw shear-force and bending-moment diagrams for the uprights.

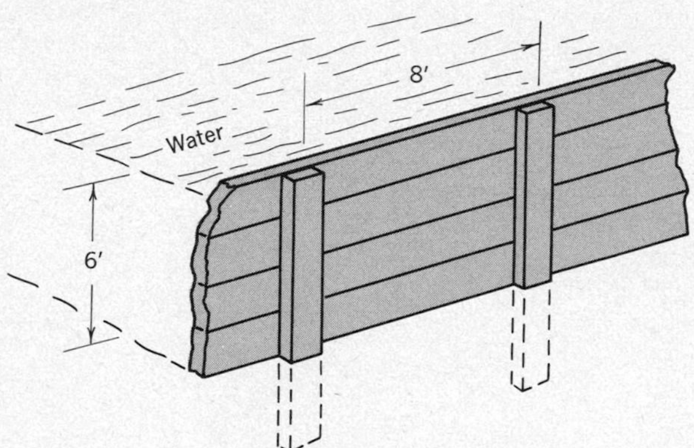

Prob. 3.34

3.35. A hollow metal sphere with a diameter of 20 in. and a weight of 160 lb is used as a valve to close a 12-in.-diameter hole in the freshwater tank shown. What force F is required to just open the valve?

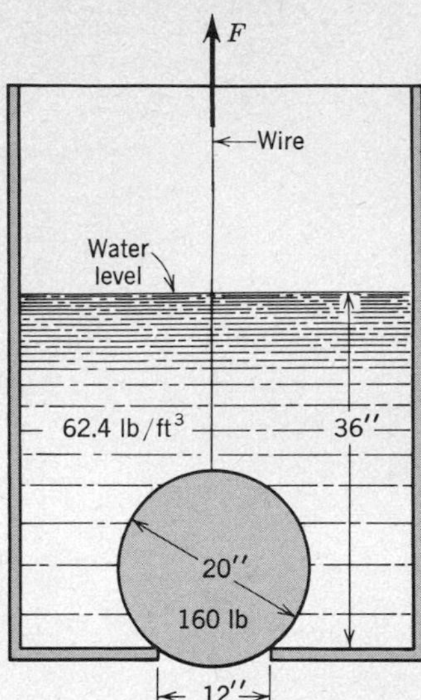

Prob. 3.35

3.36. The gate AB blocks an 8-ft-wide opening in a freshwater reservoir. Find the reaction force F on the support at A.

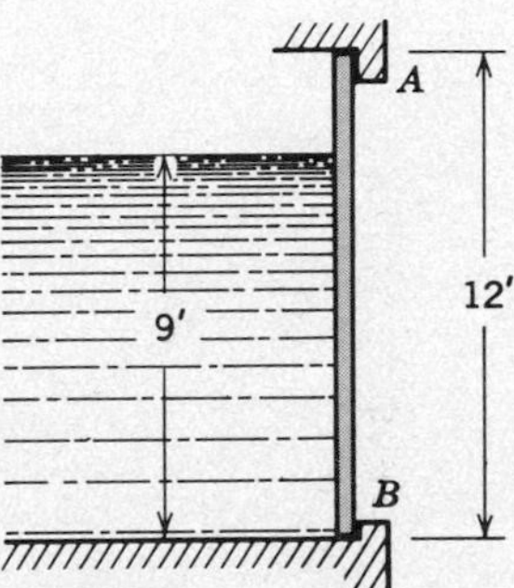

Prob. 3.36

3.37. The shaft AD is supported in bearings at A and D and has pulleys attached at B and C. The pulley at B is 8-in. diameter while that at C is 12-in. diameter. The shaft transmits a maximum of 25 hp at 1,750 rpm. The belt tensions are adjusted so that

$$\frac{T_1}{T_2} = \frac{T_3}{T_4} = 3$$

Sketch the shear-force, bending-moment, and twisting-moment diagrams for AD, labeling important values. (*Note:* A horsepower is 33,000 ft-lb/min. Rotational power is the product of torque times angular velocity in radians per unit time.)

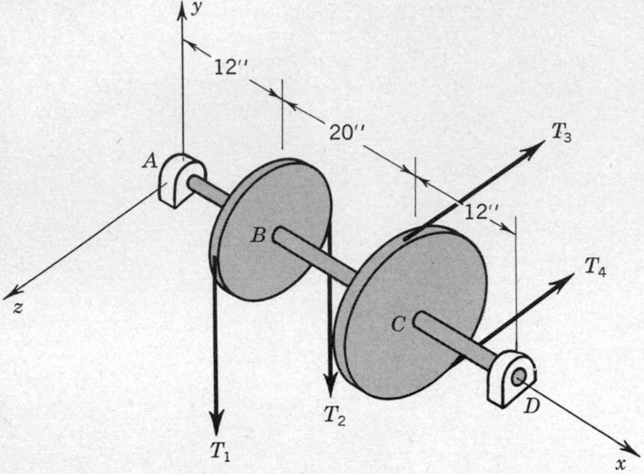

Prob. 3.37

3.38. A shaft is supported in bearings A and B, has a crank attached to one end, and transmits a moment M_o. Sketch the shear-force, bending-moment, and twisting-moment diagrams for the shaft.

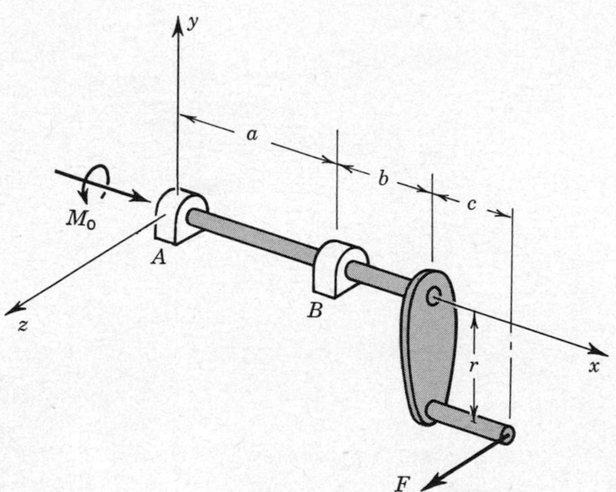

Prob. 3.38

3.39. A 5-in.-diameter 20° spur gear is attached to the end of a 5-in.-long cantilevered shaft. A smaller gear (ratio 3:1) transmits a 100 ft-lb torque. Sketch the shear-force, bending-moment, and twisting-moment diagrams for the cantilevered shaft, labeling important values.

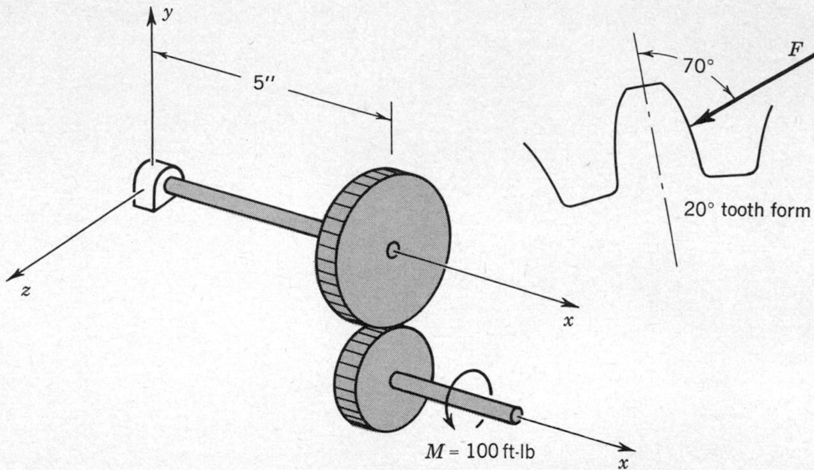

Prob. 3.39

3.40. A crankshaft for a single-cylinder engine is shown mounted in bearings at each end. It is in equilibrium under the action of the connecting-rod force and the shaft torque M_o. The engine has:

Bore 2½ in.
Stroke 3 in.
Connecting-rod length 5 in.

Show diagrams for shear, bending moment, and twisting moment for the two end sections of the crankshaft.

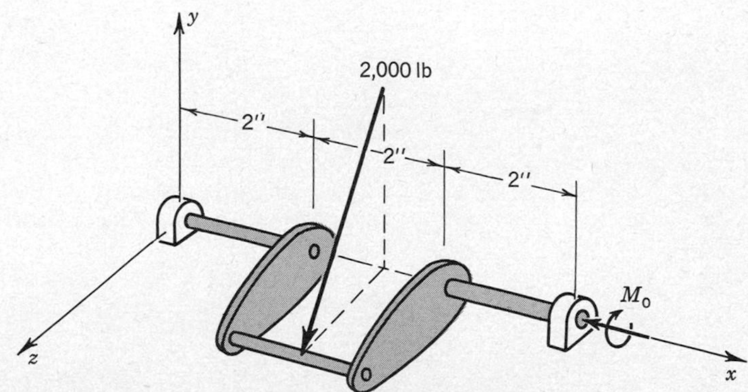

Prob. 3.40

3.41. A closely wound coil spring of radius R can have the "ends" in either of the positions shown. Calculate the shear forces, bending moments, and twisting moments for the two cases.

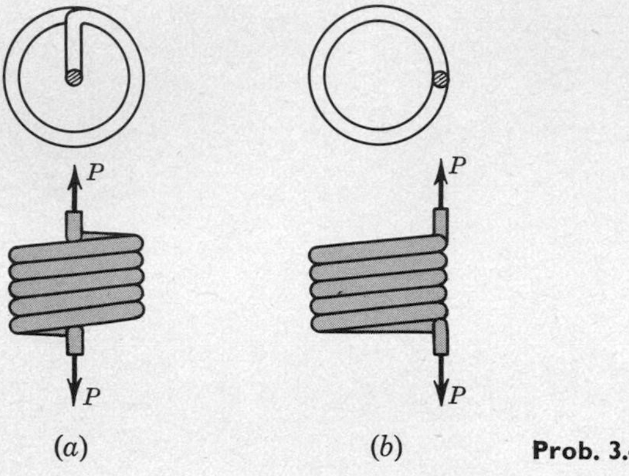

Prob. 3.41

3.42. Compression springs may buckle under an axial load. Part of the study of this phenomenon requires the effect of the bending moment M on the moments and forces at the points A and B. Determine the local shear force, bending moment, and twisting moment at a section through the wire at point A. Do the same for B.

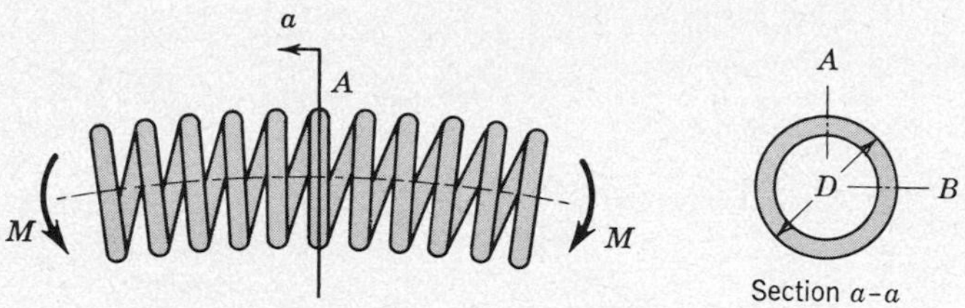

Prob. 3.42

3.43. A straight slender member along the x axis is subjected to only point loads along its length which are all parallel to the yz plane. Show that the diagrams for M_{by} and M_{bz} must be made up of a number of straight-line segments connected at the points of loading, and the maximum of either component must always occur at a loading point. Now consider the resultant bending moment

$$M_b = \sqrt{M_{by}^2 + M_{bz}^2}$$

Show that if

$$M_{by} = Ax + B$$
$$M_{bz} = Cx + D$$

where A, B, C, and D are arbitrary constants in a certain segment of the member, then

$$\frac{d^2M_b}{dx^2} \geqq 0$$

and therefore that the resultant moment curve is either straight or concave outward.

3.44. Consider the loading function $q_{-1}(x,u)$ defined by

$$q_{-1}(x,u) = \begin{cases} 0 & \text{for } -\infty < x < 0 \\ \dfrac{x}{u^2} & \text{for } 0 < x < u \\ \dfrac{2}{u} - \dfrac{x}{u^2} & \text{for } u < x < 2u \\ 0 & \text{for } 2u < x < \infty \end{cases}$$

Show that in the limit as $u \to 0$, $q_{-1}(x,u)$ approaches a unit "concentrated force" $\langle x \rangle_{-1}$ located at the point $x = 0$. The limit of the function $q_{-1}(x - a, u)$ is denoted by $\langle x - a \rangle_{-1}$. Consider a cantilever beam, free at $x = 0$ and built-in at $x = L$, subjected to the loading $q_{-1}(x,u)$. Sketch the shear-force and bending-moment diagrams and discuss their limiting forms when $u \to 0$.

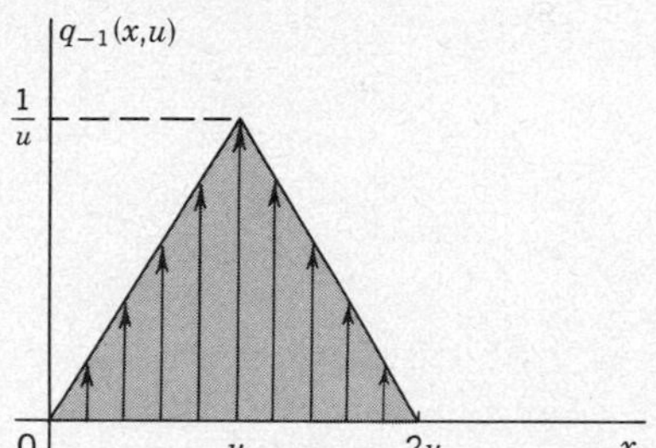

Prob. 3.44

3.45. Consider the loading function $q_{-2}(x,u)$ defined by

$$q_{-2}(x,u) = \begin{cases} 0 & \text{for } -\infty < x < 0 \\ \dfrac{1}{u^2} & \text{for } 0 < x < u \\ -\dfrac{1}{u^2} & \text{for } u < x < 2u \\ 0 & \text{for } 2u < x < \infty \end{cases}$$

Show that in the limit as $u \to 0$, $q_{-2}(x,u)$ approaches a "concentrated couple" $\langle x \rangle_{-2}$ located at $x = 0$ and with value unity. The limit of the function $q_{-2}(x - a,u)$ when located at the point $x = a$ is denoted by $\langle x - a \rangle_{-2}$. Consider a cantilever beam, free at $x = 0$ and built-in at $x = L$, subjected to the loading $q_{-2}(x,u)$. Sketch the shear-force and bending-moment diagrams and discuss their limiting forms when $u \to 0$.

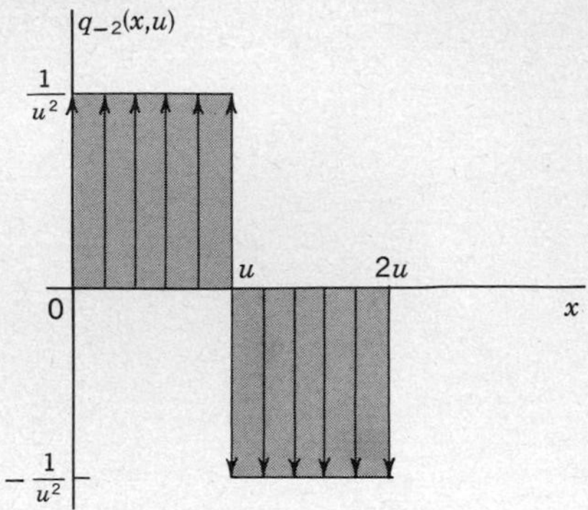

Prob. 3.45

3.46. Show that for any fixed u

$$\int_{-\infty}^{x} q_{-2}(x-a,\, u)\, dx = q_{-1}(x-a,\, u)$$

and, therefore, assuming that the limit and integration operations can be interchanged,

$$\int_{-\infty}^{x} \langle x-a \rangle_{-2}\, dx = \langle x-a \rangle_{-1}$$

4
Stress and Strain

4.1 INTRODUCTION

In Chap. 2 we considered deformable bodies in which the loadings and resulting deformations were assumed to be unidirectional. We found the relation between load and the resulting deformation in terms of the shape of the body and a property of the material (the modulus of elasticity). But we only did so under the assumption that conditions were uniform from one point to another. As we extend our study of deformable solids to more common cases where conditions are nonuniform, we shall find that it is necessary to study the behavior in differentially small elements within the body. At the same time, it is usually necessary to consider two- or three-dimensional aspects of the behavior of the material. For example, the transverse deflection of a beam in bending turns out to depend on the distribution, both along and across the beam, of the axial force per unit area. To derive the overall behavior of a body from the properties of differentially small elements within the body still requires the use of the three fundamental principles of equilibrium, geometric compatibility, and the relations between force and deformation.

In this and the following chapter we shall investigate the significance of our three basic principles when they are applied to the localized behavior of the material at a *point* within a deformable body. Equilibrium at a point and the geometry of deformation at a point are considered in the remaining sections of this chapter, and the relations between force and deformation at a point within a real material are discussed in the next chapter. Succeeding chapters will build upon this foundation in examining the action of structures of various shapes under a variety of loadings.

4.2 STRESS

Recall that in Sec. 2.2 we found that force and deformation in bars under tensile loading could be related to a material property by considering force per unit area and extension per unit length. We first consider how the concept of internal force per unit area can be extended to a more general shape and loading such as that shown in Fig. 4.1.

To examine these internal forces at some point O in the interior of the body, we pass a plane whose normal vector is $\mathbf{n}$ through the point O, as shown in Fig. 4.2. In order for the separate halves of the body to be in equilibrium, there must, in general, be internal forces transmitted across the cutting plane. If we divide the plane into a number of small areas, and we measure the forces acting on each of these, we will observe that these forces in general vary from one small area to the next, as shown in Fig. 4.2. On the small area ΔA whose normal vector is $\mathbf{n}$ centered on the point O, there will be acting a force of $\Delta\mathbf{F}$ which is inclined to the surface ΔA at some arbitrary angle (Fig. 4.3*a*). When we consider the problem of describing completely the action of the force vector $\Delta\mathbf{F}$, we have to specify the orientation and size of the face on which $\Delta\mathbf{F}$ acts, the magnitude of $\Delta\mathbf{F}$, and the orientation of $\Delta\mathbf{F}$ with respect to the face.

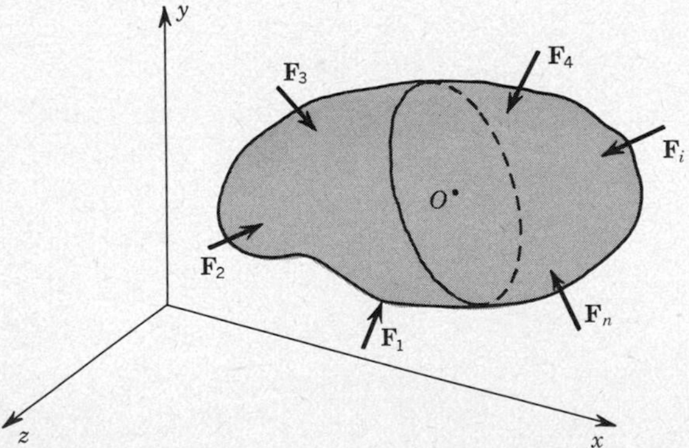

Fig. 4.1 Continuous body acted on by external forces.

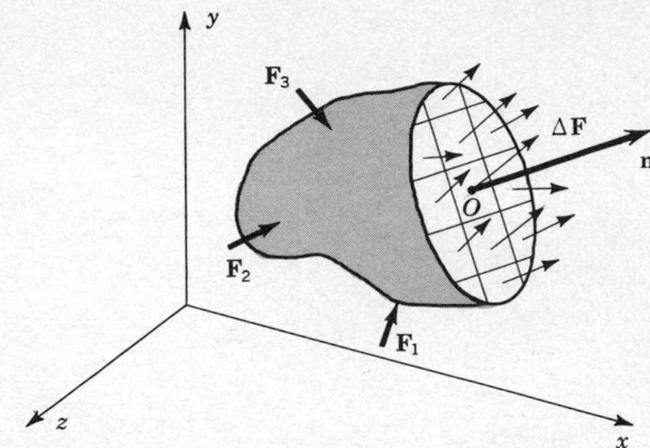

Fig. 4.2 Internal forces acting on a plane whose normal is **n**.

We now introduce the concept of the *stress vector* $\overset{(\mathbf{n})}{\mathbf{T}}$ acting at the point O on a plane whose normal is **n** passing through O. We define the stress vector as

$$\overset{(\mathbf{n})}{\mathbf{T}} = \lim_{\Delta A \to 0} \frac{\Delta \mathbf{F}}{\Delta A} \tag{4.1}$$

We see that $\overset{(\mathbf{n})}{\mathbf{T}}$ is *force intensity* or *stress* acting on a plane whose normal is **n** at the point O. We have used this notation for the force intensity to emphasize that $\overset{(\mathbf{n})}{\mathbf{T}}$ is a vector and that it acts on a plane passing through the point O whose normal is **n** (Fig. 4.3*c*); $\overset{(\mathbf{n})}{\mathbf{T}}$ does *not* act in general in the direction of **n**. In assuming the existence of the limit in (4.1), we have in effect made a *continuum hypothesis* with respect to the distribution of internal forces. This is an idealization of the behavior in a real material, where $\Delta \mathbf{F}$ decreases smoothly with ΔA only when ΔA is large in comparison with the microstructure (metallurgical or molecular) of the material (Fig. 4.3*b*).

There are four major characteristics of stress we must keep in mind: (1) the physical dimensions of stress are force per unit area, (2) stress is defined at a point upon an imaginary plane or boundary dividing the material into two parts, (3) stress is a vector equivalent to the action of one part of the material upon another, and (4) the direction of the stress vector is not restricted. We may write the stress vector in terms of its components with respect to the coordinate axes in the form

$$\overset{(\mathbf{n})}{\mathbf{T}} = \overset{(\mathbf{n})}{T_x}\mathbf{i} + \overset{(\mathbf{n})}{T_y}\mathbf{j} + \overset{(\mathbf{n})}{T_z}\mathbf{k} \tag{4.2}$$

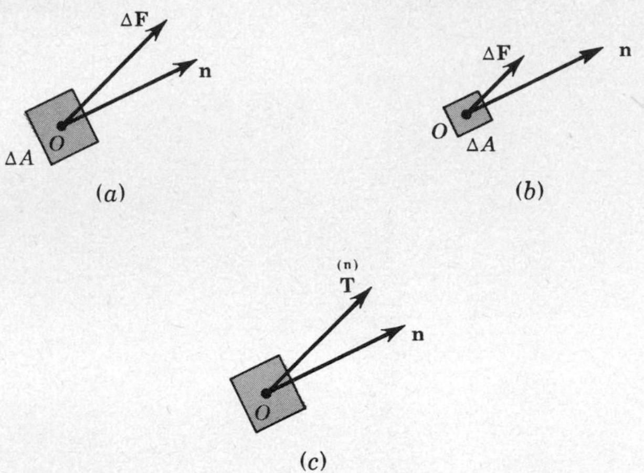

Fig. 4.3 (*a*) Force vector acting on element of area ΔA and (*b*) indication of limiting process which defines stress vector $\overset{(\mathbf{n})}{\mathbf{T}}$; (*c*) stress vector $\overset{(\mathbf{n})}{\mathbf{T}}$ acting on an element of area A whose normal is $\mathbf{n}$ at the point O.

It is convenient in what follows to express the stress vector at a point acting on a plane whose normal is $\mathbf{n}$ in terms of stress vectors acting on planes which pass through the point parallel to the coordinate planes. Once we have found the components of the stress vectors acting on the coordinate planes we will relate these to the components of $\overset{(\mathbf{n})}{\mathbf{T}}$.

Let us return to our discussion of the continuous body shown in Fig. 4.1 and now pass through the point O a plane mm parallel to the yz plane and consider the free body to the left of the plane mm. We divide the plane mm into a large number of small areas, each Δy by Δz, as shown in Fig. 4.4. We will now repeat

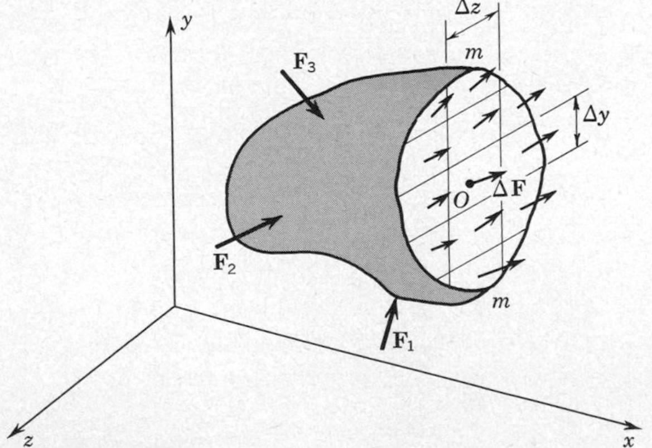

Fig. 4.4 Internal forces acting on plane mm.

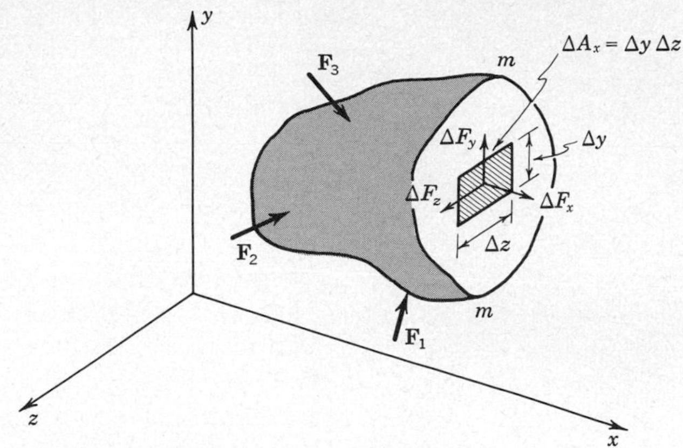

Fig. 4.5 Rectangular components of the force vector $\Delta\mathbf{F}$ acting on the small area centered on point O.

our arguments for the force vector and the force intensity for the arbitrarily defined plane with normal **n** by applying them to an element of area in the plane *mm*. On the small area ΔA centered on the point O there will be acting a force $\Delta\mathbf{F}$ which is *inclined* to the surface *mm* at some arbitrary angle. We will describe the force vector $\Delta\mathbf{F}$ in terms of a set of rectangular force components, and the most convenient set of axes will prove to be a set in which one axis is normal to the surface and the other two are parallel to the surface. Figure 4.5 shows the rectangular components of the force vector $\Delta\mathbf{F}$ referred to such a set of axes.

Before proceeding further, it will be useful to adopt a convention which allows us to identify precisely a specific area or face on the surface of a body. For example, if we cut out a parallelepiped whose edges are parallel to the x, y, z axes, as shown in Fig. 4.6, there will be six separate plane surfaces which enclose the volume, and we need a concise notation for unambiguously identifying each of these. We shall use the same convention adopted in Chap. 3 and identify a

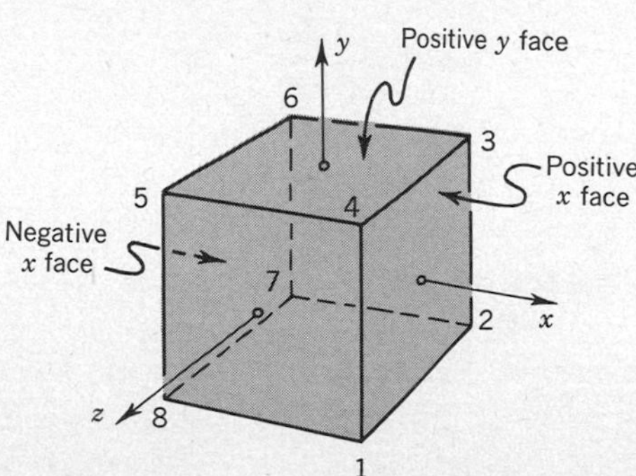

Fig. 4.6 Definition of positive and negative faces.

surface face in terms of the coordinate axis normal to the surface. A *face* will be defined as *positive* when its outwardly directed normal vector points in the direction of the positive coordinate axis, and as *negative* when its outward normal vector points in the negative coordinate direction. Thus, in Fig. 4.6, the face 1–4–5–8 is a positive z face, and the face 2–3–6–7 is a negative z face. In Fig. 4.5 the plane mm is a positive x face.

If we now return to Fig. 4.5 and take the ratios $\Delta F_x/\Delta A_x$, $\Delta F_y/\Delta A_x$, and $\Delta F_z/\Delta A_x$, we have three quantities which establish the average intensity of force on the face of area $\Delta A_x = \Delta y\,\Delta z$. In the limit as $\Delta A_x \to 0$ these ratios define the force intensity acting on the x face *at* the point O; these values of the three force intensities are defined as the *stress components associated with the x face* at point O. They are the *components* of the *stress vector* acting on the x face at the point O; that is, they are the components of the stress vector acting on an element of area whose normal points in the positive x direction.

We note that we must use *two* directions to define a stress *component*: one direction to identify the face on which the stress component acts, and a second direction to specify the force component from which the stress component is derived. The stress components parallel to the surface will be called *shear*-stress components and will be denoted by τ. The shear-stress component acting on the x face in the y direction will be identified as τ_{xy}, where the first subscript denotes the direction of the normal to the face and the second denotes the direction in which the stress component acts. The stress component perpendicular to the face will be called a *normal* stress component and will be denoted by σ. The normal stress component acting on the x face will act in the x direction and thus will be identified as σ_{xx}. Because the two subscripts on normal stress components are always identical, it has become common practice to write only one of these, with the understanding that the second subscript is implied; thus, we shall use σ_x as shorthand for σ_{xx}. Using the above notation, the stress *components* on the x face at point O are defined as follows in terms of the force intensity ratios:

$$
\begin{aligned}
\sigma_x &= \lim_{\Delta A_x \to 0} \frac{\Delta F_x}{\Delta A_x} \\
\tau_{xy} &= \lim_{\Delta A_x \to 0} \frac{\Delta F_y}{\Delta A_x} \\
\tau_{xz} &= \lim_{\Delta A_x \to 0} \frac{\Delta F_z}{\Delta A_x}
\end{aligned}
\tag{4.3}
$$

These stress components are illustrated in Fig. 4.7.

It will be convenient to adopt a sign convention for stress components. We shall again use the same convention adopted in Chap. 3 and define the resulting stress component to be *positive* when a positively directed force component acts

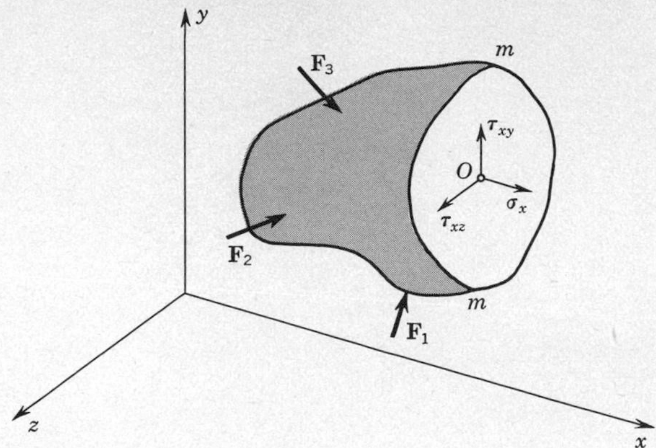

Fig. 4.7 Stress components on positive x face at point O.

on a positive face. Alternatively, a positive stress component results when a negatively directed force component acts on a negative face; by a consideration of action and reaction, we see that this definition follows from the first. When a positively directed force component acts on a negative face or a negatively directed force component acts on a positive face, the resulting stress component will be *negative*. All the stress components shown in Fig. 4.7 are positive according to this sign convention.

If we return to the continuous body shown in Fig. 4.1, we can also pass a plane parallel to the xy and xz planes through the point O. On the elements of area lying in the xy and xz plane, we define stress components in analogy with (4.3). In this way we find at the point O that the state of stress is dependent on the nine stress components:

$$\begin{matrix} \sigma_x & \tau_{xy} & \tau_{xz} \\ \tau_{yx} & \sigma_y & \tau_{yz} \\ \tau_{zx} & \tau_{zy} & \sigma_z \end{matrix} \tag{4.4}$$

We see that each line of (4.4) gives the components of the stress vector acting on the respective coordinate plane. A knowledge of the nine components is necessary in order to determine the components of the stress vector $\overset{(\mathbf{n})}{\mathbf{T}}$ acting on an arbitrary plane with normal $\mathbf{n}$. We will return to the relation between the components of $\overset{(\mathbf{n})}{\mathbf{T}}$ and the components (4.4) in Sec. 4.5.

Returning again to the continuous body shown in Fig. 4.1, suppose that in addition to the plane *mm* we pass another plane through the body, this plane being parallel to *mm* but separated from it a small distance Δx. If we similarly pass planes parallel to the xz and xy planes, we finally cut out a parallelepiped Δx by Δy by Δz. The stress components acting on the faces of this element will be as

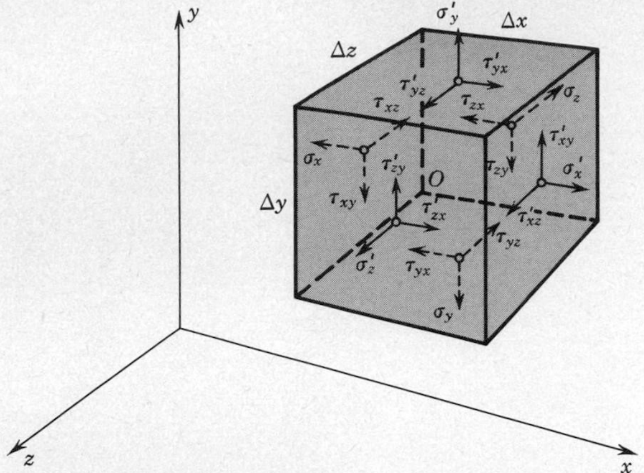

Fig. 4.8 Stress components acting on the six sides of a parallelepiped.

shown in Fig. 4.8. The primes are used to indicate that the stress components on opposite faces do not necessarily have the same magnitudes since the faces are separated by small distances. In Fig. 4.8 we assume, for example, that the stress component τ_{xy} is uniform over the negative x face, but that when we move a distance Δx to the positive x face, the stress component in the y direction has a new value τ'_{xy} which is uniform over this positive x face. Actually, the stress components in Fig. 4.8 should be thought of as average values over the respective faces of the parallelepiped.

An alternative notation called index or *indicial notation* for stress is often more convenient for general discussions in elasticity. We will briefly discuss this notation here. In indicial notation the coordinate axes x, y, and z are replaced by numbered axes, x_1, x_2, and x_3, respectively. The components of a vector such as the force $\Delta \mathbf{F}$ of Fig. 4.5 are then written as ΔF_1, ΔF_2, and ΔF_3, where the numerical subscript indicates the component with respect to the numbered coordinate axes.

The definitions of the components of stress acting on the x_1 face, (4.3), can be written in indicial form as follows:

$$\sigma_{11} = \lim_{\Delta A_1 \to 0} \frac{\Delta F_1}{\Delta A_1}$$

$$\sigma_{12} = \lim_{\Delta A_1 \to 0} \frac{\Delta F_2}{\Delta A_1}$$

$$\sigma_{13} = \lim_{\Delta A_1 \to 0} \frac{\Delta F_3}{\Delta A_1}$$

where we use the σ symbol for both normal and shear stresses. The components of stress are distinguished by two numerical subscripts, the first indicating the face on which the stress component acts and the second specifying the direction of the stress component. Similar equations can be written for components of stress associated with the other faces. The similarity of such equations suggests replacing any particular numerical subscript by an alphabetic subscript which can take on any of the three numerical values, 1, 2, or 3. All components of stress can now be defined by a single equation:

$$\sigma_{ij} = \lim_{\Delta A_i \to 0} \frac{\Delta F_j}{\Delta A_i} \tag{4.5}$$

We can therefore think of the nine components of stress in (4.4) as simply σ_{ij}, where i and j take on the values 1, 2, or 3. In Sec. 4.4 and 4.15 we shall encounter further extensions of this indicial notation.

4.3 PLANE STRESS

In many instances the stress situation is simpler than that illustrated in Fig. 4.8. For example, if we pull on a long, thin wire of uniform section and examine a small parallelepiped whose x axis coincides with the axis of the wire, then σ_x and σ'_x will be the only nonzero stress components acting on the faces of the parallelepiped (and it will be necessary for equilibrium that $\sigma'_x = \sigma_x$). Another example of practical interest is that of a thin sheet which is being pulled by forces in the plane of the sheet. If we take the xy plane to be the plane of the sheet, then σ_x, σ'_x, σ_y, σ'_y, τ_{xy}, τ'_{xy}, τ_{yx}, and τ'_{yx} will be the only stress components acting on the parallelepiped of Fig. 4.8. Furthermore, the stress components turn out to be practically constant through the thickness of the sheet. We can assume, therefore, for a thin sheet that there are no variations in the stress components in the z direction. The state of stress at a given point will only depend upon the four stress components

$$\begin{matrix} \sigma_x & \tau_{xy} \\ \tau_{yx} & \sigma_y \end{matrix} \tag{4.6}$$

in which the stress components are functions of *only* x and y. This combination of stress components is called *plane stress* in the xy plane; it is illustrated in Fig. 4.9.

There are many important problems in which the stress condition is one of plane stress, and for this reason we shall examine this special situation in some detail.

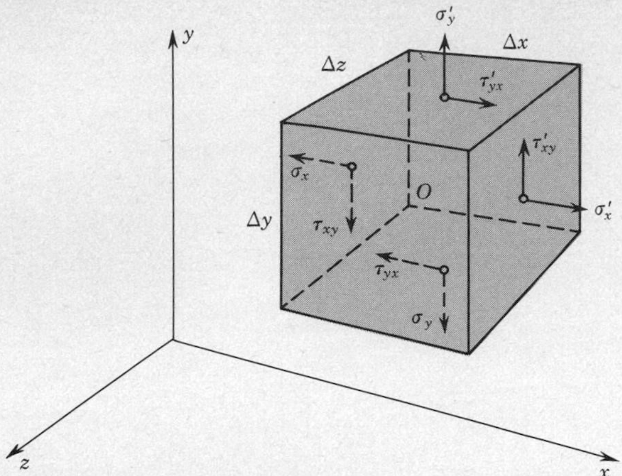

Fig. 4.9 Stress components which define a condition of plane stress in the xy plane.

4.4 EQUILIBRIUM OF A DIFFERENTIAL ELEMENT IN PLANE STRESS

If a continuous body is in equilibrium, then any isolated part of the body must be acted upon by an equilibrium set of forces. The small element shown in Fig. 4.9 represents part of a body in plane stress and therefore must be in equilibrium if the entire body is to be in equilibrium. The requirements of equilibrium will establish certain conditions which must exist between the stress components. Before determining these conditions, we shall express the stress components σ_x', τ_{xy}', σ_y', and τ_{yx}' in more convenient form, relating them to the components σ_x, τ_{xy}, σ_y, and τ_{yx}.

As we move from one point to another in the xy plane, the stress components vary in magnitude. Thus, in Fig. 4.9, as we move from the negative x face to the positive x face, the stress component τ_{xy} changes in value to τ_{xy}'. A convenient way to express the change in τ_{xy} as we move from point to point along a path in the xy plane is through use of the directional derivative of τ_{xy} with respect to distance measured along the path. This directional derivative gives the rate of change in τ_{xy} per unit distance along the path. If τ_{xy} is expressed as a function of x and y, the directional derivatives of τ_{xy} along the x and y directions are called, respectively, the *partial derivative*[1] of τ_{xy} with respect to x and the *partial derivative* of τ_{xy} with respect to y, and they are expressed in mathematical notation as follows:

$$\frac{\partial \tau_{xy}}{\partial x} \qquad \frac{\partial \tau_{xy}}{\partial y}$$

where the symbol ∂ is used in place of the usual differential operator d.

[1] For a discussion of partial derivatives, see, for example, G. B. Thomas, "Calculus and Analytic Geometry," 3d ed., p. 657, Addison-Wesley Publishing Company, Inc., Reading, Mass., 1966.

Using the concept of the partial derivative, we can approximate the amount a stress component changes between two points separated by a small distance as the product of the partial derivative in the direction connecting the two points multiplied by the distance between the points. Thus, referring to Fig. 4.9, we can express τ'_{xy} as

$$\tau'_{xy} = \tau_{xy} + \frac{\partial \tau_{xy}}{\partial x}\,\Delta x \tag{4.7}$$

and σ'_x, σ'_y, and τ'_{yx} can be expressed similarly. In Fig. 4.10 we show the element of Fig. 4.9 with the stress components expressed in this manner.

We now require that the element shown in Fig. 4.10 must satisfy the equilibrium conditions $\Sigma\mathbf{M} = 0$ and $\Sigma\mathbf{F} = 0$. $\Sigma\mathbf{M} = 0$ is satisfied by taking moments about the center of the element

$$\Sigma\mathbf{M} = \left\{(\tau_{xy}\Delta y\,\Delta z)\frac{\Delta x}{2} + \left[\left(\tau_{xy} + \frac{\partial \tau_{xy}}{\partial x}\,\Delta x\right)\Delta y\,\Delta z\right]\frac{\Delta x}{2}\right.$$
$$\left. - (\tau_{yx}\,\Delta x\,\Delta z)\frac{\Delta y}{2} - \left[\left(\tau_{yx} + \frac{\partial \tau_{yx}}{\partial y}\,\Delta y\right)\Delta x\,\Delta z\right]\frac{\Delta y}{2}\right\}\mathbf{k} = 0 \tag{4.8}$$

$\Sigma\mathbf{F} = 0$ is satisfied by the following two conditions:

$$\Sigma F_x = \left(\sigma_x + \frac{\partial \sigma_x}{\partial x}\,\Delta x\right)\Delta y\,\Delta z + \left(\tau_{yx} + \frac{\partial \tau_{yx}}{\partial y}\,\Delta y\right)\Delta x\,\Delta z$$
$$- \sigma_x\Delta y\,\Delta z - \tau_{yx}\Delta x\,\Delta z = 0 \tag{4.9}$$

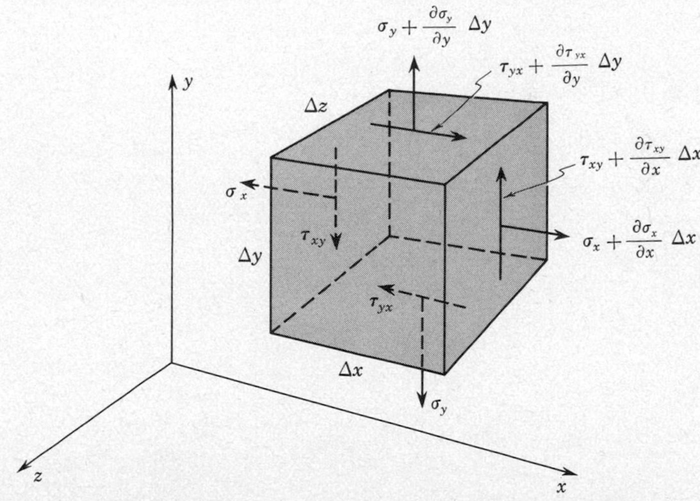

Fig. 4.10 Stress components in plane stress expressed in terms of partial derivatives.

$$\Sigma F_y = \left(\sigma_y + \frac{\partial \sigma_y}{\partial y}\,\Delta y\right)\Delta x\,\Delta z + \left(\tau_{xy} + \frac{\partial \tau_{xy}}{\partial x}\,\Delta x\right)\Delta y\,\Delta z - \sigma_y \Delta x\,\Delta z - \tau_{xy}\Delta y\,\Delta z = 0 \quad (4.10)$$

Simplifying (4.8), we obtain

$$\tau_{xy} + \frac{\partial \tau_{xy}}{\partial x}\frac{\Delta x}{2} - \tau_{yx} - \frac{\partial \tau_{yx}}{\partial y}\frac{\Delta y}{2} = 0 \quad (4.11)$$

In the limit as Δx and Δy go to zero, we see that all terms in Eq. (4.11) except τ_{xy} and τ_{yx} approach zero, and thus the equation reduces to

$$\tau_{yx} = \tau_{xy} \quad (4.12)$$

The result expressed by (4.12) is of great importance in the study of stress at a point. Equation (4.12) says that in a body in plane stress the shear-stress components on perpendicular faces must be equal in magnitude. Actually, (4.12) is true for any state of stress at a point. It can be proved in the most general case of stress at a point, as illustrated in Fig. 4.8, that to satisfy the requirement of $\Sigma\mathbf{M} = 0$ the shear-stress components acting on perpendicular faces and in directions perpendicular to the line of intersection of the faces must be equal in magnitude and directed relative to each other, as shown in Fig. 4.8. Thus, in Fig. 4.8, $\tau_{zy} = \tau_{yz}$ and $\tau_{xz} = \tau_{zx}$ in addition to the result $\tau_{yx} = \tau_{xy}$ given by (4.12).

Equation (4.12) was obtained from the free body of Fig. 4.10, where τ_{xy} and τ_{yx} are both positive according to the sign convention for stress components defined in Sec. 4.2. It is clear from Eq. (4.12) that if τ_{xy} were negative, then τ_{yx} also would have to be negative and of the same magnitude. Thus the two shear-stress components associated with the x, y set of axes must both be positive or both be negative. These two possibilities are shown in Fig. 4.11, where, because of Eq. (4.12), we use the same subscript sequence to identify the shear-stress component on both the x face and the y face.

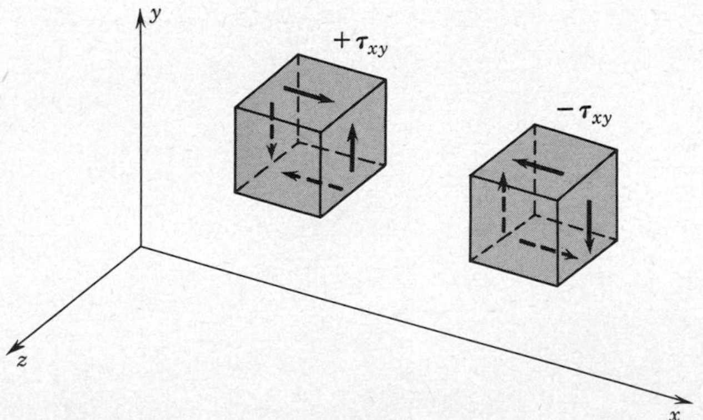

Fig. 4.11 Definition of positive and negative τ_{xy}.

If we now return to Eqs. (4.9) and (4.10), we find, using (4.12), that the requirements of $\Sigma \mathbf{F} = 0$ at a point lead to the differential equations

$$\begin{aligned} \frac{\partial \sigma_x}{\partial x} + \frac{\partial \tau_{xy}}{\partial y} &= 0 \\ \frac{\partial \tau_{xy}}{\partial x} + \frac{\partial \sigma_y}{\partial y} &= 0 \end{aligned} \tag{4.13}$$

Equations (4.13) specify the relations which must exist between the partial derivatives of the stress components on perpendicular faces at a point in order to satisfy $\Sigma \mathbf{F} = 0$ at the point.

Summarizing, we have considered the case of plane stress (Figs. 4.9 and 4.10) and have found the requirements (4.12) and (4.13) which equilibrium imposes upon the stress components acting on perpendicular faces. Moment equilibrium has the result (4.12) that the original four stress components are reduced to three independent components, while force equilibrium requires that certain relations (4.13) exist between the partial derivatives of the stress components. For a three-dimensional case of stress, moment equilibrium will reduce the original nine components of stress to six independent ones (that is, σ_x, σ_y, σ_z, τ_{xy}, τ_{yz}, τ_{zx}). The satisfaction of $\Sigma \mathbf{F} = 0$ for a three-dimensional state of stress will require three equations (see Probs. 4.1 and 4.2). In some instances it may be convenient to use a coordinate system different from the rectangular system we have discussed here; e.g., in examining the twisting of a circular shaft it is convenient to use cylindrical coordinates. The requirements of $\Sigma \mathbf{F} = 0$ can be expressed in results similar to (4.13) for stress components referred to other coordinate systems (see Probs. 4.3 and 4.4).

In indicial notation, the three-dimensional equations corresponding to Eqs. (4.13) (Prob. 4.1) take on the form, with $\sigma_{ij} = \sigma_{ji}$,

$$\begin{aligned} \frac{\partial \sigma_{11}}{\partial x_1} + \frac{\partial \sigma_{21}}{\partial x_2} + \frac{\partial \sigma_{31}}{\partial x_3} &= 0 \\ \frac{\partial \sigma_{12}}{\partial x_1} + \frac{\partial \sigma_{22}}{\partial x_2} + \frac{\partial \sigma_{32}}{\partial x_3} &= 0 \qquad \text{or} \qquad \sum_{i=1}^{3} \frac{\partial \sigma_{ij}}{\partial x_i} = 0 \\ \frac{\partial \sigma_{13}}{\partial x_1} + \frac{\partial \sigma_{23}}{\partial x_2} + \frac{\partial \sigma_{33}}{\partial x_3} &= 0 \end{aligned} \tag{4.14}$$

Note that the second form of (4.14) stands for any of three equations, depending on whether the subscript j has the value 1, 2, or 3. These three equations correspond to the three equations for equilibrium of forces in the x_1, x_2, and x_3 directions, respectively. Each equation consists of three terms, each corresponding to a gradient in stress between a pair of parallel planes. The three-dimensional indicial form of these equations makes their symmetry stand out more clearly than does the two-dimensional case with the σ–τ notation of Eqs. (4.13).

Einstein noted that in many equations of mathematical physics, the summation indicated in (4.14) is over a subscript that appears twice. As a consequence, he introduced the *summation convention* that whenever a subscript appears twice in an expression written in indicial notation, a summation over that subscript is automatically understood (unless the contrary is explicitly stated). Thus (4.14) may be stated

$$\frac{\partial \sigma_{ij}}{\partial x_i} = 0 \tag{4.15}$$

where the summation over the repeated subscript i is implied [see (4.14)].

4.5 STRESS COMPONENTS ASSOCIATED WITH ARBITRARILY ORIENTED FACES IN PLANE STRESS

In the previous section we determined the conditions which equilibrium imposes upon the stress components on perpendicular faces at a point in a stressed body. In this section we examine further the problem of equilibrium of stress at a point and determine relationships which must exist between the stress components associated with faces which are not perpendicular to each other. In particular, we will find how to express the components of the stress vector on a plane passing through a point when the normal to the plane is not parallel to one of the coordinate planes in terms of the components of stresses on the coordinate planes. We will also obtain stress components in a set of axes rotated with respect to the original axes in terms of the stress components referred to the original coordinate axes.

Let us assume that we know the values of the stress components σ_x, σ_y, and τ_{xy} at some point in a body subjected to plane stress.

We ask the first question: Do the known stress components on the x and y faces determine the *components* of the stress *vector* $\overset{(\mathbf{n})}{\mathbf{T}}$ acting on a face which passes through the point and whose normal lies in the xy plane and makes an arbitrary angle with the x axis? The question is illustrated in Fig. 4.12: Does equilibrium uniquely determine the components $\overset{(\mathbf{n})}{T}_x$, $\overset{(\mathbf{n})}{T}_y$, of the stress vector $\overset{(\mathbf{n})}{\mathbf{T}}$ in terms of σ_x, σ_y, τ_{xy}, and θ?

We answer the question by considering equilibrium of a small wedge centered on point O, as shown in Fig. 4.13. For a sufficiently small wedge we can consider the stress components to be uniform over each face. The equilibrium requirements $\Sigma\mathbf{M} = 0$ and $\Sigma\mathbf{F} = 0$ can be expressed as follows:

$$\Sigma M_z = (\tau_{xy}\,\Delta z\,\overline{MP})\frac{\overline{NP}}{2} - (\tau_{xy}\,\Delta z\,\overline{NP})\frac{\overline{MP}}{2} = 0 \tag{4.16}$$

$$\Sigma F_x = \overset{(\mathbf{n})}{T}_x\,\Delta z\,\overline{MN} - \sigma_x\,\Delta z\,\overline{MN}\cos\theta - \tau_{xy}\,\Delta z\,\overline{MN}\sin\theta = 0 \tag{4.17}$$

$$\Sigma F_y = \overset{(\mathbf{n})}{T}_y\,\Delta z\,\overline{MN} - \tau_{xy}\,\Delta z\,\overline{MN}\cos\theta - \sigma_y\,\Delta z\,\overline{MN}\sin\theta = 0 \tag{4.18}$$

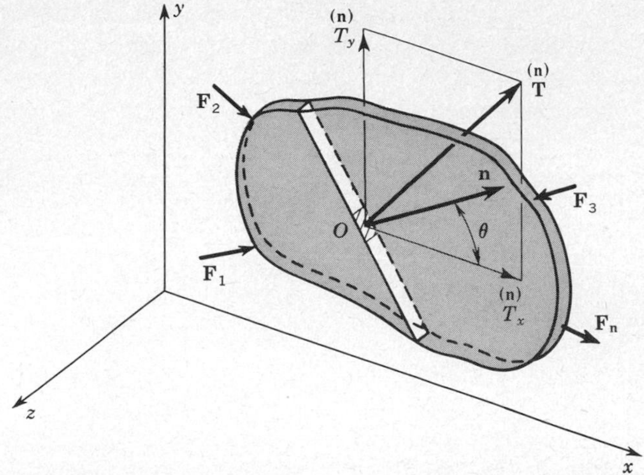

Fig. 4.12 Stress vector acting at point O on a plane whose normal is **n** for a body subjected to plane stress.

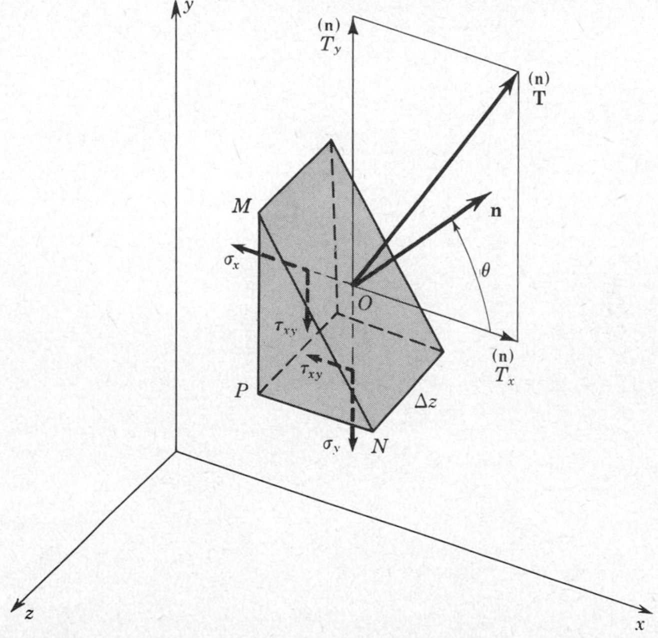

Fig. 4.13 Stress vector and stress components acting on faces of a small wedge, cut from the body of Fig. 4.12, which is in a state of plane stress in the xy plane.

We see that moment equilibrium (4.16) is satisfied identically; this is an expected result since it was the satisfaction of moment equilibrium which led to the result (4.12) which has been included in the free-body diagram of Fig. 4.13.

Upon simplification of (4.17) and (4.18) we find

$$\begin{aligned} \overset{(\mathbf{n})}{T}_x &= \sigma_x \cos\theta + \tau_{xy}\sin\theta \\ \overset{(\mathbf{n})}{T}_y &= \tau_{xy}\cos\theta + \sigma_y \sin\theta \end{aligned} \tag{4.19}$$

Equations (4.19) give the components of the stress vector acting on an element of area whose unit normal vector is

$$\mathbf{n} = \cos\theta\mathbf{i} + \sin\theta\mathbf{j} \tag{4.20}$$

Equations (4.19) are important results because a knowledge of the vector components in terms of the components on coordinate faces is often required at the boundary of a thin body. Similar results also hold for the general three-dimensional body (see Sec. 4.15).

The next question we ask is: Do the known stress components on the x and y faces determine the stress components on a face which passes through the point and whose normal x' lies in the xy plane and makes an arbitrary angle θ with the x axis? The question is illustrated in Fig. 4.14: Does equilibrium uniquely determine the stress components $\sigma_{x'}$ and $\tau_{x'y'}$ in terms of σ_x, σ_y, τ_{xy}, and θ?

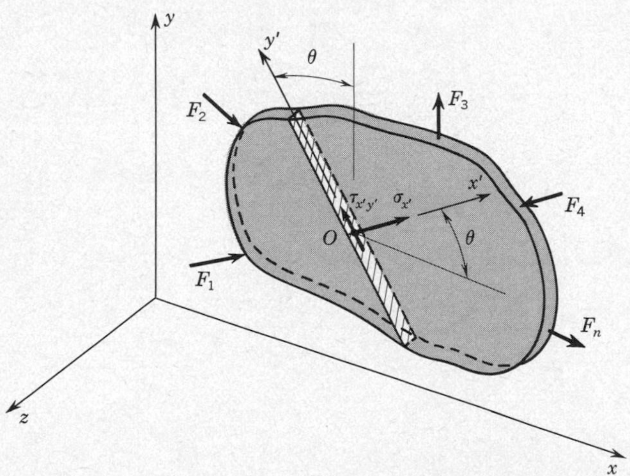

Fig. 4.14 Thin body subjected to plane stress in the xy plane. x' and y' are a set of perpendicular axes lying in the plane of the body, and $\sigma_{x'}$ $\tau_{x'y'}$ are the stress components acting on the positive x' face at point O.

We again answer the question by considering the equilibrium of a small wedge centered on point O as shown in Fig. 4.15. The equilibrium requirements $\Sigma\mathbf{F} = 0$ are now expressed in the $x'y'$ direction as follows:

$$\Sigma F_{x'} = \sigma_{x'}\,\Delta z\,\overline{MN} - (\sigma_x\,\Delta z\,\overline{MP})\cos\theta - (\tau_{xy}\,\Delta z\,\overline{MP})\sin\theta$$
$$- (\sigma_y\,\Delta z\,\overline{NP})\sin\theta - (\tau_{xy}\,\Delta z\,\overline{NP})\cos\theta = 0 \quad (4.21)$$

$$\Sigma F_{y'} = \tau_{x'y'}\,\Delta z\,\overline{MN} + (\sigma_x\,\Delta z\,\overline{MP})\sin\theta - (\tau_{xy}\,\Delta z\,\overline{MP})\cos\theta$$
$$- (\sigma_y\,\Delta z\,\overline{NP})\cos\theta + (\tau_{xy}\,\Delta z\,\overline{NP})\sin\theta = 0 \quad (4.22)$$

Using the trigonometric relations between the sides of the wedge, (4.21) and (4.22) become

$$\begin{aligned} \sigma_{x'} &= \sigma_x \cos^2\theta + \sigma_y \sin^2\theta + 2\tau_{xy}\sin\theta\cos\theta \\ \tau_{x'y'} &= (\sigma_y - \sigma_x)\sin\theta\cos\theta + \tau_{xy}(\cos^2\theta - \sin^2\theta) \end{aligned} \quad (4.23)$$

Equations (4.23) demonstrate the answer to our question: The stress components $\sigma_{x'}$ and $\tau_{x'y'}$ at the point O *are* uniquely determined by equilibrium from the stress components σ_x, σ_y, τ_{xy}, and the angle θ.

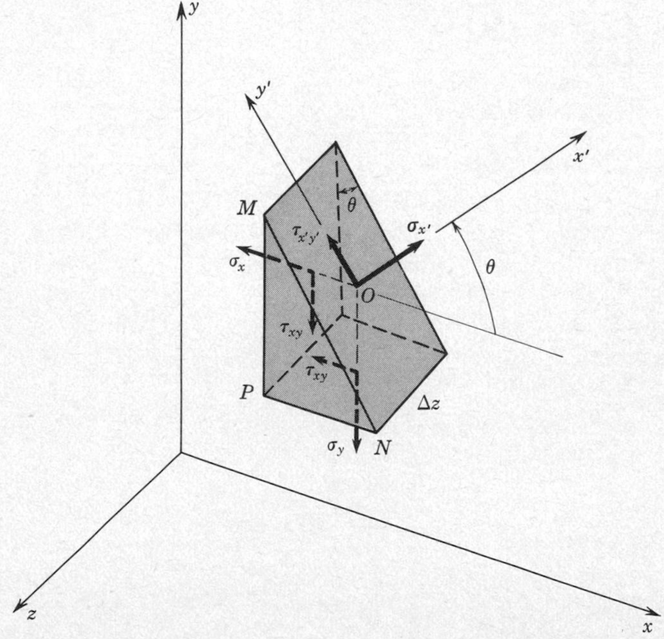

Fig. 4.15 Stress components acting on faces of a small wedge, cut from body of Fig. 4.14, which is in a state of plane stress in the *xy* plane.

From (4.23) it is evident that in plane stress if we know the stress components on any two perpendicular faces, we know the stress components on all faces whose normals lie in the plane. In particular, if we substitute $\theta + 90°$ for θ, we can obtain the normal stress $\sigma_{y'}$ acting on a face perpendicular to the y' axis.

$$\sigma_{y'} = \sigma_x \sin^2 \theta + \sigma_y \cos^2 \theta - 2\tau_{xy} \sin \theta \cos \theta \tag{4.24}$$

If we know the stress components for all possible orientations of faces through the point in question, we say that we know the *state of stress* at the point. Specification of a state of stress in plane stress thus involves knowledge of three stress components, most conveniently taken as the normal and shear components on two perpendicular faces.[1] We should be careful not to confuse a single stress component with the *state of stress* at the point.

4.6 MOHR'S CIRCLE REPRESENTATION OF PLANE STRESS

In order to facilitate application of (4.23) and (4.24), we shall make use of a simple graphical representation. To develop this representation we first rewrite (4.23) and (4.24) by introducing *double-angle* trigonometric relations

$$\begin{aligned}
\sigma_{x'} &= \frac{\sigma_x + \sigma_y}{2} + \frac{\sigma_x - \sigma_y}{2} \cos 2\theta + \tau_{xy} \sin 2\theta \\
\tau_{x'y'} &= \qquad\quad - \frac{\sigma_x - \sigma_y}{2} \sin 2\theta + \tau_{xy} \cos 2\theta \\
\sigma_{y'} &= \frac{\sigma_x + \sigma_y}{2} - \frac{\sigma_x - \sigma_y}{2} \cos 2\theta - \tau_{xy} \sin 2\theta
\end{aligned} \tag{4.25}$$

Now, in Fig. 4.16*a* we make a graphical representation of the first of (4.25) by laying out $\sigma_{x'}$ along the horizontal σ axis as follows. The first term on the right of the equation for $\sigma_{x'}$ is represented by OC. The second on the right is represented by the horizontal component of CB, and the third term is represented by the horizontal component of BA. Note that CB makes the angle 2θ with the σ axis and that BA is perpendicular to CB. The normal stress $\sigma_{x'}$ is thus represented by the abscissa of point A.

Next we turn to the second of (4.25) to obtain a graphical representation of $\tau_{x'y'}$. We note that the previous construction may also be used for this purpose if we consider shear stresses to be laid out vertically. The first term on the right of the equation for $\tau_{x'y'}$ is represented by the vertical component of CB, and the second term is represented by the vertical component of BA. The shear stress $\tau_{x'y'}$ is thus represented by the downward ordinate of point A.

[1] It can be shown that, in the general case of three-dimensional stress, specification of the state of stress at a point requires six pieces of information, most conveniently, the normal and shear-stress components associated with three mutually perpendicular faces. In a case of plane stress, at least three of these stress components are zero.

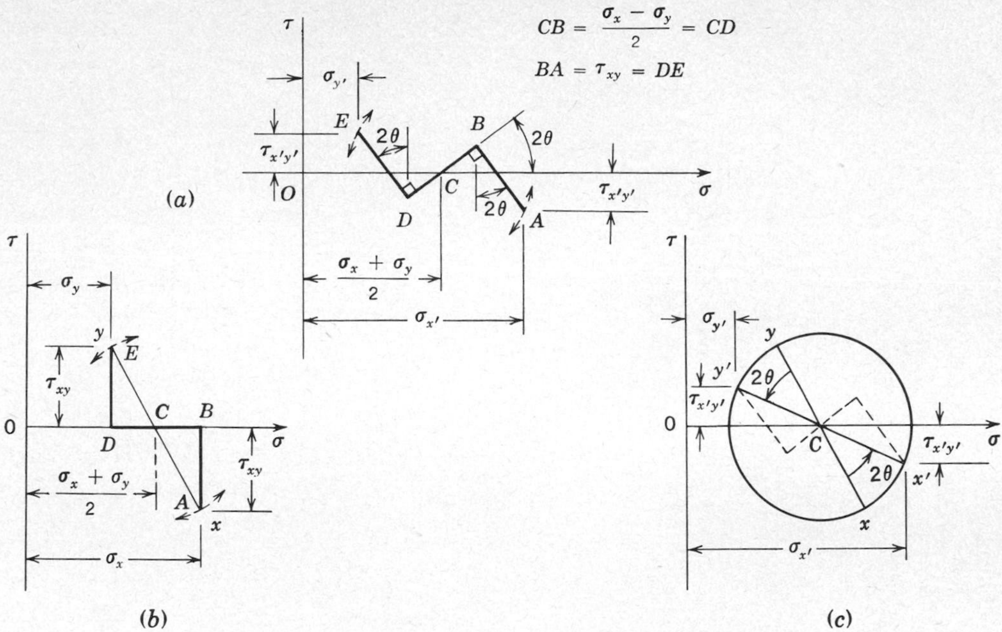

Fig. 4.16 Development of Mohr's circle for stress.

In a similar manner the construction CDE provides a representation of $\sigma_{y'}$ according to the third of (4.25) as well as a representation of $\tau_{x'y'}$ according to the second of (4.25). The normal stress $\sigma_{y'}$ is represented by the abscissa of point E, and the shear stress $\tau_{x'y'}$ is represented by the upward ordinate of point E.

In Fig. 4.16*a* the line $ABCDE$, with its two right-angle turns at B and D, is completely fixed by the original set of stress components σ_x, σ_y, and τ_{xy}. Only its orientation depends on the angle θ of Fig. 4.15. If we permit the angle θ to change, the line $ABCDE$ would change its orientation, rotating like a windmill about C. Points A and E would trace out arcs of a circle with center at C.

A particular orientation of interest is that corresponding to $\theta = 0$, as shown in Fig. 4.16*b*. Note that the common abscissa of points A and B is σ_x and that the abscissa of points D and E is σ_y. The shear stress τ_{xy} is represented by the downward ordinate of point A and by the upward ordinate of point E. We have also drawn the line ACE, labeling its end points x and y, as indicated.

In Fig. 4.16*c* a circle has been constructed using the line xy as a diameter. Also shown is the diameter $x'y'$ corresponding to the orientation of the line ACE in Fig. 4.16*a*. The angle between the xy diameter and the $x'y'$ diameter is 2θ.

The circle in Fig. 4.16*c* is called *Mohr's circle for stress.* It provides a convenient representation of the stress transformation equations (4.25). A set of stress components with respect to the xy axes is used to establish the xy diameter. Then stress components with respect to any rotated $x'y'$ axes can be determined

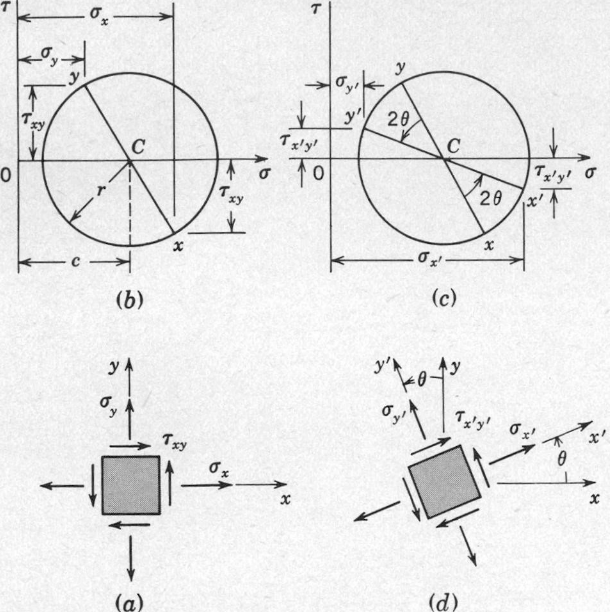

Fig. 4.17 Stress components (*a*) are used to construct Mohr's circle (*b*). Rotation of diameter through double angle in (*c*) provides stress components for inclined element (*d*).

from the corresponding $x'y'$ diameter. The manner in which the circle is constructed and used is summarized in a step-by-step list in the following paragraph. In order to follow these steps efficiently, it is necessary to understand clearly how stress components acting in a physical body are represented in the Mohr's circle diagram. The diagram is constructed in a stress plane, with normal stress σ plotted horizontally and shear stress τ plotted vertically. For normal stress, tension is *positive* and is plotted to the right of the origin of the stress plane. Compression is *negative* and is plotted to the left. For shear stress, the sign convention is complicated by the fact that we must distinguish between the x and y ends of the diameter.[1] *Positive* shear stress τ_{xy} (see Fig. 4.11) is plotted downward at x and upward at y. Negative shear stress is plotted upward at x and downward at y.

Given stress components σ_x, σ_y, and τ_{xy}, with respect to xy axes at a point O in Fig. 4.15, we can represent the state of stress at O by the rectangular element in Fig. 4.17*a*. To construct Mohr's circle in Fig. 4.17*b*, we proceed as follows:

[1] The sign convention given for the xy plane can be extended to the yz plane, the zx plane, or to any plane with rectangular axes α and β. It is only necessary to imagine the xy plane reoriented so that the positive quadrant of x and y coincides with the positive quadrant of α and β to determine which coordinate (α or β) corresponds to x and which corresponds to y.

1. Using the sign convention for stress components just given, we locate the point x with coordinates σ_x and τ_{xy} and the point y with coordinates σ_y and τ_{xy}.
2. We join points x and y with a straight line intersecting the σ axis at point C, which is to be the center of Mohr's circle. The abscissa of C is

$$c = \frac{\sigma_x + \sigma_y}{2} \tag{4.26}$$

3. With C as center and xy as diameter we draw the circle. The radius of the circle is

$$r = \left[\left(\frac{\sigma_x - \sigma_y}{2}\right)^2 + {\tau_{xy}}^2\right]^{1/2} \tag{4.27}$$

Once the circle has been constructed, it may be used to determine the stress components $\sigma_{x'}$, $\sigma_{y'}$, and $\tau_{x'y'}$ shown in Fig. 4.17*d*. These stress components apply to the same physical point O in the body but are in respect to the axes $x'y'$ which make an angle θ with the original xy axes.

4. We locate the $x'y'$ diameter with respect to the xy diameter in Mohr's circle by laying off the *double angle* 2θ in Fig. 4.17*c* in the *same sense* as the rotation θ which carries the xy axes into the $x'y'$ axes in Fig. 4.17*d*.
5. Using the sign convention for stress components in Mohr's circle, we read off the values of $\sigma_{x'}$ and $\tau_{x'y'}$ as the coordinates of point x' and the values of $\sigma_{y'}$ and $\tau_{x'y'}$ as the coordinates of point y'.

The following example illustrates how Mohr's circle is constructed and used in a specific numerical situation.

Example 4.1 We consider a thin sheet pulled in its own plane so that the stress components with respect to the xy axes are as given in Fig. 4.18*a*. We wish to find the stress components with respect to the *ab* axes which are inclined at 45° to the xy axes. Using the foregoing steps, we lay out the points x and y and construct Mohr's circle, as shown in Fig. 4.18*b*. The *ab* diameter is located at 2(45°) = 90° from the xy diameter. The stress components with respect to the *ab* axes could be read off directly from an accurately scaled diagram. Alternatively, we can use the geometry of the diagram to calculate as follows:

$$2\theta_1 = \tan^{-1}\frac{4{,}000}{3{,}000} = 53.2^\circ$$

$$r = [(3{,}000)^2 + (4{,}000)^2]^{1/2} = 5{,}000 \text{ psi}$$

$$\sigma_a = 8{,}000 + 5{,}000 \cos(90^\circ - 53.2^\circ) = 12{,}000 \text{ psi}$$

$$\sigma_b = 8{,}000 - 5{,}000 \cos(90^\circ - 53.2^\circ) = 4{,}000 \text{ psi}$$

Because point a lies above the σ axis (and point b below), the shear stress τ_{ab} is negative.

$$\tau_{ab} = -5{,}000 \sin (90^\circ - 53.2^\circ) = -3{,}000 \text{ psi}$$

These stress components are shown acting in their correct directions on the faces of the inclined element in Fig. 4.18*c*.

Mohr's circle provides a graphic overview of the state of plane stress at a point. Each possible stress-component combination $\sigma_{x'}$, $\sigma_{y'}$, $\tau_{x'y'}$ given by (4.25) is represented by some diameter of the circle. A particularly important combination is that represented by the diameter which is aligned with the normal stress axis. The end points of this diameter are designated 1 and 2 in Fig. 4.19*a*. The corresponding stress components with respect to the 1, 2 axes are shown in Fig. 4.19*b*. There are normal stresses σ_1 and σ_2, but there is *no* shear-stress component. Furthermore, σ_1 is the *maximum* possible normal stress component, and σ_2 is the *minimum* possible normal stress component at this location in the body.

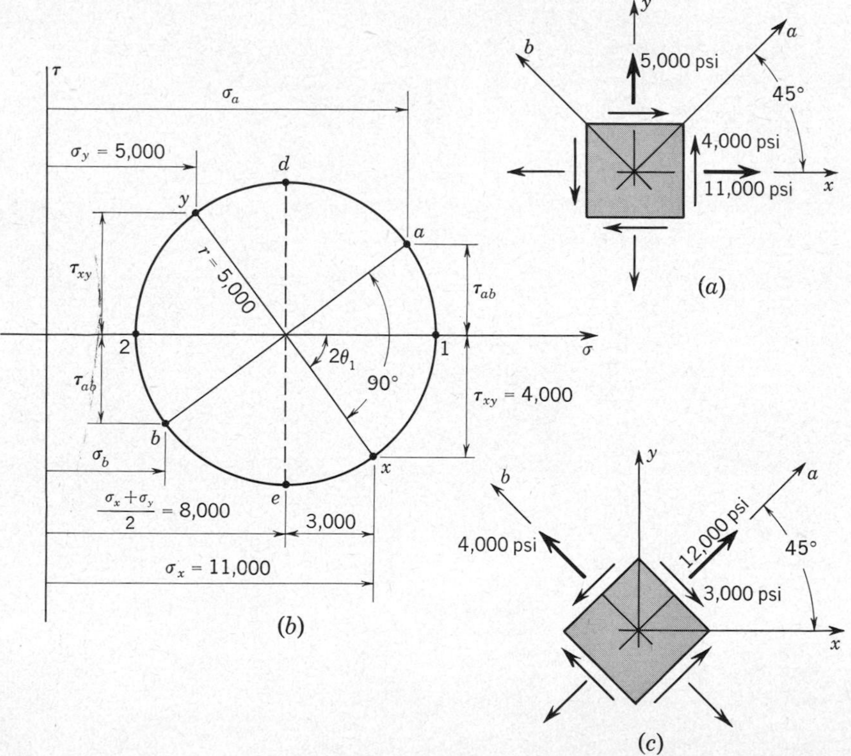

Fig. 4.18 Example 4.1.

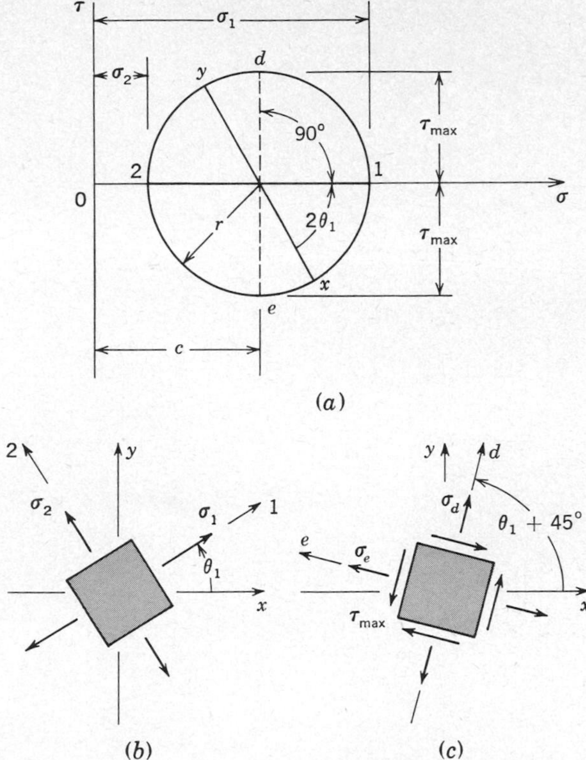

Fig. 4.19 (*a*) Principal stresses σ_1 and σ_2 and maximum shear stress τ_{max} indicated on Mohr's circle. (*b*) Element oriented along principal axes. (*c*) Element oriented along axes of maximum shear.

The stresses σ_1 and σ_2 are called *principal stresses*, and the 1 and 2 axes are called the *principal axes of stress*. In terms of the abscissa c of the center of the circle, (4.26), and the radius r, (4.27), the principal stresses are

$$\sigma_1 = c + r \qquad \sigma_2 = c - r \tag{4.28}$$

In many cases the most convenient way to describe the state of stress at a point is to give the principal axes and the corresponding principal stresses.

Another combination of stress components which is of special interest is that represented by the vertical diameter of Mohr's circle. The end points of this diameter are designated d and e in Fig. 4.19*a*, and the corresponding stress components with respect to the de axes are shown in Fig. 4.19*c*. Here the normal stresses are *equal*, and the magnitude of the shear stress is the *maximum* possible at this location. In terms of the abscissa c of the center of the circle, (4.26), and the radius r, (4.27), we have

$$\sigma_d = \sigma_e = c \qquad \tau_{\max} = r \tag{4.29}$$

The *de* axes are called the axes of *maximum shear*. The corresponding element faces, perpendicular to these axes, are said to define the *planes of maximum shear* at this location in the body. Note that the axes of maximum shear are inclined at 45° with respect to the principal axes.

Example 4.2 We reconsider the state of plane stress described in Example 4.1. It is required to (*a*) locate the principal axes and evaluate the principal stresses and (*b*) locate the axes of maximum shear and evaluate the corresponding stress components.

In Fig. 4.18*b* the principal diameter is labeled 1–2. Since $c = 8{,}000$ psi and $r = 5{,}000$ psi, the principal stresses according to (4.28) are

$$\sigma_1 = c + r = 13{,}000 \text{ psi}$$
$$\sigma_2 = c - r = 3{,}000 \text{ psi}$$

The location of the principal axes with respect to the *xy* axes is shown in Fig. 4.20*a*, and the principal stresses are indicated.

In Fig. 4.18*b* the maximum shear diameter is labeled *de*. The location of the axes of maximum shear is indicated in Fig. 4.20*b*. According to (4.29) the corresponding stress components are

$$\sigma_d = \sigma_e = 8{,}000 \text{ psi} \qquad \tau_{\max} = 5{,}000 \text{ psi}$$

These stress components are shown in Fig. 4.20*b* acting in their correct directions. Note that with respect to the *de* axes the shear stress is negative; that is, $\tau_{de} = -5{,}000$ psi.

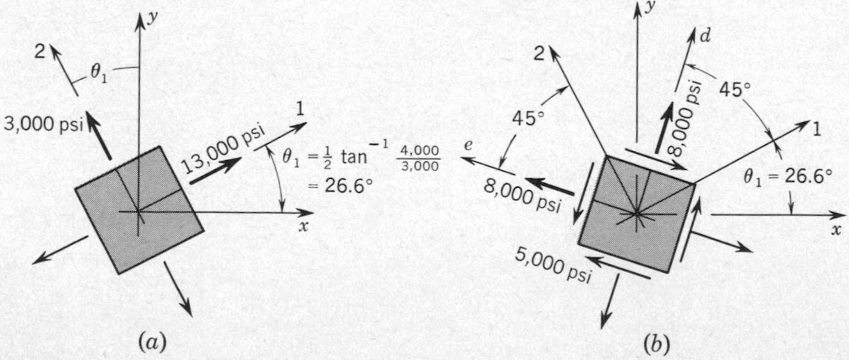

Fig. 4.20 Example 4.2

4.7 MOHR'S CIRCLE REPRESENTATION OF A GENERAL STATE OF STRESS

Thus far we have concerned ourselves with the problem of plane stress and have limited ourselves to transformation of axes within that plane. Suppose we now consider a situation in which the stress components associated with the z axis are not zero, i.e., a situation in which the state of stress may be said to be perfectly general. Further, let us suppose that we again wish to find the stress components associated with the plane whose normal x' lies in the xy plane and makes an angle of θ with the x axis. If we cut out a small wedge, similar to that of Fig. 4.15, we find that the stress components acting on the wedge are as shown in Fig. 4.21. We note that, in addition to the components $\sigma_{x'}$ and $\tau_{x'y'}$ on the $+x'$ face, we must admit the possibility of another shear-stress component $\tau_{zx'}$.

When we examine Fig. 4.21, we see that the stress components $\sigma_{x'}$ and $\tau_{x'y'}$ are unaffected by the stress components associated with the z axis. This results from the fact that for force equilibrium in the x' and y' directions the contributions of the components τ_{zx} and τ_{yz} acting on the $+z$ face of the wedge are exactly balanced by those of the components τ_{zx} and τ_{yz} acting on the $-z$ face.

If we now consider force equilibrium in the z direction for the wedge of Fig. 4.21, we find the following result:

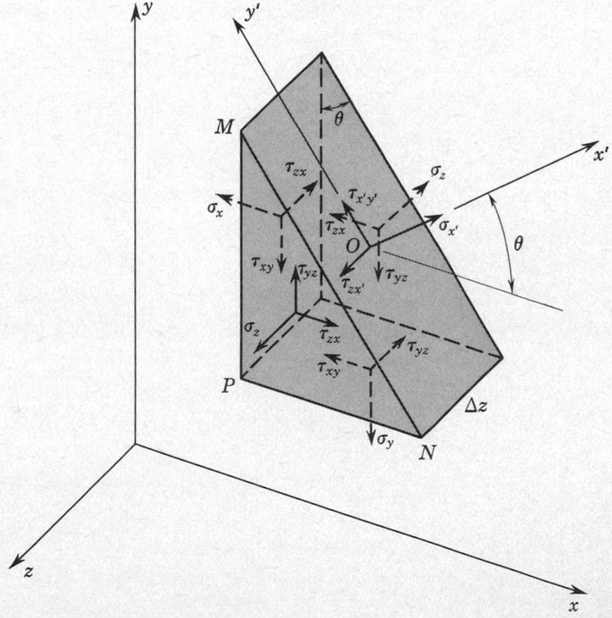

Fig. 4.21 Stress components acting on faces of a small wedge cut from a body in general state of stress.

$$\tau_{zx'}\,\Delta z\,\overline{MN} + \sigma_z \frac{\overline{NP}\,\overline{MP}}{2} - \tau_{zx}\,\Delta z\,\overline{MP} - \tau_{yz}\,\Delta z\,\overline{NP} - \sigma_z \frac{\overline{NP}\,\overline{MP}}{2} = 0$$

When the trigonometric relations between the sides $\overline{MP}$, $\overline{NP}$, and $\overline{MN}$ are used, this equilibrium requirement becomes

$$\tau_{zx'} = \tau_{zx}\cos\theta + \tau_{yz}\sin\theta \tag{4.30}$$

If we resolve the stress components τ_{zx} and τ_{yz} on the $+z$ face into components perpendicular and parallel to $\overline{MN}$, we find that the right-hand side of (4.30) is the sum of the components perpendicular to $\overline{MN}$. Thus we see that $\tau_{zx'}$ is of such magnitude as to be equal to the shear-stress component acting on the $+z$ face in the direction perpendicular to the line $\overline{MN}$.

As a consequence of our investigation of the general state of stress for the small wedge shown in Fig. 4.21, we may conclude the following:

1. The results given by Eqs. (4.25) and the Mohr's circle representation of these are correct whether or not the stress components σ_z, τ_{yz}, and τ_{zx} are zero.
2. If either τ_{yz} or τ_{zx} is nonzero, then in general there will exist a shear-stress component $\tau_{zx'}$ on the x' face in addition to $\tau_{x'y'}$. In such a case the 1 and 2 axes of Fig. 4.19 should not be called principal axes since we wish to retain the designation *principal axis of stress* for the normal to a face on which *no* shear-stress component acts.

We now consider further the case of *plane* stress and investigate the stress components referred to axes which do not lie in that plane. We begin by showing in Fig. 4.22*a* an element from the plate of Fig. 4.14 with the faces of the element oriented to coincide with the principal stress directions 1 and 2. The direction 3 is perpendicular to the 1, 2 plane (the xy plane) and thus is parallel to the z axis. Consider all planes which can be passed parallel to the 2 axis and so intersect the 1, 3 plane. The stress σ_2 would have no influence on the stress components on these planes. The stress components on one of these planes would depend only on the normal stress components along the 1 axis (σ_1) and along the 3 axis (zero), and thus we can obtain another Mohr's circle for which 1 and 3 are the principal stress directions. Finally, we can obtain a third Mohr's circle which represents stress components on planes parallel to the 1 axis; 2 and 3 are the principal stress directions. These three circles are shown in Fig. 4.22*b* for the situation where σ_1 and σ_2 both are tensile. Figure 4.22*c* shows the circles when σ_2 is compressive and Fig. 4.22*d* when both σ_1 and σ_2 are compressive.

An interesting fact illustrated by Fig. 4.22 is that plane stress does not mean that the stress components are zero on all faces except those which contain the normal to the so-called plane of stress. Figure 4.22*b* and *d* show that in some cases of plane stress the maximum shear stress at the point occurs on faces whose normal is inclined at 45° to the so-called plane of stress.

Returning again to a *general* state of stress, we assert without proof that at each point within a body there are three mutually perpendicular planes on which there are no shear-stress components acting. The normals to these three planes are called the *principal axes of stress* at the point. Figure 4.23*a* illustrates an element for which the principal axes are labeled 1, 2, and 3. In this example the 3 axis is assumed to be parallel to the z axis, and thus the axes 1 and 2 must lie in the xy plane.

If the six stress components associated with any three mutually perpendicular faces are specified, it is possible to develop equations similar to (4.23) for the normal and resultant shear-stress components on any arbitrary plane passed through the

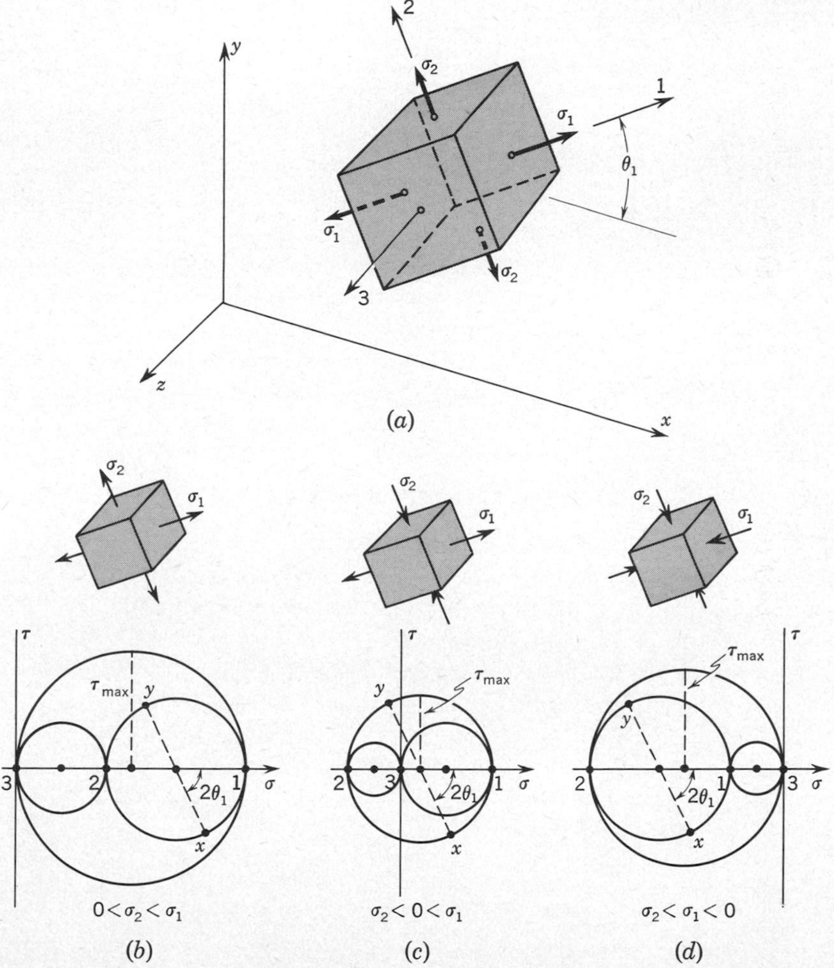

Fig. 4.22 Plane stress in xy plane.

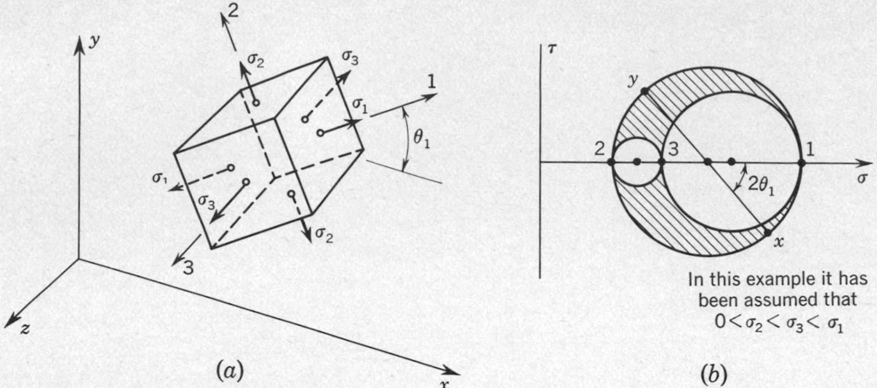

Fig. 4.23 Three-dimensional state of stress.

point. It can be shown[1] that the stress components for all possible planes are contained in the shaded area in Fig. 4.23*b* (where we have assumed a case in which $0 < \sigma_2 < \sigma_3 < \sigma_1$). In Fig. 4.23*b* the shear stress τ is the *resultant* shear-stress component acting on the plane (for example, $\sqrt{(\tau_{x'y'})^2 + (\tau_{zx'})^2}$ in Fig. 4.21). We thus see that the values of both the maximum shear-stress and the maximum normal stress components at a point are known without further calculation if the three principal stresses at the point are known. Since in most practical situations we shall be dealing with either a case of plane stress or a case in which the direction of one of the principal stress axes is known (e.g., from symmetry), we shall not examine the general state of stress beyond the level discussed here.

4.8 ANALYSIS OF DEFORMATION

Thus far in this chapter we have considered the force balance at a point within a stressed body. The results we have obtained are based only on the requirements of equilibrium and are equally true for a hypothetical "rigid" body or for a real body which deforms under the action of stress. Our concern is with real (deformable) bodies, and therefore, in addition to establishing the conditions imposed by force balance, we must determine what restrictions the requirement of *geometric compatibility* imposes upon the *deformation* of a continuous body. By a geometrically compatible deformation of a continuous body we mean one in which no voids are created in the body. This is purely a problem in the geometry of a continuum and is independent of the equilibrium requirements established in the foregoing sections of this chapter.

[1] A. J. Durelli, E. A. Phillips, and C. H. Tsao, "Introduction to the Theoretical and Experimental Analysis of Stress and Strain," p. 73, McGraw-Hill Book Company, New York, 1958; A. Nádai, "Theory of Flow and Fracture of Solids," 2d ed., p. 96, McGraw-Hill Book Company, New York, 1950.

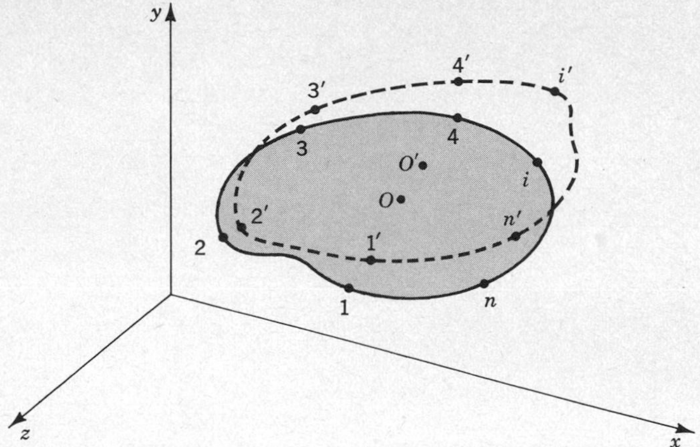

Fig. 4.24 Example of displacement of a continuous body.

We begin our study by considering the continuous three-dimensional body of Fig. 4.24 which undergoes a displacement such that point 1 goes to 1′, point 2 to 2′, etc. The displacement of an individual point is a vector quantity and, if the particles of a continuous body undergo various displacements, we can represent the displacement of each point by a displacement vector. This is illustrated in Fig. 4.25, where the displacement vectors $\mathbf{u}_1$, $\mathbf{u}_2$, $\mathbf{u}_3$, . . . , $\mathbf{u}_O$ show the displacements of the points 1, 2, 3, . . . , O. The displacement vector of any one point may be thought of as the sum of component displacements parallel to a set of suitable coordinate axes; thus, for point n of Fig. 4.25, we can write

$$\mathbf{u}_n = u_n\mathbf{i} + v_n\mathbf{j} + w_n\mathbf{k}$$

where u_n, v_n, and w_n are the xyz components of the displacement of point n.

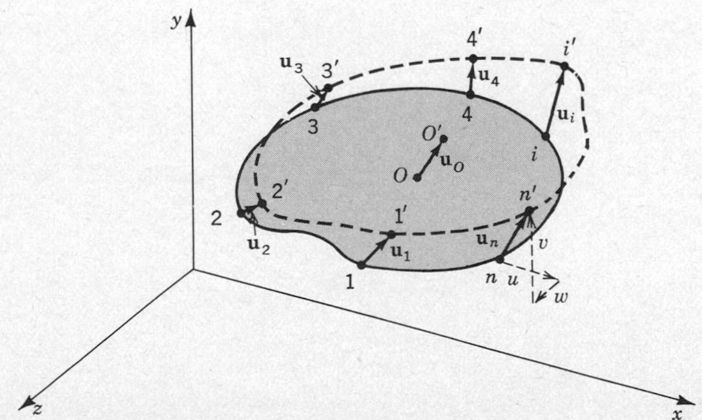

Fig. 4.25 Displacement vectors for the displacement of Fig. 4.24.

The displacement of a continuous body may be considered as the sum of two parts: (1) a translation and/or rotation of the body as a whole, and (2) a motion of the points of the body relative to each other. The translation and rotation of the body as a whole is called *rigid-body motion* because it can take place even if the body is perfectly rigid. The motion of the points of a body relative to each other is called a *deformation*. In Fig. 4.26*a* we illustrate a rigid-body translation of a triangle which is constrained to move only in the *xy* plane. In Fig. 4.26*b* a rigid-body rotation about the corner *c* is shown, and Fig. 4.26*c* illustrates a type of deformation without rigid-body motion. The displacement of Fig. 4.26*d* is the resultant of these three displacements, as may be verified by adding the displacement vectors of Fig. 4.26*a*, *b*, and *c*.

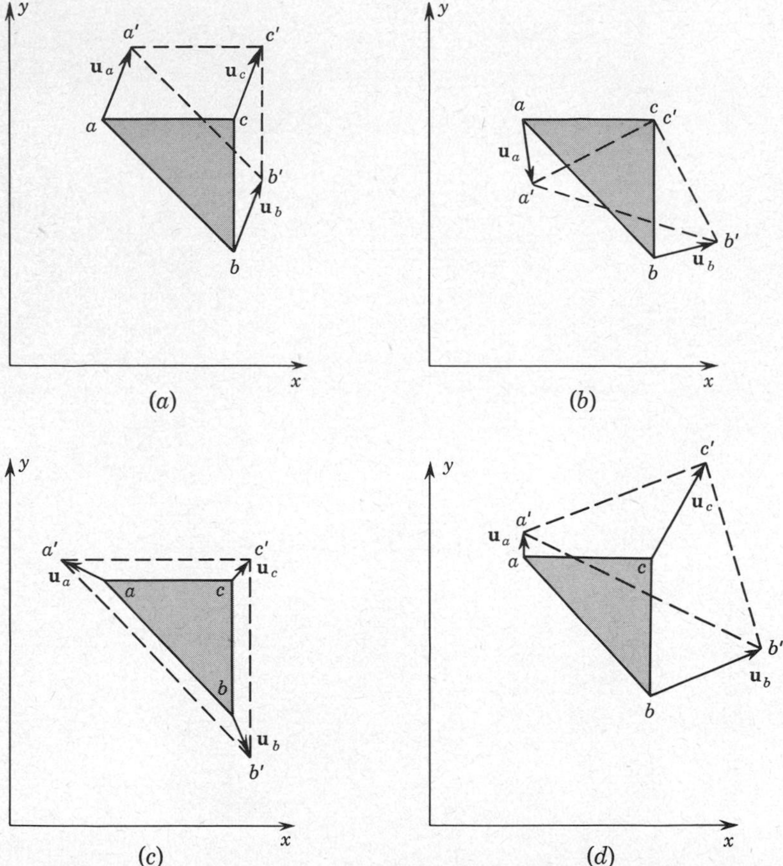

Fig. 4.26 Examples of rigid-body motion and of deformation in the *xy* plane. (*a*) Rigid-body translation. (*b*) Rigid-body rotation about *c*. (*c*) Deformation without rigid-body motion. (*d*) Sum of the displacements (*a*), (*b*), and (*c*).

The displacements associated with rigid-body motion can be either large or small, while the displacements associated with deformation usually are small. The description and analysis of rigid-body motion is important in dynamics where the forces required to produce different time rates of rigid-body motion are of interest. The description and analysis of deformation is important in our present study of the mechanics of deformable bodies where the forces required to produce different distortions are of interest. The remaining sections of this chapter will be devoted to a study of the *deformation* at a point in a continuous body.

4.9 DEFINITION OF STRAIN COMPONENTS

It will simplify our discussion if, instead of considering the general case of three-dimensional deformation, we focus our attention on a body whose particles all lie in the same plane and which deforms only in this plane. This type of deformation is called *plane strain*. We shall return to the problem of three-dimensional deformation after we have completed our study of deformation in a plane.

In Fig. 4.27 we show two examples in which a thin rubber block is deformed in its own plane. In Fig. 4.27*b* all elements in the block have been deformed the same amount; we call this a state of *uniform strain*. In Fig. 4.27*c* the elements on the right have been deformed more than those on the left; we call this a state of *nonuniform strain*.

In examining the uniform state of deformation in Fig. 4.27*b*, we note that originally straight lines are straight in the deformed state, but they may have changed their length or rotated. For example, the lines AE and CG do not rotate and line

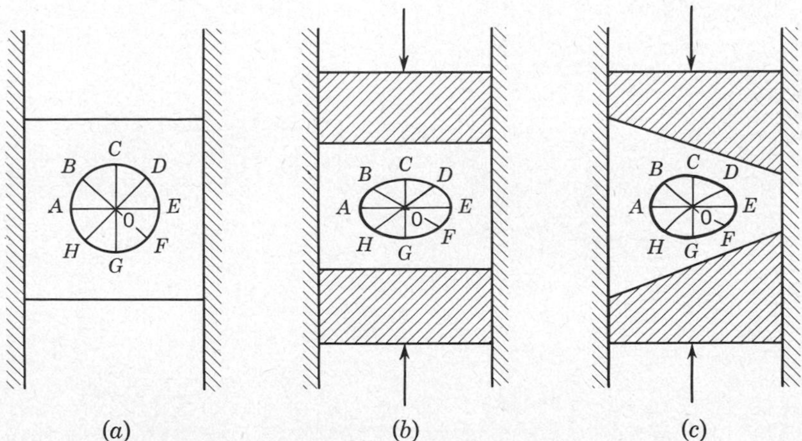

Fig. 4.27 (*a*) Undeformed block of rubber with superimposed diagram. (*b*) Rubber block of (*a*) deformed in uniform strain. (*c*) Rubber block of (*a*) deformed in nonuniform strain.

AE remains unchanged in length while CG shortens. By contrast, the lines BF and DH rotate equal and opposite amounts and both change in length by the same increment. Further examination of Fig. 4.27*b* does not reveal any other type of transformation of an originally straight line, and, in fact, it can be demonstrated rigorously that in uniform strain an originally straight line can only transform into another straight line.

Examining now the nonuniform state of deformation in Fig. 4.27*c*, we observe that originally straight lines are not necessarily straight in the deformed state. Evidently a line can remain straight, unrotated, and unchanged in length, as line AE; or it can remain straight, unrotated, but changed in length, as line CG; or it may become a continuous curve whose arc length may or may not be equal to the original length, as lines BF and DH. On a macroscopic scale, then, the deformation in nonuniform strain can be considerably more complicated than in the case of uniform strain. However, if we examine the deformation of a sufficiently small portion of the block in Fig. 4.27*c*, say a small area centered on O, we observe that over this small area the curved lines BF and DH can be replaced by their tangents, and that the deformation of the small area is then similar to that in Fig. 4.27*b*; that is, within this small area the deformation is approximately uniform. In the limit as the small area centered on O shrinks to zero this uniform deformation becomes the deformation *at* point O.

We now consider a thin, continuous body which lies entirely in the xy plane and which undergoes a *small* geometrically compatible deformation in the xy plane. If we study a small element of the body, we can, in the light of the above discussion, consider this element to be deformed in a state of uniform strain, as illustrated in Fig. 4.28. For the deformed element in Fig. 4.28, let us express the *deformation* in

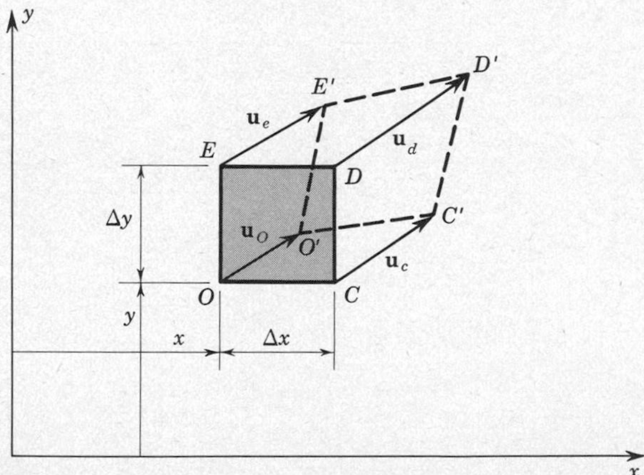

Fig. 4.28 Deformation in the xy plane of a small element of a continuous body.

the vicinity of point O quantitatively by giving the changes in length of the two lines OC and OE and the rotation of these lines relative to each other. We shall find it convenient to define dimensionless quantities which will describe these two aspects of the deformation. The first of these, which gives a measure of the elongation or contraction of a line, will be called the *normal strain* component. The second, which gives a measure of the relative rotation of two lines, will be called the *shear-strain* component.

The *normal strain* component is defined as the fractional change in the original length of a line and is designated by the symbol ϵ with a subscript to indicate the original direction of the line for which the strain is measured. Thus, from Fig. 4.28, the value of ϵ_x *at* point O is

$$\epsilon_x = \lim_{\Delta x \to 0} \frac{O'C' - OC}{OC}$$

Similarly,

$$\epsilon_y = \lim_{\Delta y \to 0} \frac{O'E' - OE}{OE}$$

From the above definition of the normal strain component it is evident that normal strain is positive when the line elongates and negative when the line contracts.

The *shear-strain* component is specified with respect to two axes which are perpendicular in the undeformed body and is designated by the symbol γ, with two subscripts to indicate these two axes. Shear strain may be defined as the tangent of the change in angle between these two originally perpendicular axes. When the axes rotate so that the first and third quadrants become smaller, the shear strain is *positive*; when the first and third quadrants get larger, the shear strain is *negative*. Using this definition, we see that for the deformation illustrated in Fig. 4.28 the shear-strain component γ_{xy} is positive. For small shear strains (those of engineering interest are mostly less than 0.01) it is adequate to define shear strain in terms of the change in angle itself (in radians) instead of in terms of the tangent of this angle change. Using this definition, we obtain from Fig. 4.28 the value of the shear strain *at* point O to be

$$\gamma_{xy} = \lim_{\substack{\Delta x \to 0 \\ \Delta y \to 0}} (\angle COE - \angle C'O'E') = \lim_{\substack{\Delta x \to 0 \\ \Delta y \to 0}} \left(\frac{\pi}{2} - \angle C'O'E'\right)$$

We have considered the deformation of the element of Fig. 4.28 in the limit as the element shrinks to zero size, and thus we can consider our strain components to define the deformation *at* the point O. In the following section we examine the connection between the strain components ϵ_x, ϵ_y, and γ_{xy} and the changes in displacement from position to position.

4.10 RELATION BETWEEN STRAIN AND DISPLACEMENT IN PLANE STRAIN

In describing the displacement of a point, it will be convenient to deal with rectangular components of the displacement vector. Thus, if we let u be the displacement in the x direction of point O in Fig. 4.28 and v the displacement in the y direction, we can express the displacement vector of point O as

$$\mathbf{u}_O = u\mathbf{i} + v\mathbf{j}$$

This relation is illustrated in Fig. 4.29 where the x and y components of the displacement of point O are indicated by u and v.

The displacement components u and v must be continuous functions of x and y to ensure that the displacement be geometrically compatible, that is, to ensure that no voids or holes are created by the displacement. Using the concept of partial derivatives, we can express the displacements of point E and C in Fig. 4.29 in terms of the displacements u and v of the point O and their partial derivatives, as illustrated in Fig. 4.29.

Using the definitions previously established for the strain components ϵ_x, ϵ_y, and γ_{xy}, we obtain the following from Fig. 4.29 under the assumption that the strains are small compared with unity.

$$\epsilon_x = \lim_{\Delta x \to 0} \frac{O'C' - OC}{OC} = \lim_{\Delta x \to 0} \frac{[\Delta x + (\partial u/\partial x)\,\Delta x] - \Delta x}{\Delta x} = \frac{\partial u}{\partial x}$$

$$\epsilon_y = \lim_{\Delta y \to 0} \frac{O'E' - OE}{OE} = \lim_{\Delta y \to 0} \frac{[\Delta y + (\partial v/\partial y)\,\Delta y] - \Delta y}{\Delta y} = \frac{\partial v}{\partial y} \tag{4.31}$$

$$\gamma_{xy} = \lim_{\substack{\Delta x \to 0 \\ \Delta y \to 0}} \left(\frac{\pi}{2} - \angle C'O'E'\right) = \lim_{\substack{\Delta x \to 0 \\ \Delta y \to 0}} \left\{\frac{\pi}{2} - \left[\frac{\pi}{2} - \frac{(\partial v/\partial x)\,\Delta x}{\Delta x} - \frac{(\partial u/\partial y)\,\Delta y}{\Delta y}\right]\right\}$$

$$= \frac{\partial v}{\partial x} + \frac{\partial u}{\partial y}$$

Not all of the relative displacement, however, involves deformation. Some of it arises from rigid-body rotation. For *small* displacement derivatives, the rotation about the z axis of the line OC, for example, is

$$(\omega_z)_{OC} = \frac{[v + (\partial v/\partial x)\,\Delta x] - v}{\Delta x} = \frac{\partial v}{\partial x}$$

Likewise, the rotation about the z axis of the line OE is

$$(\omega_z)_{OE} = \frac{-[u + (\partial u/\partial y)\,\Delta y] + u}{\Delta y} = -\frac{\partial u}{\partial y}$$

We can define an average (small) rotation of the element as a whole as the average of the rotations of the two perpendicular line segments:

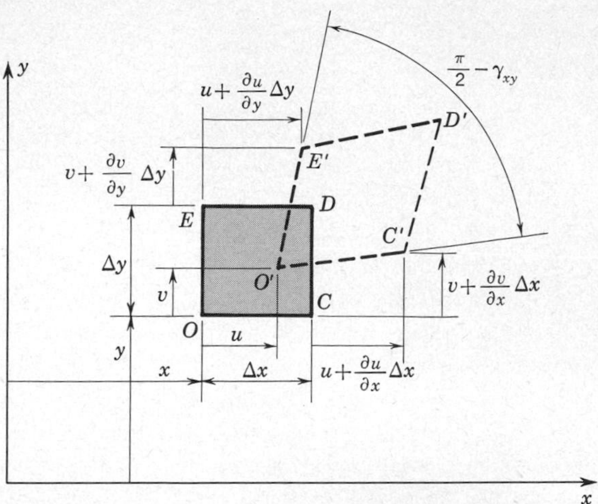

Fig. 4.29 Plane strain deformation expressed in terms of the components u and v and their partial derivatives.

$$\omega_z = \tfrac{1}{2}[(\omega_z)_{OC} + (\omega_z)_{OE}] = \frac{1}{2}\left(\frac{\partial v}{\partial x} - \frac{\partial u}{\partial y}\right) \tag{4.32}$$

The strain components in terms of the displacement derivatives can be summarized as follows:

$$\epsilon_x = \frac{\partial u}{\partial x} \qquad \epsilon_y = \frac{\partial v}{\partial y} \qquad \gamma_{xy} = \frac{\partial v}{\partial x} + \frac{\partial u}{\partial y} \tag{4.33}$$

Equations (4.33) indicate that the strain components depend *linearly* on the derivatives of the displacement components. It is important to emphasize that our derivation for the normal and shear strains is valid under the assumption of small displacement derivatives compared to unity. We think of (4.33) as the components of strain for *small strains* and *small rotations* (where both are of the same order of magnitude) compared to unity. If we have a situation in which large deformations are important, then a set of *nonlinear* strain-displacement relations are required.[1] The derivation of the nonlinear relations usually proceeds in a way different from that given above. The derivation above, however, has emphasized the geometrical meaning of the strain components. For the case of a rigid-body translation and (small) rigid-body rotation we can show as expected that the strain components vanish (Prob. 4.29).

We have considered the deformation of the element of Fig. 4.29 in the limit as the size of the element shrinks to zero. Thus, we consider the strain components

[1] See, for example, V. V. Novozhilov, "Theory of Elasticity," Pergamon Press, New York, 1961.

to define the deformation at the point O. We therefore speak of the state of plane strain at a given point in a two-dimensional body as given by the strain components

$$\begin{matrix} \epsilon_x & \gamma_{xy} \\ \gamma_{yx} & \epsilon_y \end{matrix} \qquad (4.34)$$

where we define $\gamma_{yx} = \gamma_{xy}$. The deformation at points adjacent to point O will in general differ from that at point O. The fact that two components of displacement serve to define three components of strain in (4.33) indicates that the three components of strain cannot vary arbitrarily in a field of nonuniform strain. The conditions that ensure that a single value will be found for the displacement at any point from a knowledge of the strains are called the compatibility conditions. These conditions will not be derived here.[1] Rather, for each problem we consider, we shall simply demonstrate the existence of a single-valued set of displacements from which the strain components are derived according to Eqs. (4.33).

If we now turn to the indicial notation introduced in Sec. 4.2 for stress, we may write the strain components (4.33) in the form

$$\epsilon_1 = \frac{\partial u_1}{\partial x_1} \qquad \epsilon_2 = \frac{\partial u_2}{\partial x_2} \qquad \gamma_{12} = \gamma_{21} = \frac{\partial u_2}{\partial x_1} + \frac{\partial u_1}{\partial x_2} \qquad (4.35)$$

To achieve an economy of notation parallel with that realized for stresses, we introduce an *indicial notation* e_{ij} for strains.

$$\begin{gathered} e_{ij} = \frac{1}{2}\left(\frac{\partial u_i}{\partial x_j} + \frac{\partial u_j}{\partial x_i}\right) \\ \epsilon_1 = e_{11} \qquad \epsilon_2 = e_{22} \qquad e_{12} = e_{21} = \tfrac{1}{2}\gamma_{12} \end{gathered} \qquad (4.36)$$

Note that the shear-strain component e_{12} in indicial notation is *half* the shear-strain component γ_{12} defined as the change in angle between the x_1 and x_2 axes.

For a general case of three-dimensional displacement we can describe the deformation at a point by specifying the three normal strain components and three shear-strain components associated with a set of three mutually perpendicular axes (see Prob. 4.16). In any given situation we choose the coordinate system which is most convenient for describing the deformation (see Probs. 4.17, 4.18, and 4.19).

4.11 STRAIN COMPONENTS ASSOCIATED WITH ARBITRARY SETS OF AXES

In the previous section we determined how the strain components referred to a set of perpendicular axes through a point are related to components of displacement parallel to those axes. In this section we examine further the problem of geometric

[1] See, for example, S. Timoshenko and J. N. Goodier, "Theory of Elasticity," 3d ed., p. 237, McGraw-Hill Book Company, New York, 1970.

compatibility at a point and determine relationships which must exist between the strain components associated with axes which are not perpendicular to each other.

We begin by considering the situation illustrated in Fig. 4.30, in which the deformation of a small element is shown in terms of the displacement components u' and v' parallel to the x' and y' set of axes. Analogous to Eqs. (4.31) which were developed for the element of Fig. 4.29, the strain components $\epsilon_{x'}$, $\epsilon_{y'}$, and $\gamma_{x'y'}$ can be obtained as follows in terms of the partial derivatives of u' and v' with respect to x' and y'.

$$\epsilon_{x'} = \frac{\partial u'}{\partial x'}$$

$$\epsilon_{y'} = \frac{\partial v'}{\partial y'} \tag{4.37}$$

$$\gamma_{x'y'} = \frac{\partial v'}{\partial x'} + \frac{\partial u'}{\partial y'}$$

Our next step is to inquire as to the nature of the connection between the strains $\epsilon_{x'}$, $\epsilon_{y'}$, and $\gamma_{x'y'}$ of Eqs. (4.37) and the strains ϵ_x, ϵ_y, and γ_{xy} of Eqs. (4.33). We observe that, since geometric compatibility requires u' and v' in Fig. 4.30 to be continuous functions of position in the plane, we can express these displacement components either as functions of x' and y' or as functions of x and y. If we

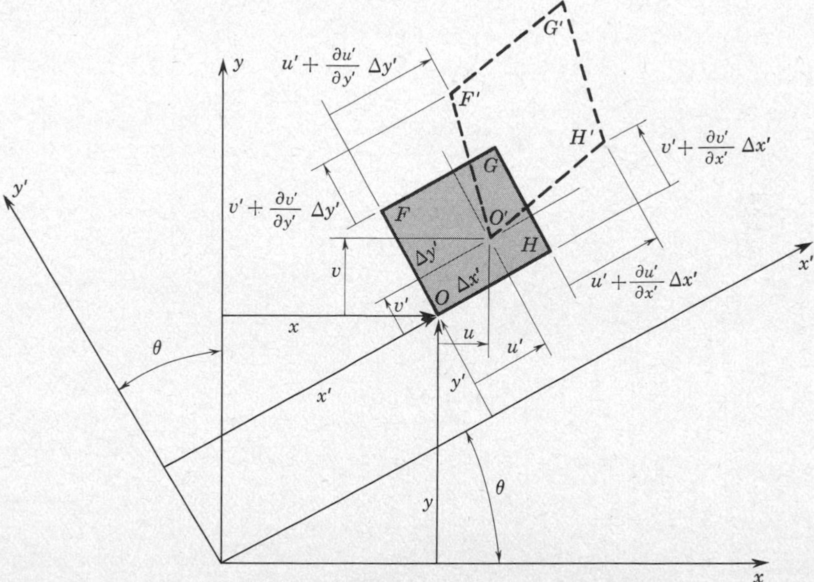

Fig. 4.30 Plane strain. Deformation of a small element with sides originally parallel to the x' and y' set of axes.

express u' and v' as functions of x and y, and if we recall the chain rule for partial derivatives, we can write the strain components (4.37) as follows:

$$
\begin{aligned}
\epsilon_{x'} &= \frac{\partial u'}{\partial x'} = \frac{\partial u'}{\partial x}\frac{\partial x}{\partial x'} + \frac{\partial u'}{\partial y}\frac{\partial y}{\partial x'} \\
\epsilon_{y'} &= \frac{\partial v'}{\partial y'} = \frac{\partial v'}{\partial x}\frac{\partial x}{\partial y'} + \frac{\partial v'}{\partial y}\frac{\partial y}{\partial y'} \\
\gamma_{x'y'} &= \frac{\partial v'}{\partial x'} + \frac{\partial u'}{\partial y'} = \left(\frac{\partial v'}{\partial x}\frac{\partial x}{\partial x'} + \frac{\partial v'}{\partial y}\frac{\partial y}{\partial x'}\right) + \left(\frac{\partial u'}{\partial x}\frac{\partial x}{\partial y'} + \frac{\partial u'}{\partial y}\frac{\partial y}{\partial y'}\right)
\end{aligned}
\tag{4.38}
$$

From the geometry of Fig. 4.30 or from Prob. 1.2 we can obtain the following relations:

$$
\begin{aligned}
x &= x' \cos\theta - y' \sin\theta \\
y &= x' \sin\theta + y' \cos\theta \\
u' &= u \cos\theta + v \sin\theta \\
v' &= -u \sin\theta + v \cos\theta
\end{aligned}
\tag{4.39}
$$

Substituting (4.39) into the first of (4.38), we have

$$
\begin{aligned}
\epsilon_{x'} &= \left(\frac{\partial u}{\partial x}\cos\theta + \frac{\partial v}{\partial x}\sin\theta\right)\cos\theta + \left(\frac{\partial u}{\partial y}\cos\theta + \frac{\partial v}{\partial y}\sin\theta\right)\sin\theta \\
&= \frac{\partial u}{\partial x}\cos^2\theta + \frac{\partial v}{\partial y}\sin^2\theta + \left(\frac{\partial v}{\partial x} + \frac{\partial u}{\partial y}\right)\sin\theta\cos\theta
\end{aligned}
$$

Finally, substituting (4.33), we find

$$
\epsilon_{x'} = \epsilon_x \cos^2\theta + \epsilon_y \sin^2\theta + \gamma_{xy}\sin\theta\cos\theta \tag{4.40}
$$

Proceeding in a similar manner with the other two equations of (4.38), and using the trigonometric relations for double angles, we get

$$
\begin{aligned}
\epsilon_{x'} &= \frac{\epsilon_x + \epsilon_y}{2} + \frac{\epsilon_x - \epsilon_y}{2}\cos 2\theta + \frac{\gamma_{xy}}{2}\sin 2\theta \\
\epsilon_{y'} &= \frac{\epsilon_x + \epsilon_y}{2} - \frac{\epsilon_x - \epsilon_y}{2}\cos 2\theta - \frac{\gamma_{xy}}{2}\sin 2\theta \\
\frac{\gamma_{x'y'}}{2} &= -\frac{\epsilon_x - \epsilon_y}{2}\sin 2\theta + \frac{\gamma_{xy}}{2}\cos 2\theta
\end{aligned}
\tag{4.41}
$$

Equations (4.41) define the conditions which geometric compatibility imposes on the strain components associated with different sets of perpendicular axes in plane strain: if the strain components ϵ_x, ϵ_y, and γ_{xy} and the angle θ are specified, the strain components $\epsilon_{x'}$, $\epsilon_{y'}$, and $\gamma_{x'y'}$ are completely determined.

4.12 MOHR'S CIRCLE REPRESENTATION OF PLANE STRAIN

When we compare Eqs. (4.41) with Eqs. (4.25), we see that if in (4.41) we make the substitution ϵ for σ and $\gamma/2$ for τ, we obtain (4.25). Thus it must be possible to represent (4.41) by a Mohr's circle.

In the Mohr's circle for strain (see Fig. 4.31) the normal strain components ϵ are plotted positive to the right, and *half* the shear-strain component, that is, $\gamma/2$, is plotted vertically. If the shear strain is *positive*, the point representing the x axis is plotted a distance $\gamma/2$ below the ϵ axis *and* the point representing the y axis a distance $\gamma/2$ above the ϵ axis. If γ is *negative*, the x-axis point is plotted a distance $\gamma/2$ above the ϵ-axis *and* the y-axis point a distance $\gamma/2$ below the ϵ axis. As in the case of the stress circle, relative angular positions are the same in the real body and in the Mohr's circle, but the angles are *doubled* in the Mohr's circle. Figure 4.31 illustrates a situation where both γ_{xy} and $\gamma_{x'y'}$ are positive.

We note that the sets of axes I, II and DE are of special interest. The I, II set is the only set for which the shear-strain component is zero. Also, the normal

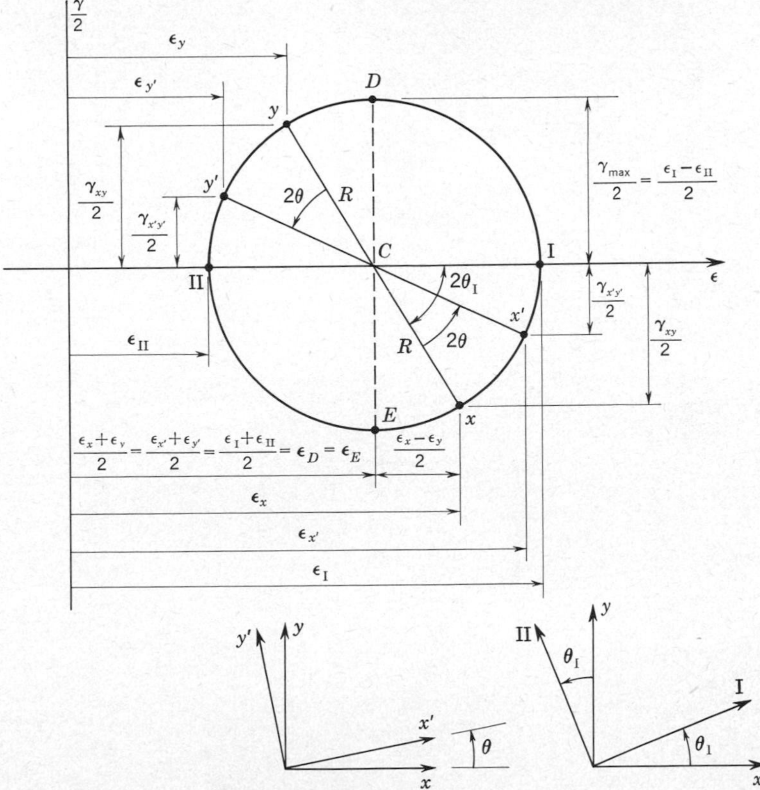

Fig. 4.31 Mohr's circle for plain strain.

strain component has its maximum and minimum values on these axes. These axes are referred to as *principal axes of strain*, and the normal strain components ϵ_{I} and ϵ_{II} for these axes are called *principal strains*. The *DE* set of axes is inclined at 45° to the principal axes of strain and is the set for which the shear-strain component is a maximum. For this reason these axes are referred to as the *axes of maximum shear strain*. Note that the normal strain components for these axes are equal.

To illustrate the use of Mohr's circle for strain, we consider the following numerical example.

Example 4.3 A sheet of metal is deformed uniformly in its own plane so that the strain components related to a set of axes xy are[1]

$$\epsilon_x = -200 \times 10^{-6}$$

$$\epsilon_y = 1000 \times 10^{-6}$$

$$\gamma_{xy} = 900 \times 10^{-6}$$

We wish to find the strain components associated with a set of axes $x'y'$ inclined at an angle of 30° clockwise to the xy set, as shown in Fig. 4.32. Also, we wish to find the principal strains and the direction of the axes on which they exist.

Figure 4.33 shows the Mohr's circle laid out on the basis of the given strains ϵ_x, ϵ_y, and γ_{xy}. Point x' lies at a relative angular position twice that existing in the actual body, i.e., at a position 60° clockwise from x on the Mohr's circle. In a manner similar to the calculations for the stress circle of Examples 4.1 and 4.2, we find

$$2\Phi_{\mathrm{I}} = \tan^{-1} 450/600 = 36.8°$$

[1] Strains are *dimensionless* quantities, although they frequently are referred to in units such as *inches per inch* or *microinches per inch*.

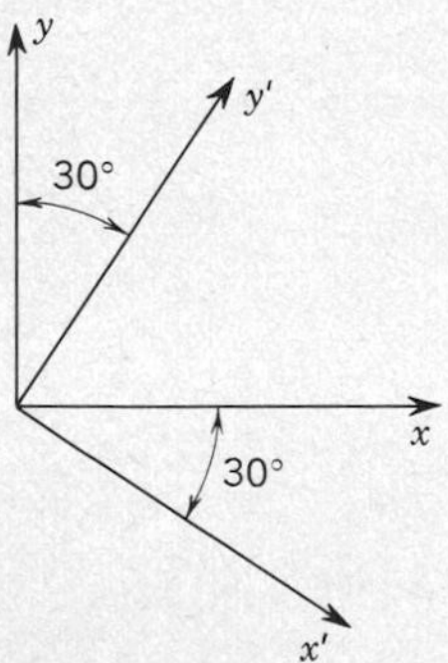

Fig. 4.32 Example 4.3. Location of x', y' axes.

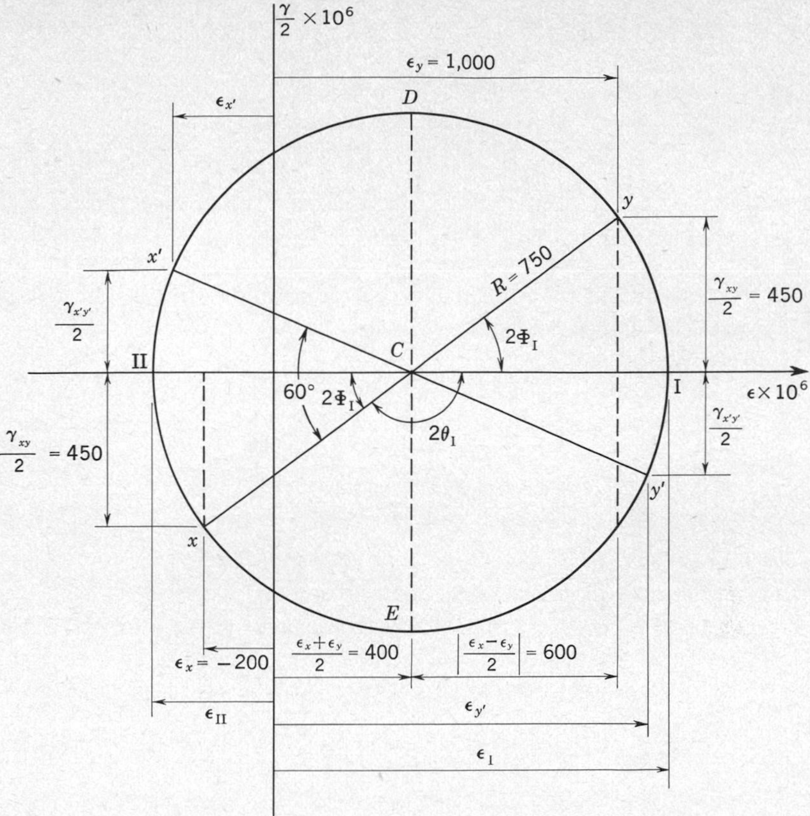

Fig. 4.33 Example 4.3. Mohr's circle.

$$R = \sqrt{600^2 + 450^2} = 750$$

$$\epsilon_{x'} = (400 \times 10^{-6}) - (750 \times 10^{-6}) \cos (60° - 36.8°) = -290 \times 10^{-6}$$

$$\epsilon_{y'} = (400 \times 10^{-6}) + (750 \times 10^{-6}) \cos (60° - 36.8°) = 1{,}090 \times 10^{-6}$$

Because point x' lies above the ϵ axis (and point y' below), the shear strain $\gamma_{x'y'}$ is negative.

$$\frac{\gamma_{x'y'}}{2} = -(750 \times 10^{-6}) \sin (60° - 36.8°) = -295 \times 10^{-6}$$

$$\gamma_{x'y'} = -590 \times 10^{-6}$$

The principal strains are

$$\epsilon_{I} = (400 \times 10^{-6}) + (750 \times 10^{-6}) = 1{,}150 \times 10^{-6}$$

$$\epsilon_{II} = (400 \times 10^{-6}) - (750 \times 10^{-6}) = -350 \times 10^{-6}$$

The directions of the principal axes of strain are shown in Fig. 4.34.

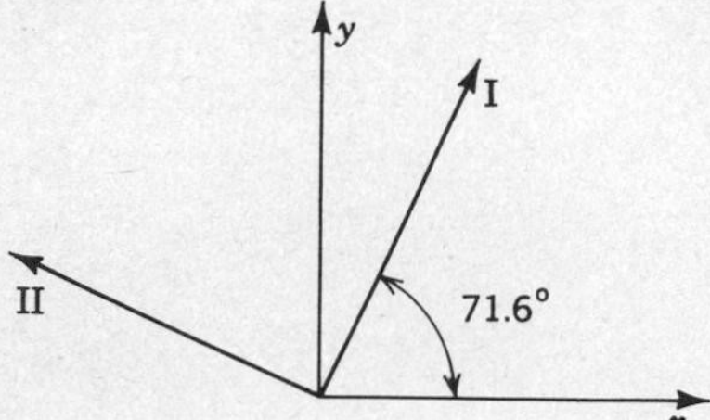

Fig. 4.34 Example 4.3. Orientation of principal axes of strain.

To invert the statement of Example 4.3, we can say that if we take a sheet of metal and extend it in direction I with a strain of $1{,}150 \times 10^{-6}$ and compress it in a perpendicular direction II with a strain of 350×10^{-6}, then on a set of axes xy oriented as shown in Fig. 4.34 the strain components will be as given in the original statement of Example 4.3.

4.13 MOHR'S CIRCLE REPRESENTATION OF A GENERAL STATE OF STRAIN

Thus far we have considered only deformation in a single plane. Suppose we now consider a situation in which the strains associated with the z direction are not zero, i.e., a situation in which the state of strain may be said to be perfectly general. Further, let us suppose that we are again interested in the strains associated with sets of axes originally lying in the xy plane. We illustrate this situation in Fig. 4.35

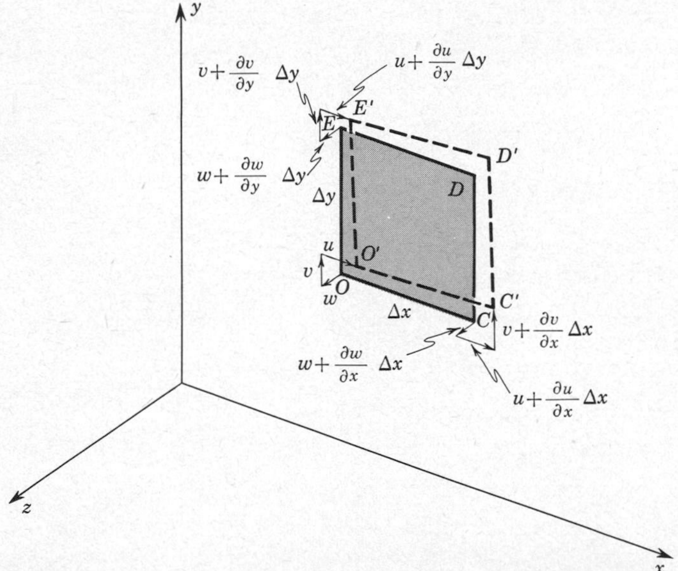

Fig. 4.35 General displacement of a small element originally parallel to the xy plane.

where the undeformed element $OCDE$, which is parallel to the xy plane, is given a general displacement in which the z displacement is designated by w. Figure 4.35 is to be compared with Fig. 4.29 in which the element $OCDE$ displaces only in the xy plane ($w = 0$).

In our consideration of plane strain (see Fig. 4.29), the effect of $(\partial v/\partial x)\,\Delta x$ on the length $O'C'$ was neglected when the strains were small compared to unity. We see in Fig. 4.35 that the displacements $(\partial v/\partial x)\,\Delta x$ and $(\partial w/\partial x)\,\Delta x$ play the same type of role in their effect on the length $O'C'$, and thus we conclude that for small strains the out-of-plane displacement w has no effect on ϵ_x or on the normal strain associated with any axis originally lying in the xy plane. To the same degree of approximation the displacements $(\partial u/\partial x)\,\Delta x$, $(\partial w/\partial x)\,\Delta x$, $(\partial v/\partial y)\,\Delta y$, and $(\partial w/\partial y)\,\Delta y$ have negligible effect on the angle $C'O'E'$, and we conclude that for small strains the shear strain associated with any set of axes originally lying in the xy plane is not influenced by out-of-plane deformations.

As a consequence of these characteristics of the general deformation illustrated in Fig. 4.35, we may state the following for *small strains*:

1. The results given by Eqs. (4.41) and the Mohr's circle representation of these are correct whether or not ϵ_z, γ_{yz}, or γ_{zx} is zero.
2. If w exists and also varies with respect to any coordinate direction in the xy plane, then with any arbitrarily oriented axis x' in the xy plane there will, in general, be associated a shear-strain component $\gamma_{zx'}$ in addition to $\gamma_{x'y'}$. In such a case the I and II axes of Fig. 4.31 should not be called principal axes, since we wish to retain the designation *principal axis of strain* for an axis with which there is associated *no* shear strain.

To complete the picture for a general three-dimensional state of strain, we state here without proof that at each point in a deformed body there are three mutually perpendicular axes which remain perpendicular after deformation. These axes are called the *principal axes of strain*, and they determine three principal planes of strain with three associated Mohr's circles illustrated by the circles I–II, II–III, and III–I of Fig. 4.36*b*. Analogous to the case of stress, the normal and the resultant shear-strain components associated with any arbitrary axis will be defined by the coordinates of some determinable point[1] within the shaded area in Fig. 4.36*b* (where we have assumed a case in which $\epsilon_{\text{II}} < \epsilon_{\text{III}} < 0 < \epsilon_{\text{I}}$). By the resultant shear-strain component we mean the maximum value associated with each axis. It can be shown that for an axis x' with which are associated shear strains $\gamma_{x'y'}$ and $\gamma_{zx'}$, the maximum shear strain between x' and any perpendicular axis is of magnitude $\sqrt{(\gamma_{x'y'})^2 + (\gamma_{zx'})^2}$. Since in most practical cases we shall be dealing

[1] A. Nádai, "Theory of Flow and Fracture of Solids," 2d ed., p. 115, McGraw-Hill Book Company, New York, 1950.

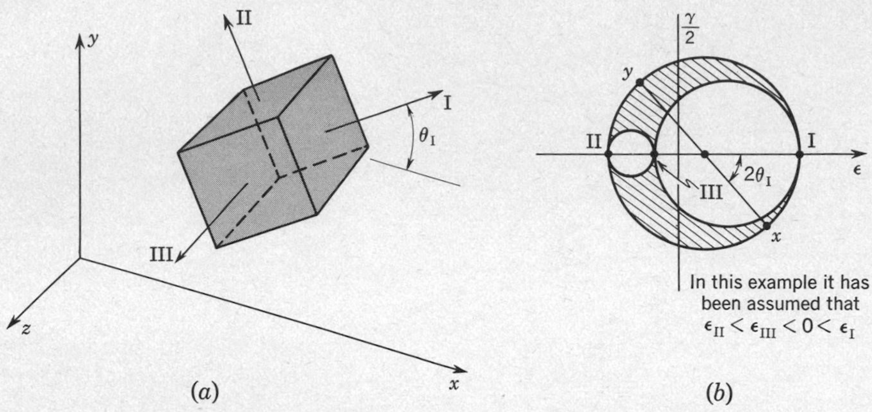

Fig. 4.36 Three-dimensional state of strain.

with strain components in a plane which contains two of the principal axes of strain, we shall not examine the general state of strain beyond the level discussed here.

4.14 MEASUREMENT OF STRAINS

In order to determine whether a theoretical model really predicts the stress distribution in a machine part or a structure with the desired degree of accuracy, it is common practice to conduct an experimental investigation to establish the actual stress condition as felt by the object when subjected to load. In many cases, however, the situation is so complicated that we are unable to develop a workable theoretical model for calculating the stresses. Instead, stresses can only be determined from measurements on the structure, or a part thereof, under actual service conditions.

One must appreciate that, with the exception of some cases involving contact stresses on the outside surface of a body, it is impossible to measure stress directly. What is usually done in practice is to measure strain at various locations, and then to compute the corresponding stress values from the quantitative relations between stress and strain which we will find in the next chapter.

The three most widely used methods for measuring strain are

1. Photoelasticity
2. Brittle coatings
3. Strain gages (of the bonded electrical resistance type)

Photoelasticity began as a special type of model testing in which models were fabricated from flat sheets of suitable (birefringent) transparent materials, subjected to loading in their own plane (plane stress), and examined in polarized light with the path of the light perpendicular to the plane of the model. The characteristic of the materials from which the models are made is such that the light is transmitted through the material with velocities which depend upon the magnitudes of both principal strains.

When the models are examined in a polariscope (or between two crossed pieces of Polaroid), the effect of the strain is to produce a pattern which can be interpreted both qualitatively and quantitatively. When white light is used, the pattern will contain a series of colored bands and, if monochromatic (one-color) light is employed, it will consist of sharply defined black lines on uniform background, as shown in Fig. 4.37 which shows a ring in diametral compression. The lines in the diagram are loci of constant difference between the principal stresses, or constant maximum shear stress.

Although the photoelastic method is basically suited to plane stress problems, methods have been found for extending it to three-dimensional applications which have been carried out with great success on models of pressure vessels with complicated shapes, such as those employed in nuclear reactors.

More recently, very successful techniques have been developed for applying photoelastic coatings to flat or curved surfaces. This has been an enormous step

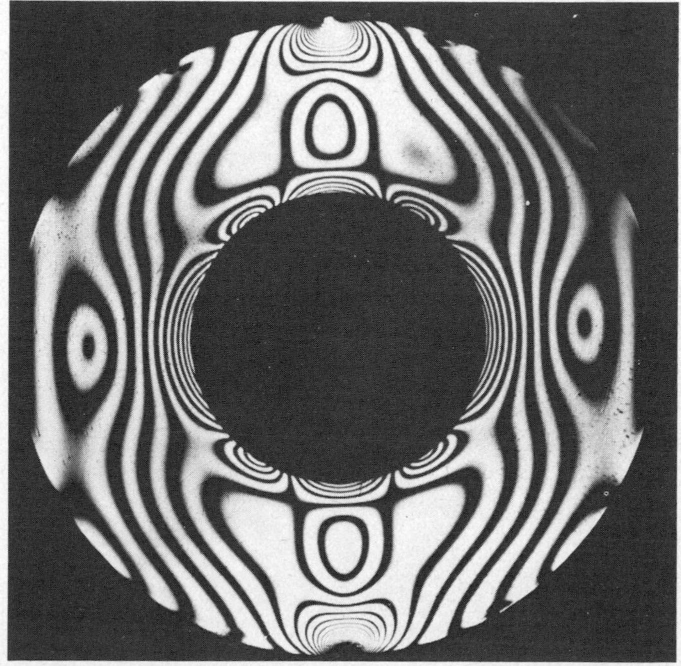

Fig. 4.37 Ring of photoelastic material in diametral compression. (*Courtesy Prof. W. M. Murray, MIT*)

Fig. 4.38 Test setup with photoelastic coating. Aluminum-filled epoxy model of nose landing gear of Boeing 747 Superjet. (*Courtesy S. S. Redner, Photolastic, Inc.*)

forward, as it has enabled the observation of the entire strain distribution over the coated surface of the structure, or a part thereof. This development has extended the use of photoelasticity from a laboratory device to a very practical method of investigating stress conditions under actual service or operating conditions. Figure 4.38 shows a study being conducted on the landing gear for a large airplane.

The brittle coating technique involves covering the test specimen with a thin layer (usually less than 10 one-thousandths of an inch thick) of material which sets or hardens in a brittle condition. Two types of coating material are in general use.

One is a lacquer which sets and is used at room temperature (65 to 80°F), and the other is a ceramic which has to be fired onto the specimen but which can be used at temperatures above those which are suitable for the lacquer.

When the test specimen is loaded, strains are developed in the surface and in the coating. When these strains are large enough, the coating will crack in the direction perpendicular to the algebraically larger principal strain. A typical example is shown in Fig. 4.39 which illustrates the crack pattern in the coating on an engine crankshaft. In conducting tests, it is customary practice to apply loads by increments and to observe the initiation of new cracks in the coating following each increment of load. In this manner a pattern can be developed to present an indication of relative magnitudes of strain. By calibration of the coating, it is possible to evaluate the strain magnitudes. Although this method is not as precise

Fig. 4.39 Stresscoat pattern on crankshaft in bending. (*By W. T. Bean. Courtesy Experimental Mechanics, Oct.* 1966, *and Magnaflux Corp.*)

as photoelasticity and strain gages, it is relatively inexpensive and it gives an overall picture of what is going on. For a preliminary test prior to a detailed strain-gage study, the use of brittle coatings can effect considerable economy by indicating where, and in what orientations, strain gages should be mounted to detect the largest tensile strains.

Bonded electric resistance strain gages are probably the most extensively used instruments for the measurement of normal strains on the surfaces of bodies. They may be grouped in three different categories: wire, foil, or semiconductor, according to the nature of the sensing element. In most cases the sensing elements are mounted on some form of very thin carrier which permits easy installation at the location at which the strain is to be measured.

For the wire gages, the sensing element consists of a grid of fine wire (about 0.001 in. in diameter, or slightly less); foil gages have a comparable grid which is produced either by photoetching or die-cutting methods from metal foil approximately 0.0002 in. thick. In the United States, silicon is the favored material for semiconductor gages whose sensing elements are made from small pieces cut from a single crystal into which a controlled amount of additive has been introduced to obtain certain specified characteristics. In some cases involving transducers, the additive is diffused into a small beam or diaphragm of silicon so that the sensing element and its carrier are all in one piece.

All three varieties of the strain gage operate upon the basic principle that certain metals, and silicon, exhibit a change in electrical resistance with change in mechanical strain. Within certain limiting conditions, the metals chosen for sensing elements exhibit essentially a linear relationship between change in strain and unit change in electric resistance. For silicon, the corresponding relationship is more complicated and basically nonlinear. However, for small strain excursions the relation can be taken as approximately linear for practical purposes.

The gages are bonded to the surface on which strain is to be measured and are usually connected to a Wheatstone-bridge circuit for determining the unit change in resistance. Both static and dynamic indications can be obtained (in the latter case, to very high frequencies). Present-day instrumentation allows for signal processing so that the readout instruments indicate directly the strain or some other quantity such as force, torque, moment, etc.

In order to provide a complete specification of the state of strain at any location on a free (unloaded) surface, it is necessary to know two perpendicular components of normal strain and the corresponding shear strain, or the equivalent in terms of three normal strains. Since the bonded resistance strain gages only measure normal strains, it is customary (unless some auxiliary information is available) to make three observations of normal strain in three independent directions as indicated in Fig. 4.40*a*. Such an arrangement of axes is called a *rosette* and the calculation of the principal strains and the direction of the principal axes from the three observed strains is known as *rosette analysis*. A general rosette is shown in Fig. 4.40*a* and two widely used special rosettes in Fig. 4.40*b* and *c*.

There are many techniques for deducing the state of strain from the strain-

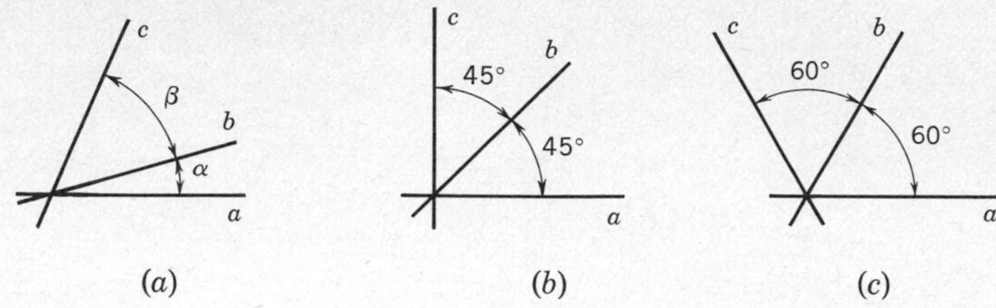

Fig. 4.40 Strain-gage rosettes. (*a*) General rosette; (*b*) 45° rosette; (*c*) 60° rosette.

rosette readings.[1] A convenient way to describe the state of strain is to obtain its Mohr's circle representation. To indicate the general nature of the problem, we shall illustrate how the Mohr's circle can be constructed from the three normal strain readings of the 45° rosette of Fig. 4.40*b*.

Example 4.4 The strain readings ϵ_a, ϵ_b, and ϵ_c have been obtained from the strain gages in a 45° rosette, and we wish to construct the Mohr's circle.

We begin by laying off three vertical lines at distances, ϵ_a, ϵ_b, and ϵ_c from the $\gamma/2$ axis. This is shown in Fig. 4.41*a* where it has been assumed that $\epsilon_a < \epsilon_c < \epsilon_b$. We next locate the point D which is midway between the two vertical lines representing the strains for the *perpendicular* axes a and c. We now, in Fig. 4.41*b*, lay off $Aa = DB$ above the ϵ axis and $Cc = DB$ below the ϵ axis. Then, on D as center we draw a circle through a and c, the circle cutting the vertical through B in the point b. The right triangles DAa, DCc, and bBD are similar since two sides of each triangle are identical in length with the corresponding two sides in the other triangles. As a consequence, in the Mohr's circle, b is situated 90° from a and c, that is, double the 45° angles in the rosette. When we examine the *relative* positions of the points a, b, and c in the Mohr's circle of Fig. 4.41*b*, however, we find that the positions are the reverse of those in the rosette, and thus this circle cannot represent the state of strain measured by the rosette. We proceed again, in Fig. 4.41*c*, and this time we lay off $Aa = BD$ *below* the ϵ axis and $Cc = DB$ *above* the ϵ axis and construct a circle on D as center; this is correct since a, b, and c have the same relative orientation in this circle as in the rosette.

When the Mohr's circle has been established, the strain components for any particular set of axes can be obtained from the geometry of the circle in the manner illustrated in Example 4.3.

[1] See, for example, M. Hetenyi, "Handbook of Experimental Stress Analysis," p. 390, John Wiley & Sons, Inc., New York, 1950; G. Murphy, *Trans. ASME*, vol. 67, p. A209, 1945; and F. A. McClintock, *Proc. Soc. Exp. Stress Anal.*, vol. 9, p. 209, 1951; J. W. Dally and W. F. Riley, "Experimental Stress Analysis," chap. 16, McGraw-Hill Book Company, New York, 1965; C. C. Perry and H. R. Lissner, "The Strain Gage Primer," 2d ed., chap. 7, McGraw-Hill Book Company, New York, 1962.

A relatively recent technique for the measurement of surface strains beyond those mentioned above is holographic interferometry.[1] With this technique, a special photograph known as a hologram is taken of the object while it is in the unstrained state. After the loads are applied and the object is deformed, a second hologram is taken. When the two holograms are observed together, interference fringes are formed and these may be interpreted to yield the strain at various points on the surface of the object. The technique requires a laser beam as a light source, the use of high-resolution film and vibration-free supports to produce satisfactory holograms. However, the technique is believed to be a highly promising one because it is extremely general and can handle objects as complex as, and with such irregular surfaces as, printed circuit boards and growing plants.

4.15 INDICIAL NOTATION

As mentioned in Sec. 4.2, it is frequently convenient, especially in general proofs for three dimensions, to describe components of stress in indicial notation. The convenience also applies as we saw to relations involving strains and displacements (4.36), and also will apply as we shall see in the next chapter to the stress-strain relations. In this section we shall briefly review stress and strain and the relations involving each in terms of indicial notation. The notation is very concise and can be advantageous in permitting one to keep general principles in mind without getting lost in the details. On the other hand, because so much meaning is implied by every symbol, it requires considerable experience before one fully appreciates the complete physical significance of deceptively simple manipulations in indicial notation. We do not emphasize the use of indicial notation in this book. The discussion in this section is provided as a brief introduction to the notation to be found in more advanced treatments of the mechanics of solids and fluids.

We will not rederive all the important equations of this chapter in indicial notation. Instead we will show how the indicial form of the equations follow from our previous derivations. The results will be motivated for two dimensions. Once the equations are obtained in two dimensions, we can by analogy obtain the corresponding results in three dimensions. For example, in plane stress the state of stress at a point is given by (4.6) which we may think of as:

$$\sigma_{ij} \qquad i, j = 1, 2 \tag{4.42}$$

However, (4.42) applies as well when $i, j = 1, 2, 3$ to the state of stress in three dimensions given by (4.4). We will normally omit the range of values assumed by the subscripts in subsequent equations.

The equilibrium equations for stress upon use of the *summation convention* were written in the form

$$\frac{\partial \sigma_{ij}}{\partial x_i} = 0 \tag{4.15}$$

[1] See, for example, Holographic Instrumentation Applications, NASA SP-248, National Aeronautics and Space Administration, Washington, D.C., 1970.

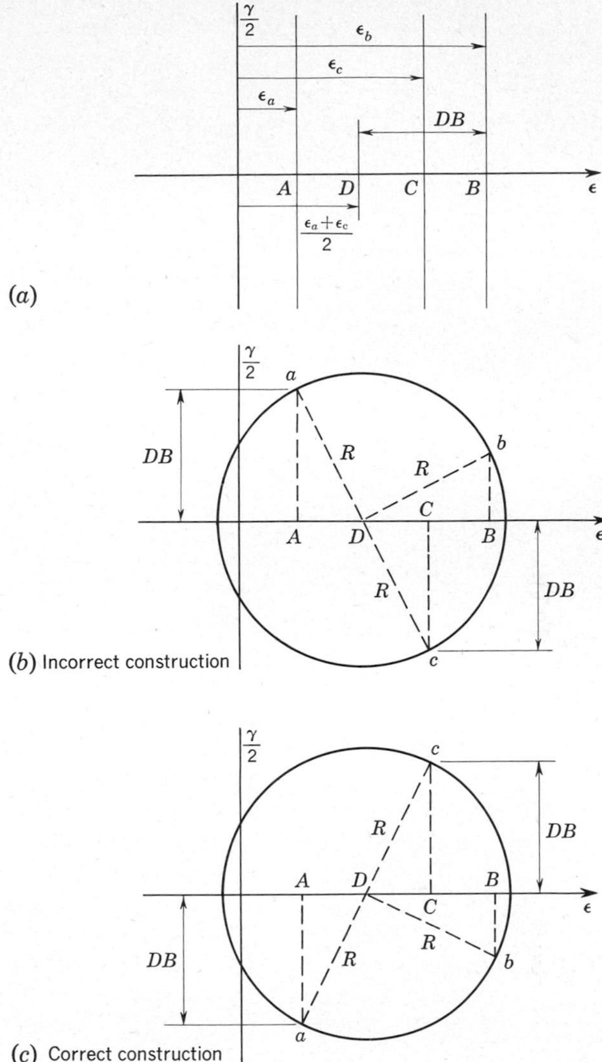

Fig. 4.41 Example 4.3. Construction of Mohr's circle from readings of 45° strain rosette (for case where $0 < \epsilon_a < \epsilon_c < \epsilon_b$).

A still further simplification in writing the equation is obtained by indicating partial differentiation of any function f by a comma preceding the corresponding subscript:

$$\frac{\partial f}{\partial x_i} = f_{,i}$$

Thus the three-dimensional equilibrium equation (4.15) may be written

$$\sigma_{ij,i} = 0 \tag{4.43}$$

If you are in doubt about the meaning of such an abbreviated notation, write out the equations in detail for each numerical value of each subscript.

The expression for a vector in terms of the indicial notation takes on a very compact form. For example, the stress vector, (4.2), can be written in the form

$$\overset{(\mathbf{n})}{\mathbf{T}} = \overset{(\mathbf{n})}{T_i}\mathbf{e}_i \tag{4.44}$$

where $\mathbf{e}_i$ are the unit vectors in the x_1, x_2, and x_3 directions.

Let us now consider the expressions for the components of the stress vector acting on a plane whose normal is $\mathbf{n}$ in terms of the stress components on the coordinate faces, (4.19). We have for the unit normal vector $\mathbf{n}$, (4.20),

$$\mathbf{n} = \cos\theta\mathbf{i} + \sin\theta\mathbf{j} = n_i\mathbf{e}_i$$

where the n_i are the *direction cosines.* Equations (4.19) now take the form

$$\overset{(\mathbf{n})}{T_1} = \sigma_{11}n_1 + \sigma_{21}n_2$$

$$\overset{(\mathbf{n})}{T_2} = \sigma_{12}n_1 + \sigma_{22}n_2$$

or

$$\overset{(\mathbf{n})}{T_j} = \sigma_{ij}n_i \tag{4.45}$$

The compactness of (4.45) should be compared with (4.19).

We now wish to express the *transformation law* for *stress* and *strain* in indicial notation. In Fig. 4.42 we define the angle between the *new* x_i' axes and the *old* x_j axes by θ_{ij}. The direction cosines of the new axes with respect to the old axes are

$$l_{ij} = \cos\theta_{ij} \tag{4.46}$$

For example, from Fig. 4.42

$$l_{11} = \cos\theta \qquad l_{12} = \cos\theta_{12} = \cos\left(\frac{\pi}{2} - \theta\right) = \sin\theta$$

$$l_{22} = \cos\theta \qquad l_{21} = \cos\theta_{21} = \cos\left(\frac{\pi}{2} + \theta\right) = -\sin\theta$$

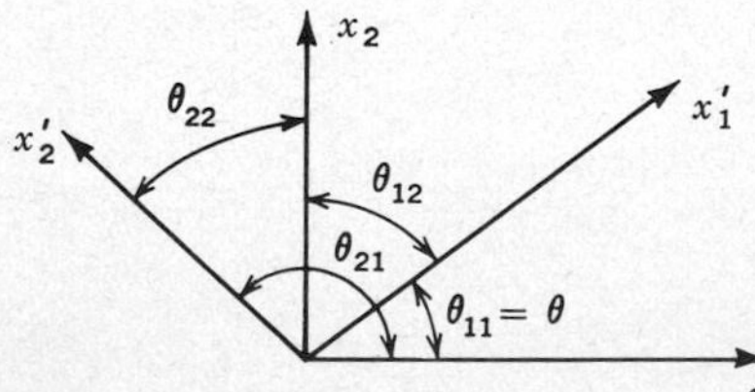

Fig. 4.42 Rotation of axes; angles.

The advantage of the use of the direction cosines is that we can write transformation relations upon rotation of axes in a simple manner.

From Prob. 1.2 we found the transformation for the components of a vector in the form

$$F'_x = F_x \cos\theta + F_y \sin\theta$$
$$F'_y = -F_x \sin\theta + F_y \cos\theta$$

where the components of the vector in the rotated x'_i axes are indicated by primes. In indicial notation these become upon use of (4.46) and the summation convention:

$$F'_1 = F_1 l_{11} + F_2 l_{12} = F_j l_{1j}$$
$$F'_2 = F_1 l_{21} + F_2 l_{22} = F_j l_{2j}$$

or in a compact form for both equations:

$$F'_i = l_{ij} F_j \tag{4.47}$$

If we now turn to the stress transformation formula of (4.23)

$$\sigma'_x = \sigma_x \cos^2\theta + \sigma_y \sin^2\theta + 2\tau_{xy} \sin\theta \cos\theta$$

we see that in view of the definition of direction cosines this may be written in the form

$$\sigma'_{11} = \sigma_{11} l_{11} l_{11} + \sigma_{22} l_{12} l_{12} + \sigma_{12} l_{11} l_{12} + \sigma_{21} l_{12} l_{11}$$

or

$$\sigma'_{11} = l_{1i} l_{1j} \sigma_{ij}$$

The general transformation law for the stress components may then be written as

$$\sigma'_{ij} = l_{ip} l_{jq} \sigma_{pq} \tag{4.48}$$

In exactly the same way, the strain transformation law from (4.40) can be written

$$e'_{ij} = l_{ip} l_{jq} e_{pq} \tag{4.49}$$

Both the stress transformation and strain transformation formula are identical in form. A quantity whose components transform in this manner upon rotation of axes is called a *tensor*. Mohr's circle is thus a graphical representation of a tensor transformation.

All the formulas we have written in indicial notation apply to three dimensions.

PROBLEMS

4.1. Show that for a general state of stress at a point, such as illustrated in Fig. 4.8, the requirement that $\Sigma\mathbf{F}=0$ leads to the following three equations:

$$\frac{\partial\sigma_x}{\partial x}+\frac{\partial\tau_{xy}}{\partial y}+\frac{\partial\tau_{zx}}{\partial z}=0$$

$$\frac{\partial\tau_{xy}}{\partial x}+\frac{\partial\sigma_y}{\partial y}+\frac{\partial\tau_{yz}}{\partial z}=0$$

$$\frac{\partial\tau_{zx}}{\partial x}+\frac{\partial\tau_{yz}}{\partial y}+\frac{\partial\sigma_z}{\partial z}=0$$

4.2. Show that if the particles of a solid are acted on by "body forces" which are distributed over the volume with intensities X, Y, and Z per unit volume, then the requirement of $\Sigma\mathbf{F}=0$ leads to

$$\frac{\partial\sigma_x}{\partial x}+\frac{\partial\tau_{xy}}{\partial y}+\frac{\partial\tau_{zx}}{\partial z}+X=0$$

$$\frac{\partial\tau_{xy}}{\partial x}+\frac{\partial\sigma_y}{\partial y}+\frac{\partial\tau_{yz}}{\partial z}+Y=0$$

$$\frac{\partial\tau_{zx}}{\partial x}+\frac{\partial\tau_{yz}}{\partial y}+\frac{\partial\sigma_z}{\partial z}+Z=0$$

4.3. Show that if a state of plane stress is to be described in terms of polar coordinates, the requirement that $\Sigma\mathbf{F}=0$ leads to the following two equations:

$$\frac{\partial\sigma_r}{\partial r}+\frac{1}{r}\frac{\partial\tau_{r\theta}}{\partial\theta}+\frac{\sigma_r-\sigma_\theta}{r}=0$$

$$\frac{\partial\tau_{r\theta}}{\partial r}+\frac{1}{r}\frac{\partial\sigma_\theta}{\partial\theta}+2\frac{\tau_{r\theta}}{r}=0$$

Note that the length of the curved boundary on the outer edge of the element is $(r+\Delta r)\,\Delta\theta$.

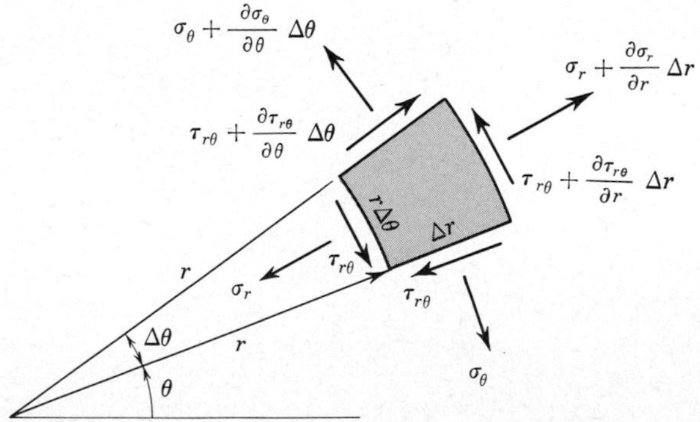

Prob. 4.3

4.4. Show that if a general state of stress is to be described in cylindrical coordinates, the requirement that $\Sigma \mathbf{F} = 0$ leads to the following three equations:

$$\frac{\partial \sigma_r}{\partial r} + \frac{1}{r}\frac{\partial \tau_{r\theta}}{\partial \theta} + \frac{\partial \tau_{zr}}{\partial z} + \frac{\sigma_r - \sigma_\theta}{r} = 0$$

$$\frac{\partial \tau_{r\theta}}{\partial r} + \frac{1}{r}\frac{\partial \sigma_\theta}{\partial \theta} + \frac{\partial \tau_{\theta z}}{\partial z} + 2\frac{\tau_{r\theta}}{r} = 0$$

$$\frac{\partial \tau_{zr}}{\partial r} + \frac{1}{r}\frac{\partial \tau_{\theta z}}{\partial \theta} + \frac{\partial \sigma_z}{\partial z} + \frac{\tau_{zr}}{r} = 0$$

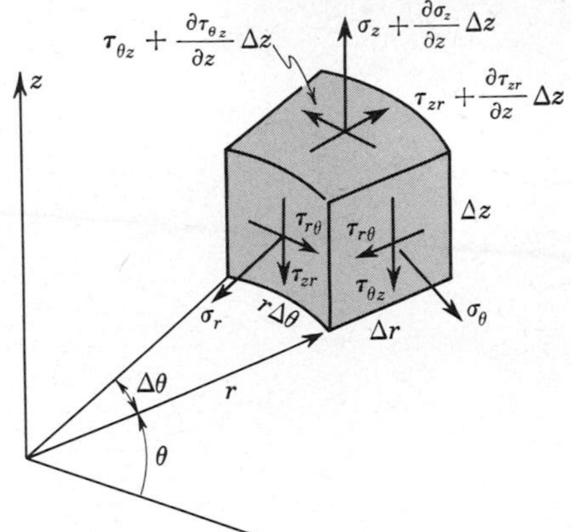

Prob. 4.4

4.5. Assume that in Example 4.1 (Fig. 4.18*a*) the direction of the $+a$ axis is reversed, as shown in the accompanying sketch. Show that the Mohr's circle for this orientation of the a and b axes is identical with that for the orientation given in Fig. 4.18*a*. Show also that when the stress components are taken from the Mohr's circle and drawn on the element in the accompanying sketch, the direction of the shear-stress components will be the same as in Fig. 4.18*c*. This demonstrates that our sign convention for shear stresses in the Mohr's circle has the necessary characteristic that the physical results are independent of the choice of positive coordinate direction in the stressed body.

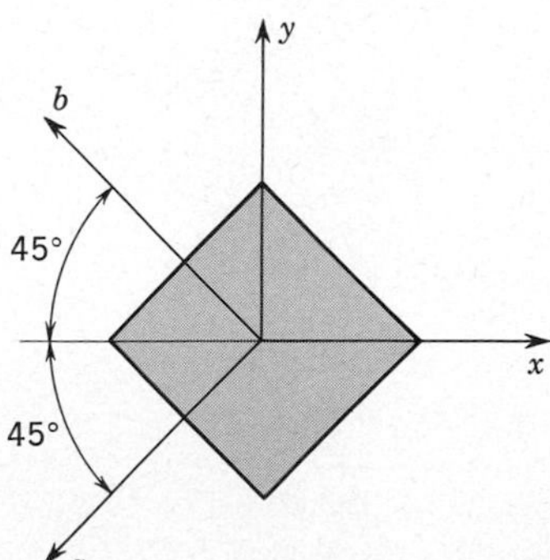

Prob. 4.5

4.6. Sketch the Mohr's circle for stress for each of the following cases of plane stress.

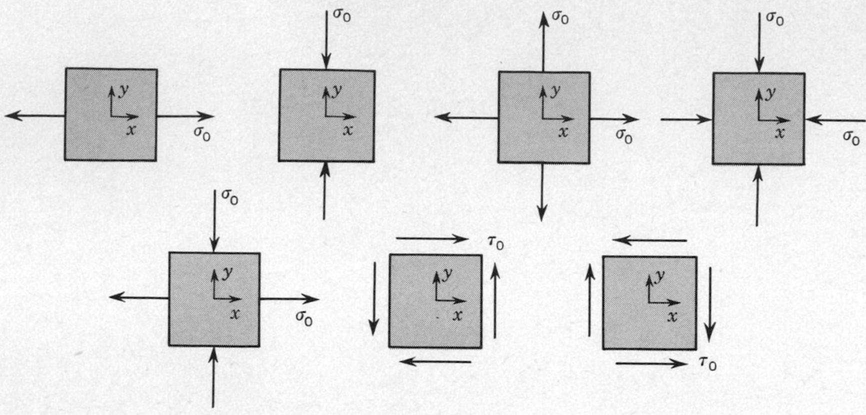

Prob. 4.6

4.7. Find the principal stresses and the orientation of the principal axes of stress for the following cases of plane stress.

(*a*) $\sigma_x = 4{,}000$ psi
$\sigma_y = 0$
$\tau_{xy} = 8{,}000$ psi

(*b*) $\sigma_x = 14{,}000$ psi
$\sigma_y = 2{,}000$ psi
$\tau_{xy} = -6{,}000$ psi

(*c*) $\sigma_x = -12{,}000$ psi
$\sigma_y = 5{,}000$ psi
$\tau_{xy} = 10{,}000$ psi

(*d*) $\sigma_x = 10{,}000$ psi
$\sigma_y = -4{,}000$ psi
$\tau_{xy} = 8{,}000$ psi

(*e*) $\sigma_x = -10{,}000$ psi
$\sigma_y = 20{,}000$ psi
$\tau_{xy} = -6{,}000$ psi

4.8. If the minimum principal stress is $-1{,}000$ psi, find σ_x and the angle which the principal stress axes make with the *xy* axes for the case of plane stress illustrated.

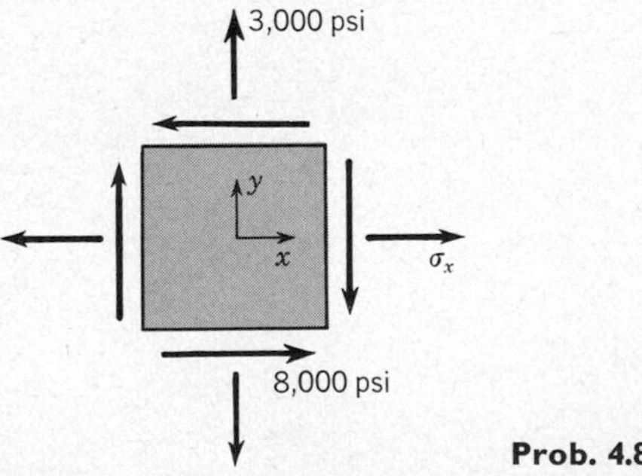

Prob. 4.8

4.9. For the state of stress given in Prob. 4.7(*c*), find the stress components on an element inclined at 30° to the *xy* axes.

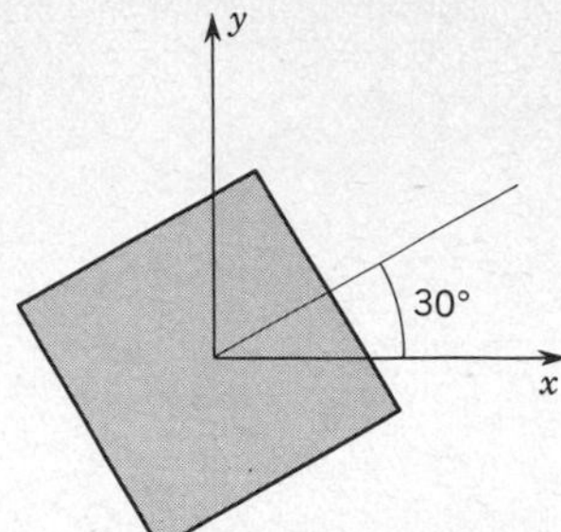

Prob. 4.9

4.10. Consider a thin-walled cylinder of internal radius r and thickness t. If the cylinder is subjected to an internal pressure p and an axial force F, show that the r, θ, z directions are the principal stress directions. Show also that if the wall is so thin that $t/r \ll 1$, then the stresses in the pipe wall are given approximately by

$$\sigma_r = 0$$

$$\sigma_\theta = \frac{pr}{t}$$

$$\sigma_z = \frac{F}{2\pi rt}$$

Prob. 4.10

4.11. For the thin-walled cylinder of Prob. 4.10, what should be the relation between F and p if the maximum shear-stress component in the wall is to have the same magnitude as the maximum normal stress component? What should be the relation between F and p if the maximum shear-stress component in the wall is to be half the magnitude of the maximum normal stress component?

4.12. Consider a thin-walled cylindrical shell of internal radius r and thickness t, with ends which will contain pressure. Show that the principal stresses in the cylinder wall are given approximately by the following when the cylinder contains an internal pressure p:

$$\sigma_r = 0$$

$$\sigma_\theta = \frac{pr}{t}$$

$$\sigma_z = \frac{pr}{2t}$$

4.13. Show that in a closed-end, thin-walled cylinder subjected to internal pressure the maximum shear-stress component in the θz plane is one-quarter the maximum normal stress component in that plane.

4.14. Show that in a thin-walled sphere of internal radius r and thickness t subjected to internal pressure p the principal stresses are given approximately by the following:

$$\sigma_r = 0$$

$$\sigma_\theta = \frac{pr}{2t}$$

$$\sigma_\phi = \frac{pr}{2t}$$

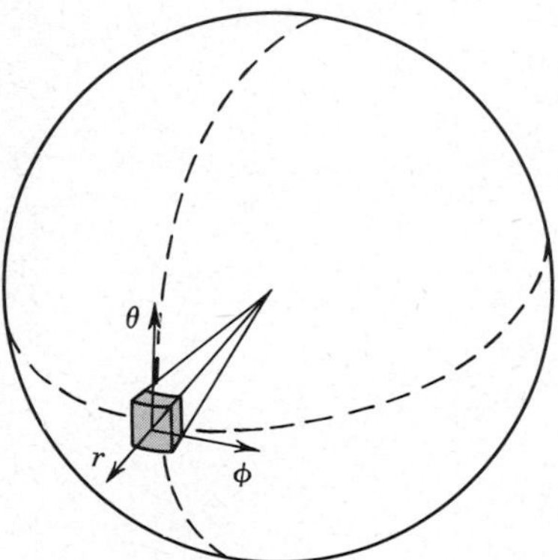

Prob. 4.14

4.15. A rectangular plate is under a uniform state of plane stress in the xy plane. It is known that the maximum tensile stress acting on any face (whose normal lies in the xy plane) is 10,000 psi. It is also known that on a face perpendicular to the x axis there is acting a compressive stress of 2,000 psi and no shear stress. No explicit information is available as to the values of the normal stress σ_y, and the shear stress τ_{yx} acting on the face perpendicular to the y axis.

Find the stress components acting on the faces perpendicular to the a and b axes which are located as shown in the lower sketch. Report your results in an unambiguous sketch.

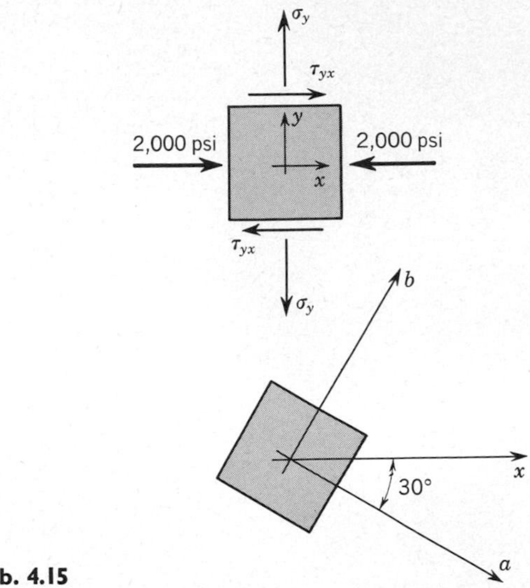

Prob. 4.15

4.16. Show in a general three-dimensional displacement that if the displacement components in the x, y, and z directions are u, v, and w, respectively, the strain components referred to the xyz axes are

$$\epsilon_x = \frac{\partial u}{\partial x} \qquad \epsilon_y = \frac{\partial v}{\partial y} \qquad \epsilon_z = \frac{\partial w}{\partial z}$$

$$\gamma_{xy} = \frac{\partial v}{\partial x} + \frac{\partial u}{\partial y} \qquad \gamma_{yz} = \frac{\partial w}{\partial y} + \frac{\partial v}{\partial z} \qquad \gamma_{zx} = \frac{\partial u}{\partial z} + \frac{\partial w}{\partial x}$$

4.17. In a case of plane strain in which each point displaces radially in a rotationally symmetric fashion about the origin O, the displacement can be expressed by a single displacement component u in the radial direction. Show that the strain components referred to the radial, tangential (r,θ) set of axes are

$$\epsilon_r = \frac{du}{dr} \qquad \epsilon_\theta = \frac{u}{r} \qquad \gamma_{r\theta} = 0$$

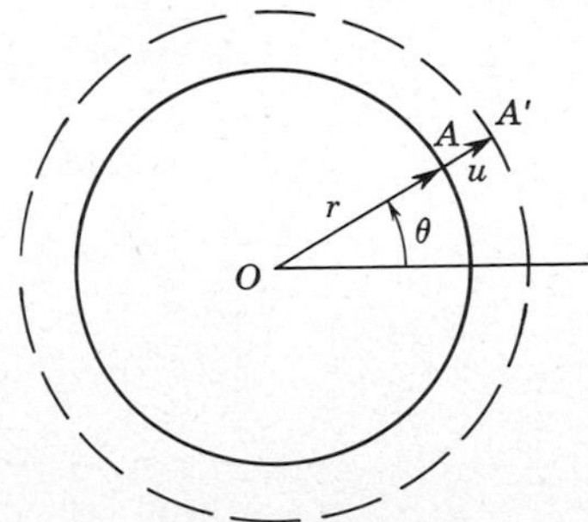

Prob. 4.17

4.18. A general deformation in plane strain can be described in polar coordinates by expressing the displacement of each point as the vector sum of a radial component u and a tangential component v. Show that in such a case the strain components referred to the r, θ set of axes are

$$\epsilon_r = \frac{\partial u}{\partial r}$$

$$\epsilon_\theta = \frac{1}{r}\frac{\partial v}{\partial \theta} + \frac{u}{r}$$

$$\gamma_{r\theta} = \frac{\partial v}{\partial r} + \frac{1}{r}\frac{\partial u}{\partial \theta} - \frac{v}{r}$$

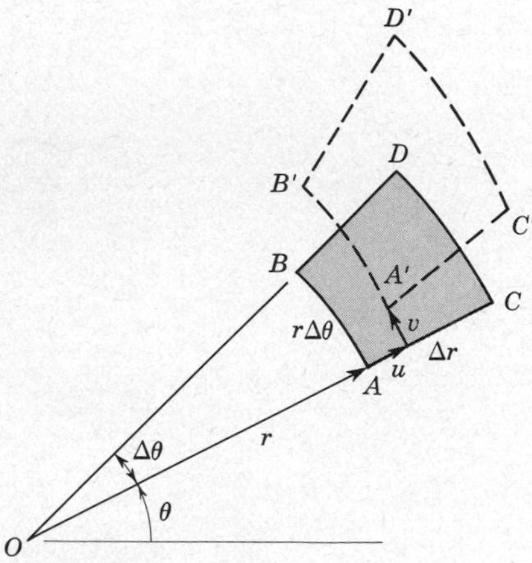

Prob. 4.18

4.19. Using the results of Prob. 4.18, show that if a general three-dimensional deformation is to be described in cylindrical coordinates r, θ, and z in which the displacement components are u, v, and w, respectively, the strain components referred to the r, θ, z axes are

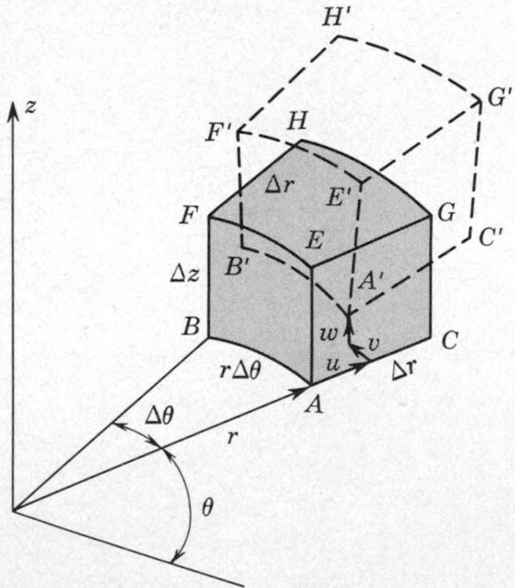

Prob. 4.19

$$\epsilon_r = \frac{\partial u}{\partial r} \qquad \epsilon_\theta = \frac{1}{r}\frac{\partial v}{\partial \theta} + \frac{u}{r} \qquad \epsilon_z = \frac{\partial w}{\partial z}$$

$$\gamma_{r\theta} = \frac{\partial v}{\partial r} + \frac{1}{r}\frac{\partial u}{\partial \theta} - \frac{v}{r} \qquad \gamma_{\theta z} = \frac{1}{r}\frac{\partial w}{\partial \theta} + \frac{\partial v}{\partial z} \qquad \gamma_{zr} = \frac{\partial u}{\partial z} + \frac{\partial w}{\partial r}$$

4.20. In a state of plane strain in the xy plane the strain components associated with the xy axes are

$$\begin{aligned} \epsilon_x &= 800 \times 10^{-6} \\ \epsilon_y &= 100 \times 10^{-6} \\ \gamma_{xy} &= -800 \times 10^{-6} \end{aligned}$$

Find the magnitude of the principal strains and the orientation of the principal strain directions

4.21. At a point in a body the principal strains are

$$\begin{aligned} \epsilon_{\text{I}} &= 700 \times 10^{-6} \\ \epsilon_{\text{II}} &= 300 \times 10^{-6} \\ \epsilon_{\text{III}} &= -300 \times 10^{-6} \end{aligned}$$

What is the maximum shear-strain component at the point? What is the orientation of the axes which experiences the maximum shear strain?

4.22. The readings of a 45° strain rosette (Fig. 4.40*b*) are

(*a*) $\epsilon_a = 100 \times 10^{-6}$ (*b*) $\epsilon_a = 1{,}200 \times 10^{-6}$

$\epsilon_b = 200 \times 10^{-6}$ $\epsilon_b = 400 \times 10^{-6}$

$\epsilon_c = 900 \times 10^{-6}$ $\epsilon_c = 60 \times 10^{-6}$

Find the magnitude of the principal strains in the plane of the rosette.

4.23. A body is in plane strain in the xy plane. The strain components associated with the xy axes are

$$\begin{aligned} \epsilon_x &= -800 \times 10^{-6} \\ \epsilon_y &= -200 \times 10^{-6} \\ \gamma_{xy} &= -600 \times 10^{-6} \end{aligned}$$

Show in a suitable sketch the location of the axes with which the maximum shear strain is associated. Show also the deformed shape of an element which originally was a parallelepiped with its faces parallel to these axes.

4.24. Using the results of Prob. 4.19, show that if a body undergoes a displacement $v = Cr$, there will be no strains resulting from this displacement. Since there are no strains, this displacement must describe a rigid-body motion. What is this motion?

4.25. A long, cylindrical pressure vessel with closed ends is to be made by rolling a strip of plastic of thickness t and width w into a helix and making a continuous fused joint, as illustrated. It is desired to subject the fused joint to a tensile stress only 80 percent of the maximum in the parent plastic. What is the maximum allowable width w of the strip?

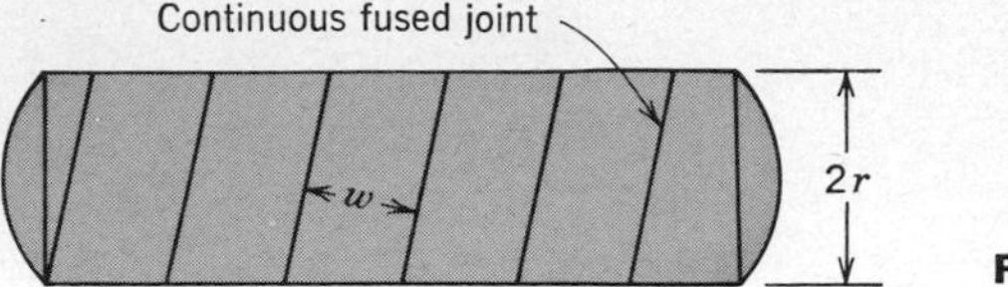

Prob. 4.25

4.26. In each of the two cases find the principal stress directions if the stress at a point is the sum of the two states of stress illustrated.

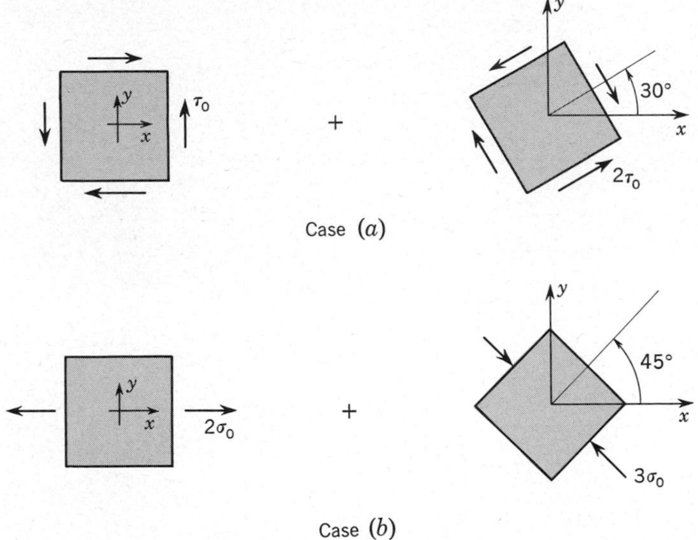

Prob. 4.26

4.27. An open-ended, thin-walled cylinder, $r = 10$ in. and $t = 0.1$ in., is acted on by an internal pressure p and an axial force F. Find the values of p and F acting in each of the following two situations:

(a) $\sigma_m = 15{,}000$ psi $\quad \sigma_n = 5{,}000$ psi $\quad \tau_{mn} = ?$

(b) $\sigma_m = 15{,}000$ psi $\quad \sigma_n = 15{,}000$ psi $\quad \tau_{mn} = ?$

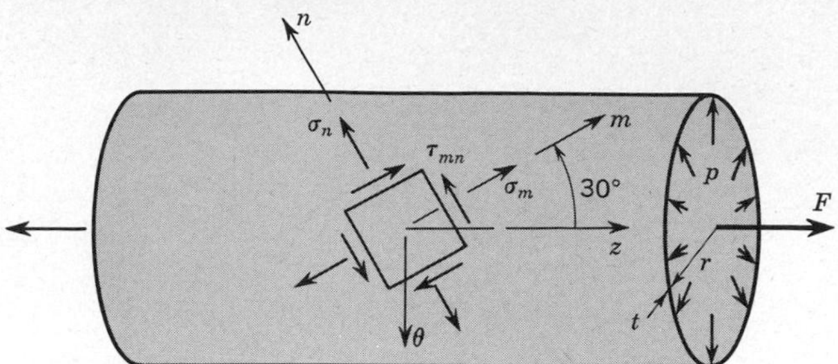

Prob. 4.27

4.28. At a point in a body in plane stress the shear stress on the xy faces is as sketched. Also, it is known that the principal stresses at this point are

$$\sigma_1 = 3{,}000 \text{ psi} \qquad \sigma_2 = -7{,}000 \text{ psi}$$

Complete the stress picture for the xy faces; that is, calculate σ_x and σ_y and show them in proper direction and magnitude in a suitable sketch.

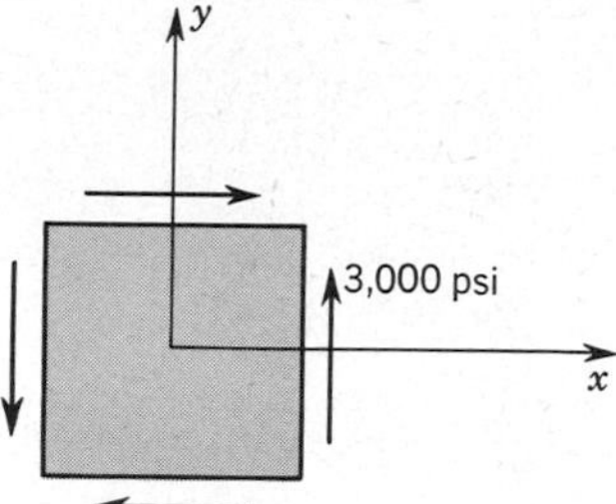

Prob. 4.28

4.29. The displacements for a rigid-body rotation through an angle β about an axis may be described by

$$u(x,y) = (\cos \beta - 1)x - \sin \beta y$$
$$v(x,y) = \sin \beta x + (\cos \beta - 1)y$$

Determine the rotation ω_z (4.32) and the strains (4.33); do the strains vanish? What are the values of the strains and the value of ω_z when the rotation β is small?

4.30. If the boundary portion AB of a structure under plane stress is *stress-free* as illustrated, the stress *vector* acting on the portion AB must be zero. Express the condition that the components of the stress vector must vanish in terms of the stress components with respect to the coordinate axes and the angle α.

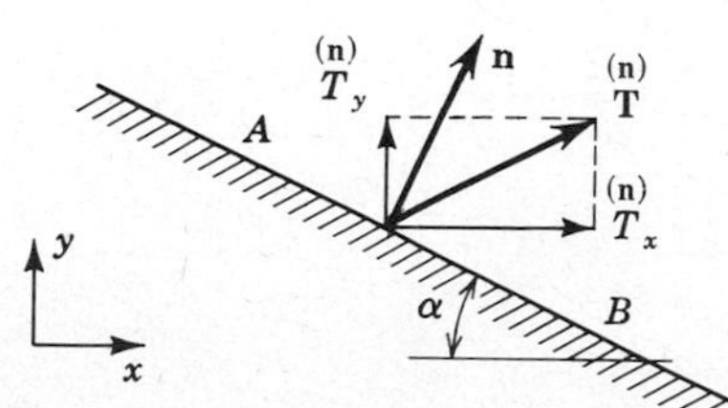

Prob. 4.30

4.31. Show from (4.25) that the angles which make $\sigma_{x'}$ a maximum or minimum correspond to the principal directions.

4.32. Lightweight pressure vessels often use glass filaments for resisting tensile forces and use epoxy resin as a binder. Find the angle of winding, α, of the filaments when the ends of the vessel are closed such that the tensile forces in the filaments are equal (see Prob. 4.12).

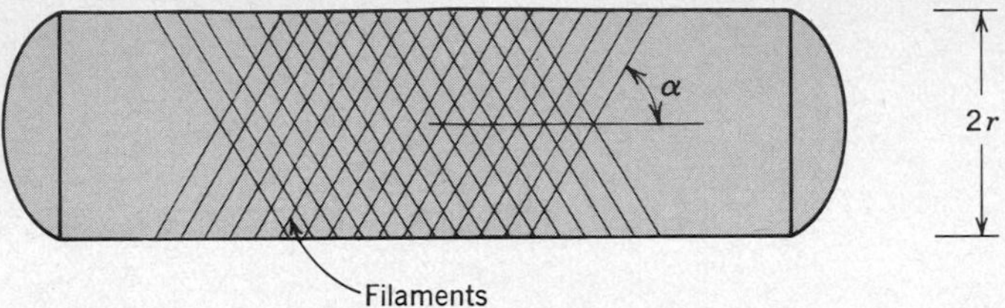

Prob. 4.32

4.33. Recall from Chap. 3 that the equilibrium equations for a slender beam (3.11) and (3.12) are

$$\frac{dV}{dx} + q = 0 \qquad \frac{dM_b}{dx} + V = 0$$

Show that an integration of the equilibrium equations (4.13) across the thickness of a beam in plane stress reduces to the above equations where

$$V = \int_{-h/2}^{h/2} \tau_{xy}\, dy \qquad M = -\int_{-h/2}^{h/2} y\sigma_x\, dy$$

$$q = \sigma_y\left(y = \frac{h}{2}\right) - \sigma_y\left(y = -\frac{h}{2}\right)$$

4.34. A sea urchin egg, which may be considered a thin-walled sphere, has a volume of 35×10^{-8} cm^3 and wall thickness $= 10^{-4}$ cm. Micropuncture techniques give values for the internal pressure of approximately 150 mm Hg. Calculate the membrane stresses in the egg wall.

5
Stress-Strain-Temperature Relations

5.1 INTRODUCTION

In Chap. 4 the ideas of stress and strain at a point were developed separately, using only geometry and the physical concepts of equilibrium of forces and continuity of displacements. We did not in Chap. 4 specify the nature of the material of the body. The presence of only three equations of equilibrium for the six components of stress and the addition of three components of displacement in the six equations relating strain to displacement indicates that further relations are needed before the equations can be solved to determine the distributions of stress and strain in a body; i.e., the distribution of stress and strain will depend on the material behavior of the body. In this chapter we shall concern ourselves with the relations *between* stress and strain. Two avenues of approach suggest themselves: (1) relations based on experimental evidence at the atomic level with theoretical extension to the macroscopic level or (2) relations based on experimental evidence at the macroscopic level. Although there has been much progress in the physics of solids during the past 35 years, the subject has not yet developed to the point where very much of the quantitative information required by engineers can be predicted from atomic data. We shall turn, therefore, to experimental data at the macroscopic level, obtaining the stress-strain relations we need by generalizing these data with the aid of physical and mathematical arguments.

In this chapter we describe the stress-strain behavior of a wide variety of structural materials, including metals, wood, polymers, and composite materials. Elastic, plastic, and viscoelastic behavior are discussed and various mathematical models are established to describe elasticity and plasticity. For application in subsequent chapters, we develop in some detail the theory of linear isotropic elasticity. More briefly we describe, and give design criteria for, yielding of ductile materials, fracture of brittle materials, and fatigue under repeated loading. An attempt is made to give an introduction[1] to most of the aspects of material behavior which are important for engineering applications.

A few preliminary experiments may serve to illustrate the kinds of load-deformation behavior that materials exhibit. Straighten out a paper clip and clamp it between a book and the edge of a table, as shown in Fig. 5.1. Press on it lightly; release the load and notice that it returns to its original shape. This illustrates *elastic deformation*, which is defined as the deformation that disappears on release of load. Reload the wire and note that greater deformations require greater loads. Increase the load and note that finally a value of the load is reached such that when this load is released the wire does not return all the way to its original shape but remains partly bent. Hang on the end of the wire a fixed load large enough to cause permanent bending, and note that the amount of deformation does not increase with time. These phenomena characterize *plastic deformation*, which is defined as the deformation which depends on the applied load, is independent of time, and remains on release of load. Bend the wire further, keeping the force perpendicular to the wire. The force required to bend the wire 90° is greater than that required to initiate plastic deformation. This increase in the load required for further plastic deformation is termed *strain-hardening*. Note that as the load increases, the elastic deformation continues to increase, as is shown by the increased springback upon release of the force. Try to break the wire by bending it sharply. You probably will be unable to do so. A *ductile* structure is defined as one for which the plastic deformation before fracture is much larger than the elastic

[1] For more complete discussions of material behavior, see F. A. McClintock and A. Argon (eds.), "Mechanical Behavior of Materials," Addison-Wesley Publishing Company, Inc., Reading, Mass., 1966. An interesting *film* depicting different material behavior is "Behavior of Structural Materials," Film No. 2 in the McGraw-Hill Film Series on The Mechanics of Structures and Materials, McGraw-Hill Book Company, New York, 1969.

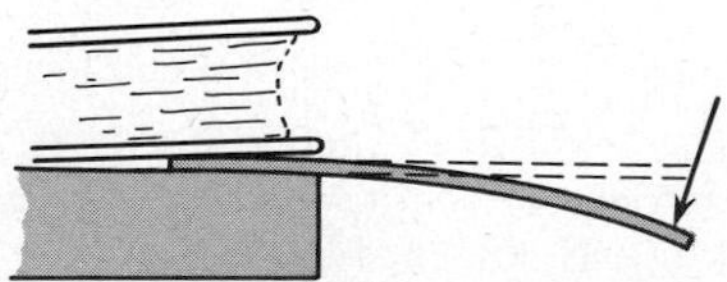

Fig. 5.1 Bending experiment.

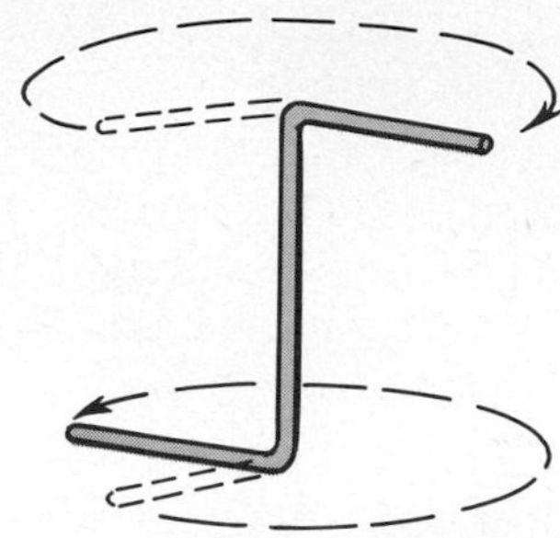

Fig. 5.2 Twisting experiment.

deformation. To get an idea of the full ductility of the wire, bend it into the shape of a U and twist it as shown in Fig. 5.2. The plastic deformation up to fracture may be a hundred times the elastic deformation. Finally, note that the wire can be broken by repeated bending back and forth. This progressive fracture under repeated load is termed *fatigue*.

Repeat the bending experiments with a piece of "lead" from a mechanical pencil. We define as *brittle* a structure which exhibits little deformation before fracture. Whether brittle or ductile behavior occurs depends not only on the material involved but also on the temperature. For example, a steel paper clip under bending is ductile at room temperature but is brittle at temperatures below about $-100°$F. Thus, the classification of a structure as being ductile or brittle depends upon the temperature of test. Unless specifically indicated otherwise, room temperature is implied when these terms are used.

Furthermore, ductility depends on the size of the structure. The cleavage of a grain or the growth of a hole from a nonmetallic inclusion are the fundamental mechanisms by which cracks grow, leading to fracture of the structure. These processes require a certain strain over regions of the order of a micron in size. To get such large strains in turn requires some plastic strain in a surrounding region perhaps a hundred to a thousand times as large. Thus in the presence of an accidental crack or flaw, a small portion of a structure may deform plastically before appreciable localized fracture occurs. Such a portion would be ductile even in the presence of the crack. A larger part, however, might still have a large elastic region surrounding the plastic zone when the crack started to grow so that the overall deformation of the part would only be little more than the elastic deformation. Such a part or structure would be called *notch-brittle*. Of course, in the absence of any notches or cracks the plastic strain required to nucleate such defects may be large enough so that even relatively large structures are ductile. This is the case with normal structural steels, but the higher-strength alloys, coming into more common use, tend to be notch-brittle.

Another important behavior of materials becomes evident at absolute temperatures which are about half the melting point of the material; at these temperatures the deformation under constant load increases with time. This time-dependent part of the deformation is called *creep*. For a steel paper clip the deformation due to creep becomes significant at a temperature of about 900°F, whereas for aluminum the corresponding temperature is about 400°F.

Another time-dependent phenomenon is the slow recovery of shape which some materials undergo when they are unloaded. This may be observed by folding a piece of paper and then releasing it. Note that there is first an immediate elastic springback followed by a relatively slow unfolding; this is called *elastic aftereffect*, or *recovery*.

Many plastics (long-chain polymers) exhibit a mixture of creep and elastic aftereffects at room temperature. The general term for such behavior is *viscoelasticity*, which includes the viscous behavior of liquids in the limiting case.

These simple experiments serve to indicate some idealized kinds of material behavior under applied loads. In the following sections we shall examine some of these aspects more fully and discuss the quantitative relationships which are at present commonly used to describe these phenomena.

5.2 THE TENSILE TEST

In common with all branches of science we wish to develop a theory which will allow us to predict behavior in a general situation from the results of an experiment made in a very simple situation. The most simple loading situation we can imagine is one in which a relatively slender member is pulled in the direction of its axis. Such a test is called a *tensile test*. Our aim is to use tensile-test data to formulate quantitative stress-strain relations which, when incorporated with equilibrium and compatibility requirements, will produce theoretical predictions in agreement with the experimental results in complicated situations.

To perform a tensile test, a piece of material is cut in the form of a cylindrical test specimen, such as that shown in Fig. 5.3. The ends are moved apart by a

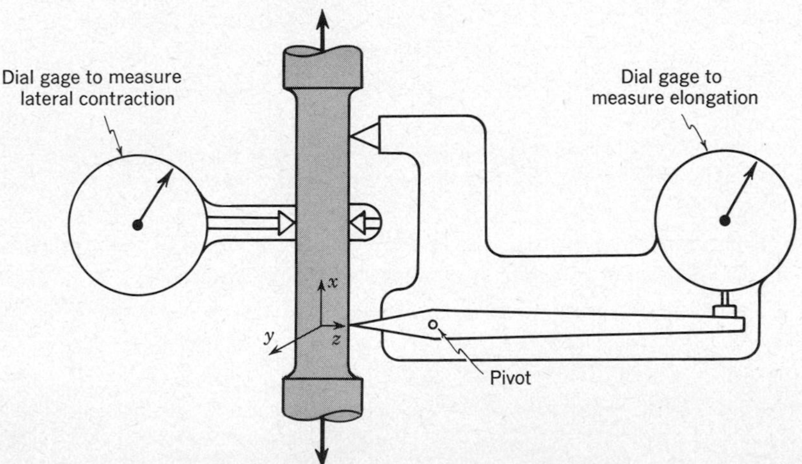

Fig. 5.3 A tensile-test specimen and gages for measuring deformation.

testing machine which indicates the load required at each stage. The elongation and lateral contraction are also noted as the test proceeds. Since we expect thicker specimens to carry higher loads, and longer ones to stretch farther, as indicated in Fig. 2.4, we convert the results of such tests from load and elongation to stress and strain.

Taking axes x, y, z such that y and z lie in the cross section and x is coincident with the axis of the tensile specimen, the only component of stress present is the axial normal component σ_x which is found by dividing the load by the cross-sectional area.

Usually the only component of strain reported from a tensile test is the axial normal component. To obtain this strain component, two gage marks can be made on the specimen, and the displacement of one measured relative to another, as shown in Fig. 5.4. Taking the lower gage mark as a point of zero displacement, the displacement vectors of various points on the specimen will be as in Fig. 5.4. If the displacements vary uniformly over the gage length L, one may write

$$u = \frac{x}{L} \Delta L$$

For small strains, application of the definition of strain (4.31) yields

$$\epsilon_x = \frac{\partial u}{\partial x} = \frac{\Delta L}{L} \tag{5.1}$$

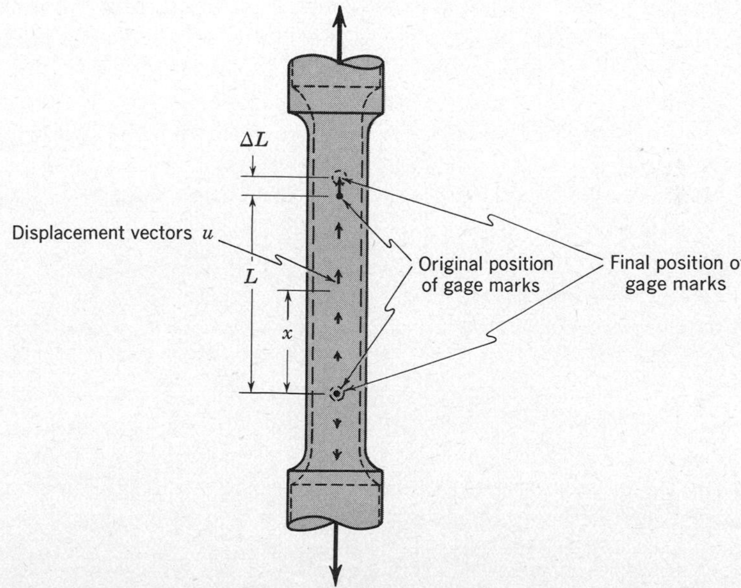

Fig. 5.4 Displacements in a tensile test.

The results of room-temperature tests on a variety of materials are shown in Fig. 5.5. These tests were carried out to strains of only a few percent; in the cases of steel, aluminum, and the aluminum alloys this was far from the strains at which fracture would occur (see Fig. 5.39). As a matter of interest, the 1020 CR curve of Fig. 5.5*a* represents a close approximation to the behavior of the steel from which paper clips are made.

There are a number of features which many of the stress-strain curves of Fig. 5.5 have in common. At first there is a region where the stress is very nearly proportional to the strain. The *proportional limit* is defined as the greatest stress for which the stress is still proportional to the strain. The *elastic limit* is defined as the greatest stress which can be applied without resulting in any permanent strain on release of stress. Figure 5.6*a* illustrates the loading and unloading behavior of a material which is elastic at stresses greater than the proportional limit. For the materials shown in Fig. 5.5, the proportional and elastic limits coincide. For these materials, if the stress is increased beyond the elastic limit and then removed, the stress-strain curve has the shape shown in Fig. 5.6*b*. The slopes of the unloading and reloading curves are nearly equal to the slope in the initial elastic region. It is convenient to think of the total strain OB under a given stress as being made up of an elastic part AB and a plastic part OA.

Neither the proportional nor the elastic limits can be determined precisely, for they deal with the limiting cases of *zero* deviation from linearity and of *no* permanent set. Since plastic deformations of the order of the elastic strains are often unimportant, instead of reporting the elastic limit it has become standard practice to report a quantity called the *yield strength*, which is the stress required to produce a certain arbitrary plastic deformation. The yield strength is determined by drawing through the point on the abscissa corresponding to the arbitrary plastic strain, usually 0.2 percent, a line which is parallel to the initial tangent to the stress-strain curve; the intersection of this line with the stress-strain curve defines the yield strength. This construction is illustrated in several cases in Fig. 5.5. Note that, because of the method of determination, these yield strengths are more sharply defined than are the proportional limits.

For many of the common steels the plastic deformation begins abruptly, resulting in an increase of strain with no increase, or perhaps even a decrease, in stress. For such materials a *yield point* is defined as a stress level, less than the maximum attainable stress, at which an increase in strain occurs without an increase in stress. The stress at which such plastic deformation first begins is called the *upper yield point*; subsequent plastic deformation may occur at a lower stress, called the *lower yield point*. (See curve for 1020 HR in Fig. 5.5*a*.) The upper yield point is very sensitive to rate of loading and accidental bending stresses or irregularities in the specimen, so the lower yield point should be used for design purposes. Unfortunately, it is the upper yield point which is often tabulated, without being labeled as such, in tables of properties.

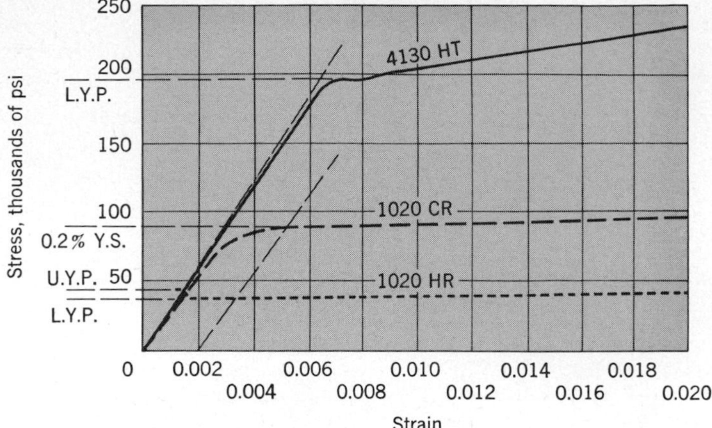

Fig. 5.5*a* Stress-strain curves for three steels.
---- Mild steel, hot-rolled (1020 HR)
— — Mild steel, cold-rolled (1020 CR)
—— 0.3% C, 0.5% Mn, 0.25% Si, 0.9% Cr, balance Fe (4130 HT)
Heat treatment: Oil quenched from 1600°F, tempered at 600°F

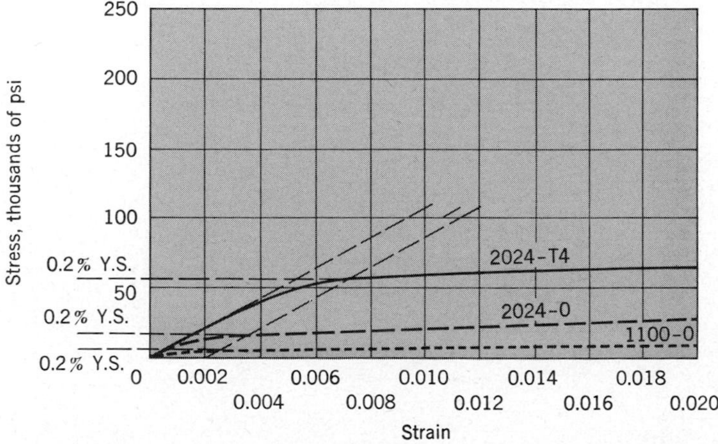

Fig. 5.5*b* Stress-strain curves for aluminum and two aluminum alloys.
---- Commercially pure aluminum, annealed (1100-0)
— — 4.6% Cu, 1.5% Mg, 0.7% Mn, balance Al, annealed (2024-0)
—— 4.6% Cu, 1.5% Mg, 0.7% Mn, balance Al (2024-T4)
Water quenched from 915°F, aged 24 hr at 250°F

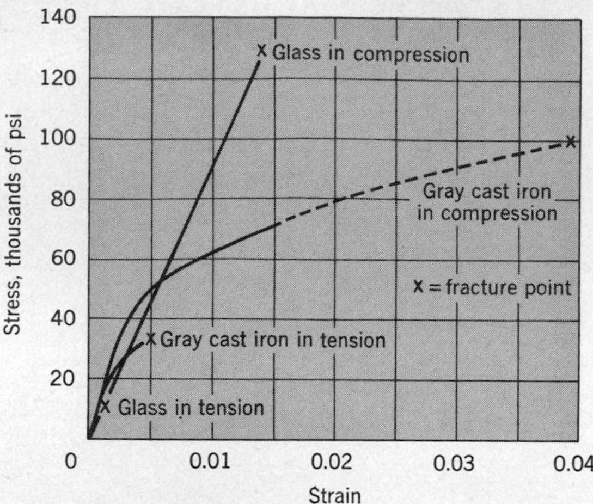

Fig. 5.5c Stress-strain curves for cast iron (*From L. F. Coffin, Jr., The Flow and Fracture of a Brittle Material, J. Appl. Mechanics, vol. 17, pp. 233–248, 1950; and S. H. Ingberg and P. D. Sale, Compressive Strength and Deformation of Structural Steel and Cast Iron Shapes at Temperatures up to 950°C, Proc. ASTM, vol. 26, part 2, pp. 33–51, 1926*) and glass (*From G. W. Morey, "Properties of Glass," Reinhold Publishing Corporation, New York, 1954.*)

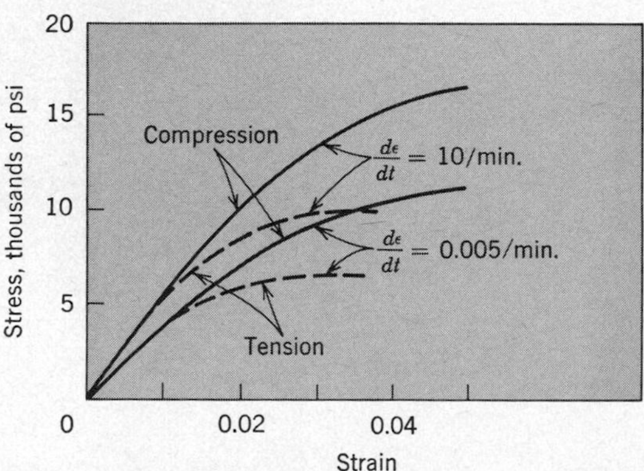

Fig. 5.5*d* Stress-strain curve for commercial polymethyl methacrylate (*From A. G. H. Dietz, W. J. Gailus, and S. Yurenka, Effect of Speed of Test upon Strength Properties of Plastics, Proc. ASTM, vol. 48, pp. 1160–1190, 1948.*)

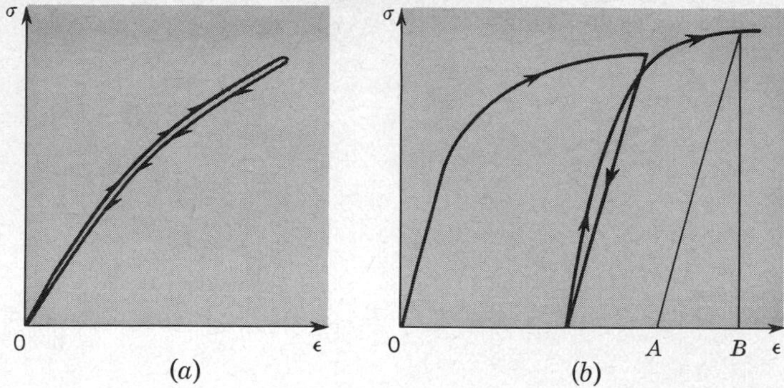

Fig. 5.6 Effect of unloading and reloading. (*a*) Elastic; (*b*) plastic.

As plastic deformation is continued, the stress required for further plastic flow, termed the *flow strength*, rises. This characteristic of the material in which further deformation requires an increase in the stress usually is referred to as *strain-hardening* of the material. Other developments as the test is carried to large plastic deformation are discussed in Sec. 5.12.

Note, in Fig. 5.5*c*, that for glass, stress is proportional to strain up to the point of fracture. Its behavior is entirely elastic; there is no evidence of plastic deformation in either tension or compression. Also, the stress at which fracture occurs is much greater in compression than in tension; this is a usual characteristic of brittle materials.

Some materials exhibit time-dependent behavior, as shown in Fig. 5.5*d*. At room temperature this *viscoelastic* behavior is commonly found in polymers whose structure consists of long-chain molecules with primary chemical bonds along the chain and secondary bonds between chains. The secondary bonds can be broken by thermal activation. At higher temperatures, metals can also exhibit time-dependent effects (Sec. 5.18). For the polymethyl methacrylate of Fig. 5.5*d*, note that the time effect is rather small: changing the strain rate by a factor of 200 changes the stress at a given strain by only 30 percent. In linearly viscous liquids, on the other hand, a 30-percent change in stress would result from only a 30-percent change in strain rate. In metals at room temperature, the strain-rate effect can be observed but is much less; typically a 30-percent change in stress would require changing the strain rate by a factor of a million or more. Thus strain-rate effects can often be neglected, although they may be important in comparing the behavior of the structure under loads over a period of 20 years to that under loads occurring during a collision or explosion.

For most ductile materials the stress-strain curves for tension and compression are nearly the same for strains small compared to unity, and in the following theoretical developments we shall assume that they are identical.

5.3 IDEALIZATIONS OF STRESS–STRAIN CURVES

In any problem in the mechanics of deformable bodies, we need to know the physical relation between stress and strain. This stress-strain relation, together with the equilibrium equations (4.13) and the strain-displacement relations (4.33), must be satisfied at every point in a deformable body in equilibrium. These relations for each differential element comprise the three steps (2.1) on which the solution must be based. From the foregoing discussion of the tensile test, it is evident that different materials often have quite dissimilar stress-strain relations, and that, with the exception of glass, no simple mathematical equation can fit the entire stress-strain curve of one of the materials in Fig. 5.5. Because we wish the mathematical part of our analysis to be as simple as possible, consistent with physical reality, we shall idealize the stress-strain curves of Fig. 5.5 into forms which can be described by simple equations. The appropriateness of any such idealization will depend on the magnitude of the strains being considered, and this in turn will depend upon whatever practical problem is being studied at the moment. To decide what idealizations of the stress-strain curves are needed, we must turn to the applications of mechanics in which these idealizations are used.

Sometimes we must design structures so as to accommodate or produce certain desired deformations. Examples of such applications are the design of springs, safety valves, bumpers, crash panels, shear pins, and blowout diaphragms. Some such elements, e.g., springs, must accommodate the desired deformations repeatedly and reproducibly. In such cases the material must operate below the elastic limit, and a linear approximation to the stress-strain curve will be required. Crash panels and automobile bumpers should not deform permanently under normal usage, but should deform plastically and so limit the deceleration in case of an accident. Here an approximation is needed for both the elastic and plastic regions. Shear pins and blowout diaphragms are intended to fracture completely at certain loads, and for such structures the elastic deformations may be of no importance at all.

Another use of the mechanics of deformable bodies is in the design of metal-forming and cutting processes. In metal forming, the strains usually are relatively small compared to unity, and elastic springback may be a problem. In metal cutting, the strains may be even greater than unity, and elastic effects often are negligible.

Perhaps the most important use of the mechanics of deformable bodies occurs in designing elements so that failure will not occur. *Excessive deformation* is one mode of failure. Machine parts often must fit closely and reliably, and this requirement may not be satisfied if plastic deformation occurs. Even in the absence of plastic deformation, rather small strains will lead to large elastic deflections in long, thin members subjected to bending and twisting. Therefore, there are many members in which the allowable strains are limited to those found in the elastic region. For these members, linear-elastic relations will be sufficient for design against excessive deformation.

There are other conditions under which some plastic deformation is allowable. A little plastic deformation around an oil hole or a fillet may not be noticeable. In other situations, if dimensional changes of a few percent can be accommodated, then the structure may still be serviceable even though some general plastic deformation has occurred. The most common instance of this is the design of structures against extreme overloads such as those caused by earthquakes, bomb blasts, or storms, and in the design of transportation equipment against bad roads, minor accidents, or storms. For calculations intended to assure safety under these conditions, it is necessary to have a relation between stress and strain which holds for values of strain up to, say, 0.05. In such cases one can see from Fig. 5.5 that for some materials it is reasonable to assume that the stress does not change after yielding has begun. This assumption greatly simplifies the mathematical treatment of the problem and, at the same time, does not lead to stresses which are in error by more than approximately 10 percent.

Fracture is the most dangerous mode of failure. Brittle structures are those that fracture with little plastic deformation compared with the elastic deformation, and so we may base all our calculations for these materials on a linear relation between stress and strain. For ductile structures there is as yet no quantitative theory which will predict fracture. Some of the difficulty in our understanding of this phenomenon probably arises from our lack of knowledge of the distributions of stress and strain in the plastic region in front of a crack, and some from our lack of knowledge of strain around the holes that grow from inclusions and coalesce to cause fracture. To study either of these strain distributions it will be necessary to have available stress-strain relations which are reasonable approximations for large plastic strains. In some large structures fracture by hole growth can spread from a sharp notch even though the average stress is below the yield strength. In such cases the elastic strains also must be taken into account.

Another form of fracture arises when stresses, perhaps even less than the yield strength, are applied repeatedly, say thousands or millions of times. These repetitions of stress eventually produce fine cracks which grow very slowly at first and then extend rapidly across the entire part, giving very little warning of the impending failure. This process is known as *fatigue*. (See Sec. 5.15 for a more complete discussion of fatigue.) Since fatigue can occur even if the stresses are below the yield strength, it is sufficient for most practical design purposes to know the relation between the stress and the strain within the elastic region. On the other hand, plastic yielding certainly does occur at the tip of the growing crack, and if we wish really to understand the physical mechanism of the growth of fatigue cracks, we must use relations between stress and strain which take plasticity into account.

Failure by *corrosion* can be greatly accelerated by the presence of stress. A small corrosion pit will cause a local stress concentration which will in turn create an electromotive force between the highly stressed and the less stressed regions. This electromotive force in turn accelerates the corrosion, and the process can lead to the development of cracks and final fracture of the part. This problem is most serious when it is least expected, for example, in normally noncorroding materials such as brass and stainless steel. As in fatigue, the phenomenon may occur when stresses are below the yield strength so that the elastic stress-strain assumptions are of practical use.

Perhaps the most common form of mechanical failure is by *wear*. The laws governing the overall friction and wear between two surfaces seem to depend primarily on the total force transmitted across the two surfaces rather than on the local distribution of the force. For this reason, in considering the overall effects, the local distributions of stress and strain are unimportant, and one may assume that the two bodies in contact are perfectly rigid. In studying the details of the actual mechanism of wear and friction, however, one must take into account the extremely small areas of actual load contact between two bodies and the elastic and plastic deformations in these regions. Another factor which must be considered in studying the detailed mechanism is the surface condition of the metal, since this condition may be such that these local points of contact behave in a manner quite different from that of the same material in bulk form.

The preceding discussion of problems arising in the mechanics of solids shows that there is a need for a variety of stress-strain relations, depending on the problem at hand. Since for most materials it is impossible to describe the entire stress-strain curve with a simple mathematical expression, in any given problem the behavior of the material is represented by an *idealized stress-strain curve* which emphasizes those aspects of the behavior which are most important in that particular problem. Six such idealized models of material behavior are described below and illustrated in Fig. 5.7.

A *rigid* material (Fig. 5.7*a*) is one which has no strain regardless of the applied stress. This idealization is useful in studying the gross motions and forces on machine parts to provide for adequate power and for resistance to wear.

A linearly *elastic* material (Fig. 5.7*b*) is one in which the strain is proportional to the stress. This idealization is useful when we are designing for small deformations, for stiffness, or to prevent fatigue or fracture in brittle structures.

A *rigid-plastic* material is one in which elastic and time-dependent deformations are neglected. If the stress is released, the deformation remains. Strain-hardening may be neglected (Fig. 5.7*c*), or a relation for the strain-hardening may be assumed (Fig. 5.7*d*); in the former case, the material is termed *perfectly plastic*. Such idealizations are useful in designing structures for their maximum loads, in studying many machining and metal-forming problems, and in some detailed studies of fracture.

An *elastic-plastic* material is one in which both elastic and plastic strains are present; strain-hardening may or may not be assumed to be negligible (Figs. 5.7*f* and *e*). These idealizations are useful in designing against moderate deformations and when carrying out detailed studies of the mechanisms of fracture, wear, and friction.

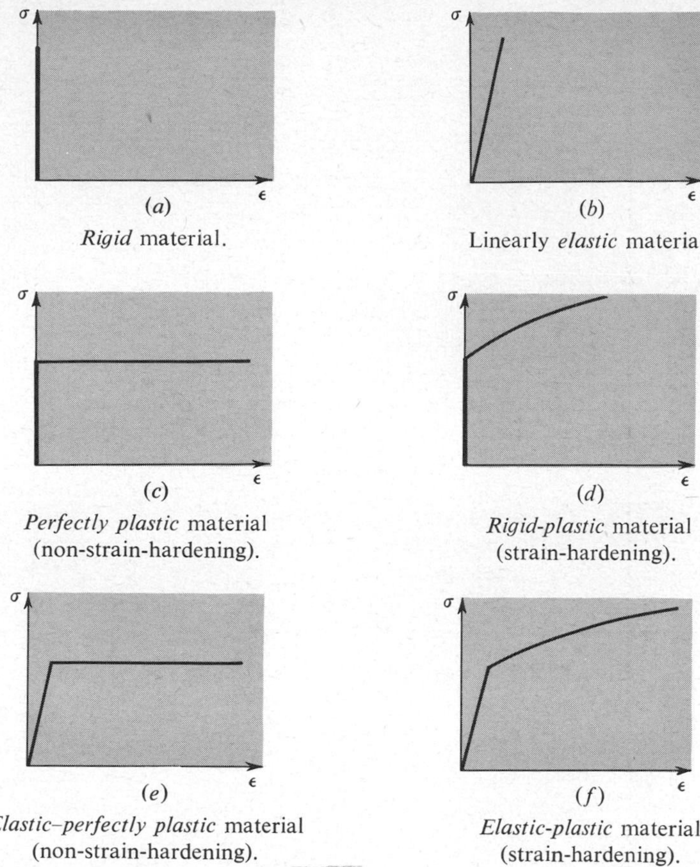

(*a*)
Rigid material.

(*b*)
Linearly *elastic* material.

(*c*)
Perfectly plastic material (non-strain-hardening).

(*d*)
Rigid-plastic material (strain-hardening).

(*e*)
Elastic–perfectly plastic material (non-strain-hardening).

(*f*)
Elastic-plastic material (strain-hardening).

Fig. 5.7 Idealized models of material behavior.

Other idealizations could be made, but these models are those of most practical use from the standpoint of mathematical simplicity. To illustrate the construction of an idealized stress-strain curve of a material and also to illustrate the use of such a curve, we consider the following example.

Example 5.1 Two coaxial tubes, the inner one of 1020 CR steel and cross-sectional area A_s, and the outer one of 2024-T4 aluminum alloy and of area A_a, are compressed between heavy, flat end plates, as shown in Fig. 5.8. We wish to determine the load-deflection curve of the assembly as it is compressed into the plastic region by an axial force P. Our first step is to construct an idealized model of the situation, which we show in Fig. 5.9 (note similarity to idealized model of Example 2.1). We assume in Fig. 5.9 that the end plates are so stiff that both tubes are shortened exactly the same amount. We now apply the steps (2.1) to the idealized model of Fig. 5.9.

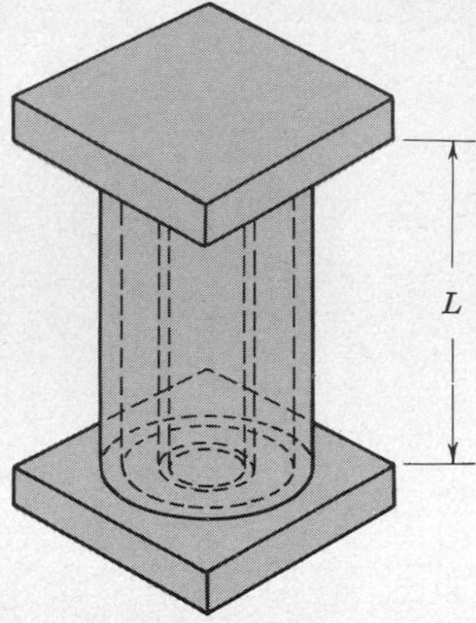

Fig. 5.8 Example 5.1.

GEOMETRIC COMPATIBILITY

From Fig. 5.9 we see that, because of geometric compatibility, we must have the following relation between the strains:

$$\epsilon_s = \epsilon_a = \epsilon = \frac{\delta}{L} \tag{a}$$

STRESS-STRAIN RELATIONS

We wish to carry the test through the elastic region to the point where both materials are in the plastic region. Looking at the curves for 1020 CR steel and 2024-T4 aluminum alloy in Fig. 5.5*a* and *b*, we conclude that we can, with reasonable accuracy, idealize both these curves as being of the elastic–perfectly plastic type of Fig. 5.7*e*, although the 1020 CR steel is described by this model somewhat better than is the 2024-T4 aluminum alloy. These idealized stress-strain curves are shown in Fig. 5.10. From Fig. 5.10 we see that there are three regions of strain which are of interest as we compress the assembly.

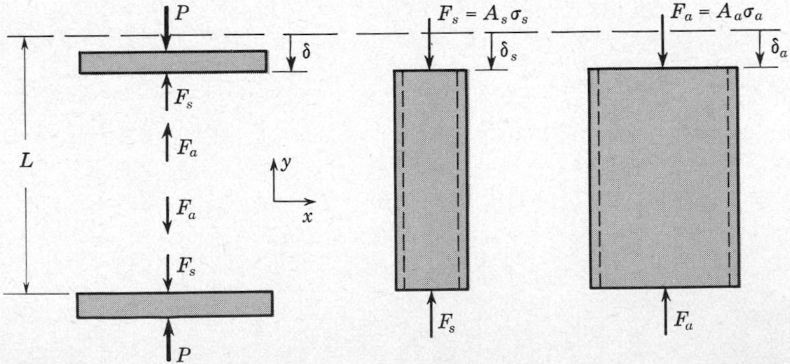

Fig. 5.9 Idealized model for Example 5.1.

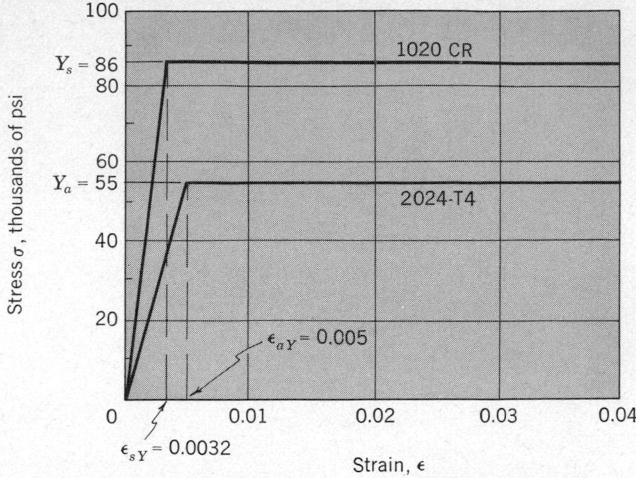

Fig. 5.10 Idealized stress-strain curves for Example 5.1.

In the range $0 \leqq \epsilon \leqq 0.0032$,

$$\begin{aligned} \sigma_s &= E_s\epsilon_s = E_s\epsilon \\ \sigma_a &= E_a\epsilon_a = E_a\epsilon \end{aligned} \tag{b}$$

where

$$E_s = \frac{86{,}000}{0.0032} = 27 \times 10^6 \text{ psi}$$

$$E_a = \frac{55{,}000}{0.005} = 11 \times 10^6 \text{ psi}$$

In the range $0.0032 \leqq \epsilon \leqq 0.005$,

$$\begin{aligned} \sigma_s &= Y_s = 86{,}000 \text{ psi} \\ \sigma_a &= E_a\epsilon_a = E_a\epsilon \end{aligned} \tag{c}$$

In the range $0.005 \leqq \epsilon$,

$$\begin{aligned} \sigma_s &= Y_s = 86{,}000 \text{ psi} \\ \sigma_a &= Y_a = 55{,}000 \text{ psi} \end{aligned} \tag{d}$$

EQUILIBRIUM

In Fig. 5.9 the top plate is in equilibrium when

$$\Sigma F_y = \sigma_s A_s + \sigma_a A_a - P = 0 \tag{e}$$

where A_s and A_a are the cross-sectional areas of the metal in the steel and aluminum tubes.

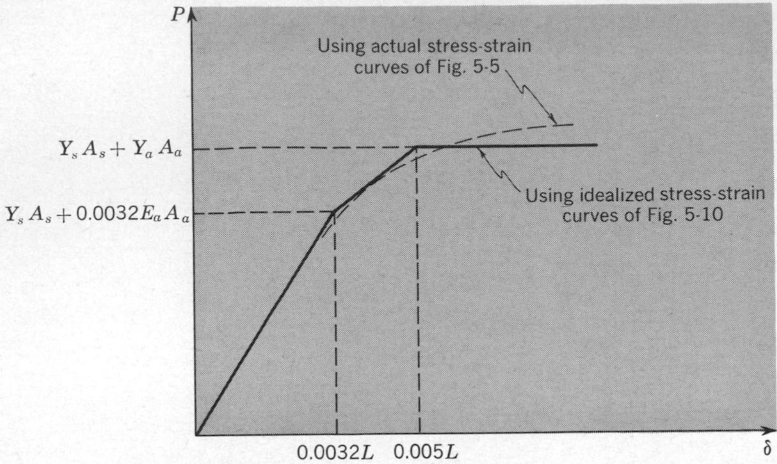

Fig. 5.11 Load-deformation curve for Example 5.1.

Combining (*e*) with (*b*), (*c*), and (*d*) in succession, we obtain the load-deformation curve of Fig. 5.11. It should be borne in mind that this is the load-deformation curve for the idealized materials of Fig. 5.10. If we repeat the analysis replacing (*b*), (*c*), and (*d*) by the actual stress-strain curves of the materials from Fig. 5.5*a* and *b*, we will obtain the dotted curve shown in Fig. 5.11, which is only slightly different from the result using the idealized curves.

The idealized models introduced above represent the behavior of material as observed in the tensile test. We now turn to the generalization of these idealized uniaxial stress-strain relations for application to more general situations, where any or all components of stress and strain may be present.

5.4 ELASTIC STRESS-STRAIN RELATIONS

In the preceding sections stress-strain relations were considered for the special case of *uniaxial* loading. Only one component of stress, the axial normal component, was present, and only the axial normal component of strain was considered. In this section we shall generalize the *elastic* behavior in the tension test to arrive at relations which connect all six components of stress with all six components of elastic strain. We shall restrict ourselves to materials which are *linearly elastic*, corresponding to the idealized model of Fig. 5.7*b*. We shall use as our definition of strains the one developed in Sec. 4.10 and expressed in Eqs. (4.33), and thus we also restrict ourselves to strains small compared to unity. These assumptions are not a serious limitation except for materials such as rubber, in which large nonlinear elastic strains may occur.

As an aid to formulating general stress-strain relations from uniaxial test data, we shall consider some of the *physical* aspects of materials. All solids have some regularity in the arrangement of the atoms of which they are composed. Metals have the atoms arranged in a regular crystal lattice. Plastics consist of long chain molecules. Even glass has some order in the tetrahedral arrangement of silicon and oxygen atoms. These structural elements have different stiffnesses in different directions. In many materials these structural elements are arranged in a random fashion and are so small that there are millions of them, sometimes billions of billions, in a cubic inch. Even a cube of such material which appears small to the naked eye will have thousands of structural elements with just about every conceivable orientation relative to the axes of the cube. If the orientations are indeed random, then two cubes cut out of the material at different angles will have the same statistical distribution of orientations of the structural elements relative to the cube axes, as illustrated in Fig. 5.12. Thus, the average stiffnesses of the two cubes will be equal—in fact, the average elastic properties will be the same for all possible orientations of coordinate axes within either cube. An *isotropic* material is defined as one whose properties are independent of orientation. Materials made up of randomly oriented structural elements may be thought of as being statistically isotropic.

The general state of stress is described by three normal components of stress and three shear components of stress. The most general state of strain likewise can be described by three normal components of strain and three shear components of strain. The general relations between stress and strain can therefore be stated in six equations, each giving the dependence of one component of strain on the various components of stress. We shall consider the various components of stress one at a time and add all their strain effects to get the resulting strain in the presence of all components of stress for a material which is assumed to be isotropic.

First, consider an element on which there is only one component of normal stress acting, as shown in Fig. 5.13. As discussed in the previous section on the tensile test, this normal component of stress will produce a corresponding normal

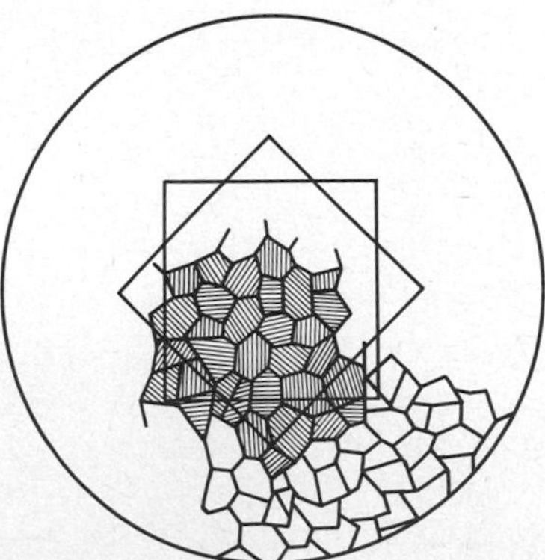

Fig. 5.12 Statistically isotropic material.

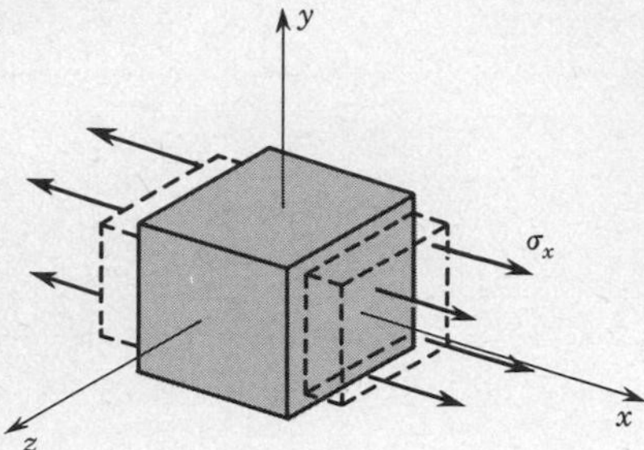

Fig. 5.13 Uniaxial normal stress.

component of strain. We are considering here only materials in which the strain will be directly proportional to the stress. This relation can be expressed in mathematical terms as

$$\epsilon_x = \frac{\sigma_x}{E}$$

This is another form of Hooke's law for uniaxial loading; we first discussed this type of behavior in Sec. 2.2, where, for a bar of area A and length L subjected to a load P, the above relation was expressed in the form of Eq. (2.2). The modulus of elasticity E is numerically equal to the slope of the line in Fig. 5.7*b*.

In addition to the normal component of strain in the x direction, our experience with rubber bands, if nothing else, leads us to expect that there will be a lateral contraction when a bar is elongated. Detailed measurements made during the tensile test bear out this supposition, and the lateral compressive strain is found to be a fixed fraction of the longitudinal extensional strain. Furthermore, tests in uniaxial compression show a lateral extensional strain which is this same fixed fraction of the longitudinal compressive strain. This fixed fraction is known as *Poisson's ratio* and is given the symbol ν. For the uniaxial stress condition illustrated in Fig. 5.13, the lateral strains ϵ_y and ϵ_z must be equal because neither the material, being isotropic, nor the mode of stressing favors either direction. For a linear-elastic material these strains can be expressed as

$$\epsilon_y = \epsilon_z = -\nu\epsilon_x = -\nu\frac{\sigma_x}{E}$$

Now, consider the possibility of shear strains resulting from the normal stress σ_x. Again the isotropy of the material simplifies the relations. Suppose that a shear strain were present, as shown in Fig. 5.14*a*. A 180° rotation of the element about the x axis would appear to give a shear strain in the opposite sense, as shown by Fig. 5.14*b*. But the material is assumed to be isotropic, so its stress-strain behavior should be independent of a 180° rotation. This contradiction is avoided only if the shear strain due to a normal stress component vanishes. Similar arguments show that the other two components of shear strain also must vanish, and that, consequently, a normal stress produces only normal strains.

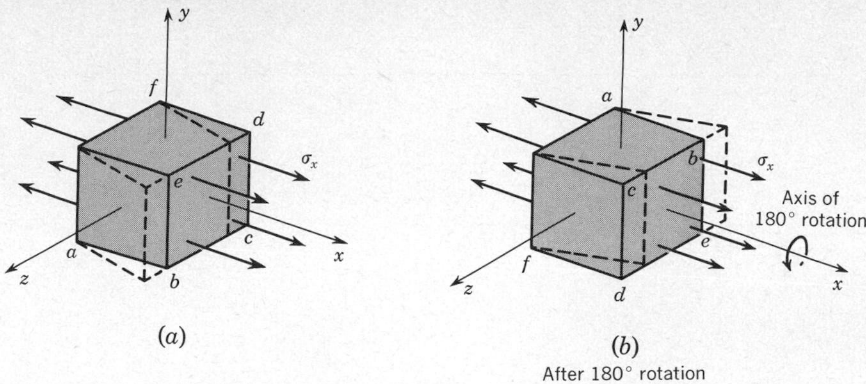

Fig. 5.14 Hypothetical shear strain due to normal stress.

Suppose, now, that a second normal component of stress σ_y is present. Because of the linearity of the stress-strain relation, an increment of stress will always produce the same increment of strain regardless of the level of stress before the increment was added. The strains resulting from σ_y are linearly related to σ_y and are directly additive to the strains due to σ_x. Furthermore, because of the isotropy, the strains due to σ_y will be

$$\epsilon_y = \frac{\sigma_y}{E}$$

$$\epsilon_x = \epsilon_z = -\nu\epsilon_y = -\nu\frac{\sigma_y}{E}$$

where the constants E and ν are the same as those appearing in the expressions for the strains due to σ_x. Analogous results are obtained for the strains due to σ_z.

Next, we show that a shear-stress component will produce only the corresponding shear-strain component. Suppose a shear stress did produce a normal strain component, as shown in Fig. 5.15*a*. If the cube were rotated 180° about a 45° axis into the position shown in Fig. 5.15*b*, the shear stress would appear the same and the material would appear the same, because it is isotropic, but the strain component would appear different. This contradiction is avoided only if the normal strain component does not exist. Similar symmetry arguments show that equal normal strains in all directions cannot be present and that the other two shear-strain components cannot exist. Thus, a shear-stress component produces only its corresponding shear-strain component, and for a linear-elastic material we may express this behavior in the following way

$$\gamma_{zx} = \frac{\tau_{zx}}{G} \qquad \gamma_{xy} = \frac{\tau_{xy}}{G} \qquad \gamma_{yz} = \frac{\tau_{yz}}{G}$$

where the constant of proportionality G is called the *shear modulus.*

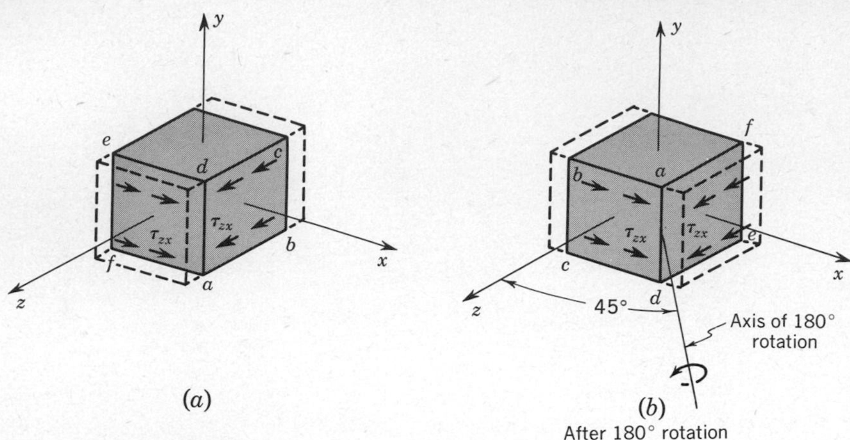

Fig. 5.15 Hypothetical normal strain due to shear stress.

Considering now a linear-elastic isotropic material with all components of stress present, we can summarize the above arguments by writing down the following stress-strain relations which are applicable to this case:

$$
\begin{aligned}
\epsilon_x &= \frac{1}{E}[\sigma_x - \nu(\sigma_y + \sigma_z)] \\
\epsilon_y &= \frac{1}{E}[\sigma_y - \nu(\sigma_z + \sigma_x)] \\
\epsilon_z &= \frac{1}{E}[\sigma_z - \nu(\sigma_x + \sigma_y)] \\
\gamma_{xy} &= \frac{\tau_{xy}}{G} \\
\gamma_{yz} &= \frac{\tau_{yz}}{G} \\
\gamma_{zx} &= \frac{\tau_{zx}}{G}
\end{aligned}
\tag{5.2}
$$

These equations are known as the *generalized Hooke's law*.

As may be seen from the above equations for the shear strains, a consequence of isotropy is the fact that the *principal axes of strain* at a point of a stressed body *coincide* with the *principal axes of stress* at that point. The angular relations in the Mohr's circles for stress and for strain are therefore identical, and in determining the location of the principal axes corresponding to a given state of stress, one may use either the Mohr's circle for stress or that for strain.

There is one more piece of information which the isotropy of the material can give. So far, in developing the stress-strain equations, we have only made use of the fact that the properties of the material are the same in the three orthogonal directions x, y, and z. A further relation between the elastic constants can be found from the fact that the material has the same properties referred to *any* set of coordinate axes. To illustrate this, we begin by considering a state of pure shear as shown in Fig. 5.16. Using (5.2) we obtain the following result for the shear strain:

$$\gamma_{xy} = \frac{\tau}{G} \tag{a}$$

It is possible to obtain another expression for γ_{xy} in the manner indicated in Fig. 5.16. To do this we proceed as follows: The stress component τ is equivalent to a principal state of stress with components $\sigma_1 = \tau$, $\sigma_2 = -\tau$ referred to the 1,2 coordinate axes. The principal strains referred to these axes are

$$\epsilon_1 = \frac{\sigma_1}{E} - \nu\frac{\sigma_2}{E} = \frac{\tau(1+\nu)}{E}$$

$$\epsilon_2 = \frac{\sigma_2}{E} - \nu\frac{\sigma_1}{E} = -\frac{\tau(1+\nu)}{E}$$

Now, upon use of the strain transformation formulas, we can express the shearing strain with respect to the xy axes in terms of the principal strains in the form

$$\gamma_{xy} = \epsilon_1 - \epsilon_2 = \frac{2(1+\nu)}{E}\tau \tag{b}$$

The expressions (a) and (b) for γ_{xy} must be equal, and this identity requires the following relation between the elastic constants:

$$G = \frac{E}{2(1+\nu)} \tag{5.3}$$

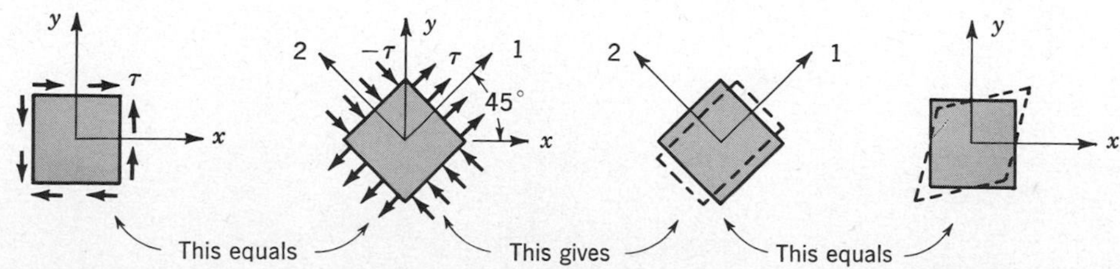

Fig. 5.16 Equivalent states of stress and strain.

Table 5.1 Elastic constants for isotropic materials at room temperature

Material	*Composition*	*Modulus of elasticity* E, 10^6 psi	*Poisson's ratio,* ν	*Shear modulus* G, 10^6 psi	*Coeff. of linear expansion* α, $10^{-6}/°F$	*Density,* lb/in.^{3}
Aluminum[1]	Pure and alloy	9.9–11.4	0.32–0.34	3.7–3.85	11.1–13.4	0.096–0.104
Brass[1,2]	60–70% Cu, 40–30% Zn	14.5–15.9	0.33–0.36	5.3–6.0	11.0–11.6	0.302–0.307
Copper[1,2,3]		17–18	0.33–0.36	5.8–6.7	9.2–9.4	0.323–0.324
Iron, cast[2,3]	2.7–3.6% C	13–21	0.21–0.30	5.2–8.2	5.8	0.251–0.265
Steel[1,2]	Carbon and low alloy	28–32	0.26–0.29	11.0–11.9	5.5–7.1	0.279–0.284
Stainless steel[3,7]	18% Cr, 8% Ni	28–30	0.30	10.6	8.3–9.4	0.276–0.286
Titanium[1,2]	Pure and alloy	15.4–16.6	0.34	6.0	4.9	0.163
Glass[4]	Various	7.2–11.5	0.21–0.27	3.8–4.7	3.3–5.3	0.086–0.14
Methyl methacrylate[5]		0.35–0.5	0.35	0.15	50	0.042
Polyethylene[5]		0.02–0.055	0.45	0.017	100	0.033
Rubber[6]		0.00011–0.00060	0.50	0.00005–0.00020	70–110	0.036–0.045

[1] C. J. Smithells, "Metals Reference Book," Interscience Publishers, Inc., New York, 1955.
[2] L. S. Marks, "Mechanical Engineers Handbook," McGraw-Hill Book Company, New York, 1958.
[3] "Metals Handbook," American Society for Metals, Cleveland, 1948.
[4] G. W. Morey, "Properties of Glass," p. 16, Reinhold Publishing Corporation, New York, 1954.
[5] *Modern Plastics*, Encyclopedia Issue, vol. 34, 1956.
[6] U.S. Rubber Co., "Engineering Properties of Rubber," Fort Wayne, Ind., 1950.
[7] C. L. Mantell, "Engineering Materials Handbook," McGraw-Hill Book Company, New York, 1958.

It is true, although it will not be proved here, that no other choice of coordinate axes gives any added information about the elastic constants, and thus for an isotropic material there are just *two independent* elastic constants. Typical values for various materials are given in Table 5.1.

A number of elastic materials are not isotropic but have a basic structural orientation which extends throughout the material. Examples of these materials and their stress-strain relations are discussed in Sec. 5.10.

5.5 THERMAL STRAIN

In the elastic region the effect of temperature on strain appears in two ways: first, by causing a modification in the values of the elastic constants, and second, by directly producing a strain even in the absence of stress. The effect on the elastic constants for many materials is small for temperature changes of a few hundred degrees Fahrenheit and will not be further considered. The strain due to temperature change in the absence of stress is called *thermal strain* and is denoted by the superscript t on the strain symbol thus: ϵ^t. For an isotropic material, symmetry arguments show that the thermal strain must be a pure expansion or contraction with no shear-strain components referred to any set of axes. The thermal strains are not exactly linear with temperature change, but for temperature changes of one or two hundred degrees Fahrenheit we can closely describe the actual variation by a linear approximation. We then obtain the following thermal strains due to a change in temperature from T_o to T.

$$\begin{aligned} \epsilon_x{}^t &= \epsilon_y{}^t = \epsilon_z{}^t = \alpha(T - T_o) \\ \gamma_{xy}{}^t &= \gamma_{yz}{}^t = \gamma_{zx}{}^t = 0 \end{aligned} \tag{5.4}$$

The quantity α is called the *coefficient of linear expansion*. Average values of α for temperature variations in the vicinity of room temperature are tabulated in Table 5.1 for a variety of common materials.

The total strain at a point in an elastic body is the sum of that due to stress and that due to temperature. Denoting the elastic strain due to stress by ϵ^e and the thermal part by ϵ^t, the total strain derived from the displacements is given by

$$\epsilon = \epsilon^e + \epsilon^t \tag{5.5}$$

For example, if the material is rigidly restrained so that no strain is possible, the elastic part of the strain will be the negative of the thermal strain, since the total strain is zero.

5.6 COMPLETE EQUATIONS OF ELASTICITY

The *theory of elasticity* is the name given to that body of knowledge which deals with the distribution of stress and strain in elastic bodies subjected to given loads, displacements, and distributions of temperature. We now are in a position to state completely the foundations of the theory of elasticity. The problem is to find distributions of stress and strain which meet the prescribed loads and displacements on the boundary and which at every point within satisfy the equilibrium equations, the stress-strain-temperature relations, and the geometrical conditions associated with the definition of strain and the concept of continuous displacements. The problem was outlined previously in broad generality by the three steps given in (2.1). For convenience we summarize below, under the three steps of (2.1), explicit[1] equations which must be satisfied at each point of a nonaccelerating, *isotropic*, linear-elastic body subject to small strains.

EQUILIBRIUM

On the surface the stress components must be in equilibrium with the given external loads, and within the body they must satisfy the following equilibrium equations:

$$\begin{aligned} &\frac{\partial \sigma_x}{\partial x} + \frac{\partial \tau_{xy}}{\partial y} + \frac{\partial \tau_{zx}}{\partial z} + X = 0 \\ &\frac{\partial \tau_{xy}}{\partial x} + \frac{\partial \sigma_y}{\partial y} + \frac{\partial \tau_{yz}}{\partial z} + Y = 0 \\ &\frac{\partial \tau_{zx}}{\partial x} + \frac{\partial \tau_{yz}}{\partial y} + \frac{\partial \sigma_z}{\partial z} + Z = 0 \end{aligned} \tag{5.6}$$

where X, Y, and Z are *body forces* which are distributed over the volume with intensities X, Y, and Z per unit volume. Equations (5.6) are generalizations of the two-dimensional equilibrium equations given in (4.13) (see Prob. 4.2).

GEOMETRIC COMPATIBILITY

The displacements must match the geometrical boundary conditions and must be continuous functions of position with which the strain components are associated, as follows:

$$\begin{aligned} \epsilon_x &= \frac{\partial u}{\partial x} & \gamma_{xy} &= \frac{\partial v}{\partial x} + \frac{\partial u}{\partial y} \\ \epsilon_y &= \frac{\partial v}{\partial y} & \gamma_{yz} &= \frac{\partial w}{\partial y} + \frac{\partial v}{\partial z} \\ \epsilon_z &= \frac{\partial w}{\partial z} & \gamma_{zx} &= \frac{\partial u}{\partial z} + \frac{\partial w}{\partial x} \end{aligned} \tag{5.7}$$

where u, v, and w are the displacement components in the x, y, and z directions. These equations are three-dimensional extensions of (4.33) (see Prob. 4.16).

[1] These equations may be written in abbreviated form by using indicial notation. See Prob. 5.49.

STRESS-STRAIN-TEMPERATURE RELATIONS

In addition to the relations between stress and strain components, we must include the effect of temperature on the strain components. Both of these effects are included in the following relations:

$$\begin{aligned}
\epsilon_x &= \frac{1}{E}[\sigma_x - \nu(\sigma_y + \sigma_z)] + \alpha(T - T_o) \\
\epsilon_y &= \frac{1}{E}[\sigma_y - \nu(\sigma_z + \sigma_x)] + \alpha(T - T_o) \\
\epsilon_z &= \frac{1}{E}[\sigma_z - \nu(\sigma_x + \sigma_y)] + \alpha(T - T_o) \\
\gamma_{xy} &= \frac{\tau_{xy}}{G} \\
\gamma_{yz} &= \frac{\tau_{yz}}{G} \\
\gamma_{zx} &= \frac{\tau_{zx}}{G}
\end{aligned} \tag{5.8}$$

The equilibrium equations (5.6), the strain-displacement equations (5.7), and the strain-stress-temperature relations (5.8) provide 15 equations for the six components of stress, the six components of strain, and the three components of displacement. These 15 equations are the foundation for what is commonly called *linear elasticity theory*. The equations are *linear* because of the linear material behavior assumed in (5.8) and also because of the restriction to small strains in (5.7). An additional consequence of (5.7) is that the strains (and hence the stresses) are associated with the undeformed configuration. This in turn means that (5.6) represents an application of the equilibrium requirements in the undeformed configuration. Thus the complete equations (5.6), (5.7), and (5.8) apply to deformations of isotropic, linearly elastic solids which involve small strains and for which it is acceptable to apply the equilibrium requirements in the undeformed configuration.

In order to obtain solutions to these equations it is generally necessary, because of the derivatives which appear in (5.6) and (5.7), to perform integrations. To fix the solution for a particular elastic body it is necessary to prescribe *boundary conditions* at every point on the surface of the body. Most commonly, either the *displacement* vector of the boundary point is specified, or the *stress* vector (4.2) applied by an external load is specified. It can be shown[1] that, if at every point on the surface of an elastic body, either the displacement vector or the surface stress

[1] See, for example, S. Timoshenko and J. N. Goodier, "Theory of Elasticity," 3d ed., p. 269, McGraw-Hill Book Company, New York, 1970; or C. Wang, "Applied Elasticity," p. 38, McGraw-Hill Book Company, New York, 1953.

vector is prescribed, then there exists a *unique* solution which satisfies (5.6), (5.7), and (5.8) throughout the interior and which meets the prescribed boundary conditions on the surface.

Uniqueness proofs also have been obtained for certain types of plastic material behavior.[1] The three factors of equilibrium, geometric compatibility, and force-deformation behavior of the material form the central core on which these proofs of existence and uniqueness are based. This is the basis for stating that the three steps of (2.1) contain all the principles which need to be considered in problems in the mechanics of deformable solids.

Our aim in presenting the complete equations of elasticity in this introductory text is to acquaint the reader with the essentially simple nature of the rigorous foundations of an important field in the mechanics of solids. In spite of the simplicity of the general problem statement, it has been difficult to obtain exact solutions to particular cases. There is a growing store of known solutions, but in most engineering applications it is still necessary to use results which are in some sense approximate. For example, a common approximation is to replace the actual boundary conditions for a physical system by simpler ones in an idealized model. In other cases approximate solutions are constructed which satisfy some, but not all, of the interior equations; for instance, the equilibrium requirements might be satisfied exactly at every interior point, but the geometric compatibility requirements would be satisfied only in some gross overall average sense. Numerical methods often involve approximating the body by finite elements which satisfy the governing equations in an overall manner but may violate equilibrium, compatibility, or boundary conditions at points along their boundaries. The degree of approximation which is acceptable will depend upon the circumstances of each situation; the deflection of the hairspring of a good watch would obviously have to be determined with greater accuracy than that of the wind-up spring of a child's cheap toy. The question of whether or not an approximation is good can be answered by comparing several different approximations which bound the actual situation or by making the ultimate test of any physical theory, which is to compare the theoretical results predicted by an analysis with experimental results obtained from the actual physical situation.

In this introductory study of the mechanics of solids we shall primarily be concerned with the three steps of (2.1), expressed not in the infinitesimal formulation of (5.6), (5.7), and (5.8) but expressed, instead, on a macroscopic level in terms of rods, shafts, and beams. We do consider, however, two simple examples of complete solutions to the infinitesimal formulation in this and the following section. Furthermore, we use the formulation of (5.6), (5.7), and (5.8) to make critical evaluations of the macroscopic force-deflection relations which are developed in subsequent chapters.

[1] A discussion of this point at the graduate level is given by R. Hill, A General Theory of Uniqueness and Stability in Elastic-plastic Solids, *J. Mech. Phys. Solids*, vol. 6, pp. 236–249, 1958.

To illustrate how a relatively simple problem situation can be idealized into a theoretical model and, further, to illustrate the application of the equations of elasticity to this model, we shall consider the following problem.

Example 5.2 A long, thin plate of width b, thickness t, and length L is placed between two rigid walls a distance b apart and is acted on by an axial force P, as shown in Fig. 5.17*a*. We wish to find the deflection of the plate parallel to the force P. We idealize the situation in Fig. 5.17*b*. In constructing this model we have assumed:

1. The axial force P results in an axial normal stress uniformly distributed over the plate area, including the end areas.
2. There is no normal stress in the thin direction. (Note that this implies a case of *plane stress* in the xy plane.)
3. There is no deformation in the y direction, that is, $\epsilon_y = 0$. (Note that this implies a case of *plane strain* in the xz plane.)
4. There is no friction force at the walls (or, alternatively, it is small enough to be negligible).
5. The normal stress of contact between the plate and wall is uniform over the length and width of the plate. We now satisfy the requirements (2.1) for the idealized model of Fig. 5.17*b*.

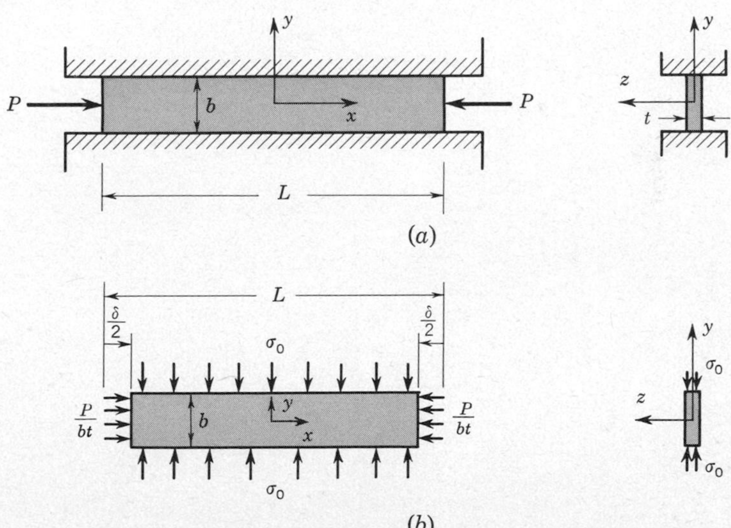

Fig. 5.17 Example 5.2. (*a*) Actual problem; (*b*) idealized model.

EQUILIBRIUM

Equilibrium with the external loads is satisfied when the stresses existing in the plate are

$$\sigma_x = -\frac{P}{bt} \qquad \sigma_y = -\sigma_0 \qquad \sigma_z = 0$$
$$\tau_{xy} = \tau_{yz} = \tau_{zx} = 0 \tag{a}$$

These stresses also satisfy the equilibrium equations (5.6), and therefore we assume them to be the stresses acting throughout the plate.

GEOMETRIC COMPATIBILITY

Since the walls are rigid, the plate cannot expand in the y direction, and therefore

$$\epsilon_y = 0 \tag{b}$$

Also, in terms of δ, we can write

$$\epsilon_x = -\frac{\delta}{L} \tag{c}$$

STRESS-STRAIN RELATIONS

In view of (*a*) and because the temperature is constant, Eqs. (5.8) reduce to

$$\epsilon_x = \frac{1}{E}(\sigma_x - \nu\sigma_y) \qquad \epsilon_y = \frac{1}{E}(\sigma_y - \nu\sigma_x) \qquad \epsilon_z = -\frac{\nu}{E}(\sigma_x + \sigma_y)$$
$$\gamma_{xy} = \gamma_{yz} = \gamma_{zx} = 0 \tag{d}$$

Solving the system of equations (*a*), (*b*), (*c*), and (*d*), we find

$$\sigma_y = \nu\sigma_x = -\nu\frac{P}{bt}$$
$$\delta = (1 - \nu^2)\frac{PL}{Ebt} \tag{e}$$
$$\epsilon_z = \nu(1 + \nu)\frac{P}{Ebt} = \frac{\nu}{1 - \nu}\frac{\delta}{L}$$

We note that the presence of the rigid walls reduces the axial deflection of the plate by the factor $(1 - \nu^2)$.

STRAIN-DISPLACEMENT RELATIONS

If we consider the origin of coordinates to be in the center of the plate and assume that this point does not move in either the x or the z direction, then by substituting the strains into Eqs. (5.7) and integrating all six relations, it can be shown that the proper displacements for this plate are

$$u = -\frac{\delta}{L}x$$

$$v = 0 \tag{f}$$

$$w = \frac{\nu}{1-\nu}\frac{\delta}{L}z$$

We have thus obtained a rigorous and exact solution to the elasticity problem for the idealized model of Fig. 5.17*b*. It should be emphasized that we have *not* obtained an exact solution to the actual problem of Fig. 5.17*a*, where there is a concentrated force rather than a uniformly distributed stress acting on the ends of the plate. The problem of Fig. 5.17*a* was itself an approximation to a more realistic loading over a small region of contact. On the basis of experiments in similar situations, it is probable that the overall deflection of the actual plate is near that estimated from the idealized model. Also, it is probable that away from the ends the stress distribution for the actual plate is quite close to that for the model, even though the stress distributions are quite different in the vicinity of the ends. We shall return to this point in Sec. 5.7 in connection with the discussion of St. Venant's principle. This problem is an illustration from a class of situations in which it is very difficult to get an exact solution to the real problem, but in which it is relatively easy to get an exact or nearly exact solution to an idealized approximation of the real problem.

5.7 COMPLETE ELASTIC SOLUTION FOR A THICK-WALLED CYLINDER

In this section we consider a further example of an exact solution within the theory of elasticity for homogeneous isotropic solids. The problem in this case arises in many engineering applications so that the solution is, in itself, of considerable technical importance. Furthermore, the solution exhibits several characteristics which are representative of a wide class of more complicated problems. Much insight into the nature of elastic solutions can thus be gained from a study of this one example.

Consider the cylinder with inner radius r_i and outer radius r_o shown in Fig. 5.18*a*. We shall determine the distribution of stress within the cylinder when it is subjected to the external loadings indicated in Fig. 5.18*b*. There is uniform inner pressure p_i, uniform outer pressure p_o, and uniform axial tensile stress σ_o. This configuration provides a model for several practical problems. For example, the cylinder might be a thick-walled pressure vessel where the important load is the inner pressure or a submersible hull where the important load is the outer pressure.

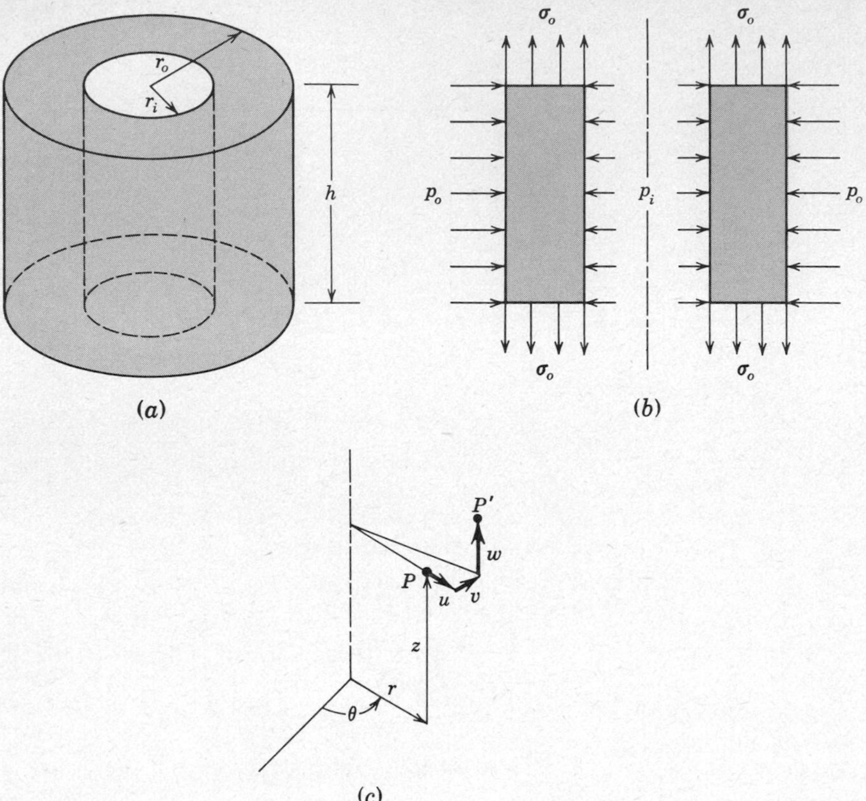

Fig. 5.18 Thick-walled cylinder (*a*) subjected to inner and outer pressures and axial tension (*b*). Cylindrical coordinates and displacement components (*c*).

Alternatively, if the height h were small compared with the radii, the cylinder might be a plate or a disk, and the important load might be the inner pressure which arises from a "shrink-fit" attachment to a shaft.

To take advantage of the cylindrical symmetry, we use the cylindrical coordinates r, θ, and z shown in Fig. 5.18*c*. Also indicated in Fig. 5.18*c* are the components u, v, and w in the r, θ, and z directions of the displacement vector in cylindrical coordinates. The 15 equations of elasticity in cylindrical coordinates consist of the three *equilibrium* equations of Prob. 4.4, six *strain-displacement* equations of Prob. 4.19, and the six *stress-strain* equations in cylindrical coordinates corresponding to (5.2). The boundary conditions are

$$\sigma_r = -p_i \qquad \tau_{rz} = 0 \qquad \tau_{r\theta} = 0 \tag{a}$$

on the inner surface $r = r_i$,

$$\sigma_r = -p_o \qquad \tau_{rz} = 0 \qquad \tau_{r\theta} = 0 \tag{b}$$

on the outer surface $r = r_o$, and

$$\sigma_z = \sigma_o \qquad \tau_{rz} = 0 \qquad \tau_{\theta z} = 0 \tag{c}$$

on the top and bottom surfaces, where $z = h$ and $z = 0$. The problem is to determine the stresses which together with the strains and displacements satisfy the 15 interior equations and the three boundary conditions at each point on the boundary surface.

The general problem is greatly simplified in the present case because of the radial symmetry of loading. Based on symmetry, we shall look for a solution in which v, the θ component of displacement, vanishes everywhere and in which all stresses, strains, and displacements are independent of θ. If we find such a solution, we know from the uniqueness principle that it is indeed *the* solution. We also make the tentative hypothesis, based on the uniformity of the axial loading, that $\sigma_z = \sigma_o$ throughout the interior and that all stresses and strains are independent of z. With these assumptions the problem collapses to manageable size. The shear stresses $\tau_{r\theta}$, $\tau_{\theta z}$, τ_{rz} and the corresponding strains $\gamma_{r\theta}$, $\gamma_{\theta z}$, γ_{rz} vanish everywhere. What remains are two displacements u and w, two stresses σ_r and σ_θ, and three strains ϵ_r, ϵ_θ, and ϵ_z, with a single *equilibrium* equation,

$$\frac{d\sigma_r}{dr} + \frac{\sigma_r - \sigma_\theta}{r} = 0 \tag{d}$$

three *strain-displacement* equations,

$$\epsilon_r = \frac{du}{dr} \qquad \epsilon_\theta = \frac{u}{r} \qquad \epsilon_z = \frac{dw}{dz} \tag{e}$$

and three *stress-strain* equations,

$$\begin{aligned}
\epsilon_r &= \frac{1}{E}[\sigma_r - \nu(\sigma_\theta + \sigma_z)] \\
\epsilon_\theta &= \frac{1}{E}[\sigma_\theta - \nu(\sigma_z + \sigma_r)] \\
\epsilon_z &= \frac{1}{E}[\sigma_z - \nu(\sigma_r + \sigma_\theta)]
\end{aligned} \tag{f}$$

together with the boundary conditions

$$\begin{aligned}
\sigma_r &= -p_i \quad \text{at} \quad r = r_i \\
\sigma_r &= -p_o \quad \text{at} \quad r = r_o
\end{aligned} \tag{g}$$

We now outline a possible systematic procedure for obtaining a solution to (*d*), (*e*), and (*f*) which meets the boundary conditions (*g*). Starting from an unknown radial displacement $u(r)$, we use (*e*) to express the transverse strains ϵ_r and ϵ_θ as functions of u. Then from the first two equations of (f) we solve for the transverse stresses σ_r and σ_θ in terms of ϵ_r and ϵ_θ and thus obtain the stresses also as functions of u. Finally, substituting the stresses into (d) leads to the following differential equation for $u(r)$

$$\frac{d^2u}{dr^2} + \frac{1}{r}\frac{du}{dr} - \frac{u}{r^2} = 0 \tag{h}$$

The general solution to (h) is

$$u = Ar + \frac{B}{r} \tag{i}$$

where A and B are constants of integration. We now retrace our steps, substituting (i) in place of u to obtain expressions for the strains and the stresses which are functions of r and the constants A and B. The constants are then evaluated by requiring the radial stress to meet the boundary conditions (g). Finally, after substituting back for A and B, we obtain the following expressions for the transverse stresses

$$\begin{aligned}\sigma_r &= -\frac{p_i[(r_o/r)^2 - 1] + p_o[(r_o/r_i)^2 - (r_o/r)^2]}{(r_o/r_i)^2 - 1} \\ \sigma_\theta &= \frac{p_i[(r_o/r)^2 + 1] - p_o[(r_o/r_i)^2 + (r_o/r)^2]}{(r_o/r_i)^2 - 1}\end{aligned} \tag{5.9}$$

The axial strain is obtained by substituting these stresses together with $\sigma_z = \sigma_o$ into the third equation of (f),

$$\epsilon_z = \frac{\sigma_o}{E} - \frac{2\nu}{E}\frac{p_i r_i^{\,2} - p_o r_o^{\,2}}{r_o^{\,2} - r_i^{\,2}} \tag{5.10}$$

Note that ϵ_z is independent of position within the cylinder. The axial displacement w thus varies linearly with z. It can be shown that in obtaining these results we have exactly satisfied the original set of 15 equations and the boundary conditions of (a), (b), and (c). This justifies our tentative hypotheses concerning the symmetry of the solution. We turn now to a discussion of the significance of the results (5.9) and (5.10).

We note that the transverse stresses σ_r and σ_θ both vary with the radial coordinate r and depend linearly on the pressure loadings p_i and p_o. The transverse stresses, however, are independent of the axial loading σ_o. The uniform axial strain ϵ_z depends on the axial loading σ_o and on the pressure loadings (through the action of Poisson's ratio).

To study the behavior of the transverse stresses (5.9), we consider several special cases. When the inner and outer pressures are both equal (that is, $p_i = p_o = p$), we find that $\sigma_r = \sigma_\theta = -p$ throughout the interior. When the outer pressure is absent ($p_o = 0$), we note that an inner pressure p_i results in a compressive radial stress which varies from $\sigma_r = -p_i$ at the inner wall to $\sigma_r = 0$ at the outer wall. The tangential stress σ_θ is tensile throughout with peak magnitude occurring at the inner wall $r = r_i$. The manner in which σ_r and σ_θ vary with r is indicated in Fig. 5.19*a* for a cylinder in which $r_o = 2r_i$. Figure 5.19*b* displays the corresponding distributions of stress due to an outer pressure p_o when the inner pressure is absent ($p_i = 0$). Note that the numerically greatest stress in both Fig. 5.19*a* and Fig. 5.19*b* is the tangential stress σ_θ at the inner wall of the cylinder.

When the cylinder wall-thickness $t = r_o - r_i$ becomes small in comparison with r_i, the solution (5.9) approaches the thin-walled-tube approximation of Prob. 4.10 (see Prob. 5.47).

The axial stress $\sigma_z = \sigma_o$ and the axial strain (5.10) do not vary in the interior of the cylinder. Two important special cases occur commonly in applications. When the axial stress vanishes ($\sigma_o = 0$), the cylinder is said to be subject to a *plane stress* distribution. In this case the axial strain ϵ_z is generally not zero (unless $p_i r_i^2 = p_o r_o^2$). When the axial strain vanishes ($\epsilon_z = 0$), the cylinder is said to be

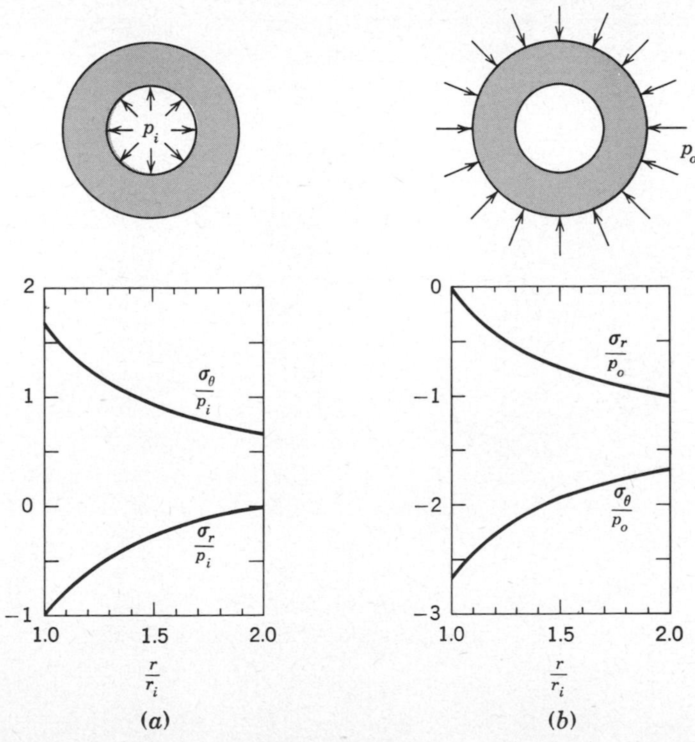

Fig. 5.19 Distribution of radial stress $\sigma_r(r)$ and tangential stress $\sigma_\theta(r)$ in cylinder with $r_o = 2r_i$ due to (*a*) inner pressure p_i and (*b*) outer pressure p_o.

subject to a *plane strain* distribution. In this case the axial stress is generally not zero (again, unless $p_i r_i^2 = p_o r_o^2$).

Two characteristics of the solution just obtained for a circular cylinder extend to isotropic elastic cylinders of arbitrary contour (but constant height h) subjected to transverse loads which vary arbitrarily around the contour (but which are uniform with respect to z). In such cases it can be shown that the transverse stress distribution is independent of any uniform axial stress or axial strain. In particular, the transverse stress distribution in plane stress is identical with the stress distribution in plane strain if the transverse boundary conditions are identical.

A second characteristic of the solution (5.9) is that, although it depends on the material's being homogeneous, isotropic, and linearly elastic, the stresses are *independent* of the actual magnitudes of the elastic parameters E and ν. This result extends to cylinders of arbitrary contour,[1] provided that the resultants of the transverse boundary stresses acting on each closed contour are separately zero (e.g., in Fig. 5.18 the resultant of the pressure forces acting on the inner wall alone is zero, and the resultant of the pressure forces acting on the outer wall alone is also zero). This condition would not be met, for example, if, after the cylinder of Fig. 5.18 was shrunk onto a shaft, the shaft was pressed or twisted so as to exert a net force or moment on the cylinder. The fact that the stress distributions in these circumstances are independent of the elastic parameters is of great importance for experimental stress analysis. It permits predicting the stress distribution in one material from measurements on a geometrically similar specimen of another material. In particular, photoelastic analysis using birefringent polymer specimens can be used to predict stress distributions in metals, even though the elastic moduli are of quite different order of magnitude (see Sec. 4.14).

The exact solution (5.9) permits us to verify a very general qualitative principle in elasticity known as *St. Venant's principle*.[2] Consider two sets of boundary stress vectors which are to be applied to the same elastic body. Let these loadings be identical over all the boundary surface except over a certain small region R where they differ. The resulting internal stress distributions will, in general, be different throughout the interior, but St. Venant's principle asserts that significant differences in internal stress will be localized in the immediate neighborhood of R if the two loadings over the region R are *statically equivalent*. No general statement can be made as to precisely how large the neighborhood of significant difference will be. This depends on the size, shape, and location of the small region R, as well as on the nature of the different loadings over R. If ϵ represents a representative lineal dimension of the region R, then a rough rule of thumb is

[1] See, for example, N. I. Muskhelishvilli, "Some Basic Problems of the Mathematical Theory of Elasticity," 2d English ed., p. 160, P. Noordhoff, Ltd., Groningen, Netherlands, 1963.

[2] See, for example, S. Timoshenko and J. N. Goodier, "Theory of Elasticity," 3d ed., p. 39, McGraw-Hill Book Company, New York, 1970.

that, for engineering purposes, the differences in internal stress become an insignificant fraction of the surface stress differences at distances from the surface that are more than two or three times greater than ϵ.

To verify this we consider the cylinder of Fig. 5.18 in the special case where the inner radius is very small in comparison with the outer radius ($r_i \ll r_o$). Let one transverse loading system consist of no load at all (that is, $p_i = 0$, $p_o = 0$), and let the other loading consist of an inner pressure $p_i = p$, with $p_o = 0$. Here the loadings are identical except over the region R, which consists of the inner wall of small radius r_i. The representative dimension ϵ of the region R may be taken as r_i. On this surface the two loadings are statically equivalent since the resultant of the inner pressure forces is zero. Now the interior stresses for the first loading system vanish everywhere, while for the second loading system the internal stresses as given by (5.9) are approximately

$$\sigma_r = -p\left(\frac{r_i}{r}\right)^2 \qquad \sigma_\theta = p\left(\frac{r_i}{r}\right)^2 \tag{j}$$

where $r_i \ll r_o$. On the inner boundary the stresses for the two loading systems differ in magnitude by p. Equations (j) indicate that the differences in internal stresses decay to less than 10 percent of this magnitude at interior points for which r is slightly greater than $3r_i$. At such points the distance to the boundary is slightly greater than $2r_i$. This provides rough confirmation of the rule of thumb given in connection with St. Venant's principle. In Example 5.2 the nature of the stress distribution to be expected in the actual plate away from the ends was obtained by appealing to St. Venant's principle. In that case the region R is the end surface of the bar, and the representative dimension ϵ is the width of the bar b.

Finally, we can use the exact result (5.9) to illustrate the concept of *stress concentration*. When an elastic body with a local geometrical irregularity, such as an oil hole, a keyway, or a notch, is stressed, there usually is a localized variation in stress state in the immediate neighborhood of the irregularity. The peak stress levels at the irregularity may be several times larger than the nominal stress levels in the bulk of the body. Under these circumstances the irregularity is said to cause a stress concentration. As an illustration consider again the cylinder of Fig. 5.18 for the case where $r_i \ll r_o$, but now let the transverse loading be the outer pressure only, that is, $p_i = 0$, $p_o = p$. The small hole of radius r_i can be considered as a geometrical irregularity in an otherwise solid cylinder. The exact stress distribution in the interior is given by (5.9). It can be shown (see Prob. 5.48) that away from the small hole the stress state is very nearly biaxial compression, with $\sigma_r = \sigma_\theta = -p$, while at the surface of the hole the stress state is simple tension, with $\sigma_r = 0$ and σ_θ very nearly equal to $2p$. The peak stress level is thus nearly twice that in the bulk of the cylinder. The change in stress state is concentrated in the near vicinity of the hole: approximately 90 percent of the change takes place within a radius of $r = 3r_i$. Further discussion of stress concentrations is given in Sec. 5.9.

5.8 STRAIN ENERGY IN AN ELASTIC BODY

In Sec. 2.6 the concept of elastic energy was introduced in terms of springs and uniaxial members. Here we extend the concept to arbitrary linearly elastic bodies subjected to small deformations. In (2.10) the elastic energy U stored in a *linear* spring is given in three forms: in terms of the deflection δ, in terms of the force F, or in terms of the deflection δ *and* the force F. We shall find the latter form

$$U = \tfrac{1}{2}F\delta \tag{5.11}$$

most convenient for our present purposes. Because of the linearity, force and deflection grow *in proportion* during the loading process, and thus the total work done is just *one-half* the product of the final force and the final deflection.

Let us apply this concept to an infinitesimal element within a linearly elastic body. Figure 5-20*a* shows a uniaxial stress component σ_x acting on a rectangular element, and Fig. 5-20*b* shows the corresponding deformation including the elongation due to the strain component ϵ_x. The elastic energy stored in such an element is commonly called *strain energy*. In this case the force $\sigma_x\,dy\,dz$ acting on the positive x face does work as the element undergoes the elongation $\epsilon_x\,dx$. In a linearly

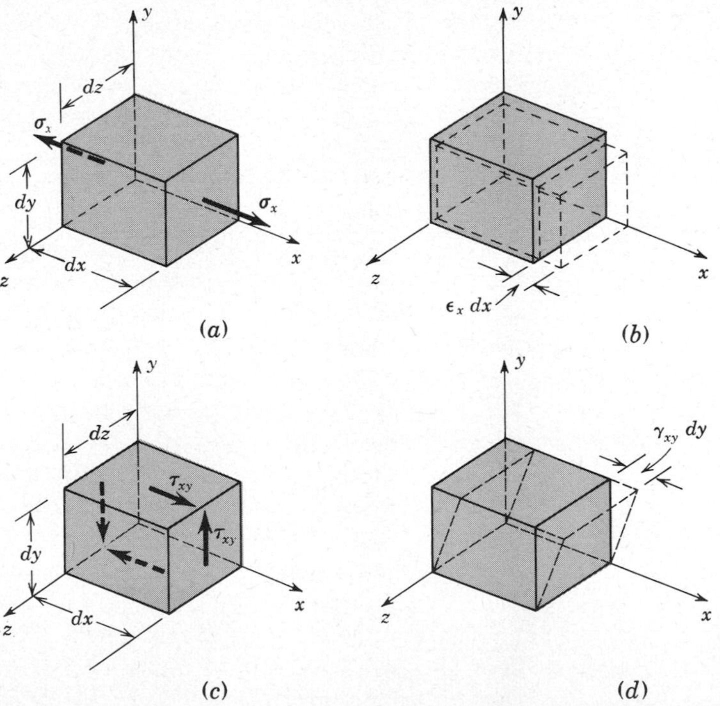

Fig. 5.20 Infinitesimal element subjected to: uniaxial tension (*a*), with resulting deformation (*b*); pure shear (*c*), with resulting deformation (*d*).

elastic material, strain grows in proportion to stress. Thus the strain energy dU stored in the element, when the final values of stress and strain are σ_x and ϵ_x, is

$$dU = \tfrac{1}{2}(\sigma_x\, dy\, dz)(\epsilon_x\, dx) = \tfrac{1}{2}\sigma_x\epsilon_x\, dV \tag{5.12}$$

where $dV = dx\, dy\, dz$ is the volume of the element. If an elastic body of total volume V is made up of such elements, the total strain energy U is obtained by integration

$$U = \tfrac{1}{2}\int_V \sigma_x\epsilon_x\, dV \tag{5.13}$$

As an elementary application of (5.13) we reconsider the linear uniaxial member of Figs. 2.4 and 2.5. In this case the stress and strain can be taken to be uniform through the volume of the member. With $\sigma_x = P/A$ and $\epsilon_x = \delta/L$, we have

$$U = \tfrac{1}{2}\left(\frac{P}{A}\right)\left(\frac{\delta}{L}\right)\int_V dV = \tfrac{1}{2}P\delta \tag{5.14}$$

since $V = LA$. Note that (5.14) agrees with the result (2.11) obtained in Sec. 2.6.

Consider next the shear-stress component τ_{xy} acting on the infinitesimal element in Fig. 5.20*c*. The corresponding deformation due to the shear-strain component γ_{xy} is indicated in Fig. 5.20*d*. In this case the force $\tau_{xy}\, dx\, dz$ acting on the positive y face does work as that face translates through the distance $\gamma_{xy}\, dy$. Because of linearity, γ_{xy} and τ_{xy} grow in proportion as the element is deformed. The strain energy stored in the element, when the final values of strain and stress are γ_{xy} and τ_{xy}, is

$$dU = \tfrac{1}{2}(\tau_{xy}\, dx\, dz)(\gamma_{xy}\, dy) = \tfrac{1}{2}\tau_{xy}\gamma_{xy}\, dV \tag{5.15}$$

Results analogous to (5.12) and (5.15) can be written for any other pair of stress and strain (components for example, σ_y and ϵ_y or τ_{yz} and γ_{yz}) whenever the stress component involved is the only stress acting on the element.

Finally, we consider a general state of stress in which all six stress components are present. The corresponding deformation will in general involve all six strain components. The individual strain components may depend on more than one stress component [e.g., see the first three equations of (5.2)], but we assume that the dependence is linear. Thus, if we imagine a gradual loading process in which all stress components maintain the same relative magnitudes as in the final stress state, the strain components will also grow in proportion, maintaining the same relative magnitudes as in the final strain state. During this process in which all stresses and strains are growing, a single stress component such as σ_x will do work *only* on the deformation due to its corresponding strain ϵ_x (this can be seen in Fig. 5.20 by observing that, for small strains, none of the other strains ϵ_y, ϵ_z, γ_{xy}, γ_{yz}, or γ_{xz} involve a change in the distance between the x faces of the element). The total strain energy stored in the element when the final stresses are σ_x, σ_y, σ_z, τ_{xy}, τ_{yz}, τ_{zx} and the final strains are ϵ_x, ϵ_y, ϵ_z, γ_{xy}, γ_{yz}, γ_{zx} is thus

$$dU = \tfrac{1}{2}(\sigma_x\epsilon_x + \sigma_y\epsilon_y + \sigma_z\epsilon_z + \tau_{xy}\gamma_{xy} + \tau_{yz}\gamma_{yz} + \tau_{zx}\gamma_{zx})\,dV \tag{5.16}$$

In general, the final stresses and strains vary from point to point in the body. The strain energy stored in the entire body is obtained by integrating (5.16) over the volume of the body

$$U = \tfrac{1}{2}\int_V (\sigma_x\epsilon_x + \sigma_y\epsilon_y + \sigma_z\epsilon_z + \tau_{xy}\gamma_{xy} + \tau_{yz}\gamma_{yz} + \tau_{zx}\gamma_{zx})\,dV \tag{5.17}$$

This formula for strain energy applies to small deformations of any linearly elastic body. For isotropic materials the stress-strain relations (5.2) can be inserted in (5.17) to obtain formulas for the strain energy completely in terms of strain components, or completely in terms of stress components (see Prob. 5.51). In the case of plane stress or plane strain parallel to the xy plane, (5.17) reduces to

$$U = \tfrac{1}{2}\int_V (\sigma_x\epsilon_x + \sigma_y\epsilon_y + \tau_{xy}\gamma_{xy})\,dV \tag{5.18}$$

In Chaps. 6 and 7 we shall use these results to develop special formulas for strain energy in torsion and bending.

5.9 STRESS CONCENTRATION

We noted in Sec. 5.7 that when an elastic body with a local geometrical irregularity, such as an oil hole, a keyway, or a notch, is stressed, there usually is a localized variation in the stress state in the immediate neighborhood of the irregularity. The maximum stress levels at the irregularity may be several times larger than the nominal stress levels in the bulk of the body. This increase in stress caused by the irregularity in geometry is called a *stress concentration*. Where the stress concentration cannot be avoided by a change in design, it is important to base the design on the *local* value of the stress rather than on an average value. The usual procedure in design is to obtain the local value of the stress by use of a *stress concentration factor*. The stress concentration factor K_t is defined as the ratio of the maximum stress σ_{max}, in the presence of a geometric irregularity or discontinuity, to the nominal stress σ_{nom}, which would exist at the point if the irregularity were not there

$$K_t = \sigma_{\text{max}}/\sigma_{\text{nom}}$$

The magnitude of this factor depends upon the particular geometry and loading involved, but factors of 2 or more are common.

The determination of the stress concentration factor in practical engineering situations is a difficult problem, and only a few stress concentrations have been studied exactly.[1] What has been done is to obtain a few exact solutions, check them against results from photoelastic stress analysis, and from them make estimates of what the stress concentration factor should be in a variety of other practical cases. A typical stress concentration factor curve is shown in Fig. 5.21. Similar curves may be found in the literature. In the absence of a stress concentration curve for a particular design application, it may be possible to estimate the stress concentration factor by an empirical equation.[2]

Consider the stress concentration shapes shown in Fig. 5.22. In each of these, there is a radius of curvature a at the root of a notch and a relevant dimension which may be taken to be either the half-thickness of the remaining material, or the height of a shoulder. Let the least of such dimensions be called c.

[1] For more complete discussions, see R. E. Peterson, "Stress Concentration Factors in Design," John Wiley & Sons, Inc., New York, 1953; G. N. Savin, "Stress Concentration Around Holes," Pergamon Press, New York, 1961 (translated from the Russian edition).

[2] More complete discussion of the use of the empirical equation may be found in F. A. McClintock and A. Argon, *op. cit.*, chap. 11.

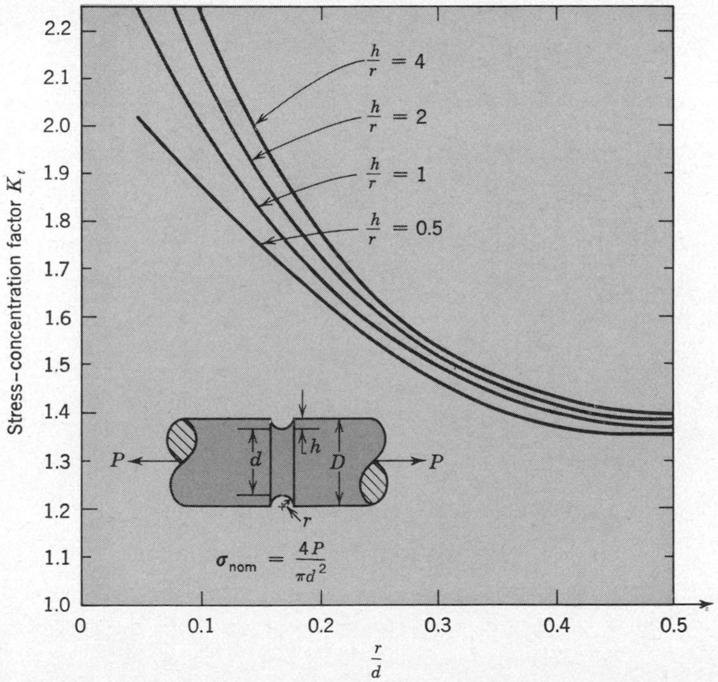

Fig. 5.21 Stress concentration factor K_t for a circular groove in a solid circular shaft with tensile force P. (*From C. Lipson and R. Juvinall, "Handbook of Stress and Strength," The Macmillan Company, New York, 1963.*)

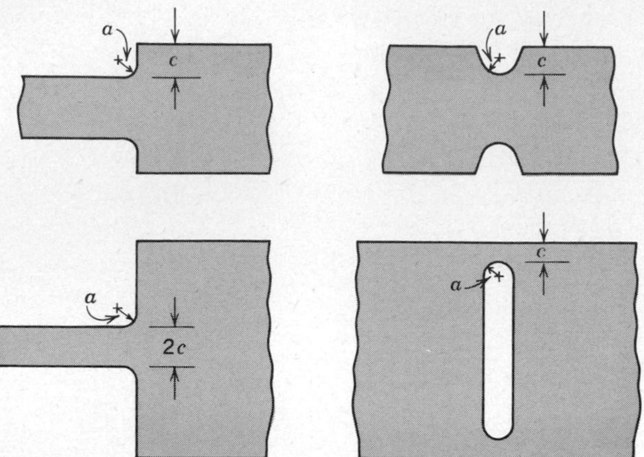

Fig. 5.22 Examples of stress concentration configurations.

Then the strain or stress concentration factor is given approximately by

$$\frac{\epsilon_{\text{max}}}{\epsilon_{\text{nom}}} \approx \frac{\sigma_{\text{max}}}{\sigma_{\text{nom}}} \approx 1 + (0.3 \text{ to } 2)\sqrt{\frac{c}{a}} \tag{5.19}$$

In Sec. 5.7 we noted that the stress concentration factor for a small hole in a large cylinder under outer pressure was approximately equal to 2. The same result may be obtained from (5.19) by taking both c and a to equal the hole radius r_i and by choosing the intermediate value of unity for the numerical coefficient. In using (5.19), the numerical coefficient should be chosen at the low end of the range for fillets with generous radii and for bending and torsion, and it should be chosen at the high end for tension. While Eq. (5.19) appears to possess some vagueness, it does provide a handy working approximation to elastic stress and strain concentrations.

Usually the stress concentration factor is defined in terms of maximum stress. We shall find that other aspects of the state of stress may be important; e.g., for yielding it is often the maximum difference of principal stress components that is critical. In cases involving plastic flow and ductile fracture, the *strain* concentration rather than the stress concentration may be the important variable. In elasticity the stress and strain concentrations are more or less in proportion, whereas in plasticity the strain concentrations can be much higher. The plastic stress concentrations are limited to at most a few times the yield strength. However, discussion of plastic stress and strain concentrations is beyond the scope of this text.[1] In the remainder of this book we will discuss a number of situations where the effect of stress concentrations may play a role. However, in the interests

[1] See further discussion in F. A. McClintock and A. Argon, *op. cit.*, chaps. 11 and 18.

of simplicity we do not usually consider the evaluation and use of stress concentration factors. In practical engineering situations, however, the possibility of stress concentrations should *always* be considered.

5.10 COMPOSITE MATERIALS AND ANISOTROPIC ELASTICITY

Fiber-reinforced composite materials are of rapidly growing practical importance; they are also very old. When bricks were simply dried mud, it is mentioned in the Bible that the Egyptian pharaoh punished the Israelites by making them gather their own straw for brickmaking. The importance of fiber length was known even then, for we are told that the Israelites had to gather stubble instead of straw. Other composites are rope and fabrics, especially coated fabrics. Steel-reinforced concrete was introduced in the 19th century. We shall consider reinforced concrete later in connection with the bending of beams, but here we turn to the newer materials, consisting of fine, extremely high-strength fibers embedded in a matrix. For room-temperature service a polymer matrix, typically epoxy, is most useful. Metal matrices for high-temperature service have not yet proven practical.

Properties of the currently most promising fibers are shown in Table 5.2. Glass has the advantage of least cost, but has relatively low stiffness. The boron filament is commercially available, but it has the disadvantage that it is produced by plating onto a tungsten substrate which is inherently expensive. Furthermore, the relatively large diameter of the boron filaments means that much of the available strength is lost in bending the fibers to form curved parts. Graphite filaments are the most promising because of their very high stiffness and strength, combined with low weight. They are formed by heating fine monofilaments in a vacuum under

Table 5.2 Properties of high-strength fibers

	Glass[1]		*Boron*[2]	*Graphite*	
Property	*Type* E	*Type* S		*Stiff*	*Strong*
Diameter, in.	0.0002–0.0008		0.004	0.00027–0.00035	
Specific gravity	2.54	2.50	2.63	1.96	1.74
Modulus of elasticity, 10^6 psi	10.5	12.6	55	>55	>35
Tensile strength, 10^3 psi	450–550	650	450	>200	>350
Cost per pound in epoxy tape 1968				$595	
1969				$495	
1970	$5		$150	$380	

[1] L. J. Broutman and R. H. Krock (eds.), "Modern Composite Materials," Addison-Wesley Publishing Company, Inc., Reading, Mass., 1967.

[2] Manufacturers' data in this and the following tables include those from Celanese, 3M, PPG, and Whittaker.

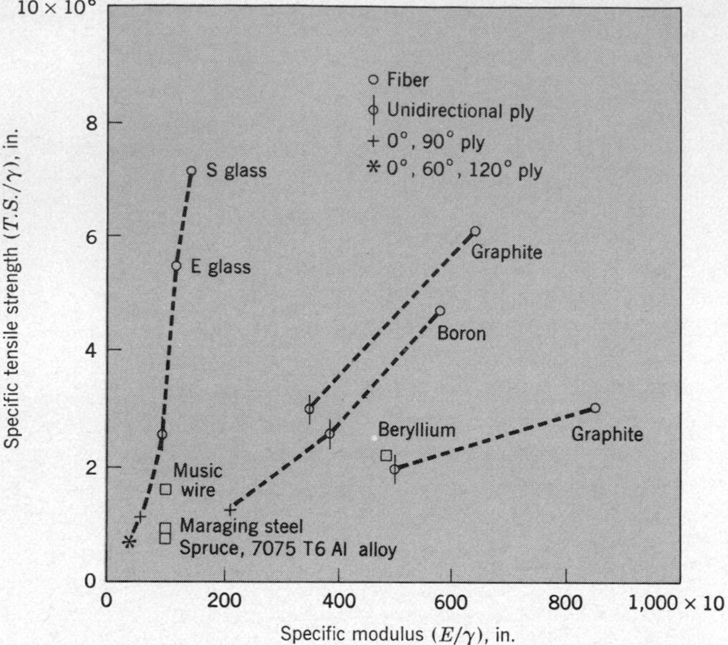

Fig. 5.23 Specific tensile strength (tensile strength per unit weight density γ) and specific modulus of high-strength materials. Metals are in sheet form and isotropic.

tension to drive off the hydrogen and to form graphite crystals aligned with the fiber axis. Variations in the pyrolyzing temperature give essentially two different grades of graphite, one stiffer but weaker than the other. The chief problem is the cost. If it continues to drop to the order of \$100/lb there will be a very wide range of uses, not only in aircraft and space but also in high-speed machinery where stiffness is important to control vibrations. Examples are machine tools, circular saws, printing presses, and textile machinery. Other uses include springs, bearings, and pressure vessels. Where weight is important, it is interesting to compare the fibrous materials with more conventional ones on the basis of tensile strength per unit density and modulus of elasticity per unit density, as shown in Fig. 5.23. Beryllium is the most attractive solid metal, but its brittleness prevents its use in many cases and makes its processing quite expensive (it is also highly toxic).

Fibers require a matrix to hold them in place during normal handling. Much of the advantage of the fibers themselves is lost in this process, as indicated in Table 5.3 and Fig. 5.23. In general the compressive modulus should be the same as the tensile value. The compressive strength is limited by buckling of fibers within the matrix, whereas the tensile strength is determined by fracture from flaws. The bending properties are included for reference.

Table 5.3 Uniaxial properties of fiber laminates. Parallel fibers in epoxy matrix

Property		*Glass*	*Boron*	*Graphite*	
Volume, % fiber		64 (wt.)	50	55	
Specific gravity		1.8		1.5–1.6	
Tensile modulus, 10^6 psi,	72°F	5.7	30–32	25–33	18–22
	300°F			25–32	18–20
Tensile strength, 10^3 psi,	72°F	160	190–230	105–120	150–200
	300°F	50		105–115	150–190
Compressive modulus, 10^6 psi,	72°F		35–36		
Compressive strength, 10^3 psi,	72°F		440–460	75–110	120–160
Flexural modulus, 10^6 psi,	72°F	5.3	28	23–32	18–21
	300°F	1.0	24	20–30	17–18
Flexural strength, 10^3 psi,	72°F	165	245	120–140	185–250
	300°F	30	215	95–105	155–220
Short beam, shear strength, 10^3 psi,	72°F		17	7–9	13–16
	300°F		9	5	8–9

Naturally the properties of a composite material will be different across the fibers from what they are along them. Materials with different properties in different directions are called *anisotropic*. Examples of other anisotropic materials are shown in Fig. 5.24. The most frequently observed example is wood. Flat-sawed wood cut from the outer sections of a tree will have the growth rings or cylinders aligned in approximately parallel planes. The stiffness of the wood is markedly different across these growth planes from what it is parallel to them. A single crystal of metal has the atoms arranged in a regular array. Since the atoms are more densely packed in certain directions than others, it is not surprising that the stiffnesses are different in different directions. Even polycrystalline metals, if rolled, are likely to have a preferred orientation of structure because the plastic deformation tends to line up the crystals in certain directions and tends at the same time to elongate and flatten them. In discussing materials such as these, where there is a basic structure, it is convenient to choose coordinate axes related to that structure. In the examples shown in Fig. 5.24 the structures all appear identical after a 180° rotation about any one of the three orthogonal coordinate axes. Such materials are called *orthotropic*.

An orthotropic material is a special case of an anisotropic material. An example of more general anisotropy is the crystal structure shown in Fig. 5.25, in which the atomic spacings in the three crystallographic directions are unequal and the crystallographic directions are not orthogonal. No rotations short of 360° give the same geometrical configuration. Here there is an interrelation between all components of stress and strain. If we assume that the strains in an elastic anisotropic material are linearly related to the stresses, then the stress-strain relations are given in (5.20), where the elastic constants appearing in the stress-strain

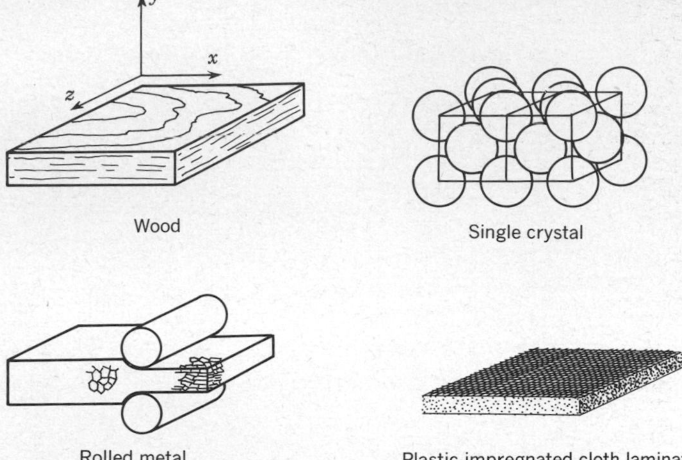

Fig. 5.24 Orthotropic materials.

relations are denoted by double subscripts, the first referring to the strain component and the second to the stress component:[1]

$$\begin{aligned}
\epsilon_x &= S_{11}\sigma_x + S_{12}\sigma_y + S_{13}\sigma_z + S_{14}\tau_{xy} + S_{15}\tau_{yz} + S_{16}\tau_{zx} \\
\epsilon_y &= S_{21}\sigma_x + S_{22}\sigma_y + S_{23}\sigma_z + S_{24}\tau_{xy} + S_{25}\tau_{yz} + S_{26}\tau_{zx} \\
\epsilon_z &= S_{31}\sigma_x + S_{32}\sigma_y + S_{33}\sigma_z + S_{34}\tau_{xy} + S_{35}\tau_{yz} + S_{36}\tau_{zx} \\
\gamma_{xy} &= S_{41}\sigma_x + S_{42}\sigma_y + S_{43}\sigma_z + S_{44}\tau_{xy} + S_{45}\tau_{yz} + S_{46}\tau_{zx} \\
\gamma_{yz} &= S_{51}\sigma_x + S_{52}\sigma_y + S_{53}\sigma_z + S_{54}\tau_{xy} + S_{55}\tau_{yz} + S_{56}\tau_{zx} \\
\gamma_{zx} &= S_{61}\sigma_x + S_{62}\sigma_y + S_{63}\sigma_z + S_{64}\tau_{xy} + S_{65}\tau_{yz} + S_{66}\tau_{zx}
\end{aligned} \tag{5.20}$$

Actually, the elastic constants with unequal subscripts are the same when the order of the subscripts is reversed: $S_{12} = S_{21}$, $S_{45} = S_{54}$, etc. This can be proved from

[1] In many books, 4 denotes yz, 5 denotes zx, and 6 denotes xy.

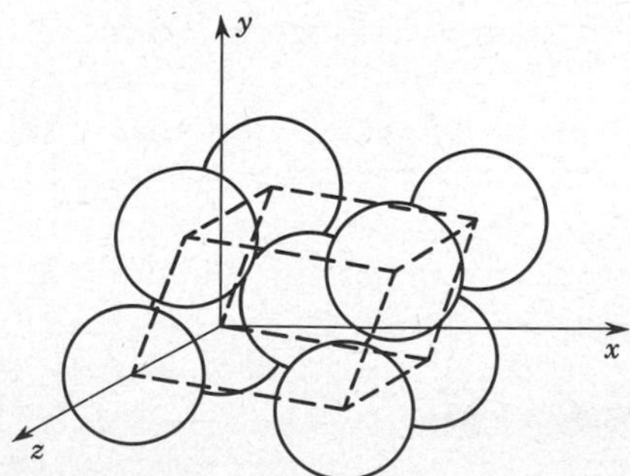

Fig. 5.25 An anisotropic crystal structure.

the fact that the net work on loading and subsequent unloading of the material should be zero, regardless of the order of application or removal of the various components of stress.

The symmetry of an orthotropic material requires that there be no interaction between the various shear components or the shear and normal components when the x,y,z axes are chosen parallel to the axes of structural symmetry. Thus the stress-strain relations reduce to:

$$\begin{aligned} \epsilon_x &= S_{11}\sigma_x + S_{12}\sigma_y + S_{13}\sigma_z & \gamma_{xy} &= S_{44}\tau_{xy} \\ \epsilon_y &= S_{21}\sigma_x + S_{22}\sigma_y + S_{23}\sigma_z & \gamma_{yz} &= S_{55}\tau_{yz} \\ \epsilon_z &= S_{31}\sigma_x + S_{32}\sigma_y + S_{33}\sigma_z & \gamma_{zx} &= S_{66}\tau_{zx} \end{aligned} \tag{5.21}$$

where $S_{12} = S_{21}$, $S_{13} = S_{31}$, and $S_{23} = S_{32}$.

Table 5.4 gives the orthotropic elastic constants for several fiber-reinforced epoxy materials. Complete data are difficult to find in the literature, and data are often given in terms of different constants. Note, as also shown in Fig. 5.23, that where transverse strength is desired, the cross-ply construction reduces the specific stiffness and strength to values that are little if any in excess of those corresponding to high-strength alloys. Table 5.5 gives orthotropic elastic constants for four types of wood. Table 5.6 shows the variation of modulus of elasticity with orientation as a function of angle for several materials. In general, these results can be obtained from the elastic coefficients referred to the structure of the material,

Table 5.4 Orthotropic elastic constants for fiber-epoxy materials

Fiber axis x, sheet normal z

						Coeff. of linear expansion, $10^{-6}/°F$	
Fiber	*Directions*	S_{11}, 10^{-6} in.^{2}/lb	$\dfrac{S_{22}}{S_{11}}$	$\dfrac{S_{12}}{S_{11}}$	$\dfrac{S_{44}}{S_{11}}$	α_x	α_y
Isotropic		$1/E$	1	$-\nu$	$2(1+\nu)$		
Glass A[1]	0°	0.178	3.25	−(0.28–0.30)	15.0		
Glass B[1]	0°	0.175				4.8	12.3
	0°, 90°	0.27	1			7.1	7.1
	±45°	0.62	1			7.1	7.1
	0°, 60°, 120°	0.35–0.38	1			8.4	8.4
Boron	0°	0.031–33	8.2–9.4	−(0.17–0.20)	16.2–17.6		
	0°, 90°	0.055–58	1	−0.05			
	±45°	0.26	1	−0.85			

[1] Glasses A and B are from two different manufacturers.

Table 5.5 Orthotropic elastic constants for various woods
Axes defined in Fig. 5.24

Material	*Moisture,* %	*Density,* lb/ft³	S_{11}, 10^{-6} in.²/lb	$\frac{S_{22}}{S_{11}}$	$\frac{S_{33}}{S_{11}}$	$\frac{S_{12}}{S_{11}}$	$\frac{S_{23}}{S_{11}}$	$\frac{S_{31}}{S_{11}}$	$\frac{S_{44}}{S_{11}}$	$\frac{S_{55}}{S_{11}}$	$\frac{S_{66}}{S_{11}}$
Balsa	9	4–14	1–8	20	70	−0.3	−15	−0.5	18	200	29
Yellow birch	12	40	0.48–0.50	13	20	−0.5	−9	−0.5	14	60	15
Douglas-fir	12	27–30	0.51–0.69	15	20	−0.4	−7	−0.5	16	140	13
Sitka spruce	12	22–25	0.59–0.64	13	23	−0.4	−6	−0.5	16	20	16

Note: Thermal coefficients of expansion are not given because moisture effects are much more important than thermal effects under normal conditions.
Source: "Wood Handbook," U.S. Department of Agriculture Handbook no. 72, 1955.

as discussed below for cubic metals. Alternatively, explicit expressions can be found for the elastic constants as a function of angle.[1]

If a material has equal properties in three orthogonal directions, it is said to have a *cubic structure*. In this case many of the elastic constants in Eqs. (5.21) are identical, and those equations reduce to:

$$\begin{aligned} \epsilon_x &= S_{11}\sigma_x + S_{12}(\sigma_y + \sigma_z) \qquad & \gamma_{xy} &= S_{44}\tau_{xy} \\ \epsilon_y &= S_{11}\sigma_y + S_{12}(\sigma_z + \sigma_x) \qquad & \gamma_{yz} &= S_{44}\tau_{yz} \\ \epsilon_z &= S_{11}\sigma_z + S_{12}(\sigma_x + \sigma_y) \qquad & \gamma_{zx} &= S_{44}\tau_{zx} \end{aligned} \tag{5.22}$$

The isotropy condition (equal properties in *all* directions) which resulted in Eq. (5.3) is not in general satisfied, so there remain three independent elastic constants. The elastic constants for several common cubic metals are listed in Table 5.7. Although some progress has been made in calculating these constants from quantum mechanics, the values are more accurately found by experiment.

Even with cubic symmetry, the stiffness of a crystal depends markedly on the orientation. Suppose we wish to determine the effects of a single normal component of stress referred to noncrystallographic axes, such as the a, b, c axes of Fig. 5.26. Using Mohr's circle for stress, we can determine the components of stress referred to the crystallographic axes x, y, and z. Then, from the stress-strain relations expressed relative to the crystallographic directions, Eqs. (5.22), we can determine the strains referred to the crystallographic axes. Finally, using Mohr's circle for strain, the state of strain can be described in terms of the specimen axes a, b, and c. The ratio between the normal stress component σ_a and the normal strain component ϵ_a gives the modulus of elasticity in the a direction. This modulus of elasticity may differ from that in one of the crystallographic directions by a large

[1] See, for example, McClintock and Argon, *op. cit.*, pp. 74, 78.

Table 5.6 Modulus of elasticity for orthotropic materials in sheet form
For various directions in the plane of the sheet as a function of angle with principal structural direction in plane of sheet

Material	*Principal structural direction*	*Angle* 0° E, 10^6 psi	45° E, 10^6 psi	90° E, 10^6 psi
Cold-rolled iron[1]	Direction of rolling	32.8	29.3	39.1
Cold-rolled copper[2]	Direction of rolling	19.8	15.5	20.0
Cold-rolled copper, recrystallized[2]	Direction of rolling	10.0	17.5	9.5
Glass-fiber-reinforced polyester[3]	Direction of warp	2.0–2.7	1.2–1.8	1.7–2.4

[1] E. Goens and E. Schmid, On the Elastic Anisotropy of Iron, *Naturwissenschaften*, vol. 19, pp. 520–524, 1931.
[2] J. Weertz, Elasticity of Copper Sheet, *Z. Metallk.*, vol. 25, pp. 101–103, 1931.
[3] "Plastics for Aircraft," part I, "Reinforced Plastics," ANC 17 Panel, Civil Aeronautics Board, U.S. Department of Commerce, 1955.

amount. In the case of a single crystal of iron, for example, the stiffness is about twice as great in the direction of the body diagonal of the cube as it is in the direction of the edge of the cube.

A more striking feature of the elastic behavior of cubic or orthotropic materials is that a normal stress component will produce a shear-strain component, and vice versa, when the axes of the specimen do not correspond to a symmetry axis of the structure of the material. In this case, the stress-strain relations referred to the specimen axes take on their most general anisotropic form, as given in Eqs. (5.20).

Table 5.7 Elastic constants for cubic materials

Material	S_{11}, 10^{-7} in.^{2}/lb	S_{12}, 10^{-7} in.^{2}/lb	S_{44} 10^{-7} in.^{2}/lb	$S_{11} - S_{12} - \frac{1}{2}S_{44}$, 10^{-7} in.^{2}/lb
Al	1.10	−0.40	2.43	0.28
Cu	1.03	−0.43	0.92	1.00
Fe	0.522	−0.195	0.595	0.419
Pb	6.43	−0.29	4.80	4.32
W	0.178	−0.050	0.455	0.000
95% Al, 5% Cu	1.04	−0.48	2.56	0.024
72% Cu, 28% Zn	1.34	−0.58	0.96	0.96

Source: F. Seitz and T. A. Read, Theory of the Plastic Properties of Solids, I, *J. Appl. Phys.*, vol. 12, pp. 100–118, 1941.

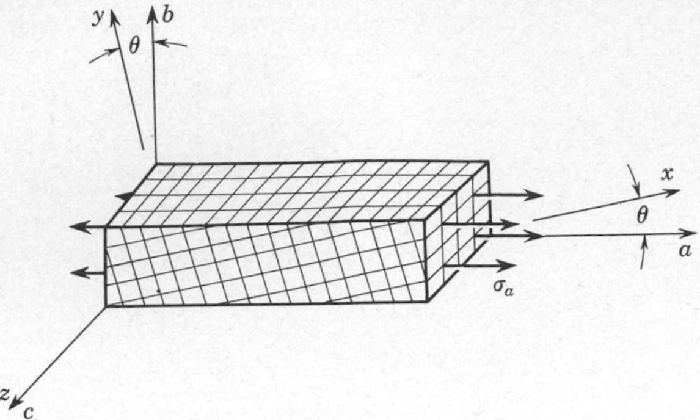

Fig. 5.26 Tensile test with specimen axes not coincident with crystallographic axes.

When making a tensile test on a specimen whose axis is not a symmetry axis of the structure, the occurrence of the shear-strain component makes it difficult to design specimens and grips which produce purely tensile stresses without secondary bending and shearing stresses.

In orthotropic materials the coefficients of thermal expansion will, in general, be different in the different crystallographic directions. For a material with cubic symmetry, it can be shown by symmetry arguments that the thermal strains referred to the axes of cubic symmetry consist only of three equal normal strains. From Mohr's circle for strain it then can be seen that there are no shear-strain components, and thus the thermal strain in a material with cubic symmetry also must be one of equal expansion or contraction in all directions.

5.11 CRITERIA FOR INITIAL YIELDING

We now turn to the problem of what happens when, in a general state of stress, the material is stressed to the point where it no longer behaves in a linearly elastic manner. For some materials, such as rubber, the deformation is still elastic, even after the proportional limit has been passed. For viscoelastic materials, the nonlinear stress-strain curve in the tension test may be due to viscous flow. For most materials, however, including metals, the deviation from proportionality in a uniaxial tensile test is an indication of the beginning of plastic flow (yielding). In this section we shall consider the problem of determining the conditions for yielding when any or all of the stress components are present; that is, using the data obtained in a simple tensile test, we shall look for relations which will predict the onset of yielding in a general state of stress. Again we shall restrict ourselves to polycrystalline materials which are at least statistically isotropic.

A brief introduction to the physical phenomena occurring in a metal being deformed elastically and plastically may help one to understand the ideas which will be proposed to correlate quantitatively the yielding phenomenon. During elastic deformation of a crystal, there is a uniform shifting of whole planes of atoms relative to each other, as shown in Fig. 5.27*a*, where the solid circles represent the atoms in their deformed position. Plastic deformation, on the other hand, depends on the motion of individual imperfections in the crystal structure. One kind of imperfection, called an *edge dislocation*, is shown in cross section in Fig. 5.27*b*. The dislocation consists of a series of planes (parallel to the paper) in each of which there is an area, shown enclosed in the circle, where three atoms are located above two atoms. Under the presence of a shear stress this dislocation will tend to migrate, as shown in the series of sketches, until there has been a displacement of the upper part of the crystal relative to the lower by approximately one atomic spacing. These dislocations can move in a variety of directions on a number of different crystallographic planes. By a combination of such motions, plastic strain can be produced. If this slip, or plastic strain, takes place on one crystallographic plane, very low stresses, of the order of 100 psi, are required to produce it. The relatively high stresses observed in engineering metals arise from the interactions of dislocations with each other, with alloying atoms or phases, and with the boundaries of the multitude of crystals making up the polycrystalline

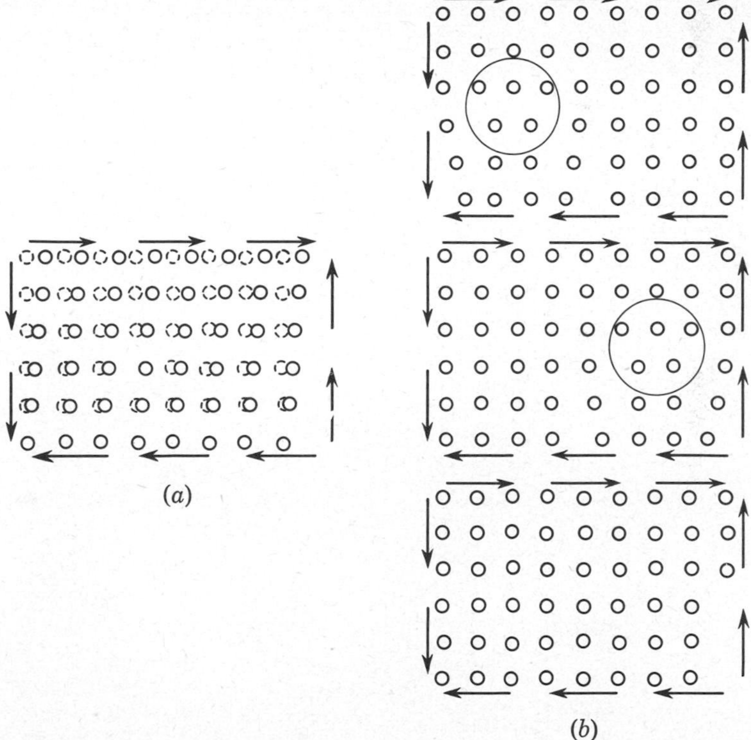

Fig. 5.27 Deformation of a crystal lattice. (*a*) Elastic deformation; (*b*) plastic deformation.

aggregate. There may be millions of these crystals and a million miles of dislocations in a cubic inch of plastically deformed metal. It is important to note that a consequence of this simple model is that shear stress is the dominant agent in the migration of these dislocations. A hydrostatic state of stress (equal normal components in three directions) would not tend to move the dislocation. Further, volume changes are small, since net changes in the crystal-lattice spacing are restricted to localized areas around the dislocation, and while some intersecting dislocations can leave strings of vacancies or holes behind them, this effect is small compared with the other deformations involved.

Not too surprisingly, little quantitative understanding of plastic deformation has yet been obtained from atomic physics. We shall derive the relations between stress and plastic-strain increments from symmetry arguments and from empirical evidence gathered from large-scale (not atomic-scale) experiments, coupled with the observations from the above model that density changes are unimportant and that pure pressure does not produce plastic strain.

While some very fine-scale plastic flow takes place by occasional motion of dislocations in a few individual crystals at low stress levels, for most engineering metals the readily observable plastic deformation takes place rather abruptly. In the uniaxial tensile test, the condition for the beginning of flow was described by the yield strength, giving the axial normal component of stress at which practically important plastic deformation was observed. When several components of stress are present, yielding must depend on some particular combination of these components. For example, consider a thin-walled cylinder of internal radius r and wall thickness t with an internal pressure and axial load, as discussed in Prob. 4.10. The radial stress is small compared with the tangential stress (by the ratio t/r), and thus we may consider a small element of this shell as being in plane stress with the principal stress components indicated in Fig. 5.28*b*. Experiments have been carried out on such thin-walled tubes with various amounts of axial load applied to determine under what combinations of these two normal components of stress the material will yield. The results of such experiments are shown in Fig. 5.29.

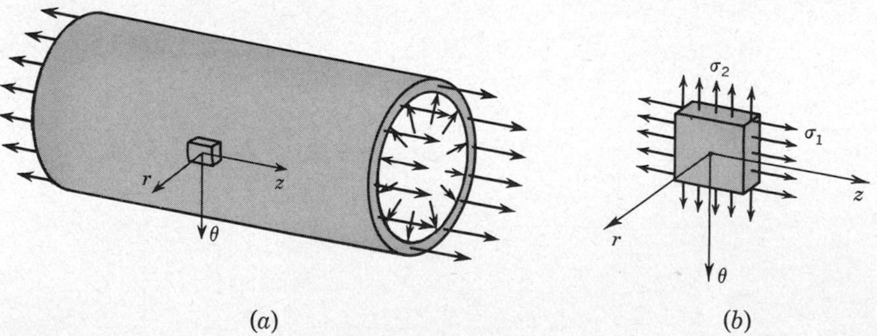

Fig. 5.28 Example of biaxial stress in a thin-walled cylinder.

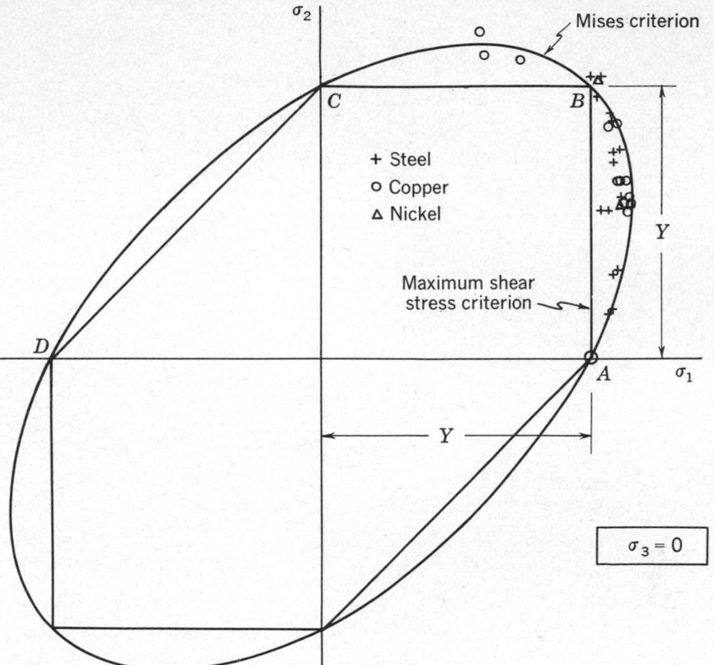

Fig. 5.29 Yielding of thin-walled tubes under combined stress. (*From W. Lode, Versuche uber den Einfluss der mittleren Hauptspannung auf das Fliessen der Metalle Eisen, Kupfer, und Nickel, Z. Physik, vol. 36, pp. 913–939, 1926.*)

There is at present no theoretical way of calculating what the relation between the stress components should be to correlate yielding in a three-dimensional state of stress with yielding in the uniaxial tensile test. Two empirical equations have been proposed which are reasonably simple and, at the same time, reasonably descriptive of the data. Each is based on the following two general considerations. First, the state of stress can be described completely by giving the magnitude and orientation of the principal stresses, but, since we are considering only isotropic materials, the orientation of the principal stresses is unimportant, and thus these criteria for yielding are based only on the magnitude of the principal stresses. Second, since experimental work[1] has substantiated the expectation from dislocation theory that a hydrostatic state of stress does not affect yielding, the two criteria are based not on the absolute magnitude of the principal stresses but rather on the magnitude of the *differences* between the principal stresses.

[1] P. W. Bridgman, "Studies in Large Plastic Flow and Fracture," McGraw-Hill Book Company, New York, 1952.

The first of these criteria assumes that yielding can occur in a three-dimensional state of stress when the root mean square of the differences between the principal stresses reaches the same value which it has when yielding occurs in the tensile test. Letting Y denote the stress at which yielding begins in the simple tensile test, the principal stresses are $\sigma_1 = Y$, $\sigma_2 = \sigma_3 = 0$. Thus, in the tensile test, yielding occurs when the root mean square of the differences between the principal stresses is

$$\sqrt{\tfrac{1}{3}[(\sigma_1 - \sigma_2)^2 + (\sigma_2 - \sigma_3)^2 + (\sigma_3 - \sigma_1)^2]}$$
$$= \sqrt{\tfrac{1}{3}[(Y - 0)^2 + (0 - 0)^2 + (0 - Y)^2]} = \sqrt{\tfrac{2}{3}}\, Y$$

Carrying the factor $\sqrt{\tfrac{2}{3}}$ over to the left side for convenience, this criterion can be expressed as follows: For a general state of stress, yielding can occur when

$$\sqrt{\tfrac{1}{2}[(\sigma_1 - \sigma_2)^2 + (\sigma_2 - \sigma_3)^2 + (\sigma_3 - \sigma_1)^2]} = Y \tag{5.23}$$

where, again, Y is the stress at which yielding begins in the tensile test. This criterion is known by the names of a number of men who independently conceived it: Maxwell, Mises, Huber, Hencky. It is also known as the distortion-energy criterion, or the octahedral shear-stress criterion, for yielding. We shall refer to it as the *Mises yield criterion*. When a stress state is known in terms of stress components with respect to nonprincipal axes, it is convenient to use the following transformation of the Mises yield criterion:[1]

$$[\tfrac{1}{2}\{(\sigma_x - \sigma_y)^2 + (\sigma_y - \sigma_z)^2 + (\sigma_z - \sigma_x)^2\} + 3\tau_{xy}^{\ 2} + 3\tau_{yz}^{\ 2} + 3\tau_{zx}^{\ 2}]^{1/2} = Y \tag{5.24}$$

A geometrical interpretation of the criterion (5.23) can be visualized by considering a space in which the coordinates of a point are given by the principal stresses σ_1, σ_2, and σ_3. The criterion (5.23) then is represented in this space by a right-circular cylinder of radius $\sqrt{\tfrac{2}{3}}\, Y$ whose axis makes equal angles with the σ_1, σ_2, and σ_3 coordinate axes, as illustrated in Fig. 5.30. Yielding occurs for any state of stress which lies on the surface of this circular cylinder. When we have a state of plane stress ($\sigma_3 = 0$), the Mises criterion is represented by an ellipse on a diagram for which σ_1 and σ_2 are the coordinates, as shown in Fig. 5.29.

The second empirical criterion assumes that yielding occurs whenever the maximum shear stress reaches the value it has when yielding occurs in the tensile test. The maximum shear stress (see Fig. 4.23) is one-half the difference between the maximum and minimum principal stresses, and it occurs on faces inclined at 45° to the faces on which the maximum and minimum principal stresses act. In

[1] This can be shown, for example, from W. Prager and P. G. Hodge, "Theory of Perfectly Plastic Solids," pp. 16, 22, 23, John Wiley & Sons, Inc., New York, 1951; or A. Mendelson, "Plasticity, Theory and Application," pp. 37, 77, The Macmillan Company, New York, 1968.

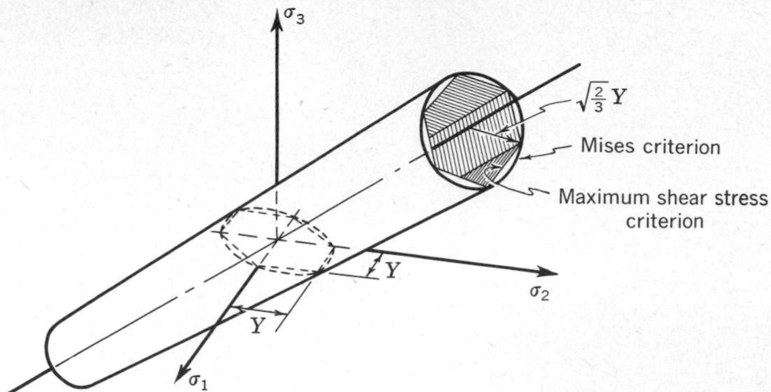

Fig. 5.30 Geometrical representation in principal stress space of the Mises and maximum shear-stress yield criteria.

the tensile test the maximum shear stress is $Y/2$, so this criterion says that yielding occurs when

$$\tau_{\max} = \frac{\sigma_{\max} - \sigma_{\min}}{2} = \frac{Y}{2} \tag{5.25}$$

This criterion is known as the Tresca, the Guest, or the maximum shear-stress criterion. We shall refer to it as the *maximum shear-stress criterion.*

A geometrical interpretation also can be made for the maximum shear-stress criterion. In the above-mentioned σ_1, σ_2, σ_3 space, the criterion (5.25) can be represented by a hexagonal cylinder inscribed within the right-circular cylinder of the Mises criterion, as illustrated in Fig. 5.30. Yielding occurs for any state of stress which lies on the surface of this hexagonal cylinder. For plane stress ($\sigma_3 = 0$) the maximum shear-stress criterion is represented by the six-sided polygon shown inscribed within the Mises ellipse in Fig. 5.29.

The application of the maximum shear-stress criterion to the problem of yielding under combined stress, such as that occurring in the thin-walled cylinder of Fig. 5.28, is not as straightforward as the application of the Mises criterion because some judgment must be exercised in determining which difference between the principal stresses is the maximum. For example, consider the tube to be under primarily axial tension with just a little internal pressure. The stress on an element of the tube wall will be as shown in Fig. 5.31. The maximum difference in

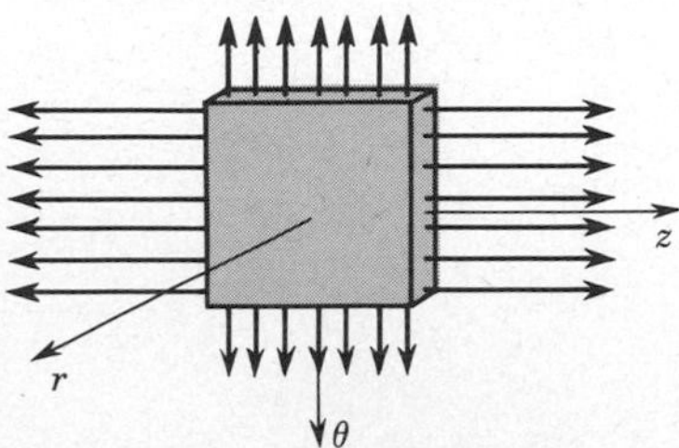

Fig. 5.31 State of stress in cylinder wall of Fig. 5.28*a* when σ_z and σ_r determine the maximum shear stress.

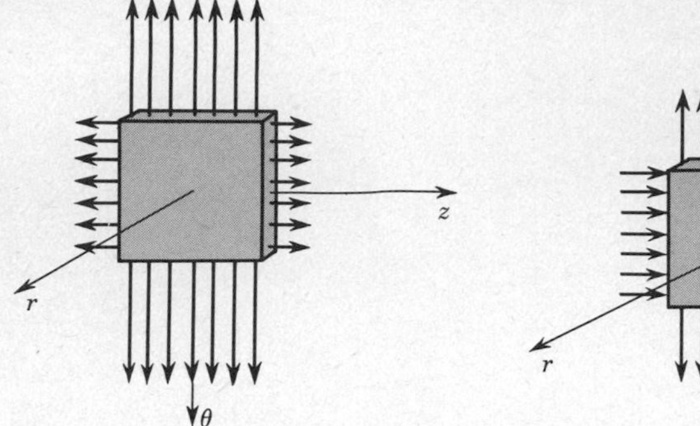

Fig. 5.32 State of stress in cylinder wall of Fig. 5.28*a* when σ_θ and σ_r determine the maximum shear stress.

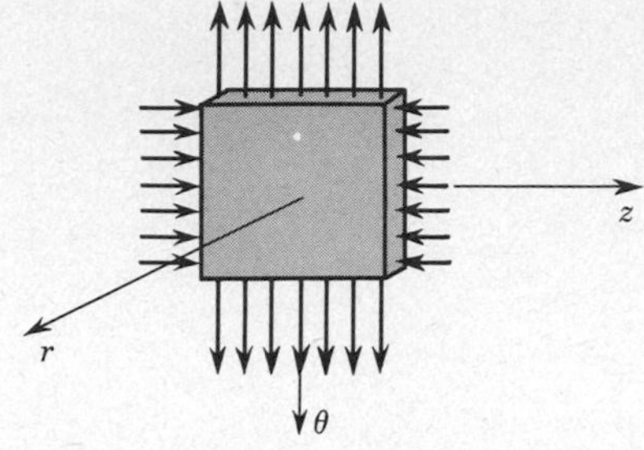

Fig. 5.33 State of stress in cylinder wall of Fig. 5.28*a* when σ_θ and σ_z determine the maximum shear stress.

principal stresses occurs between the z and r components, the latter component being essentially zero (σ_r is smaller than σ_θ by a factor of t/r and thus may be neglected for a thin-walled tube). Increasing the internal pressure does not change either the σ_r (approximately) or the σ_z components of stress, so the addition of the internal pressure does not increase the tendency to yield. This corresponds to proceeding along the straight line from A toward B in Fig. 5.29. When the internal pressure is great enough so that the σ_θ component of stress equals the σ_z component of stress, the shear stress on the plane at 45° to the θ and r axes becomes equal to the shear stress on the plane at 45° to the z and r axes. This situation corresponds to the point B of Fig. 5.29. Further increases in the ratio of internal pressure to axial load result in the maximum difference in principal stresses occurring between the σ_θ and the σ_r components so that now the axial load does not contribute to the tendency to yielding. This situation is sketched in Fig. 5.32, and if we keep the internal pressure constant and decrease the axial load, this corresponds to proceeding along the straight line from B toward C in Fig. 5.29. If we now start to compress the tube in the axial direction, we have the condition shown in Fig. 5.33. Now σ_r is the intermediate stress, and Eq. (5.25) shows that any increase in the axial compressive stress must be accompanied by an equal decrease in the tangential tensile stress; this corresponds to proceeding along the straight line from C to D in Fig. 5.29. Thus the different sides of the polygon in Fig. 5.29 correspond to different components of the stress being the contributing ones in the shear-stress yield criterion.

The experimental results plotted in Fig. 5.29 are intermediate between the maximum shear-stress and the Mises criteria but closer to the latter. This is generally the case for initial yielding of annealed isotropic metals.

5.12 BEHAVIOR BEYOND INITIAL YIELDING IN THE TENSILE TEST

When the tensile test of a ductile material is carried beyond the point of initial yielding, we can consider unloading and subsequent loading of the specimen, or we can carry the test further by increasing the load until the specimen fractures. Let us first consider the case of loading and unloading (see Fig. 5.6*b*). The following description is an idealized description of the behavior of a real material during loading and unloading beyond initial yielding; a more complete description of the factors contributing to postyielding plastic behavior in a general state of stress will be given in Sec. 5.16. It is assumed that the stress-strain curve obtained from a uniaxial tensile test is as shown in Fig. 5.34*a*. A fresh specimen of the material (Fig. 5.34*b*) is stretched in tension to point A, where the plastic extensional strain is $\frac{1}{3}\bar{\epsilon}_B{}^p$ and the stress is $\bar{\sigma}_A$. The load is released, bringing the specimen to point C, and then reapplied as compression. Further yielding begins when the stress $-\bar{\sigma}_A$ is reached at point A'. As the compressive load is increased, yielding continues along the curve $A'B'$, which has the same shape as the curve AB in Fig. 5.34*a*. When the point B' is reached, a compressive plastic strain of $\frac{2}{3}\bar{\epsilon}_B{}^p$ has occurred between A' and B', and the stress required to cause further yielding has reached the value $-\bar{\sigma}_B$. If the load is now released, the material returns to D'. A reapplication of the tensile load will cause the material to move along the curve $D'B'F'$, which is identical with the curve DBF in Fig. 5.34*a*. Thus, in constructing the curve of Fig. 5.34*b*, the assumption has been made that all the plastic-strain increments along the loading path have contributed in a positive manner to the strain-hardening so that the material in state D' has been strain-hardened the same amount as the material in state D in Fig. 5.34*a*.

We shall, in the following chapters, consider applications in which only very simple models of plastic behavior are required. As an illustration of a case where both elastic and plastic strains are of importance, we consider the following problem.

Example 5.3 Returning to Example 5.1, we ask, what will happen if we remove the load P after we have strained the combined assembly so that both the steel and the aluminum are in the plastic range, that is, beyond a strain of 0.005?

We can again use the model of Fig. 5.9, and the equilibrium relation (*e*) and geometric compatibility relation (*a*) still remain valid. We need new stress-strain relations which will be valid during unloading.

STRESS-STRAIN RELATIONS

As indicated in Fig. 5.34, unloading is an elastic process, no further plastic deformation occurring until the stress has been completely reversed and reaches the current value of the flow stress. If we let δ_o be the deflection when the assembly is loaded by P and δ the deflection after the load has

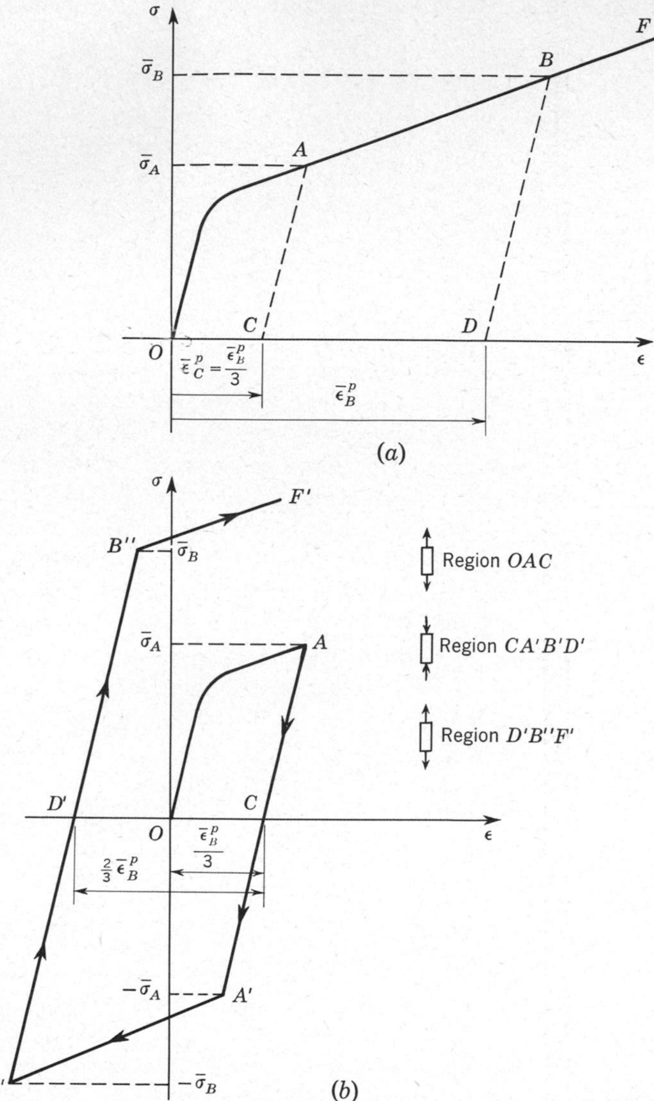

Fig. 5.34 Example of simple loading path. (*a*) Stress-strain curve in uniaxial tensile test; (*b*) stress-strain behavior in alternate uniaxial tension and compression.

been decreased somewhat, then, taking into account the compatibility relation (*a*), the stress-strain behavior of the material will be as illustrated in Fig. 5.35, where *S* and *A* represent the states of the steel and aluminum under the load *P*, and *S'* and *A'* represent the states after the load has been

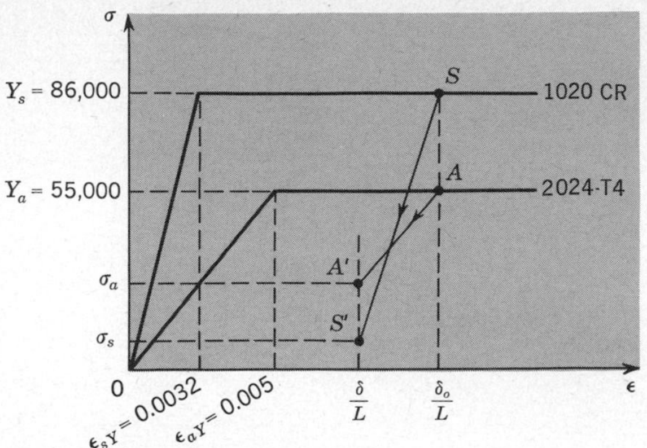

Fig. 5.35 Example 5.3. Stress-strain behavior of the assembly of Example 5.1 when the load P is decreased after both the steel and aluminum alloy have been strained plastically.

decreased somewhat. From Fig. 5.35 we obtain the following stress-strain relations describing the unloading curves SS' and AA':

$$\sigma_s = Y_s - E_s \frac{\delta_o - \delta}{L}$$
$$\sigma_a = Y_a - E_a \frac{\delta_o - \delta}{L} \qquad (f)$$

Substituting (f) into Eq. (e) of Example 5.1 and setting $P = 0$, we obtain

$$A_s\left(Y_s - E_s \frac{\delta_o - \delta}{L}\right) + A_a\left(Y_a - E_a \frac{\delta_o - \delta}{L}\right) = 0 \qquad (g)$$

From this we find

$$\frac{\delta_o - \delta}{L} = \frac{A_s Y_s + A_a Y_a}{A_s E_s + A_a E_a} \qquad (h)$$

Substituting (h) into (f), we find the residual stresses which remain in the assembly after the load has been removed.

$$\sigma_{s_{\text{residual}}} = Y_s \frac{1 - \dfrac{Y_a/E_a}{Y_s/E_s}}{1 + E_s A_s/E_a A_a} = Y_s \frac{1 - \dfrac{\epsilon_{aY}}{\epsilon_{sY}}}{1 + E_s A_s/E_a A_a}$$
$$\sigma_{a_{\text{residual}}} = Y_a \frac{1 - \dfrac{Y_s/E_s}{Y_a/E_a}}{1 + E_a A_a/E_s A_s} = Y_a \frac{1 - \dfrac{\epsilon_{sY}}{\epsilon_{aY}}}{1 + E_a A_a/E_s A_s} \qquad (i)$$

The algebraic signs of σ_s and σ_a were defined in Fig. 5.9; a positive value in (i) indicates compression and a negative value indicates tension.

We see that, as a result of the yielding, there are residual stresses "locked into" the assembly upon release of the load. It is worthwhile both as a check of the equations (i) and as an aid to the development of judgment to consider several limiting cases. We note from (i) that the residual stresses will be zero only when the initial yield strains $\epsilon_{sY} = Y_s/E_s$ and $\epsilon_{aY} = Y_a/E_a$ are equal. Since in the present case $\epsilon_{aY} > \epsilon_{sY}$, the equations ($i$) show that the steel will be in tension and the aluminum in compression. We also note from the first of (i) that since $\epsilon_{aY} < 2\epsilon_{sY}$, the residual stress in the steel cannot reach the value of the flow stress, whatever the ratio A_s/A_a, and thus the equations (i) are valid for any value of A_s/A_a. Finally, we note that if the area of steel is reduced to zero, we would expect the steel to exert a negligibly small force on the aluminum through the end plates; the second of Eqs. (i) predicts this result.

Let us now consider the case again when a tensile test of a ductile material is carried beyond the point of initial yielding. The behavior is as illustrated in Fig. 5.36: as the specimen is deformed plastically, there is an increase in the load required for further elongation, but eventually the load reaches a maximum from which it steadily decreases until the specimen fractures. To describe the behavior of the *material* during this process—as distinct from the overall behavior of the *specimen*—it is necessary to distinguish between the stress based on the original cross-sectional area of the specimen and the stress based on the actual cross-sectional area at any stage of the elongation. Since the cross-sectional area decreases as the specimen is elongated, the stress based on the actual area is greater than that based on the original area.

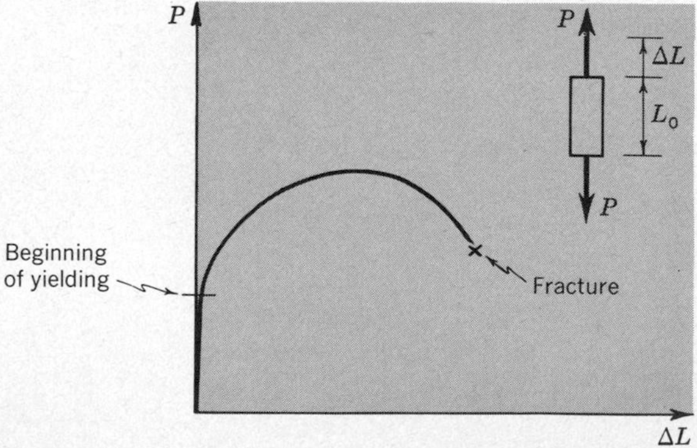

Fig. 5.36 Complete load-elongation curve of a ductile material in a tensile test.

The intensity of load per unit of *actual* area is called the *true stress*; this stress describes the load intensity the material actually experiences. The intensity of load per unit of *original* area is called the *engineering stress.* For strains small compared to unity, the fractional decrease in cross-sectional area is small, and the true stress and engineering stress are essentially equal; even when the axial strain has reached the relatively large (for engineering purposes) value of 0.05, the true stress is only about 5 percent greater than the engineering stress.

Returning to the tensile-test behavior, as plastic deformation is continued, the load required for further plastic flow increases. Finally a point is reached where the increase in flow strength no longer compensates for the decrease in cross-sectional area, and the load required to cause further elongation begins to decrease. At this point the load has passed a maximum and, consequently, so also has the engineering stress. This maximum value of the engineering stress is termed the *tensile strength.* The phrase "tensile strength" is something of a misnomer, for the true stress at this point is already higher than the tensile strength, and will continue to rise as the test proceeds. Due to unavoidable variations some one particular section of the specimen will arrive at the condition where the increase in flow stress will not compensate for the decrease in area, while the other sections of the specimen will still be able to carry higher loads. From this stage on, the plastic deformation will be concentrated in that section which first reaches the condition of maximum load. As the load decreases, governed by the product of area times flow stress in the weakest section, the other sections remain at essentially constant area but with decreasing stress to maintain equilibrium with the decreasing load. This process of nonuniform deformation is called *necking*, and examples[1] of it are shown in Fig. 5.37. The tensile test reaches its conclusion when a small crack develops at the center of the neck and spreads outward to complete the fracture.

In describing the strain in the tensile test, two different approaches have been used. The first of these uses Eq. (5.1) to define the strain as the ratio of the change in length to the *original* length of the specimen. Thus for a specimen of original length L_o the strain corresponding to the length L_f is defined as

$$\epsilon_x = \frac{\Delta L}{L_o} = \frac{L_f - L_o}{L_o} \tag{5.26}$$

This is analogous to the engineering stress defined in terms of the original area and is called an *engineering strain.* The second approach to strain regards the total strain as being the sum of a number of *increments* of strain, thus

$$\epsilon_x = \Sigma\Delta\epsilon_x = \Sigma\frac{\Delta L}{L}$$

where L is the current length of the specimen when the increment of elongation ΔL occurs. If L_o is the original length of the specimen, then in the limit as $\Delta L \to 0$

[1] Further discussion of the phenomenon of necking of a tensile specimen is given in Sec. 9.7.

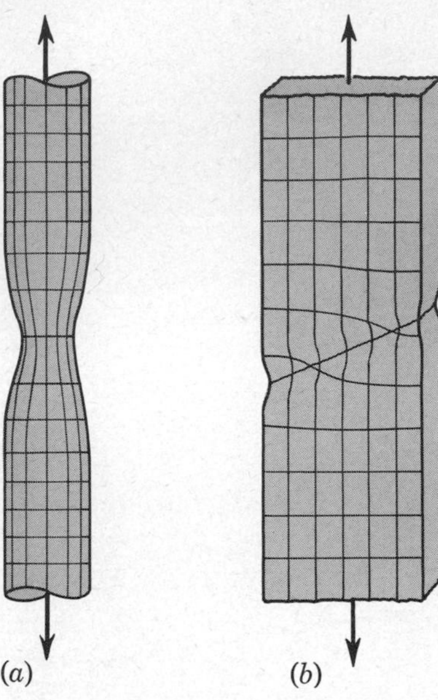

Fig. 5.37 Examples of necking.

the strain ϵ_x corresponding to the length L_f is given by the following integral:

$$\epsilon_x = \int_{L_o}^{L_f} \frac{dL}{L} = \ln \frac{L_f}{L_o} \tag{5.27}$$

This strain, obtained by adding up the increments of strain which are based on the current dimensions, is called a *true strain*. Sometimes true strain is called logarithmic strain because of the form of (5.27); however, in some situations the idea of redefining the coordinates for each incremental deformation does not lead to a logarithmic form, and hence the term *true strain* is preferred.

When deciding which definition of strain to use in describing the behavior beyond initial yielding in the tensile test, the balance is in favor of using *true* strain. One reason for this is that most of the dislocation processes are more conveniently described by an incremental concept of strain. A second reason is that when a ductile metal is tested both in tension and in compression, the true-stress and true-strain curves practically coincide, whereas the two curves are quite different when engineering strain is used; from the standpoint of dislocation theory one would expect the curves to coincide.

For strains small compared to unity it is not quantitatively important which definition is used, since the numerical results will be substantially equal. For example, if the engineering strain is 0.05, the true strain is 0.0488, a difference of only about 2 percent. It should be noted that even where the total strain is small compared to unity, the plastic strain can be many times larger than the elastic strain. For instance, when the total strain in the 1020 HR steel in Fig. 5.5*a* is

0.01, the plastic strain is about 8 times the elastic part. Thus even for very small strains it is a good approximation to assume for our tensile specimen that

$$A_o L_o = A_f L_f$$

since dislocation theory suggests and experiment verifies that the volume remains nearly constant during plastic deformation. Substituting the above relation into (5.27), we obtain

$$\epsilon_x = \ln \frac{A_o}{A_f} = 2 \ln \frac{D_o}{D_f} \tag{5.28}$$

where D_o is the original diameter and D_f is the diameter corresponding to ϵ_x. If we now measure the change in length during the early portion of the tensile test (say up to a strain of 0.05) and calculate ϵ_x from (5.27) [or, for that matter, from (5.26)], and then measure the minimum diameter of the specimen during the latter part of the test and calculate ϵ_x from (5.28), we shall determine throughout the entire tensile test the true strain of the most highly strained part of the material. The advantage of (5.28) over (5.27) is that the latter allows us to measure the strain of a specimen in the necked region where, because of the nonuniform strain, one must use in effect a zero gage length.

Figure 5.38 illustrates the results obtained when the data from uniaxial tensile and compression tests are plotted on an *engineering* basis and on a *true* basis. In compression, as the specimen is shortened, the cross-sectional area increases, and thus the true stress is less than the engineering stress.

The complete tensile true-stress–true-strain curves for the materials of Fig. 5.5*a* and *b* are shown in Fig. 5.39*a* and *b*. In addition to the yield point (or 0.2 percent yield strength) the *tensile strength* (T.S.) is indicated for each material.

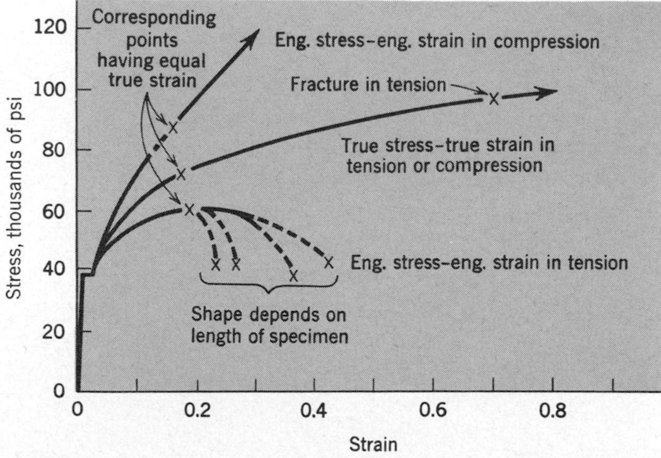

Fig. 5.38 Stress-strain curves for hot-rolled low-carbon steel (1020 HR).

Essentially the tensile strength tells the designer the maximum load that a long, thin part can carry in tension. However, before the tensile strength is reached the part may become unserviceable due to excessive deformation.

In addition to its strength, another important property of a material is its ductility before fracture. The ductility of a material can be described by the *reduction of area* (R.A.), defined as the ratio of the decrease in area to the initial area. One can show that for round bars the reduction of area is related to the true strain at fracture ϵ_f by the equation

$$\text{R.A.} = 1 - e^{-\epsilon_f}$$

where e denotes the base of natural logarithms. As shown in Fig. 5.37*b*, the necking process is much more complicated in the case of sheets. A more frequently quoted measure of the ductility of the material is the *elongation*, defined as the change in gage length to final fracture divided by the original gage length (i.e., the engineering strain at fracture). As a measure of ductility of the material, the elongation has the disadvantage that it is an engineering, rather than a true, strain, and furthermore it consists of some sort of weighted average of the uniform

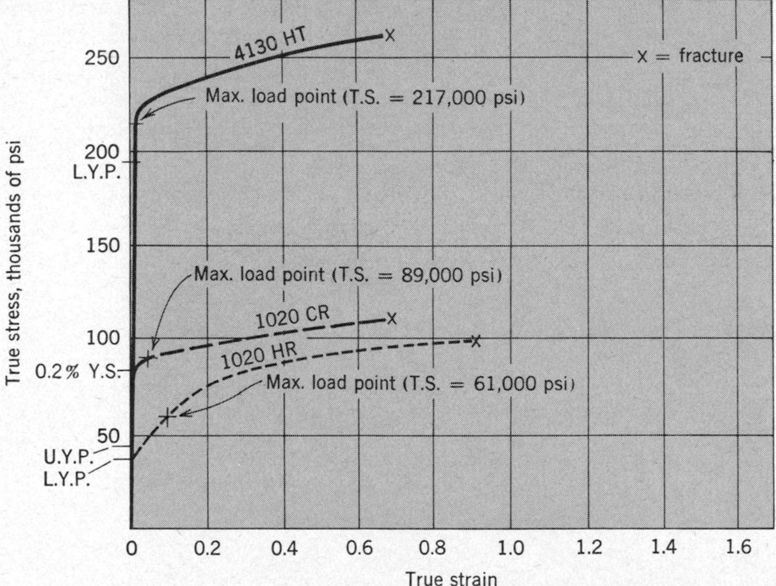

Fig. 5.39*a* Complete stress-strain curves from tensile tests of three steels.
- - - - - Mild steel, hot-rolled (1020 HR)
— — Mild steel, cold-rolled (1020 CR)
——— 0.3% C, 0.5% Mn, 0.25% Si, 0.9% Cr, balance Fe (4130 HT)
Oil quenched from 1600°F, tempered at 600°F

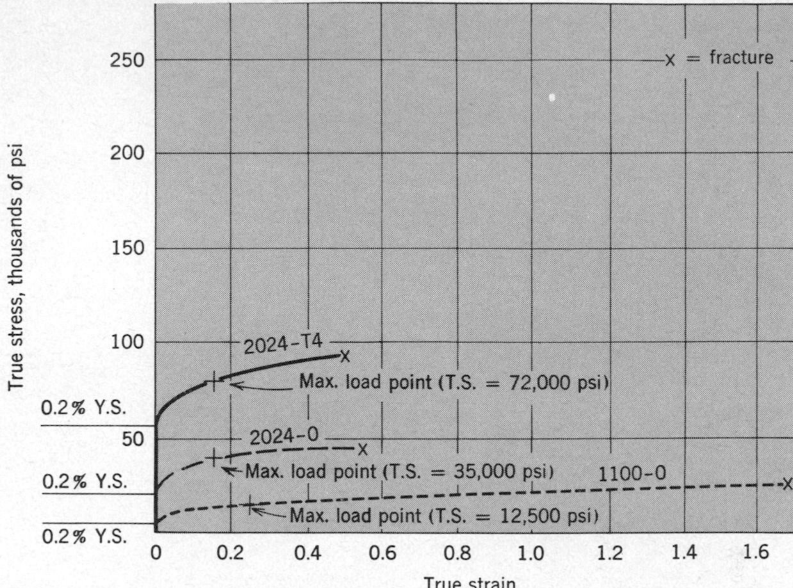

Fig. 5.39*b* Complete stress-strain curves from tensile tests of aluminum and two aluminum alloys.

- - - - Commercially pure aluminum, annealed (1100-0)
— — 4.6% Cu, 1.5% Mg, 0.7% Mn, balance Al, annealed (2024-0)
——— 4.6% Cu, 1.5% Mg, 0.7% Mn, balance Al (2024-T4)
Water quenched from 915°F, aged 24 hr at 250°F

strain in the un-necked portion and the higher strain in the necked region of the specimen. As such, it is very dependent on the length, as indicated in Fig. 5.38, as well as on the cross-sectional dimensions of the specimen.

5.13 FRACTURE OF DUCTILE SPECIMENS AND STRUCTURES

The question might be asked as to whether information on the ductile behavior of a tensile-test specimen can be used to predict fracture in structures under a more general state of stress. The answer is not yet known. For one thing, even fracture in a tensile test is complicated by the fact that the curvature at the neck of the tensile specimen alters the state of stress in the neck. Bridgman[1] has given an approximate method of correcting for the effect of the curvature, and the curves of Fig. 5.39 include this correction. Recent work, however, casts some doubt on this correction without providing a firm alternative. Another problem of predicting the fracture of ductile structures under general loadings from the

[1] P. W. Bridgman, *op. cit.*, chap. 1.

tensile-test behavior is that compressive behavior is different from tensile behavior. For instance, in "compression" testing, many materials fracture only after sufficient barreling has occurred to produce a transverse tension. Bridgman has further shown, in uniaxial tensile tests in the presence of large hydrostatic pressures, that the stress-strain curve at small strains is nearly identical to that for ordinary (atmospheric) pressure, but that fracture occurs at much higher strains. Thus, fracture depends not only on the differences of the principal stresses, as does yielding, but also on their absolute values as well.

The actual fracture process of a ductile specimen or structure can occur by several different metallurgical mechanisms, the most important of which are cleavage (separation normal to the direction of maximum tensile stress) and the growth of holes by plastic deformation around inclusions (which depends on the mean normal tension and on the shear strain). In other words, a ductile structure can fracture by either a cleavage or a hole-growth mechanism. Conversely, a large enough structure with a crack in it can be brittle, whether it fails by the cleavage or the hole-growth mechanism. It is unwise, therefore, even though the terms are used frequently, to speak of brittle and ductile *materials*. Rather, one should speak of brittle or ductile *structures*. Either kind of structure may fracture by either the cleavage or the hole-growth mechanism. In fact, depending on the size of the structure, as well as on the temperature and the strain rate, steel can undergo an abrupt transition from the hole-growth mechanism to the cleavage mechanism. Although progress is being made,[1] there is not yet a comprehensive theory nor a quantitative correlation of these various phenomena of fracture.

5.14 FRACTURE OF BRITTLE SPECIMENS AND STRUCTURES

At first glance, there are some materials, such as glass and ceramics, which seem to fracture with no prior plastic deformation. However, even glass, when cut with a sharp tungsten carbide cutter, exhibits microscopic curved chips similar to the larger ones produced in machining metals. The apparent brittleness of glass and of structures of large sections made of normally ductile steels and aluminum alloys arises from the presence of cracks which concentrate the stress and strain so that the cleavage or hole-growth mechanisms of fracture can occur before the entire specimen or structure becomes plastic. We first consider specimens that are large enough, for the materials and microcracks involved, so that design can be based on an average stress. To design against fracture in such specimen-structure combinations (e.g., glass and ceramics), we must seek a criterion for fracture rather than for yielding. A criterion can be derived from the assumption that fracture occurs when the local stress around the worst crack in the material reaches a critical value. The problem, then, is to find the highest stress on the worst

[1] See, for example, McClintock and Argon, *op. cit.*, chap. 16; or H. Liebowitz (ed.), "Treatise on Fracture," Academic Press, Inc., New York, 1969–1970.

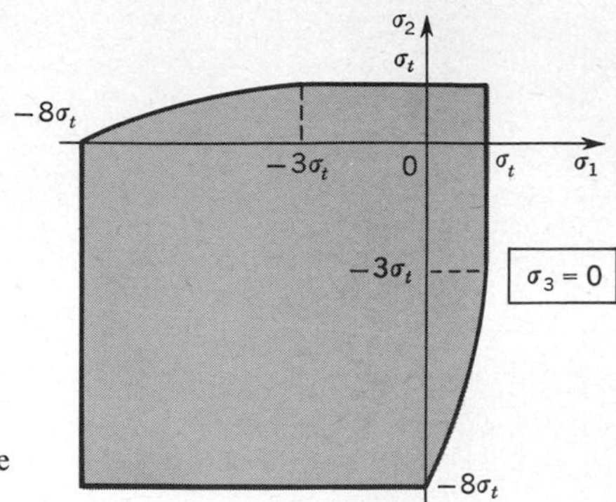

Fig. 5.40 Griffith criterion for brittle fracture under plane stress, $\sigma_3 = 0$.

crack under the average macroscopic stress. When this highest stress reaches the same value it would have for fracture in a tensile loading of the material, the crack grows. Griffith[1] developed these ideas into the fracture criterion shown in Fig. 5.40 (for the case in which one of the principal stresses is zero, analogous to that shown in Fig. 5.29 for plastic yielding). Note that tensile stresses are far more serious than compressive stresses. When both principal stresses are tensile, the fracture criterion is simply that failure will occur when the larger principal stress becomes the same as the fracture stress in a uniaxial tensile test.

When *all three* principal stresses are compressive, as is the case with rocks deep in the ground, Griffith's theory seems to predict lower strengths than are actually obtained. The presence of normal or shear forces transmitted across the faces of cracks, a possibility not considered by Griffith, is part of the cause of this discrepancy, but more appears to be due to the fact that cracks in compression are initially stable, and that macroscopic fracture only occurs when a number of cracks interact in some unknown statistical fashion.

When the specimen is not large enough in relation to the critical crack sizes involved, it is necessary to carry out tests on precracked specimens rather than on smooth specimens. The analysis and procedures for such testing have been developed within the last 15 years.[2] The further problem of predicting the behavior of large brittle structures from tests on small ductile specimens is as yet unsolved.

5.15 FATIGUE

As illustrated by the experiment with a paper clip in Sec. 5.1, Fig. 5.2, loads or deformations which will not cause fracture in a single application can result in fracture when applied repeatedly. Fracture may occur after a few cycles, as in the

[1] See, for example, E. Orowan, Fracture and Strength of Solids, *Repts. Progr. in Phys.*, Phys. Soc. London, vol. 12, pp. 185–232, 1949.

[2] See, for example, McClintock and Argon, *op. cit.*, or Liebowitz, *op. cit.*

paper-clip experiment, or after millions of cycles. This process of fracture under repeated loading is called *fatigue*. Fatigue is important in that it is one of the three common causes of mechanical failure, the others being wear and corrosion. At the present time the mechanisms of fatigue failure are not well understood, and design procedures to avoid fatigue are not precise. In this section we hope to provide an awareness of the problems encountered and an introduction to current methods of solution.

Consider a situation in which the stress at a point in a body varies with time, as shown in Fig. 5.41. Experiments show that the alternating stress component σ_a is the most important factor in determining the number of cycles of load a material can withstand before fracture, while the mean stress level σ_m is less important, particularly if σ_m is negative (compressive). For a given machine part the fatigue life is strongly influenced by the quality of the surface finish, the possible residual stresses within the part, the presence of surface or subsurface cracks, the presence of stress concentrations, the chemical nature of the environment, and the material itself.

The results of tests on an aluminum alloy are shown in Fig. 5.42*a*. Here the mean stress σ_m was zero. The stress amplitude σ_a is plotted against (the logarithm of) the number of cycles N required to cause failure. Notice that the life increases very rapidly with decrease in stress, especially after a life of about 10^7 cycles has been reached. It is customary to designate the stress which can be withstood for some specified number of cycles as the *fatigue strength* of the material. For example, from Fig. 5.42*a* it can be said that the fatigue strength of 2024-T4 aluminum alloy, when tested as unnotched bars, is about 30,000 psi for 10^6 cycles and 24,000 psi for 10^7 cycles. Some materials, of which ferrous metals are an outstanding example, have the property of a stress level which the material can withstand for an indefinite number of cycles without failure; this stress level, corresponding to essentially infinite life, is called the *endurance limit*.

It may be seen from Fig. 5.42*a* that when tests are run on notched specimens, the number of cycles to failure is substantially less than that for smooth specimens having the same diameter at the root of the notch. In plotting the data in Fig. 5.42*a*, it has been assumed that the stresses are determined only by the minimum

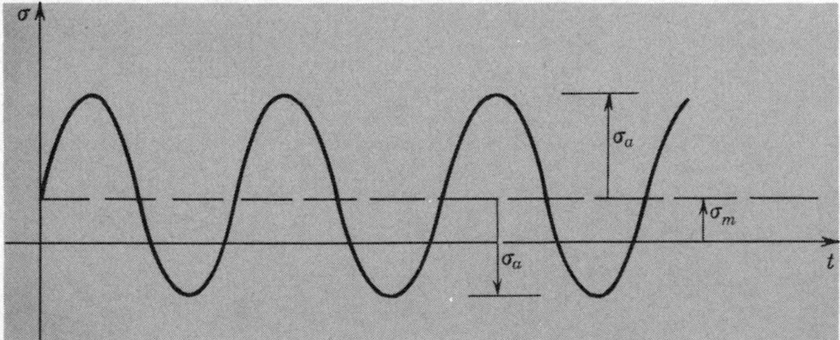

Fig. 5.41 Time-varying stress.

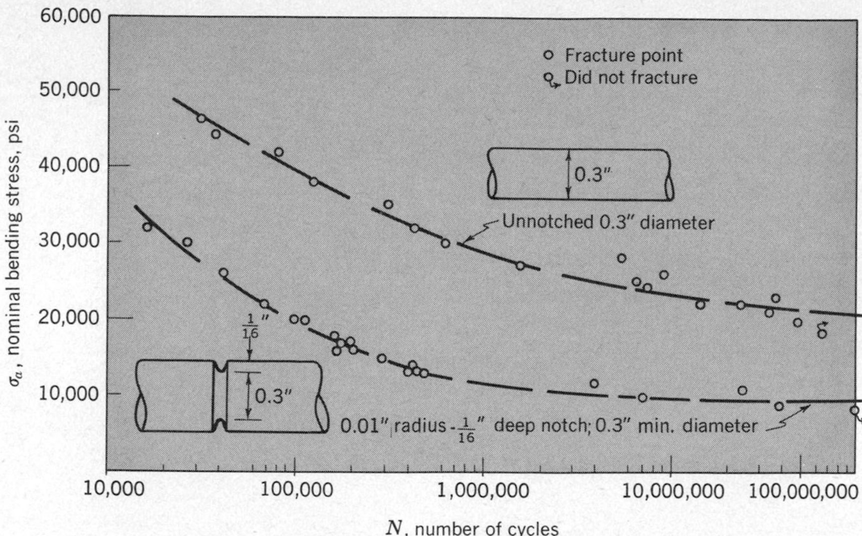

Fig. 5.42a Fatigue curve for 2024-T4 aluminum alloy. [*From C. W. McGregor and N. Grossman, The Effects of Cyclic Loading on the Mechanical Behavior of 24S-T4 and 75S-T6 Aluminum Alloys and SAE 4130 Steel, Nat. Advisory Comm. Aeronaut., Tech. Notes, No. 2812, October, 1952.* (*24S-T4 is the former designation of the alloy now designated as 2024-T4*).]

diameter, and thus, that the stress in the minimum cross section of the notched bar is equal to that in the smooth specimen when equal loads are applied to each. Actually, in a notched specimen the stress distribution is complex, and in a small region close to the notch the stresses are larger than in a smooth specimen of the same diameter. This increase in stress is caused by the abrupt change in the geometry of the specimen in the vicinity of the notch and is a stress concentration (Sec. 5.9). If the *actual* stress at the root of the notch had been used in plotting the data for the notched bar in Fig. 5.42*a*, there would be much better agreement between the two sets of data.

Fatigue cracks are most likely to form and grow from locations where holes or sharp reentrant corners (e.g., notches) cause stress concentrations. In designing parts to withstand repeated stresses, it is important to avoid stress concentrations. Keyways, oil holes, and screw threads are potential sources of trouble and require special care in design in order to prevent fatigue failures.

Figure 5.42*b* shows fatigue curves for a number of different materials. To determine each curve may require many tests, because of the flat slope and the scatter due to variations in surface finish (Fig. 5.42*c*). Each curve also represents only one state of mean stress (most often zero), one surface finish, one environment, etc. The problems associated with obtaining a complete set of fatigue data for a wide range of conditions are readily apparent. In the remainder of this section, we shall briefly consider effects of fatigue variables and simple design criteria.

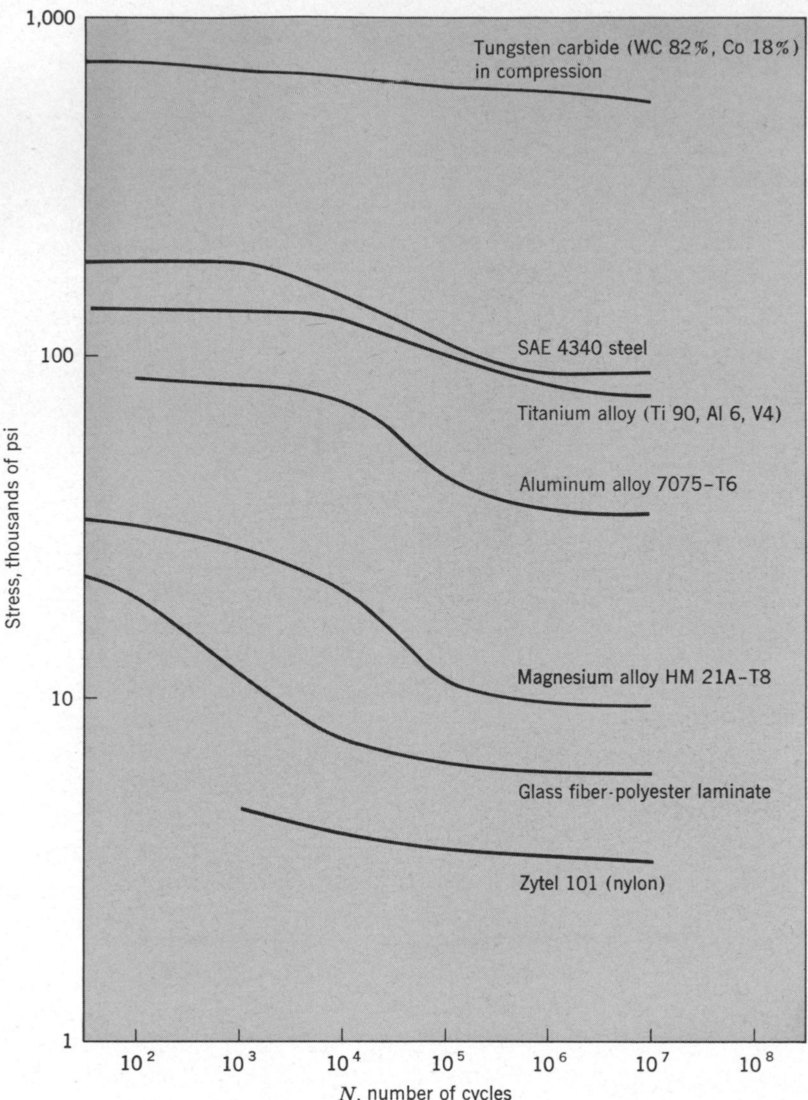

Fig. 5.42b Fatigue data for different materials. (*Data obtained from Carl C. Osgood, "Fatigue-Design," Interscience Publishers, New York, 1971; F. A. McClintock and A. Argon, op. cit.; I. Johansson, G. Persson, and R. Hiltscher, Determination of static and fatigue compressive strength of hard-metals, Powder Met., vol. 13, no. 26, p. 449, 1970; C. R. Smith, Small specimen data for predicting life of full scale structures, in "Symposium on Fatigue Tests of Aircraft Structures: Low-Cycle, Full Cycle, and Helicopters," ASTM Special Technical Publication 338, 1962, American Society for Testing and Materials; "Metals Handbook," 8th ed., vol. 1, Properties and Selection of Metals, American Society for Metals.*)

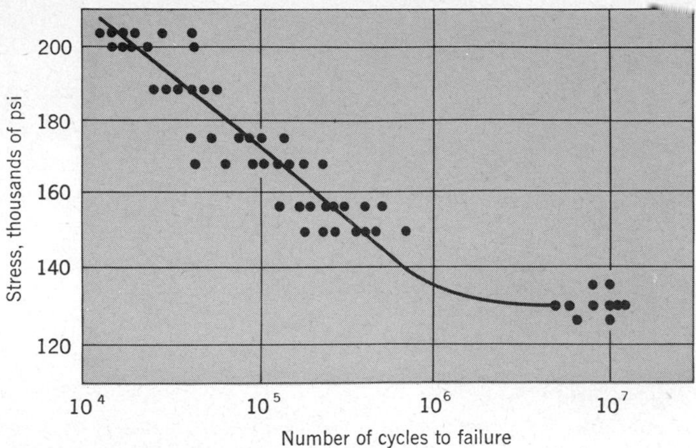

Fig. 5.42c Fatigue data for 5 percent Cr-Mo-V steel bars heat-treated to 240,000 to 265,000 psi tensile strength. Note that at a typical stress level there is a range of almost a factor of 10 between the shortest and longest fatigue life, while at any fatigue life, there is a range of about 20 percent between the highest and lowest stress level.

Endurance limit when mean stress is zero There is a tremendous amount of fatigue data in the literature, and one can usually find test data appropriate to the particular problem. However, as a crude "rule of thumb" we find that, for reversed bending tests ($\sigma_m = 0$) on *smooth polished* specimens *in air*, the approximate endurance limit stress σ_e is related to the ultimate tensile strength σ_{ult} as follows:[1]

Ferrous materials:	$\sigma_e/\sigma_{ult} \approx 0.4$ (infinite life)
Nonferrous materials:	$\sigma_e/\sigma_{ult} \approx 0.25$ (10^8 cycles)

If the ultimate strength is not known, the yield stress can be used and the result will be conservative.

Stress concentration From Fig. 5.42*a* we see that a stress concentration can greatly reduce the allowable alternating stress component. For design purposes, we search for fatigue data, under appropriate conditions, with proper stress concentrations. In the absence of such data, we require a criterion for adjusting the allowable endurance limit for different stress concentrations. Unfortunately, this is not a simple, nor precise, problem.

In Sec. 5.9 we discussed the magnitude of the stress concentration factor K_t, in *elastic* media; the subscript "t" in K_t refers to the fact that this is a *theoretical*

[1] In this section we shall let σ_e represent either the endurance limit, when such a limit exists, or the fatigue strength at a specified very large number of cycles, namely, 10^8 cycles. This is not consistent with general practice wherein these two quantities are given different symbols.

ratio of maximum stress to nominal stress σ_{nom} if the material remains purely elastic, that is,

$$K_t = \left(\frac{\sigma_{\text{max}}}{\sigma_{\text{nom}}}\right)_{\text{elastic, theoretical}}$$

It has become common practice to regard a stress concentration as reducing the *strength* of a material rather than increasing the *stress*. That is, the endurance limit in terms of the nominal stress for a material with no size effect and no yielding would be regarded as being reduced by the factor K_t. However, the effect of the stress concentration is not as severe as would appear from K_t. Yielding complicates the picture by decreasing the stress concentration while increasing the strain concentration, and it is the local microplastic strain that initiates fatigue cracks. Practically these effects are all combined into a *fatigue-strength reduction factor* K_f,

$$K_f \equiv \frac{\text{unnotched fatigue strength}}{\text{notched fatigue strength}} \qquad (1 < K_f < K_t)$$

Values of K_f are tabulated for different materials and geometries.[1] For large specimens or structures, K_f tends to approach K_t. For very high values of K_t corresponding to preexisting cracks, there is hardly any definite endurance limit, and it is more useful to correlate fatigue crack growth rates with a stress intensity factor.[2]

Endurance limit in the presence of mean stress Experimental data for infinite fatigue life of unnotched specimens under a mean stress σ_m and an alternating stress σ_a tend to follow the curved line shown in the diagram of Fig. 5.43*a*. These experimental results are limited by the fact that, when the maximum stress (mean stress plus alternating stress) is equal to the yield strength σ_Y, the specimen may begin to fail by deformation even without fatigue. In this case an elastic stress analysis will no longer apply. The limitation on the stress levels because of yielding is shown by the dashed lines in Fig. 5.43*a*. Also, it is found experimentally that a compressive mean stress has no detrimental (nor beneficial) effect on fatigue life. These observations have given rise to a simple design criterion for infinite life in the presence of a mean and alternating stress for *uniaxial* states of stress. We note first that in the absence of any alternating stress, the specimen will not fail if the mean stress is less than the ultimate tensile stress. This is point B in Fig. 5.43*a*. Similarly, in the absence of mean stress the specimen will not fail if the alternating stress is less than the endurance limit as determined above. This is point A in Fig. 5.43*a*. The straight line joining A and B is called the *Goodman-Soderberg criterion*. It is a simple approximate representation for the combinations of σ_m and σ_a which

[1] See, for example, G. Sines and J. Waisman (eds.), "Metal Fatigue," McGraw-Hill Book Company, New York, 1959.

[2] See, for example, H. Liebowitz (ed.), *op. cit.*

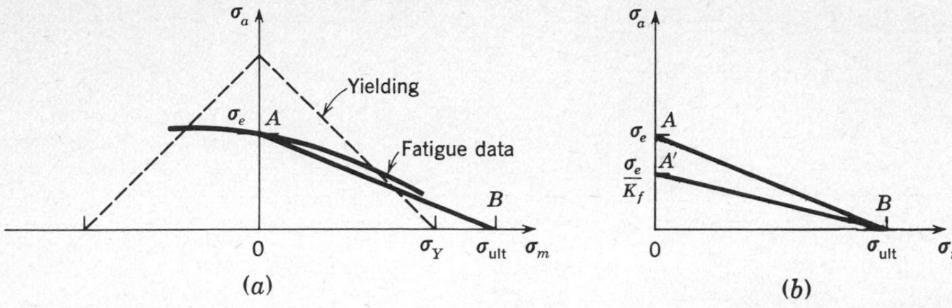

Fig. 5.43 Diagram for the prediction of fatigue life. (*a*) Goodman-Soderberg line *AB*; (*b*) effect of fatigue-strength reduction factor.

provide infinite fatigue life. This criterion for the prediction of fatigue life is generally conservative, since most data lie above the straight line *AB*.

As we discussed above, when $\sigma_m = 0$, the presence of a stress concentration reduces the nominal endurance limit by the fatigue-strength reduction factor K_f. Because of plastic flow in ductile materials, the effect of the mean stress level σ_m on fatigue life is not appreciably altered by the stress concentration. As a consequence, the Goodman-Soderberg criterion is extended to include the effects of stress concentration by taking the straight line *A'B* in Fig. 5.43*b* to represent infinite life combinations of mean and alternating stresses. For ductile ferrous materials, $K_f \leq 3$ for threads, holes, fillets, keyways, etc. For aluminum, magnesium, and titanium, unless one has data to the contrary it is best to assume that $K_f = K_t$.

Surface condition In most instances (not all) fatigue cracks originate at the surface of a part at an imperfection or other stress concentration. Thus the physical character of the surface is most important. Contrary to common belief, a smooth, unstressed, virgin surface is *not* best for fatigue strength. Such a surface, which is produced by electrochemical machining or electrochemical polishing, will generally give an endurance limit as low as ½ of that obtainable from a specimen having high surface compressive stresses produced by gentle grinding, shot peening, etc. On the other hand, a cracked or tensile stressed surface can give an endurance limit ½ that of the virgin surface; such a surface is generally obtained from various "hot" processes such as electrodischarge machining, abrasive grinding, laser cutting, etc.[1] Thus a 4:1 variation can be obtained through process variations. Table 5.8 shows typical data for Inconel 718.

Environment: thermal and chemical The effect of temperature on the endurance limit appears closely related to the effect of temperature on the ultimate tensile strength. Thus, as temperature changes, we alter σ_e in proportion to the change

[1] "Surface Integrity of Machined Structural Components," Air Force Materials Laboratory Report AFML-TR-70-11, March, 1970.

Table 5.8 Effect of surface finishing method on endurance limit–Inconel 718

	Process	*Endurance limit,* psi
A.	Gentle surface grind	60,000
B.	Electrochemical machine	39,000
C.	Electrodischarge machine	22,000
D.	B Plus shot peen	78,000
E.	C Plus shot peen	66,000

in σ_{ult}; i.e., if σ_{ult} is reduced by 30 percent, we assume a 30-percent reduction in σ_e. Generally, within useful temperature ranges, this effect is small.

The effect of the chemical environment, on the other hand, can be extremely large. Embrittling media, such as liquid metals, can reduce the material strength to zero (as in the case of aluminum immersed in liquid gallium at 100°F). Corrosive media generally have strong effects; the fatigue strength of steel, in a salt-water-spray environment, may be reduced to 10 to 20 percent of its nominal value. A freshwater environment may have half this effect.

Example 5.4 A rod is to be loaded axially with a mean tensile stress σ_0 plus an alternating axial stress $\sigma_a = \frac{1}{2}\sigma_0$. The rod has a small radial hole such that $K_t = 3$. We wish to have a factor of safety of 2. The material is a titanium alloy with an ultimate tensile strength of 120,000 psi and an endurance limit ($\sigma_m = 0$) of 60,000 psi. How large can σ_0 be?

Let us plot a Goodman-Soderberg-type diagram as shown in Fig. 5.44.

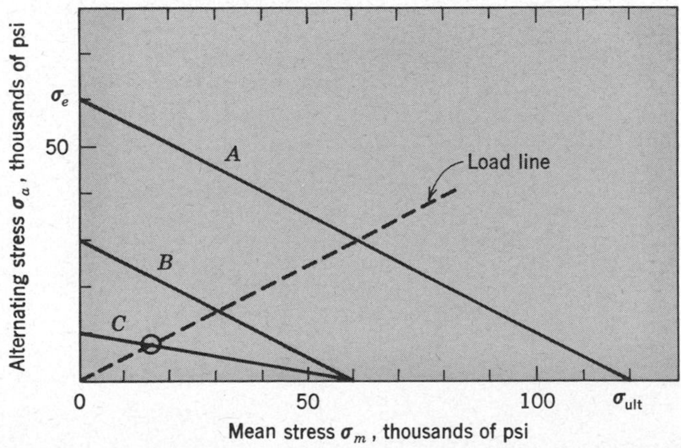

Fig. 5.44 Example 5.4. Goodman-Soderberg diagrams: *A*, for basic material; *B*, for material with a safety factor of 2; *C*, for material with a safety factor of 2 plus allowance for stress concentration factor $K_t = 3$.

The upper line represents the basic material; the next line shows stress combinations which will provide a safety factor of 2; the lowest line incorporates the fatigue strength reduction factor K_f of 3 (not applied to σ_m). The dashed line is the load line $\sigma_a = \frac{1}{2}\sigma_m$. The required value of $\sigma_0 = 15{,}000$ psi is read off as the abscissa of the point of intersection of the latter two lines.

5.16 CRITERIA FOR CONTINUED YIELDING

The next question of interest is how the tendency for further yielding in a *general* state of stress varies as the material is plastically deformed. In answering this question for materials which yield initially according to the Mises criterion, it has been assumed that the tendency for further yielding can be measured by an *equivalent stress* $\bar{\sigma}$ which is defined as follows:

$$\bar{\sigma} = \sqrt{\tfrac{1}{2}[(\sigma_1 - \sigma_2)^2 + (\sigma_2 - \sigma_3)^2 + (\sigma_3 - \sigma_1)^2]} \tag{5.29}$$

To illustrate the use of this equivalent stress, we would say that initial yielding can occur when $\bar{\sigma} = Y$.

If the tendency to yield is given by (5.29), then our original question can be reduced to asking: How does $\bar{\sigma}$ vary with the plastic deformation? This is a question which required many years and much experimentation to answer, but with the advantage of hindsight we can develop a logical answer to this question in the short argument which follows.

If the material does not strain-harden in the tensile test, the equivalent stress presumably remains constant at the value Y. If the material strain-hardens, however, the equivalent stress rises. On what parameter of the deformation does this increase depend? Since elastic deformations do not affect the dislocation structure of the material, we expect the increase to depend only on the plastic strain. Just as the equivalent stress depends on more than one component of the stress, so the amount of strain-hardening would be expected to depend on more than one component of the plastic strain. If we assume that the material remains reasonably isotropic, we can base the description of the plastic strains on the principal components of plastic strain without regard to their orientation. Furthermore, since the volume of the metal remains nearly constant, we expect the strain-hardening to depend on some function of the differences between the plastic-strain components. Finally, if we bend a piece of soft wire and then straighten it to its original shape, it becomes more resistant to further plastic deformation—but the total strain is zero and therefore the strain-hardening cannot depend directly on the total strains. It seems reasonable, therefore, to suggest that the strain-hardening depends on some function of the plastic-strain *increments*, a function which is always increasing since we do not usually observe strain-softening. These conditions are all met by assuming that the equivalent stress depends upon the *equivalent plastic strain*, defined by

$$\bar{\epsilon}^p = \int \sqrt{\tfrac{2}{9}[(d\epsilon_1{}^p - d\epsilon_2{}^p)^2 + (d\epsilon_2{}^p - d\epsilon_3{}^p)^2 + (d\epsilon_3{}^p - d\epsilon_1{}^p)^2]} \tag{5.30}$$

where the integral is taken over the entire loading path. The factor $\frac{2}{9}$ is introduced so that the equivalent strain is equal to the axial strain in a tensile test (see Prob. 5.35).

Equation (5.30) postulates a model of material behavior in which compressive plastic-strain increments have the same effect on strain-hardening as do extensional plastic-strain increments. Also, this material has the property that if it will yield under a given set of stresses it will also yield if all the stresses are reversed. These properties of material behavior were illustrated in Fig. 5.34, which illustrated the physical reasoning underlying (5.30). In constructing the curve of Fig. 5.34*b*, the basic assumptions underlying Eq. (5.30) have been followed, in that *all* the plastic-strain increments along the loading *path* have contributed in a positive manner to the strain-hardening so that the material in state D' has been strain-hardened the same amount as the material in state D in Fig. 5.34*a*.

Whether or not (5.29) and (5.30) are good descriptions of the yielding tendency of a material depends upon how well $\bar{\sigma}$ and $\bar{\epsilon}^p$ correlate with the available data; Fig. 5.45*a* shows the equivalent stress $\bar{\sigma}$ plotted versus the equivalent plastic strain $\bar{\epsilon}^p$ for thin-walled tubes with various amounts of internal pressure and axial load. This type of correlation is fairly satisfactory when the ratios of the principal stresses

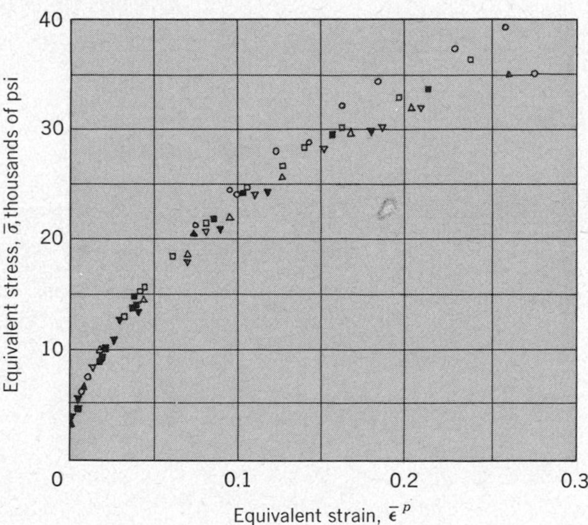

Fig. 5.45*a* Tests on plastic deformation of copper tubes with different fixed ratios of stress components correlated on the basis of equivalent stress and equivalent plastic strain. (Points represent seven different ratios of σ_1/σ_2, varying from 0 to 1.) (*From E. A. Davis, Increase of Stress with Permanent Strain and Stress-strain Relations in the Plastic State for Copper under Combined Stresses, Trans. ASME, vol. 65, pp. A187–A196, 1943.*)

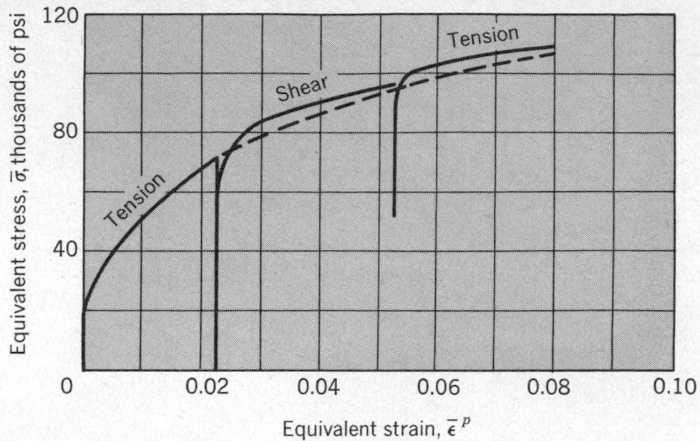

Fig. 5.45*b* Alternating tension and shear correlated on the basis of equivalent stress and equivalent plastic strain. (*W. Sautter, A. Kochendorfer, and U. Dehlinger, The Laws of Plastic Deformation, Z. Metallk., vol. 44, pp. 553–565, 1953.*)

remain constant during the test. The correlation is put to a more severe test when the kind of stressing is changed during the test, as, for example, when first tensile, then shear, and then tensile stresses are applied. The results of such tests are given in Fig. 5.45*b*. When the change in stress during the test is a complete reversal, the correlation is less satisfactory, as illustrated in Fig. 5.45*c*. The lowered elastic limit observed on the reversals of load in Fig. 5.45*c* is called the *Bauschinger* effect. It has been found that the correlation of equivalent stress with equivalent plastic strain is not good when appreciable anisotropy is developed during the straining, since the equivalent strain "remembers" only a sort of averaged total amount of plastic strain and not the distribution of components of this total with respect to various axes within the material.

For materials which yield initially according to the maximum shear-stress criterion, it has been found that the tendency for further yielding can be measured by an *equivalent shear stress* $\bar{\tau}$, defined as follows:

$$\bar{\tau} = \frac{\sigma_{max} - \sigma_{min}}{2} \tag{5.31}$$

Using this concept of yielding, we would say that initial yielding can occur when $\bar{\tau} = Y/2$.

As may be seen from Fig. 4.36, the maximum shear strain is numerically equal to the difference between the maximum and the minimum normal strains. Further, since the principal axes of stress and of strain are coincident in a body, the maximum shear strain occurs on the same set of axes with which is associated the maximum shear stress.

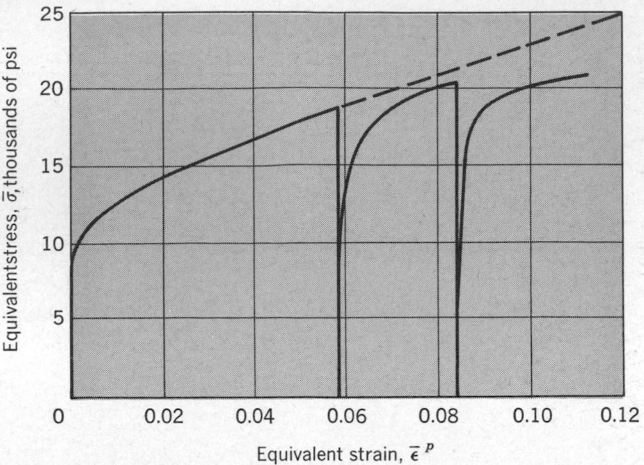

Fig. 5.45c Tests involving complete reversal of torsion of thin-walled copper tubes, correlated on the basis of equivalent stress and equivalent plastic strain. (99.99% pure copper, annealed at 600°F, 1,000 grains/mm².) (*From J. A. Meyer, unpublished research at M.I.T., 1957.*)

In determining how $\bar{\tau}$ will vary with the plastic deformation, there is some theoretical evidence[1] for again choosing the equivalent strain to depend on the strain components in the same way that the equivalent stress depends on the stress components. Thus, the *equivalent plastic shear strain* has been taken to be the integral of the maximum shear component of the plastic-strain increments, as follows:

$$\bar{\gamma}^p = \int[(d\epsilon^p)_{\max} - (d\epsilon^p)_{\min}] \tag{5.32}$$

where, again, the integral is to be taken over the entire loading path. When the experimental results plotted in Fig. 5.45*a* and *b* are replotted on this basis, the results shown in Fig. 5.46*a* and *b* are obtained. The correlation by the shear-stress criterion is usually not quite as good as with the Mises criterion, but Figs. 5.45*a* and 5.46*a* show that the shear-stress criterion gives better correlation at high strains for these tests on copper tubes. The choice as to which correlation to use in a given situation is often governed by mathematical convenience, although, as we shall see below, the components of strain derived from the Mises yield criterion are much closer to physical observations than are those found from the maximum shear-stress criterion.

[1] R. Hill, "The Mathematical Theory of Plasticity," p. 52, Oxford University Press, New York, 1950; D. C. Drucker, A More Fundamental Approach to Plastic Stress-strain Relations, *Proc. First U.S. Natl. Congr. Appl. Mech.*, pp. 487–491, 1951.

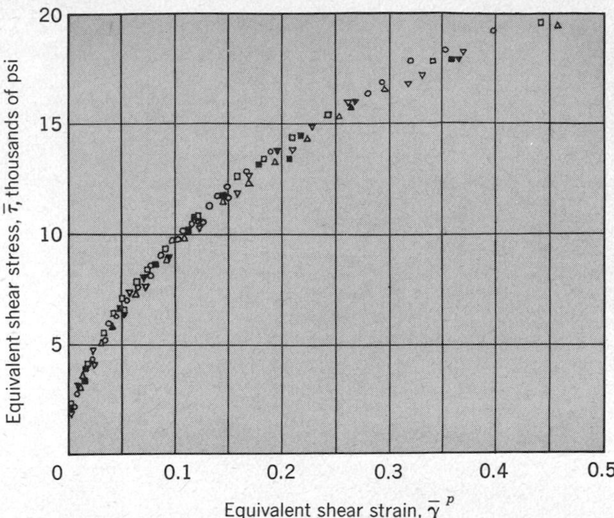

Fig. 5.46a Tests on plastic deformation of copper tubes, with different fixed ratios of stress components correlated on the basis of equivalent shear stress and equivalent plastic shear strain. (Same data as in Fig. 5.45*a*.)

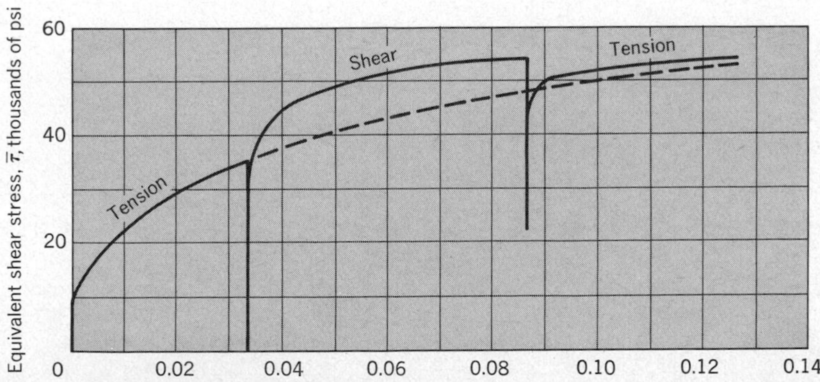

Fig. 5.46b Alternating tension and shear correlated on the basis of equivalent shear stress and equivalent plastic shear strain. (Same data as in Fig. 5.45*b*.)

It should be noted that the data in Figs. 5.45*a* and 5.46*a* extend up to very large strains, and in plotting the data true stress and true strain (based on incremental definitions) were used. Most practical stress analysis—i.e., cases where analysis can yield meaningful numerical results—is limited to strains less than about 0.05 to 0.10, and in such cases it is sufficiently accurate to use engineering stress and strain in the analysis. In metal-working operations, where the shape must be changed, however, the true stress and strain are needed.

5.17 PLASTIC STRESS-STRAIN RELATIONS

In the previous section we discussed how the data from the tensile test could be generalized through the use of an equivalent stress and an equivalent plastic strain to predict when plastic deformation would occur under a general state of stress. However, this correlation gave no indication of how the individual components of plastic strain depend on the stress components. In this section we shall discuss relations which give the plastic-strain components as functions of the stress components. We shall discuss more than one such set of relations because, as was indicated by the experimental correlations discussed in the previous section, no one simple plastic-deformation theory has been found to give superior agreement with experiment in all situations. In the developments which follow we shall use plausibility arguments to lead us rather directly to the stress-strain relations; the reader should be aware that in so doing we compress into a few pages the contributions of many men over a number of years.[1]

As a start let us take the elastic stress-strain relations (5.2) and attempt to modify them to fit the plastic case, assuming that yielding occurs according to the Mises criterion. As discussed in Sec. 5.11, plastic straining occurs with essentially no volume change, so we first modify (5.2) to fit this experimental fact. It can be shown (see Prob. 5.2) that the percentage change in volume is given by

$$\frac{\Delta V}{V} = \epsilon_x + \epsilon_y + \epsilon_z$$

Substituting (5.2) in this equation, we obtain

$$\frac{\Delta V}{V} = \frac{1-2\nu}{E}(\sigma_x + \sigma_y + \sigma_z)$$

and thus to ensure zero volume change for the plastic components of strain, we must take $\nu = \frac{1}{2}$.

A second modification of (5.2) follows from the fact that plastic strain is not linearly proportional to stress. In place of E we must use a factor which varies in some manner with the amount of plastic deformation. It can be shown that if

[1] See A. Mendelson, "Plasticity: Theory and Application," The Macmillan Company, New York, 1968, for further discussion and references.

we use the ratio of the equivalent stress to the equivalent plastic strain[1] of the Mises criterion, that is, $\bar{\sigma}/\bar{\epsilon}^p$, we obtain a set of stress-strain relations which have the required symmetry implied by (5.2) and which predict the correct result for the tensile test. With these modifications in the elastic equations (5.2) we obtain the Hencky[2] stress-strain relations

$$\begin{aligned} \epsilon_x{}^p &= \frac{\bar{\epsilon}^p}{\bar{\sigma}}[\sigma_x - \tfrac{1}{2}(\sigma_y + \sigma_z)] & \gamma_{xy}{}^p &= 3\frac{\bar{\epsilon}^p}{\bar{\sigma}}\tau_{xy} \\ \epsilon_y{}^p &= \frac{\bar{\epsilon}^p}{\bar{\sigma}}[\sigma_y - \tfrac{1}{2}(\sigma_z + \sigma_x)] & \gamma_{yz}{}^p &= 3\frac{\bar{\epsilon}^p}{\bar{\sigma}}\tau_{yz} \\ \epsilon_z{}^p &= \frac{\bar{\epsilon}^p}{\bar{\sigma}}[\sigma_z - \tfrac{1}{2}(\sigma_x + \sigma_y)] & \gamma_{zx}{}^p &= 3\frac{\bar{\epsilon}^p}{\bar{\sigma}}\tau_{zx} \end{aligned} \tag{5.33}$$

These equations are referred to as representing a *deformation* theory of plasticity. While these equations are applicable in many situations, there are some problems to which they definitely are not applicable because a deformation theory of plasticity overlooks an important physical fact connected with plastic deformation. To gain further insight into this problem, let us return to some more paper-clip experiments. As before, open up the clip into the shape of a U. Twist it elastically. Then change the loading from twisting to bending, again being careful not to cause any plastic deformation. Note that the deformation is now the same as if you had omitted the twisting and simply bent it in the first place. That is, in the elastic region the deformation depends only on the final load and is independent of the path by which the load was reached.

Now carry out a similar experiment with plastic deformation. Twist the base of the U through 90° and then gradually apply a bending moment while releasing the twisting moment; continue bending until a little plastic bending deformation has occurred. Note that the plastic twisting deformation still is present, even after the load has been changed to bending. Thus we see that in plastic deformation the total plastic strains depend not only on the final load but also on the path by which the load was reached.

Try another experiment. Twist the wire plastically a little, and unload. Note the twisting deformation. Now bend it plastically a little. Note that the initial twisting deformation remains and there is added to it some bending deformation. Now twist it some more. Note the increase in twisting deformation, while the bending deformation remains. From this experiment we can conclude that in

[1] To be theoretically consistent in this development, we should obtain $\bar{\epsilon}^p$ by eliminating the integral sign in (5.30) and using $\epsilon_1{}^p, \ldots,$ instead of $d\epsilon_1{}^p \ldots,$ under the square root sign.

[2] H. Hencky, Zur Theorie Plastischer Deformationen und der hierdurch im Material Hervorgerufenen Nebenspannungen, *Proc. First Intern. Congr. Appl. Mech.*, pp. 312–317, Delft, 1924.

plastic deformation the kind of load determines the *increment* of strain but not its total value; the total value of a plastic-strain component is given by the integral of the increments of the plastic-strain component over whatever loading history the material has undergone. When Hencky's equations are modified to include the idea that the current stress components determine only the current *increments* of the plastic-strain components, the Levy-Mises equations[1] are obtained.

$$
\begin{aligned}
d\epsilon_x{}^p &= [\sigma_x - \tfrac{1}{2}(\sigma_y + \sigma_z)]\frac{d\bar{\epsilon}^p}{\bar{\sigma}} & d\gamma_{xy}{}^p &= 3\tau_{xy}\frac{d\bar{\epsilon}^p}{\bar{\sigma}} \\
d\epsilon_y{}^p &= [\sigma_y - \tfrac{1}{2}(\sigma_z + \sigma_x)]\frac{d\bar{\epsilon}^p}{\bar{\sigma}} & d\gamma_{yz}{}^p &= 3\tau_{yz}\frac{d\bar{\epsilon}^p}{\bar{\sigma}} \\
d\epsilon_z{}^p &= [\sigma_z - \tfrac{1}{2}(\sigma_x + \sigma_y)]\frac{d\bar{\epsilon}^p}{\bar{\sigma}} & d\gamma_{zx}{}^p &= 3\tau_{zx}\frac{d\bar{\epsilon}^p}{\bar{\sigma}}
\end{aligned}
\tag{5.34}
$$

These relations generally are referred to as the Mises equations and represent an *incremental* theory of plasticity.

To illustrate how the equations (5.34) can be used, if a body is at the point of incipient plastic deformation and a stress increase is applied, giving an increase in equivalent stress of $d\bar{\sigma}$, the increment in equivalent strain, $d\bar{\epsilon}^p$, is found from the equivalent stress-strain curve of the material, e.g., Fig. 5.45*a*. Then the increments in the plastic-strain components can be found from (5.34). For materials with a horizontal stress-strain curve in the plastic region, i.e., perfectly plastic materials, no increase in yield strength can occur, and the extent of the plastic deformation, and hence the magnitude of $d\bar{\epsilon}^p$, is determined by external conditions. For example, in twisting a paper clip nearly constant torque is required from the time plastic flow begins until fracture, and the extent of deformation rather than the torque controls the process.

There is theoretical and experimental evidence[2] for using the Mises equations (5.34) only with the Mises yield criterion (5.29). The same theoretical ideas suggest that when using the maximum shear-stress yield criterion (5.31), the plastic deformation should be assumed to be a pure shear strain of the coordinate axes with which are associated the maximum shear stress. This pure shear is in fact the equivalent plastic shear-strain increment $d\bar{\gamma}^p$ defined by (5.32), and from a consideration of the Mohr's circle for the plastic-strain increments, it may be deduced that this deformation implies that the principal plastic-strain increments are

[1] M. Levy, *Compt. rend.*, vol. 70, p. 1323, 1870; and R. von Mises, Mechanik der festen Körper in plastisch-deformablen Zustand, *Ges. Wiss. Göttingen, Nachs.*, mathphys. Klasse, 1913, pp. 582–592.

[2] R. Hill, "The Mathematical Theory of Plasticity," p. 52, Oxford University Press, New York, 1950; D. C. Drucker, A More Fundamental Approach to Plastic Stress-strain Relations, *Proc. First U.S. Natl. Congr. Appl. Mech.*, pp. 487–491, 1951; McClintock and Argon, *op. cit.*, pp. 283–288.

$$\begin{aligned}(d\epsilon^p)_{max} &= \frac{d\bar{\gamma}^p}{2}\\ (d\epsilon^p)_{int} &= 0 \\ (d\epsilon^p)_{min} &= -\frac{d\bar{\gamma}^p}{2}\end{aligned} \tag{5.35}$$

where max, int, and min refer, respectively, to the axes having the (algebraically) maximum, intermediate, and minimum principal stresses. Equations (5.35) satisfy the requirements that there be zero volume change and that the plastic-strain *increments* be determined by the current stresses (which define the principal stress directions). When $\bar{\tau}$ and $d\bar{\tau}$ are known, the increment in equivalent plastic-strain $d\bar{\gamma}^p$ is found from a plot such as Fig. 5.46*a*. For a perfectly plastic material the value of $d\bar{\gamma}^p$ is determined by the boundary conditions, which must prescribe the magnitude of the deformation.

The choice between using the Mises yield criterion (5.29) and the Mises flow rule (5.34) or the shear-stress yield criterion (5.31) and the shear flow rule (5.35) is a moot one. The Mises formulation fits most experimental data more accurately. The shear formulation is sometimes simpler but is ambiguous at other times, for example, when two of the principal stresses are equal. The Hencky equations (5.33) will not predict correct results in a situation involving a complicated loading path, but in a situation where the ratios of the stresses are kept constant and the stresses are continuously increased, and never decreased, the results are identical with the Mises incremental theory (5.34). There are many situations where these conditions are approximately fulfilled, and the use of (5.33) in such cases may considerably simplify the mathematics and predict results which are within a few percent of the exact ones.

From a strictly theoretical viewpoint, when considering plastic behavior, one should also include the elastic effects (e.g., most real materials are elastic-plastic, as illustrated in Fig. 5.7*e* and *f*). Thus the elastic strains given by (5.8) should be added to the plastic strains given by an appropriate set of plastic stress-strain relations such as (5.33), (5.34), or (5.35). These combined stress-strain relations, together with the equilibrium equations (5.6) and the geometrical conditions (5.7), constitute the complete equations of plasticity.

Much of the present theoretical work in plasticity is concerned with cases where the elastic strains are neglected; i.e., the material is considered to be rigid-plastic, as illustrated in Fig. 5.7*c* and *d*. This assumption reduces the complexity of the problem greatly and provides a fairly good approximation for many important practical problems. Using this type of material model, approximate methods have been developed[1] for estimating the load-carrying capacity of reasonably complex structures. In most of this work—and certainly in elastic-plastic analysis—the

[1] See, for example, J. F. Baker, M. R. Horne, and J. Heyman, "The Steel Skeleton," vol. II, "Plastic Behaviour and Design," Cambridge University Press, New York, 1956; and P. G. Hodge, Jr., "Plastic Analysis of Structures," McGraw-Hill Book Company, New York, 1959.

strains are small enough (less than 0.05 to 0.10) so that the engineering and true definitions of stress and strain yield essentially equal results, and the choice of which to use is governed by convenience.

The theory of plasticity is of more recent origin than the theory of elasticity, and, for this reason, as well as because of the obviously more complicated nature of the plastic stress-strain relations, relatively few problems have been solved. The most common ones involve plane strain of a nonhardening material; some of these results are strikingly different from what might be guessed from elasticity theory. Also, plasticity theory has given a much better understanding of some of the observed failures of structures and machines.

In presenting the foregoing detailed discussion of yielding and plastic behavior, it has been our aim to acquaint the reader with the basic concepts which are currently useful in describing these phenomena. There is little doubt that this area will continue to be one of the most active fields of research in the mechanics of solids since so many problems of scientific and engineering interest involve plastic behavior.

5.18 VISCOELASTICITY

Thus far we have discussed time-independent behavior of materials. In this section we will briefly discuss the time-dependent behavior of engineering metals at high temperature and of plastics (polymers).

At ordinary temperatures the stress-strain relations of most engineering metals are independent of the duration of the loading. At elevated temperatures, however, the strain resulting from a fixed stress continues to grow with time. A typical plot of strain versus time is sketched in Fig. 5.47. The intercept of the curve at $t = 0$ gives the elastic (and possibly also plastic) strain which occurs almost immediately on application of the load. There follows a period of decreasing rate of strain (primary creep), then a period of uniformly increasing strain (secondary creep), and finally a period of accelerating strain rate as the specimen necks or cracks just before the fracture (tertiary creep). The uniform rate of creep during secondary creep has been experimentally determined[1] for many materials from uniaxial tensile creep tests at constant load.

[1] See, for example, Report on the Elevated-temperature Properties of Chromium-molybdenum Steels, *ASTM Spec. Tech. Publ.* 151, 1953; Report on the Elevated-temperature Properties of Selected Super-strength Alloys, *ASTM Spec. Tech. Publ.* 160, 1954.

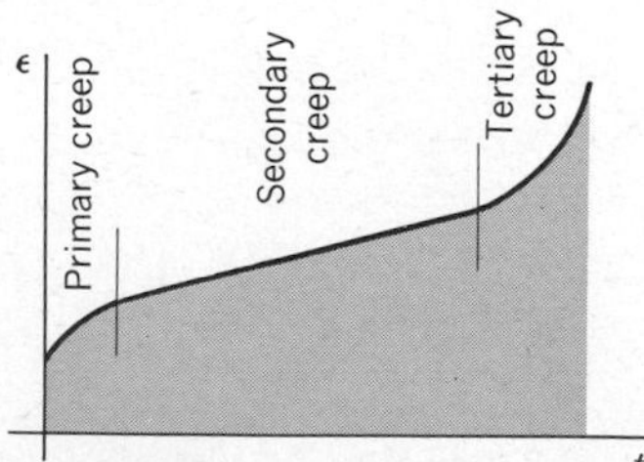

Fig. 5.47 Creep curve.

In using this uniaxial creep data to predict creep rates in situations under combined stress, the ideas of equivalent stress and strain used in plasticity have proved useful. Assuming that the equivalent creep stress $\bar{\sigma}$ is defined by the Mises criterion (5.29), the following equations for the creep rates have been found to give reasonable agreement with experiment:

$$\frac{d\epsilon_x^{\,c}}{dt} = \left[\sigma_x - \frac{1}{2}(\sigma_y + \sigma_z)\right]\frac{d\bar{\epsilon}^c/dt}{\bar{\sigma}} \tag{5.36}$$

with similar equations for $d\epsilon_y^{\,c}/dt$ and $d\epsilon_z^{\,c}/dt$, and

$$\frac{d\gamma_{xy}^{\,c}}{dt} = \tau_{xy}\,\frac{3d\bar{\epsilon}^c/dt}{\bar{\sigma}} \tag{5.37}$$

with similar equations for $d\gamma_{yz}^{\,c}/dt$ and $d\gamma_{zx}^{\,c}/dt$. In these equations the effective creep rate $d\bar{\epsilon}^c/dt$ is defined by the quantity under the integral sign in (5.30) with creep rates substituted for the plastic-strain increments, i.e., with $d\epsilon_1^{\,c}/dt$ used in place of $d\epsilon_1^{\,p}$, etc.

For metals, the coefficient

$$\frac{3d\bar{\epsilon}^c/dt}{\bar{\sigma}}$$

depends very strongly on both stresses and temperature. In the case of liquids, however, the ratio is relatively independent of stress. When the coefficient of viscosity μ is introduced for the reciprocal of this ratio, (5.36) and (5.37) become the stress-strain rate equations for a viscous incompressible fluid. This can be seen if we express the strain rates in (5.36) and (5.37) in terms of the velocity components $\dot{u}$, $\dot{v}$, and $\dot{w}$:

$$\sigma_x - \tfrac{1}{3}(\sigma_x + \sigma_y + \sigma_z) = 2\mu\,\frac{\partial \dot{u}}{\partial x} \tag{5.38}$$

$$\tau_{xy} = \mu\left(\frac{\partial \dot{u}}{\partial y} + \frac{\partial \dot{v}}{\partial x}\right) \tag{5.39}$$

The negative of the mean normal stress $\tfrac{1}{3}(\sigma_x + \sigma_y + \sigma_z)$ is the pressure p. Equations (5.38) and (5.39), together with the corresponding equations for σ_y, σ_z, τ_{yz}, and τ_{xz}, are the constitutive equations for viscous incompressible flow.[1] When combined with momentum and continuity requirements they give the Navier-Stokes equations of fluid mechanics.

Another form of time-dependent behavior appears in polymers. Polymers consist of long-chain molecules in which the links along the chain are held together by primary chemical bonds, typically between carbon atoms. The binding between the chains is due to secondary chemical bonds, which have low enough

[1] See, for example, I. Shames, "Mechanics of Fluids," pp. 277–279, McGraw-Hill Book Company, New York, 1962.

energy so that thermal motion is continually breaking and rearranging the bonds. An applied stress will bias such rearrangements, resulting in gradual flow or deformation with time. At low stress levels the effects are *linearly* proportional to the applied stress. Such mechanisms give rise to creep under constant applied stress, to the elastic aftereffect observed in the gradual unfolding of a freshly folded piece of paper, and to the strain-rate effect in the stress-strain curve of polymethyl methacrylate, shown in Fig. 5.5*d*. Such time-dependent phenomena are called *viscoelasticity*.

The linear dependence of viscoelastic strain on stress turns out to mean that whenever the stress distribution does not depend on the elastic constants E and ν, which is true for almost all of the problems of plane elasticity discussed in this book, the viscoelastic *stress* distribution is constant with time and identical to the elastic stress distribution. With constant applied loads, the resulting time-dependent *deformation* can be found from the creep curve of the material, which gives strain as a function of time under constant applied stress[1] (Fig. 5.47).

The linear dependence of strain at a given time on the stress history allows creep curves for all stress levels within the linear range to be correlated by plotting the strain per unit applied stress, called the creep compliance, as a function of time. A typical plot is shown in Fig. 5.48. At high stress levels the flow of polymer chains past each other is produced directly by the stress, without waiting for thermal motion, and the strain has a greater-than-linear dependence on stress. As a rough estimate, linear behavior usually extends to 1 or 2 percent of immediately recoverable elastic strain, which includes the maximum strains of interest in most structural applications.

As might be expected from the importance of thermal motion in breaking secondary bonds, temperature has a strong effect on the creep rate. To show the effect over a reasonable range of times and temperatures, it is necessary to use a logarithmic scale. At the same time, for comparison with the modulus of elasticity, the stress divided by the time-dependent strain, called the creep modulus, is plotted as in Fig. 5.49*a*. Note that a change in temperature seems to correspond to a shift in the logarithm of the time. This correlation is usually but not always valid. It is often used to extrapolate laboratory tests to the long times needed to predict deflections in structures. The effect of temperature is shown even more dramatically in Fig. 5.49*b*, although the time-temperature shift is not evident in this form.

[1] For examples of the calculation of deformations see, for example, D. C. Drucker, "Introduction to Mechanics of Deformable Solids," McGraw-Hill Book Company, New York, 1967.

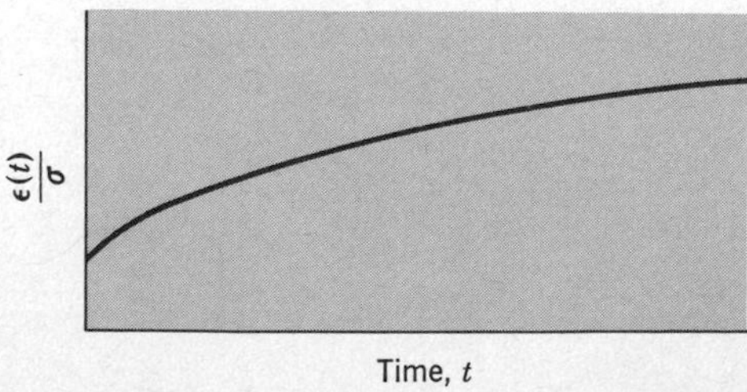

Fig. 5.48 Creep compliance as a function of time.

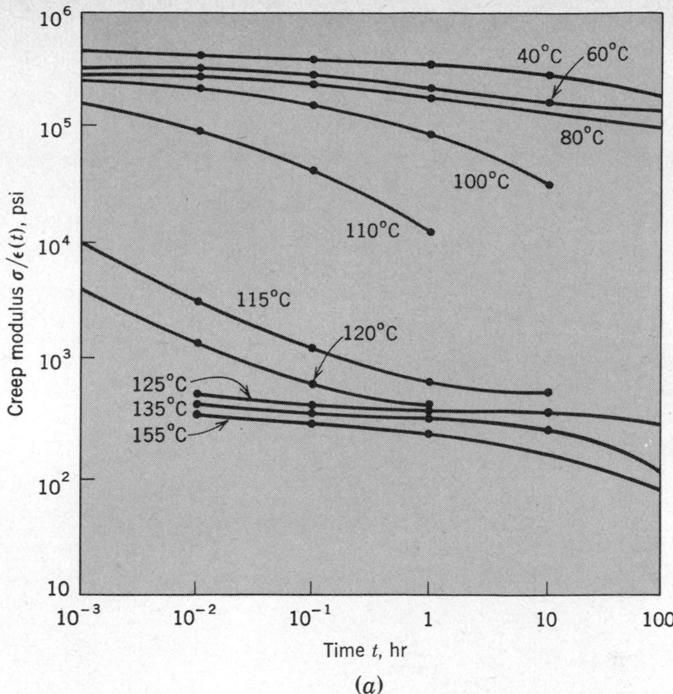

(*a*)

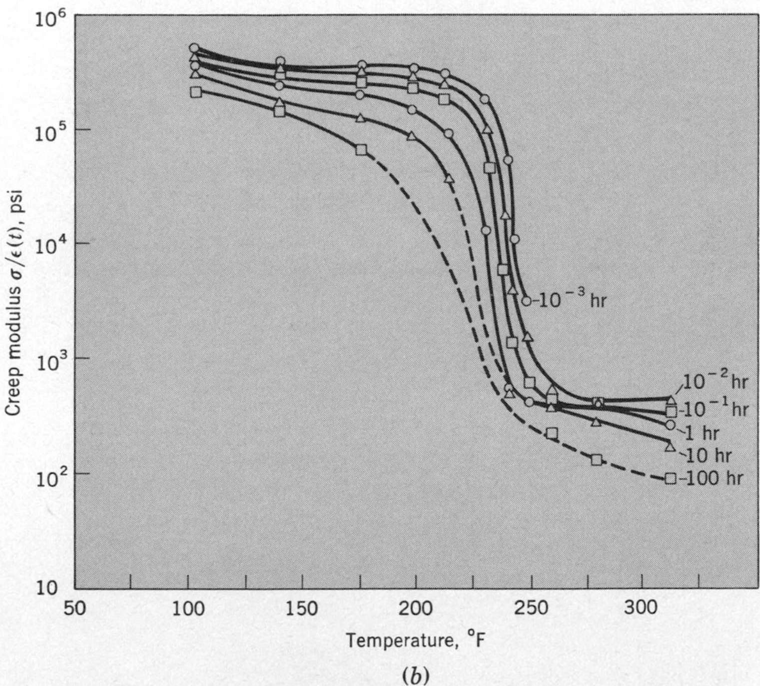

(*b*)

Fig. 5.49 Creep in polymethyl methacrylate. Calculated from data of McLoughlin and Tobolsky, *op. cit.*, using formula of Ferry, *op. cit.*

The time-temperature behavior of viscoelasticity makes it difficult to present data in tabular form. Most data are available in the form of curves derived from tests in which a constant strain is applied and the resulting relaxation of stress with time is noted. These tests have the advantage that shape changes are negligible, which may not be the case for creep in the relatively soft, viscous regimes. Figures 5.49*a* and 5.49*b* were actually obtained from such data, using the conversion scheme described by Ferry.[1] Further data may be found in the literature.[2]

When the history of stress is not constant, the history of strain can still be obtained by numerically superimposing the results of creep, relaxation, constant strain rate, or other tests.[3]

PROBLEMS

5.1. An elastic material with modulus of elasticity E and Poisson's ratio ν originally fills a square cavity of sides $2a$ and height L in a rigid block. A rigid cap is placed on top of the elastic material, and when there is a force F_o acting on the cap the height of the elastic material is observed to have decreased by an amount c. Calculate the magnitude of the force F_o.

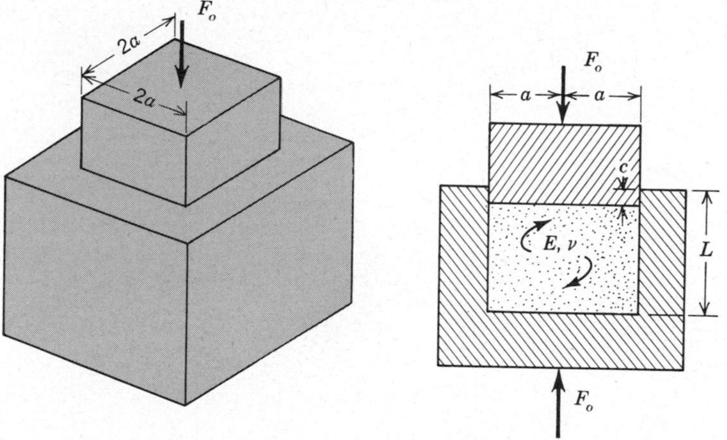

Prob. 5.1

[1] J. D. Ferry, "Viscoelastic Properties of Polymers," chap. 4, John Wiley & Sons, Inc., New York, 1961.

[2] J. R. McLoughlin and A. V. Tobolsky, The Viscoelastic Behavior of Polymethylmethacrylate, *J. Colloid. Sci.*, vol. 7, pp. 555–568, 1952; A. E. Moehlenpah, O. Ishai, and A. T. DiBenedetto, The Effect of Time and Temperature on the Mechanical Behavior of a "Plasticized" Epoxy Resin under Different Loading Modes, *J. Appl. Polymer Sci.*, vol. 13, pp. 1231–1245, 1969; J. T. Seitz and C. F. Balazs, Application of Time-temperature Superposition Principles to Long-term Engineering Properties of Plastic Materials, *Polymer Eng. Sci.*, pp. 151–160, April, 1968; M. Takahashi, M. C. Shen, R. B. Taylor, and A. V. Tobolsky, Master Curves for Some Amorphous Polymers, *J. Appl. Polymer Sci.*, vol. 8, pp. 1549–1561, 1964; J. B. Yannas and A. V. Tobolsky, Approximate Master Curves for Amorphous Polymers from Modulus-temperature Data, *J. Macromol. Chem.*, vol. 1, pp. 399–402, 1966.

[3] See, for example, McClintock and Argon, *op. cit.*, pp. 247–248.

5.2. Show that for small strains the fractional change in volume is the sum of the normal strain components associated with a set of three perpendicular axes.

5.3. (*a*) Using the result of Prob. 5.2, prove that for an isotropic, linearly elastic material,

$$\frac{\Delta V}{V} = \frac{(\sigma_x + \sigma_y + \sigma_z)(1 - 2\nu)}{E}$$

(*b*) The *bulk modulus* of a linearly elastic material is defined as the ratio of the hydrostatic pressure to the fractional decrease in volume.

$$B = \frac{p}{(-\Delta V/V)}$$

Show that for an isotropic material the bulk modulus is given by

$$B = \frac{E}{3(1 - 2\nu)}$$

5.4. With the aid of the results of Prob. 5.3, discuss the stability under hydrostatic pressure of a hypothetical material with a Poisson's ratio greater than ½.

5.5. Prove that in a linear isotropic material a τ_{xy} component of shear stress cannot produce a γ_{yz} component of shear strain.

5.6. Prove that in a linear isotropic material a τ_{xy} component of shear stress cannot produce a uniform expansion or contraction consisting of three equal normal components of strain.

5.7. Invert the stress-strain-temperature relations (5.8) to obtain the stresses in terms of the strains.

5.8. The stresses in a flat steel plate in a condition of plane stress are

$$\begin{aligned}\sigma_x &= 13{,}000 \text{ psi}\\ \sigma_y &= -7{,}000 \text{ psi}\\ \tau_{xy} &= 8{,}000 \text{ psi}\end{aligned}$$

Find the magnitude and orientation of the principal strains in the plane of the plate. Find also the magnitudes of the third principal strain (perpendicular to the plane of the plate).

5.9. In a flat steel plate which is loaded in the xy plane, it is known that

$$\begin{aligned}\sigma_x &= 21{,}000 \text{ psi}\\ \tau_{xy} &= 6{,}000 \text{ psi}\\ \epsilon_z &= -3.6 \times 10^{-4}\end{aligned}$$

What is the value of the stress σ_y?

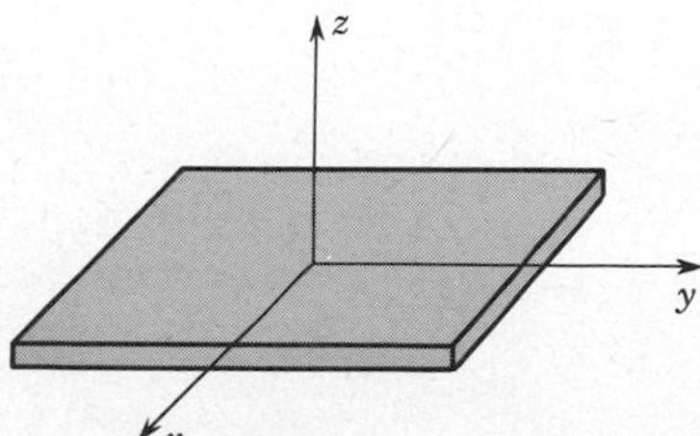

Prob. 5.9

5.10. The principal strains in the plane of a flat aluminum plate which is loaded in its plane are

$$\epsilon_1 = 3.2 \times 10^{-4}$$
$$\epsilon_2 = -5.4 \times 10^{-4}$$

Find the stresses σ_x, σ_y, and τ_{xy}, where the x, y axes are located as shown in the sketch.

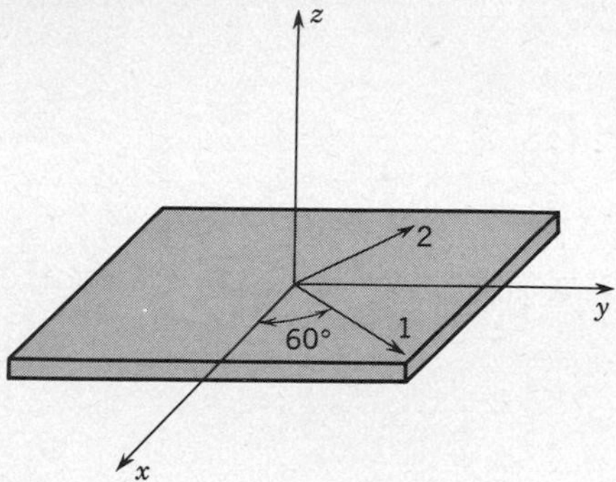

Prob. 5.10

5.11. Let A_o be the original area of a tensile-test specimen and let A be the value at some point during a test. If the test has progressed to a point where the elastic strains can be neglected compared to the plastic strains, show that the reduction of area of the specimen is given by

$$\text{Reduction of area} = \frac{A_o - A}{A_o} = 1 - e^{-\epsilon_x}$$

where ϵ_x is the true strain [defined by Eq. (5.27)] and e is the base of natural logarithms.

5.12. A long, thin-walled cylindrical tank has a radius r and a wall thickness t. Its ends are closed, and when a pressure p is put in the tank a strain gage mounted on the outside surface in a direction parallel to the axis of the tank measures a strain of ϵ_o. What is the pressure in the tank?

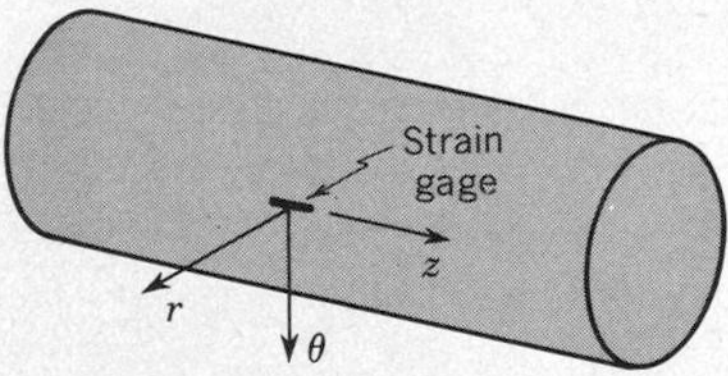

Prob. 5.12

5.13. A sheet of metal in the form of a circle has a small circular hole cut out of it, as indicated in the sketch. If the sheet is initially free of stress and is not restrained in any fashion, what general shape will the originally circular boundaries assume if the sheet is heated uniformly to a temperature T above its original temperature?

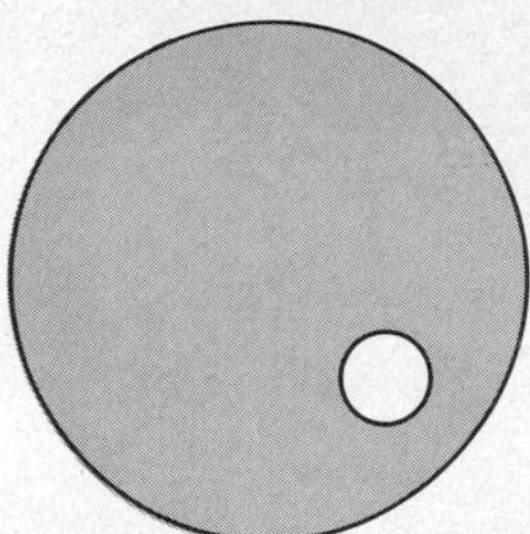

Prob. 5.13

5.14. A small experimental pressure vessel is made from a 9-in.-long brass cylinder of 6-in. mean diameter and $\frac{1}{16}$-in. wall thickness, and two $\frac{1}{2}$-in.-thick steel plates held together by three $\frac{1}{4}$-in.-diameter steel bolts set on a 7-in.-diameter bolt circle. The vessel is put together with the nuts on the three bolts brought up snug, and then each nut is tightened one-half turn additional. Estimate the internal pressure at which the vessel is *certain* to leak.

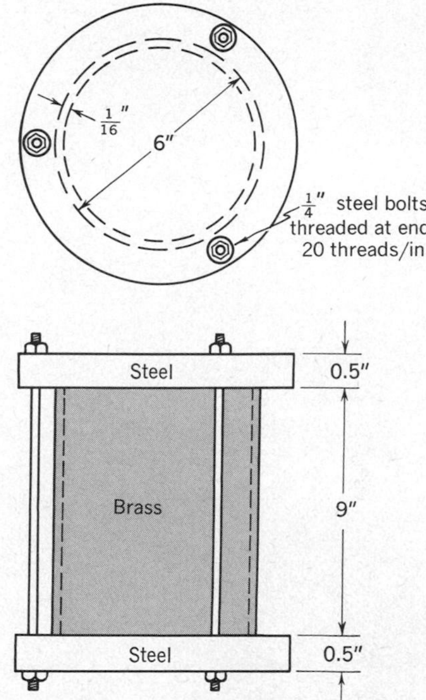

Prob. 5.14

5.15. Discuss stress changes which may be found in a continuous railroad rail due to temperature changes during a 24-hr period. What forces must be applied to hold the rail in place? Where must these forces be applied and how do they depend upon the length of the rail?

5.16. If a steel plate is clad with a thin layer of soft aluminum on both sides, how hot can the assembly be heated without causing plastic flow? Assume that the aluminum does not slide on the steel and that the stress-strain curves are independent of temperature.

5.17. It is desired to produce a tight fit of a steel shaft in a steel pulley. The internal diameter of the hole in the pulley is 0.998 in., while the outside diameter of the shaft is 1.000 in. The pulley will be assembled on the shaft by either heating the pulley or cooling the shaft and then putting the shaft in the pulley hole and allowing the assembly to reach a uniform temperature. Is it more effective to heat the pulley or to cool the shaft? What temperature change would be required in each case to produce a clearance of 0.001 in. for easy assembly?

5.18. An aircraft fuselage is constructed with circular ribs connected by longitudinal stringers and an outer skin riveted to the stringers but not in contact with the ribs. If insulation is placed between the skin and the ribs, the temperature of the skin and stringers may become quite different from that of the ribs. Find the state of stress and strain in the skin due to cooling it and the stringers 50°F below the temperature of the ribs. Note that the skin is free to expand longitudinally, but that lateral expansion is at least partly constrained by the ribs. Neglect the effects of curvature. For simplicity, consider two extreme cases, the first in which the ribs are so heavy as to prevent any lateral strain, and the second in which the ribs are so light that they offer practically no constraint at all. Consider the material to be 2024-T4 aluminum alloy.

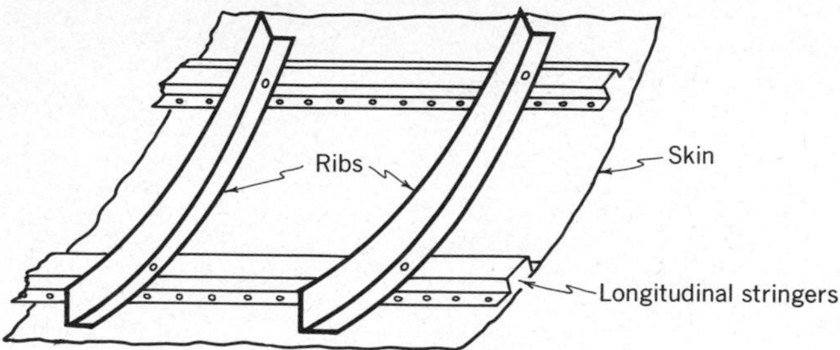

Prob. 5.18

5.19. A cantilever bridge is built out from piers at either end and is to be joined in the middle. Discuss with a numerical example the problem of lining up the two halves of the span if the truss on one side of the roadway is in the sun and that on the other side is in the shade.

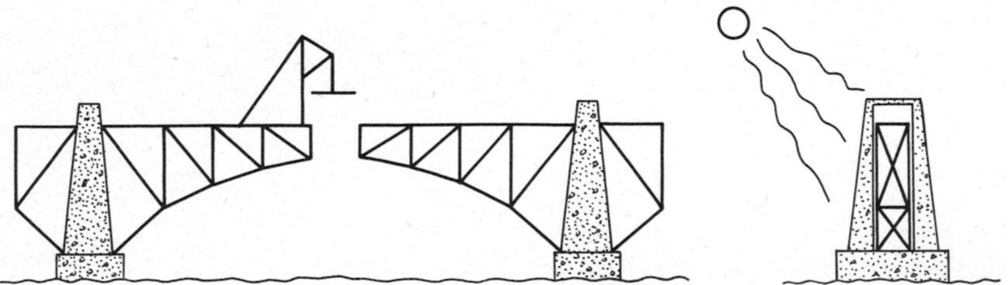

Prob. 5.19

5.20. A long, thin-walled *cylindrical tank* of length L just fits between two rigid end walls when there is no pressure in the tank. Estimate the *force* exerted on the rigid walls by the tank when the pressure in the tank is p and the material of which the tank is made follows Hooke's law.

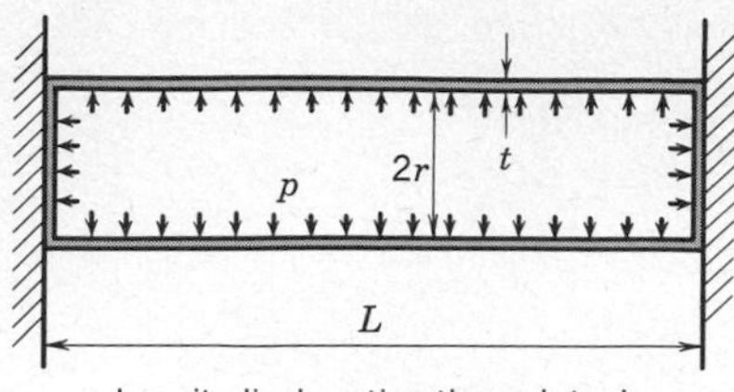

Longitudinal section through tank and rigid end walls

Prob. 5.20

5.21. A long, thin-walled cylinder tank just fits in a rigid cylindrical cavity when there is no pressure in the tank. Estimate the tangential stress in the cylindrical-tank wall when the pressure in the tank is p and the material of which the tank is made follows Hooke's law.

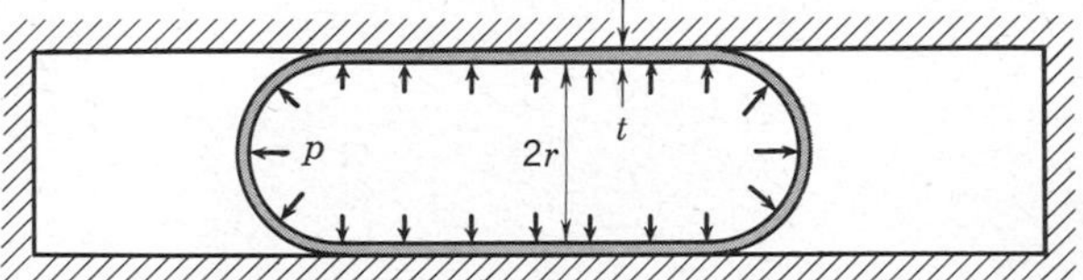

Longitudinal section through tank and rigid cavity

Prob. 5.21

5.22. A thin-walled cylindrical tank with its ends closed by thick plates is put under internal pressure. Derive an expression for the ratio of the change in length to the change in diameter.

5.23. A thin-walled cylinder with closed ends and a thin-walled sphere of the same diameter and wall thickness are put under the same internal pressure. Find the ratio of the change in diameter of the cylinder to the change in diameter of the sphere.

5.24. Find expressions for the elastic displacements in a uniform bar under tensile loading. Show that your solution satisfies the 15 equations of the theory of elasticity.

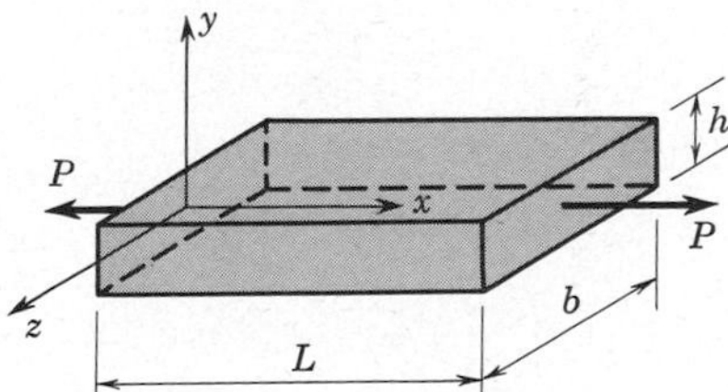

Prob. 5.24

5.25. If the clad plate of Prob. 5.16 is heated 100°F above the temperature which initiates yielding and then is cooled, what is the resulting state of stress and strain? Assume that the clad plate was free of stress before it was heated.

5.26. A composite hoop (the same hoop as in Prob. 2.31) consists of a brass hoop of 10-in. internal radius and 0.1-in. thickness, and a steel hoop of 10.1-in. internal radius and 0.2-in. thickness. Both hoops are 2 in. wide normal to the plane of the hoop. If the composite hoop is heated uniformly to a temperature 150°F above room temperature, estimate the stresses in the brass and steel hoops.

5.27. A batch of 2024-T4 aluminum alloy yields in uniaxial tension at the stress $\sigma_o = 48{,}000$ psi. If this material is subjected to the following state of stress, will it yield according to (*a*) the Mises criterion, and (*b*) the maximum shear-stress criterion?

$$\sigma_x = 20{,}000 \text{ psi} \qquad \tau_{xy} = 20{,}000 \text{ psi}$$
$$\sigma_y = -10{,}000 \text{ psi} \qquad \tau_{yz} = 0$$
$$\sigma_z = 0 \qquad \tau_{zx} = 0$$

5.28. It is proposed to check the safety of a thin-walled, cylindrical pressure vessel made of hot-rolled low-carbon steel by measuring changes in length and circumference as the internal pressure is increased. How much change in length and circumference would occur before the material yielded?

5.29. Calculate the ratio of axial to tangential strain when the pressure inside a long, thin-walled, closed-end cylinder is increased to the point where enough plastic flow has occurred so that elastic strains may be neglected.

5.30. Alclad sheet consists of an aluminum alloy covered with a thin layer of purer aluminum for corrosion resistance. For example, 16-gage 2024-T4 alclad has a total thickness of 0.051 in., of which 0.005 in. on either side is the pure aluminum cladding. What are the residual stresses in the cladding and in the core as a narrow strip of this material is stretched 2 percent and then released? Consider two approximations, in both cases using an elastic–perfectly plastic model to describe the data for the two materials in Fig. 5.5*b*.

(*a*) First make a simple analysis in which it is assumed that the cladding is so thin that it exerts negligible influence on the core.

(*b*) Repeat (*a*), taking into account the finite thickness of the cladding and its restraining effect on the core.

5.31. A chain hoist is attached to the ceiling through two tie rods at an angle θ to the vertical, as shown in the sketch. The tie rods are made of cold-rolled steel with yield strength Y, and each has an area A.

(*a*) What is the load at which both rods become plastic, so that large-scale plastic deformation begins?

(*b*) By how much would this load be increased if a third rod of area A were added, as shown by the dotted line?

(*c*) What is the load-deflection relation when the deflections are elastic in all three rods? (*Hint:* For an assumed vertical deflection δ, find the loads carried by the pair of diagonal bars and by the vertical strut separately, and then add to get the total load as a function of deflection.)

(*d*) If the three-member frame is loaded until all three rods become fully plastic, and then the load is released, find the residual stress in the central rod. What result would you expect for the extreme cases of $\theta = 0$ and $\theta = \pi/2$? Does your analysis give this?

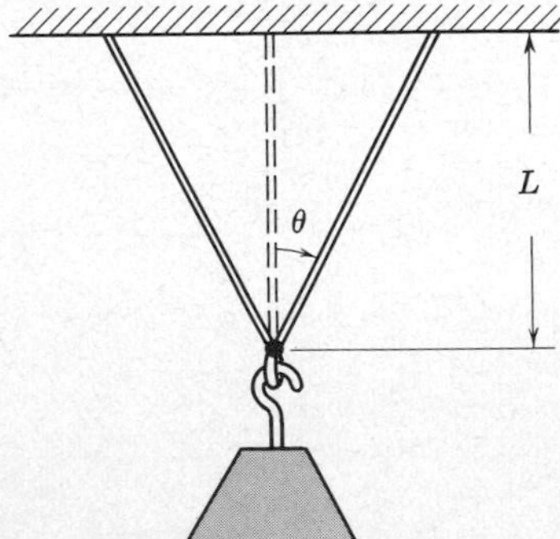

Prob. 5.31

5.32. A stiff horizontal bar 012 is hinged at one end. Two rods of equal length and area are attached to the bar as shown. The material of the rods is elastic–perfectly plastic. The load P is increased until both rods have become plastic, and then it is removed. Find the residual stresses in the two rods. Find also the change in angle which the bar 012 makes with the horizontal when the load is removed.

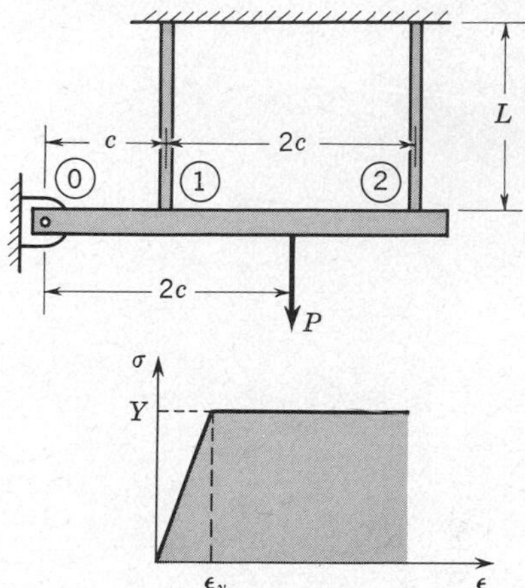

Prob. 5.32

5.33. If a very long, thin bar of metal is stretched to fracture, almost all the elongation will be due to uniform strain, since the necked region is so short relative to the length of the piece. On this basis calculate the change in length before fracture in rods ¼-in. diameter by 50 in. long of cold-rolled and hot-rolled mild steel. Contrast the ratio of these changes in length with the ratio of the true strains at fracture for the two materials.

5.34. The most commonly quoted results of a tensile test are the yield point or yield strength, the tensile strength, the percentage elongation, and the percentage of reduction of area. How can one tell from these results whether or not any necking occurred before fracture?

5.35. Show that, in a tensile test where the plastic stress-strain relations (5.34) govern the material behavior, the equivalent plastic strain defined by Eq. (5.30) is equal to the axial normal component of strain.

5.36. In the solution for the displacement in a thick-walled cylinder, Eq. (*i*) of Sec. (5.7), evaluate the constants of integration A and B from the boundary conditions to obtain an expression for the displacement in terms of the outer and inner pressures.

5.37. There are many practical situations where it is desirable to shrink-fit an external member on a shaft. The inner diameter of the external member is usually made slightly less than the outer diameter of the shaft. The external member is then expanded by heating, slipped over the shaft, and allowed to cool. A steel shaft with an outer diameter of 18 in. and an inner diameter of 12 in. has a steel tube 3-in.-thick shrunk-fit onto it. The inner diameter of the tube is machined to be 0.05 in. less than the outer diameter of the shaft. Determine the expressions for the stresses in the shaft.

5.38. Show that if the x, y, z and a, b, c axes are related as shown in Fig. 5.26, then the shear strain γ_{ab} resulting from the normal stress σ_a is

$$\frac{\gamma_{ab}}{\sigma_a} = +(S_{11} - S_{22} - \tfrac{1}{2}S_{44}) \sin 2\theta \cos 2\theta$$

Find the maximum value this ratio will have for iron.

5.39. Show by symmetry arguments that in linear orthotropic materials a shear-stress component referred to the structural axes will produce only the corresponding shear-strain component referred to the structural axes.

5.40. A pipe is held by two fixed supports as shown in the figure. When mounted, the temperature of the pipe was 70°F. In use, however, cold fluid moves through the pipe, causing it to cool considerably. If we assume that the pipe has a uniform temperature of 0° and if we take the coefficient of linear expansion to be 6.5×10^{-6}/°F for this temperature range, determine the state of stress and strain in the pipe as a result of this cooling. Neglect all body forces and the surface forces on the lateral surfaces of the pipe.

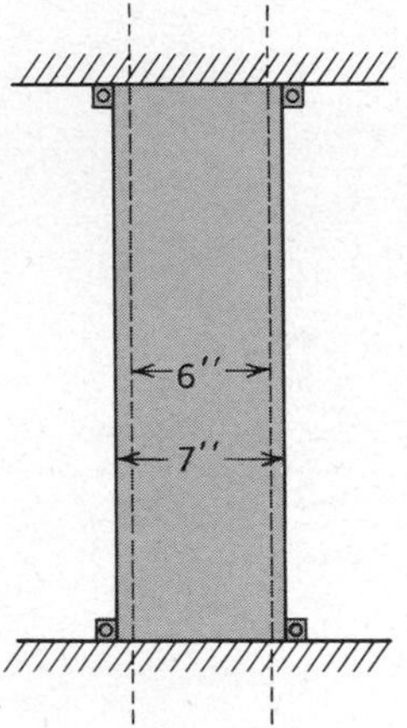

Prob. 5.40

5.41. Shown is a steel bolt and nut and an aluminum sleeve. The bolt has 16 threads per inch and, when the material is at 60°F, the nut is tightened one-quarter turn. The temperature is then raised from 60 to 100°F. Determine the stresses in both bolt and sleeve.

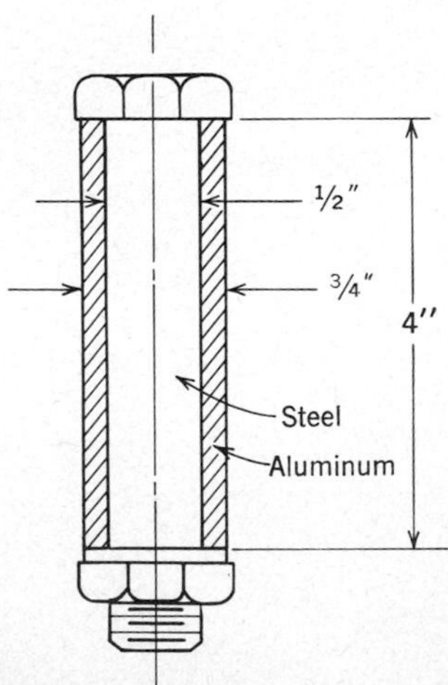

Prob. 5.41

5.42. Blowing up a spherical toy balloon becomes easier once the balloon "starts" to inflate. A typical experimental curve of internal pressure versus balloon diameter is shown. Attempt to formulate a *simple* explanation of these phenomenon under the assumption that the rubber of the balloon is an incompressible linearly elastic material. You should be able to obtain an analytical expression for the curve of pressure in the balloon versus diameter which is similar to the *first* portion of the experimental curve.

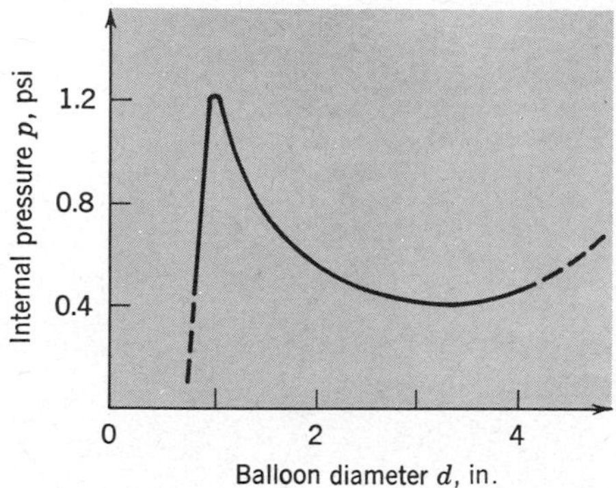

Prob. 5.42

5.43. In the structure shown, all three bars have the same cross-sectional area A and are compelled to have the same length L, although this common length is free to expand or contract as the temperature changes. The bar materials have unequal thermal expansion coefficients and elastic moduli:

$$\alpha_1 = \alpha \qquad \alpha_2 = 2\alpha$$
$$E_1 = E \qquad E_2 = 2E$$

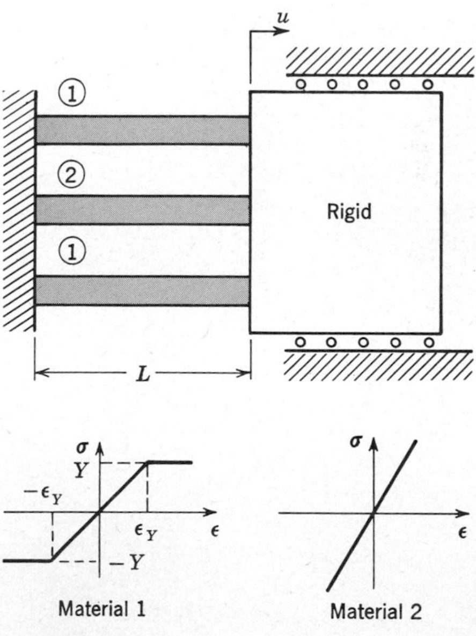

Prob. 5.43

Material 1 is elastic-plastic, being ideally plastic beyond a strain of ϵ_Y, while material 2 can be taken to remain elastic throughout the excursions described below. The system is assembled at temperature $T=0$ with no stress in the bars. The problem is to analyze the behavior of the structure as the temperature T is increased. Specifically, answer the following questions:

(*a*) If T is small enough, the entire system is elastic and the displacement u returns to zero when T returns to zero. Find the limiting temperature T_Y for which, as soon as $T > T_Y$, some yielding occurs in material 1.

(*b*) When $T = T_Y$, what is the corresponding displacement u?

(*c*) If now the temperature is raised to $T = 2T_Y$, what is the displacement u at this temperature?

(*d*) When the temperature is $T = 2T_Y$, what is the plastic strain in the material 1?

5.44. A bolt-sleeve-washer combination is tightened with a torque wrench to produce an axial strain in the bolt of 0.0005. The assembly is then heated to 200°F. What will be the compressive stress in the washer at the new temperature? The bolt is steel and has cross-sectional area $A_s = 0.5$ in.^{2}. The sleeve is aluminum and has cross-sectional area $A_a = 1.0$ in.^{2} and length $L = 6$ in. The washer is steel and has cross-sectional area $A_c = 1.0$ in.^{2} and thickness $h = 0.06$ in.

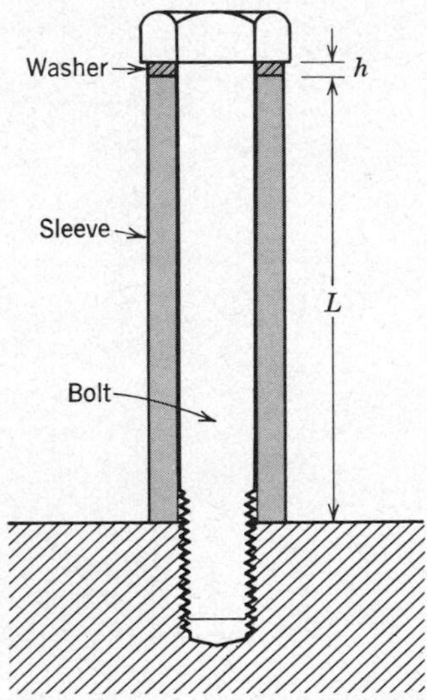

Prob. 5.44

5.45. A materials test is performed by pressurizing the chamber shown. The specimen is machined to have cross-sectional area A at the ends and area kA in the test section $(0 < k < 1)$. What is the stress state in the test section when the pressure (above atmospheric) in the chamber is p? Under atmospheric conditions, the material yields in simple tension at $\sigma_Y = 40{,}000$ psi. How large must the pressure p be to produce yielding? For this case, does it make any difference whether you use the Mises or the maximum shear-stress criterion?

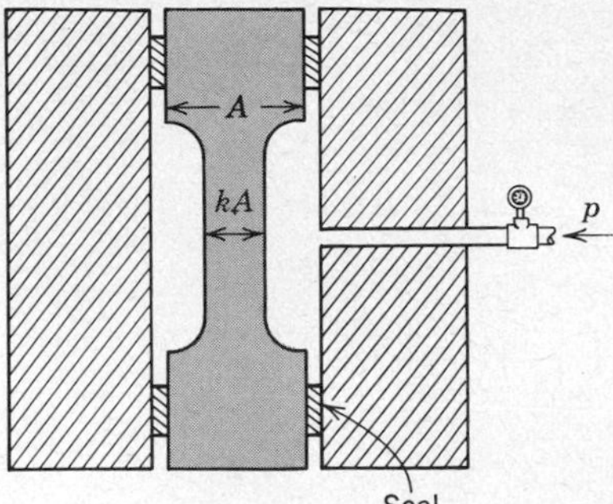

Prob. 5.45

5.46. We wish to electroplate copper onto sheet steel to improve corrosion resistance. However, when copper is plated under normal conditions, the resulting copper layer has a residual tensile stress of 3,000 psi, which is harmful from a fatigue point of view. In order to obtain a copper layer which is either stress-free or under compressive stress, it is suggested that we either:

(*a*) Alter the temperature of the plating bath such that at room temperature the layer will be stress-free, or

(*b*) Carry out the plating while the steel sheet is subject to a uniaxial tensile loading, such that when the load is removed the layer will be stress-free.

Analyze both methods, determining the bath temperature for the first case and the tensile stress for the second. Would both methods provide the same resultant stress configuration? Which would *you* recommend?

	Thickness	α, (°F)$^{-1}$	σ_Y, psi
Steel	0.010 in.	6×10^{-6}	100,000
Copper	0.001 in.	9×10^{-6}	10,000

5.47. Show that the stresses (5.9) for $p_o = 0$, $p_i = p$ reduce to the results for a thin-walled tube (Prob. 4.10) as t/r_i becomes small. Plot σ_r and σ_θ for $t/r_i = 0.05$ from the exact solution and from the thin-walled-tube approximation.

5.48. A circular plate (or short cylinder) of outer radius a has a small central hole of radius c, where $c \ll a$. The plate is subjected to an outer pressure p as shown; there is no axial force.

(*a*) Calculate the maximum normal stress existing in the plate in the limit as $c/a \to 0$, that is, as the hole becomes microscopic in size.

(*b*) Calculate the ratio of this maximum stress to the maximum stress which would exist in a solid circular plate of radius a loaded with an outer pressure p.

This ratio gives the *stress concentration* factor for a small hole in a plate which is subjected to *hydrostatic* (equal in all directions) stress in the plane of the plate.

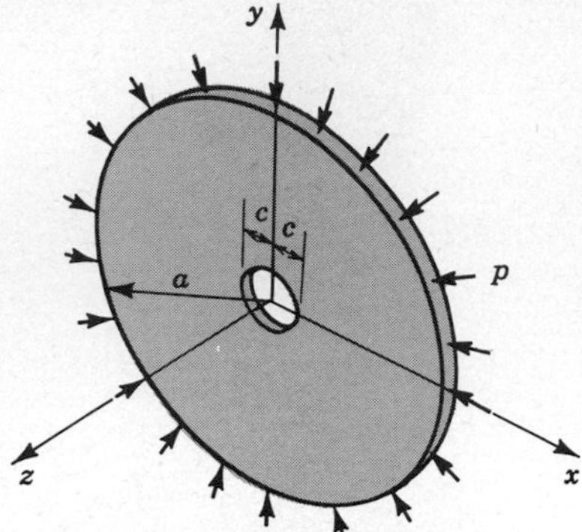

Prob. 5.48

5.49. Consider the indicial form (see Sec. 4.15) of the complete equations of elasticity in Sec. 5.6. Show that (5.6)–(5.8) can be written in the form

$$\sigma_{ij,i} + X_j = 0$$

$$e_{ij} = \tfrac{1}{2}(u_{i,j} + u_{j,i})$$

$$e_{ij} = \frac{1+\nu}{E}\sigma_{ij} - \frac{\nu}{E}\delta_{ij}\theta + \alpha\delta_{ij}\,\Delta T$$

where δ_{ij} is called the *Kronecker delta* and is equal to 1 when $i = j$ and equal to zero when $i \neq j$, and where

$$\theta = \sigma_{11} + \sigma_{22} + \sigma_{33} = \sigma_{ii}$$

5.50. Show that the strain energy stored in a body (5.17) can be written in indicial notation in the form

$$U = \tfrac{1}{2}\int(\sigma_{ij}e_{ij})\,dV$$

5.51. Show that the strain-energy expression (5.17) for an isotropic material can be written in terms of stresses in the form

$$U = \int\left[\frac{1}{2E}(\sigma_x^2 + \sigma_y^2 + \sigma_z^2) - \frac{\nu}{E}(\sigma_x\sigma_y + \sigma_y\sigma_z + \sigma_z\sigma_x) + \frac{1}{2G}(\tau_{xy}^2 + \tau_{yz}^2 + \tau_{zx}^2)\right]dV$$

or in terms of the strain

$$U = \int\left\{\frac{E(1-\nu)}{2(1+\nu)(1-2\nu)}(\epsilon_x + \epsilon_y + \epsilon_z)^2 - \frac{E}{(1+\nu)}[\epsilon_x\epsilon_y + \epsilon_y\epsilon_z + \epsilon_z\epsilon_x - \tfrac{1}{4}(\gamma_{xy}^2 + \gamma_{yz}^2 + \gamma_{zx}^2)]\right\}dV$$

5.52. A composite material is made by aligning continuous fibers of boron $d = 0.002$-in. diameter and bonding them together in a linear array with an epoxy resin as shown in the sketch. The modulus of elasticity of the boron fibers is 5×10^7 psi and that of the epoxy resin 5×10^5 psi.

(*a*) What are Young's moduli of the composite material in the 1 and 2 directions for a volume fraction of 40 percent of the boron fibers (the densities of the boron and resin may be assumed equal)?

(*b*) If a second composite layer with fibers lined up parallel to the 2 direction is glued on top of the first layer, what would the new moduli be in the 1 and 2 directions?

(*c*) Does the structure described in part (*b*) possess isotropy in the plane?

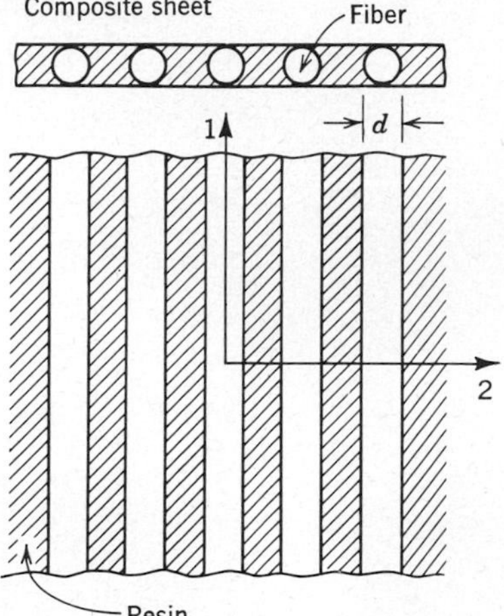

Prob. 5.52

5.53. Show that Eq. (5.20) in indicial notation can be written in the form

$$\epsilon_j = S_{ji}\tau_i \qquad S_{ij} = S_{ji} \qquad = 1, \ldots, 6$$

where

$$\epsilon_j = \epsilon_x, \ldots, \gamma_{zx} \qquad \tau_i = \sigma_x, \ldots, \tau_{zx}$$

5.54. A threaded steel rod is subject to a mean tensile stress of 30,000 psi. The ultimate strength for the material is 130,000 psi, and K_f for the threaded section is 2.6. Estimate the maximum alternating stress that can be applied?

5.55. A closed-end cylindrical tank of 10-in. diameter is made of 0.05-in.-thick steel having an endurance limit of 40,000 psi. The tank is supplied with air from a pump in such a fashion that there are alternating pressure pulses, equal in amplitude to 15 percent of the mean pressure. For a safety factor of 3, what maximum mean pressure would you recommend? (Assume that due to fittings and so forth, K_f may approach 3.)

5.56. A magnesium alloy rod (HM21A-T8) is to be used for a *finite* life application. The mean stress is 12,000 psi and the alternating stress is 17,000 psi. Estimate by using a suitable modification of the Goodman-Soderberg diagram and Fig. 5.42*b* the number of cycles to failure of the rod.

6
Torsion

6.1 INTRODUCTION

In Chap. 2 we saw how the three steps of (2.1) formed the basis for the analysis of simple problems in the mechanics of solids. In the intervening chapters we have studied each of the three steps separately, extending considerably our understanding of these steps. We are now ready to use them in the analysis of problems of major engineering importance.

In this chapter we shall consider the problem of *twisting*, or *torsion*. A slender element subjected primarily to twist is usually called a *shaft*. Twisted shafts play an important part in many mechanisms. A major use of such shafts is in the transfer of mechanical power from one point to another; Figs. 6.1 and 6.2 show familiar examples of this use of shafts in torsion. In other situations a twisted shaft can be used to provide a spring with prescribed stiffness with respect to rotation; examples of this are the torsion-bar spring system on automobiles (see Prob. 6.34), and, on a different scale, the measurement of extremely small forces by an instrument which uses a very fine wire in torsion as the basic "spring" (see Prob. 6.25).

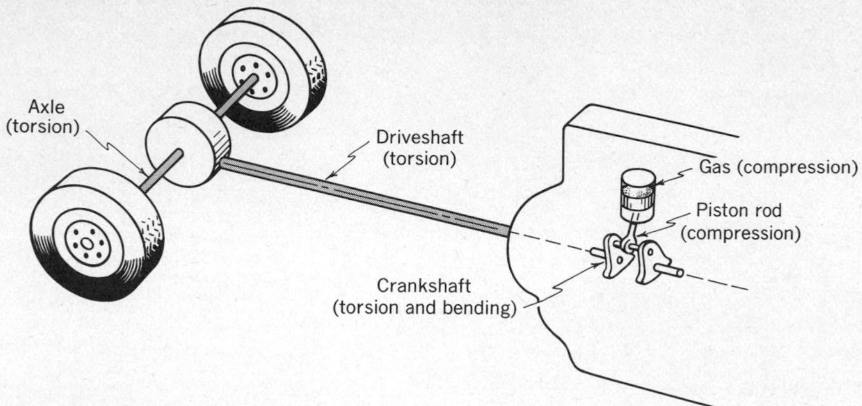

Fig. 6.1 Schematic representation of power transmission in an automobile. Combustion in the cylinder compresses gas which exerts compressive force on the piston rod. Piston rods tend to twist (and bend) the crankshaft. The torsion is transmitted by the drive shaft to the rear gear box, which in turn exerts torsion on the rear axles.

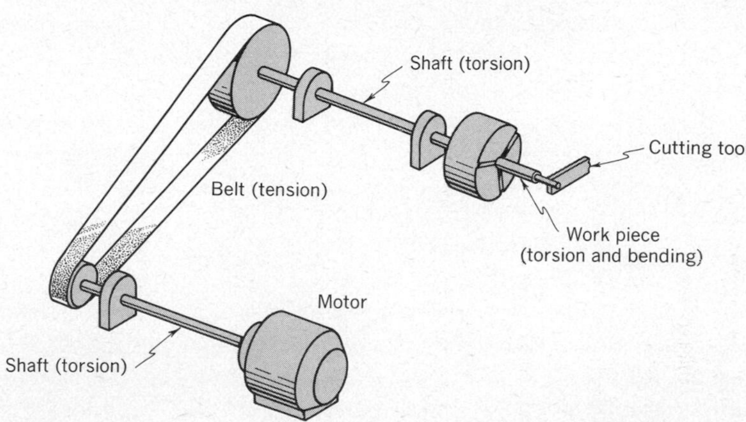

Fig. 6.2 Schematic representation of power transmission in a lathe. The torque generated electromagnetically in the motor is transmitted to a pulley by a shaft in torsion. A belt in tension transfers power to another shaft in torsion which drives the work piece.

In the transmission of power by a shaft in torsion, we are interested primarily in the twisting moment which can be transmitted by the shaft without damage to the material, and hence we wish to know what the stresses are in the shaft. In the use of a shaft as a torsional spring, we are interested primarily in the relation between the applied twisting moment and the resulting angular twist of the shaft. In order to obtain this *overall force-deformation* relation, we shall have to consider the distribution of stress and strain throughout the entire member. This will give us an opportunity to study a complete nontrivial solution in the theory of elasticity. We shall also consider simple cases involving plasticity. The analyses considered in this chapter furnish the foundation for most engineering calculations involving shafts.

We have said that a complete solution to a problem in the mechanics of solids can be obtained by following the three steps of (2.1). In a new nonroutine problem it is not always clear in what order these steps should be executed. In many cases it is necessary to come back to each step several times before a solution is obtained. For example, one might consider all the obvious relations under step 1 and then proceed to step 2, but in carrying out step 2 it might become clear that certain aspects of step 1 had been overlooked. In the torsion problem it proves most convenient to begin our analysis by examining the geometric behavior of a twisted shaft and from this to construct a plausible model for the deformation. Next the stress-strain relations are incorporated, and then, finally, the conditions of equilibrium are applied.

6.2 GEOMETRY OF DEFORMATION OF A TWISTED CIRCULAR SHAFT

We shall consider uniform circular shafts made of isotropic material. Figure 6.3*a* illustrates a circular shaft loaded only by twisting couples M_t at its ends. The cylindrical coordinates r, θ, z, which we shall employ in the subsequent development, are shown. Figure 6.3*b* illustrates the fact that equilibrium requires that each cross section of the shaft must transmit the twisting moment M_t.

Let us start our consideration of possible modes of deformation by isolating from the shaft a slice Δz in length with faces originally plane and normal to the axis of the shaft, as shown in Fig. 6.4*a*. We take this slice from somewhere near the middle of the shaft so that we are away from any possible end effects. Before proceeding further, we observe that, since any other slice taken from this region will have an identical original shape and will be subject to the same twisting moment, we can expect it to have the same deformation; i.e., the pattern of deformation will not vary along the length of the shaft. Our aim is to deduce a plausible mode of deformation for such a slice. In this process we shall appeal to one of the most powerful arguments of science, the argument of *symmetry*. Because of its simple shape, the circular shaft is particularly vulnerable to symmetry arguments.

Suppose, in Fig. 6.4*a*, that the originally straight radius OA deformed into the curved line OA'. Since the material is isotropic and the slice has full geometric circular symmetry about the z axis, we must then conclude that *all* radii are deformed into identical curved lines, as illustrated by the radius OF in Fig. 6.4*b*. Furthermore, these curved lines must all lie in a *plane*; i.e., the ends of the slice remain plane during the deformation. We can demonstrate this by a simple symmetry argument. Suppose that one end of the slice bulged out or dished in. Symmetry would then require that the other end do the same: but it would be

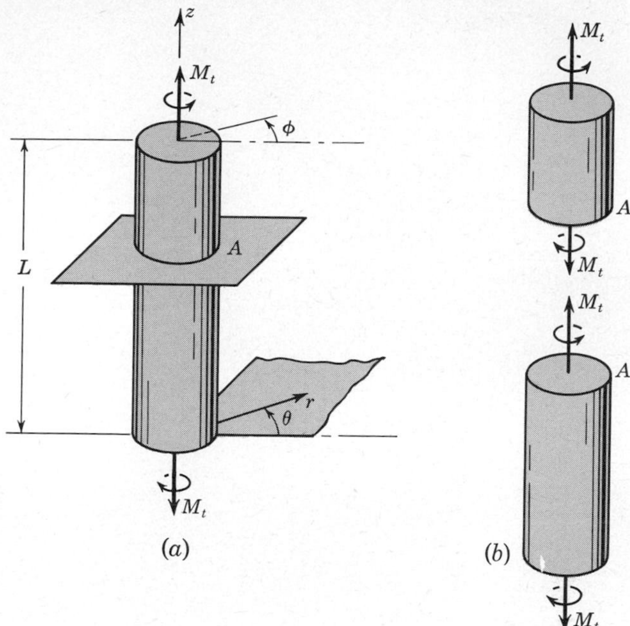

Fig. 6.3 (*a*) Circular shaft loaded at ends by twisting moments M_t which cause a relative rotation ϕ between the ends; (*b*) illustrating that every cross section A is acted on by a twisting moment M_t.

impossible to fit a number of such deformed slices together to form a complete continuous shaft. Thus we conclude that when a circular shaft is twisted, its cross sections *must remain plane.*

We consider next the slice shown in the upper part of Fig. 6.5. We focus our attention on the section *HOABCJ*, which is planar before deformation. Now, when the twisting moment M_t is applied, this section will deform into some

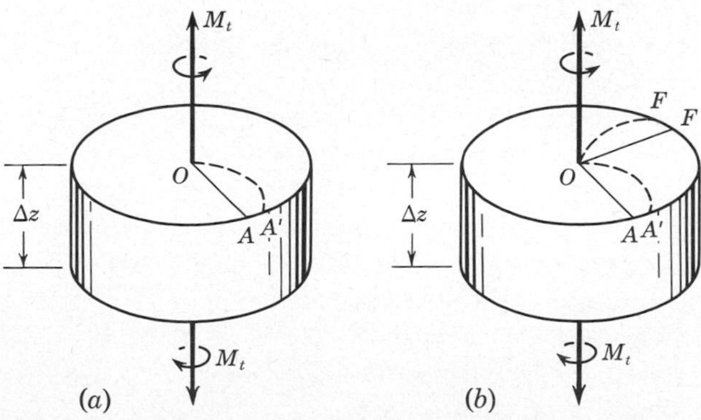

Fig. 6.4 Thin slice showing hypothetical deformation.

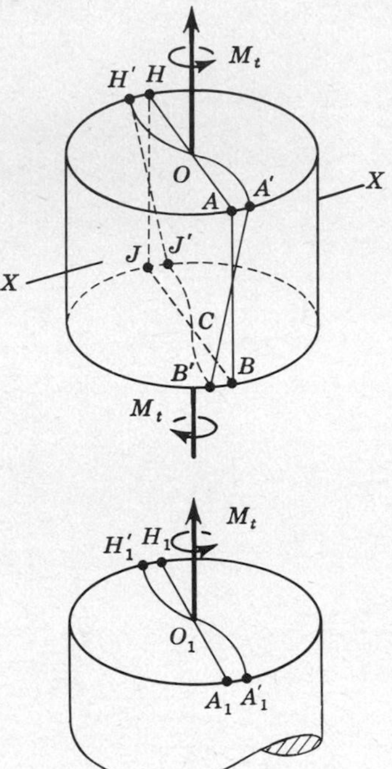

Fig. 6.5 The assumed shape of $B'CJ'$ does not match $A_1'O_1H_1'$.

distorted shape $H'OA'B'CJ'$. Let us tentatively assume that the curvatures of the lines $H'OA'$ and $B'CJ'$ are as indicated in Fig. 6.5. We can show, however, that this assumption leads to a contradiction by considering in the lower part of Fig. 6.5 the very next slice of the shaft. Since this element is subjected to the same deformation as the element above, we expect the diameter $A_1O_1H_1$ to deform into a curved line $A_1'O_1H_1'$ with the same shape as $A'OH'$ above. Such a deformation, however, would violate geometric compatibility since the curvatures of $A_1'O_1H_1'$ and $B'CJ'$ have opposite sense. It would be impossible to fit these two elements together when deformed as shown.

Since the curvature assumed for $B'CJ'$ in Fig. 6.5 has led to a contradiction, we next make the tentative hypothesis that $B'CJ'$ has the *opposite* curvature, as indicated in Fig. 6.6*a*. This would permit matching adjacent elements. Nevertheless, this pattern of deformation also leads to a contradiction. To see this, we rotate the element in Fig. 6.6*a* about the axis XX which is perpendicular to the element $HOABCJ$. After a rotation of 180° the element is upside down, as shown in Fig. 6.6*b*. Now we compare Fig. 6.6*a* with 6.6*b*. The elements are of identical shape and material and are subjected to identical loadings. Therefore they should have identical deformations. The curvatures of diameters in the two elements, however, are of opposite sense. Thus the assumption that diametral lines deform

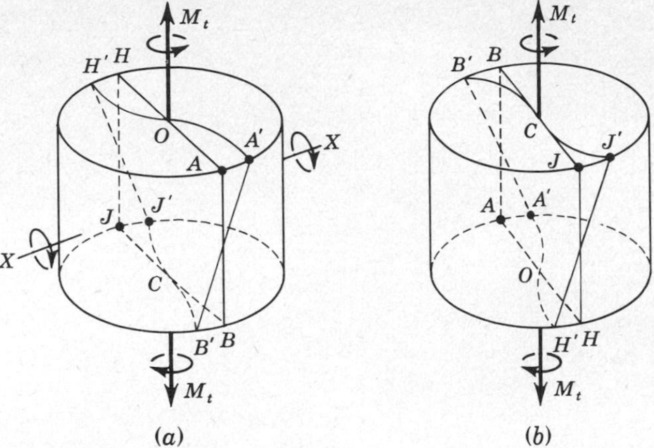

Fig. 6.6 Rotating (*a*) about *X*-*X* through 180° yields (*b*), which has undergone different deformation even though the twisting moment and geometry are the same.

into curved lines is ruled out by symmetry, and we are forced to the conclusion that the deformation pattern must be as indicated in Fig. 6.7. Straight diameters are carried into straight diameters by the twisting deformation.

To summarize, our repeated application of the argument of symmetry has established that the circular shaft must deform such that each plane cross section originally normal to the axis remains plane and normal and does not distort within its own plane. Symmetry of deformation has not ruled out a symmetrical

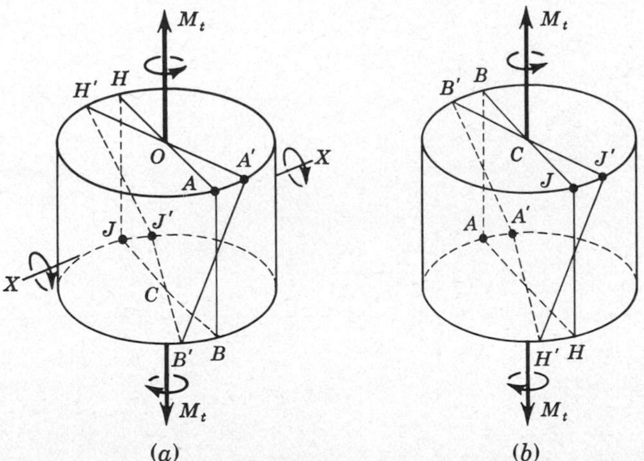

Fig. 6.7 If the diameter *HA* remains straight during deformation, then rotation of (*a*) about *X*-*X* produces (*b*) which is identical in terms of deformation.

expansion or contraction of the circular cross section or a lengthening or shortening of the cylinder. It does not seem plausible, however, that such *dilational* deformations would be an important part of the deformation due to a twisting moment. We shall make the tentative *assumption* that the extensional strains all vanish.

$$\epsilon_r = \epsilon_\theta = \epsilon_z = 0$$

It will turn out that on the basis of this assumption we shall arrive at a consistent theory which meets all the requirements of the theory of elasticity, providing the amount of twist is small. For most structural materials the amount of twist is small. For materials like rubber, where large twists are possible, our theory is too simple, and the above assumptions must be reexamined.[1]

With the assumption that extensional strains vanish, the only remaining possible mode of deformation is one in which the cross sections of the shaft remain undeformed but rotate relative to each other. This deformation is pictured in Fig. 6.8. A slice of length Δz is shown before twisting in Fig. 6.8*a*. After twisting, the bottom section has rotated through the angle ϕ and the top section has rotated through the angle $\phi + \Delta\phi$, as shown in Fig. 6.8*b*. The relative rotation causes rectangular elements *EFGH* to *shear* into parallelograms $E_1F_1G_1H_1$. The originally right angle *EHG* is sheared into the acute angle $E_1H_1G_1$. With reference

[1] See, for example, R. S. Rivlin and D. W. Saunders, Large Elastic Deformations of Isotropic Materials. VII. Experiments on the Deformation of Rubber, *Phil. Trans. Roy. Soc.* (*London*) *A*, vol. 243, pp. 251–288, 1951.

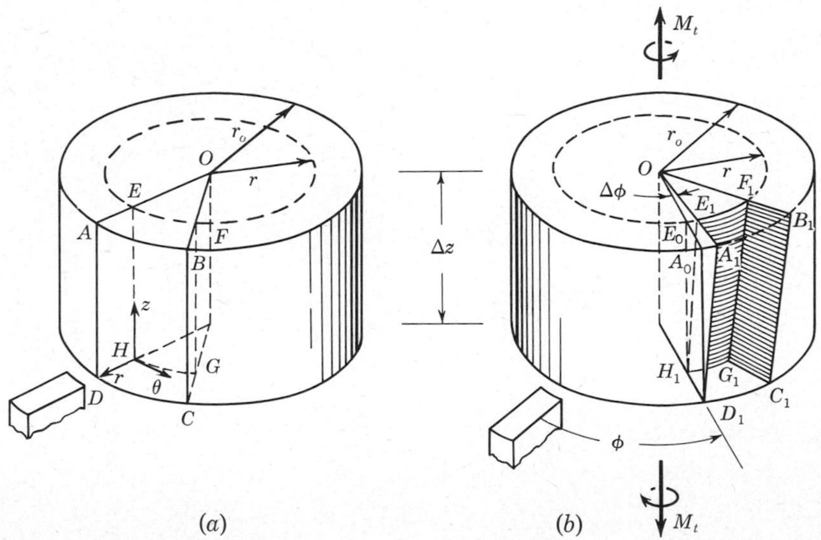

Fig. 6.8 Analysis of deformation of a slice of circular shaft subjected to torsion.

to the r, θ, z system of coordinates shown in Fig. 6.8*a*, this kind of shear deformation is denoted by the symbol $\gamma_{\theta z}$. The magnitude of $\gamma_{\theta z}$ is given by the limiting value of the angle $E_0H_1E_1$ in Fig. 6.8*b* when the size of the element approaches zero. The geometry of the triangles $E_0H_1E_1$ and OE_0E_1 permits us to relate the shear strain $\gamma_{\theta z}$ to the twist angle ϕ.

$$\gamma_{\theta z} = \lim_{\Delta z \to 0} \frac{E_0E_1}{H_1E_0} = \lim_{\Delta z \to 0} \frac{r\Delta\phi}{\Delta z} = r\frac{d\phi}{dz} \tag{6.1}$$

It is important to emphasize that this states that the shear strain varies in direct proportion to the radius, from no shear at the center to a greatest shear at the outside, where $r = r_o$ (i.e., the element $A_1B_1C_1D_1$ in Fig. 6.8 has this greatest shear strain).

We have already noted that each slice of length Δz deforms in the same way as any other, so that we can conclude that $d\phi/dz$ is a *constant* along a uniform section of shaft subjected to twisting moments at the ends. We call $d\phi/dz$ the *twist per unit length*, or the *rate of twist*.

Returning to Fig. 6.8, we can see that our assumption regarding no distortion within the plane of the cross section implies that the right angle DHG goes over into a right angle $D_1H_1G_1$ and hence that $\gamma_{r\theta} = 0$. Also the right angle EHD goes over into the right angle $E_1H_1D_1$ (at least for small deformation) because of our assumption that plane cross sections remain plane. This implies that $\gamma_{rz} = 0$.

Thus, from symmetry and the plausible assumption that the extensional strains are zero, we have arrived at the following distribution of strains:

$$\begin{aligned} &\epsilon_r = \epsilon_\theta = \epsilon_z = \gamma_{r\theta} = \gamma_{rz} = 0 \\ &\gamma_{\theta z} = r\frac{d\phi}{dz} \end{aligned} \tag{6.2}$$

These strains were derived from the geometrically compatible deformation of Fig. 6.8 by simple geometry. An alternative procedure leading to (6.2) is to describe the *displacements* mathematically in terms of cylindrical coordinates and to obtain the strains from the differential equations relating strains to displacements. This procedure is outlined in Prob. 6.36 at the end of the chapter.

We next turn to a consideration of the force-deformation relations of the shaft material.

6.3 STRESSES OBTAINED FROM STRESS-STRAIN RELATIONS

It should be noted that in deducing the strains (6.2) in the previous section, the only restriction placed on the material was that it must be isotropic, but within this restriction it could be elastic or plastic, linear or nonlinear. In this section we wish to consider the stresses for a material which follows Hooke's law.

Using Hooke's law in cylindrical coordinates, we find that the stress components related to the strain components given by (6.2) are

$$\sigma_r = \sigma_\theta = \sigma_z = \tau_{r\theta} = \tau_{rz} = 0$$

$$\tau_{\theta z} = G\gamma_{\theta z} = Gr\,\frac{d\phi}{dz} \tag{6.3}$$

where G is the shear modulus. This state of stress is illustrated in Fig. 6.9*a*, where $\tau_{\theta z}$ is the only stress component acting on the small element referred to the cylindrical coordinates. The stress components acting on a cross section of the shaft are shown in Fig. 6.9*b*; the only component acting is the tangential shear-stress component $\tau_{\theta z}$, whose magnitude varies linearly with radius as given by (6.3).

6.4 EQUILIBRIUM REQUIREMENTS

Analysis of the deformations and stress-strain relations have led us to a proposal for the form of the strain and stress distributions in a circular shaft subjected to torsion. Both the shear strain $\gamma_{\theta z}$ and the shear stress $\tau_{\theta z}$ are proportional to the rate of twist $d\phi/dz$, which is still unknown. We next require that our stresses meet the conditions of equilibrium.

First of all, we note that the stress distribution given by (6.3) and shown in Fig. 6.9*b* leaves the external cylindrical surface of the shaft free of stress, as it should. Inside the shaft each element, such as that shown in Fig. 6.9*a*, is in equilibrium because the shear stress $\tau_{\theta z}$ does not change in the θ direction (because of symmetry) nor in the z direction (because of the uniformity of the deformation

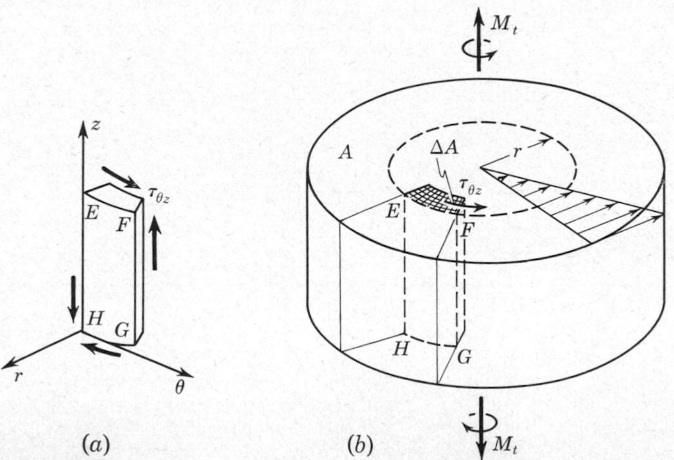

Fig. 6.9 (*a*) Stress components acting on a small element; (*b*) distribution of shearing stress on cross section.

and stress pattern along the length of the shaft). The shearing stress is therefore the same on each z and θ face of the element in Fig. 6.9*a*, and thus the element is in equilibrium. We can also show that the stresses (6.3) satisfy the equilibrium equations in cylindrical coordinates (Prob. 4.4).

On each cross section of the shaft the resultant of the stress distribution must be equal to the applied twisting moment M_t. Because of the rotational symmetry of the stress distribution shown in Fig. 6.9*b*, it is clear that the force resultant must be zero. The only resultant of the stress distribution in Fig. 6.9*b* is, therefore, the moment

$$\int_A r(\tau_{\theta z}\, dA) = M_t \tag{6.4}$$

where dA is the element of area and the integral is to be taken over the cross-sectional area A of the shaft.

6.5 STRESS AND DEFORMATION IN A TWISTED ELASTIC CIRCULAR SHAFT

In the preceding three sections we have considered the geometry of deformation, the stress-strain relations, and the equilibrium requirements of a twisted circular shaft made of isotropic material which obeys Hooke's law. The results of this analysis are summarized in Eqs. (6.2), (6.3), and (6.4). In this section we combine these physical results in a manner which will produce the interrelations which are of interest to us: the relation between $\tau_{\theta z}$ and M_t and the relation between ϕ and M_t.

If we substitute $\tau_{\theta z}$ from (6.3) into the equilibrium relation (6.4), we obtain

$$M_t = G\frac{d\phi}{dz}\int_A r^2\, dA = G\frac{d\phi}{dz} I_z \tag{6.5}$$

where $I_z = \int_A r^2\, dA$ is called the *polar moment of inertia* of the cross-sectional area about the axis of the shaft. For a shaft of radius r_o and diameter d, we can readily perform the integration to obtain

$$I_z = \frac{\pi r_o{}^4}{2} = \frac{\pi d^4}{32} \tag{6.6}$$

From (6.5) we obtain the rate of twist, $d\phi/dz$, in terms of the applied twisting moment

$$\frac{d\phi}{dz} = \frac{M_t}{GI_z} \tag{6.7}$$

For a shaft of length L which is loaded at the ends by twisting moments, as illustrated in Fig. 6.3*a*, the total *angle of twist* between the ends is obtained by integrating (6.7):

$$\phi = \int_0^L \frac{M_t}{GI_z}\, dz = \frac{M_t L}{GI_z} \tag{6.8}$$

where ϕ is given in *radians*.

When we substitute $d\phi/dz$ from (6.7) into (6.3), we obtain the stress in terms of the applied twisting moment.

$$\tau_{\theta z} = \frac{M_t r}{I_z} \tag{6.9}$$

Recapitulating, we have considered the problem of the twisting of a solid circular shaft and have obtained a relation (6.8) between the applied twisting moment and the resulting angle of twist and a relation (6.9) between the twisting moment and the resulting stress distribution. The stress and strain distributions (6.2) and (6.3) on which these were based do, in fact, satisfy the fundamental equations of elasticity so that for every small element the requirements of equilibrium, of geometric compatibility, and of Hooke's law are satisfied (see Prob. 6.36). Furthermore, the boundary condition of no stress on the outside cylindrical surface is satisfied. The only possible shortcoming of our solution occurs at the ends of the shaft. If our solution is to be valid at the ends, then the externally applied twisting moment must actually be distributed according to the pattern of Fig. 6.9*b*. In many practical cases (e.g., when M_t is applied by the jaws of a wrench), the actual stress distribution at the end, although statically equivalent to M_t, is widely different from that of Fig. 6.9*b*. According to St. Venant's principle, we expect that our solution provides an excellent approximation to the actual stress and strain distribution in the central portion of the shaft and that the approximation is probably adequate to within a diameter or two of the ends. This means that if the shaft is reasonably long our estimate (6.8) of the total twist is probably not very much affected by the manner of loading at the ends. We cannot, however, use (6.9) to predict the local stresses at the ends.

It is unusual to encounter a situation in which the shape of the body, the loading imposed on it, and the resulting deformations are so simple that an exact solution according to the theory of elasticity can be obtained as straightforwardly as in this case. In most engineering situations we have to be content with a "solution" which does not satisfy all the conditions of elasticity exactly.

When we consider a length of circular shaft to be a structural element, the relation (6.8) becomes the load-deformation relation to be used in the third step of (2.1). It is often convenient to state (6.8) in the form

$$\frac{M_t}{\phi} = \frac{GI_z}{L} \tag{6.10}$$

which gives the twisting moment per radian of twist. This ratio is analogous to a spring constant which gives tensile force per unit length of stretch. The ratio in (6.10) is called the *torsional stiffness* of the shaft and is often denoted by the symbol k or c.

The complete solution to any problem involving a twisted elastic shaft consists of establishing equilibrium between the internal twisting moments and the external loads, of satisfying the conditions of geometric restraint on the rotation of the shaft, and of satisfying the load-deformation relations (6.8) or (6.10). In almost all practical cases we have to consider the stresses in the shaft as an independent design consideration in addition to the angle of twist, and we use (6.9) to evaluate the stresses.

We illustrate the application of (2.1) to the behavior of twisted circular shafts in the following examples.

Example 6.1 Two small lathes are driven by the same motor through a ½-in.-diameter steel shaft, as shown in Fig. 6.10*a*. We wish to know the maximum shear stress in the shaft due to twisting and the angle of twist between the two ends of the shaft.

We begin the analysis by idealizing the situation, as shown in Fig. 6.10*b*. Here we represent each pulley loading by its static equivalent of a force of

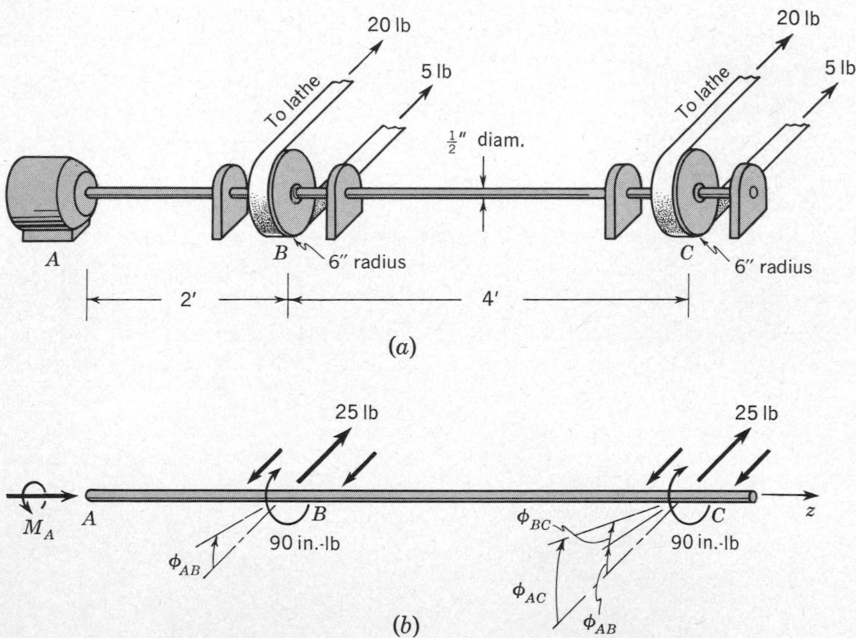

Fig. 6.10 Example 6.1.

25 lb through the axis of the shaft and a couple about the z axis of 6(20 − 5) = 90 in.-lb. Because each pulley is supported by a *pair* of immediately adjacent bearings, we make the idealization that the 25-lb transverse forces are balanced by the bearing reactions in such a way that there is negligible shear force and bending moment transmitted beyond the bearings. In this case it is only necessary for the motor to supply a torque M_A, as shown. We now apply the steps of (2.1) to the model of Fig. 6.10*b*.

EQUILIBRIUM

Establishing moment equilibrium, we have, since all moment vectors are parallel to z, $\Sigma \mathbf{M}_A = 0$ if

$$\begin{aligned} M_A - 90 - 90 &= 0 \\ M_A &= 180 \text{ in.-lb} \end{aligned} \qquad (a)$$

The twisting moments in sections AB and BC of the shaft are then clearly

$$M_{AB} = 180 \text{ in.-lb} \qquad M_{BC} = 90 \text{ in.-lb} \qquad (b)$$

GEOMETRIC COMPATIBILITY

We wish to find the angle ϕ_{AC} which describes the rotation of the end C with respect to the end A. From the sketch in Fig. 6.10*b* we see that

$$\phi_{AC} = \phi_{AB} + \phi_{BC} \qquad (c)$$

LOAD-DEFORMATION RELATION

From Eq. (6.8) we have

$$\phi_{AB} = \frac{M_{AB}L_{AB}}{GI_z} \qquad \phi_{BC} = \frac{M_{BC}L_{BC}}{GI_z} \qquad (d)$$

where, from Table 5.1, $G = 11.5 \times 10^6$ psi. Combining Eqs. (*b*), (*c*), and (*d*), and having proper regard for units, we find

$$\phi_{AC} = 0.123 \text{ rad} = 7.0^\circ \qquad (e)$$

The maximum shear stress occurs at the outside of the shaft in section AB. Using (6.9) we find

$$(\tau_{\theta z})_{\max} = \frac{M_{AB}r_o}{I_z} = 7{,}300 \text{ psi} \qquad (f)$$

Example 6.2 A couple of 600 in.-lb is applied to a 1-in.-diameter 2024-0 aluminum-alloy shaft, as shown in Fig. 6.11*a*. The ends A and C of the shaft are built-in and prevented from rotating, and we wish to know the angle through which the center cross section O of the shaft rotates.

We idealize the situation in Fig. 6.11*b*. The shaft is statically indeterminate since we cannot determine M_A and M_C from equilibrium considerations

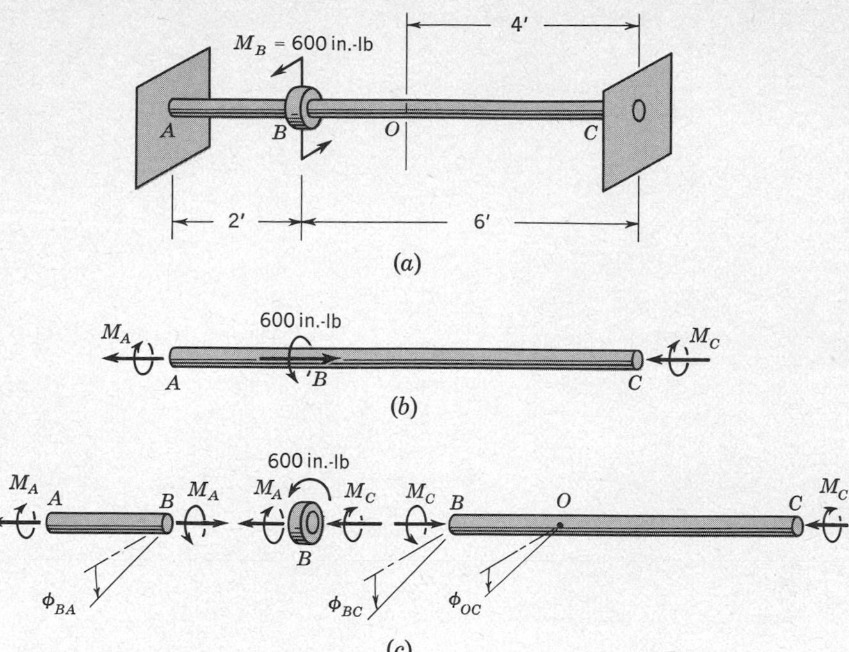

Fig. 6.11 Example 6.2.

alone. In Fig. 6.11c we show isolated free bodies of three sections of the shaft. Applying (2.1), we obtain:

EQUILIBRIUM

Satisfying moment equilibrium for the complete shaft shown in Fig. 6.11*b* (or for the middle segment in Fig. 6.11*c*) yields

$$M_A + M_C - 600 = 0 \tag{a}$$

GEOMETRIC COMPATIBILITY

Continuity of the shaft at the point B requires that

$$\phi_{BC} = \phi_{BA} \tag{b}$$

LOAD-DEFORMATION RELATION

From Eq. (6.8) we have

$$\phi_{BA} = \frac{M_A L_{AB}}{GI_z} \qquad \phi_{BC} = \frac{M_C L_{BC}}{GI_z}$$
$$\phi_{OC} = \frac{M_C L_{OC}}{GI_z} \tag{c}$$

Combining Eqs. (*a*), (*b*), and (*c*), we find

$$M_A = \frac{L_{BC}}{L_{AC}} M_B = 450 \text{ in.-lb}$$

$$M_C = \frac{L_{AB}}{L_{AC}} M_B = 150 \text{ in.-lb} \tag{d}$$

$$\phi_{OC} = 0.020 \text{ rad} = 1.1^\circ$$

6.6 TORSION OF ELASTIC HOLLOW CIRCULAR SHAFTS

If we examine the arguments which were used in developing Eqs. (6.8) and (6.9) for the solid circular shaft, we shall find the arguments apply with equal validity[1] to a circular shaft with a concentric hole. The only difference is that the integral in (6.4) now extends over an annulus instead of a complete circle. Thus (6.8) and

[1] These arguments apply with only minor alterations to any case in which there is *circular symmetry* of the shaft, e.g., a shaft made up of two concentric tubes (see Prob. 6.19) or a composite shaft in which a tube of one material is bonded to a core of another material (see Prob. 6.3).

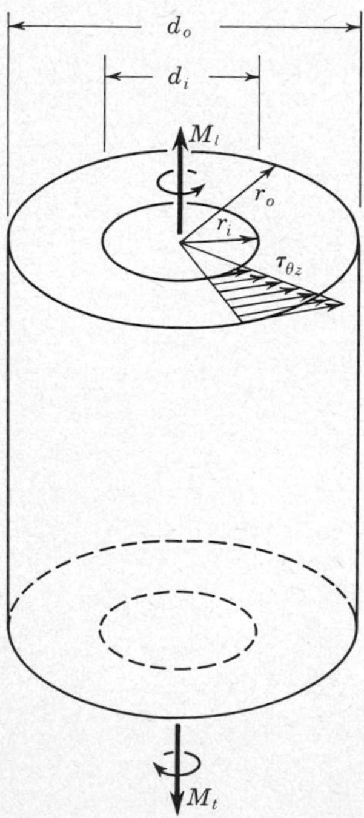

Fig. 6.12 Stress distribution in elastic hollow circular shaft.

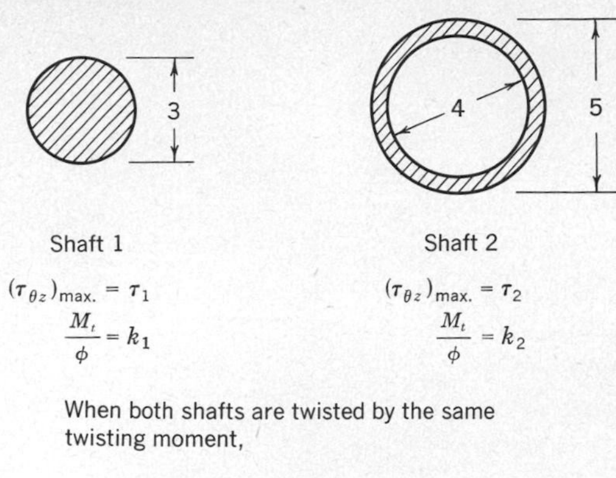

Fig. 6.13 Illustration of advantages of hollow shaft over solid shaft of *same* cross-sectional area.

(6.9) describe the behavior of the hollow circular shaft of Fig. 6.12, providing the polar moment of inertia is taken as

$$I_z = \frac{\pi r_o^4}{2}\left(1 - \frac{r_i^4}{r_o^4}\right) = \frac{\pi d_o^4}{32}\left(1 - \frac{d_i^4}{d_o^4}\right) \tag{6.11}$$

Making a concentric hole in a shaft does not reduce the torsional stiffness in proportion to the amount of material removed. An element of material near the center of the shaft has a low stress and a small moment arm and thus contributes less to the twisting moment than an element near the outside of the shaft. More precisely, the torsional stiffness (6.10) for a given length of given material depends only on the polar moment of intertia I_z, and from (6.11) it is clear that the size of the hole enters only in the fourth power of the ratio d_i/d_o. The maximum shear stress (6.9) for a given twisting moment also depends in the same manner on the size of the hole. To dramatize this behavior, consider the two shafts in Fig. 6.13 which have the *same* cross-sectional area but markedly different maximum stresses and deformation. It is apparent that a given amount of material is used most efficiently in torsion when it is formed into a hollow shaft. When construction of a hollow shaft requires the extra labor of boring out a solid shaft of the correct outside diameter, it is not worthwhile to make a hollow shaft except in applications where weight is critical. There is also a limit on the increase in effectiveness that can be obtained by increasing the diameter and decreasing the wall thickness. If the wall is made too thin, the cylinder wall will buckle[1] due to compressive stresses which act in the wall on surfaces inclined at 45° to the axis of the cylinder.[2]

[1] See Sec. 9.3.

[2] See Sec. 6.7.

6.7 STRESS ANALYSIS IN TORSION; COMBINED STRESSES

It is sometimes of interest to determine the stress components related to axes other than the r, θ, z set. A convenient way to determine these stress components is to use Mohr's circle for stress. Figure 6.14 shows the stress components related to the θ, z axes (we may use the two-dimensional Mohr's circle because there is no stress in the r direction), and Fig. 6.15 shows the resulting Mohr's circle. From the Mohr's circle we see that the principal stresses are σ_1 (tensile) and σ_2 (compressive) with magnitudes $|\sigma_1| = |\sigma_2| = |\tau_{\theta z}|$. The orientation of the principal diameter in the Mohr's circle is 90° from the θ, z diameter. The principal directions in the shaft must then be 45° from the θ, z directions. The principal stress components are sketched in Fig. 6.14.

In certain special cases the existence of these principal stress components and their principal directions can be demonstrated directly. For example, if a piece of chalk (which is a brittle material with a low tensile strength and much larger strength in compression and shear) is twisted, the chalk will fracture along a spiral line normal to the direction of maximum tension (e.g., along the line AB in Fig. 6.14). To give another example, as was mentioned in Sec. 6.6 a very thin-walled, hollow cylinder will buckle in the direction of maximum compression; if a

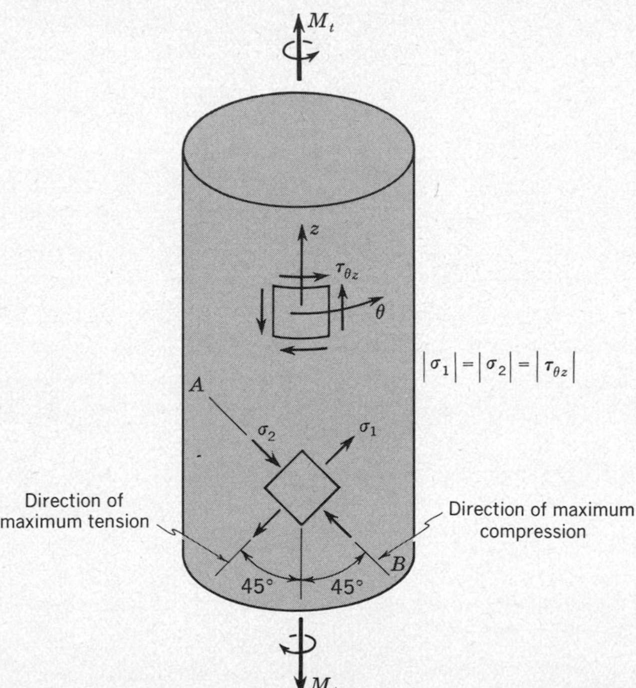

Fig. 6.14 The principal stresses in torsion are equal tension and compression acting on faces inclined at 45° to the axis of the shaft.

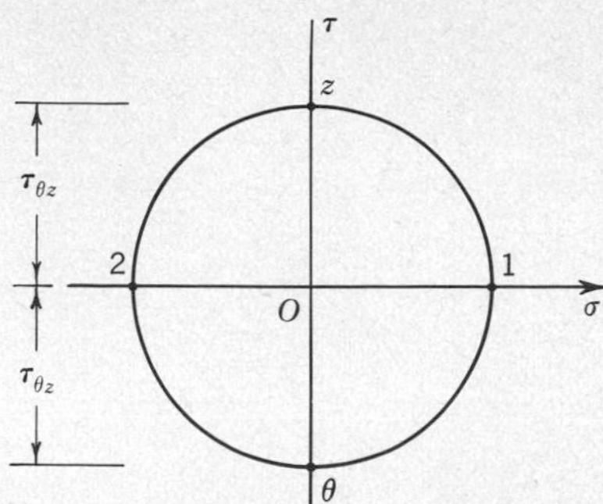

Fig. 6.15 Mohr's circle for stress for element of shaft in torsion.

piece of paper is rolled into a cylinder and twisted, this mode of failure will become evident.

A circular shaft is often subjected to longitudinal and bending deformations in addition to torsion. At this stage in our discussion we have not yet considered the distribution of stresses in bending, but we are familiar with simple axial (tensile or compressive) loadings on slender members. We are thus equipped to consider the *combined-stress* problem resulting from axial loading combined with torsion. The following example illustrates how the resultant state of stress is obtained by *superposition* of the individual effects. The justification for superposition lies in the *linearity* of Eqs. (5.6), (5.7), and (5.8) underlying the theory of elasticity. The stresses and strains contributed by one form of loading are not altered by the presence of another kind of loading.

Example 6.3 In Fig. 6.16*a* a uniform, homogeneous, circular shaft is shown subjected simultaneously to an axial tensile force P and a twisting moment M_t. In Fig. 6.16*b* the individual stress distributions are sketched for the separate loads. Due to the twisting moment, we have the distribution of Fig. 6.9*b* which is given analytically by (6.9).

$$\tau_{\theta z} = \frac{M_t r}{I_z} \tag{a}$$

Due to the tensile force, we show a uniform stress distribution in Fig. 6.16*b*. A demonstration of the validity of this distribution can be given along the same lines used at the beginning of this chapter (see also Prob. 5.24). Symmetry considerations lead to the postulate that plane cross sections remain plane but displace uniformly under tensile load. This implies a uniform distribution of axial strain and hence a uniform distribution of axial stress. In order to be in equilibrium with P the magnitude of the axial stress must be

$$\sigma_z = \frac{P}{\pi r_o^{\,2}} \tag{b}$$

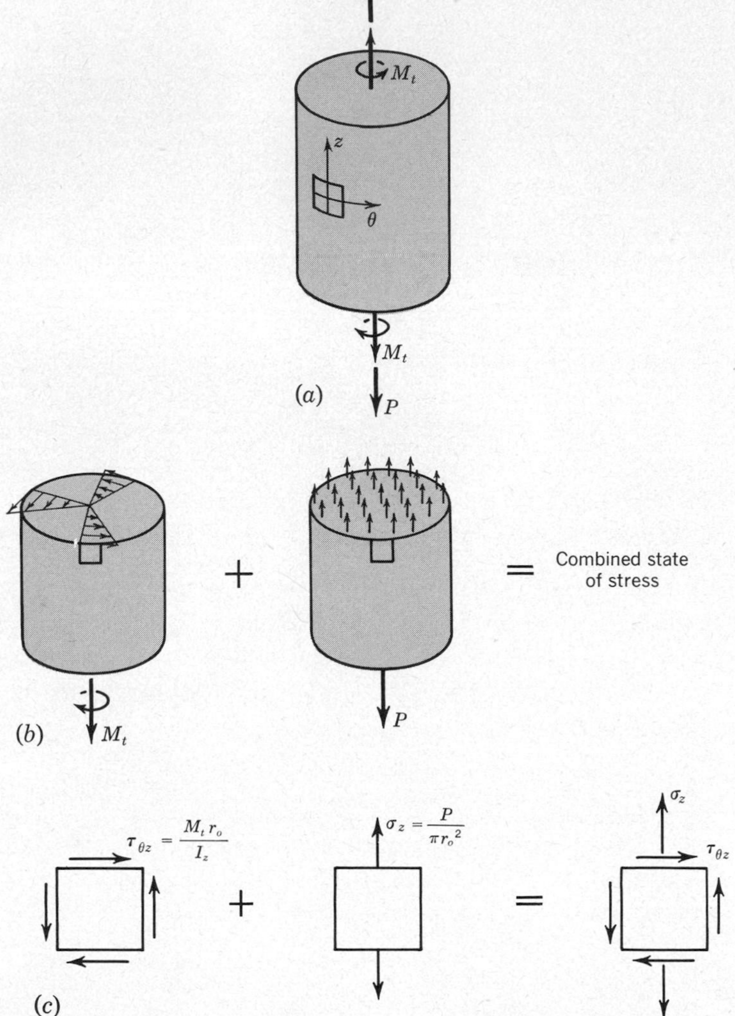

Fig. 6.16 Example 6.3. Combined stresses due to torsion and tension.

In Fig. 6.16*c* these stresses are shown acting on a small element on the surface of the shaft where $r = r_o$. The individual stresses are first shown separately and then superposed to represent the combined-stress state.

The most convenient method of describing the combined-stress state is to use the principal stress components. The Mohr's circle diagram used to obtain the principal stresses is sketched in Fig. 6.17*a*, and the principal directions indicated in Fig. 6.17*b*. Note that this element is in a state of plane stress, i.e., the third principal stress σ_3 is zero.

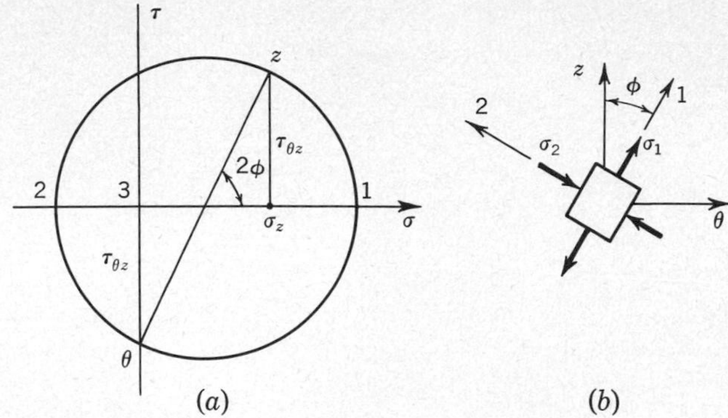

Fig. 6.17 Example 6.3. Principal directions and principal stresses.

6.8 STRAIN ENERGY DUE TO TORSION

In Sec. 2.6 the concept of elastic energy is introduced, and its application in the calculation of elastic deflections by the use of Castigliano's theorem is demonstrated. In Sec. 5.8 a formula for the strain energy in a linearly elastic body subjected to an arbitrary distribution of stress and strain is developed. In this section we apply that result specifically to the case of torsion of circular members and consider an example of Castigliano's theorem applied to torsional deformation.

In torsion of an isotropic elastic shaft of circular cross section, the only nonvanishing stress and strain components are $\tau_{\theta z}$ and $\gamma_{\theta z}$, according to (6.3) and (6.2). The total strain energy (5.17) thus reduces to

$$U = \tfrac{1}{2} \int_V \tau_{\theta z} \gamma_{\theta z} \, dV \tag{6.12}$$

where the integration is over the volume of the shaft. Setting $\gamma_{\theta z} = \tau_{\theta z}/G$ and introducing (6.9), we find

$$U = \tfrac{1}{2} \int_V \frac{1}{G} \left[\frac{M_t r}{I_z} \right]^2 dV = \tfrac{1}{2} \int_L \frac{M_t^2}{G I_z^2} \, dz \int_A r^2 \, dA$$

where the integrations are over the length L and cross-sectional area A of the shaft. Since the latter integral is just the polar moment of inertia I_z, the torsional strain energy is

$$U = \int_L \frac{M_t^2}{2 G I_z} \, dz \tag{6.13}$$

This formula may also be derived by considering each differential slice of thickness dz of the shaft to act as a torsional spring. If the final values of the twisting moment and twisting angle are M_t and $d\phi$, respectively, the work done during a loading process in which these grow in proportion is

$$dU = \tfrac{1}{2}M_t\, d\phi = \tfrac{1}{2}M_t \frac{d\phi}{dz}\, dz \tag{6.14}$$

When (6.7) is substituted for $d\phi/dz$ and the result is integrated over the length of the shaft, we obtain (6.13) again.

We recall that Castigliano's theorem states that if the total elastic energy in a system is expressed in terms of the external loads, the in-line deflection δ_i of the point of application of a particular load P_i is given by the partial derivative

$$\delta_i = \frac{\partial U}{\partial P_i}$$

We illustrate the application of this to a torsional system in the following example.

Example 6.4 Consider a closely wound coil spring of radius R loaded by a force P (Fig. 6.18*a*). The spring consists of n turns of wire with wire radius r. We wish to find the deflection of the spring and hence the spring constant.

First we find the internal forces and moments acting on a section of the spring (Prob. 3.41). From the free-body diagram in Fig. 6.18*b*, we see that the twisting moment M_t is independent of position on the spring and is equal to PR. The strain energy associated with the twisting moment is

$$U = \int_L \frac{P^2R^2}{2GI_z}\, dz = \int_0^{2\pi n} \frac{P^2R^2}{2GI_z} R\, d\theta = \frac{P^2R^3}{2GI_z} 2\pi n \tag{a}$$

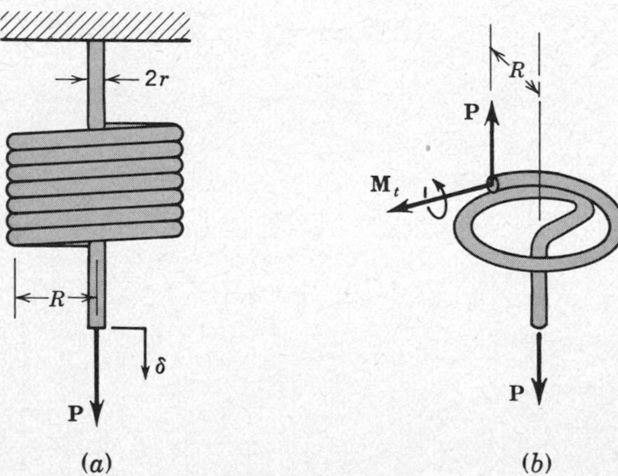

Fig. 6.18 Example 6.4.

There is additional strain energy in the spring due to the transverse shear force P. It can be shown, however (see Prob. 7.27), that the ratio of strain energy due to transverse shear to strain energy due to torsion is proportional to $(r/R)^2$ and hence is small for springs of usual design. For simplicity we shall neglect the contribution of the transverse force and consider (*a*) to represent the total strain energy in the spring.

Therefore the deflection in the direction of P is

$$\delta = \frac{\partial U}{\partial P} = \frac{PR^3}{GI_z} 2\pi n \tag{b}$$

and the spring constant becomes

$$k = \frac{P}{\delta} = \frac{GI_z}{2\pi n R^3} \tag{c}$$

Upon substituting for the moment of inertia I_z in (*c*), we find that

$$k = \frac{Gr^4}{4nR^3}$$

where r is the radius of the wire. We see that the spring constant is inversely proportional to the number of coils n and directly proportional to the fourth power of the wire radius. For example, if we increase the wire radius by 19 percent, the spring constant is doubled.

In this example, Castigliano's theorem has provided a simple means of evaluating an elastic deflection in a system of some geometric complexity. It is possible to obtain the same result by direct application of (6.7), but the analysis is considerably less simple (see Prob. 6.30).

6.9 THE ONSET OF YIELDING IN TORSION

In Chap. 5 two criteria for the initiation of yielding of metals were described in general terms. We return to these briefly for the special case of torsion.

In order to apply either criterion to a particular material it is necessary to obtain (experimentally) the yield stress Y in uniaxial tension. Then, to decide whether yielding will occur in a general state of stress, we compute the equivalent or effective stress $\bar{\sigma}$ (or $\bar{\tau}$) according to the criterion employed and compare with Y.

The principal stresses acting on an element of a shaft in torsion were obtained in Sec. 6.7 (see Fig. 6.15).

$$\sigma_1 = \tau_{\theta z} \qquad \sigma_2 = -\tau_{\theta z} \qquad \sigma_3 = 0 \tag{6.15}$$

Using the *Mises criterion* (5.23), the effective stress $\bar{\sigma}$ is

$$\begin{aligned}\bar{\sigma} &= \sqrt{\tfrac{1}{2}[(2\tau_{\theta z})^2 + (-\tau_{\theta z})^2 + (-\tau_{\theta z})^2]} \\ &= \sqrt{3}\tau_{\theta z} \end{aligned} \tag{6.16}$$

and thus an element of a shaft in torsion would be expected to begin yielding when

$$\tau_{\theta z} = \frac{1}{\sqrt{3}} Y = 0.577\, Y \tag{6.17}$$

according to the Mises criterion.

Using the *maximum shear-stress criterion* (5.25), the equivalent shear stress $\bar{\tau}$ is simply

$$\bar{\tau} = \tau_{\theta z} \tag{6.18}$$

and thus an element of a shaft in torsion would be expected to begin yielding when

$$\tau_{\theta z} = \tfrac{1}{2} Y = 0.500\, Y \tag{6.19}$$

according to the maximum shear-stress criterion.

Torsion involves a stress state which gives rise to the *maximum* discrepancy between the two criteria (see Fig. 5.29). As can be seen from (6.17) and (6.19), this discrepancy is about 15 percent. From the point of view of the designer trying to avoid yielding, it is more conservative to design on the basis of (6.19).

Since the shear stress $\tau_{\theta z}$ is proportional to the radius r in an elastic shaft, it is clear that according to either criterion the elements on the *outer* surface of the shaft will reach the yield condition first. In the following section we investigate what happens if the shaft is twisted beyond this point.

6.10 PLASTIC DEFORMATIONS

We shall consider a solid circular shaft twisted into the plastic range. It is important to remember that in passing from elastic to plastic behavior there is no alteration in the conditions of equilibrium or in the conditions of geometric compatibility. The only change is in the stress-strain relation. The symmetry arguments at the beginning of this chapter which led to the conclusion that the only nonvanishing strain component was $\gamma_{\theta z}$ remain valid whether the material is elastic or plastic. What *will* be different is the relation between $\gamma_{\theta z}$ and $\tau_{\theta z}$.

One way in which the relation between $\tau_{\theta z}$ and $\gamma_{\theta z}$ for a given material can be obtained is by direct experiment in which the material is subjected to uniform pure *shear* (e.g., torsion of a thin-walled tube). A simpler, if less exact, procedure is to make use of *tension* test data and to predict the relation between $\tau_{\theta z}$ and $\gamma_{\theta z}$ in torsion by using one of the *plastic flow rules* (5.34) or (5.35).

In this chapter we shall confine our analytical treatment to the *elastic–perfectly plastic* material whose stress-strain curve is shown in Fig. 5.7*e*. For this material the problem of predicting the relation between torsional shear stress and shear strain then becomes trivial. In the elastic range the constant of proportionality between $\tau_{\theta z}$ and $\gamma_{\theta z}$ is the shear modulus G. Once the material yields at a particular point there is no strain-hardening, so the equivalent or effective stress $\bar{\sigma}$ (or $\bar{\tau}$) remains constant at that point for all further plastic strain. This means that $\tau_{\theta z}$ remains constant, as shown in Fig. 6.19. The only difference between the Mises and the maximum shear-stress flow rules is in the yield stress values, which are given by (6.17) and (6.19), respectively. In order to include both possibilities we have called the yield-point shear stress simply τ_Y in Fig. 6.19. The corresponding shear strain is called γ_Y.

We now turn to the analysis of the twisting of a solid circular shaft of material whose stress-strain relation in pure shear is given by Fig. 6.19. As long as the shaft remains elastic the results of Sec. 6.5 still apply. As the shaft is twisted further, plane cross sections continue to rotate with respect to one another, and the variation of shear *strain* $\gamma_{\theta z}$ still remains linearly proportional to r, as indicated in Fig. 6.8. Because of Fig. 6.19, the shear *stress* $\tau_{\theta z}$ will be distributed as shown in Fig. 6.20.

To obtain quantitative representations of the sketches in Fig. 6.20, we proceed as follows. The elastic relations (6.8) and (6.9) apply until the yield-point situation in Fig. 6.20*b* is reached. Let us call the twisting moment and twisting angle associated with this stress distribution T_Y and ϕ_Y, respectively. Then from (6.8) and (6.9) we have

$$\begin{aligned} T_Y &= \frac{\tau_Y I_z}{r_o} = \frac{\pi}{2} \tau_Y r_o^3 \\ \phi_Y &= \frac{\tau_Y L}{G r_o} \end{aligned} \tag{6.20}$$

where L is the length of the shaft. Now as the shaft is twisted further the shear strain at the outer radius becomes larger than γ_Y. At some intermediate radius r_Y the strain will be just equal to γ_Y. We still have the geometric relation (6.1) between shear strain and twist angle

$$\gamma_{\theta z} = r \frac{d\phi}{dz} = r \frac{\phi}{L} \tag{6.21}$$

from which we can solve for r_Y when $\phi > \phi_Y$.

$$r_Y = \frac{L \gamma_Y}{\phi} \tag{6.22}$$

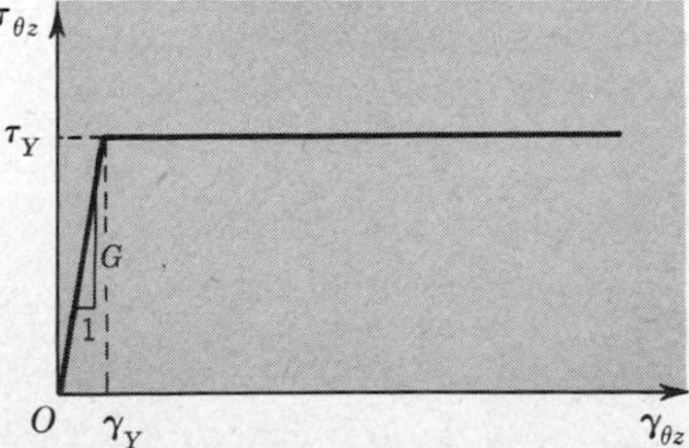

Fig. 6.19 Shear-stress–shear-strain curve for elastic–perfectly plastic material.

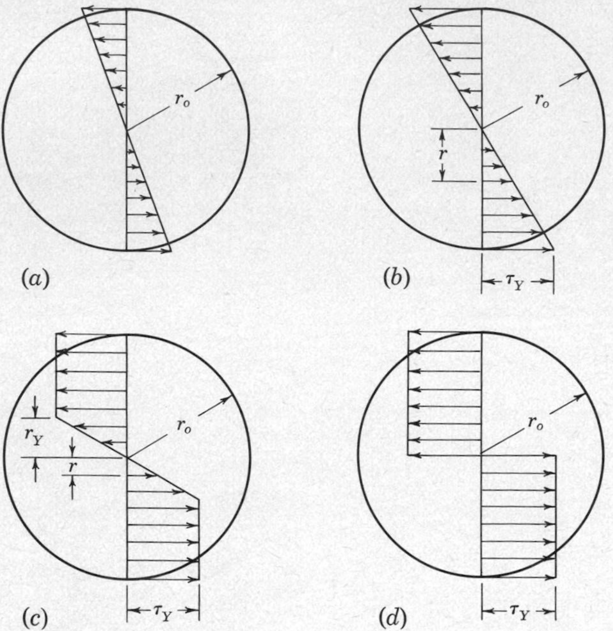

Fig. 6.20 Shear-stress distributions in a twisted shaft of material having the stress-strain curve of Fig. 6.19. (*a*) Entirely elastic; (*b*) onset of yield; (*c*) partially plastic; (*d*) fully plastic.

Using the fact that $\tau_Y = G\gamma_Y$ and introducing the second of (6.20), we find

$$r_Y = \frac{L\tau_Y}{G\phi} = r_o \frac{\phi_Y}{\phi} \tag{6.23}$$

Next, we obtain a quantitative representation for the stress distribution $\tau_{\theta z}$ corresponding to the strain distribution $\gamma_{\theta z}$ of (6.21) by using the stress-strain relation of Fig. 6.19. In the inner elastic core $O < r < r_Y$,

$$\begin{aligned} \tau_{\theta z} &= G\gamma_{\theta z} \\ &= G\frac{\phi}{L}r = \tau_Y \frac{r}{r_Y} \end{aligned} \tag{6.24}$$

In the outer plastic region $r_Y < r < r_o$,

$$\tau_{\theta z} = \tau_Y \tag{6.25}$$

The stress distribution defined by (6.24) and (6.25) is sketched in Fig. 6.20*c*.

Finally, we use the equilibrium requirement that the stress distribution of Fig. 6.20*c* should be equivalent to the applied twisting moment M_t.

$$\begin{aligned} M_t &= \int_A r\tau_{\theta z}\, dA \\ &= \int_0^{r_Y} r\left(\frac{r}{r_Y}\tau_Y\right)2\pi r\, dr + \int_{r_Y}^{r_o} r\tau_Y 2\pi r\, dr \\ &= \frac{\pi}{2} r_Y{}^3\tau_Y + \frac{2\pi}{3}(r_o{}^3 - r_Y{}^3)\tau_Y \\ &= \frac{2\pi}{3}\tau_Y r_o{}^3\left(1 - \frac{1}{4}\frac{r_Y{}^3}{r_o{}^3}\right) \end{aligned} \tag{6.26}$$

This result can be put into a more useful final form by introducing the yield-point twisting moment from (6.20) and the twisting angle from (6.23).

$$M_t = \frac{4}{3} T_Y\left(1 - \frac{1}{4}\frac{\phi_Y{}^3}{\phi^3}\right) \tag{6.27}$$

This *nonlinear* relationship is valid when $\phi > \phi_Y$. For smaller angles of twist the connection between twisting moment and twisting angle is given by the *linear* relationship (6.8). The resulting curve made up from (6.8) and (6.27) is sketched in Fig. 6.21. The *limit* or *fully plastic twisting moment* T_L corresponding to the stress distribution of Fig. 6.20*d* is $\frac{4}{3}T_Y$ and is theoretically approached only in the limit as $\phi \to \infty$. As Fig. 6.21 shows, this limit is approached very rapidly (for example, $M_t = 1.32T_Y$ when $\phi = 3\phi_Y$).

6.11 RESIDUAL STRESSES

The result incorporated in Fig. 6.21 is valid for a steadily increasing twisting moment. If we assume that the material of the shaft unloads *elastically* after it has been strained plastically (see Fig. 5.6), then if at any stage the twisting moment were to be decreased, the twisting-moment–twisting-angle curve would trace out a straight line parallel to the original elastic relation of (6.8), as sketched in Fig. 6.22. The justification for this lies in the fact that the geometric and

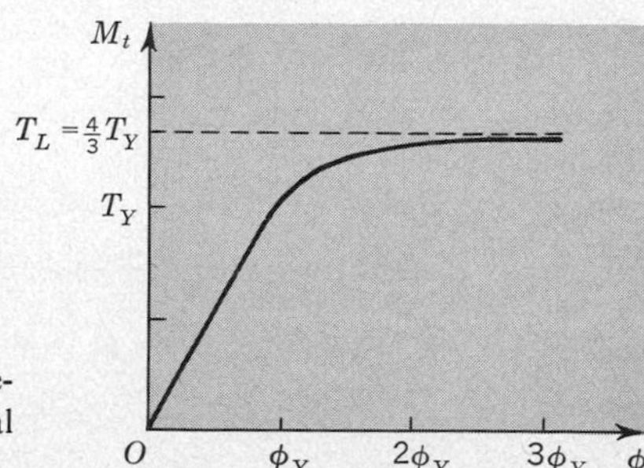

Fig. 6.21 Twisting-moment–twisting-angle relationship for solid circular shaft made of material with stress-strain curve of Fig. 6.19.

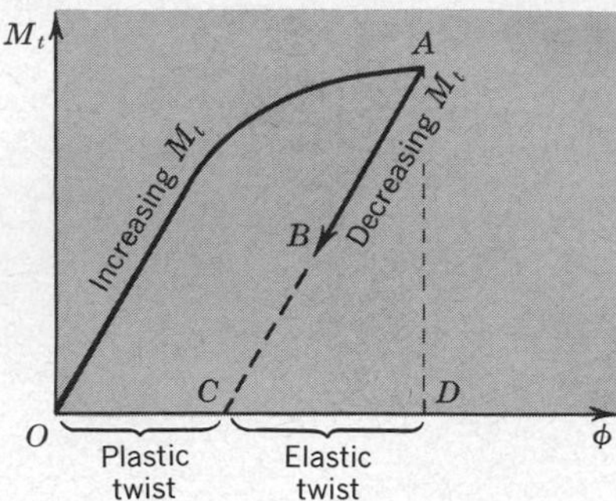

Fig. 6.22 Unloading a plastically deformed shaft.

equilibrium requirements for torsion remain unchanged while the stress-increment–strain-increment relation is now elastic for the entire shaft.

If the twisting moment were to be decreased to zero (point C in Fig. 6.22), the shaft would be left with the permanent twist OC. It is interesting to note that although there is no external load on the shaft in this condition, there is a distribution of self-balancing internal stresses in the shaft. These internal stresses which are "locked in" the material by the plastic deformation are called *residual* stresses.

The distribution of residual stresses can be found by using superposition. The plastic deformation under M_t leads to the stress distribution of Fig. 6.23*a* which corresponds to increasing the twist until point A of Fig. 6.22 is reached. This process is *nonlinear*, but the elastic unloading along ABC in Fig. 6.22 is linear. The *hypothetical* distribution of elastic stresses required to develop a twisting moment of $-M_t$ is shown in Fig. 6.23*b*. When parts *a* and *b* of Fig. 6.23 are superposed, we end up with no external twisting moment but with a distribution of residual stresses, as shown in Fig. 6.23*c*. The outer part of the shaft carries shearing stresses of the opposite sense to that imposed by the original

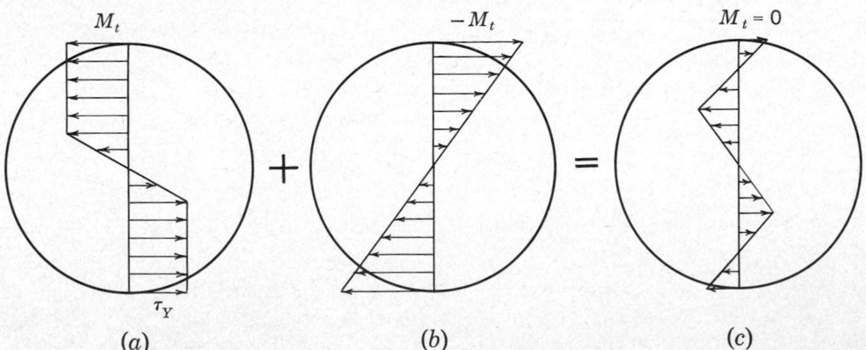

Fig. 6.23 Residual shear-stress distribution in a shaft which has been twisted into the plastic region and unloaded.

application of the load, while the inner part carries stresses of the same sense as those originally imposed.

Under some circumstances the reversed stresses obtained in this manner might be larger than the yield stress in the opposite direction. In this case simple linear superposition would not be applicable (see Prob. 6.41).

6.12 LIMIT ANALYSIS

In the preceding two sections we have seen the complications which arise when we try to follow the details of a deformation which involves both elastic and plastic strains. For the simple case of pure torsion we are actually able to perform the analysis. In more complicated structures the corresponding analysis is often so forbiddingly complex as to make exact analysis almost hopeless in practical cases.

For design purposes there exists an extraordinarily simple approximate analysis which includes the basic yield phenomenon of plasticity but omits all detailed considerations of intermediate stress and strain distributions. This procedure is called *limit analysis* and is used by designers of complex structures constructed from metals having a pronounced yield point. We shall introduce the ideas of limit analysis here in connection with torsion. In Chap. 8 we shall apply the same technique to the bending of beams.

The basic idea underlying limit analysis is that any member made of a material with a pronounced yield point will have a fairly well-defined *limit load* such that the deformations never become large until the load approaches the limit load. The deformations, even though partially plastic, will be small as long as the load remains appreciably less than the limit load. Whenever the member undergoes a large deformation, the load that it carries is approximately the limit load. This is just the behavior that we have obtained for a simple shaft of elastic–perfectly plastic material in torsion. The twisting angle is finite and small (see Fig. 6.21) as long as the twisting moment is appreciably less than the limit moment. For large twisting angles the twisting moment approaches the limit moment asymptotically.

When such a shaft is built into a statically indeterminate structure, it often happens that large deformations of the shaft are only possible when there are large deformations in *other parts of the structure.* The structure as a whole may not suffer large deformation until *several interconnected parts simultaneously undergo large deformations.*

The process of limit analysis consists of studying the geometry of a structure to determine which combinations of parts must undergo large deformations simultaneously in order to cause large deformations of the structure and then of *studying the equilibrium requirements to determine what external loads correspond to these deformations.* This last step is much simpler than usual in the mechanics of solids because now we know that each individual part undergoing large deformation must be carrying its fully plastic or limiting load. The external limit loading is obtained from the condition that it be in equilibrium with the limit loads on the individual parts.

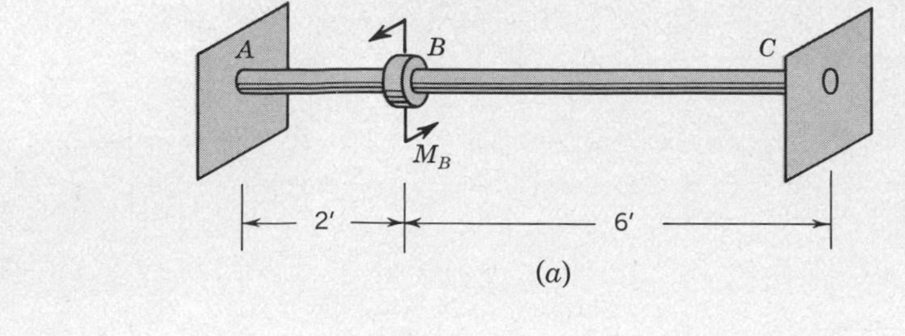

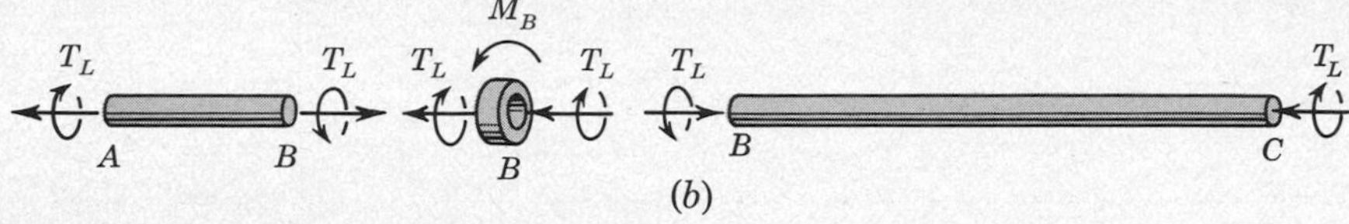

Fig. 6.24 Example 6.5. Limit analysis of a statically indeterminate torsion member.

Example 6.5 We return to the system of Example 6.2, shown here in Fig. 6.24*a*. Instead of asking for the deformation under a given moment M_B when the shaft remains elastic, we ask what is the limiting value for M_B which results in large angular displacements at B.

In order for there to be large rotation at B, there must be large twisting of *both* AB and BC. This occurs, as shown in Fig. 6.24*b*, when both AB and BC carry the fully plastic or limiting twisting moment T_L. Equilibrium at B requires

$$(M_B)_L = 2T_L \qquad (a)$$

The simplicity of this limit analysis as compared with the elastic analysis of Example 6.2 should be noted. Note that the three basic considerations (equilibrium, geometric compatibility, and force-deformation relation) are used, but in a simplified form. It is *geometric compatibility* which requires that *both* AB and BC have large deformations. The fact that both shafts carry T_L is due to a *force-deformation* relation; in order to have large deformation the twisting moment must be fully plastic. Finally, *equilibrium* at B was used to establish (*a*).

For comparison with (*a*) we might use the elastic analysis to find the value of M_B which first caused yielding anywhere in the system. From Example 6.2 it is clear that the largest stress in the system will occur in AB. The twisting moment at A will be equal to T_Y when the loading M_B takes the value $(M_B)_Y$. The relation between these can be obtained from (*d*) in Example 6.2 without repeating the analysis.

$$(M_B)_Y = \frac{L_{AC}}{L_{BC}} T_Y$$

$$= \tfrac{4}{3} T_Y \qquad (b)$$

To compare with (*a*), we use the result of Fig. 6.21 that $T_L = \frac{4}{3}T_Y$ to obtain

$$(M_B)_L = 2(M_B)_Y \tag{c}$$

Thus the loading at B which causes unlimited twisting is twice that which first causes yielding anywhere in the shaft.

In the above example there was only one possible mechanism resulting in large deformations. In more complicated systems there are often two or more combinations which can result in large deformation. When this is so, it is necessary to evaluate the limit load for each possible combination. The actual combination which will occur is that corresponding to the lowest limit load.

6.13 TORSION OF RECTANGULAR SHAFTS

In some situations slender members with other than circular cross sections are subjected to torsion. The analysis at the beginning of this chapter, which was based on symmetry arguments, breaks down when applied to such sections. For example, if a shaft having a *square* cross section is twisted, symmetry arguments like those in Sec. 6.2 can be used to show that in the element pictured in Fig. 6.25 the four dotted lines remain straight (see Probs. 6.20 and 6.21). However, all other lines in the cross section can deform when the shaft is twisted without violating the requirements of symmetry.

If we rule a grid of small squares on the surface of a square shaft, the resulting deformation after twist is illustrated in Fig. 6.26. A corner element can have no stresses acting on it and hence is undistorted in the deformed shaft. The originally plane cross sections have deformed or *warped* out of their own planes. In general, torsion of shafts which do not possess circular symmetry produces deformations that involve rigid-body rotations of one cross section with respect to another accompanied by warping out of the original planes of the cross sections.

Exact solutions for the torsion of rectangular shafts using the equations developed in Chap. 5 are available, both for elastic[1] and elastic–perfectly plastic[2]

[1] See, for example, S. Timoshenko and J. N. Goodier, "Theory of Elasticity," 3d ed., chap. 10, McGraw-Hill Book Company, New York, 1970.

[2] See, for example, W. Prager and P. G. Hodge, Jr., "The Theory of Perfectly Plastic Solids," p. 67, John Wiley & Sons, Inc., New York, 1951.

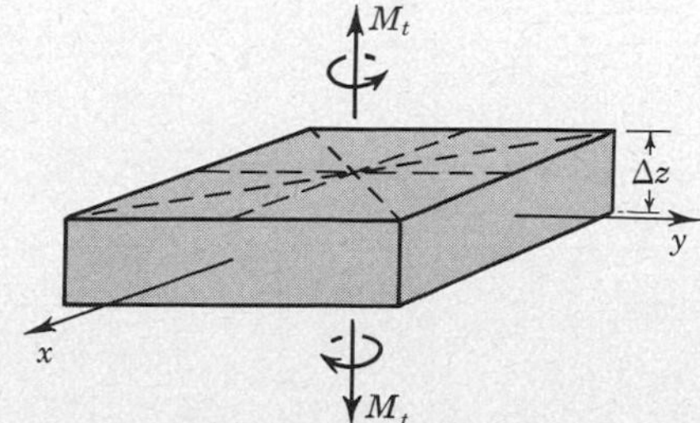

Fig. 6.25 In torsion of a square shaft, symmetry requires that the dotted lines remain straight and perpendicular to the z axis. All other lines in the cross section can deform.

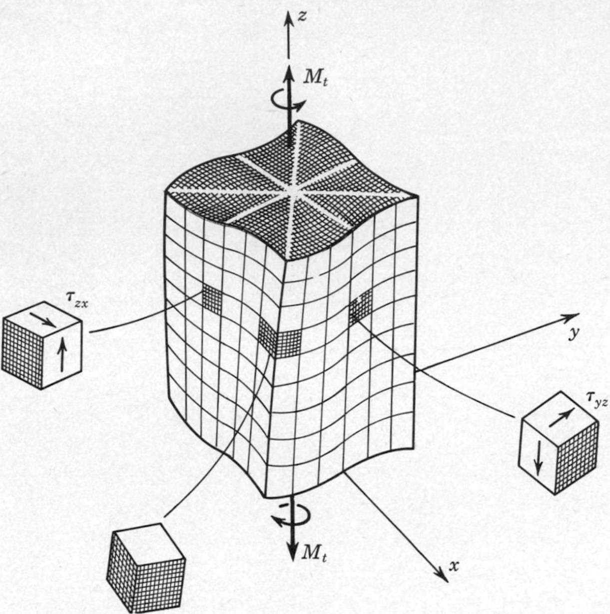

Fig. 6.26 Deformation of a square shaft in torsion. The originally plane cross sections have warped out of their own planes.

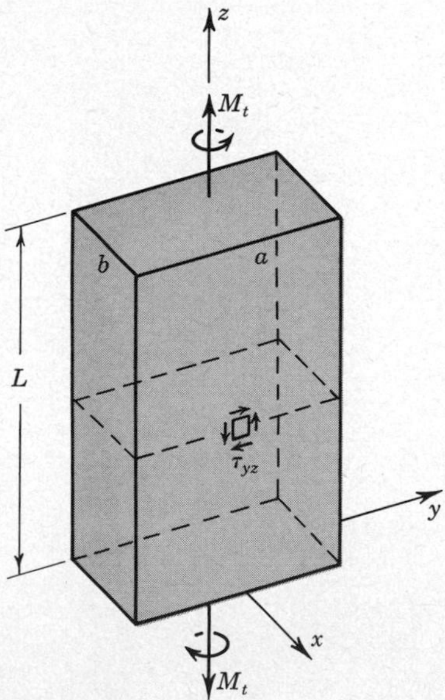

Fig. 6.27 Rectangular shaft subject to twisting moments.

Table 6.1 Coefficients for torsion of rectangular shafts

a/b	c_1	c_2
1	4.81	0.141
1.5	4.33	0.196
2	4.06	0.229
3	3.74	0.263
5	3.44	0.291
10	3.20	0.312

materials, but relatively sophisticated mathematical techniques are required. We shall not pursue the theoretical development any further but will simply list some of the results of the elastic analysis.

For a long, rectangular shaft with cross-section dimensions a and b with $b \leqq a$, the *maximum shearing stress* neglecting end effects occurs in the middle of the side a, as shown in Fig. 6.27, and has the magnitude

$$(\tau_{yz})_{\max} = c_1 \frac{M_t}{ab^2} \tag{6.28}$$

where c_1 is given in Table 6.1 as a function of the ratio a/b. The *torsional stiffness* of a long rectangular shaft is

$$\frac{M_t}{\phi} = c_2 \frac{Gab^3}{L} \tag{6.29}$$

where G is the shear modulus and c_2 is given in Table 6.1.

6.14 TORSION OF HOLLOW, THIN-WALLED SHAFTS

In the case of torsion of circular shafts, we were able to obtain a mathematically exact solution by elementary methods. For rectangular shafts exact solutions have been obtained but by fairly advanced methods. In this section we discuss a class of shafts for which a simple approximate analysis is available.

An exact analysis requires the simultaneous consideration of equilibrium, geometric compatibility, and the force-deformation relationship. Our approximate analysis uses equilibrium, but instead of carefully examining the other two requirements, we substitute a plausible guess as to the nature of the stress distribution.

We consider a long, hollow, cylindrical shaft of noncircular section and with a wall thickness t which need not be constant around the circumference but which is small compared with the overall dimensions of the cross section. A small element of length Δz of such a shaft is shown subjected to twisting moments M_t in Fig. 6.28*a*. The n, s, z axes identify the normal, the tangential, and the axial directions.

We begin by making a plausible assumption regarding the type of stress distribution which results from the twisting. Our previous experience with torsion suggests that there will certainly be shearing stresses τ_{sz}. What about other stress components? The normal stress σ_n is zero on the inside and outside surfaces. Since the wall thickness is small, it is not unreasonable to assume that σ_n is zero throughout. A similar argument applies to the components τ_{nz} and τ_{ns}. The components σ_z and σ_s are not ruled out by thinness of the wall. We can, however, rule out the existence of σ_s by considering the equilibrium of an element such as that in Fig. 6.28*b* (see Prob. 6.22). Regarding σ_z, we can say that since there is no axial force, the *resultant* of the σ_z distribution must be zero. This does not guarantee that σ_z is everywhere zero, but it does suggest that σ_z may not be an important stress component. On this basis we make the idealizing assumption that σ_z is zero. We thus have made the plausible assumption that τ_{sz} is the *only* nonzero stress component in the shaft of Fig. 6.28.

In Fig. 6.28*b* we show a free-body sketch of an arbitrary element 1-2 taken from the cross section of Fig. 6.28*a*. The shear stress τ_{sz} will be distributed along the top and bottom and on the vertical faces 1 and 2. A convenient concept to introduce at this time is the *shear flow q*, defined as follows:

$$q = \int_{-t/2}^{t/2} \tau_{sz}\, dn \tag{6.30}$$

The shear flow is the shear force per unit length. It is the resultant, across the thickness, of the shear-stress distribution. It is convenient to use the shear flow in discussions where the precise distribution of τ_{sz} across the thickness is unknown or unimportant.

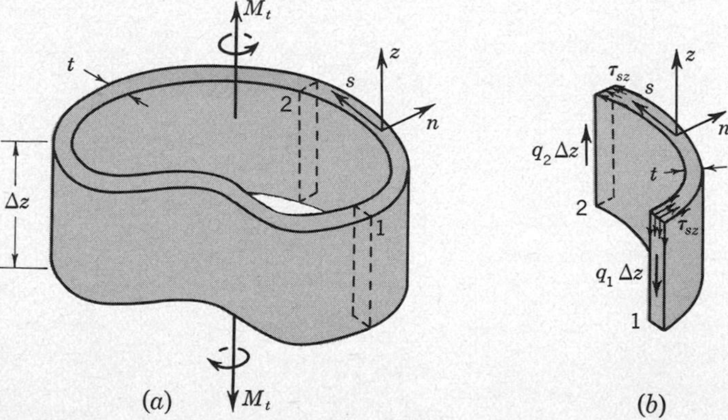

Fig. 6.28 Shear flow in a thin-walled tube.

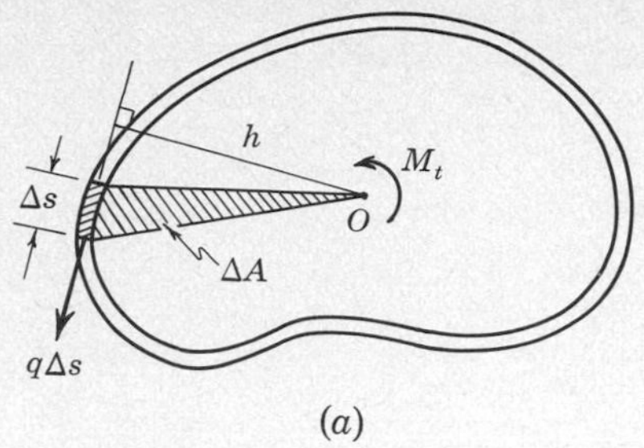

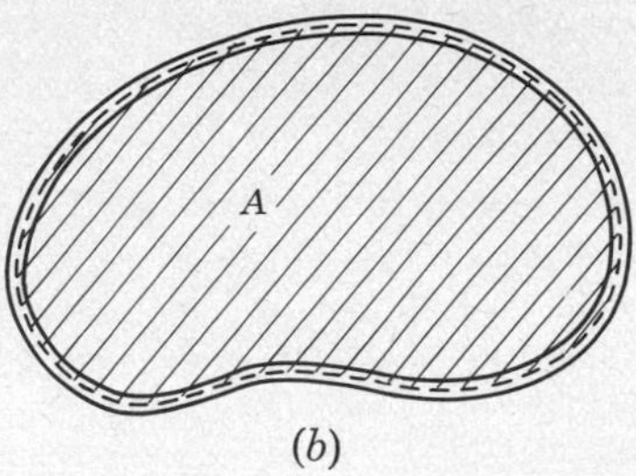

Fig. 6.29 Evaluation of the twisting moment.

The vertical sides 1 and 2 in Fig. 6.28*b* have length Δz so the total vertical forces on these faces are $q_1 \, \Delta z$ and $q_2 \, \Delta z$, as indicated. The next step is to use the fact that the element in Fig. 6.28*b* is in *equilibrium*. For balance of vertical forces we must have

$$q_1 = q_2 \tag{6.31}$$

Thus, the shear flows at two arbitrary positions must be equal. This means that the *shear flow is constant around the cross section of the shaft*. It is this fact which accounts for the designation shear *flow*. There is a clear analogy between shear flow in the thin-walled shaft and the flow of an incompressible fluid around a closed channel.

It remains to evaluate the shear flow in terms of the applied twisting moment M_t. In Fig. 6.29*a* the force on the element of length Δs is $q \, \Delta s$ and its lever arm about O is h. The contribution of this element to the twisting moment is

$$\Delta M_t = q \, \Delta sh \tag{6.32}$$

When Δs is small, we can approximate Δsh by twice the area ΔA of the shaded triangle in Fig. 6.29. The total twisting moment is the sum of all such contributions or simply

$$M_t = 2qA \tag{6.33}$$

where A is the total area enclosed by the shaft. Optimum accuracy is obtained by extending A to the mid-thickness of the wall as shown in Fig. 6.29*b*; however, our approximation is based on the thinness of the shaft wall so that use of the inner or outer area should not make much quantitative difference. If there is a large difference between these, we are probably outside the realm of validity of our approximation.

The result (6.33) links the shear flow with the applied twisting moment. It was obtained on the basis of equilibrium considerations alone after making the plausible assumption that only the τ_{sz} stress component was involved. If it is valid at all, it should be valid independently of the material behavior, e.g., elastic or plastic. More complete investigations[1] have shown that this approximation is in fact very good for thin-walled, hollow shafts.

In the elastic range it is plausible to postulate further that the stress τ_{sz} is *uniformly* distributed across the thickness. In this case (6.30) yields $q = \tau_{sz}t$, and, on substituting in (6.33), we obtain

$$\tau_{sz} = \frac{M_t}{2At} \tag{6.34}$$

as the shearing stress in a thin-walled, hollow shaft due to a twisting moment M_t.

The approximate theory just presented is useful but incomplete. We have said nothing about the deformations and are still unable to do so without further analysis.[2] The result (6.33) is valid for plastic behavior but (6.34) is not (see Prob. 6.27). This fact is not of great practical significance because in a thin-walled shaft the difference between the twisting moment which first initiates yielding and that which corresponds to the fully plastic condition is very small. (The difference approaches zero as the wall thickness t approaches zero.)

Example 6.6 Compare the stresses in a uniform thin-walled *circular* shaft as predicted by the approximate theory of this section and as predicted by the exact theory of Sec. 6.6.

Figure 6.30 shows the cross section of the shaft. According to the exact theory, the shear stress $\tau_{\theta z}$ varies linearly with the radius and has its maximum value at the outer radius. Using (6.9) and (6.11), we have

$$(\tau_{\theta z})_{\max} = \frac{M_t r_o}{\frac{\pi}{2}(r_o^4 - r_i^4)}$$

$$= \frac{M_t r_o}{\frac{\pi}{2}(r_o^2 + r_i^2)(r_o + r_i)(r_o - r_i)} \tag{a}$$

[1] See, for example, S. Timoshenko and J. N. Goodier, "Theory of Elasticity," 3d ed., p. 332, McGraw-Hill Book Company, New York, 1970.

[2] See, for example, J. T. Oden, "Mechanics of Elastic Structures," p. 46, McGraw-Hill Book Company, 1967. See also Prob. 6.42.

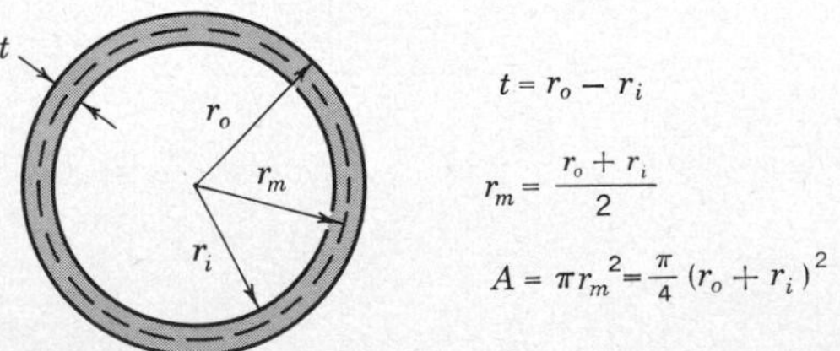

Fig. 6.30 Example 6.6.

According to the approximate theory of this section, the shear stress τ_{sz} is uniformly distributed across the wall thickness. Using (6.34) and the quantities in Fig. 6.30, we obtain

$$\tau_{sz} = \frac{M_t}{2At}$$

$$= \frac{M_t}{\frac{\pi}{2}(r_o + r_i)^2(r_o - r_i)} \qquad (b)$$

The percentage difference between (*a*) and (*b*) would be computed from the ratio

$$\frac{\tau_{sz} - (\tau_{\theta z})_{\max}}{(\tau_{\theta z})_{\max}} = -\frac{r_i}{r_o}\frac{r_o - r_i}{r_o + r_i} \qquad (c)$$

For example, when $r_i/r_o = 0.9$, there is only 4.7 percent difference between (*a*) and (*b*); when $r_i/r_o = 0.75$, there is 10.7 percent difference.

PROBLEMS

6.1. If a twisting moment of 10,000 in.-lb is applied to the end of a 2-in.-diameter steel shaft, what is the maximum shearing stress developed and the angle of twist in a 5-ft length of the shaft?

6.2. A hollow steel shaft 8 ft long must transmit a torque of 20,000 ft-lb. The total angle of twist over the length of the shaft is not to exceed 2.0° and the maximum allowable shearing stress is 12,000 psi. Find the dimensions of the shaft, i.e., its inside and outside diameters.

6.3. A composite shaft is made up of an inner circular cylinder of elastic material with shear modulus G_1 and an outer circular annulus of elastic material with shear modulus G_2. The

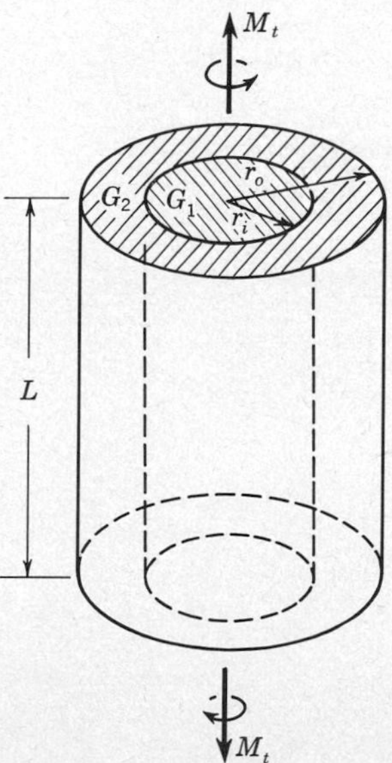

Prob. 6.3

materials are bonded securely at the interface r_i. Using the text derivation in Secs. 6.2 to 6.5 as a model, derive formulas for the twist angle ϕ and for the shear stress $\tau_{\theta z}$ which result from the application of the twisting moment M_t.

6.4. A 4130 HT steel shaft of $1\frac{1}{4}$-in. diameter and of length 2 ft transmits the maximum torque which does not cause plastic deformation. Calculate the angle of twist between the ends of the shaft.

6.5. A hollow steel shaft of 2-in. outside diameter is made of 1020 CR steel. What is the maximum internal diameter of the shaft which will just allow it to transmit without any yielding a torque of (*a*) 2,500 ft-lb, and (*b*) 5,000 ft-lb?

6.6. A torque M_t of 100 in.-lb is applied as shown to the steel shafts geared together. Calculate the angle of twist at the point where the torque is applied.

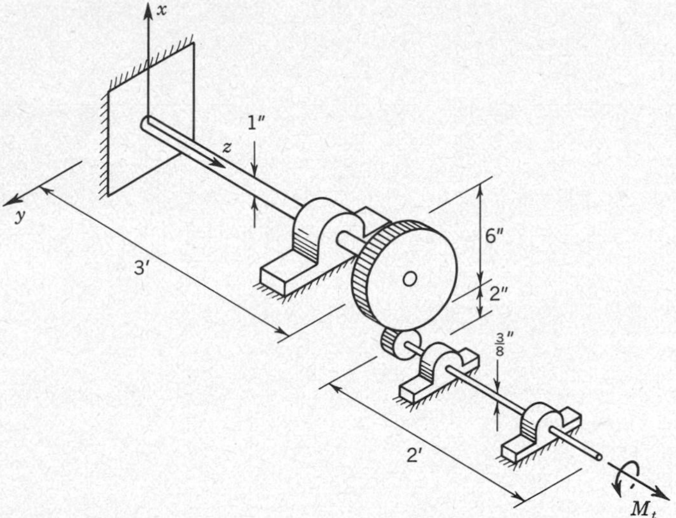

Prob. 6.6

6.7. For the system of Prob. 6.6, what maximum torque M_t may be applied before a shear stress of 40,000 psi is reached in either shaft?

6.8. In testing the fatigue behavior of gears, the illustrated system of gears and shafts is frequently employed. The parts are made so that in assembly it is necessary to hold one of the 3-in. gears stationary and rotate the corresponding 5-in. gear through a 3° angle in order to get the gears to mesh. There is a "locked-in" torque in the system, and when the system is driven by application of an external torque to one of the gears, the system of gears and shafts constitutes a power loop in which the average power is much larger than the externally supplied power. Calculate the maximum shear stresses in the two shafts after they are assembled but before external torque is applied. The shafts are made of 1020 CR steel.

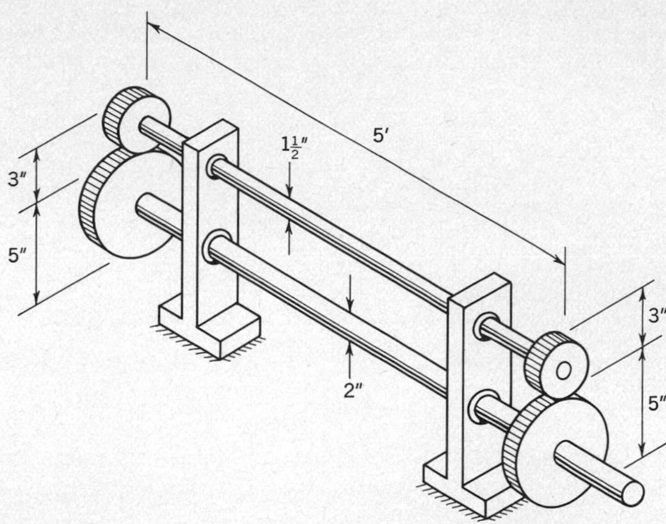

Prob. 6.8

6.9. When the system of shafts and gears as shown is assembled, it is found that, to get the gears to mesh, the 3-in.-diameter gear must be held stationary and the 5-in.-diameter gear rotated through 3°. Calculate the maximum shear stresses in the two shafts after assembly but before external torque is applied. The shafts are made of 1020 CR steel.

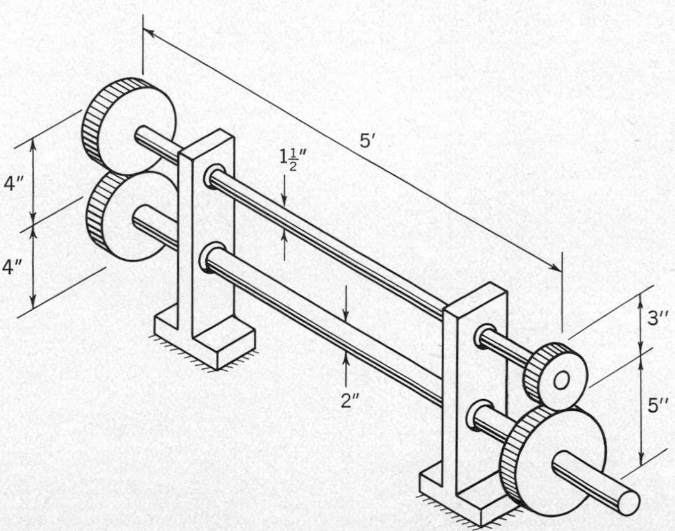

Prob. 6.9

6.10. Two shafts AB and BC of the same material but different diameter are welded together at point B. Ends A and C are fastened securely so that the shafts cannot rotate at these points. An external twisting couple M_o is applied to the shafts at point B. Find the twisting couples exerted on the ends of the shafts at A and C.

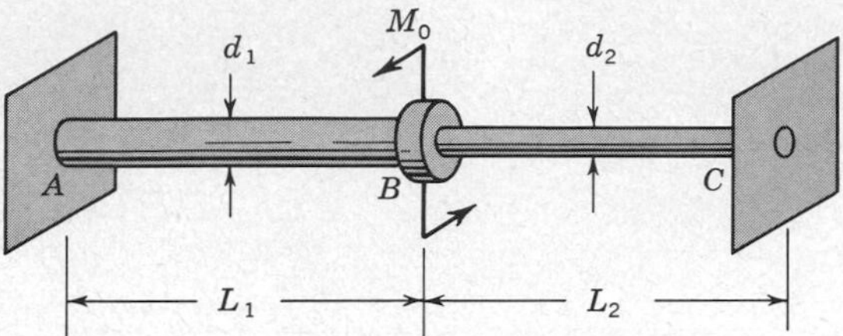

Prob. 6.10

6.11. For the diagram of the system in Prob. 6.10, assume that the two rods are made of different materials. It is found that the twisting couples at A and at C are equal. What relation must hold between the diameters, lengths, and moduli of the two materials?

6.12. A plate is riveted to a fixed member by means of six $\frac{3}{4}$-in. rivets as shown. Show how the theory of torsion can be used to estimate the shear stress in the rivets. How much additional force P could be exerted before the average shear stress exceeded 10,000 psi in any of the rivets?

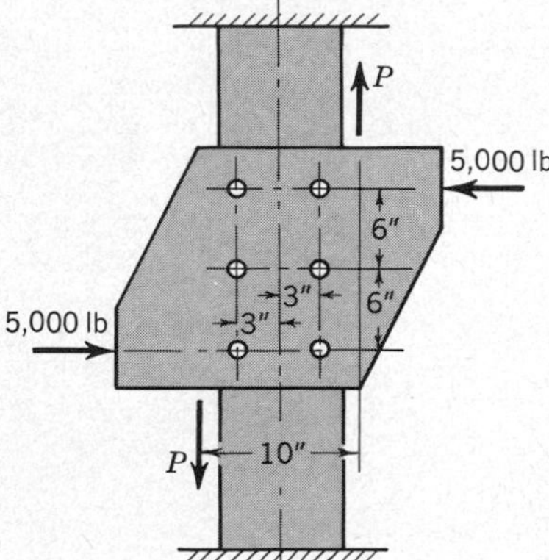

Prob. 6.12

6.13. A composite shaft is made of a 1-in.-diameter core of 2040-0 aluminum alloy and a 4130 steel (tempered 600° F) case of outside diameter 1.10 in. What maximum torque can be applied to a 5-ft length of the shaft (*a*) before plastic yielding starts, and (*b*) before the shaft fails through excessive plastic deformation? At what angle of twist does plastic deformation commence?

6.14. Consider the steel shaft shown in the figure which has applied to its surface a linearly varying twisting-moment distribution. Take the distributed twisting moment to have a resultant moment M_t. Determine the angle of twist at the end of the shaft.

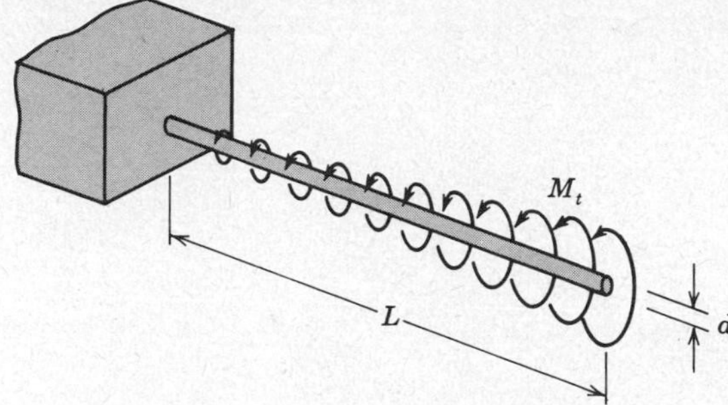

Prob. 6.14

6.15. If the free end of the shaft of Prob. 6.14 is now fixed into a rigid wall support, find the twisting moment at the walls. Take $L = 3$ m and $d = 8$ cm.

6.16. Find the distribution of twisting moment and angle of twist distribution along the steel shaft shown.

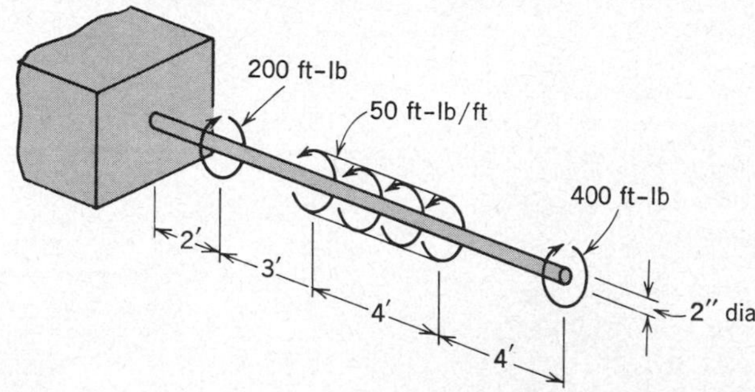

Prob. 6.16

6.17. For a hollow shaft whose outside diameter is twice its inside diameter, derive the relation between the horsepower that may be transmitted, the rpm, the maximum allowable shear stress, and the outside diameter d_o.

6.18. A hollow shaft of 1020 HR steel has an outside diameter of 5 in. and a bore diameter of 4.5 in., the centers of the two circles being eccentric by 0.1 in. Estimate the maximum torque that the shaft will carry without the occurrence of yielding.

6.19. A torsion member is fabricated from two concentric thin-walled tubes as shown. At the ends the tubes are welded to rigid disks so that both tubes are forced to twist as a unit. Let the shaft length be L, and let the material in both shafts be elastic–perfectly plastic with shear modulus G and yield stress τ_Y in shear.

(*a*) Find the twisting moment T_Y and twisting angle ϕ_Y corresponding to the first occurrence of yielding in the assembly.

(*b*) Find the limiting twisting moment T_L when the assembly becomes fully plastic. At what twisting angle is T_L reached?

(*c*) After reaching the fully plastic condition of (*b*), the applied twisting moment is removed. What will be the elastic springback angle, and what will be the residual stress distribution?

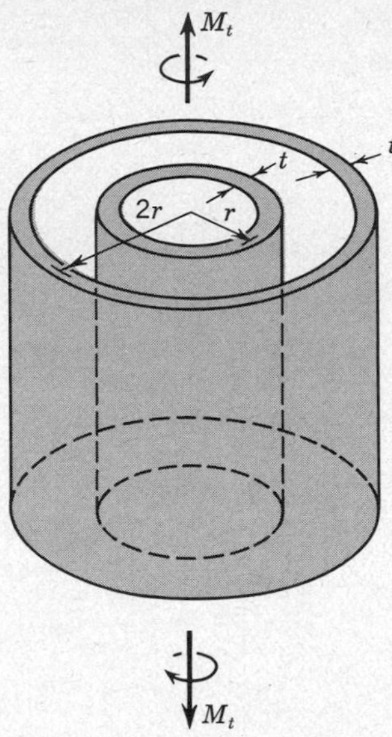

Prob. 6.19

6.20. A square isotropic shaft is being twisted. The sketches show proposed deformations of the radii OM and ON in the plane of the cross section. Use a symmetry argument to show that one of these is an impossible mode of deformation.

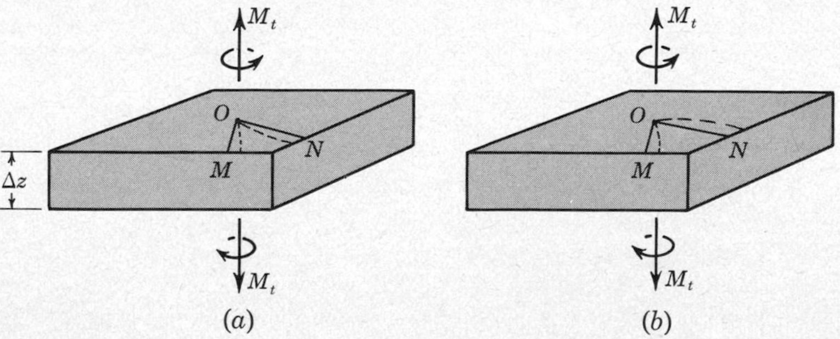

Prob. 6.20

6.21. Consider the two possible patterns of displacement in the z direction for the points M and N shown in the square shaft subjected to torsion. Use a symmetry argument to show that one of these is an impossible mode of deformation.

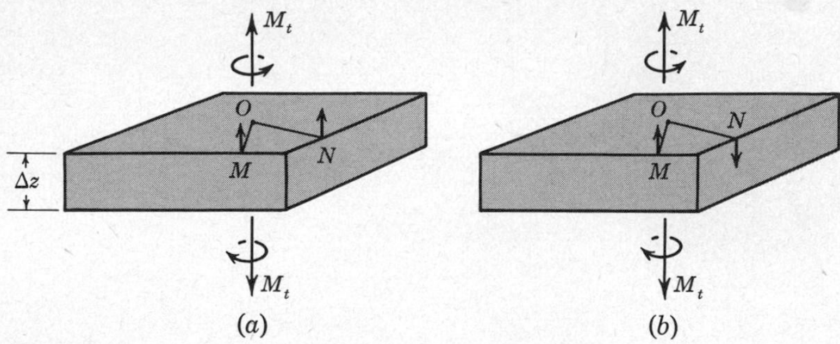

Prob. 6.21

6.22. Refer to Sec. 6.14 and Fig. 6.28. Assume that $\sigma_n = \tau_{nz} = \tau_{ns} = 0$ everywhere. Use the element sketched to deduce that σ_s must also vanish.

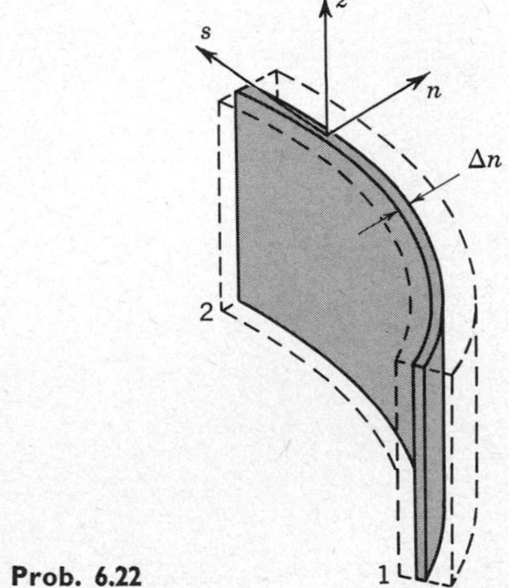

Prob. 6.22

6.23. An elevator system consists of an electric motor, a drive shaft, a wheel, and the elevator cage and counterweight as shown. When a certain number of people step into the cage, it moves down through 0.2 in. Assuming that the shaft and the cable are both made of steel, that the rotor of the motor does not move, and that there are sufficient bearings to prevent bending of the shaft, calculate the weight of the passengers that stepped in.

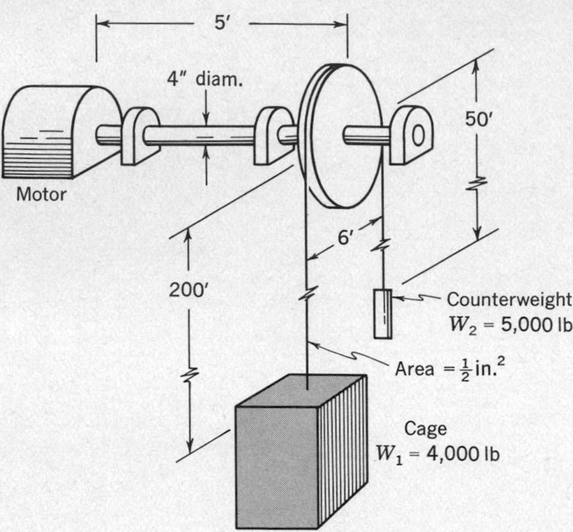

Prob. 6.23

6.24. A flexible shaft consists of a $\frac{1}{8}$-in.-diameter steel wire in a flexible hollow tube which imposes a frictional torque of 0.1 in.-lb per inch. The shaft is to be used for applying a torque of 3 in.-lb to actuate a switch. What is the maximum length of shaft that may be used without exceeding the elastic limit in shear of the wire, 40,000 psi? What will then be the "play" at the knob end if the shaft is used to turn the switch first in one direction and then in the other?

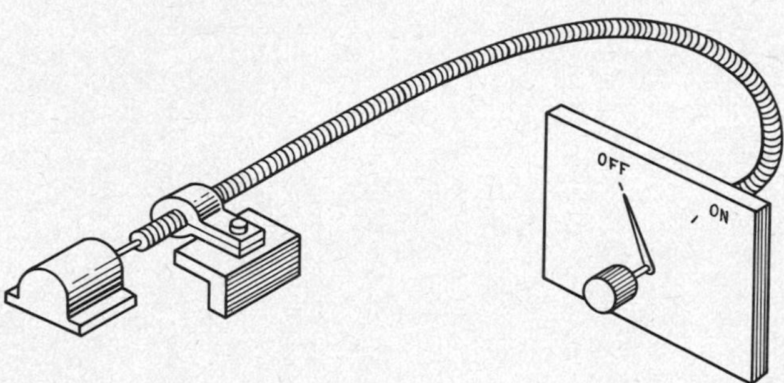

Prob. 6.24

6.25. An adsorption microbalance consists of a horizontally mounted quartz fiber AB of diameter 0.001 cm and length 8 cm which is put under enough tension so that it is nearly straight. To its middle C is glued a thin rod DE, a metal foil of dimensions 3 cm $\times$ 2 cm being attached to D and a counterbalance of negligible surface area to E, so that DE is horizontal in dry air. If the microbalance is placed in moist air so that a monolayer of water (thickness 10^{-8} cm) forms on all surfaces, by how much must the pointer at B be turned to bring DE back to the horizontal position? ($G_{\text{quartz}} = 2.7 \times 10^8$ g/cm^2).

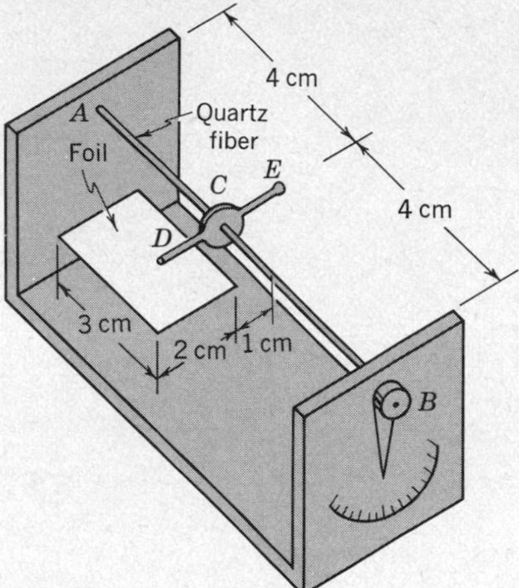

Prob. 6.25

6.26. An extruded 2024-T4 aluminum-alloy tube has a thin-wall rectangular cross section and is subjected to a twisting moment M_t as shown. Neglecting the stress concentrations at the corners, estimate the value of M_t which first causes yielding in the section.

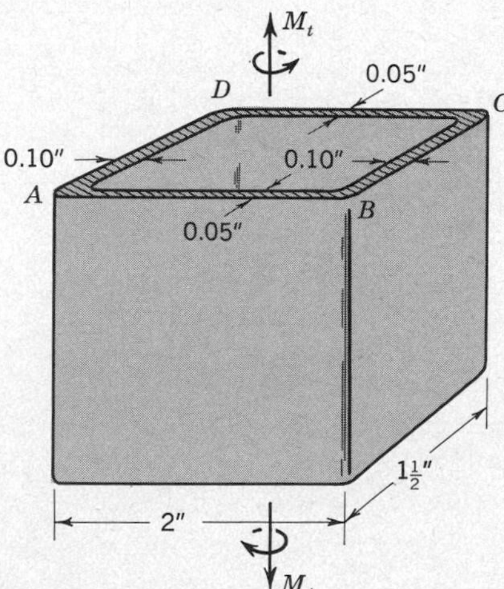

Prob. 6.26

6.27. The figure gives the fully plastic stress distribution for the tube of Prob. 6.26, assuming the material can be taken as elastic–perfectly plastic. Find the magnitude of x in order that there should be the same shear flow in (a) as in (b). Neglecting complications at the corners, evaluate the limiting twisting moment T_L and compare it with T_Y, the twisting moment at the first occurrence of yielding.

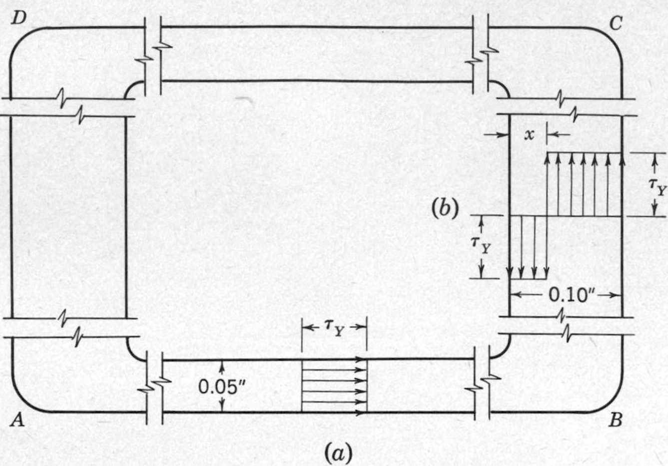

Prob. 6.27

6.28. The steering system of an automobile consists of a steering wheel, of a steering column which is a shaft of ¾-in. diameter, and of a linkage which gives a 20 : 1 reduction in angular rotation between the steering wheel and the tires. Each of the front wheels carries 1,000 lb of the weight of the car, and the tires are inflated to a pressure of 30 psi. If the coefficient of friction between rubber and the ground is 0.6, calculate the maximum stress set up in the steering column while the wheels are being turned. (*Note:* To keep the calculation simple, assume that the contacting region between tire and ground is circular, and that the pressure over it is uniformly 30 psi.)

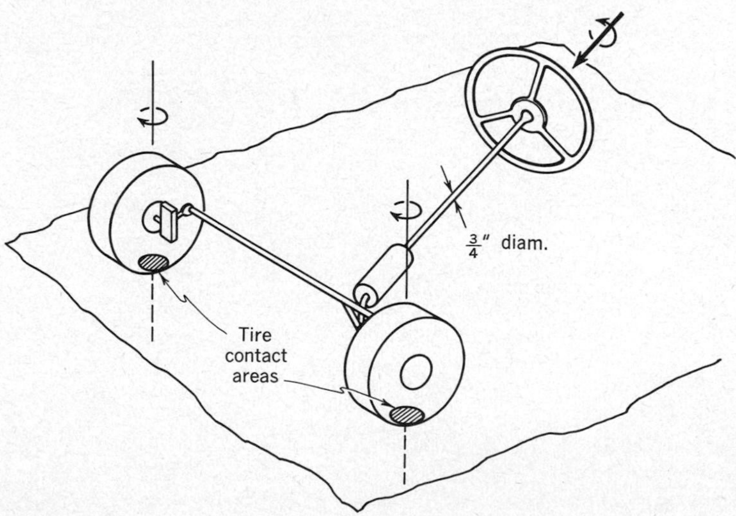

Prob. 6.28

6.29. A solid shaft of radius r_o, made of an elastic–perfectly plastic material, is twisted through an angle ϕ_1, which is sufficient to cause plastic deformation in those parts of the shaft with $r > r_o/2$. If the twisting moment is then removed, calculate the permanent twist angle ϕ_2 which remains in the shaft.

6.30. The close-coiled spring shown is made of an elastic material with shear modulus G. Show that an element of length $R\,\Delta\theta$ is principally subjected to a twisting moment and a shearing force. Assume that the deflection of the spring is due primarily to the twist. Study the geometry to determine the contribution to the total deflection δ which comes from twisting the element of length $R\,\Delta\theta$. Use integration to show that if there are n complete coils in the spring, then

$$\delta = \frac{4PR^3n}{Gr^4}$$

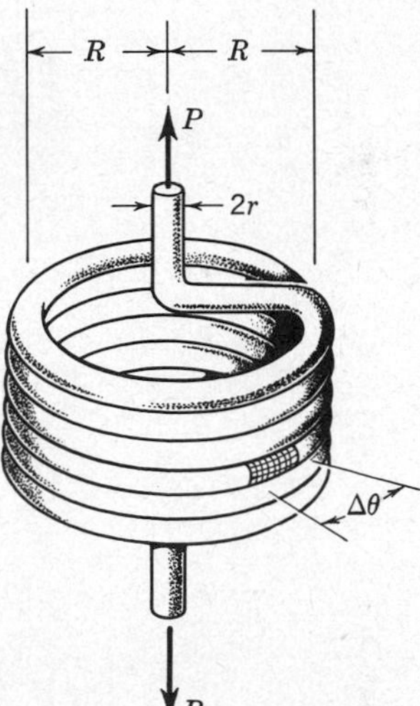

Prob. 6.30

6.31. A handyman's screw driver has a shaft of ¼-in. diameter. Estimate the compressive force and twisting torque that he might apply in tightening a screw. What is the order of magnitude of the corresponding compressive stress in the shaft compared with the shear stress due to torsion?

6.32. The center section of a T-handle socket wrench is made of alclad tubing (2024-T4 aluminum alloy with two thin layers of 1100-0 pure aluminum on the outside and inside for added corrosion resistance). The mean diameter of the tubing is ¾ in., the thickness of the alloy is ⅛ in., and the pure aluminum layers are each 1⁄32-in. thick. If a 100-lb force is applied to each end of the T handle as shown, and the forces are then removed, estimate the residual stresses which remain in the pure aluminum and in the aluminum alloy.

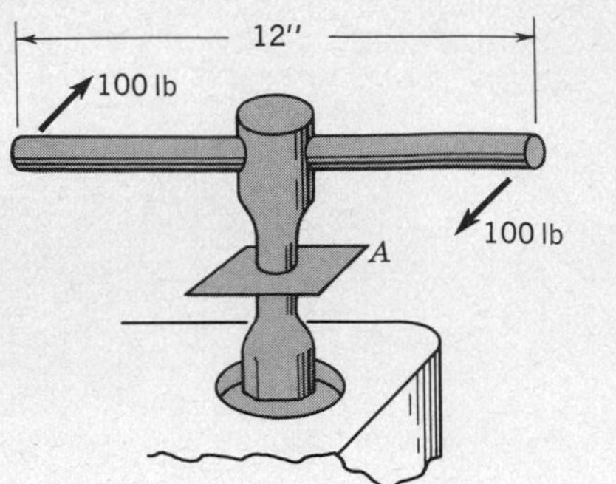

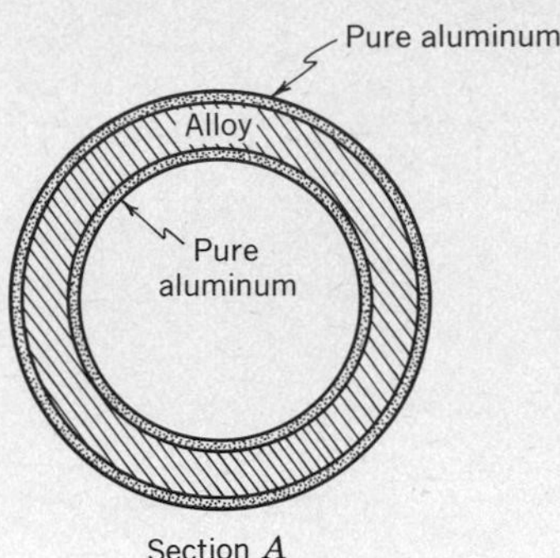

Prob. 6.32

6.33. A diamond-drill boring machine consists of a steel bit set with diamonds which is rotated by means of hollow steel rods coupled by screw joints. Water is forced through the hollow rods by a pump, and returns to the surface through the annular space between the rods and the walls of the hole, carrying the cuttings with it. Enough tension is maintained at the top of the shaft so that the compressive force between the bit and the bottom of the hole is small. If a torque of 2,000 ft-lb is applied to the top of the shaft, how deep would the hole be when yielding begins in the shaft? The yield stress in simple tension is $Y = 50{,}000$ psi. Use either the Mises or maximum shear yield criterion.

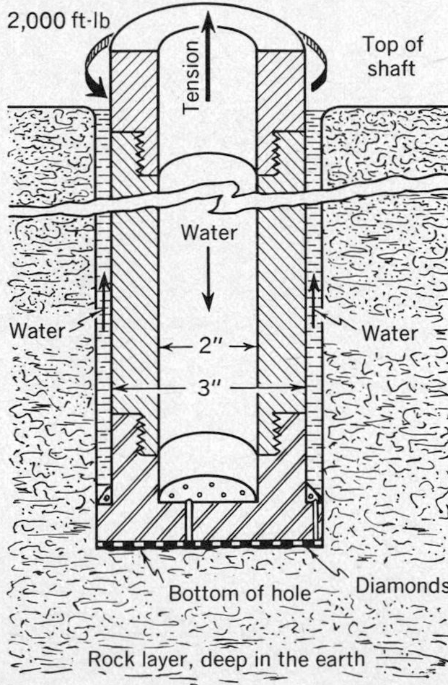

Prob. 6.33

6.34. It is proposed to use torsion-spring suspensions for an automobile's front wheels. The conditions under which each steel torsion spring, consisting of a solid, circular shaft, must operate, are as follows:

(1) It must take up the static force at the wheel, 1,000 lb.

(2) It must provide a spring constant at the wheel of 125 lb/in.

(3) Deflections of the wheel, up or down, of up to 6 in. must be possible without a shear stress exceeding 50,000 psi being set up in the shaft.

(4) The length L of the shaft should not exceed 120 in.

(5) The length x of the arm on the shaft cannot be larger than 30 in.

Derive a relation between x and L and see if you can design a suspension which meets these conditions. What will be the shaft diameter?

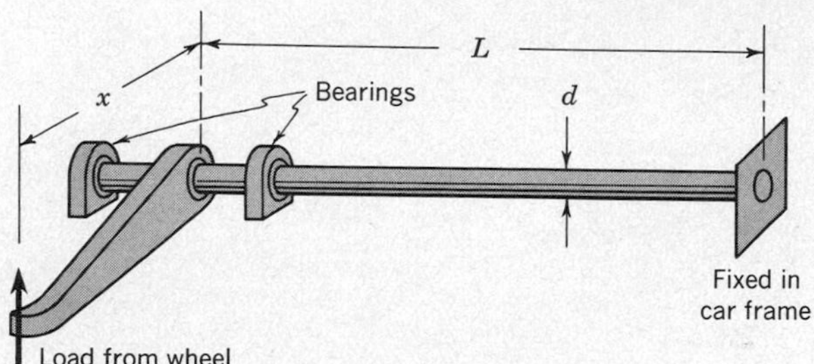

Prob. 6.34

6.35. The system shown consists of two horizontal shafts, A and B, clamped at M and N and passing through frictionless bearings at P and Q. The shafts are of length L and diameters d_A and d_B, respectively. Two stiff horizontal beams, C and D, are attached perpendicularly to the ends of the shafts, and contact one another at E. When a weight W is suspended from C at a distance $a/2$ from the end, what will be the angles of twist ϕ_A and ϕ_B, and what will be the force F transmitted at point E? The shear modulus of the shafts is G.

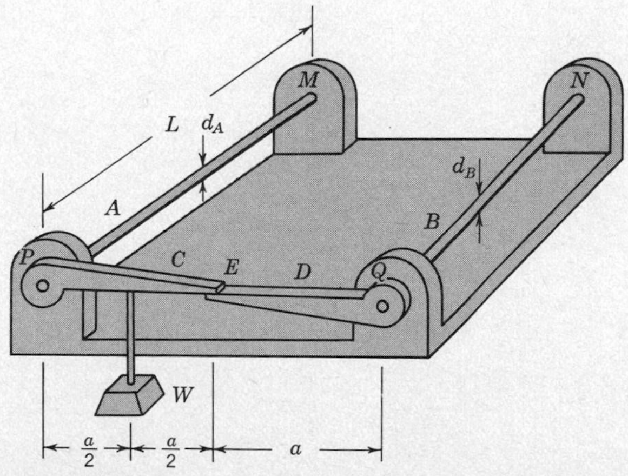

Prob. 6.35

6.36. Let u, v, and w be the displacements in the r, θ, and z directions of cylindrical coordinates. Using the results of Prob. 4.19, show that the strains of (6.2) are compatible with the displacements

$$u = 0$$
$$v = r\phi$$
$$w = 0$$

Verify that the solution of the torsion problem obtained in Sec. 6.5 also satisfies Hooke's law in cylindrical coordinates and the equilibrium equations in cylindrical coordinates (given in Prob. 4.4) and that therefore the solution is an exact solution within the framework of the theory of elasticity.

6.37. Show that by assuming only that ϵ_r and ϵ_θ vanish, the development of the text in Secs. 6.2 to 6.5 can be repeated and that in order for the solution to satisfy the 15 equations of the theory of elasticity it is necessary for ϵ_z to be zero everywhere.

6.38. A series of rods of fresh compact bone from a human femur were tested in torsion. If the samples had a cross-sectional diameter of approximately 1.9 mm and four tests yielded breaking torques of 1,000, 1,250, 1,000, and 1,100 g-cm, find the average torsional shear stress at breaking.

6.39. A given piece of metal will store more energy the more highly stressed it is below the yield point of the material. The "efficiency" of a spring can be defined as the ratio of the strain energy in the spring when the maximum stress is equal to the yield stress, to the strain energy in the spring when all the spring material is at the yield stress. Calculate the efficiency of a torsional spring, i.e., a shaft of length L with a twisting moment T.

6.40. Obtain expressions for the maximum stress and the angle of twist as a function of twisting moment for a circular shaft if the material follows the stress-strain law $\gamma = k\tau^2$.

6.41. A solid circular shaft of material whose stress-strain law is shown in Fig. 6.19 has been given a very large pretwist in one direction. Determine the twisting-moment–twisting-angle curve for this shaft if it is now twisted in the opposite direction. Show that this curve is initially linear but begins to depart from linearity when the magnitude of the reverse twisting moment is $\frac{2}{3} T_Y$. This illustrates the fact that in general the force-deformation relation of a solid depends on the past history of the solid. Only when the material remains in the linear-elastic range can we be sure that the force-deformation relation is independent of the previous history.

6.42. Consider Figs. 6.28 and 6.29 and Eq. (6.34). We wish to calculate the angle of twist of a thin-walled tube using Castigliano's theorem. The energy stored in the tube may be written in the form

$$U = \frac{1}{2G} \iiint \tau_{sz}^2 \, dn \, ds \, dz$$

Show that the angle of twist is given by

$$\phi = \frac{dU}{dM_t} = \frac{M_t L}{4A^2 G} \oint \frac{ds}{t}$$

6.43. The compact torsion-bar spring sketched below consists of an inner shaft of radius R_i and a sleeve whose outer radius is R_o. There is a very small clearance between the shaft and the inner surface of the sleeve. The material has an elastic shear modulus G and a yield stress in shear of τ_Y.

(*a*) Determine the torsional spring constant of the spring under the action of the twisting moment M_t.

(*b*) In a well-designed spring the outer sleeve will yield under the same twisting moment as the inner shaft. Develop an equation for determining the ratio R_o/R_i in order that this will occur.

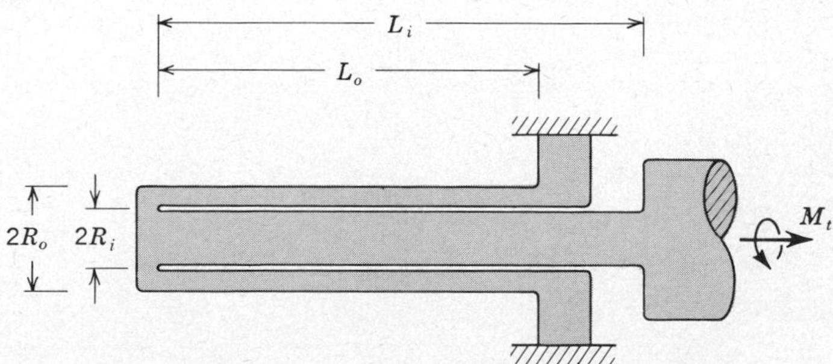

Prob. 6.43

6.44. A circular shaft AE of length $5a$, diameter d, and shear modulus G is welded to a support at end A and has a rigid arm of length b welded to its other end, E. A pointer is fastened to the shaft at C. When the shaft is unloaded, the pointer and the rigid arm are both horizontal, as shown in (*a*). A weight W is now hung on the end of the rigid arm and, at the same time, moments M_B and M_D are applied to the shaft at points B and D such that the pointer and the rigid arm remain horizontal, as shown in (*b*). Find the values of M_B and M_D required to keep the pointer and rigid arm horizontal when the weight W is hung on the arm.

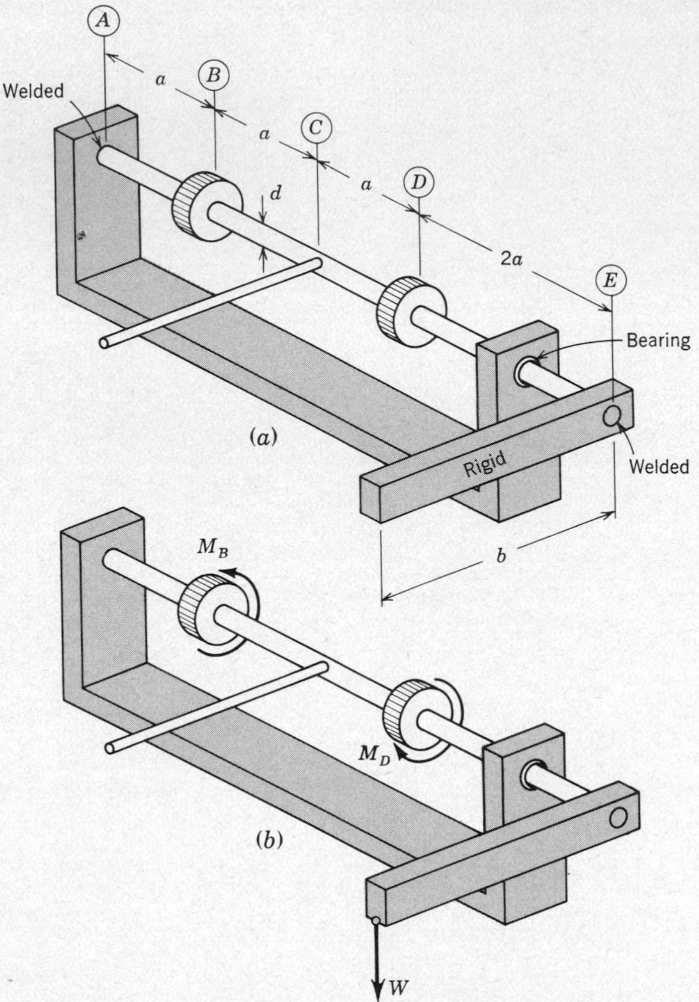

Prob. 6.44

6.45. It is proposed to measure Newton's gravitational constant by hanging a 10-m length of Mylar tape from a rod in a high ceiling and fastening the end to a 2-m-length aluminum rod. On each end of the rod a cast-iron cannon ball of mass 5 kg is fastened. Take the Mylar to have a tension modulus of 1.4×10^9 N/m^2, a Poisson's ratio of 0.40, and cross-sectional dimensions 25.4 by 0.254 mm. Verify that hanging the weights on the tape will not overload the Mylar, which has a maximum elastic strain capacity of 10 percent. When the whole device has stopped shaking and oscillating, the position of the balls is carefully determined, and then an aluminum bucket filled with lead shot is placed near each of the balls as shown. The distance between the center of the iron balls and the center of the buckets is 40 cm, and each bucket is filled with 100 kg of lead shot. Estimate the expected deflection of each cannon ball if the gravitational attraction between two masses each of 1 kg separated by 1 m is 6.67×10^{-11} N.

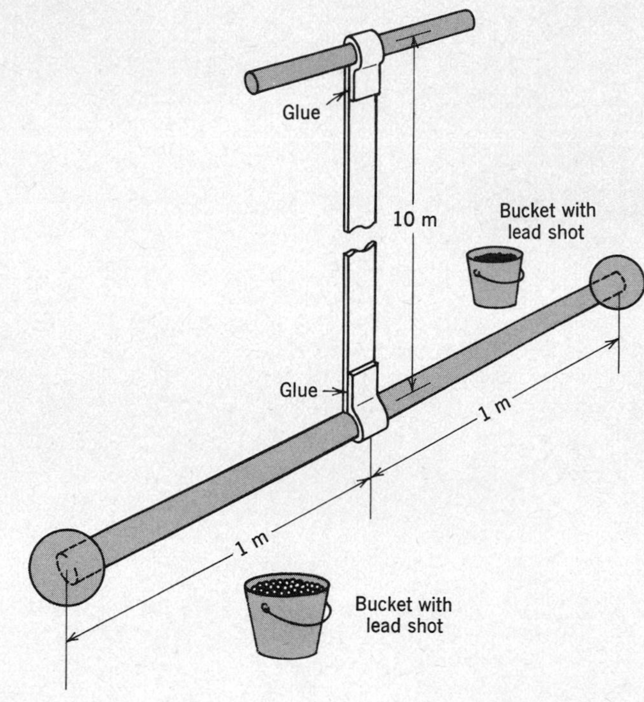

Prob. 6.45

7
Stresses Due to Bending

7.1 INTRODUCTION

When a slender member is subjected to transverse loading, we say it acts as a beam. We can find many examples of beam action in our immediate environment: the horizontal members in buildings (commonly called joists and girders) act as beams in transferring the vertical floor loading to the columns and foundation walls; the leaf springs of an automobile suspension transfer the body weight to the axle through beam action; the wings of an airplane act as beams in supporting the weight of the fuselage.

In Chap. 3 we found that if we sectioned a transversely loaded member, as illustrated in Fig. 7.1, a shear force and a bending moment would in general have to act on the cross section in order to maintain equilibrium. Our aim in this chapter is to determine the distributions of stresses which have the shear force V and the bending moment M_b as their resultant. Although we shall, of necessity, consider the nature of the deformation of beams, we shall postpone until Chap. 8 a detailed study of the deflection of beams under various conditions of support and loading.

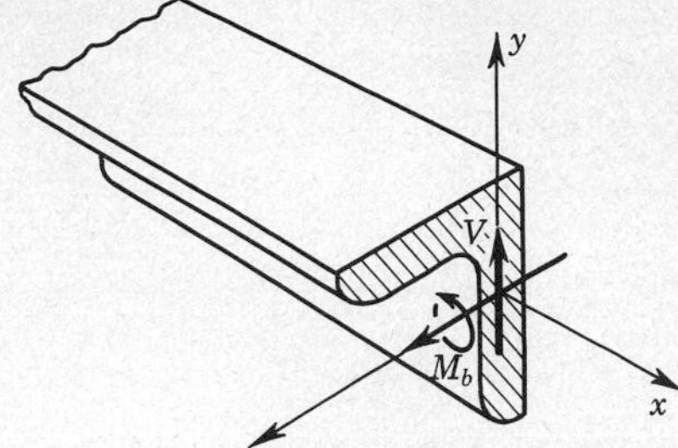

Fig. 7.1 Cross section of a beam showing shear force and bending moment resulting from loading in the xy plane.

Our method of approach will be similar to that followed in the investigation of torsion in Chap. 6, and to a certain extent our results will be similar. In Chap. 6 we were able to satisfy the requirements of the three steps of (2.1) at each point in a circular shaft; i.e., we were able to obtain the exact solution according to the theory of elasticity for this cross section. For the case of thin-walled, hollow shafts we obtained a useful approximate solution on the basis of equilibrium conditions alone. In this chapter we shall also obtain an exact solution within the theory of elasticity for the special case of a beam subjected to pure bending. For more general cases we shall obtain approximate distributions of stresses on the basis of equilibrium considerations.

7.2 GEOMETRY OF DEFORMATION OF A SYMMETRICAL BEAM SUBJECTED TO PURE BENDING

We begin by considering an originally straight beam which is uniform along its length, whose cross section is symmetrical about the plane of loading, as illustrated in Fig. 7.2, and whose material properties are constant along the length of the beam

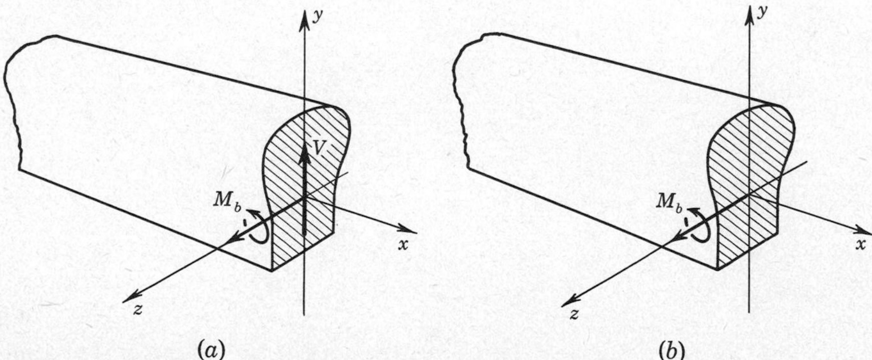

Fig. 7.2 Symmetrical beam loaded in its plane of symmetry. (*a*) In general, both shear force and bending moment are transmitted; (*b*) in *pure bending* there is no shear force, and a constant bending moment is transmitted.

and symmetrical with respect to the plane of loading. We further restrict our immediate attention to the case where such a beam transmits a bending moment which is constant along its length, as shown in Fig. 7.2*b*. A beam which transmits a constant bending moment is said to be in *pure bending*. We select this simple case as a starting point because, as will become evident, there is sufficient symmetry in this situation so that the deformation pattern can be fixed by symmetry arguments alone. After establishing the nature of the deformation, we shall introduce the stress-strain relations and then complete the three steps (2.1) by requiring that the resulting stress distribution have the resultant M_b which is required for equilibrium of the beam as a whole.

Since our originally straight beam will deform into some curved shape, it is a useful preliminary step to introduce the concept of *curvature*. The curvature of a plane curve is defined as the rate of change of the slope angle of the curve with respect to distance along the curve. In Fig. 7.3 we illustrate a curve AD whose curvature is in the xy plane. The normals to the curve at B and C intersect in the point O'. The change in the slope angle between B and C is $\Delta\phi$. When $\Delta\phi$ is small, the arc Δs is approximately $O'B\,\Delta\phi$. In the limit as point C approaches B, that is, as $\Delta s \to 0$, the curvature at point B is defined as

$$\frac{d\phi}{ds} = \lim_{\Delta s \to 0} \frac{\Delta\phi}{\Delta s} = \lim_{\Delta s \to 0} \frac{1}{O'B} = \frac{1}{\rho} \tag{7.1}$$

where $\rho = OB$ is the *radius of curvature* at point B.

We turn now to the geometry of deformation of pure bending in Fig. 7.4*a*, where AD, BE, and CF represent three equidistant plane sections, all perpendicular to the axis of an initially straight beam. In Fig. 7.4*b* we show the beam bent by bending moments M_b applied at the ends in the plane of symmetry, and in Fig. 7.4*c* we show the two deformed elements formed by the surfaces A_1D_1, B_1E_1, and C_1F_1.

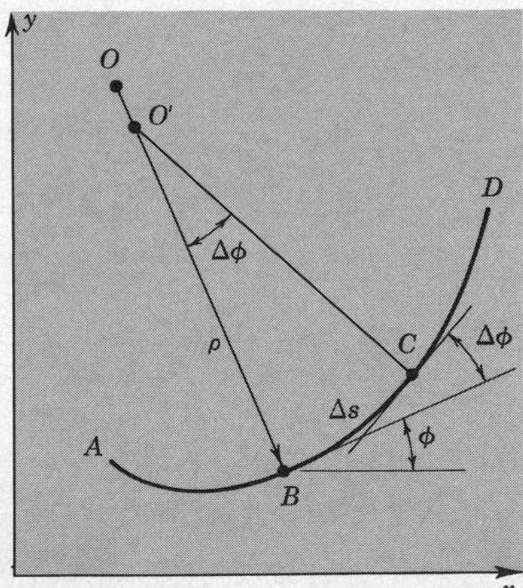

Fig. 7.3 The line AD has *curvature* $d\phi/ds = 1/\rho$ at point B, where $\rho = OB$ is the *radius of curvature* at point B.

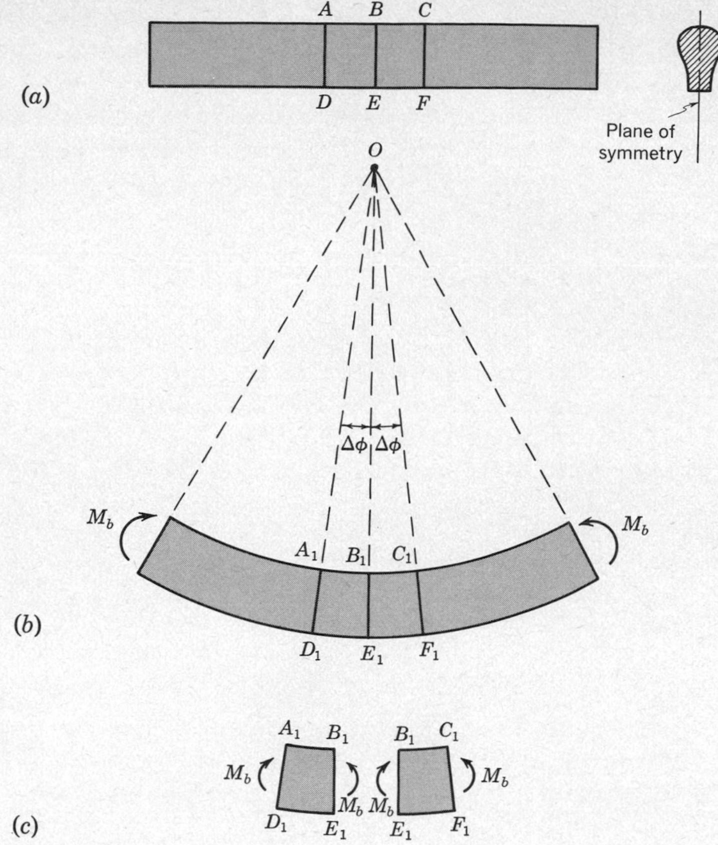

Fig. 7.4 Overall deformation of a symmetrical beam subjected to pure bending in its plane of symmetry.

Because each of these elements is loaded in its plane of symmetry, we can argue that the deformation of each will be symmetrical about its plane of symmetry. Further, since in their undeformed shape these elements were identical and since they are subjected to identical bending moments, it is reasonable to suppose (at least where these elements are far from the ends) that their deformed shapes will be identical. If, for instance, the surface A_1D_1 of element $A_1D_1E_1B_1$ bulged out, we would expect the corresponding surface B_1E_1 of element $B_1E_1F_1C_1$ to bulge out by the same amount. However, the latter action requires the surface B_1E_1 of element $A_1D_1E_1B_1$ to be dished in, and this destroys the end-to-end symmetry of deformation which this element must possess. We conclude, therefore, that the surfaces A_1D_1, B_1E_1, and C_1F_1 must be plane surfaces perpendicular to the plane of symmetry. Thus in pure bending in a plane of symmetry *plane cross sections remain plane.* Furthermore, the fact that each element deforms identically means that the initially parallel plane sections now must have a common intersection, as illustrated by point O in Fig. 7.4*b*, and that the beam bends into the arc of a circle centered on this intersection.

It should be noted that the above arguments have not ruled out the possibility of deformation of a plane section within its own plane. Such deformation does in fact occur, the only restriction on it being that it must be symmetrical with respect to the plane of symmetry. We shall postpone detailed consideration of this deformation of the cross section until Sec. 7.5.

We now pursue further the geometry of deformation in order to obtain the distribution of strain which is implied by our conclusion that plane cross sections remain plane. In Fig. 7.5*a* we show a segment of the beam before deformation. In Fig. 7.5*b* the deformed trace of the beam in the plane of symmetry is shown. While cross sections have remained plane, the originally straight longitudinal lines have become arcs of circles. Some of these lines have elongated and some of them have shortened in the deformation. There is one line in the plane of symmetry

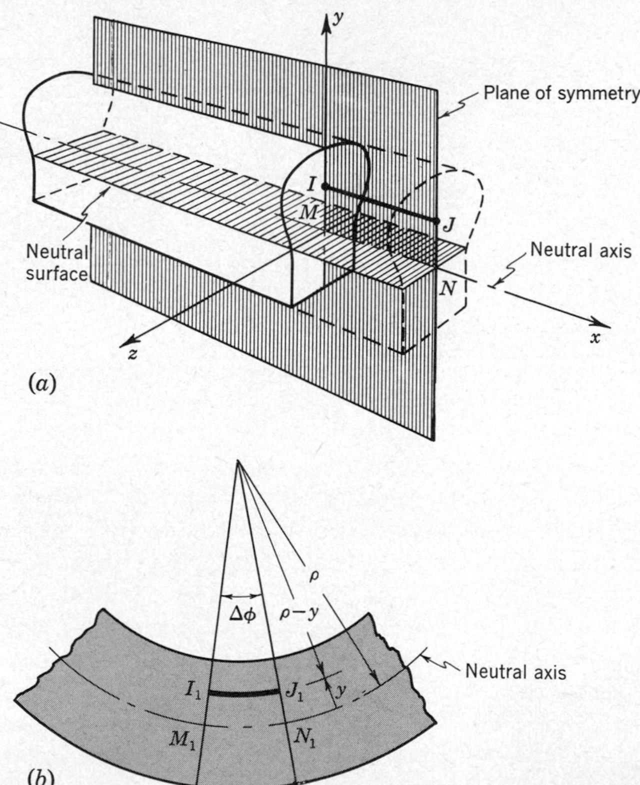

Fig. 7.5 (*a*) Undeformed beam; (*b*) trace of deformed beam in the plane of symmetry.

which has not changed in length. Although we do *not yet know its precise location*, we call this line the *neutral axis* and set up our coordinate system in the undeformed beam so that the x axis coincides with the neutral axis. The xy plane is the plane of symmetry, and the xz plane is called the *neutral surface*.

Now, although we admit that there may be deformation of the cross section in its own plane, we make the following assumption. We *assume* that the deformation will be sufficiently small so that we can use the coordinates of a point in the undeformed cross section to provide an adequate approximation to the location of the point after deformation. Thus if IJ and MN, which are separated by the distance y in the undeformed beam of Fig. 7.5*a*, are deformed into concentric circular arcs I_1J_1 and M_1N_1 in Fig. 7.5*b*, we assume that the difference between their radii of curvature can still be taken as y. We use the symbol ρ for the radius of curvature of the deformed neutral axis M_1N_1. The radius of curvature of I_1J_1 is then $\rho - y$.

Since $IJ = MN = M_1N_1$ from the definition of neutral axis, the strain of I_1J_1 is

$$\epsilon_x = \frac{I_1J_1 - IJ}{IJ}$$

$$= \frac{I_1J_1 - M_1N_1}{M_1N_1} \tag{7.2}$$

The circular arcs in Fig. 7.5*b* can be expressed in terms of the angle $\Delta\phi$.

$$M_1N_1 = \rho\,\Delta\phi \qquad I_1J_1 = (\rho - y)\,\Delta\phi \tag{7.3}$$

Inserting these in (7.2) and using the definition of curvature (7.1), we obtain

$$\epsilon_x = -\frac{y}{\rho} = -\frac{d\phi}{ds}\,y \tag{7.4}$$

as the distribution of longitudinal strain in the plane of symmetry of the beam. We see that the strain varies linearly with y; the minus sign indicates that there is shortening above the neutral axis and lengthening below. The derivation of (7.4) applies strictly only to the *plane of symmetry*, but we shall *assume* that (7.4) describes the longitudinal strain at all points in the cross section of the beam. When we have obtained our solution, we shall discuss the validity of all our assumptions.

In addition to (7.4) we can conclude from the symmetry arguments which require plane sections to remain plane that

$$\gamma_{xy} = \gamma_{xz} = 0 \tag{7.5}$$

for all points in the cross section of the beam. We can make no quantitative statements about the strains ϵ_y, ϵ_z, and γ_{yz} beyond the remark that they must be symmetrical with respect to the xy plane.

7.3 STRESSES OBTAINED FROM STRESS-STRAIN RELATIONS

The arguments which led to (7.4) and (7.5) are independent of the beam material (or materials), provided only that the material properties do not change along the length of the beam and that these properties are symmetrical with respect to the xy plane. Within these limitations the material can be nonisotropic, linear or nonlinear, elastic or plastic. In this section we shall restrict ourselves to beams made of *linear isotropic elastic* material, i.e., to materials which follow Hooke's law (5.2). For the analysis of composite elastic beams see Probs. 7.46 and 7.51.

The strain components of (7.4) and (7.5) are related to the stress components by (5.2), as follows:

$$\epsilon_x = \frac{1}{E}[\sigma_x - \nu(\sigma_y + \sigma_z)] = -\frac{y}{\rho}$$

$$\gamma_{xy} = \frac{\tau_{xy}}{G} = 0 \tag{7.6}$$

$$\gamma_{xz} = \frac{\tau_{xz}}{G} = 0$$

Thus we learn that the shear-stress components τ_{xy} and τ_{xz} must vanish in pure bending.

7.4 EQUILIBRIUM REQUIREMENTS

In this section we turn to the third of the steps of (2.1) and consider the requirements of equilibrium. Equilibrium requires that the resultant of the stress distribution over the cross section of the beam should equal the bending moment M_b, as indicated in Fig. 7.6. Letting $\sigma_x \, \Delta A$ be the force on the elemental area ΔA of Fig. 7.7,

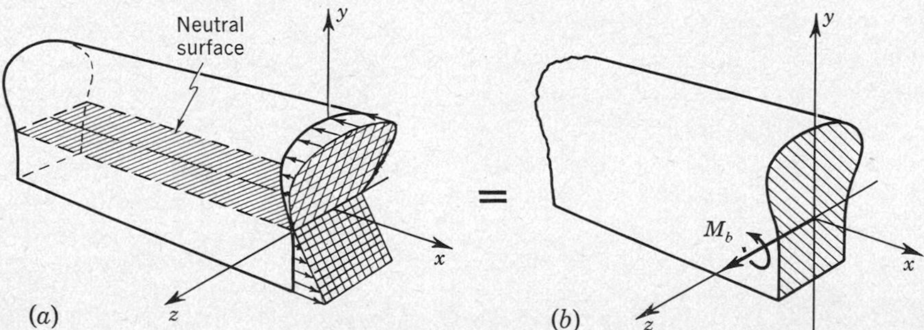

Fig. 7.6 The resultant of the stress distribution in pure bending must be the bending moment M_b.

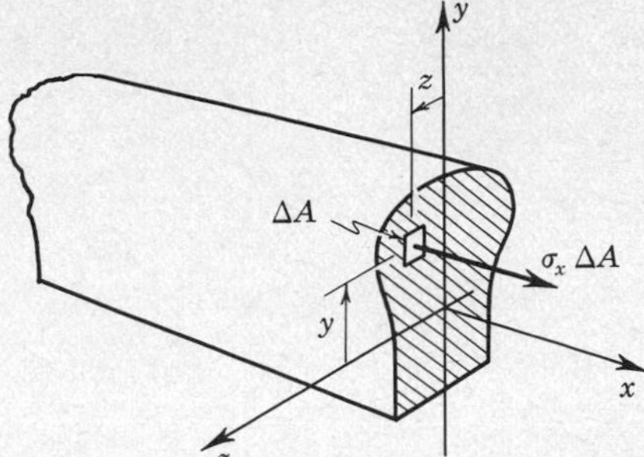

Fig. 7.7 Force acting on an elemental area ΔA of the beam of Fig. 7.6*a*.

we can express the equilibrium requirements as follows:

$$\begin{aligned} \Sigma F_x &= \int_A \sigma_x \, dA = 0 \\ \Sigma M_y &= \int_A z\sigma_x \, dA = 0 \\ \Sigma M_z &= -\int_A y\sigma_x \, dA = M_b \end{aligned} \tag{7.7}$$

where the integrals are to be taken over the total area A of the cross section. Here again, we make the fundamental assumption that the deformation of the cross section is sufficiently small so that we can use the undeformed coordinates to locate points in the deformed cross section; i.e., although the stresses must be accompanied by the deformation, we shall assume that for the purposes of equilibrium we can associate the stress at a point with the position of that point in the undeformed beam.

7.5 STRESS AND DEFORMATION IN SYMMETRICAL ELASTIC BEAMS SUBJECTED TO PURE BENDING

We have considered the problem from the standpoint of all three of the steps of (2.1). Our aim now is to draw out the solution which meets the requirements of (7.6) and (7.7). Examining (7.6) and (7.7), we note that the requirements (7.7) involve only σ_x while the first of (7.6) also includes the transverse normal stresses σ_y and σ_z. We could eliminate σ_y and σ_z by using the second and third relations of Hooke's law (5.2) to introduce the transverse normal strains ϵ_y and ϵ_z, but this would not get us very far since we can say nothing quantitative about these strains. An impasse has been reached, and in order to proceed it is necessary to make some assumption about the *transverse* behavior. In searching for a reasonable assumption, we observe that the external surfaces of the elemental slice of thickness Δx in Fig. 7.8 are free of normal and shear stresses. The slenderness of the beam suggests the plausibility of assuming that the transverse stresses σ_y, σ_z, and τ_{yz}

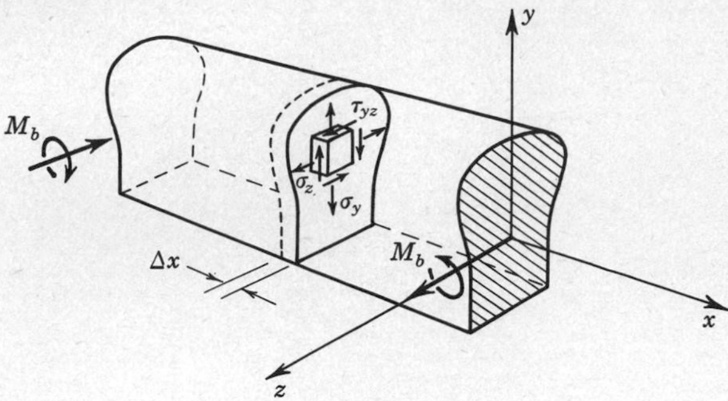

Fig. 7.8 The transverse stresses σ_y, σ_z, and τ_{yz} are assumed to be zero.

remain zero throughout the interior of the beam. We shall proceed on the basis of this assumption; i.e., we shall *assume*

$$\sigma_y = \sigma_z = \tau_{yz} = 0 \tag{7.8}$$

With the assumption (7.8) we are left with only one nonzero stress component. From the first of (7.6) we obtain

$$\sigma_x = -E\frac{y}{\rho} = -E\frac{d\phi}{ds}y \tag{7.9}$$

as the form of the longitudinal *normal stress* distribution in *pure bending* of a beam whose *material follows Hooke's law.* This linear variation of stress with distance from the neutral surface was used for the stress distribution illustrated in Fig. 7.6*a*. It remains to locate the position of the neutral surface.

Continuing now to draw out the solution, we substitute (7.9) into the first of (7.7).

$$\Sigma F_x = \int_A \sigma_x \, dA = -\int_A E\frac{y}{\rho}\, dA = -\frac{E}{\rho}\int_A y\, dA = 0 \tag{7.10}$$

This result tells us that when a linearly elastic beam of constant modulus E bends (i.e., when E/ρ does not vanish), the first moment of the cross-sectional area about the neutral surface must be zero. Stated otherwise, *the neutral surface must pass through the centroid of the cross-sectional area.*

It should be noted that for linear-elastic beams made of more than one material or for beams whose material behaves in a nonlinear fashion, the neutral surface can still be located by setting $\Sigma F_x = 0$, but in general in such cases, the neutral surface will *not* pass through the centroid of the cross-sectional area (see Probs. 7.46 and 7.50).

Substituting (7.9) into the second of (7.7), we have

$$\Sigma M_y = \int_A z\sigma_x \, dA = -\int_A E\frac{y}{\rho} z \, dA = -\frac{E}{\rho}\int_A yz \, dA = 0 \tag{7.11}$$

The integral on the right in (7.11) is zero because of the symmetry of the cross section with respect to the xy plane, and thus the second of (7.7) is satisfied.

Substituting (7.9) into the last of (7.7), we have

$$\Sigma M_z = -\int_A y\sigma_x \, dA = \int_A yE\frac{y}{\rho} \, dA = \frac{E}{\rho}\int_A y^2 \, dA = M_b \tag{7.12}$$

The integral on the right in (7.12) is known as the *second moment* of the beam cross-sectional area or as the *moment of inertia* of the area about the neutral axis. It can be computed once the specific shape of the cross section is known. We shall denote this integral by I_{yy}.

$$I_{yy} = \int_A y^2 \, dA \tag{7.13}$$

Important properties of the moment of inertia are summarized in Probs. 7.5 and 7.9.

Substituting (7.13) in (7.12), we obtain the following expression for the curvature as a function of the bending moment:

$$\frac{d\phi}{ds} = \frac{1}{\rho} = \frac{M_b}{EI_{yy}} \tag{7.14}$$

When the bending moment is positive, the curvature is positive, that is, concave upward.

Finally, by substituting (7.14) in our earlier expressions for strain and stress, namely, (7.4) and (7.9), we get the results

$$\epsilon_x = -\frac{M_b y}{EI_{yy}} \tag{7.15}$$

and

$$\sigma_x = -\frac{M_b y}{I_{yy}} \tag{7.16}$$

which express the longitudinal strain and stress in terms of the applied bending moment. Again we note that the stress distribution is linear and from (7.16) that the fibers on the top surface of the beam are in compression while the fibers on the bottom surface are in tension (Fig. 7.6*a*).

To complete our solution, we substitute (7.8) and (7.16) into Hooke's law (5.2) to obtain the transverse strain components.

$$\epsilon_y = \nu \frac{M_b y}{EI_{yy}} = -\nu\epsilon_x$$

$$\epsilon_z = \nu \frac{M_b y}{EI_{yy}} = -\nu\epsilon_x \tag{7.17}$$

$$\gamma_{yz} = 0$$

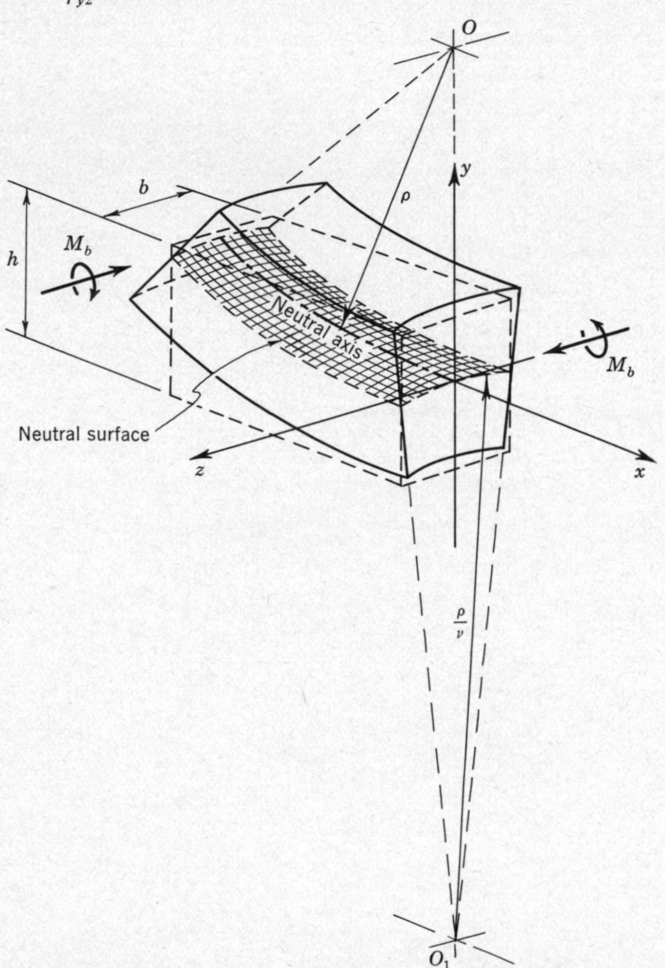

Fig. 7.9 Deformed shape of an originally rectangular beam subjected to pure bending in a plane of symmetry.

Thus there is a deformation of the cross section. The normal strains in the plane of the cross section are proportional to the axial normal strain but of opposite sense. Since the axial normal strain is compressive at the top of the beam and tensile at the bottom, the top of the cross section expands while the bottom of the cross section contracts. The deformed shape of an originally rectangular beam is shown in Fig. 7.9. Our result for ϵ_z implies that lines in the cross section originally parallel to the z axis have deformed into arcs of circles and, in particular, that the trace of the neutral surface on the cross section has become an arc with curvature $-\nu(1/\rho)$, as illustrated in Fig. 7.9. This transverse curvature of the beam is called *anticlastic curvature*; it can be seen quite easily if a rubber eraser is bent between the thumb and forefinger in the manner indicated in Fig. 7.10. As a result of the anticlastic curvature, the deformed neutral surface is a surface of double curvature (Fig. 7.9). A further result of the anticlastic curvature is that the neutral axis is the only line in the deformed neutral surface whose curvature is in a plane parallel to the original plane of symmetry of the beam.

In arriving at our solution, we have had to make several assumptions: it was assumed that the locations of stresses and strains in the deformed beam of Fig. 7.9 could be approximated by the corresponding locations in the undeformed beam of Fig. 7.5*a*; it was assumed that the strain variation (7.4) applied to the entire cross section and not just to the plane of symmetry; and, finally, it was assumed that the transverse stresses were zero. Nevertheless, our solution can be shown to satisfy all the requirements of the theory of elasticity for small deformations of slender members, providing the externally applied bending moments are applied by a distribution of stresses conforming to (7.16). The strains (7.5), (7.15), and (7.17) are geometrically compatible (see Prob. 7.32); the stresses (7.6), (7.8), and (7.16) satisfy the differential equations of equilibrium; and at every point the stresses and

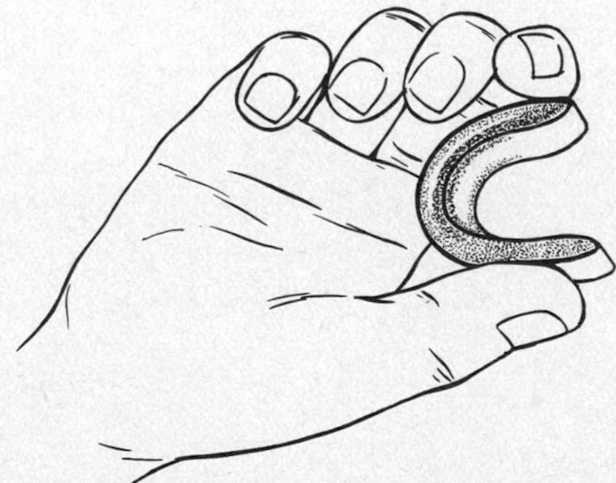

Fig. 7.10 Illustrating how a rubber eraser can be bent to demonstrate the presence of anticlastic curvature.

strains satisfy Hooke's law (5.2). More extensive studies have shown[1] that when the externally applied moments are applied in a manner which differs from (7.16), our solution is still very accurate in the central portion of the beam in accord with St. Venant's principle and only becomes appreciably in error near the ends. The length of these transition regions at the ends are of the order of the depth of the beam cross section.

Before illustrating the foregoing theory by applying it to some examples, it is interesting to observe the similarity between the bending-stress distribution illustrated in Fig. 7.6*a* and the shearing-stress distribution in torsion of circular shafts shown in Fig. 6.9*b*. Also of interest is the correspondence between Eqs. (7.14) and (6.7) and between (7.16) and (6.9).

It should be mentioned also that the above discussion of initially straight beams may be carried over to the pure bending of curved beams (see Prob. 7.33). It has been found from the analysis of curved beams that formula (7.16) for the distribution of stress across the beam thickness is reasonably accurate for curved beams when the radius of the curved beam is greater than about 5 times the thickness of the beam.[2]

Example 7.1 A steel beam 1 in. wide and 3 in. deep is pinned to supports at points *A* and *B*, as shown in Fig. 7.11*a*, where the support *B* is on rollers and free to move horizontally. When the ends of the beam are loaded with 1,000-lb loads, we wish to find the maximum bending stress at the mid-span of the beam and also the angle $\Delta\phi_o$ subtended by the cross sections at *A* and *B* in the deformed beam.

We attack this problem, as usual, within the framework of (2.1). Our first step is to determine the bending moment which is required to satisfy equilibrium at each point along the beam; this is shown in the diagram in Fig. 7.11*b*. From this diagram we see that the central portion *AB* is one of constant bending moment, and thus this part of the beam is in a state of pure bending.

To calculate the stress from (7.16), we must locate the coordinate axes and calculate I_{yy}. The centroid lies at the mid-height of the cross section, as shown in Fig. 7.11*c*, and using (7.13) we find

$$I_{yy} = \int_{-h/2}^{h/2} y^2 b\, dy = \frac{bh^3}{12} \qquad (a)$$

which on substituting $b = 1$ in. and $h = 3$ in. yields

$$I_{yy} = 2.25 \text{ in.}^4 \qquad (b)$$

[1] G. Horvay, The End Problem of Rectangular Strips, *Trans. ASME*, vol. 75, pp. 87–94, 576–582, 1953.

[2] Formulas for the calculation of maximum stress in curved members are given in R. J. Roark, "Formulas for Stress and Strain," 4th ed., p. 164, McGraw-Hill Book Company, New York, 1965.

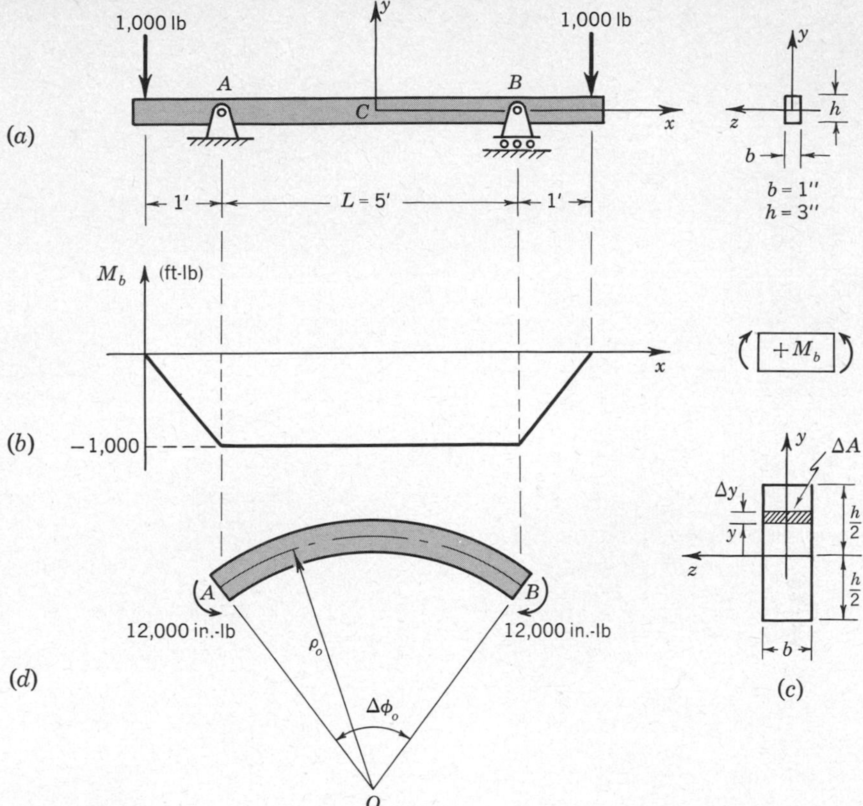

Fig. 7.11 Example 7.1.

The maximum bending stress occurs at the distance farthest from the neutral surface. At the mid-span the bending stress at the top of the beam is found from (7.16) to be

$$\sigma_x = -\frac{(-12{,}000 \text{ in.-lb})(1.5 \text{ in.})}{2.25 \text{ in.}^4} = 8{,}000 \text{ psi} \tag{c}$$

If we use $y = -1.5$ in., we obtain a numerically equal compressive stress at the bottom of the beam.

To obtain the angle change $\Delta\phi_o$, we begin by observing that the "force-deformation" relation which is applicable to this situation is the moment-curvature relation (7.14). Using $E = 30 \times 10^6$ psi, we find the curvature in the segment AB to be

$$\frac{d\phi}{ds} = \frac{M_b}{EI_{yy}} = \frac{-12{,}000 \text{ in.-lb}}{(30 \times 10^6 \text{ psi})(2.25 \text{ in.}^4)}$$
$$= -0.000178 \text{ rad/in.}$$
$$= -0.00213 \text{ rad/ft} \qquad (d)$$

The total angle change between A and B is found by integration of the curvature relation (d).

$$\phi_B - \phi_A = \int_A^B d\phi = \int_{-L/2}^{L/2} \frac{d\phi}{ds}\, ds$$
$$= -0.00213 \text{ rad/ft} \times 5 \text{ ft}$$
$$= -0.0106 \text{ rad}$$
$$= -0.61^\circ \qquad (e)$$

The magnitude of the angle labeled $\Delta\phi_o$ in Fig. 7.11d is thus 0.61°. Note that this angle has been exaggerated in the figure. As a matter of interest, the radius of curvature of the section AB can be obtained from (7.1).

$$\rho = \frac{1}{d\phi/ds} = \frac{1}{-0.00213 \text{ rad/ft}} = -470 \text{ ft} \qquad (f)$$

This is indicated in Fig. 7.11d, where $\rho_o = -\rho$.

Example 7.2 We wish to find the maximum tensile and compressive bending stresses in the symmetrical T beam of Fig. 7.12a under the action of a constant bending moment M_b.

Since we have the relation (7.16) available, our task in this problem centers around the location of the neutral surface and the evaluation of I_{yy}. As a first step we must locate the z axis in the centroid of the cross section. (A review of the calculation for the location of the centroid of an area is given in Prob. 7.1.) In Fig. 7.12b we consider the beam to be made up of the rectangle 1 of dimensions b by $2h$ and the rectangle 2 of dimensions $6b$ by $h/2$, and we let $\bar{y}$ represent the distance from the base to the centroid of the cross section. Then (see Prob. 7.1),

$$\bar{y} = \frac{\sum_i \bar{y}_i A_i}{\sum_i A_i} = \frac{\tfrac{3}{2}h(2bh) + (h/4)(3bh)}{2bh + 3bh} = \frac{3}{4}h \qquad (a)$$

The location of the axes in the cross section is shown in Fig. 7.12c. We calculate the moment of inertia for the rectangle 1 by use of the *parallel-axis theorem* (see Prob. 7.5).

$$(I_{yy})_1 = \frac{b(2h)^3}{12} + 2bh(\tfrac{3}{4}h)^2 = \tfrac{43}{24}bh^3 \qquad (b)$$

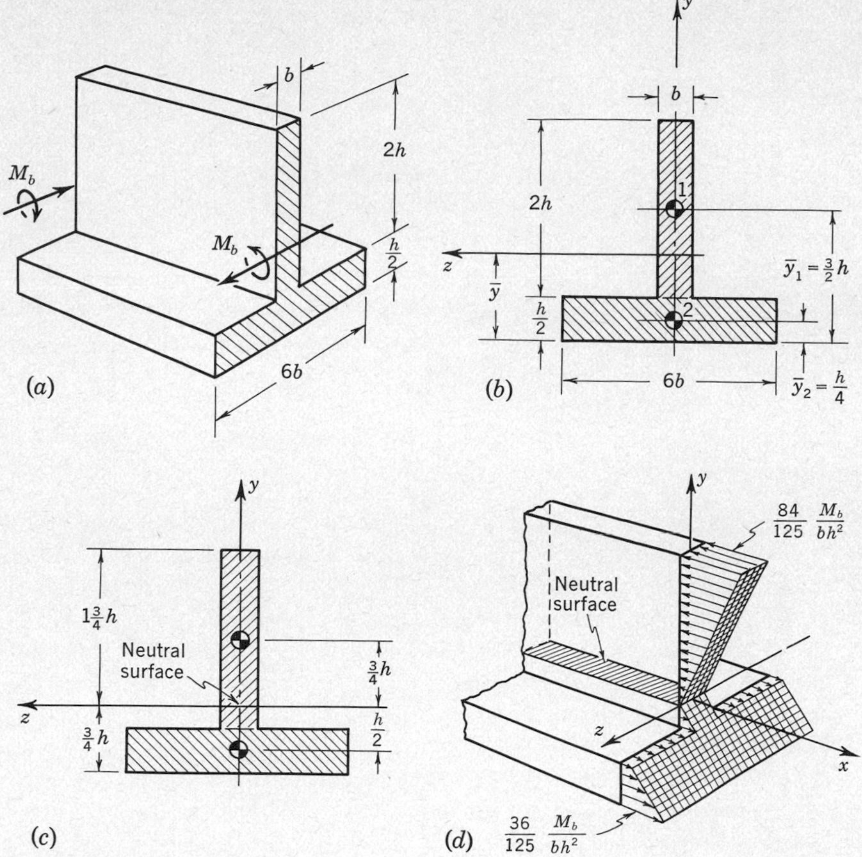

Fig. 7.12 Example 7.2.

Similarly, for the rectangle 2 we obtain

$$(I_{yy})_2 = \frac{6b(h/2)^3}{12} + 3bh(h/2)^2 = {}^{13}\!/_{16}bh^3 \tag{c}$$

Then, for the entire cross section

$$I_{yy} = (I_{yy})_1 + (I_{yy})_2 = {}^{43}\!/_{24}bh^3 + {}^{13}\!/_{16}bh^3 = {}^{125}\!/_{48}bh^3 \tag{d}$$

Now, substituting (d) in (7.16) together with $y = -{}^3\!/_4h$, we find the maximum tensile bending stress

$$\sigma_x = -\frac{M_b(-{}^3\!/_4h)}{{}^{125}\!/_{48}bh^3} = \frac{36}{125}\frac{M_b}{bh^2} \tag{e}$$

The maximum compressive bending stress occurs at $y = +1\frac{3}{4}h$,

$$\sigma_x = -\frac{M_b(\frac{7}{4}h)}{\frac{125}{48}bh^3} = -\frac{84}{125}\frac{M_b}{bh^2} \tag{f}$$

The stress distribution in the beam is illustrated in Fig. 7.12*d*. We see that the maximum compressive stress is approximately 2.3 times greater than the maximum tensile stress.

This example illustrates that the determination of the stress distribution in beams requires the location of the centroidal axis and the evaluation of the moment of inertia about this axis. For the more common structural cross sections, such as I beams, this information is available in tables.[1]

7.6 STRESSES IN SYMMETRICAL ELASTIC BEAMS TRANSMITTING BOTH SHEAR FORCE AND BENDING MOMENT

Pure bending is a relatively uncommon type of loading for a beam. As we have seen in Chap. 3, it is much more common for a shear force to be present, as illustrated in Fig. 7.2*a*. It is more difficult to obtain an exact solution to this problem since the presence of the shear force means that the bending moment *varies* along the beam and hence many of the symmetry arguments of Sec. 7.2 are no longer applicable. Exact solutions within the theory of elasticity are available for certain types of load variation along the beam,[2] but their development is beyond the mathematical scope of this book.

In this section we shall describe what is frequently referred to as the *engineering* theory of the stresses in beams—to distinguish it from the *elasticity* theory mentioned in the preceding paragraph. To develop this engineering theory, we make the *assumption that the bending-stress distribution* (7.16) *is valid even when the bending moment varies along the beam, i.e., when a shear force is present.* Thus, the engineering theory of beams assumes that the longitudinal or bending stress distribution at a location x is given by (7.16). Then, by requiring that equilibrium be satisfied for certain well-selected free bodies, we can estimate the stress distribution which has the shear force as its resultant. Since we do not include the satisfaction of geometric compatibility and satisfaction of the stress-strain relations in our analysis, we have no a priori certainty that the results are accurate. However, experimental evidence and comparison with some of the aforementioned solutions from the theory of elasticity show that the estimates of the stress distribution are satisfactory for most engineering purposes.

Beginning our analysis, we show in Fig. 7.13*a* a length Δx of a beam which is subjected to both bending and shear. We take the case where there is no external

[1] See, for example, "The American Institute of Steel Construction Manual of Steel Construction."
[2] S. Timoshenko and J. N. Goodier, "Theory of Elasticity," 3d ed., chap. 11, McGraw-Hill Book Company, New York, 1970.

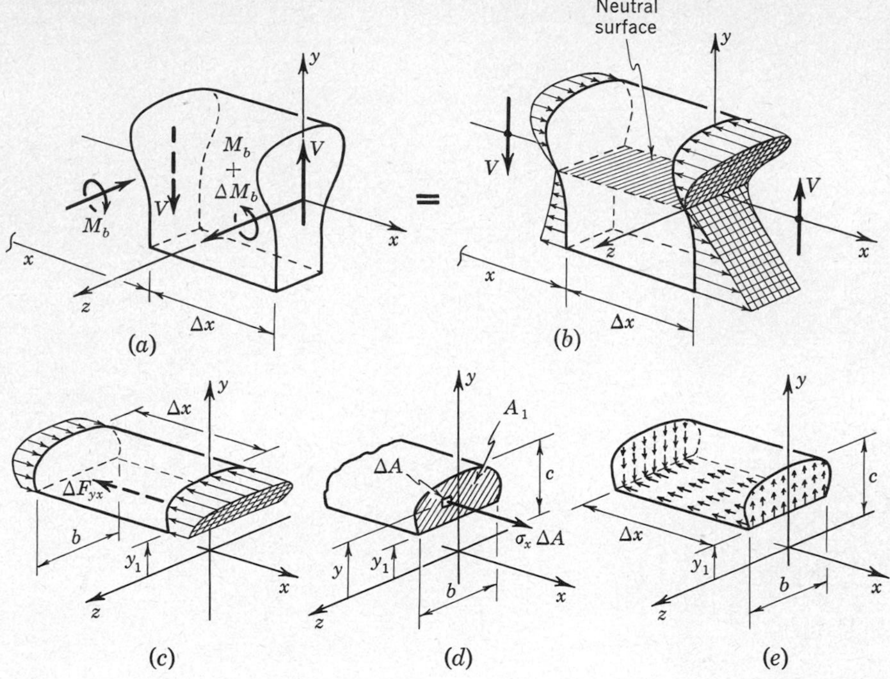

Fig. 7.13 Calculation of shear stress τ_{xy} in a symmetrical beam from equilibrium of a segment of the beam.

transverse load acting on the element so that the transverse shear force V is independent of x. A variation in bending moment with x is represented by the increment ΔM_b. We assume that the bending stresses are given by (7.16). As indicated diagrammatically in the sketch in Fig. 7.13*b*, due to the increase ΔM_b in the bending moment over the length Δx, the bending stresses acting on the positive x face of the beam element will be somewhat larger than those on the negative x face. We next consider the equilibrium of the segment of the beam shown in Fig. 7.13*c*, which we obtain by isolating that part of the beam element of Fig. 7.13*b* above the plane defined by $y = y_1$. Due to the unbalance of bending stresses on the ends of this segment, there must be a force ΔF_{yx} acting on the negative y face to maintain force balance in the x direction. We show this positive shear force ΔF_{yx} in the negative x direction consistent with the fact that the face on which it acts is a negative y face.

Expressing this equilibrium requirement in quantitative form, we have

$$\Sigma F_x = \left[\int_{A_1} \sigma_x \, dA\right]_{x+\Delta x} - \Delta F_{yx} - \left[\int_{A_1} \sigma_x \, dA\right]_x = 0 \tag{7.18}$$

where the integrals are to be taken over the shaded area A_1 in Fig. 7.13*d*, that is, over the range from $y = y_1$ to $y = c$. Substituting (7.16) in (7.18), we find

$$\Delta F_{yx} = -\int_{A_1} \frac{(M_b + \Delta M_b)y}{I_{yy}}\,dA + \int_{A_1} \frac{M_b y}{I_{yy}}\,dA$$

$$= -\frac{\Delta M_b}{I_{yy}} \int_{A_1} y\,dA \tag{7.19}$$

Dividing both sides of (7.19) by Δx and taking the limit, we obtain

$$\frac{dF_{yx}}{dx} = \lim_{\Delta x \to 0} \frac{\Delta F_{yx}}{\Delta x} = -\frac{dM_b}{dx}\frac{1}{I_{yy}} \int_{A_1} y\,dA \tag{7.20}$$

Substituting the moment equilibrium requirement (3.12) for the beam element of Fig. 7.13*a*,

$$\frac{dM_b}{dx} = -V \tag{3.12}$$

we get the result

$$\frac{dF_{yx}}{dx} = \frac{V}{I_{yy}} \int_{A_1} y\,dA \tag{7.21}$$

This result may be written more concisely by introducing the following abbreviations:

$$q_{yx} = \frac{dF_{yx}}{dx}$$
$$Q = \int_{A_1} y\,dA \tag{7.22}$$

The quantity q_{yx}, which is the total longitudinal shear force transmitted across the plane defined by $y = y_1$ per unit length along the beam, is called the *shear flow*. This is the same sense in which the term shear flow was used in the discussion of torsion of hollow, thin-walled shafts in Sec. 6.14. The integral Q is simply the first moment of the shaded area A_1 in Fig. 7.13*d* about the neutral surface. Introducing (7.22) into (7.21), we obtain as the relation for shear flow due to bending

$$q_{yx} = \frac{VQ}{I_{yy}} \tag{7.23}$$

If in the presence of a shear force the bending stresses are in fact given by (7.16), then (7.23) is an exact expression for the shear flow. The shear flow q_{yx} obviously is the resultant of a shear stress τ_{yx} distributed across the width b of the beam. Our derivation gives us no information as to the nature of this distribution, but if we make the *assumption* that the shear stress is uniform across the beam, we can estimate the shear stress τ_{yx} at $y = y_1$ to be

$$\tau_{yx} = \frac{q_{yx}}{b} = \frac{VQ}{bI_{yy}} \tag{7.24}$$

Finally, making use of the moment equilibrium requirement (4.12), we estimate the shear stress on the x faces to be uniform across the width of the beam and of magnitude

$$\tau_{xy} = \tau_{yx} = \frac{VQ}{bI_{yy}} \tag{7.25}$$

In Fig. 7.13*e* we show on the faces of the segment the nature of the shear-stress distribution given by (7.25). The stress has a constant value along the negative y face, and on the x faces it varies from this constant value at the bottom to zero at the top.

The foregoing theory can be proved to be internally consistent in that it can be shown that for a beam of arbitrary cross section the resultant of the stress distribution (7.25) over the cross section is in fact the shear force V (see Prob. 7.43).

In summary, we have assumed that the bending stresses in a beam with a variable moment distribution are given by (7.16). On the basis of this assumption and from consideration of equilibrium, we found the shear-flow distribution (force per unit length) given by (7.23). Finally, we assumed the shear stress to be uniform across the width of the beam to obtain the shear-stress distribution on the x face given by (7.25) (where both Q and b are functions of y_1).

SHEAR-STRESS DISTRIBUTION IN RECTANGULAR BEAMS

An alternative procedure for obtaining the shear-stress distribution in a beam of *rectangular* cross section is possible if we assume from the start that the shear stresses are distributed uniformly across the width. The stress distribution in Fig. 7.14 is then a case of plane stress and the equilibrium equations (4.13) apply.

$$\begin{aligned} \frac{\partial \sigma_x}{\partial x} + \frac{\partial \tau_{xy}}{\partial y} &= 0 \\ \frac{\partial \tau_{xy}}{\partial x} + \frac{\partial \sigma_y}{\partial y} &= 0 \end{aligned} \tag{4.13}$$

If we deal with a case where the shear force does not vary with x, the shear stress also will be independent of x, and the second of (4.13) is automatically satisfied since the normal stress σ_y has been assumed to be zero [cf. Eq. (7.8)]. Substituting (7.16) into the first of (4.13) and using (3.12), we obtain

$$-\frac{\partial \tau_{xy}}{\partial y} = \frac{\partial}{\partial x}\left(-\frac{M_b y}{I_{yy}}\right) = \frac{V}{I_{yy}}\,y \tag{7.26}$$

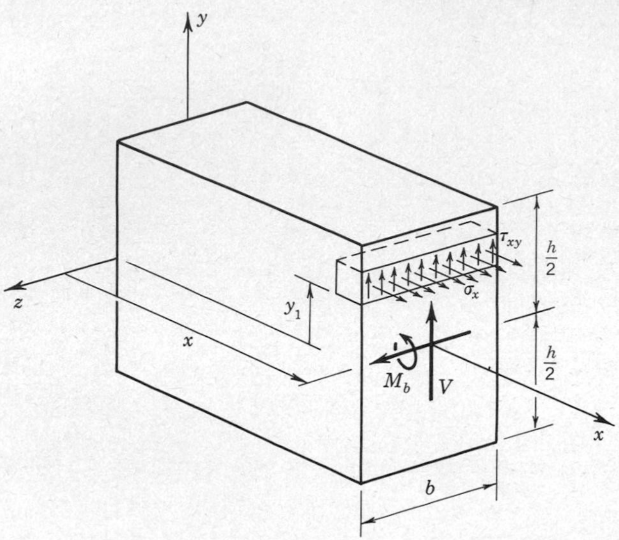

Fig. 7.14 Calculation of shear stress τ_{xy} in a rectangular beam from the equilibrium equations (4.13).

We now integrate (7.26) with respect to y from the level $y = y_1$, where τ_{xy} is to be evaluated, to the level $y = h/2$, which defines the top surface of the beam.

$$-\int_{y_1}^{h/2} \frac{\partial \tau_{xy}}{\partial y}\, dy = \frac{V}{I_{yy}} \int_{y_1}^{h/2} y\, dy$$

$$-\Big[\tau_{xy}\Big]_{y_1}^{h/2} = \frac{V}{I_{yy}} \left[\frac{y^2}{2}\right]_{y_1}^{h/2}$$

Because there is no shear stress on the exposed top surface of the beam, τ_{xy} is zero at $y = h/2$. Thus, when the limits are substituted, we obtain the following result for the shear stress at $y = y_1$:

$$\tau_{xy} = \frac{V}{2I_{yy}} \left[\left(\frac{h}{2}\right)^2 - y_1^{\,2}\right] \tag{7.27}$$

The shear stress is a *maximum* at the *neutral* surface and falls off parabolically, as illustrated in Fig. 7.15. The reader should verify that, in the case of a rectangular cross section, (7.25) reduces to the same distribution as (7.27).

If we calculate the shear strain in a rectangular beam by substituting the stress distribution (7.27) into Hooke's law (5.2), we find that the shear strain γ_{xy} also varies parabolically across the section, from a maximum at the neutral surface to zero at the top and bottom. This implies that the originally plane cross sections distort in the manner illustrated in Fig. 7.16. We note that if the shear force is constant along the length of the beam, any longitudinal line IJ does not change its

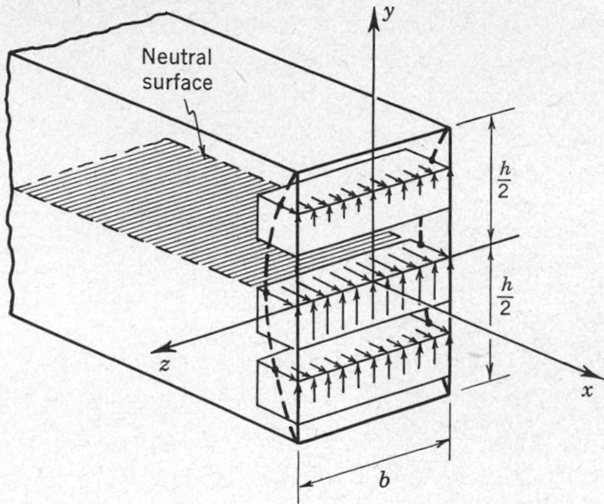

Fig. 7.15 Illustration of parabolic distribution of shear stress τ_{xy} in a rectangular beam.

length as it deforms into the position I_1J_1. From this we would suppose that the presence of a constant shear force would have little effect on the bending-stress distribution (7.16). This is, in fact, the case; the exact solution from the theory of elasticity shows that (7.14) and (7.16) are still correct when there is a constant shear force. As stated previously, this means that the expression (7.23) for the shear flow is also exact for the case of constant shear force. Both (7.14) and (7.16) are in error when the shear force *varies* along the beam, but the magnitude of error is small for long, slender beams and, consequently, (7.23) represents a good estimate

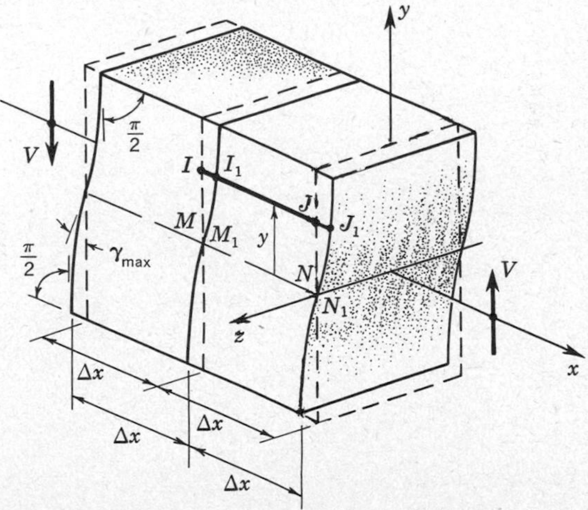

Fig. 7.16 Distortion of rectangular beam due to shear force which is constant along the length of the beam.

even in the presence of a *varying shear force*. Thus at each section of a beam in the *engineering theory of elastic beams*, we take the bending stresses to be distributed according to (7.16) and the transverse shear stresses to be distributed according to (7.25) independently of how $M_b(x)$ and $V(x)$ vary along the length of the beam.

SHEAR-STRESS DISTRIBUTION IN I BEAMS

If we examine the I beam in Fig. 7.17, we gain further insight into the shear-stress distribution in beams. In Fig. 7.17*b* we show a small segment which has been cut from the top flange by a vertical plane through BC. We see that here there must be a shear force ΔF_{zx} on the *positive* z face to maintain equilibrium in the x direction. If we carry through an analysis similar to that which led to (7.23), we obtain for the shear flow on the positive z face the result

$$q_{zx} = -\frac{VQ}{I_{yy}} \tag{7.28}$$

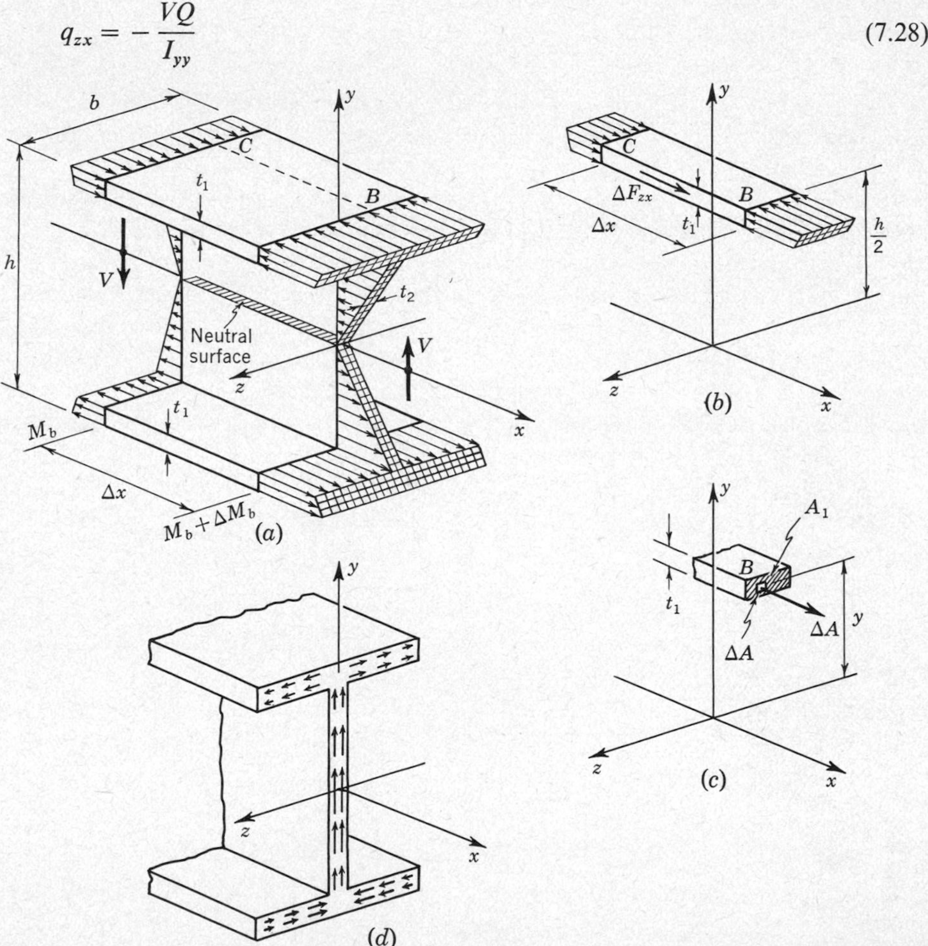

Fig. 7.17 Calculation of shear stress in an I beam.

where Q is the first moment of the shaded area A_1 in Fig. 7.17*c* about the z axis. If we make the assumption that the shear stress is uniform across the thickness t_1 of Fig. 7.17*b* (which becomes a better and better approximation as the section is thinner), we can estimate the shear stress at the point B in the flange to be

$$\tau_{xz} = \tau_{zx} = \frac{q_{zx}}{t_1} = -\frac{VQ}{t_1 I_{yy}} \tag{7.29}$$

The shear stress τ_{xy} in the web can be estimated from (7.25). In Fig. 7.17*d* we show the shear-stress distribution over the cross section of the beam; in each flange the stress τ_{xz} varies linearly from a maximum at the junction with the web to zero at the edge, while in the web the stress τ_{xy} has a parabolic distribution. There also are τ_{xy} stresses in the flanges, but they are small compared with the τ_{xz} stresses illustrated in the sketch. The stress distribution at the junction of the web and flange is quite complicated; standard rolled I beams are provided with generous fillets at these points to reduce the stress concentration.

It is worthwhile to emphasize again that in the engineering theory of beams the *shear flows in bending* are obtained simply from the *equilibrium requirement of force balance along the axis of the beam*. Another illustration of this calculation is given in the following example.

Example 7.3 In making the brass beam of Fig. 7.18*a*, the box sections are soldered to the ¼-in. plate, as indicated in Fig. 7.18*b*. If the shear stress in the solder is not to exceed 1,500 psi, what is the maximum shear force which the beam can carry?

The solder has to carry the unbalanced bending stress acting over the cross-sectional area of the box shown in Fig. 7.18*c*. Assuming that the shear flow is equal in each solder joint, the shear flow q_{zx} in each joint is given by (7.28)

$$2q_{zx} = \frac{VQ}{I_{yy}} \tag{a}$$

where Q is the first moment of the shaded area in Fig. 7.18*c* about the z axis. If we approximate this area by the complete square annulus, we have

$$Q = 5 \text{ in.}[(2 \text{ in.})^2 - (1.75 \text{ in.})^2] = 4.7 \text{ in.}^3 \tag{b}$$

Assuming that the shear stress has the constant value of 1,500 psi across each solder joint, the shear flow in each joint is

$$q_{zx} = (1{,}500 \text{ psi})(\tfrac{1}{8} \text{ in.}) = 188 \text{ lb/in.} \tag{c}$$

Substituting these values into (*a*), we obtain

$$V = \frac{2q_{zx}I_{yy}}{Q} = \frac{2(188 \text{ lb/in.})(75 \text{ in.}^4)}{4.7 \text{ in.}^3} = 6{,}000 \text{ lb}$$

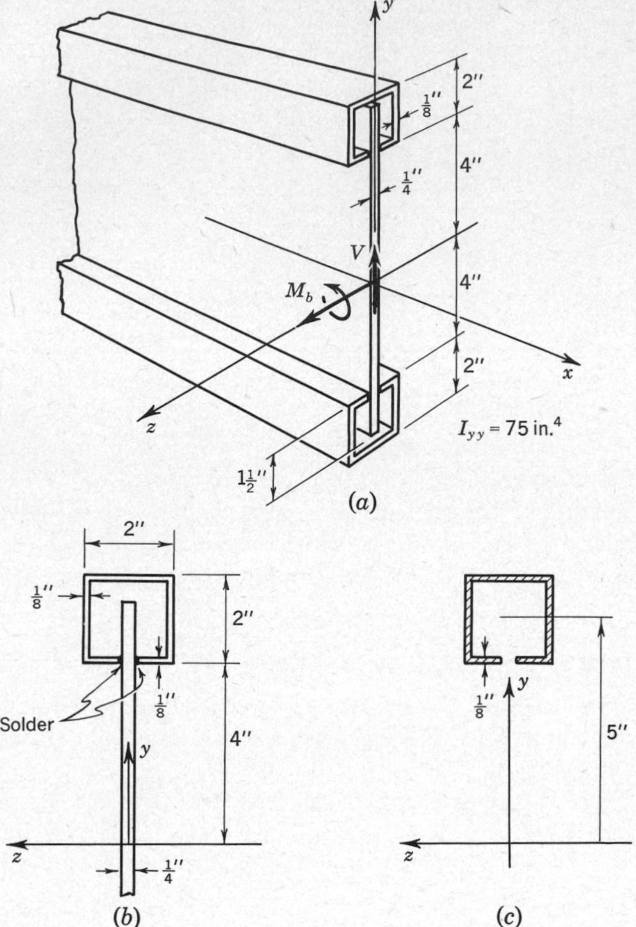

Fig. 7.18 Example 7.3.

as the maximum shear force the beam can carry without exceeding an average shear stress of 1,500 psi in the solder joints.

COMPARATIVE MAGNITUDES OF BENDING AND SHEAR STRESSES

It is of interest to investigate the comparative magnitudes of the bending stresses and the shear stresses in beams. The ratio between the maximum values of these stresses will be different for different beams and for different types of loadings on the same beam, but we can get an idea of the important factors which affect this ratio by investigating a specific example.

Example 7.4 A rectangular beam is carried on simple supports and subjected to a central load, as illustrated in Fig. 7.19. We wish to find the ratio of the maximum shear stress $(\tau_{xy})_{max}$ to the maximum bending stress $(\sigma_x)_{max}$.

The maximum bending stress occurs at mid-span where the bending moment has its maximum value of

$$M_b = \frac{PL}{4} \tag{a}$$

The bending stresses are of equal magnitude on the top and the bottom of the beam (compression on the top and tension on the bottom). From Example 7.1 we have

$$I_{yy} = \frac{bh^3}{12} \tag{b}$$

Substituting (*a*) and (*b*) in (7.16), we find the bending stress at the bottom ($y = -h/2$) to be

$$(\sigma_x)_{max} = \frac{-(PL/4)(-h/2)}{bh^3/12} = \frac{3}{2}\frac{PL}{bh^2} \tag{c}$$

The shear force has the constant magnitude $P/2$ between the load and each support. The shear stress is a maximum at the neutral surface, i.e., at the mid-height of the beam, as illustrated in Fig. 7.19. Substituting $y_1 = 0$ in (7.27), we find

$$(\tau_{xy})_{max} = \frac{P/2}{2(bh^3/12)}\left[\left(\frac{h}{2}\right)^2 - 0^2\right] = \frac{3}{2}\frac{P/2}{bh} = \frac{3}{4}\frac{P}{bh} \tag{d}$$

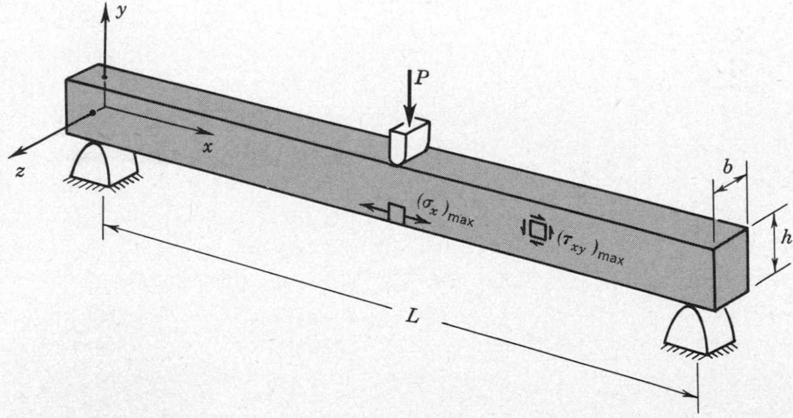

Fig. 7.19 Example 7.4. Rectangular beam on simple supports and with a central load.

We note that (d) states that the *maximum* shear stress in a rectangular beam is one and one-half times the *average* shear stress.

Combining (c) and (d), we get the ratio of the maximum shear stress to the maximum bending stress in the beam of Fig. 7.19.

$$\frac{(\tau_{xy})_{\max}}{(\sigma_x)_{\max}} = \frac{1}{2}\frac{h}{L} \tag{e}$$

Thus the bending and shear stresses are of comparable magnitude only when L and h are of the same magnitude. Since L is much greater than h in most beams (say, $L > 10h$), it may be seen from (e) that the shear stresses τ_{xy} will usually be an order of magnitude smaller than the bending stresses σ_x.

If a different loading is put on the beam in Fig 7.19, the ratio of the maximum stresses will again be found to depend upon the *ratio of the depth to the length of the beam*, although, of course, the factor of proportionality will differ from that just found (see Prob. 7.44). If beams of other cross-sectional shape are investigated. similar results are obtained. The factor of proportionality does, however, depend importantly on the shape of the section; e.g., the factor of ½ in (e) can be as large as 3 or 4 for I beams with thin webs (see Prob. 7.45).

LOCALIZED BUCKLING IN I BEAMS

From the point of view of reducing bending stress, it is apparent from (7.16) that for a given cross-sectional area of beam it is best to distribute that area so that I_{yy} is as large as practical, i.e., to concentrate the area as far as possible from the centroid. Rolled steel beams, of which the I beam of Fig. 7.17 is an example, are designed to have this feature. If the design is pushed too far in this direction, however, buckling phenomena will be encountered. For instance, if the cross-sectional area of the I beam of Fig. 7.17 was kept constant while the depth was increased at the expense of a decrease in the flange thickness, the beam might fail by a buckling of the compression flange at a stress level well below that at which the material would yield. On the other hand, if an increase in beam depth was accomplished at the expense of a decrease in web thickness, the compressive stresses resulting from the transmission of shear along the beam might cause buckling of the web. The orientation of these compressive stresses for an element at the neutral surface is shown in Fig. 7.20 (see Example 7.5). It is possible to make reasonable estimates[1] of the onset of localized buckling in these cases, but we shall not pursue the problem further here because the details of these analyses are beyond the scope of this book. We shall, however, return in Chap. 9 to a general discussion of buckling phenomena.

[1] F. R. Shanley, "Strength of Materials," pp. 618, 624, McGraw-Hill Book Company, New York, 1957.

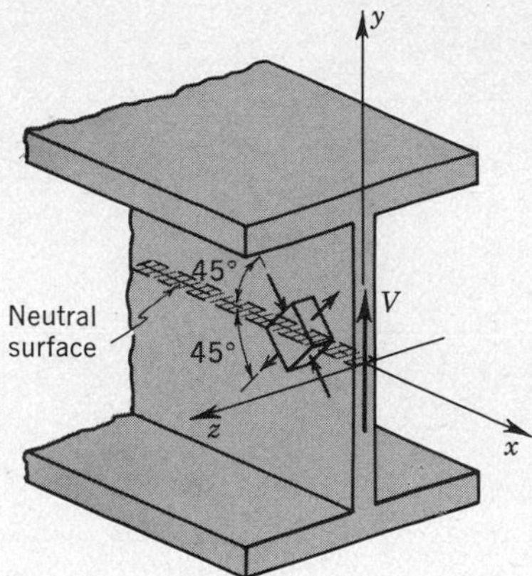

Fig. 7.20 Illustration of compressive and tensile stresses acting on an element at the neutral surface in the web of an I beam transmitting a shear force (see Example 7.5).

7.7 STRESS ANALYSIS IN BENDING; COMBINED STRESSES

If one is interested in the possibility of yielding, or fracture, or localized buckling of a beam, it is necessary to examine the stress distribution in detail and to determine the *state of stress* at critical points in the beam. To illustrate the type of analysis involved when a beam transmits both shear force and bending moment, we consider the following example.

Example 7.5 It is desired to investigate the state of stress at points B and C in the top flange and web of the I beam of Fig. 7.21*a* when a shear force V and a bending moment M_b are being transmitted.

The bending stress σ_x is given by (7.16), and the shear stress τ_{xy} is given by (7.25). The magnitude of the shear stress τ_{xz} is given by (7.29); its sense is shown in Fig. 7.17*d*. These stress components are shown in Fig. 7.21*a*. The Mohr's circles for stress drawn in Fig. 7.21*b* and *c* are constructed from these components and lead to the principal stresses σ_1 and σ_2 which are indicated in Fig. 7.21*d*. To determine the most critical point in the cross section, it would be necessary to have a criterion for comparison (e.g., a yield criterion, a fracture criterion, or a buckling criterion) and then to compare the states of stress at B and C as the points take on all possible positions in the flange and the web. This can be a lengthy analysis. In long, slender beams, as pointed out in Example 7.4, the greatest shear stress τ_{xy} or τ_{xz} is an order of magnitude smaller than the greatest bending stress σ_x. Hence for many practical purposes we can neglect the shear-stress contribution, except, as we noted previously, for consideration of possible buckling.

A frequent type of loading condition is one in which a beam transmits a longitudinal force and/or a twisting moment in addition to a shear force and a bending moment. When there is a longitudinal force in addition to shear and

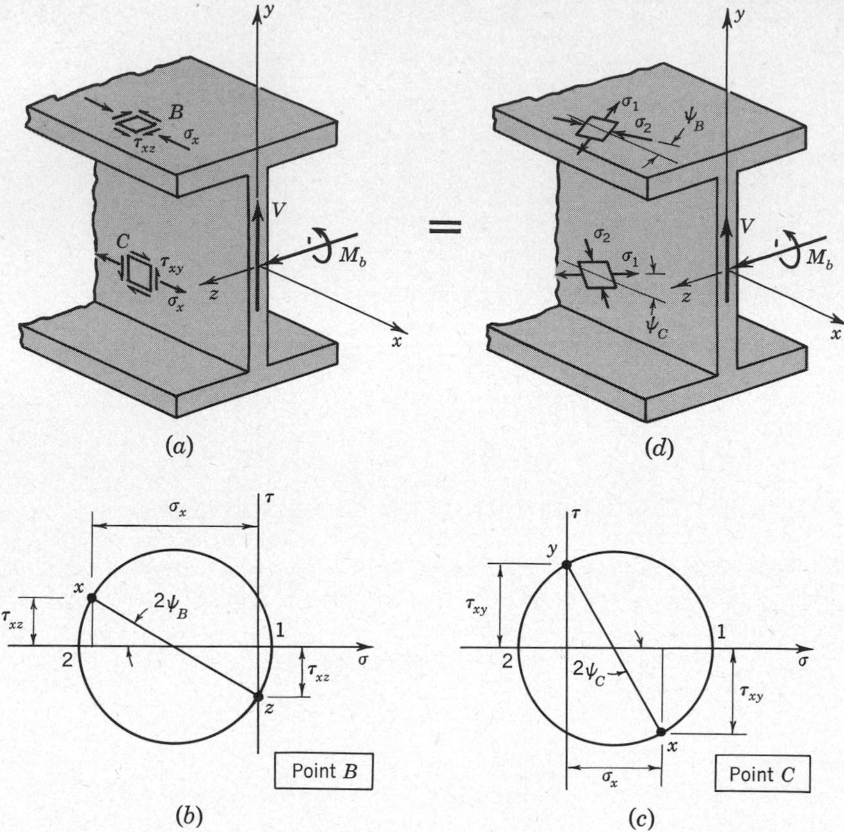

Fig. 7.21 Example 7.5. Combined stresses in an I beam due to shear and bending.

bending, the analysis is similar to that outlined in Example 7.5, with the addition of a uniform axial stress due to the longitudinal force. In the following example we consider the combination of bending moment, twisting moment, and longitudinal force.

Example 7.6 In Fig. 7.22*a* an elastic circular shaft is shown transmitting simultaneously a bending moment M_b, an axial tensile force P, and a twisting moment M_t. We wish to study the state of combined stress.

This problem represents the addition of a bending moment to the loading condition considered in Example 6.3, and our method of approach is exactly parallel. In Fig. 7.22*b* the individual stress distributions are sketched for the separate loads. Due to the bending moment, we have the distribution

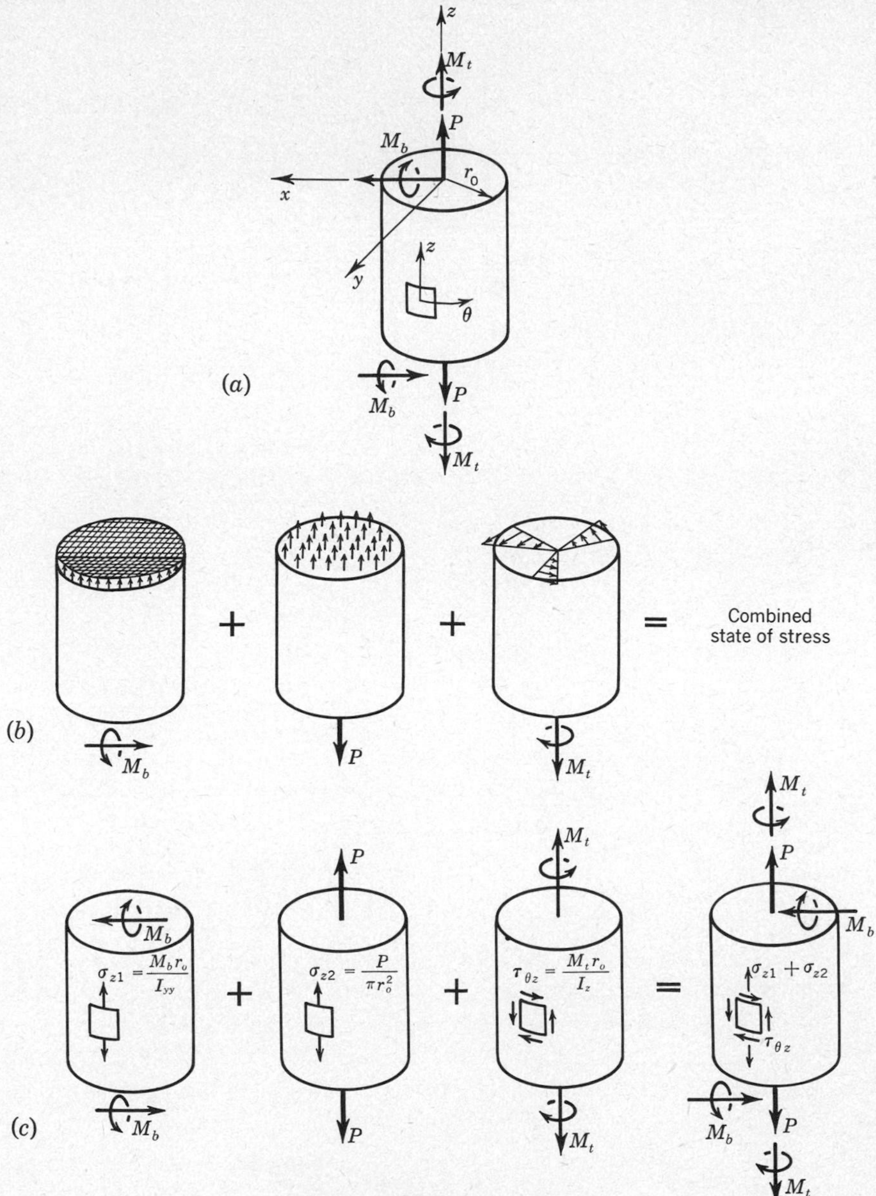

Fig. 7.22 Example 7.6. Combined stresses in a solid circular member due to bending, tension, and torsion.

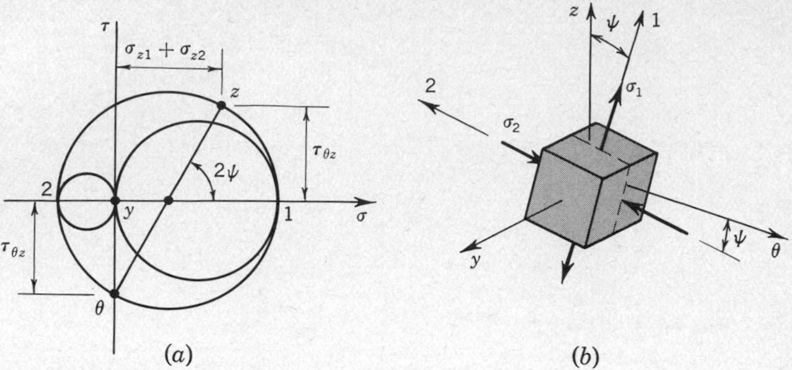

Fig. 7.23 Example 7.6. Principal directions and principal stresses at the most critically stressed point.

given by (7.16) (note that M_b in Fig. 7.22*a* is such as to cause tension when y is positive).

$$\sigma_{z1} = +\frac{M_b y}{I_{yy}} \tag{a}$$

Due to the axial tensile force P, we have a uniform axial stress.

$$\sigma_{z2} = \frac{P}{\pi r_0^{\,2}} \tag{b}$$

Finally, due to the twisting moment, we have the distribution given by (6.9).

$$\tau_{\theta z} = \frac{M_t r}{I_z} \tag{c}$$

In Fig. 7.22*c* we show the individual stresses acting on an element at $y = r_o$. (At this element the distributions (*a*) and (*c*) have their greatest magnitude.) These are superposed to give the resultant combined stress state at the right. The Mohr's circle diagram for this point is sketched in Fig. 7.23*a*, and the principal stress directions are indicated in Fig. 7.23*b*. Note that this element has an unloaded surface, and thus the state is one of *plane stress*.

7.8 STRAIN ENERGY DUE TO BENDING

The strain energy in a linearly elastic body subjected to an arbitrary distribution of stress and strain is given by (5.17). In this section we specialize that result to beams subjected to bending. We consider first the case of pure bending where the only nonvanishing stress component is the longitudinal stress (7.16). The total strain energy (5.17) thus reduces to

$$U = \tfrac{1}{2}\iiint \sigma_x \epsilon_x \, dx\, dy\, dz = \iiint \frac{\sigma_x^2}{2E}\, dx\, dy\, dz \tag{7.30}$$

where the integration is over the entire volume of the beam. Substituting from (7.16) for σ_x, we find

$$U = \iiint \frac{1}{2E}\left(\frac{M_b y}{I_{yy}}\right)^2 dx\, dy\, dz = \int_L \frac{M_b^2}{2EI_{yy}^2}\, dx \iint_A y^2\, dy\, dz$$

where the integrations are over the length L and cross-sectional area A of the beam. Since the latter integral is just the second moment of area I_{yy}, the bending strain energy is

$$U = \int_L \frac{M_b^2}{2EI_{yy}}\, dx \tag{7.31}$$

This formula may also be derived by considering each differential element of length dx to act as a bending spring. If, in Fig. 7.24, the final values of the bending moment and bending angle are M_b and $d\phi$, respectively, the work done during a loading process in which these grow in proportion is

$$dU = \tfrac{1}{2}M_b\, d\phi = \tfrac{1}{2}M_b \frac{d\phi}{dx}\, dx$$

When (7.14) is substituted for $d\phi/dx$ and the result is integrated over the length of the beam, we obtain (7.31) again.

When a beam is subjected to transverse shear in addition to bending, there are, in general, transverse shear-stress components τ_{xy} and τ_{xz} in addition to the bending stress σ_x; e.g., see Fig. 7.17*d*. The total strain energy (5.17) then becomes

$$\begin{aligned} U &= \tfrac{1}{2}\iiint (\sigma_x\epsilon_x + \tau_{xy}\gamma_{xy} + \tau_{xz}\gamma_{xz})\, dx\, dy\, dz \\ &= \iiint \frac{\sigma_x^2}{2E}\, dx\, dy\, dz + \iiint \frac{\tau_{xy}^2 + \tau_{xz}^2}{2G}\, dx\, dy\, dz \end{aligned} \tag{7.32}$$

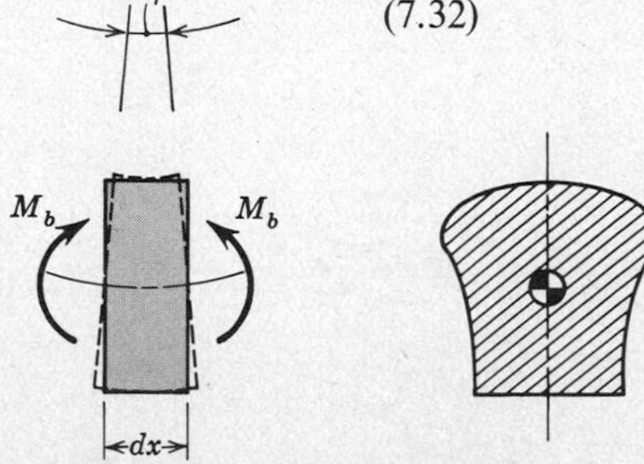

Fig. 7.24 Differential element of beam bends through angle $d\phi$ under action of bending moment M_b.

The first integral on the right is identical in form with the strain energy (7.30) in pure bending. The second integral is the contribution to the strain energy due to the transverse shear stresses. For slender members the latter contribution is almost always negligible in comparison with the former. This may be inferred from the discussion in Sec. 7.6 concerning the comparative magnitudes of the bending and shear stresses. If σ_x is an order of magnitude larger than τ_{xy} and τ_{xz}, then, since the integrals in (7.32) depend on the squares of the stresses, we see that the first integral is *two* orders of magnitude larger than the second. As a consequence, it is common to neglect the contribution to the strain energy due to the transverse shear stresses. The pure-bending formula (7.31) is then used to represent the total strain energy in a beam whether there is transverse shear or not. In any particular case the magnitude of the second integral in (7.32) can be estimated if an approximate distribution of τ_{xy} and τ_{xz} can be determined.

As an illustration we return to the rectangular beam of Fig. 7.19 to compute the total strain energy and to compare the relative magnitudes of the bending and shear contributions. The bending moment $M_b(x)$ and shear force $V(x)$ for the loading of Fig. 7.19 are

$$M_b(x) = P(x/2 - \langle x - L/2\rangle^1)$$
$$V(x) = P(-1/2 + \langle x - L/2\rangle^0)$$

for $0 < x < L$. The distribution of transverse shear stress in a beam with rectangular cross section, obtained in Sec. 7.6, is

$$\tau_{xy} = \frac{V(x)}{2I_{yy}}\left[\left(\frac{h}{2}\right)^2 - y^2\right]$$
$$\tau_{xz} = 0$$

for $0 < x < L$, $-h/2 < y < h/2$, and $-b/2 < z < b/2$. If we denote by U_b the first integral on the right of (7.32) and use (7.31) to evaluate it, we have

$$U_b = \int_0^L \frac{M_b{}^2}{2EI_{yy}}\,dx = 2\int_0^{L/2} \frac{(Px/2)^2}{2EI_{yy}}\,dx$$

because of the symmetry of the bending-moment diagram. Thus the *bending contribution* to the strain energy is

$$U_b = \frac{P^2L^3}{96EI_{yy}} = \frac{P^2L^3}{8Ebh^3} \tag{7.33}$$

where we have used $I_{yy} = bh^3/12$. If we denote by U_s the second integral on the right of (7.32), the *transverse shear contribution* to the energy is

$$U_s = \int_0^L \frac{V^2\,dx}{8GI_{yy}{}^2}\int_{-h/2}^{h/2}\left[\left(\frac{h}{2}\right)^2 - y^2\right]^2 dy \int_{-b/2}^{b/2} dz$$
$$= \frac{P^2Lbh^5}{960GI_{yy}{}^2} = \frac{3}{20}\frac{P^2L}{Gbh} \tag{7.34}$$

The total strain energy in the beam is the sum of these contributions.

$$U = U_b + U_s = \frac{P^2L^3}{8Ebh^3}\left[1 + \frac{6}{5}\frac{E}{G}\left(\frac{h}{L}\right)^2\right] \tag{7.35}$$

Note that the ratio of the two contributions is

$$\frac{U_s}{U_b} = \frac{6}{5}\frac{E}{G}\left(\frac{h}{L}\right)^2 = \frac{12}{5}(1 + \nu)\left(\frac{h}{L}\right)^2$$

Thus, for a beam with $L > 10h$ and with Poisson's ratio $\nu = 0.28$, the shear contribution is less than 3 percent of the bending contribution. For engineering purposes it would generally be permissible to neglect the transverse shear contribution in evaluating the strain energy of such a beam.

For beams with other loadings and other cross-sectional shapes, the ratio of U_s to U_b is always proportional to the square of the ratio of beam depth to beam length. The factor of proportionality does, however, depend on the loading pattern and on the shape of the cross section; e.g., the numerical factor of $6/5$ in (7.35) can be as large as 12 for I beams.

7.9 THE ONSET OF YIELDING IN BENDING

When a beam is subjected to *pure* bending, the state of stress is one in which the principal stresses are given by

$$\sigma_1 = \sigma_x \qquad \sigma_2 = \sigma_3 = 0 \tag{7.36}$$

which is the same state of stress that exists in a tension (or compression) test. Thus if the material has a yield stress Y in simple tension, the criterion for yielding in pure bending is simply that yielding will occur when

$$\sigma_x = Y \tag{7.37}$$

As soon as the state of stress becomes more complicated (e.g., by the addition of shear forces or twisting moments or longitudinal forces), the onset of yielding must include the type of combined stress analysis outlined in Sec. 7.7. There are two criteria available to signal the onset of yielding: the Mises criterion (5.23) and the maximum shear-stress criterion (5.25). Even in relatively simple structures the most critically stressed point may not be obvious, and calculations may have to be made for more than one point, as illustrated in the following example.

Example 7.7 A circular rod of radius r is bent into the shape of a U to form the structure of Fig. 7.25*a*. The material in the rod has a yield stress Y in simple tension. We wish to determine the load P that will cause yielding to begin at some point in the structure.

As indicated in Fig. 7.25*b*, there are five possible locations for the most critically stressed point. The bending and twisting moments acting at these locations are shown in Fig. 7.25*c* through *g*. From these the choice narrows down to either location B_1 or B_2; it is not obvious which of these is the more critical.

The bending and torsional shear stresses acting on an element on the top of the beam at location B_1 are indicated in Fig. 7.26*a*. Although there is a shear force *P* at this location, the corresponding shear stress τ_{xy} is zero

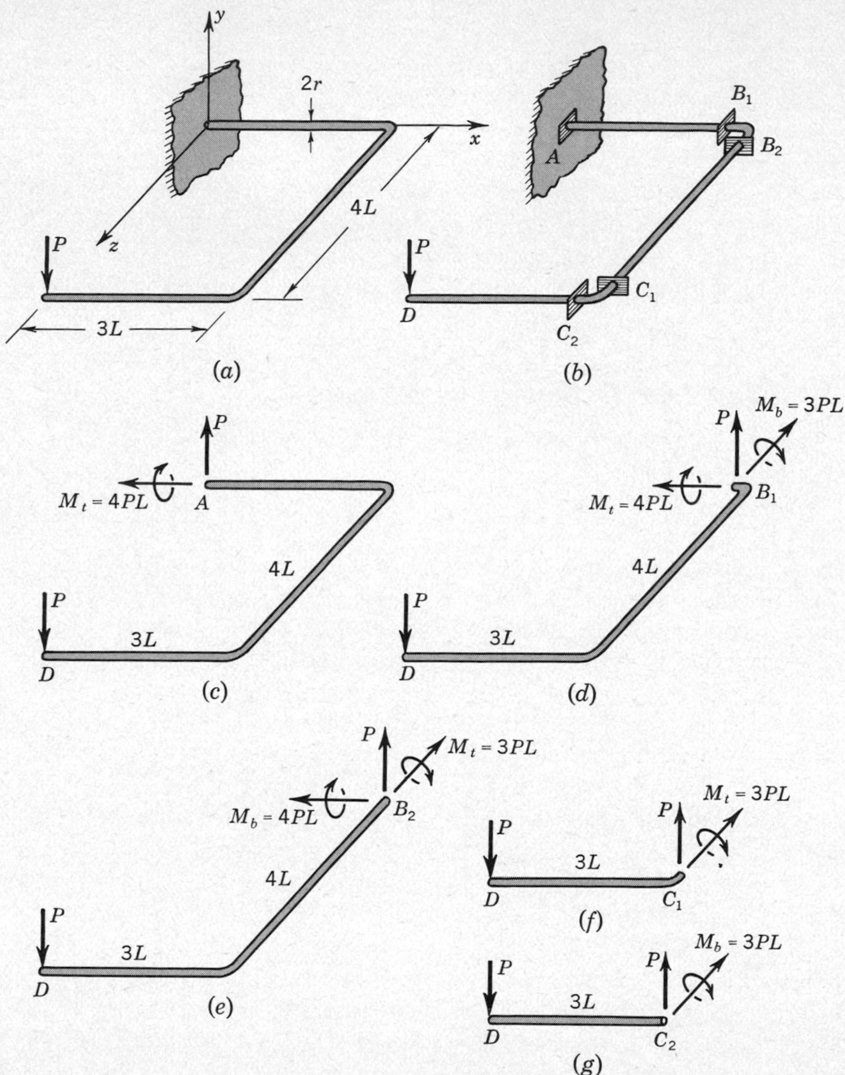

Fig. 7.25 Example 7.7. Bending and twisting moments at five critical locations in a structure.

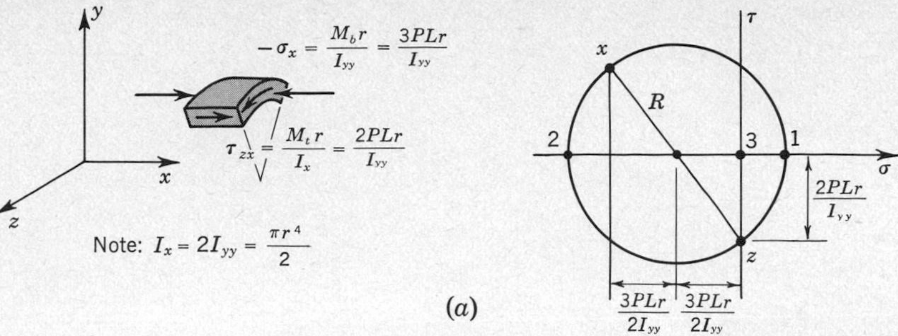

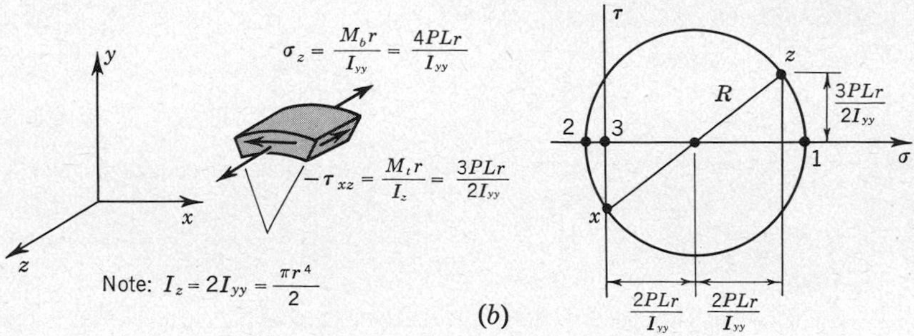

Fig. 7.26 Example 7.7. (*a*) Maximum stress condition at location B_1; (*b*) maximum stress condition at location B_2.

at the top and bottom of the beam where the bending and torsional shear stresses are maximum. The radius of the Mohr's circle for the element at B_1, shown in Fig. 7.26*a*, is

$$R = \sqrt{\left(\frac{3}{2}\frac{PLr}{I_{yy}}\right)^2 + \left(\frac{2PLr}{I_{yy}}\right)^2} = \frac{5}{2}\frac{PLr}{I_{yy}} \tag{a}$$

Using (*a*), we find the principal stresses at the point to be

$$\sigma_1 = +\frac{PLr}{I_{yy}} \qquad \sigma_2 = -4\frac{PLr}{I_{yy}} \qquad \sigma_3 = 0 \tag{b}$$

Substituting (*b*) into the Mises yield criterion (5.23),

$$\sqrt{\frac{1}{2}\left[\left(\frac{PLr}{I_{yy}} + 4\frac{PLr}{I_{yy}}\right)^2 + \left(-4\frac{PLr}{I_{yy}} - 0\right)^2 + \left(0 - \frac{PLr}{I_{yy}}\right)^2\right]} = Y \tag{c}$$

we obtain the result that yielding begins when

$$P = 0.218\frac{I_{yy}Y}{Lr} \tag{d}$$

Substituting (*b*) into the maximum shear-stress criterion (5.25),

$$\tau_{max} = \frac{1}{2}\left(\frac{PLr}{I_{yy}} + 4\frac{PLr}{I_{yy}}\right) = \frac{Y}{2} \tag{e}$$

we find that yielding is predicted when

$$P = 0.200\,\frac{I_{yy}Y}{Lr} \tag{f}$$

Note the discrepancy of 9 percent between the loads predicted by the two criteria.

Repeating the foregoing calculations for the element on top of the beam at location B_2, illustrated in Fig. 7.26*b*, we find the principal stresses to be

$$\sigma_1 = +\frac{9}{2}\frac{PLr}{I_{yy}} \qquad \sigma_2 = -\frac{1}{2}\frac{PLr}{I_{yy}} \qquad \sigma_3 = 0 \tag{g}$$

and that according to the Mises criterion, yielding occurs when

$$P = 0.210\,\frac{I_{yy}Y}{Lr} \tag{h}$$

while according to the maximum shear-stress criterion, yielding occurs when

$$P = 0.200\,\frac{I_{yy}Y}{Lr} \tag{i}$$

The maximum shear-stress criterion predicts yielding at locations B_1 and B_2 at the same load, indicating that the Mohr's circles in Fig. 7.26*a* and *b* are of equal size. The Mises criterion identifies B_2 as the critical location and predicts yielding there at a load 5 percent greater than the load for yielding according to the maximum shear-stress criterion.

7.10 PLASTIC DEFORMATIONS

We now consider the behavior of a beam in *pure* bending as the bending moment is increased beyond the value which produces the onset of yielding at the point farthest removed from the neutral surface. We shall restrict our attention to symmetrical beams. We shall further restrict our inquiry to beams in which the material has the elastic–perfectly plastic stress-strain behavior of Fig. 5.7*e*; such a stress-strain diagram is repeated in Fig. 7.27. The Mises and the maximum shear-stress criteria predict yielding at the same bending-stress level since pure bending corresponds to a uniaxial state of stress.

In Fig. 7.28 we illustrate the changes which occur in the bending-stress distribution in a rectangular beam as the curvature is increased. As has been emphasized earlier, the nature of the geometric deformation is independent of the

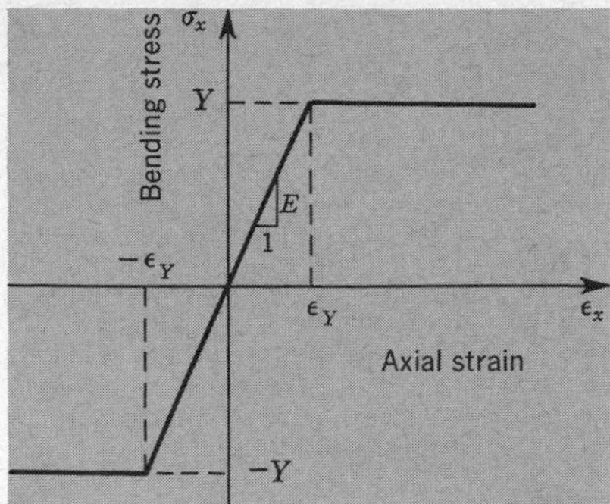

Fig. 7.27 Elastic–perfectly plastic material.

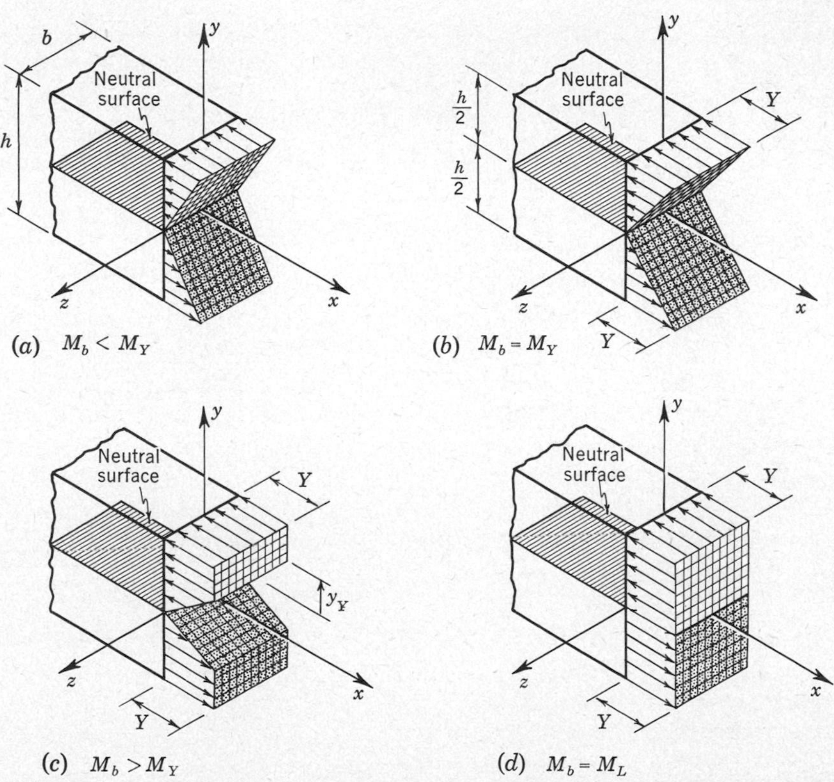

Fig. 7.28 Bending-stress distribution in a rectangular beam of elastic–perfectly plastic material as the curvature is increased until the fully plastic moment M_L is reached at infinite curvature.

stress-strain behavior of the material, and hence (7.4) describes the bending strains throughout the entire range of bending deformation of the beam

$$\epsilon_x = -\frac{y}{\rho} = -\frac{d\phi}{ds}\,y \tag{7.4}$$

In the elastic region, that is, $0 < (\sigma_x)_{\max} < Y$, the moment-curvature relation is given by

$$\frac{d\phi}{ds} = \frac{1}{\rho} = \frac{M_b}{EI_{yy}} \tag{7.14}$$

and the stress distribution is

$$\sigma_x = -\frac{M_b y}{I_{yy}} \tag{7.16}$$

as illustrated in Fig. 7.28*a*. We give the symbol M_Y to the bending moment which corresponds to the onset of yielding in the beam, as shown in Fig. 7.28*b*. M_Y corresponds to the situation where $\sigma_x = -Y$ at $y = +h/2$, and thus from (7.16) we obtain

$$M_Y = \frac{Y(bh^3/12)}{h/2} = \frac{bh^2}{6}\,Y \tag{7.38}$$

We use the notation $(1/\rho)_Y$ to indicate the curvature corresponding to M_Y. Since $\epsilon_x = -\epsilon_Y$ at $y = +h/2$, we can use (7.4) to express this curvature as

$$\left(\frac{1}{\rho}\right)_Y = \frac{\epsilon_Y}{h/2} \tag{7.39}$$

We now examine the behavior as the curvature is increased beyond $(1/\rho)_Y$. As the curvature is increased, the strain increases according to (7.4), and because of the stress-strain behavior shown in Fig. 7.27, the resulting stress distribution is as illustrated in Fig. 7.28*c*. Letting y_Y be the coordinate which defines the extent of the inner elastic region of behavior, we can describe the variation in stress above the neutral surface by

$$\begin{aligned} \sigma_x &= -\frac{y}{y_Y}\,Y \qquad \text{when } 0 < y < y_Y \\ \sigma_x &= -Y \qquad \text{when } y_Y < y < \frac{h}{2} \end{aligned} \tag{7.40}$$

The stress below the neutral surface varies in the same manner but with opposite sign. The stress distributions above and below the neutral surface will contribute equally to the bending moment, so, taking an element of area of size $\Delta A = b\,\Delta y$, we can express the bending moment as (with due regard for the sign convention for stresses)

$$M_b = -\int_A \sigma_x y \, dA$$
$$= 2\left(-\int_0^{y_Y} \sigma_x yb \, dy - \int_{y_Y}^{h/2} \sigma_x yb \, dy\right) \tag{7.41}$$

Substituting (7.40) in (7.41) and performing the integration, we obtain, after simplification, the following result for the bending moment:

$$M_b = \frac{bh^2}{4} Y\left[1 - \frac{1}{3}\left(\frac{y_Y}{h/2}\right)^2\right] \tag{7.42}$$

The strain at y_Y has the value $-\epsilon_Y$, and using this, we obtain from (7.4) the curvature corresponding to the moment given by (7.42).

$$\frac{1}{\rho} = \frac{\epsilon_Y}{y_Y} \tag{7.43}$$

Combining (7.39) and (7.43), we get

$$\frac{y_Y}{h/2} = \frac{(1/\rho)_Y}{1/\rho} \tag{7.44}$$

Finally, substituting (7.38) and (7.44) in (7.42), we find the bending moment to be given by

$$M_b = \frac{3}{2} M_Y\left\{1 - \frac{1}{3}\left[\frac{(1/\rho)_Y}{1/\rho}\right]^2\right\} \tag{7.45}$$

when the curvature $1/\rho$ is greater than $(1/\rho)_Y$.

The variation of M_b with curvature is shown in Fig. 7.29. As the curvature increases, the moment approaches the asymptotic value $\frac{3}{2}M_Y$ which we call the *fully plastic moment*, or *limit moment*, and for which we use the symbol M_L. The stress distribution which produces M_L is shown in Fig. 7.28*d*; the stress has the magnitude Y over the entire cross section. As the curvature increases, M_b approaches M_L more rapidly than y_Y approaches zero; for example, when $y_Y = \frac{1}{6}(h/2)$, the bending moment is within 1 percent of the fully plastic value. Thus, whenever the curvature of a beam is large compared with the curvature $(1/\rho)_Y$ at which yielding begins, we can assume that the bending moment transmitted is essentially the limit moment M_L. The ratio K of the fully plastic moment to the moment which causes the onset of yielding is a function of the geometry of the cross section. Values of this ratio are given in Table 7.1 for a few cross sections.

The theory just presented provides a useful basis for the engineering design of beams which will be loaded into the plastic range. It is not, however, a complete theory within the framework of the mathematical theory of plasticity.

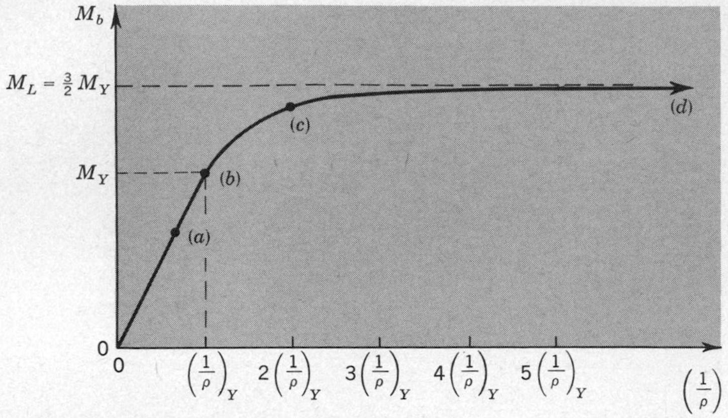

Fig. 7.29 Moment-curvature relation for the rectangular beam of Fig. 7.28. The positions (*a*), (*b*), (*c*), and (*d*) correspond to the stress distributions shown in Fig. 7.28.

When a shear force is present, i.e., when the deformation is no longer pure bending, the plastic behavior is altered somewhat since the state of stress is no longer a simple uniaxial state. However, the results of analyses which include the shear force[1] show that the effect of shear force on the value of the bending moment corresponding to fully plastic behavior is negligible in beams of reasonable length. Consequently, in the engineering theory of plastic bending it is assumed that the bending-stress distributions of Fig. 7.28 and the moment-curvature relation (7.45) are still valid when a shear force is present, i.e., when the bending moment varies along the beam. We turn next to the consideration of such a case.

In Fig. 7.30*a* a hypothetical experiment is sketched. A rectangular beam of elastic–perfectly plastic material is to be forced down at the center by a screw jack. A load cell measures the resulting force P which is transmitted to the beam by the screw jack. The bending moment in the beam varies linearly from zero at the ends to a maximum of $Pa/2$ at the center. The sketches in Fig. 7.30 show

Table 7.1 Ratio of limit bending moment to bending moment at onset of yielding

Cross section	$K = M_L/M_Y$
Solid rectangle	1.5
Solid circle	1.7
Thin-walled circular tube	1.3
Typical I beam	1.1–1.2

[1] P. G. Hodge, Jr., "Plastic Analysis of Structures," p. 213, McGraw-Hill Book Company, New York, 1959.

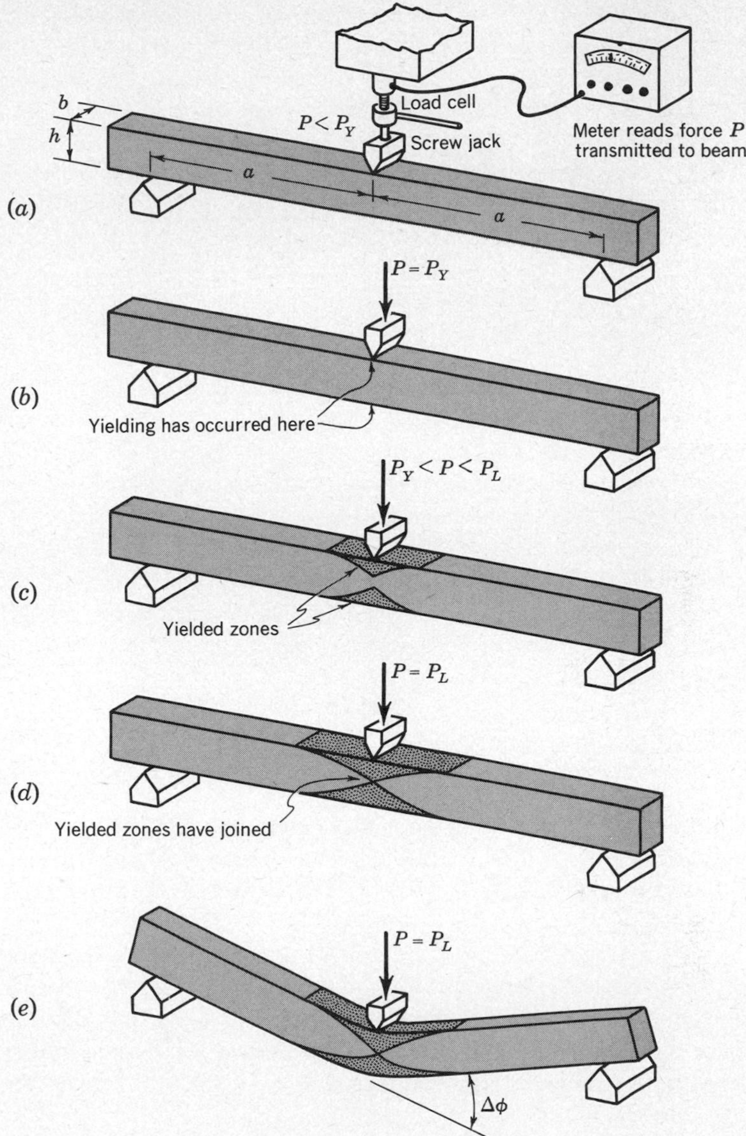

Fig. 7.30 Creation of a plastic hinge as the center of the beam is forced downward by a screw jack. The load cell measures the force P developed by the screw jack.

successive stages as the screw jack gradually depresses the center of the beam. At first the behavior is completely elastic, and the force P increases in linear proportion to the central deflection. In the following chapter we shall obtain the precise relationship between deflection and force for cases like this. Here it is sufficient to say that the deflection involved is very small.

In Fig. 7.30*b* the point has been reached where the central bending moment is M_Y, and yielding begins at the top and bottom of the cross section. The value of the force P corresponding to the onset of yielding is, using (7.38),

$$P_Y = \frac{2}{a} M_Y = \frac{bh^2}{3a} Y \tag{7.46}$$

As the beam is deflected further, two zones of yielded material begin to grow, as shown in Fig. 7.30*c*. These zones extend from the center where the bending moment is greater than M_Y out to the points where the bending moment is just equal to M_Y. The force P continues to grow during this stage of the deflection but at a less rapid rate than when the beam was entirely elastic.

The yielded zones continue to grow as the screw jack is advanced until the configuration of Fig. 7.30*d* is reached. At this point the yielded zones just touch at the center of the beam. The bending moment there is M_L with the stress distribution pictured in Fig. 7.28*d*. The value of P corresponding to the limit bending moment is

$$P_L = \frac{2}{a} M_L = \frac{bh^2}{2a} Y \tag{7.47}$$

Although the curvature at the central point is now infinite, the slope is still finite and continuous and so is the deflection. The central deflection can, in fact, be shown[1] to be only 2.2 times the central deflection corresponding to the onset of yield in Fig. 7.30*b*.

Since the central cross section is now completely plastic, any further motion of the screw jack can be accommodated by plastic flow at the center without any additional deformation of the rest of the beam and without further increase in the force P. Thus further deflection of the beam results in a finite discontinuity in slope at the center, as shown in Fig. 7.30*e*. This localized deformation is called a *plastic hinge*.

Although the hypothetical experiment just discussed was based on the engineering theory of elastic–perfectly plastic beams, it represents a surprisingly accurate model of the behavior of real materials with pronounced yield points. Experiments[2] with beams of 1020 HR steel agree very well with the theory up to

[1] See Prob. 8.63.

[2] J. F. Baker, M. R. Horne, and J. Heyman, "The Steel Skeleton," vol. II, "Plastic Behavior and Design," chap. 3, Cambridge University Press, London, 1956.

moderate values of the hinge angle $\Delta\phi$. For very large deformations the strain-hardening of the material causes deviations from the perfectly plastic theory.

If the loading on a beam is distributed along the beam, the plastic behavior of the beam will differ in some respects from that described above. As the deflection is increased, the maximum bending moment will mathematically only asymptotically approach the value M_L. However, from a practical standpoint, when the deflection is large compared to the deflection at which yielding begins, the maximum bending moment may be taken to be M_L. Large plastic strains will not be restricted to the section of the beam which has the maximum bending moment, but will occur in a localized region extending to either side of the section of maximum bending moment. This region of localized deformation, although less sharply defined than that occurring under a concentrated load, is also called a plastic hinge.

In Sec. 8.7 we shall show how plastic limit analysis can be extended to structures constructed from beams. We shall see that large deformations of a structure require the formation of one or more plastic hinges. The important result that we have developed here is that whenever a plastic hinge forms, the bending moment transmitted across the hinge may be taken to be the limit moment M_L.

In the following example we consider another aspect of the plastic behavior of beams.

Example 7.8 An originally straight rectangular bar is bent around a circular mandrel of radius $R_0 - h/2$, as shown in Fig. 7.31a. As the bar is released from the mandrel, its radius of curvature increases to R_1, as indicated in Fig. 7.31b. This change of curvature is called *elastic springback*; it becomes a factor of great importance when metals must be formed to close dimensional tolerances. Our interest here is in the amount of this springback and in the residual stresses which remain after the bar is released.

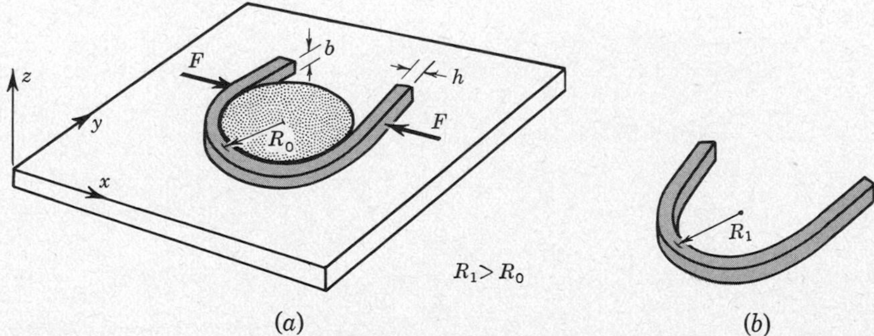

Fig. 7.31 Example 7.8. Illustration of elastic springback which occurs when an originally straight rectangular bar is released after undergoing large plastic bending deformation.

The moment-curvature behavior of the bar is sketched in Fig. 7.32. As the bar is bent around the mandrel, the curve OFA traces out the moment-curvature relation. When the bar is released, the resultant bending moment decreases elastically to zero along the line AC parallel to the original elastic portion OF. The decrease in curvature due to the springback thus is

$$\frac{1}{R_0} - \frac{1}{R_1} = \frac{3}{2}\left(\frac{1}{\rho}\right)_Y \tag{a}$$

Using (7.39), we can express $(1/\rho)_Y$ as

$$\left(\frac{1}{\rho}\right)_Y = \frac{\epsilon_Y}{h/2} = \frac{Y}{E}\frac{2}{h} \tag{b}$$

Combining (*a*) and (*b*), we find

$$\frac{1}{R_0} - \frac{1}{R_1} = \frac{Y}{E}\frac{3}{h} \tag{c}$$

The stress distribution when the bar has the curvature $1/R_0$ is shown in Fig. 7.33*a*. The bending moment corresponding to this stress distribution is

$$M_b = M_L = 3/2 M_Y \tag{d}$$

If to this stress distribution we now add the fictitious elastic-stress distribution shown in Fig. 7.33*b*, which has as its resultant a bending moment of magnitude

$$M_b = -3/2 M_Y \tag{e}$$

we have zero net bending moment corresponding to the released condition in Fig. 7.31*b*. The residual-stress distribution then existing in the bar is as illustrated in Fig. 7.33*c*. Above the neutral surface the stress varies linearly from $-Y$ at the center to $+Y/2$ at the inner radius of the bar; below

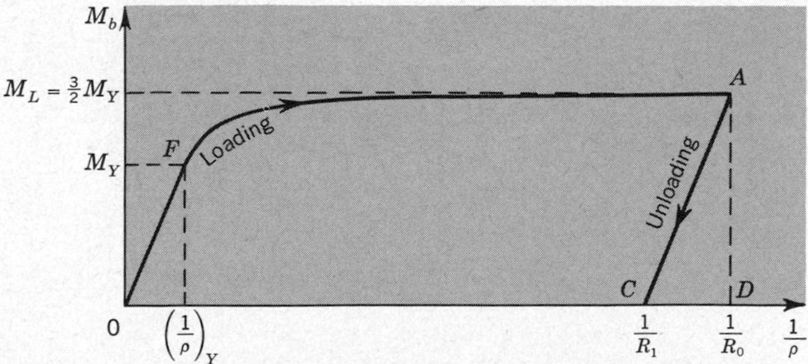

Fig. 7.32 Example 7.8. Moment-curvature relation for the complete cycle of loading and unloading the rectangular bar in Fig. 7.31.

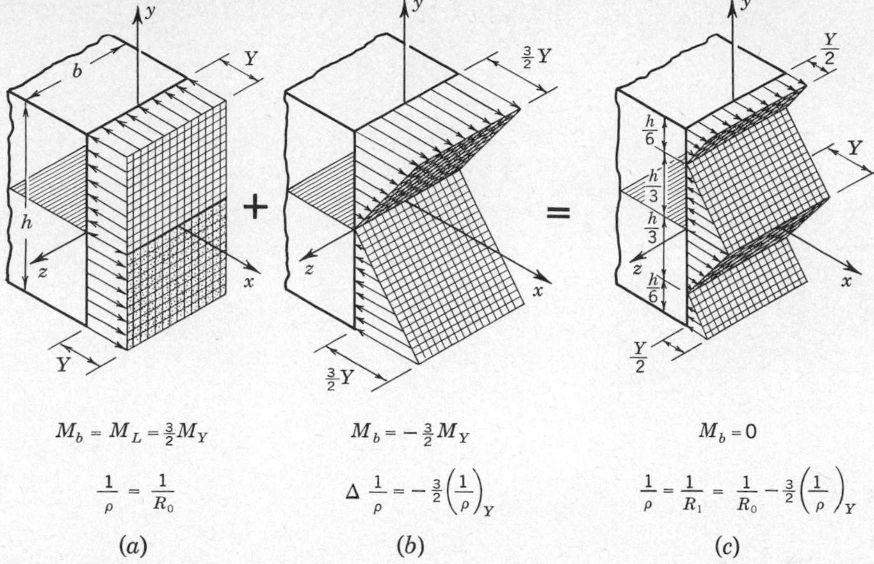

$M_b = M_L = \frac{3}{2}M_Y$ $\qquad$ $M_b = -\frac{3}{2}M_Y$ $\qquad$ $M_b = 0$

$\frac{1}{\rho} = \frac{1}{R_0}$ $\qquad$ $\Delta \frac{1}{\rho} = -\frac{3}{2}\left(\frac{1}{\rho}\right)_Y$ $\qquad$ $\frac{1}{\rho} = \frac{1}{R_1} = \frac{1}{R_0} - \frac{3}{2}\left(\frac{1}{\rho}\right)_Y$

(a) $\qquad$ (b) $\qquad$ (c)

Fig. 7.33 Example 7.8. Illustrating calculation of the residual-stress distribution in the bar of Fig. 7.31*b*.

the neutral surface the variation is linear from $+Y$ at the center to $-Y/2$ at the outer radius.

If we now added a further negative bending moment, we could decrease the curvature still further beyond the value $1/R_1$. At first, such action would be elastic, but when this additional bending moment exceeded the value

$$M_b = -\tfrac{1}{2}M_Y \tag{f}$$

there would be reversed yielding at the inner and outer radii of the bar.

7.11 BENDING OF UNSYMMETRICAL BEAMS

In this section we extend the theory of bending of elastic beams beyond the basic case of a symmetrical cross section bent in its plane of symmetry. As before, we derive the distribution of bending stress over the cross section only for the case of pure bending. According to the engineering theory of bending, we simply *assume* that the same distribution occurs in the presence of transverse shear. Our results will apply to beams of symmetrical cross section which are loaded unsymmetrically and to beams of unsymmetrical cross section. The general case is illustrated in Fig. 7.34 where a beam of arbitrary cross section (shown as triangular for aid in visualization) transmits a bending moment M_b of arbitrary orientation.

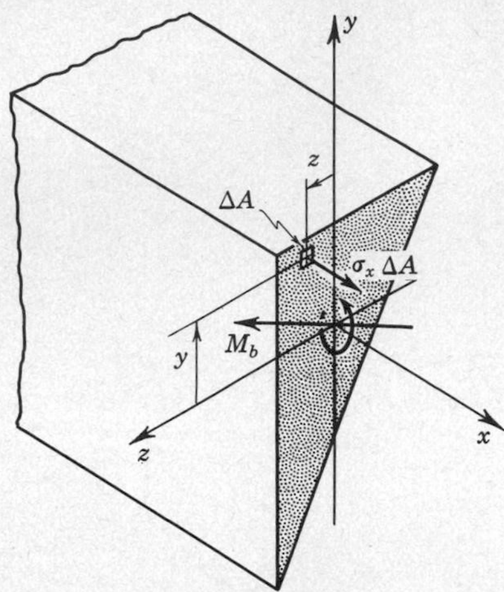

Fig. 7.34 Unsymmetrical beam subjected to pure bending.

Our procedure will be similar to that followed for symmetrical beams, although we shall take advantage of the results already obtained. We first study the geometry of a curve whose plane of curvature has arbitrary orientation. We shall show that when the slope angles are small the resultant curvature may be decomposed into component curvatures in two planes at right angles. The stresses which correspond to these component curvatures are then obtained and superposed to give the stress distribution and bending moment corresponding to the resultant curvature. This solution can also be inverted to provide the curvature when an arbitrary bending moment is given. The results we will obtain are also valid for beams subjected to transverse shear loads in addition to the bending moment.

We begin by considering the geometry of Fig. 7.35 where the curvature of the arc AC is in the xm plane which makes an angle θ with the xy plane. The projections of AC are A_1C_1 and A_2C_2 in the xy and xz planes. We shall show that when the angle ϕ is small there is a very simple relation between the curvature of AC and the curvatures of the projections A_1C_1 and A_2C_2.

In Fig. 7.36 we show the arc AC in the xm plane. By definition its curvature is

$$\frac{d\phi}{ds} = \lim_{\Delta s \to 0} \frac{\Delta\phi}{\Delta s} \tag{7.48}$$

When ϕ is small, the arc Δs is approximately equal to its projection Δx. The angle CAB between tangent and chord at A is one-half of the central angle AOC (Fig. 7.36). When ϕ and $\Delta\phi$ are small, the intercept BC is then approximately

$$BC = \Delta x \frac{\Delta\phi}{2} \tag{7.49}$$

The curvature of AC (7.48) can then be approximated as follows:

$$\frac{d\phi}{ds} \approx \lim_{\Delta x \to 0} \frac{2\overline{BC}}{(\Delta x)^2} \tag{7.50}$$

when ϕ is small. Similar results hold for the projections A_1C_1 and A_2C_2.

$$\begin{aligned} \frac{d\alpha}{ds_1} &\approx \lim_{\Delta x \to 0} \frac{2\overline{B_1C_1}}{(\Delta x)^2} \\ \frac{d\beta}{ds_2} &\approx \lim_{\Delta x \to 0} \frac{2\overline{B_2C_2}}{(\Delta x)^2} \end{aligned} \tag{7.51}$$

Turning back to Fig. 7.35, the relations between the intercepts B_1C_1, B_2C_2, and BC are

$$\begin{aligned} \overline{B_1C_1} &= \overline{BC} \cos\theta \\ \overline{B_2C_2} &= \overline{BC} \sin\theta \end{aligned} \tag{7.52}$$

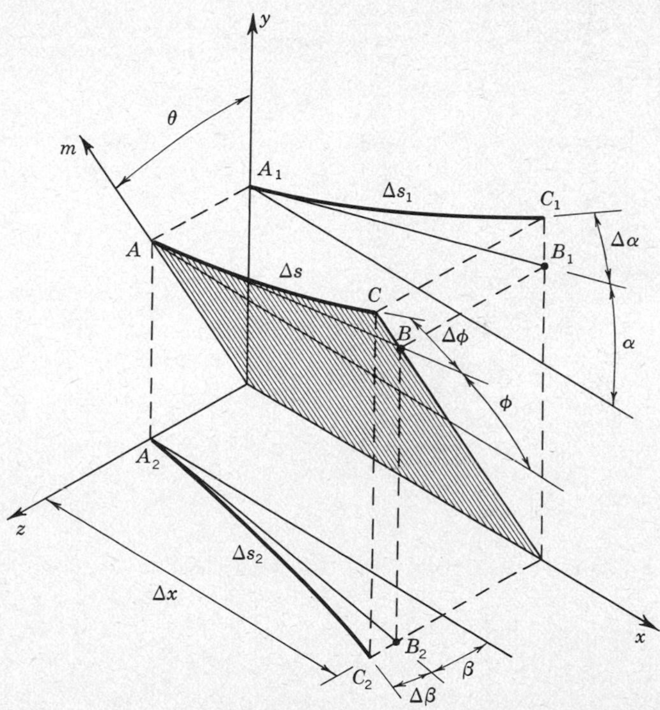

Fig. 7.35 AC is an element of an arc with curvature $d\phi/ds$ in the xm plane. A_1C_1 and A_2C_2 are projections of AC with curvature $d\alpha/ds_1$ and $d\beta/ds_2$, respectively.

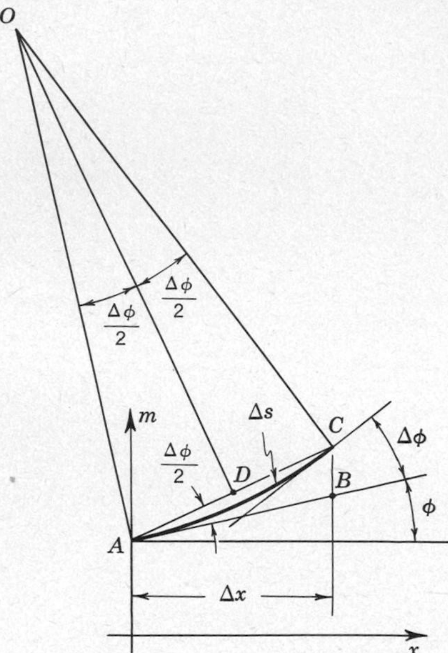

Fig. 7.36 View of xm plane in Fig. 7.35, showing the curve AC in true size.

When these are inserted in (7.51) and the results compared with (7.50), we see that within the limitations of the small-angle restriction we have established the important fact that

$$\frac{d\alpha}{ds_1} = \frac{d\phi}{ds}\cos\theta$$
$$\frac{d\beta}{ds_2} = \frac{d\phi}{ds}\sin\theta \tag{7.53}$$

The curvatures of the projected curves are simply the *components* of the curvature of the original curve.

We now apply this result to the *pure* bending of an unsymmetrical beam. We assume that the neutral axis of an originally straight beam does in fact have the curvature $d\phi/ds$ in the xm plane, as represented by the arc AC in Fig. 7.35. We shall determine the resulting stress distribution and the bending moment transmitted. It will be convenient to determine the stress distribution due to each of the two component curvatures (7.53) occurring separately and then to superpose these results to obtain the stress distribution due to the resultant curvature $d\phi/ds$ in the xm plane.

Figure 7.37 shows the stresses which result from the curvature $d\alpha/ds_1$ in the

xy plane. The argument here is based on the development at the beginning of this chapter. We assume that the longitudinal strain is still given by (7.4).

$$\epsilon_x = -\frac{d\alpha}{ds_1} y \tag{7.54}$$

Assuming as before that σ_x is the only nonzero stress component, Hooke's law (5.2) gives

$$\sigma_x = -E\frac{d\alpha}{ds_1} y \tag{7.55}$$

which is the distribution pictured in Fig. 7.37. In order for there to be no net longitudinal force resultant, it is again necessary for the neutral surface to pass through the centroid of the cross section.

In a completely similar fashion we obtain the stress distribution

$$\sigma_x = -E\frac{d\beta}{ds_2} z \tag{7.56}$$

shown in Fig. 7.38 which results from a curvature $d\beta/ds_2$ in the xz plane. Here also, the neutral surface must pass through the centroid.

We now consider the general case where there is a curvature $d\phi/ds$ in the xm plane. This curvature may be considered as the sum of the two component curvatures of (7.53). The resulting stress distribution, pictured in Fig. 7.39a, is then just the sum of the distributions (7.55) and (7.56).

$$\sigma_x = -E\left(\frac{d\alpha}{ds_1} y + \frac{d\beta}{ds_2} z\right) \tag{7.57}$$

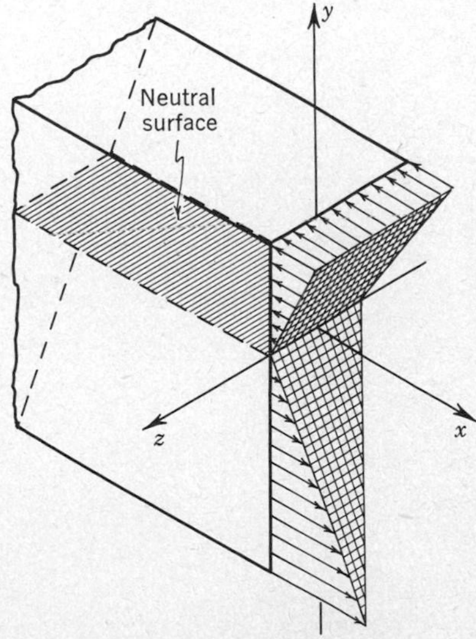

Fig. 7.37 Bending-stress distribution due to curvature $d\alpha/ds_1$ in the xy plane.

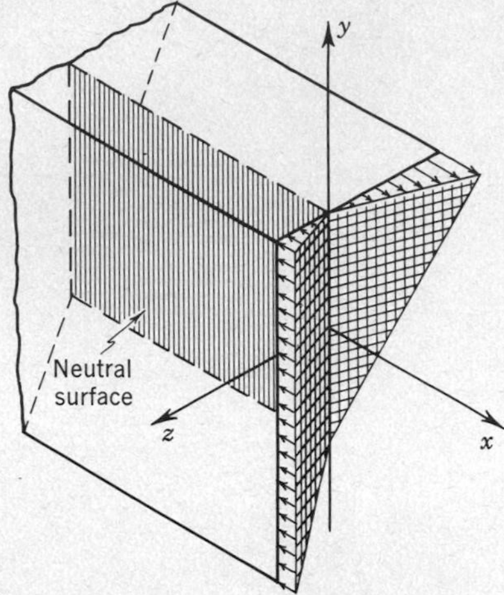

Fig. 7.38 Bending-stress distribution due to curvature $d\beta/ds_2$ in the xz plane.

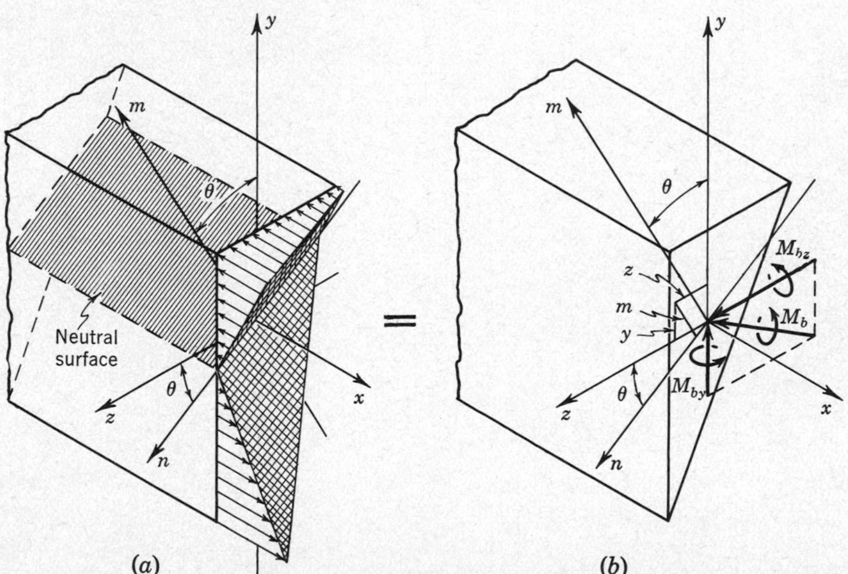

Fig. 7.39 Bending-stress distribution due to curvature $d\phi/ds$ in the xm plane.

The resultant of this distribution of stress is the bending moment M_b shown in Fig. 7.39*b* with components M_{by} and M_{bz}. These resultants are computed as follows:

$$
\begin{aligned}
M_{bz} &= -\int_A y\sigma_x\, dA = E\left(\frac{d\alpha}{ds_1}\int_A y^2\, dA + \frac{d\beta}{ds_2}\int_A yz\, dA\right) \\
M_{by} &= \int_A z\sigma_x\, dA = -E\left(\frac{d\alpha}{ds_1}\int_A yz\, dA + \frac{d\beta}{ds_2}\int_A z^2\, dA\right)
\end{aligned}
\tag{7.58}
$$

where the integrals are taken over the entire cross-sectional area A. We introduce the following symbolism for these integrals:

$$
\begin{aligned}
I_{yy} &= \int_A y^2\, dA \\
I_{zz} &= \int_A z^2\, dA \\
I_{yz} &= \int_A yz\, dA
\end{aligned}
\tag{7.59}
$$

The first two, I_{yy} and I_{zz}, are the *moments of inertia* of the cross-sectional area, and the third, I_{yz}, is called the *product of inertia* of the cross-sectional area with respect to the y and z axes. It can be shown (see Prob. 7.9) that there is always some orientation of the y, z axes, even in unsymmetrical cross sections, for which the product of inertia vanishes. Axes for which the product of inertia vanishes are called *principal axes of inertia*. Using the notation of (7.59), we can write (7.58) as follows:

$$
\begin{aligned}
M_{bz} &= E\left(I_{yy}\frac{d\alpha}{ds_1} + I_{yz}\frac{d\beta}{ds_2}\right) \\
M_{by} &= -E\left(I_{yz}\frac{d\alpha}{ds_1} + I_{zz}\frac{d\beta}{ds_2}\right)
\end{aligned}
\tag{7.60}
$$

Finally, we can restate this in terms of the resultant curvature $d\phi/ds$ of the neutral axis and the angle θ which locates the plane of resultant curvature by using (7.53).

$$
\begin{aligned}
M_{bz} &= E\frac{d\phi}{ds}(I_{yy}\cos\theta + I_{yz}\sin\theta) \\
M_{by} &= -E\frac{d\phi}{ds}(I_{yz}\cos\theta + I_{zz}\sin\theta)
\end{aligned}
\tag{7.61}
$$

This pair of equations can be considered as the generalization of the moment-curvature relation (7.14). Equations (7.61) do in fact reduce to (7.14) when the curvature is in the xy plane ($\theta = 0$) and the xy plane is a plane of symmetry ($I_{yz} = 0$).

In general the orientation of the resultant bending moment is *not* parallel to the neutral surface. It may be shown from (7.61) that the necessary and sufficient condition for the bending-moment vector to be parallel to the neutral surface is that the bending-moment vector must be parallel to a principal axis of inertia of the cross section.

For a known curvature $d\phi/ds$ with known orientation θ, we can use (7.61) to obtain the bending-moment components. The bending-stress distribution is given by (7.57), and on using (7.53) we have

$$\sigma_x = -E\frac{d\phi}{ds}[y\cos\theta + z\sin\theta] \tag{7.62}$$

which can be considered as a generalization of (7.9). Note that the bracketed term on the right of (7.62) is equal to the coordinate m in Fig. 7.39*b*. The two relations (7.61) and (7.62) yield the complete solution when the curvature is given.

To solve the opposite problem, namely, to find the curvature and stresses when the bending moment is given, we *invert* our previous results. Going back to (7.60), we consider the bending-moment components to be known and solve for the curvature components.

$$\begin{aligned}\frac{d\alpha}{ds_1} &= \frac{d\phi}{ds}\cos\theta = \frac{M_{bz}I_{zz} + M_{by}I_{yz}}{E(I_{yy}I_{zz} - I_{yz}^2)}\\ \frac{d\beta}{ds_2} &= \frac{d\phi}{ds}\sin\theta = -\frac{M_{bz}I_{yz} + M_{by}I_{yy}}{E(I_{yy}I_{zz} - I_{yz}^2)}\end{aligned} \tag{7.63}$$

This pair of expressions permits us to solve for the magnitude and orientation of the resultant curvature. (The magnitude is given by the square root of the sums of the squares of the curvature components, and the tangent of θ is given by their quotient.) We can obtain the resulting stress distribution without calculating the component curvatures by simply substituting (7.63) in (7.57) to obtain

$$\sigma_x = -\frac{(yI_{zz} - zI_{yz})M_{bz} + (yI_{yz} - zI_{yy})M_{by}}{I_{yy}I_{zz} - I_{yz}^2} \tag{7.64}$$

This is the stress distribution pictured in Fig. 7.39*a*; the bending-moment components are shown in Fig. 7.39*b*. The distribution (7.64) may be considered as the generalization of (7.16). Note that (7.64) does in fact reduce to (7.16) when $M_{by} = 0$ and when $I_{yz} = 0$. Note also that the intersection of the neutral surface and the yz plane (i.e., the n axis in Fig. 7.39) is the locus of points for which $\sigma_x = 0$ in (7.64).

In applying the above theory to practical cases, it usually is possible to orient the coordinate system so that some of the terms in the equation drop out. One approach, and probably the simplest, is to choose the y, z axes so that they are principal axes of inertia for the cross section. Another approach is to align either the y or z axis with the resultant moment vector; this also results in a simplification of the equations. The latter approach is used in the following example.

Example 7.9 A rectangular cantilever beam transmits a bending moment whose plane of action is inclined at 30° to the long axis of symmetry, as shown in Fig. 7.40*a*. We wish to determine the curvatures in the xy and xz planes and the bending stress in the beam.

The moments and product of inertia for the beam cross section are (see Prob. 7.12)

$$I_{yy} = 1.75c^4 \qquad I_{zz} = 0.75c^4 \qquad I_{yz} = 0.87c^4 \tag{a}$$

Substituting these in the first of (7.63), we find

$$\frac{d\alpha}{ds_1} = \frac{0.75c^4}{(1.75c^4)(0.75c^4) - (0.87c^4)^2}\frac{M_{bz}}{E}$$

$$= 1.36\frac{M_{bz}}{Ec^4} \tag{b}$$

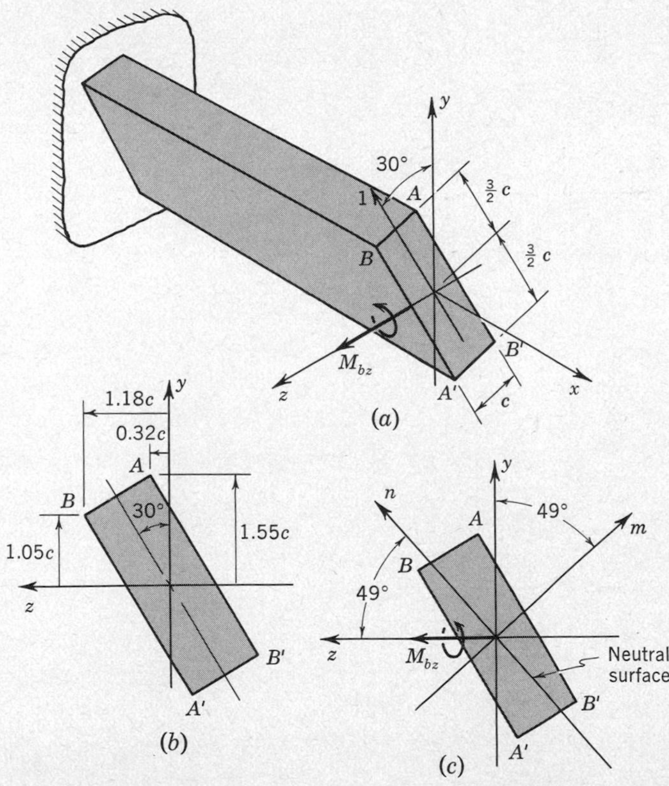

Fig. 7.40 Example 7.9.

From the second of (7.63) we then obtain

$$\frac{d\beta}{ds_2} = -\frac{I_{yz}}{I_{zz}}\left(\frac{d\alpha}{ds_1}\right) = -\frac{0.87c^4}{0.75c^4}\left(1.36\frac{M_{bz}}{Ec^4}\right) = -1.58\frac{M_{bz}}{Ec^4} \tag{c}$$

The bending stress will be a maximum at a corner. We shall investigate the corners A and B; the stresses at A' and B' will be reversed in sign. The coordinates of A and B are shown in Fig. 7.40*b*. Substituting these in (7.64), we get:

At A,

$$\sigma_x = -\frac{(1.55c)(0.75c^4) - (0.32c)(0.87c^4)}{(1.55c^4)(0.75c^4) - (0.87c^4)^2} M_{bz}$$

$$= -1.60\frac{M_{bz}}{c^3} \tag{d}$$

At B,

$$\sigma_x = -\frac{(1.05c)(0.75c^4) - (1.18c)(0.87c^4)}{(1.55c^4)(0.75c^4) - (0.87c^4)^2} M_{bz}$$

$$= +0.44\frac{M_{bz}}{c^3} \tag{e}$$

Since the stresses are of opposite sign at the corners A and B, we conclude that the neutral surface must intersect the side AB. This conclusion is verified when, from (7.63), we find

$$\theta = \tan^{-1}\left(-\frac{I_{yz}}{I_{zz}}\right) = \tan^{-1}\left(-\frac{0.87c^4}{0.75c^4}\right) = -49° \tag{f}$$

which places the neutral surface in the position shown in Fig. 7.40*c*. It is of interest to note how the beam tends to bend in its "weak plane," i.e., how closely the neutral surface coincides with the long axis of symmetry of the cross section.

7.12 SHEAR FLOW IN THIN-WALLED OPEN SECTIONS; SHEAR CENTER

The determination of shear flows in unsymmetrical beams follows the same procedure as outlined previously for symmetrical beams. When shear forces are present, the bending moment varies along the length of the beam. We *assume*, however, that the bending stresses are still distributed according to (7.64). The shear flows are then obtained from the requirement of longitudinal equilibrium applied to free bodies of elements of the beam.

We shall confine our treatment to thin-walled open sections, as shown in Fig. 7.41*a*. The term "open" is used to distinguish these sections from the thin-walled sections considered in Sec. 6.14 which were "closed." We shall assume that the resultant shear flow in the cross section at any point has the direction s of the center line of the wall; i.e., we neglect any shear flow in the direction n. This assumption is only strictly true in the limit of infinitely thin walls, but it furnishes a useful approximation whenever the wall thickness is small compared with the overall dimensions of the cross section.

In Fig. 7.41 an unsymmetrical beam with a thin-walled open section is loaded in a plane parallel to the xy plane, and thus $V_z = M_{by} = 0$. To calculate the shear flow at some point C in the cross section, we isolate the segment BC shown in Fig. 7.41*b*, where B is a free edge. The force $q_{sx}\,\Delta x$ on the positive s face must balance the unequal bending stresses acting on the ends of the element. Carrying through the analysis in the same manner as in Sec. 7.6, with the exception that in place of (7.16) we use (7.64) for σ_x, we obtain the result

$$q_{xs} = q_{sx} = \frac{-V_y}{I_{yy}I_{zz} - I_{yz}^2}\left(I_{zz}\int_{A_1} y\,dA - I_{yz}\int_{A_1} z\,dA\right) \tag{7.65}$$

where the integrals are, respectively, the first moment of the shaded area A_1 (Fig. 7.41) about the plane $y = 0$ and about the plane $z = 0$.

If we now determine the resultant of the shear flow q_{xs} acting over the face of the cross section, we shall find it to be a force V_y, but in general this force will not act through the centroid as indicated in Fig. 7.41*a*; rather, the resultant will be located some distance, say e_z, from the centroid, as illustrated in Fig. 7.42*a*.

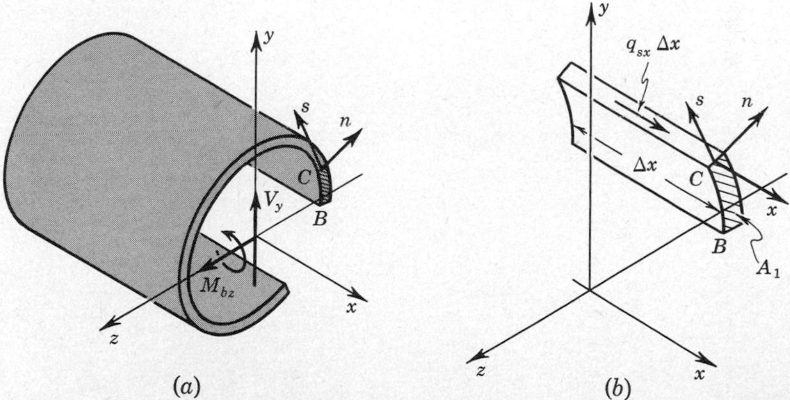

Fig. 7.41 Calculation of shear flow in an unsymmetrical beam of thin-walled, open cross section due to loading in a plane parallel to the xy plane.

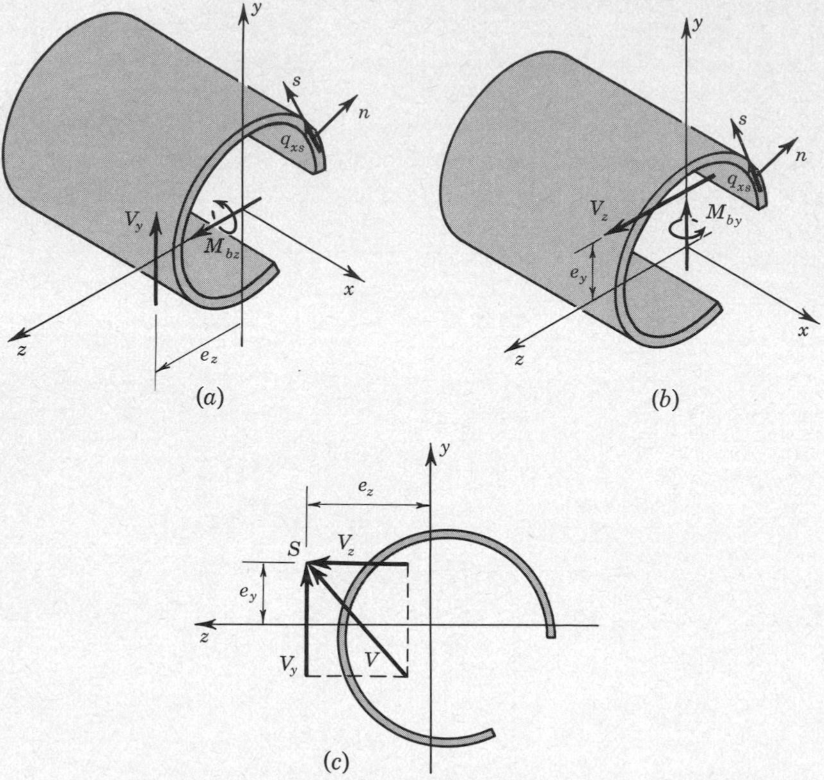

Fig. 7.42 Resultants of shear-flow distributions pass through the shear center S.

When the loading is in a plane parallel to the xz plane,[1] q_{xs} is given by

$$q_{xs} = q_{sx} = \frac{-V_z}{I_{yy}I_{zz} - I_{yz}^2}\left(I_{yy}\int_{A_1} z\,dA - I_{yz}\int_{A_1} y\,dA\right) \tag{7.66}$$

and the resultant of the shear flow is a force V_z which will in general be located some distance, say e_y, from the centroid as illustrated in Fig. 7.42*b*. The distances e_z and e_y depend only on the pattern of the shear flow and are independent of the magnitudes V_y and V_z.

When the loadings of Fig. 7.42*a* and *b* are superposed to give the general case, the resultant of the shear flows will be a force V which is the vector sum of V_y and V_z and whose line of action must pass through the point S with coordinates (e_y, e_z). This point, shown in Fig. 7.42*c*, is called the *shear center*. No matter what the magnitude or orientation of the resultant shear V, the line of action of V

[1] See Prob. 3.33.

will pass through the shear center. Note that the location of S depends on the shear-flow distributions (7.65) and (7.66), which in turn depend on the bending-stress distribution (7.64) which we have assumed to be valid for the case of varying bending moment. Every elastic beam cross section has a shear center S, although by using the engineering theory of stresses in beams, we are only able to locate S in symmetrical sections and in thin-walled sections.

An important property of the shear center is shown in Fig. 7.43, where a cantilever beam having the cross section of Fig. 7.42 is loaded through the centroid

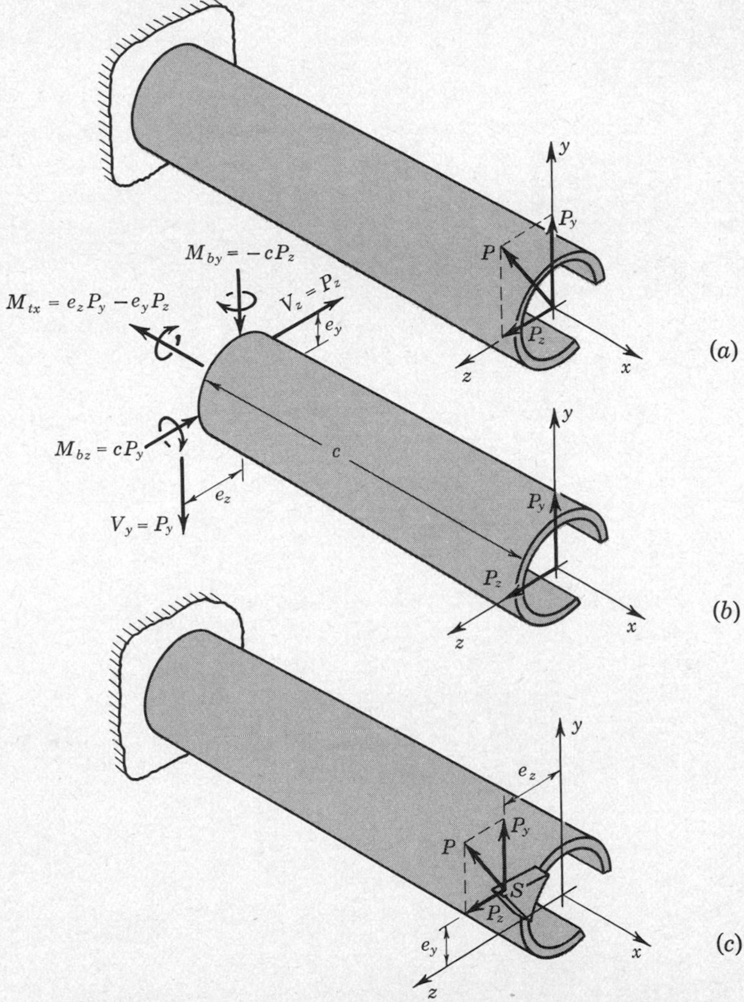

Fig. 7.43 Illustrating the presence of a twisting moment when the force P is not applied through the shear center S.

of the end cross section by a force P with components P_y and P_z. We consider in Fig. 7.43*b* a free body of the length c of the beam. We assume that the bending stresses in the section $x = -c$ are distributed according to (7.64). The shear flows will then be given by (7.65) and (7.66), and the resultant shear-force components will have the lines of action shown in Fig. 7.43*b*. For the free body of Fig. 7.43 to be in equilibrium, a twisting moment M_{tx} must be acting on the left end to maintain moment equilibrium about the x axis. There must be torsional strains and stresses in the beam in order to generate the twisting moment M_{tx}, and, as a consequence, the beam of Fig. 7.43*a* will twist as well as bend when the force P is applied. It is apparent that if the force P was applied at the shear center S in the end cross section, as illustrated in Fig. 7.43*c*, then the twisting moment would be zero and the force P would produce only bending.

Thus, if it is desired to bend a beam by transverse forces *without twisting the beam*, each transverse force should pass through the shear center of the cross section of application.

To locate the shear center, one finds the intersection of resultants of the shear-flow distributions for successive loading in two perpendicular planes, as indicated in Fig. 7.42. When the section has an axis of symmetry, the shear center lies on this axis, and thus only loading in the plane perpendicular to this axis need be considered. The following is an example of the calculation for a section without symmetry.

Example 7.10 We wish to determine the distribution of shear flows in the angle section of Fig. 7.44*a* whose centroid location and moments and product of inertia we obtain from Prob. 7.7.

$$I_{yy} = \tfrac{4}{3}a^3t \qquad I_{zz} = \tfrac{1}{4}a^3t \qquad I_{yz} = -\tfrac{1}{3}a^3t \tag{a}$$

We begin by considering loading in a plane parallel to the xy plane. For purposes of illustration the bending-stress distribution given by (7.64) is shown in Fig. 7.44*b*. The shear flow q_{xz} in the horizontal leg can be obtained from (7.65). In developing the relation (7.65) for q_{xs}, the coordinate s in Fig. 7.41 increased as we moved around the section in a counterclockwise manner. In the horizontal leg the coordinate z increases as we move in a counterclockwise manner, and hence z corresponds to s. Thus from (7.65) we can write

$$q_{xz} = q_{xs} = \frac{-V_y}{I_{yy}I_{zz} - I_{yz}^2}\left(I_{zz}\int_{A_1} y\,dA - I_{yz}\int_{A_1} z\,dA\right) \tag{b}$$

We evaluate the integrals from the sketch in Fig. 7.44*c*.

$$\begin{aligned}\int_{A_1} y\,dA &= [t(z_1 + \tfrac{5}{6}a)]\tfrac{2}{3}a \\ \int_{A_1} z\,dA &= [t(z_1 + \tfrac{5}{6}a)]\tfrac{1}{2}(z_1 - \tfrac{5}{6}a)\end{aligned} \tag{c}$$

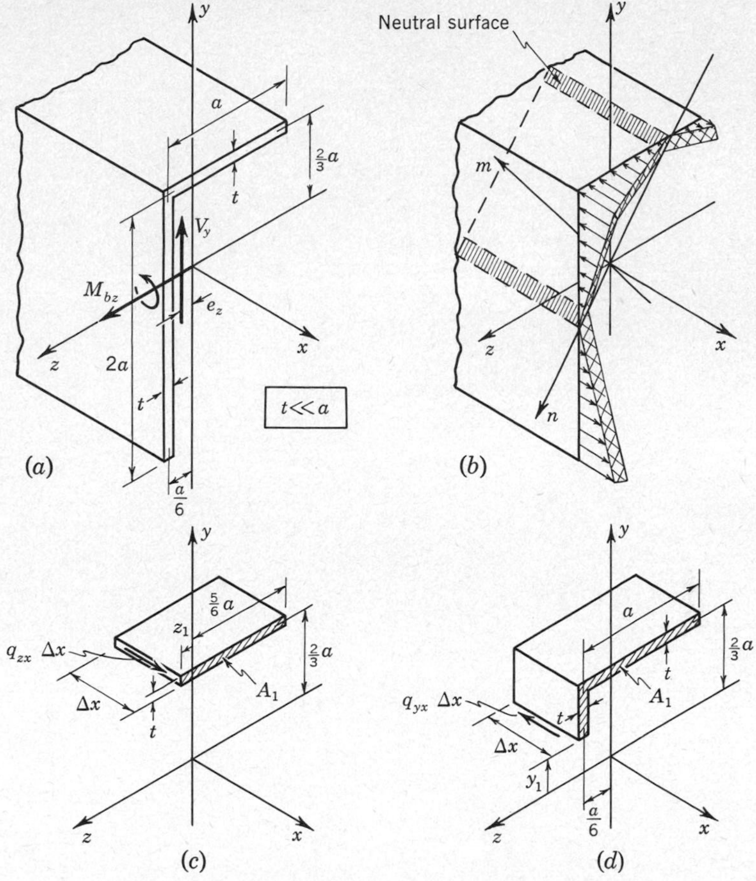

Fig. 7.44 Example 7.10. Calculation of shear-flow distribution in a thin-walled angle section due to loading in plane parallel to xy plane.

Substituting (a) and (c) in (b) and simplifying, we find

$$q_{xz} = -\frac{1}{48}\frac{V_y}{a^3}(36z_1{}^2 + 36az_1 + 5a^2) \tag{d}$$

The shear flow q_{xy} in the vertical leg also can be obtained from (7.65). As we move in a counterclockwise manner in the vertical leg, the coordinate y decreases and hence y corresponds to $-s$. Thus from (7.65) we can write

$$q_{xy} = -q_{xs} = \frac{V_y}{I_{yy}I_{zz} - I_{yz}{}^2}\left(I_{zz}\int_{A_1} y\,dA - I_{yz}\int_{A_1} z\,dA\right) \tag{e}$$

From Fig. 7.44*d* we obtain

$$\int_{A_1} y\,dA = (at)\tfrac{2}{3}a + [t(\tfrac{2}{3}a - y_1)]\tfrac{1}{2}(\tfrac{2}{3}a + y_1)$$
$$\int_{A_1} z\,dA = (at)(-\tfrac{1}{3}a) + [t(\tfrac{2}{3}a - y_1)](\tfrac{1}{6}a) \tag{f}$$

Combining (*a*), (*f*), and (*e*), we find, after simplifying,

$$q_{xy} = \frac{1}{48}\frac{V_y}{a^3}(32a^2 - 12ay_1 - 27y_1^{\,2}) \tag{g}$$

The distributions (*d*) and (*g*) are indicated in Fig. 7.45*a*. The maximum value of q_{xy} (which is positive throughout the vertical leg) and the maximum positive value of q_{xz} coincide with the points where the neutral surface cuts the vertical and horizontal legs in Fig. 7.44*b*.

As a check on the internal consistency of our analysis, we integrate across the horizontal leg to find

$$\int_{-5/6a}^{a/6} q_{xz}\,dz_1 = -\frac{1}{48}\frac{V_y}{a^3}\int_{-5/6a}^{a/6}(36z_1^{\,2} + 36az_1 + 5a^2)\,dz_1 = 0 \tag{h}$$

which agrees with the assumption in Fig. 7.44*a* that $V_z = 0$.

The resultant force in the vertical leg is found to be

$$\int_{-4/3a}^{2/3a} q_{xy}\,dy_1 = \frac{1}{48}\frac{V_y}{a^3}\int_{-4/3a}^{2/3a}(32a^2 - 12ay_1 - 27y_1^{\,2})\,dy_1 = V_y \tag{i}$$

and in Fig. 7.45*b* we show the total resultant of the shear-flow distribution to be the shear force V_y in the vertical leg.

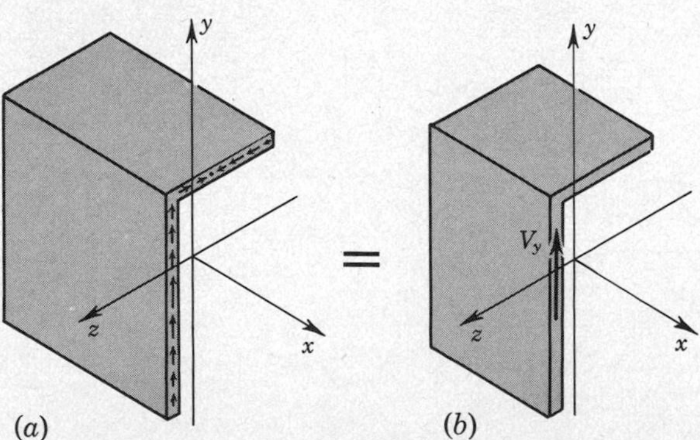

Fig. 7.45 Example 7.10. Location of resultant of shear-flow distribution.

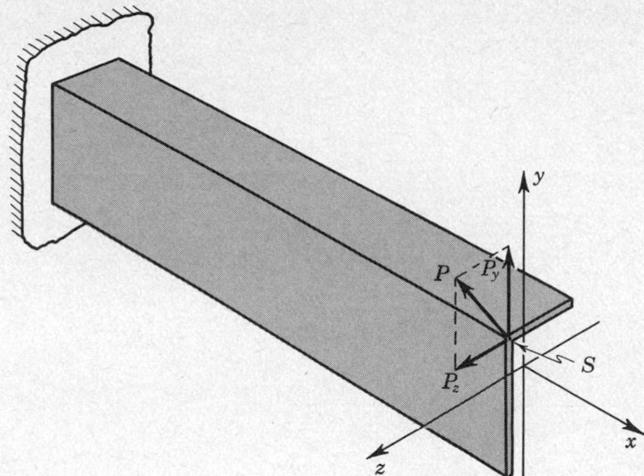

Fig. 7.46 Example 7.10. The shear center for any thin-walled angle section is at the intersection of the two legs of the angle.

A similar analysis applies to loadings in a plane parallel to the *xz* plane. It is clear that for this loading the resultant of the shear-flow distribution will be a force in the horizontal leg. The intersection of these two shear-flow resultants is at the intersection of the two legs, and this point is, therefore, the shear center for the angle. For any *thin*-walled angle section the horizontal shear flow is confined to the horizontal leg and the vertical shear flow is confined to the vertical leg. Their resultant must always pass through the intersection of the legs. Thus without further calculation we see that the shear center *S* of any thin-walled angle section must be at the intersection of the legs, as illustrated in Fig. 7.46.[1]

PROBLEMS

7.1. Demonstrate that if a complicated area is considered to be the sum of a number of simple shapes, as illustrated in the accompanying sketches, then Eq. (3.3) for the location of the centroid can be expressed in the form

$$\bar{y} = \frac{\sum_i \bar{y}_i A_i}{\sum_i A_i} \qquad \bar{z} = \frac{\sum_i \bar{z}_i A_i}{\sum_i A_i}$$

[1] Additional discussion of shear center may be found, for example, in B. Venkatraman and S. A. Patel, "Structural Mechanics with Introductions to Elasticity and Plasticity," p. 227, McGraw-Hill Book Company, New York, 1970.

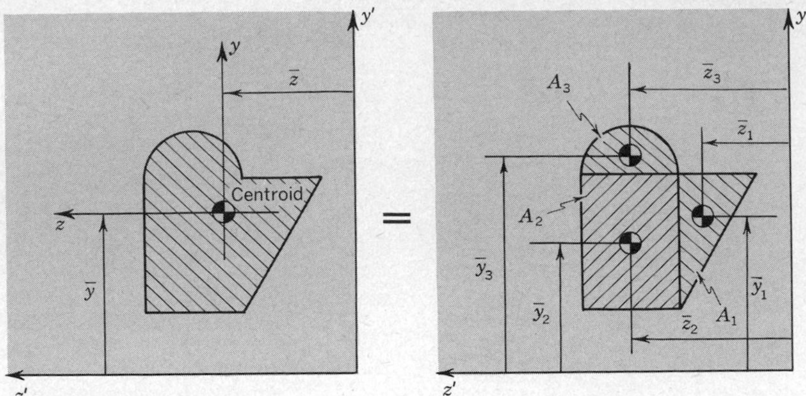

Prob. 7.1

7.2. Verify that the centroid of the angle section has the location shown.

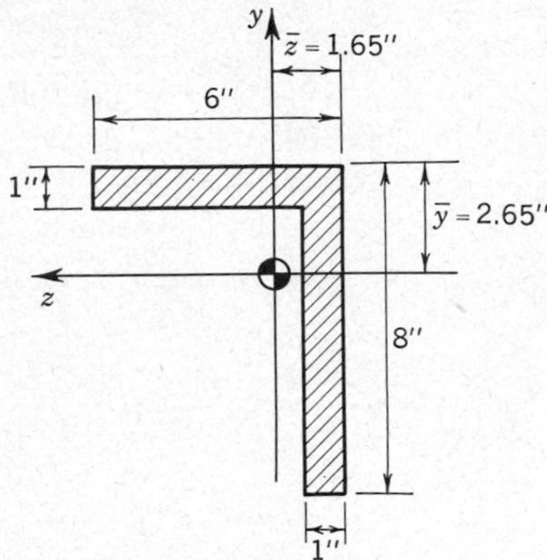

Prob. 7.2

7.3. Show that the centroid of the triangle is located as shown and, also, that for the y, z axes through the centroid the moments and products of inertia are

$$I_{yy} = \int_A y^2\, dA = \frac{bh^3}{36}$$

$$I_{zz} = \int_A z^2\, dA = \frac{b^3h}{36}$$

$$I_{yz} = \int_A yz\, dA = -\frac{b^2h^2}{72}$$

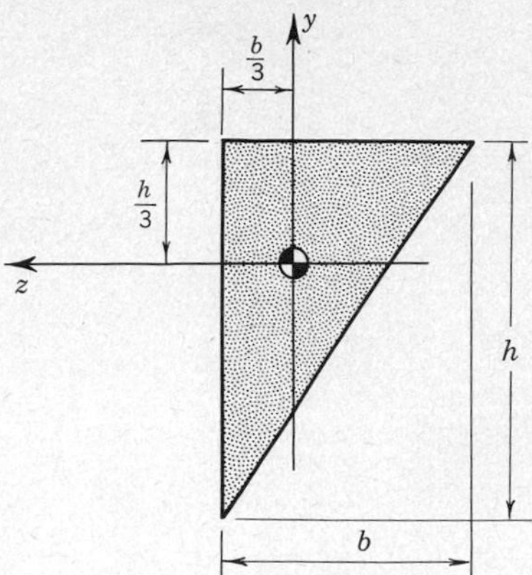

Prob. 7.3

7.4. Show that for a beam of arbitrary cross section where

$$I_x = \int_A r^2\,dA$$

the following relation holds:

$$I_{yy} + I_{zz} = I_x$$

Use this result to show that for a set of axes located in the centroid of the cross section of a solid circular shaft of radius r,

$$I_{yy} = I_{zz} = \frac{I_x}{2} = \frac{\pi r^4}{4}$$

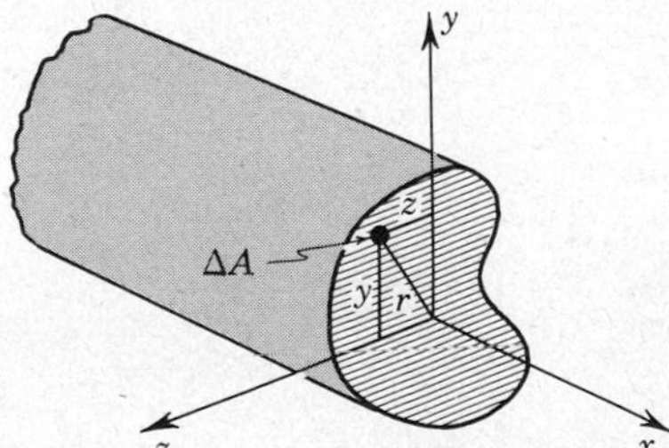

Prob. 7.4

7.5. Letting I_{yy} and I_{zz} be the moments of inertia of the area A for the y and z axes through the centroid, and letting I_{yz} be the product of inertia for the same axes, derive the following relations:

$$I_{y'y'} = I_{yy} + \bar{y}^2 A$$

$$I_{z'z'} = I_{zz} + \bar{z}^2 A$$

$$I_{y'z'} = I_{yz} + \bar{y}\bar{z}A$$

These relations illustrate the *parallel-axis theorem.*

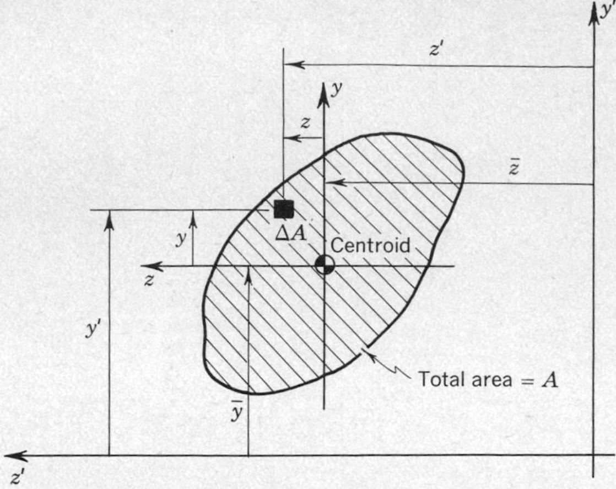

Prob. 7.5

7.6. For the angle section of Prob. 7.2, show that

$$I_{yy} = 80.8 \text{ in.}^4$$
$$I_{zz} = 38.8 \text{ in.}^4$$
$$I_{yz} = 32.3 \text{ in.}^4$$

7.7 Show that the centroid of the angle section is located at

$$\bar{y} = \frac{2}{3}a \qquad \bar{z} = \frac{a}{6}$$

and that I_{yy}, I_{zz}, and I_{yz} have the values

$$I_{yy} = \tfrac{4}{3}a^3t \qquad I_{zz} = \tfrac{1}{4}a^3t \qquad I_{yz} = -\tfrac{1}{3}a^3t$$

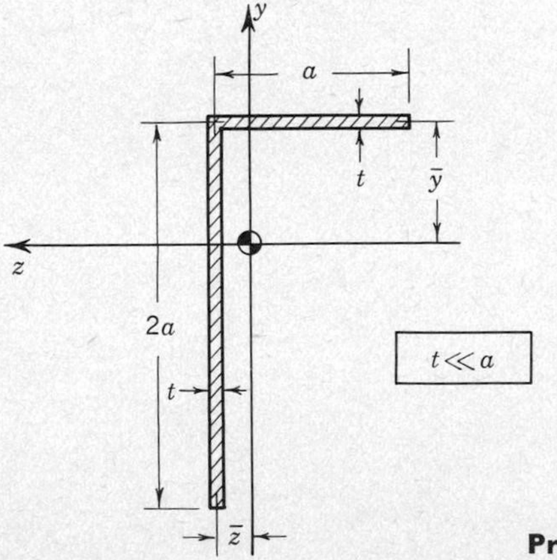

Prob. 7.7

7.8. Calculate the moment of inertia I_{yy} for the beam cross section illustrated.

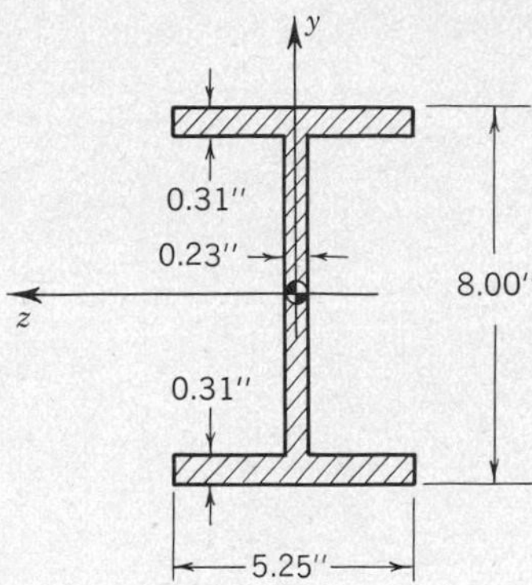

Prob. 7.8

7.9. Show that the moments of inertia

$$I_{mm} = \int_A m^2 \, dA \qquad I_{nn} = \int_A n^2 \, dA$$

and the product of inertia

$$I_{mn} = \int_A mn \, dA$$

for the m, n axes can be expressed in terms of I_{yy}, I_{zz}, and I_{yz} as follows:

$$I_{mm} = I_{yy} \cos^2 \theta + I_{zz} \sin^2 \theta + 2I_{yz} \sin \theta \cos \theta$$
$$I_{nn} = I_{yy} \sin^2 \theta + I_{zz} \cos^2 \theta - 2I_{yz} \sin \theta \cos \theta$$
$$I_{mn} = -(I_{yy} - I_{zz}) \sin \theta \cos \theta + I_{yz} (\cos^2 \theta - \sin^2 \theta)$$

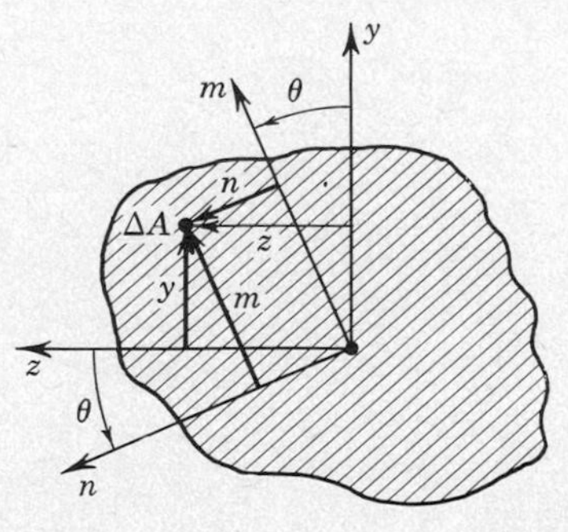

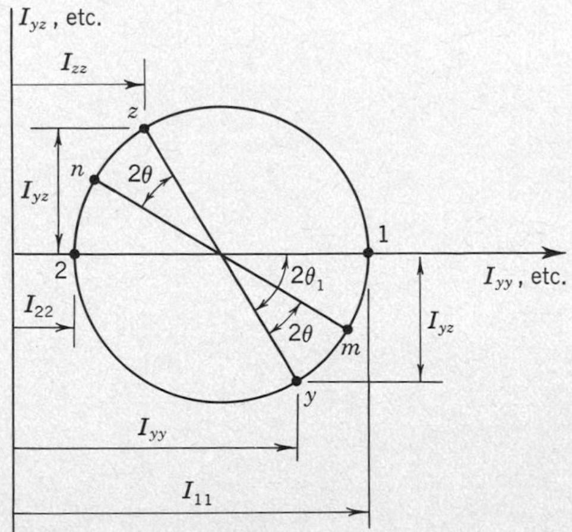

Prob. 7.9

Then, by comparing these results with Eqs. (4.23) and (4.24), show that the moments and product of inertia for various sets of axes through a point can be represented by the Mohr's circle shown. Note that 1 and 2 represent a set of axes for which the product of inertia is zero and with respect to which the moment of inertia has its maximum and minimum values; these axes are called the *principal axes of inertia* for this particular point in the cross section.

7.10. Using the results of Probs. 7.6 and 7.9, show that the principal axes of inertia through the centroid of the angle section of Prob. 7.2 are located as indicated in the accompanying sketch, and also that

$$I_{11} = 98.2 \text{ in.}^4$$

$$I_{22} = 21.4 \text{ in.}^4$$

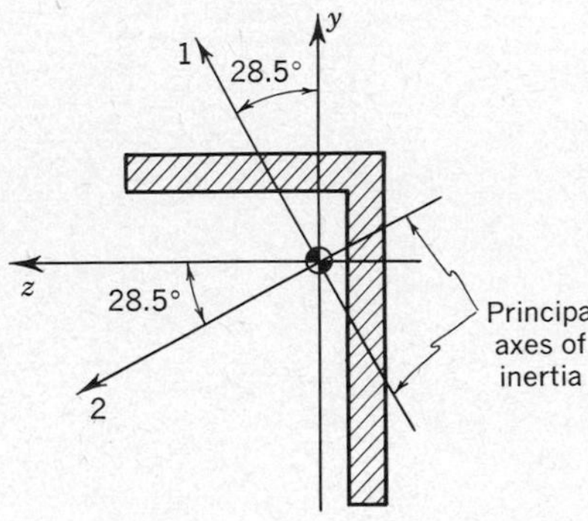

Prob. 7.10

7.11. Using the results of Prob. 7.3, show that the principal axes of inertia through the centroid of the triangle are located as drawn in the sketch, and also that

$$I_{11} = \frac{bh}{72}[(h^2 + b^2) + \sqrt{(h^2 - b^2)^2 + b^2h^2}]$$

$$I_{22} = \frac{bh}{72}[(h^2 + b^2) - \sqrt{(h^2 - b^2)^2 + b^2h^2}]$$

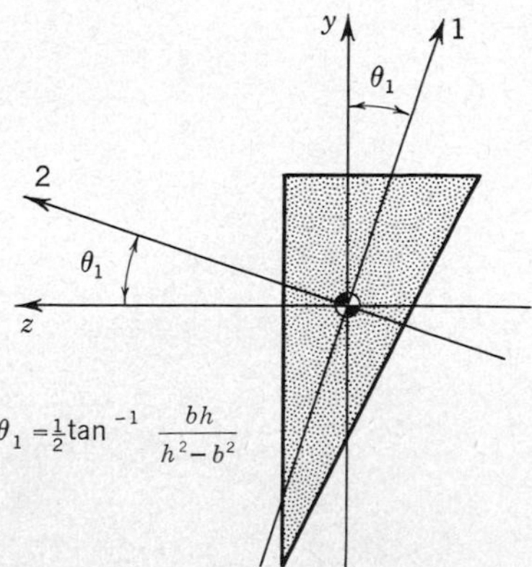

Prob. 7.11

7.12. Calculate the principal moments of inertia I_{11} and I_{22} for the rectangular beam cross section of Example 7.9. Use the Mohr's circle transformation of Prob. 7.9 to show that

$$I_{yy} = 1.75c^4 \qquad I_{zz} = 0.75c^4 \qquad I_{yz} = 0.87c^4$$

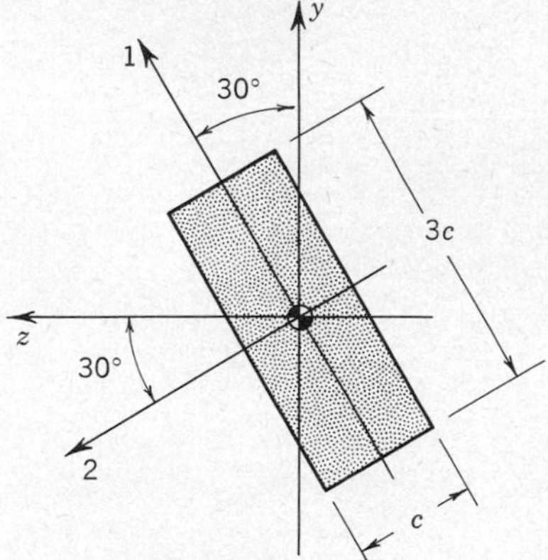

Prob. 7.12

7.13. A section of a steel beam of rectangular cross section 2 × 1 in. is loaded by a moment of 15,000 in.-lb about an axis parallel to the smallest side. Sketch the stress distribution across the beam.

7.14. A steel cantilever beam 20 ft long, whose cross section is shown in Prob. 7.8, is loaded by a 1,200-lb load. Find the maximum bending stress in the beam.

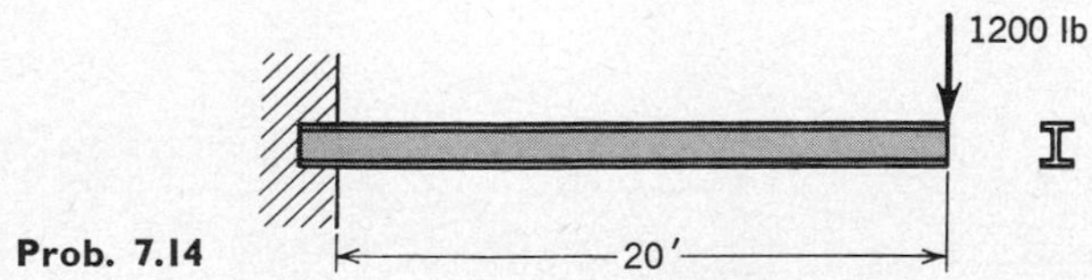

Prob. 7.14

7.15. It is proposed to use flat steel belts for a belt drive in which very precise control of the motion is required. The pulley diameter is 12 in. What is the thickest belt which can be wrapped 180° around the pulley without exceeding a stress of 40,000 psi? What would the maximum stress be if the belt thickness were halved?

7.16. A steel beam whose cross section is shown carries a uniform load per unit length (including the weight of the beam) of 2,000 lb/ft. Calculate the maximum bending stress in the beam.

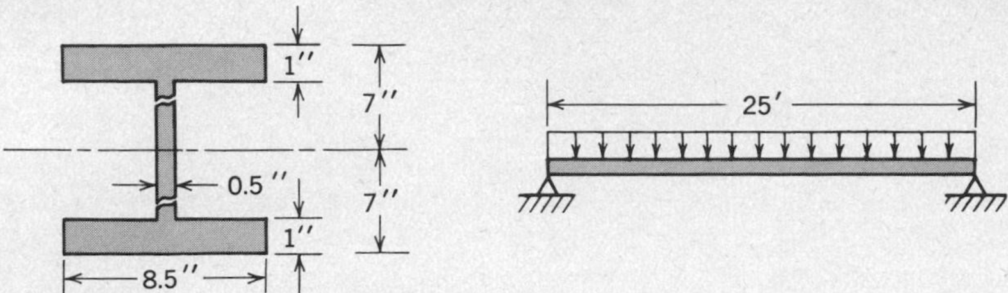

Prob. 7.16

7.17. Consider the beam shown. If the material has a maximum allowable stress of 5,000 psi in tension and 20,000 psi in compression, find the maximum value of P.

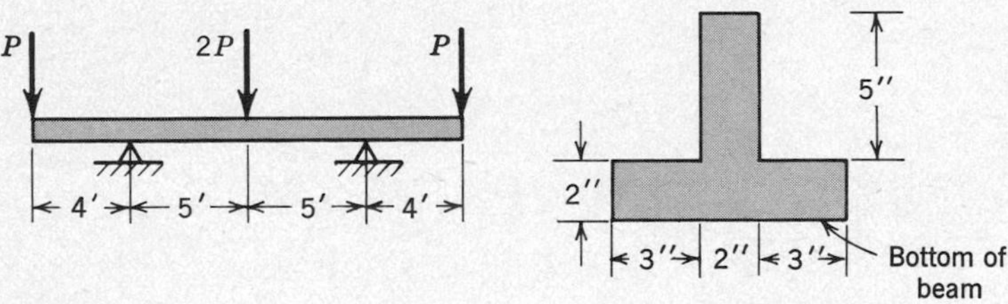

Prob. 7.17

7.18. A cast-iron T beam is to carry a distributed load of intensity w_o over a span of $10a$ as shown. The beam is of depth a and the flange width is αa. Cast iron fractures in tension at a stress approximately one-third of that for which failure occurs in compression (see Fig. 5.5*c*). Assuming that the stresses are low enough so that Hooke's law is a good assumption, find the value of α for which the maximum tensile bending stress σ_T will be one-third that of the maximum compressive bending stress σ_C. Find also the maximum load intensity w_o which can be carried for a given value of σ_T.

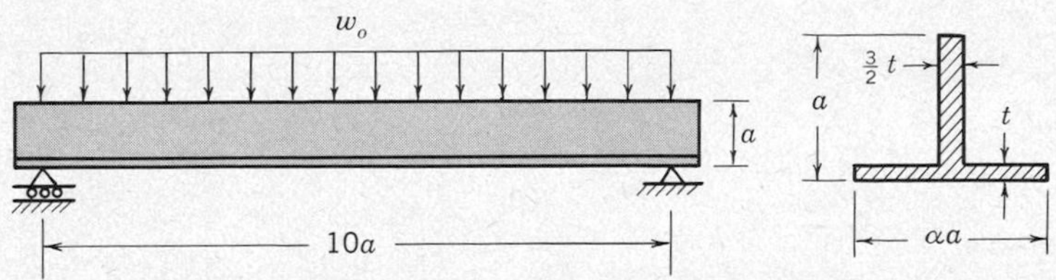

Prob. 7.18

7.19. A straight, thin steel strip of thickness t and width w is clamped to a rigid block of radius R with a length $4c$ extending from the clamp. The end of the strip is loaded with a force P sufficient to bring the strip into contact with the block over a distance c. Assuming that $c \ll R$, find the distribution of force between the strip and the block in the region BC. Find also the magnitude of the force P in terms of the dimensions of the strip and block (and any other quantities deemed necessary).

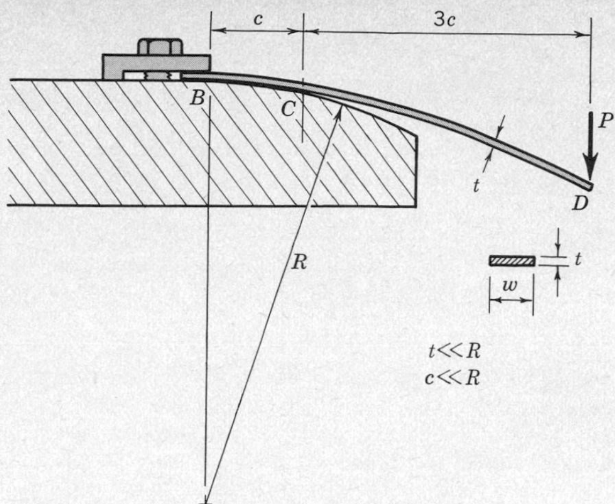

Prob. 7.19

7.20. A rough sketch of a human femur subjected to a vertical load of 100 lb is shown.

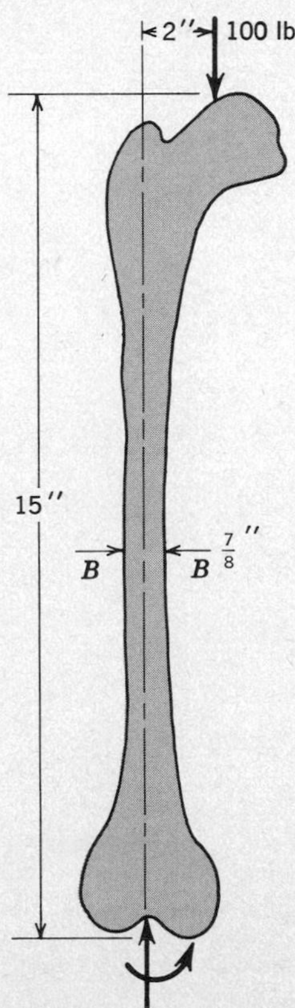

Prob. 7.20

(*a*) Determine the distribution of stress across the section *BB* assuming that the circular section is solid bone.

(*b*) Same as (*a*), except that this time assume that the inner half of the bone radius consists of "spongy" bone. Assume that the "spongy" bone does not carry appreciable stress.

(*c*) What is the percentage increase in the maximum stress of distribution (*b*) compared with the maximum stress of distribution (*a*)?

7.21. A new theory of Egyptian pyramid-building proposes that the large pyramid blocks were lifted onto sledges by the counterweighted wooden lever system shown in the figure. The sledges were then pulled up the sides of the pyramid by manpower. If the wood in the levers has an ultimate tensile stress of 7,500 psi and ultimate shear stress of 1,500 psi, find on the basis of these ultimate stresses the dimensions of the smallest square piece of timber which will support the pyramid blocks as shown.

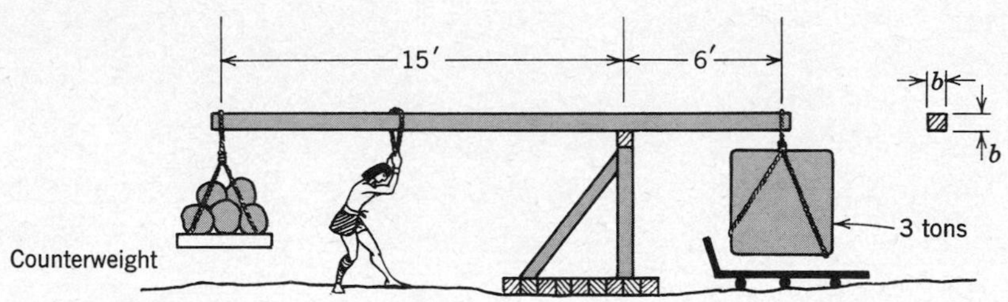

Prob. 7.21

7.22. A closed, thin-walled tube of radius r and thickness t is transmitting a bending moment M_b. Calculate the bending stress as a function of θ.

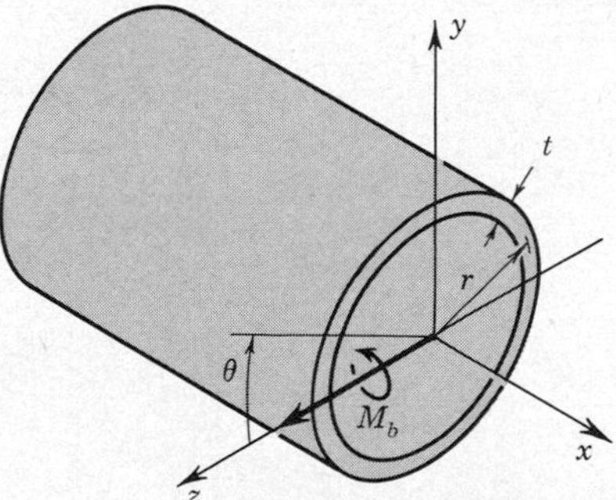

Prob. 7.22

7.23. A thin-walled cylindrical tank of radius r, thickness t, and length L is supported at its ends. It is filled with a heavy liquid which is vented to the atmosphere. If the weight of the tank is negligible compared with the weight of the liquid, show that the maximum bending stress in the tank is independent of the radius of the tank.

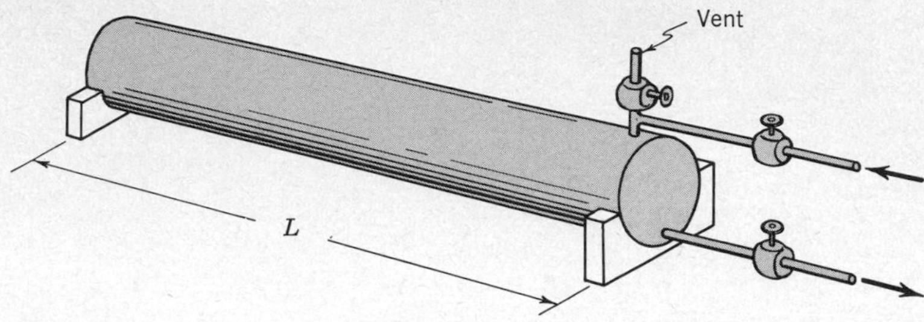

Prob. 7.23

7.24. A cantilever beam of width b and length L has a depth which tapers uniformly from d at the tip to $3d$ at the wall. It is loaded by a force P at the tip, as shown. Find the location and magnitude of the maximum bending stress.

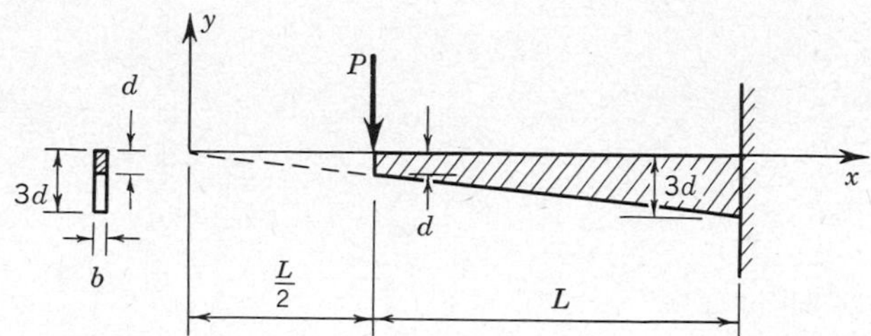

Prob. 7.24

7.25. A very thin cylindrical shell is stiffened by six equally spaced longitudinal bars welded to the cylinder. Assuming that the bending stress is carried entirely by the longitudinal bars, estimate the maximum bending stress when the bending moment is 100,000 in.-lb.

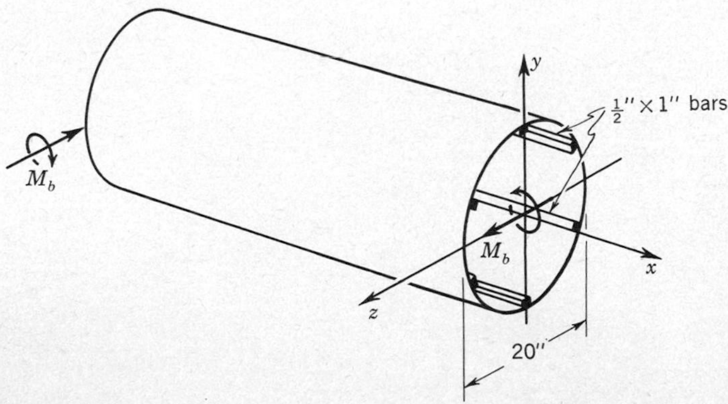

Prob. 7.25

7.26. A cross section of a cilium (see Prob. 3.21) is shown in the figure. The dark areas are fibrils which are thought to be responsible for the cilium motion. The bending moment at the base of the cilium is estimated to be 5×10^{-10} dyne-cm, and an experimental value of the radius of curvature at the base is 6 μm. Assuming that the bending forces are carried by the fibrils alone, estimate the elastic modulus of the fibrils. The total second moment I_{yy} of all the fibril cross-sectional areas is approximately 4×10^{-21} cm^4.

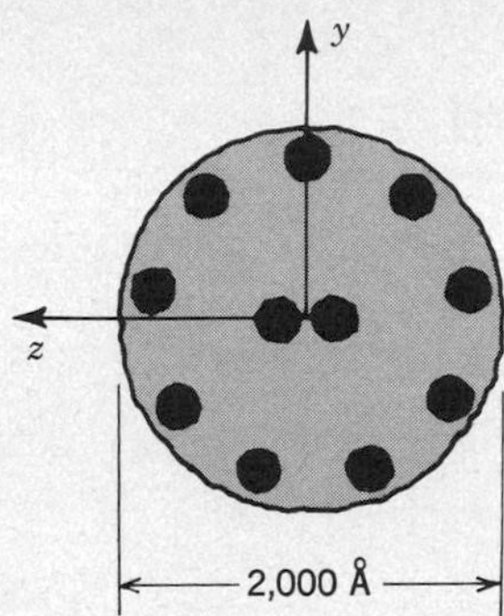

Prob. 7.26

7.27. To estimate the relative importance of transverse shear in comparison with torsion in the deformation of the tightly coiled spring of Fig. 6.18, one can proceed as follows. By arguments analogous to those in Sec. 7.6, one can derive an approximate distribution of transverse shear stress across the circular section of the wire. If the z axis is directed along the axis of the wire and the y axis parallel to the transverse force P, the transverse shear stress in a wire of radius r is distributed parabolically

$$\tau_{yz}(y_1) = \frac{4P}{3\pi r^4}(r^2 - y_1{}^2)$$

Calculate the strain energy in the spring due to this distribution of stress, and verify that the ratio of strain energy U_s due to transverse shear to the strain energy U_t due to torsion is

$$\frac{U_s}{U_t} = \frac{5}{9}\left(\frac{r}{R}\right)^2$$

7.28. Under average conditions, what is the maximum bending stress in the lead of your pencil? Make your own estimate of the geometry and the loading conditions.

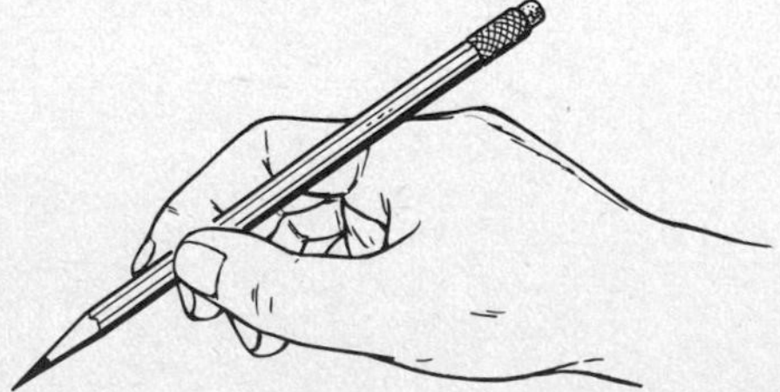

Prob. 7.28

7.29. A beam is made of two identical metal bars soldered together. What is the ratio of the stiffness

$$k_b = \frac{M_b}{d\phi/ds}$$

of this beam to the stiffness of a beam in which the two bars are not soldered and act independently? What is the ratio of the maximum bending stresses for the two cases?

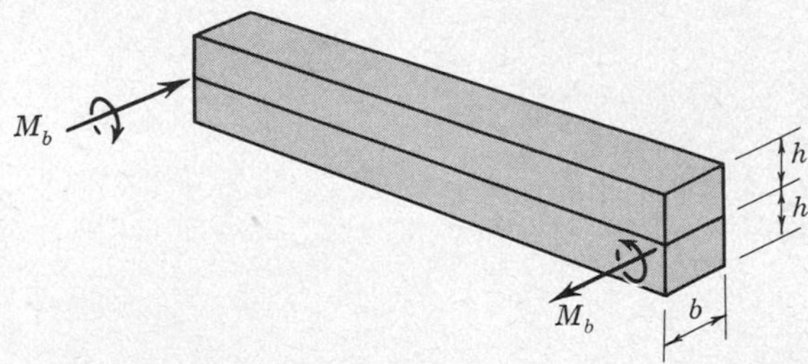

Prob. 7.29

7.30. The roof truss shown carries a central load W. Calculate the forces in the horizontal members on the top and bottom of the truss. Then calculate the *stresses* in these members and plot as a function of position along the truss. Now consider the truss to be a continuous beam in which only the top and bottom members are effective in bending. Using beam theory, calculate the stresses in the top and bottom of the beam as a function of position along the truss and plot on the previous graph for comparison. If one does consider the truss as a beam, what role do the diagonal members play in the beam action?

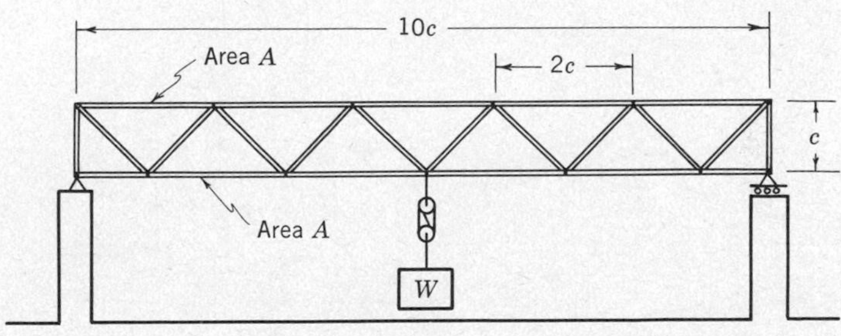

Prob. 7.30

7.31. If a rectangular beam is made of a material whose stress-strain curve in both tension and compression is well represented by $\sigma = c|\epsilon|^n$, derive an expression for the maximum bending stress in terms of the applied moment.

7.32. In the development of the theory of pure bending of symmetrical beams, the following strains were obtained:

$$\epsilon_x = -\frac{y}{\rho} \qquad \epsilon_y = \epsilon_z = \nu\frac{y}{\rho} \qquad \gamma_{xy} = \gamma_{yz} = \gamma_{zx} = 0$$

Using Eqs. (5.7), verify that these strains result from the continuous displacements (see Fig. 7.9)

$$u = -\frac{xy}{\rho}$$

$$v = \frac{x^2 + \nu(y^2 - z^2)}{2\rho}$$

$$w = \nu\frac{yz}{\rho}$$

and that, therefore, the strains are geometrically compatible. Since, for constant temperature, the solution also satisfies the equilibrium equations (5.6) and the stress-strain-temperature relations (5.8), it represents a complete solution within the theory of elasticity.

7.33. Consider a symmetrical beam that is initially curved in its plane of symmetry. Repeat the arguments of Sec. 7.2 to show that when a bending moment acts in the plane of initial curvature, plane cross sections remain plane; i.e., plane radial cross sections in the undeformed beam become plane radial cross sections in the deformed beam. Demonstrate that the *increase* in curvature of the neutral axis is

$$\frac{\Delta\phi}{R_0\phi} = \frac{1}{R_1} - \frac{1}{R_0}$$

Show by taking an appropriate free body that equilibrium requires the existence of radial normal stresses in the interior of the beam. Finally, decide whether or not the tangential strain distribution is linear across the radial depth of the beam.[1]

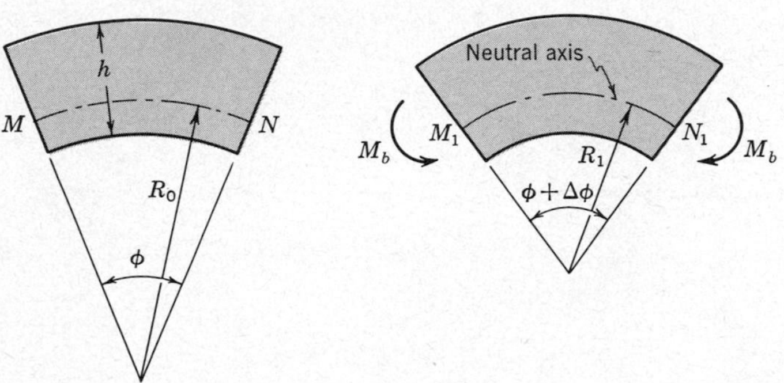

Prob. 7.33

7.34. A plate is a "beam" in which the thickness is very small compared to the width, as shown in part (*a*) of the figure. It is observed experimentally that, when a plate is bent by the application of moments to the ends, the central part forms a cylindrical surface and the anticlastic curvature is restricted to the vicinity of the edges, as shown in part (*b*) of the figure. (Bend a piece of

[1] For a detailed discussion of initially curved beams, see F. R. Shanley, "Strength of Materials," p. 322, McGraw-Hill Book Company, New York, 1957.

cardboard to verify this behavior.) Proceeding from this observed geometric behavior, develop the following relation for the curvature of the central cylindrical portion:

$$\frac{d\phi}{ds} = \frac{1}{\rho} = \frac{12(1-\nu^2)M}{Eh^3}$$

where M is the bending moment per unit width of plate in the central portion. Show also that in the central cylindrical portion the bending stress σ_x is given by

$$\sigma_x = \frac{12My}{h^3}$$

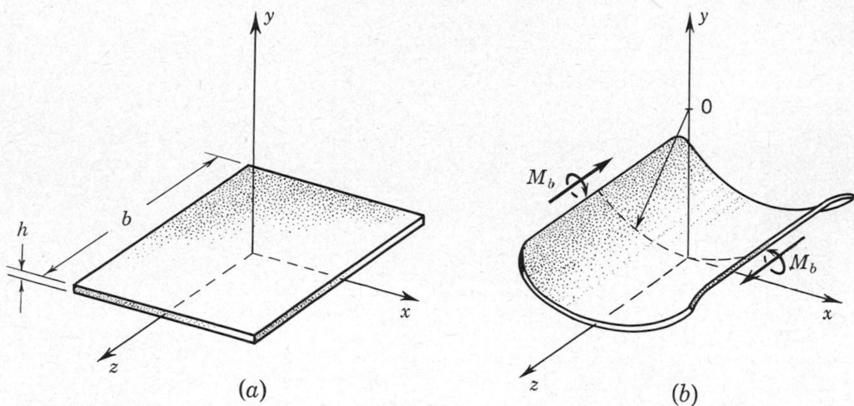

Prob. 7.34

7.35. A bookshelf is made out of ¼-in. plate glass. For long-time service, ordinary plate glass cannot safely be stressed to more than about 1,000 psi in tension. If the supports are located in the optimum position, estimate the average weight of books per unit length which can be placed along the shelf.

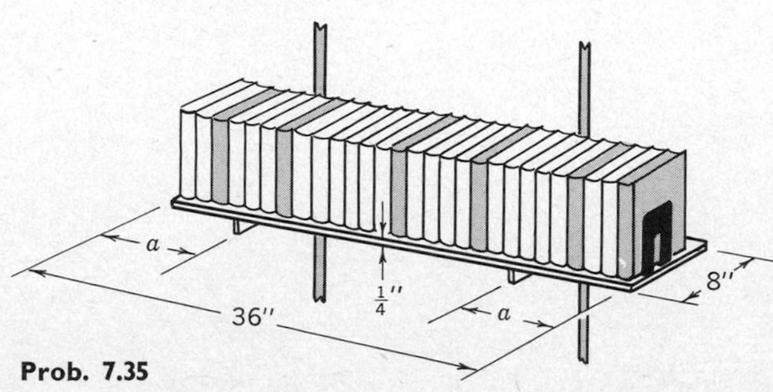

Prob. 7.35

7.36. Two 2 × 4-in. beams are glued together as shown. What is the required glue strength for the two loading directions?

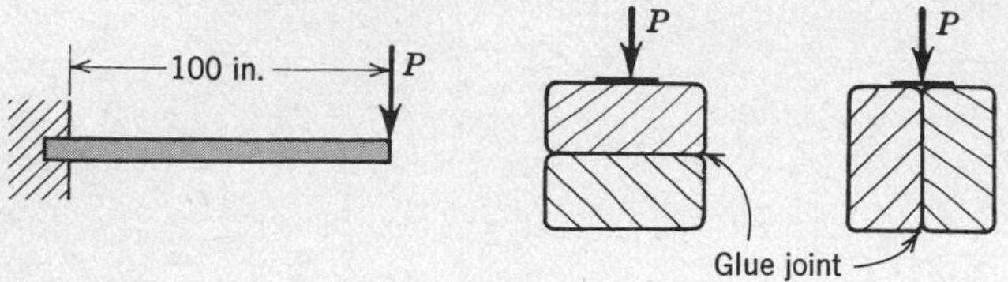

Prob. 7.36

7.37. When the shear force is 6,000 lb, calculate the shear flow q_{xy} in the vertical plate of the beam of Example 7.3 at the following locations:

(*a*) Just above the solder joint
(*b*) Just below the solder joint
(*c*) At the neutral surface

7.38. A closed, thin-walled tube of radius r and thickness t is transmitting a shear force V. Calculate the shear-flow distribution q_{xs} as a function of θ.

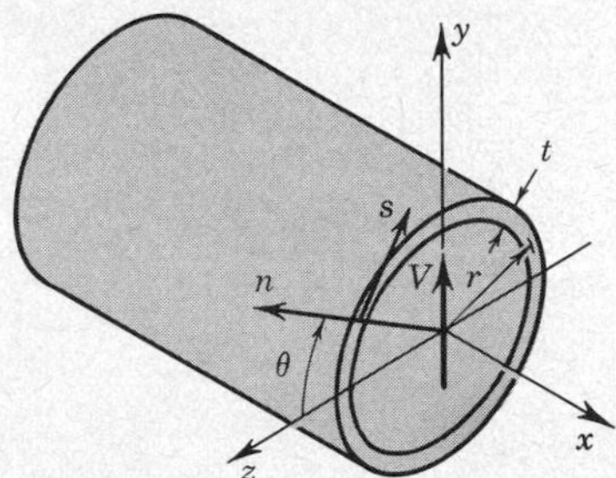

Prob. 7.38

7.39. Two designs have been suggested for building a box beam by nailing together four pieces of wood of equal thickness. The dimensions b and h and the spacing s are equal in both designs. If the beam is to carry loading in the xy plane, is one design better than the other?

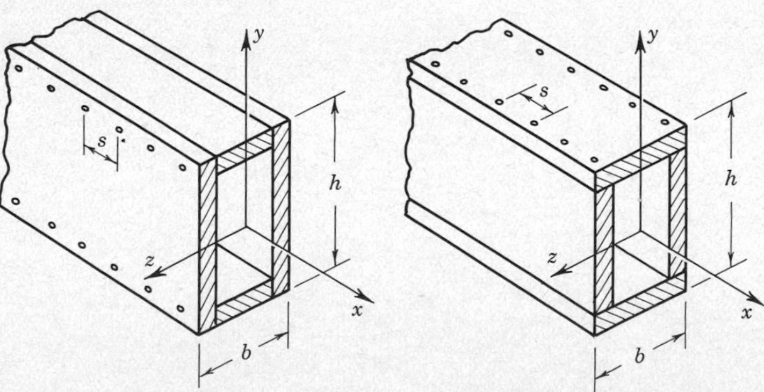

Prob. 7.39

7.40. The sketch shows the cross section of a T beam which is transmitting both a bending moment and a shear force. What is the ratio of the maximum bending stress in the stem to that in the flange? What is the ratio of the maximum average shear stress τ_{xy} in the stem to the maximum average shear stress τ_{xz} in the flange?

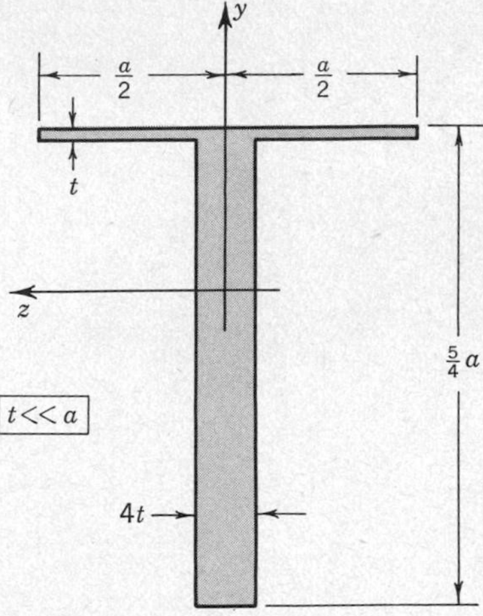

Prob. 7.40

7.41. The beam illustrated is clamped together with ¼-in. bolts with a spacing s as shown. If each bolt can safely resist a shear force across it of 400 lb, what is the bolt spacing required when the shear force V is 10,000 lb?

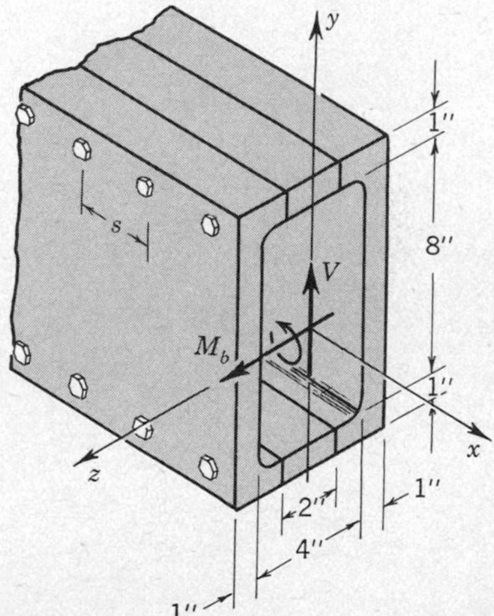

Prob. 7.41

7.42. Using (7.27) for the distribution of shear stress τ_{xy} in a rectangular beam, show that the resultant of this stress distribution is the shear force V.

7.43. With Q defined as in (7.22), show that

$$\frac{dQ}{dy_1} = -by_1$$

Using this fact and the technique of integration by parts, show that for an arbitrary symmetrical beam the resultant of the shear stress τ_{xy} given by (7.25) is the shear force V; that is, show that

$$\int_A \tau_{xy}\, dA = V$$

where A is the area of the beam cross section.

7.44. A rectangular beam on simple supports has bricks piled uniformly along its length such that there is a total weight w_o of bricks per unit length along the beam. Determine the ratio of the maximum bending stress σ_x to the maximum shear stress τ_{xy}.

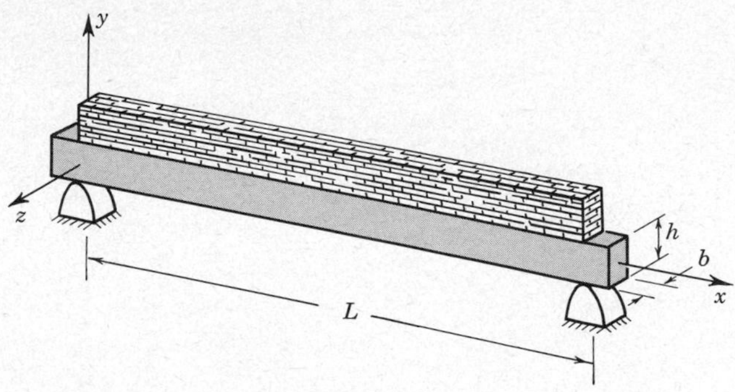

Prob. 7.44

7.45. The beam illustrated has cross-sectional proportions which are typical of "wide-flange" steel beams that are used extensively in building construction. Determine the ratio of the maximum bending stress σ_x to the maximum shear stress τ_{xy} when the beam carries a central load as indicated.

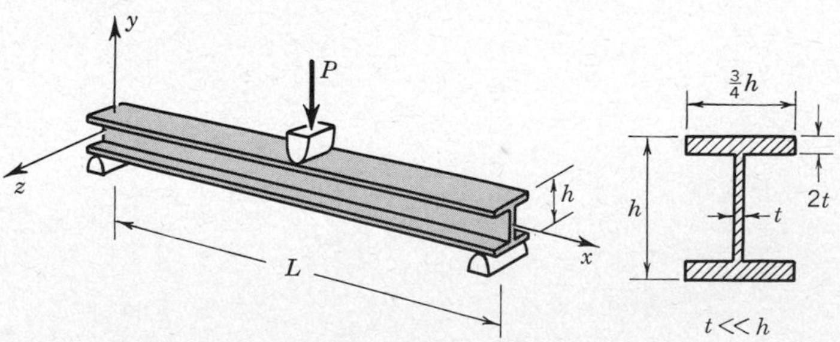

Prob. 7.45

7.46. Consider the problem of pure bending of the symmetrical *composite* beam which has been made by bonding together two materials of different elastic properties. Carry out a development parallel to that given in Secs. 7.2 to 7.5 to obtain the deformation and the stresses in the composite beam. Show that the neutral surface is located by the distance y_N pictured, where

$$(a) \qquad y_N = \frac{E_1 \bar{y}_1 A_1 + E_2 \bar{y}_2 A_2}{E_1 A_1 + E_2 A_2}$$

and that the moment-curvature relation is

$$(b) \qquad \frac{d\phi}{ds} = \frac{1}{\rho} = \frac{M_b}{E_1 (I_{yy})_1 + E_2 (I_{yy})_2}$$

where $(I_{yy})_1$ and $(I_{yy})_2$ are, respectively, the moments of inertia of the areas A_1 and A_2 about the neutral surface. Finally, show that the bending stress in the beam is given by

$$(c) \qquad (\sigma_x)_i = -E_i \frac{M_b y}{E_1 (I_{yy})_1 + E_2 (I_{yy})_2}$$

where i takes on the value of 1 or 2, depending on which material we are interested in.

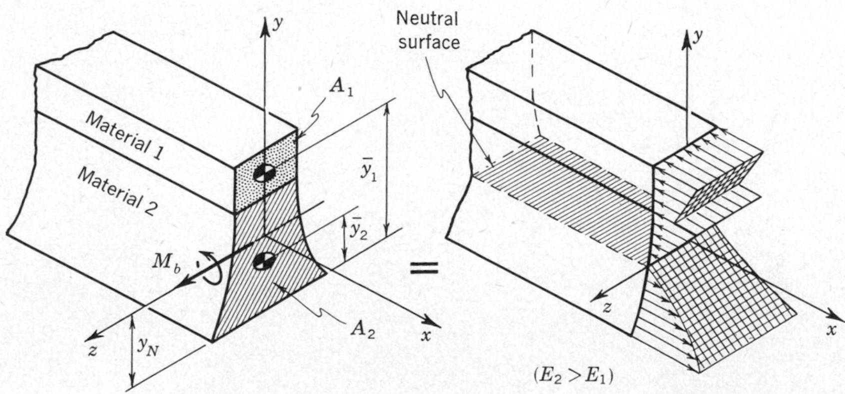

Prob. 7.46

7.47. Study the behavior in the vicinity of the joint which bonds the two materials in the composite beam of Prob. 7.46. Consider, in particular, the case when the two materials have different Poisson's ratios, and discuss the validity of the assumption that in pure bending of composite beams all stresses vanish except σ_x.

7.48. In an attempt to make a beam which combines light weight with large stiffness, ¼-in. steel plates are riveted to the top and bottom of an aluminum-alloy I beam for which $I_{yy} = 57$ in.[4]

By what ratio is the stiffness

$$k_b = \frac{M_b}{d\phi/ds}$$

of the I beam increased by the addition of the plates? In the composite beam what is the ratio of the maximum bending stress in the aluminum to that in the steel?

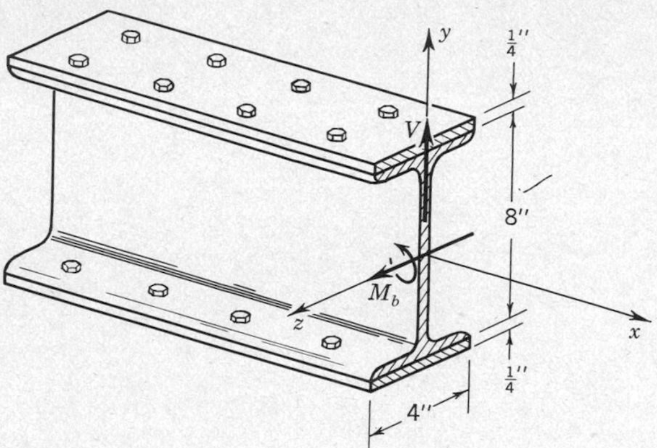

Prob. 7.48

7.49. A 1020 HR steel pipe in a chemical plant is lined with 2024-0 aluminum alloy for corrosion resistance. When the pipe is installed in the piping system, what is the maximum bending moment it can withstand without exceeding the yield stress of either the steel or the aluminum alloy?

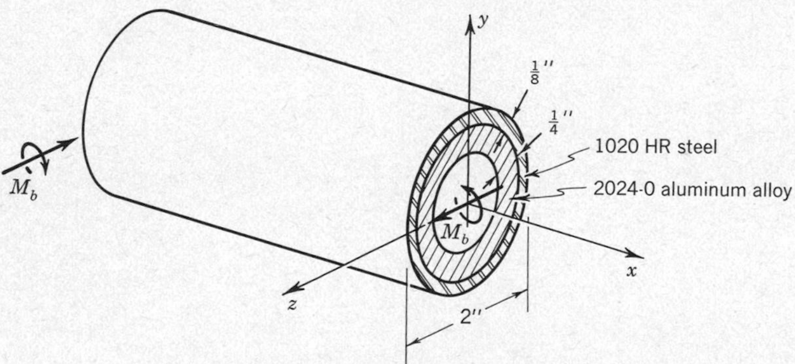

Prob. 7.49

7.50. A rectangular beam is made of a material with different properties in compression from those in tension, as shown by the curve. Find the maximum bending moment the beam can resist without exceeding the yield stress in tension or in compression.

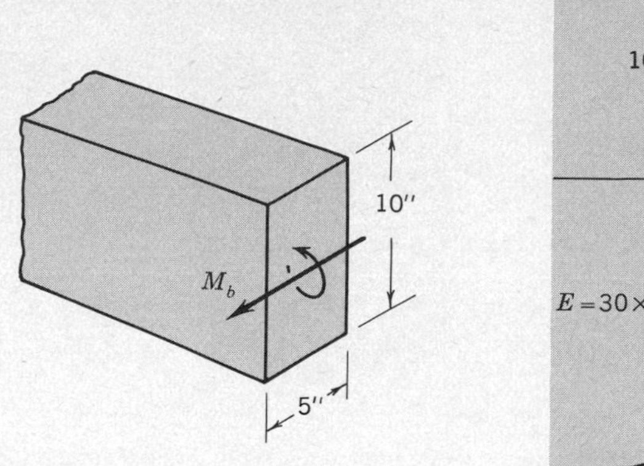

Prob. 7.50

7.51. Concrete is a brittle material which has good strength in compression but very little strength in tension. Despite its low tensile strength, economic use can be made of concrete in *reinforced-concrete* construction in which steel bars are imbedded in the concrete to provide tensile action. For a reinforced-concrete beam, carry out a development parallel to that given in Secs. 7.2 to 7.5 under the assumptions that *no* tensile stresses are carried by the concrete and that the tensile stress in the steel is uniform over the bars. Show that the neutral surface is located at a distance kd below the top of the beam, where the factor k is determined by the following quadratic equation.

$$E_s(d - kd)A_s - E_c \frac{b(kd)^2}{2} = 0$$

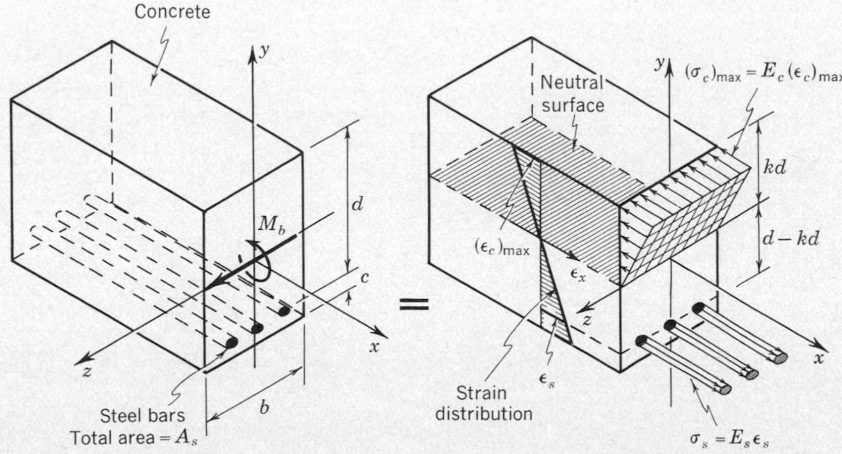

Prob. 7.51

Show also that the tensile stress in the steel and the maximum compressive stress in the concrete are given by

$$\sigma_s = \frac{M_b}{A_s d(1 - k/3)}$$

$$(\sigma_c)_{max} = \frac{2M_b}{bd^2k(1 - k/3)}$$

7.52. The reinforced-concrete beam shown in the sketch contains five ¾-in.-diameter steel bars. If the tensile stress in the steel is not to exceed 20,000 psi and the compressive stress in the concrete is not to exceed 1,350 psi, what is the maximum bending moment which the beam can transmit? Take E_c as 1.5×10^6 psi.

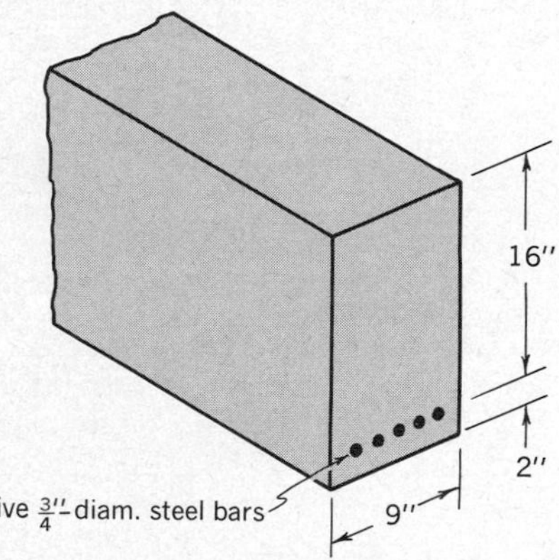

Prob. 7.52

7.53. Assume that the diameter of the steel bars in the reinforced-concrete beam of Prob. 7.52 is changed so that, when the beam is transmitting its maximum allowable bending moment, the tensile stress in the steel is 20,000 psi *and* the maximum compressive stress in the concrete is 1,350 psi. What would be the diameter of the bars in the new design, and what would be the maximum allowable bending moment? (*Note:* A beam designed so that the maximum permissible stresses in the steel and the concrete are reached simultaneously, as in the above design, is said to have *balanced* reinforcement.)

7.54. *Prestressed concrete* is a type of reinforced concrete which makes maximum use of both the compressive strength of concrete and the high tensile strength that can be obtained in cold-drawn steel wires. A prestressed-concrete beam can be made by stretching the steel reinforcing before the concrete is poured and then removing the forces on the ends of the reinforcing after the concrete has hardened and cured, thereby straining the concrete in compression. In subsequent bending, the concrete at any point can experience a tensile bending strain equal in magnitude to the compressive prestrain at the point without, in fact, experiencing a *net* tensile strain. Consider a 9×18-in. beam prestressed by seventy-six 0.15-in.-diameter cold-drawn steel wires arranged as indicated in the sketch. The tension in the wires is 130,000 psi after the ends of the wires are released. Calculate the stress and strain distribution in the concrete after the ends of the wires are released. Then calculate the maximum bending moment which this prestressed beam can transmit without

(*a*) Producing a *net* tensile strain in the concrete, *or*
(*b*) Exceeding a *net* compressive stress of 2,250 psi in the concrete, *or*
(*c*) Exceeding a *net* tensile stress of 150,000 psi in the steel wire

Take $E_c = 1.5 \times 10^6$ psi.

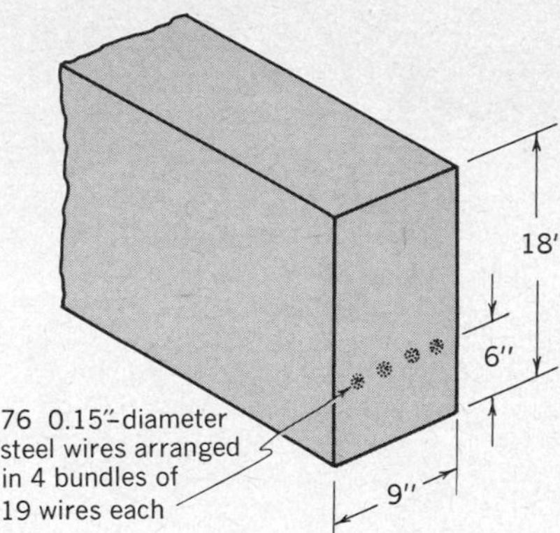

Prob. 7.54

7.55. Consider the case where a *composite* beam transmits a shear force in addition to a bending moment. Using the results of Prob. 7.46, repeat the arguments of Sec. 7.6 to show that the average shear stress τ_{xy} at a distance y_o from the neutral surface is given by

$$\tau_{xy} = \tau_{yx} = \frac{q_{yx}}{b} = \frac{V}{b[E_1(I_{yy})_1 + E_2(I_{yy})_2]} \int_{A_o} Ey\, dA$$

where the integral is to be taken over the area A_o in part (*b*) of the figure, that is, over the range from $y = y_o$ to $y = c$.

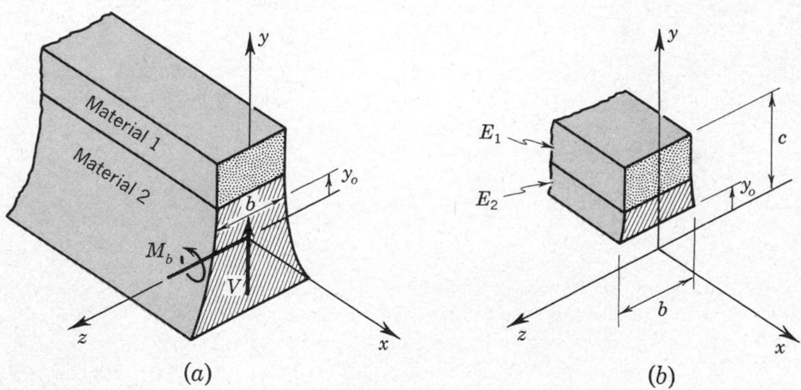

Prob. 7.55

7.56. Calculate the maximum allowable spacing of the rivets in the composite beam of Prob. 7.48 if the maximum shear force V is 6,000 lb and each rivet can safely carry a shear force across it of 500 lb.

7.57. A 4-in., 7.25-lb channel section is used for the main member of a clamp, as shown. The centroid of the channel is 0.46 in. from its base, its area is 2.12 in.^{2} and $I_{yy} = 0.44$ in.^{4} The clamping force acts at a distance of $1\frac{1}{4}$ in. above the base of the channel. What is the maximum stress in the channel if the clamping force is 3,000 lb?

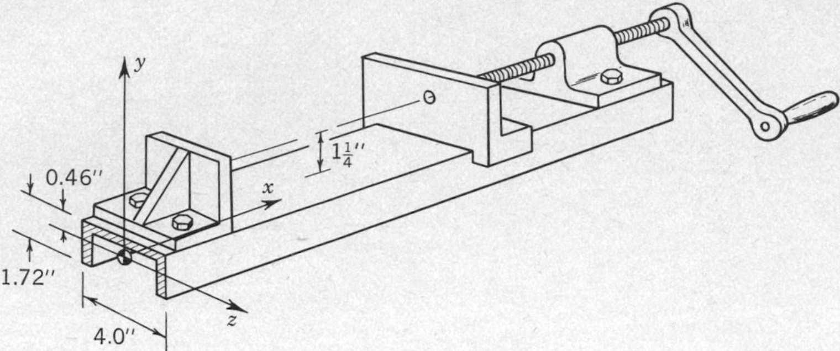

Prob. 7.57

7.58. Consider again the cylindrical tank of Prob. 7.23 for a particular case where the dimensions are fixed and $L \gg r$. Calculate the maximum bending stress when the liquid in the tank weighs γ per unit volume. Suppose now the vent is closed and the liquid is pressurized. At what pressure p will the maximum axial tensile stress in the tank wall be double the maximum axial tensile stress when the vent is open?

7.59. A ½-in. 1045 steel bar is being reduced to 0.300-in. diameter in a lathe-turning operation in which the axial feed force is 136 lb and the tangential force is 360 lb, as indicated in the sketch. If the tool face is located 3 in. from the collet which holds the workpiece, estimate the maximum shear stress existing in the workpiece at the collet.

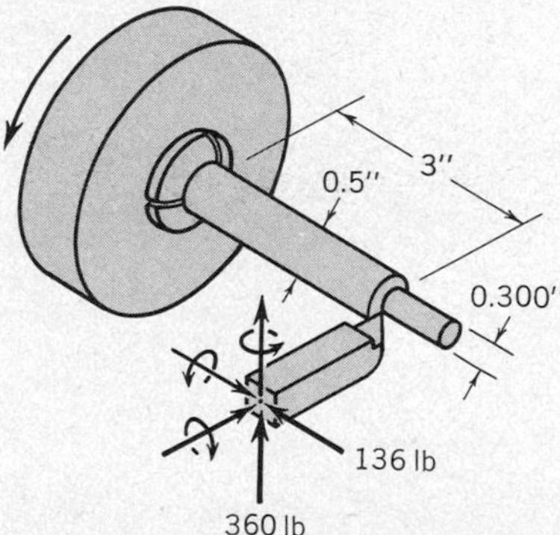

Prob. 7.59

7.60. The frame of a modern chair is made of 1020 CR steel bar stock 2 in. wide and t in. thick. If the chair is to be used by a heavy man, say 250 lb, what thickness of steel would you recommend? Make your own estimate of the loading conditions.

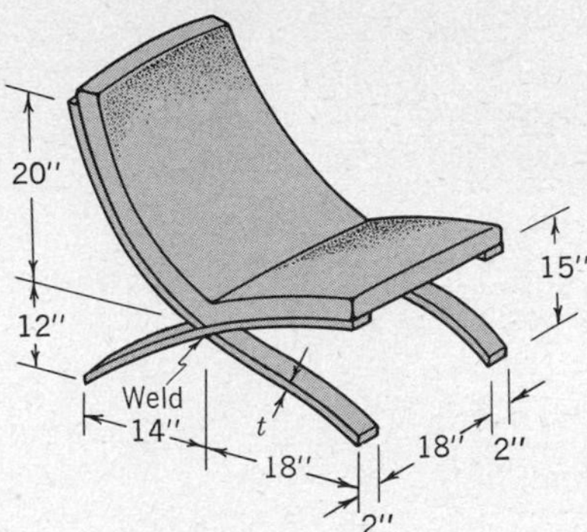

Prob. 7.60

7.61. The frame of a hacksaw is to be formed from $\frac{1}{16}$-in. 1020 CR steel sheet. If the tension in the blade will be about 75 lb, is the frame design reasonable?

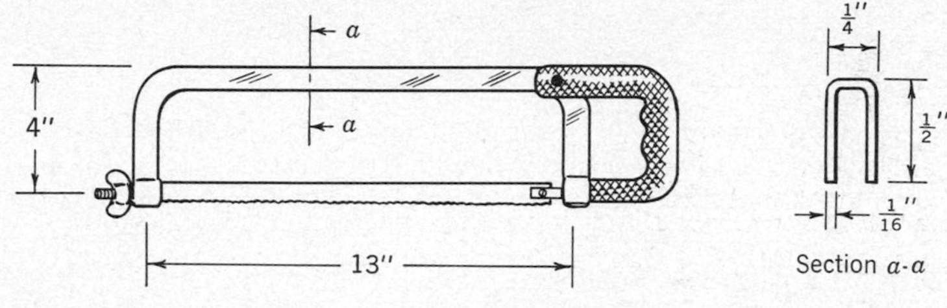

Prob. 7.61

7.62. A 3-in.-diameter 4130 HT steel rod 240 in. long is bent into the shape shown, where all the angles of the bent rod are *right* angles, and built into a wall at one end. What is the maximum force P which can be put on the free end in the direction shown without causing yielding of the rod?

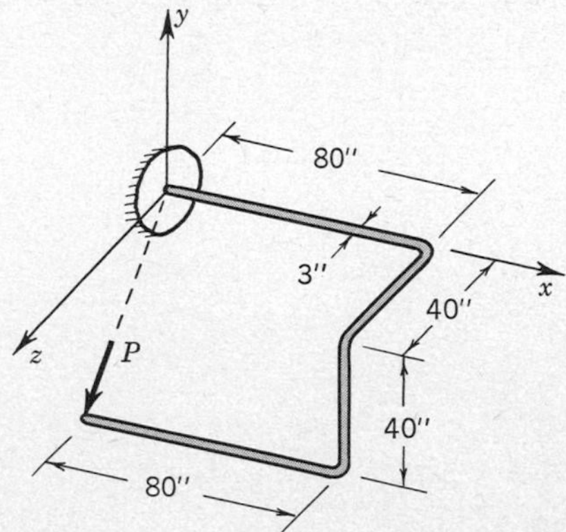

Prob. 7.62

7.63. The offset arm shown consists of two uniform members, one a rod of radius r and length a and the other a rod of radius $1.3r$ and length $2a$, and a conical connecting member of length a which tapers uniformly from a radius r to a radius $1.3r$. If the yield stress in tension is Y, determine the maximum force P which can act as shown without causing yielding anywhere.

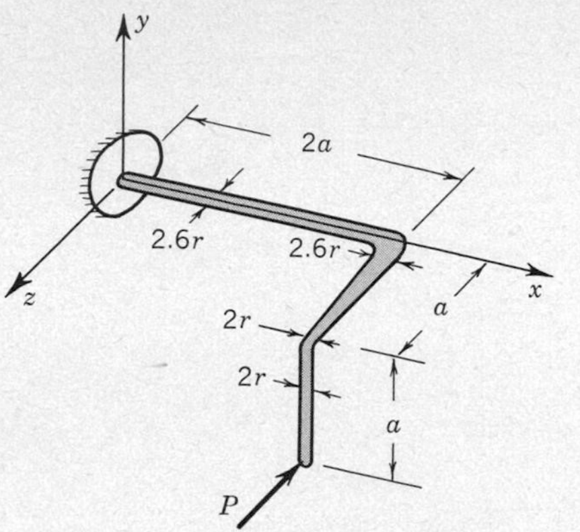

Prob. 7.63

7.64. The thin-walled tube shown is built-in at one end and is fitted with a frictionless piston at the other end. The only means of support of the tube is at the built-in end. The tube is filled with an incompressible fluid, and the piston is acted on by an axial force F. Point N is located on the top of the tube at the built-in end. For the case where $L_1/R = 5$, $L_2/R = 6$, and $R_1/R = 1/\sqrt{6}$, find the magnitudes of the principal stresses at point N in terms of F, R, and t. What is the value of F in terms of R, t, and Y for the onset of yielding at point N according to the maximum shear-stress yield criterion?

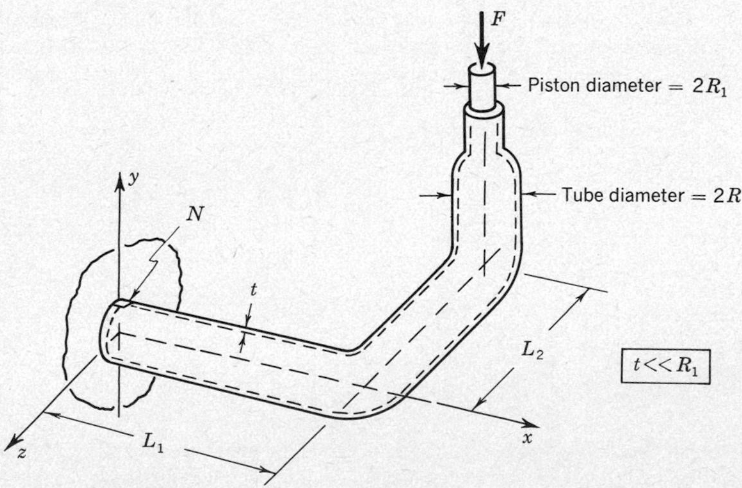

Prob. 7.64

7.65. Find the ratio $K = M_L/M_Y$ for the T beam shown.

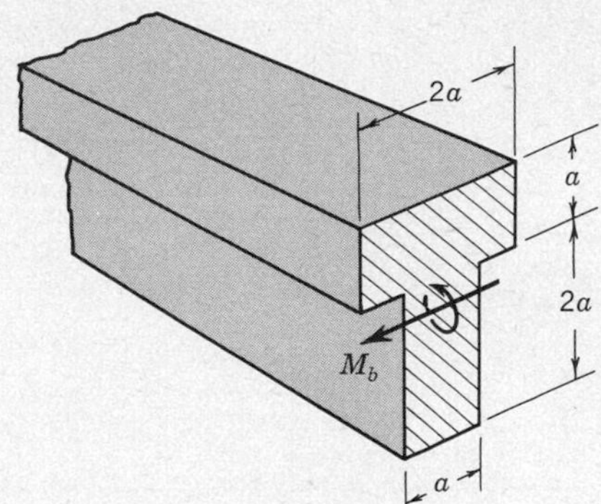

Prob. 7.65

7.66. Estimate the fully plastic moment M_L for the composite pipe of Prob. 7.49.

7.67. A beam is built of alternate layers of ⅛-in. 2024-T4 aluminum alloy and foam plastic, as shown. The foam plastic contributes very little to the bending stiffness since its modulus of elasticity is so low; its purpose is to maintain the separation of the four aluminum strips. The beam is bent with a moment of such magnitude that the strain in the bottom strip is 0.016. What are the stresses in the aluminum strips and the bending moment corresponding to this state of deformation? If the bending moment is now removed, what will be the residual stresses remaining in the aluminum strips?

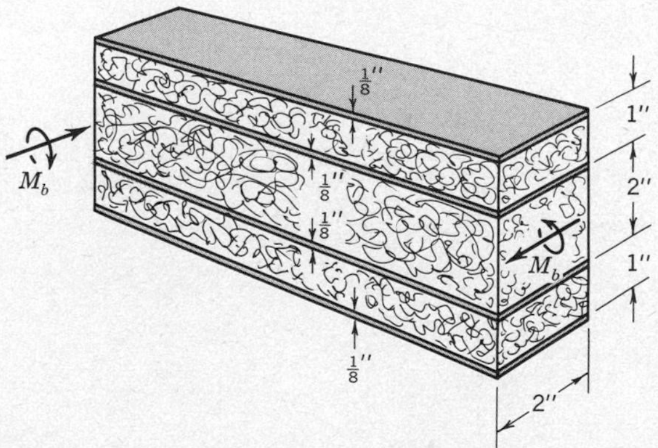

Prob. 7.67

7.68. Obtain by experiment an estimate of the fully plastic moment M_L of the wire of a paper clip. Measure the diameter of the wire and from the experimental value of M_L calculate the yield stress Y. (Note that the stress depends on the cube of the diameter, so the diameter measurement should be reasonably accurate.) Compare your result against the curves given in Fig. 5.5*a*. Which of the steels in Fig. 5.5*a* is most nearly like that of the paper clip?

7.69. The drive between two rotating cylinders is a phosphor bronze band of thickness 0.040 in. The band is made by butt-joining the two ends of a bronze strip with a silver brazing alloy [Fig. (*b*)]. During operation, the driving side of the band transmits a tensile force of up to 100 lb, and it is found that after a relatively short period of operation, cracks are found in the brazed joint and an imminent breaking of the band at that location is feared. Discuss the following proposed remedies, and choose the ones you think are suitable.

(*a*) Retain the butt-brazed joint, but increase the band thickness to 0.080 in.

(*b*) Retain the butt-brazed joint, but reduce the band thickness to 0.020 in.

(*c*) Retain the band thickness of 0.040 in., but change to a lap-brazed joint [Fig. (*c*)].

(*d*) Retain the band thickness of 0.040 in., but change to a modified lap-brazed joint [Fig. (*d*)].

(*e*) Retain the band of thickness 0.040 in. and the butt-brazed joint, but increase the band width to 3 in.

Mechanical properties

	A. Phosphor bronze	*B. Silver-brazing alloy*
Young's modulus	16,000,000 psi	11,300,000 psi
Tensile strength	81,000 psi	40,000 psi
Fatigue limit	32,000 psi	15,000 psi

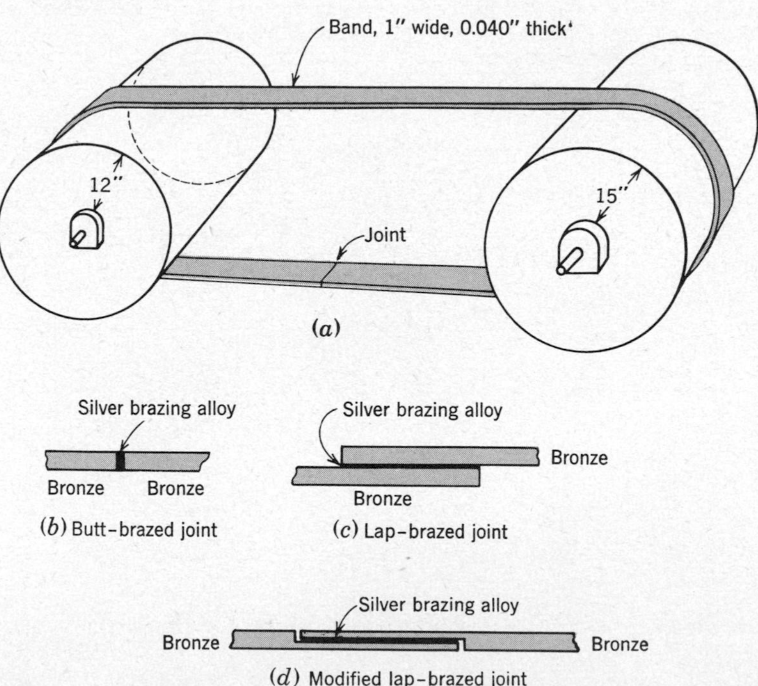

(*a*)

(*b*) Butt-brazed joint

(*c*) Lap-brazed joint

(*d*) Modified lap-brazed joint

Prob. 7.69

7.70. A strip of steel 18 in. wide, 84 in. long, and $\frac{1}{10}$ in. thick is to be bent to form a beam of rectangular cross section 6 in. deep and 3 in. wide. It is to be used to carry a central load of 4,000 lb on a simple span of 80 in. as shown in Fig. (*a*). In Figs. (*b*), (*c*), and (*d*) are shown three possible locations for the longitudinal weld to be made between the edges of the strip after the strip has been bent into the beam of rectangular shape.

(*a*) Which location of the weld is worst? Why?

(*b*) Which location of the weld is best? Why?

Calculate the numerical value of the longitudinal shear stress which will have to be carried by the weld when it is located in the *worst* position.

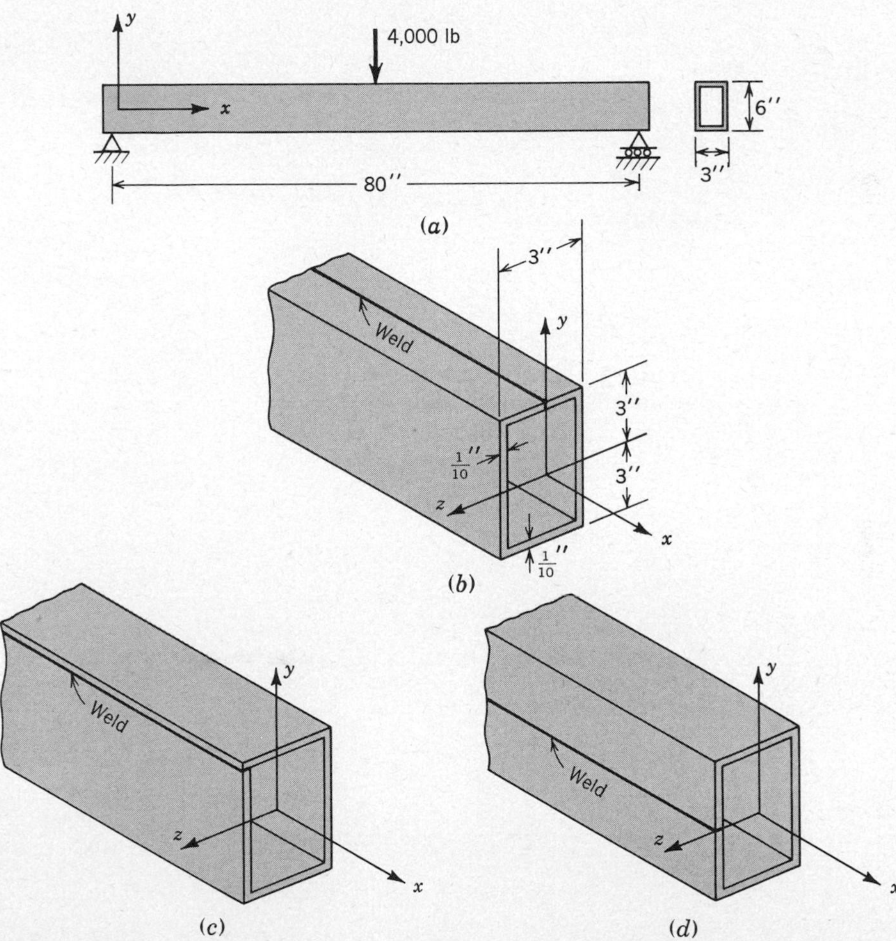

Prob. 7.70

7.71. A "metal" ski is a combination of wood and metal bonded together as shown in the cross section on the left. For purposes of analysis we will assume that the ski cross section can be modeled as shown on the right, where the modulus of elasticity of the wood is taken to be $E_w = 0.06E_m$. Using this model estimate the maximum bending stress in the ski if the ski is supported between snow hummocks and loaded centrally as illustrated.

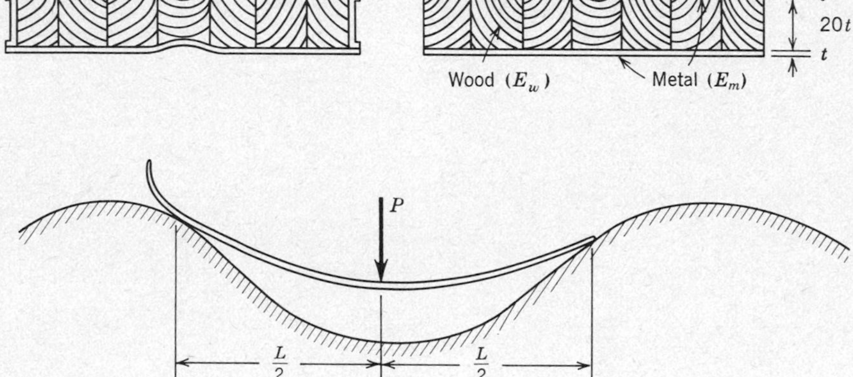

Prob. 7.71

7.72. When the central section of the beam in Fig. 7.30 is fully plastic, as indicated in Fig. 7.30*d*, what is the extent of the plastic zone along the bottom of the beam?

7.73. Calculate the curvature $d\phi/ds$ of the neutral axis of the beam in Example 7.9. Also, check the calculation in Example 7.9 for the bending stress at point *A* by calculating this stress from

$$\sigma_x = -E \frac{d\phi}{ds} m$$

7.74. The stress distribution and deformation in an unsymmetrical beam may be calculated by considering the resultant bending moment to be the vector sum of components M_{b1} and M_{b2} in the directions of the principal axes of inertia through the centroid, as indicated in the accompanying sketch. Verify that Eq. (7.64) then reduces to

$$\sigma_x = \frac{M_{b1}}{I_{22}} \xi_2 - \frac{M_{b2}}{I_{11}} \xi_1$$

where ξ_1 and ξ_2 are the position coordinates in the 1 and 2 directions, and

$$I_{11} = \int_A \xi_1{}^2 \, dA \qquad I_{22} = \int_A \xi_2{}^2 \, dA$$

Using this approach, calculate the stress at point *A* in Example 7.9.

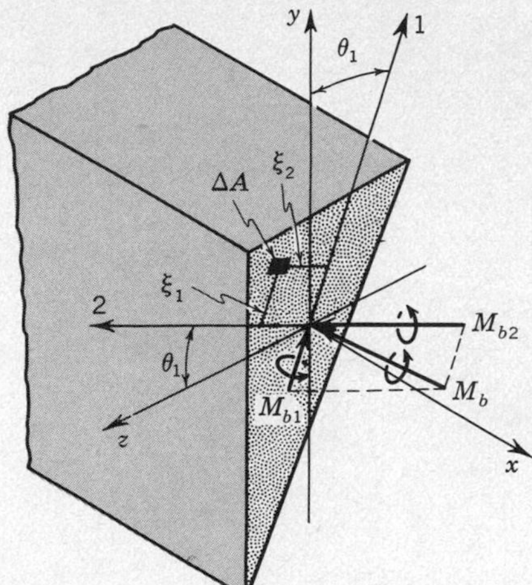

Prob. 7.74

7.75. Locate the shear center for the thin-walled channel section.

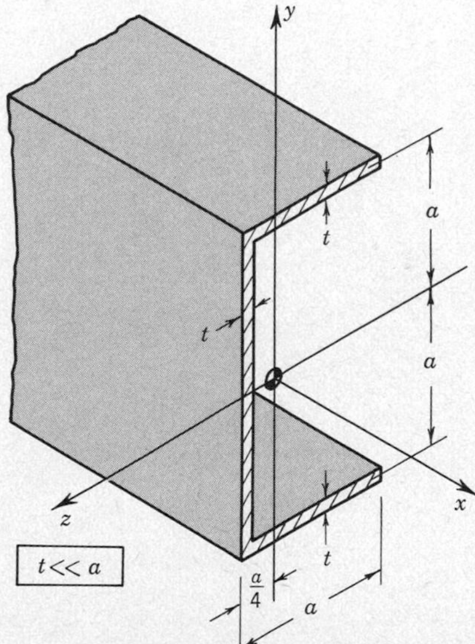

Prob. 7.75

7.76. Find the location of the shear center for the slit, thin-walled tube of radius r and thickness t.

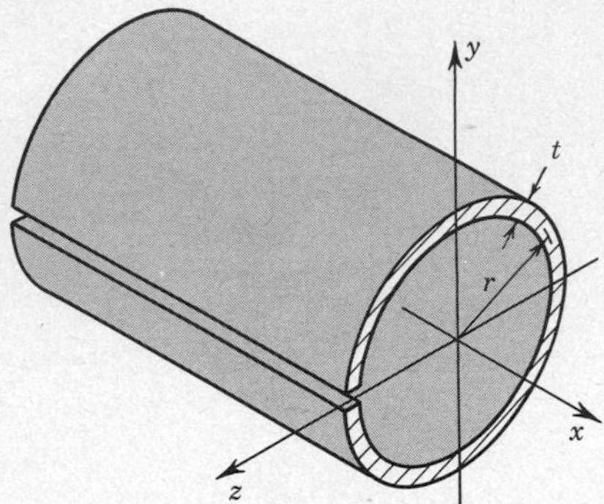

Prob. 7.76

7.77. A cantilever beam has a Z-shaped section for which

$$I_{yy} = \tfrac{8}{3}ta^3 \qquad I_{zz} = \tfrac{2}{3}ta^3 \qquad I_{yz} = -ta^3$$

Calculate the maximum bending stress in the beam when it is loaded with an end load P as shown.

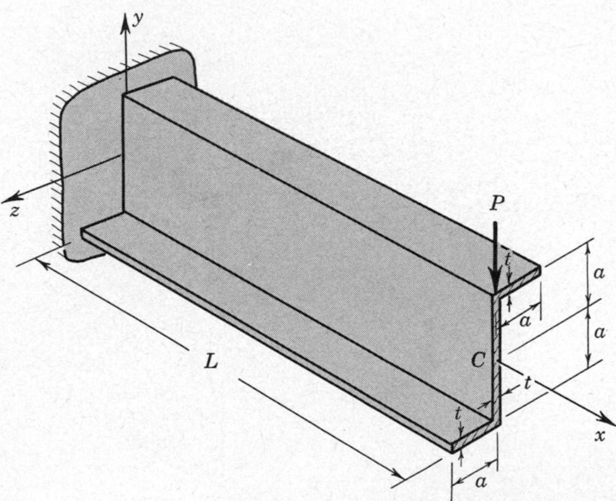

Prob. 7.77

7.78. Determine the maximum bending stress in a beam made from an angle which has the section of Probs. 7.2 and 7.10 if it transmits a bending moment of $M_{by} = 100{,}000$ in.-lb.

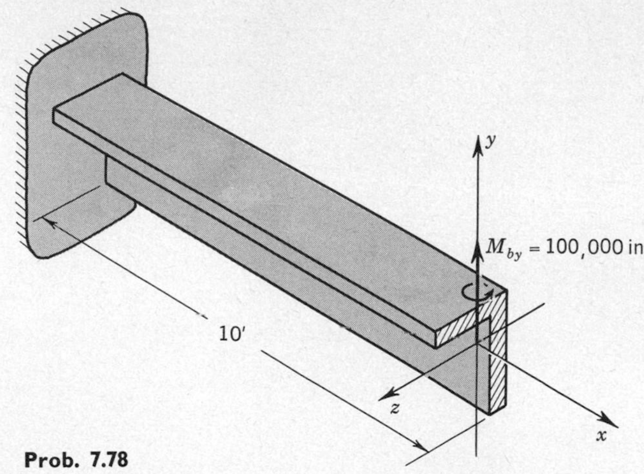

Prob. 7.78

7.79. If the beam is not to twist when the force P is applied, what should be the location of P; that is, what should be the value of e?

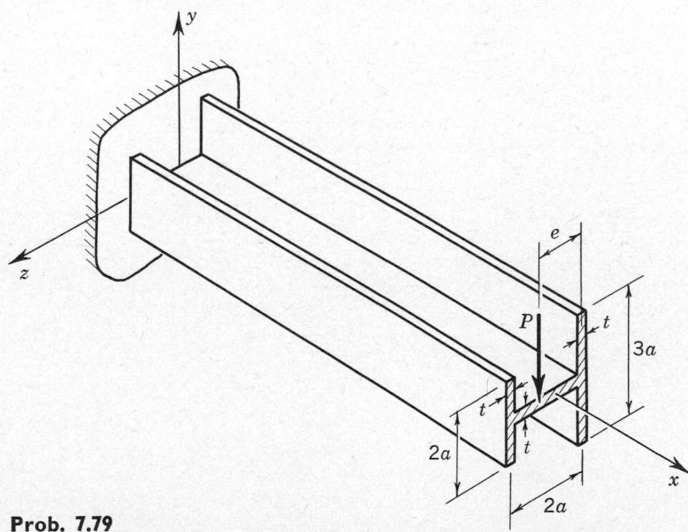

Prob. 7.79

7.80. Show that for a rectangular cantilever beam with an end load the principal stress directions are as indicated in the figure. (These two orthogonal sets of curves are called the *stress trajectories*. The principal stresses at any point are tangent to the two curves which intersect at the point. The curves indicate direction only; the magnitudes of the principal stresses vary along any given curve.)

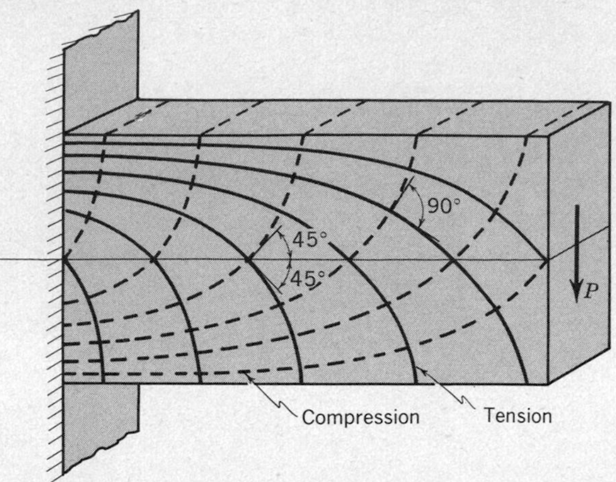

Prob. 7.80

7.81. The cross section of a solid circular shaft of radius r is acted on by a bending moment M_b and a twisting moment M_t. Show that the maximum shear stress in the shaft is given by

$$\tau_{\max} = \frac{r}{2I_{yy}} \sqrt{M_b^2 + M_t^2}$$

Show also that this shear stress acts on planes whose normals make an angle

$$\psi = \tfrac{1}{2} \tan^{-1} \frac{M_b}{M_t}$$

with the axial and tangential directions. Finally, verify that these results are valid for a hollow circular shaft (not necessarily thin-walled).

8
Deflections Due to Bending

8.1 INTRODUCTION

In this chapter we consider the deflections of slender members which transmit bending moments. There are many practical design problems in which deflection considerations are of great importance. For example, in high-speed machinery with close tolerances, excessive deflections can cause interference between moving parts; many machine elements, such as leaf springs, are designed primarily on the basis of their deflections; the failure to limit deflections in the structural framework of buildings is often indicated by the development of cracks in plastered walls and ceilings.

The determination of bending deflections, like all the problems considered in this book, involves first the selection of a model which is to represent the actual physical member. The model for the analysis of beams was first introduced in Chap. 3. In Chap. 7 the local deformation and the stress distribution across a section of a beam were discussed. We did not, however, evaluate the overall deformation of the beam. We now return to this question and develop the theory for small deflections of *elastic* beams. We then shall be able to make full use of

(2.1) in structural problems involving beams. In particular we shall treat statically indeterminate beams which require simultaneous consideration of *all three* of the steps (2.1). We also shall study mechanisms of *plastic* collapse for statically indeterminate beams.

8.2 THE MOMENT-CURVATURE RELATION

In Chap. 7 we saw that when a symmetrical, linearly elastic beam element is subjected to pure bending, as shown in Fig. 8.1, the curvature of the neutral axis is related to the applied bending moment by the equation

$$\frac{1}{\rho} = \lim_{\Delta s \to 0} \frac{\Delta \phi}{\Delta s} = \frac{d\phi}{ds} = \frac{M_b}{EI_{yy}} \tag{8.1}$$

where E is the modulus of elasticity and I_{yy} is the moment of inertia of the cross-sectional area. Throughout this chapter we shall always maintain the orientation of Fig. 8.1; i.e., the long dimension of the beam will be in the x direction, and the bending will take place in the xy plane. Under these circumstances it will cause

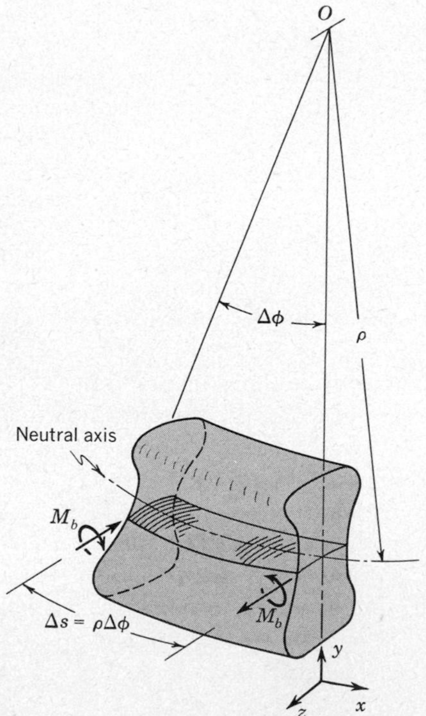

Fig. 8.1 Deformation of an element of a beam subjected to bending moments M_b.

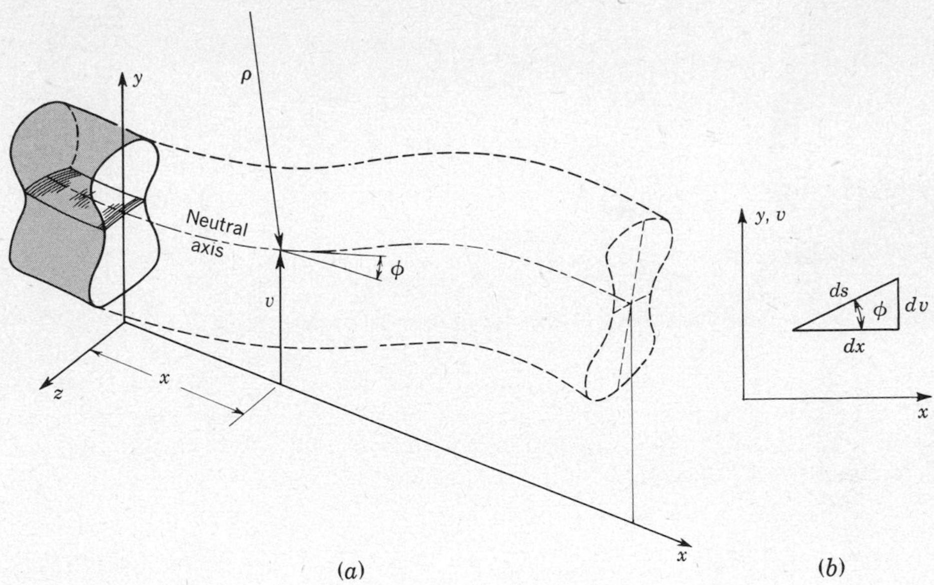

Fig. 8.2 Geometry of the neutral axis of a beam bent in the xy plane.

no ambiguity if, instead of the symbol I_{yy} for the moment of inertia about the neutral surface, we use the abbreviation I.

The curvature of the neutral axis completely defines the deformation of an element in pure bending. To extend this to the case of *general bending* where the bending moment varies along the length of the beam, we make a simplifying assumption. We *assume* that the shear forces which necessarily accompany a varying bending moment do not contribute significantly[1] to the overall deformation. Thus we assume that the deformation is still defined by the curvature and that the curvature is still given by (8.1). If we know how the bending moment varies along the length of the beam, we will then know how the curvature varies.

To determine the bent shape of the beam, we thus need to be able to deduce the deflection of the neutral axis from a knowledge of its curvature. To facilitate this, we first derive a differential equation relating the curvature $d\phi/ds$ to the deflection $v(x)$.

We start with the definition of the *slope* of the neutral axis in Fig. 8.2*a*,

$$\frac{dv}{dx} = \tan \phi$$

[1] Some additional deformation due to shear does of course occur; however, for long, slender beams this additional deformation is negligible in comparison with the bending. See Probs. 8.41 and 8.42, and also S. Timoshenko and J. N. Goodier, "Theory of Elasticity," 3d ed., pp. 46, 49, and 121, McGraw-Hill Book Company, New York, 1970.

Next, differentiation with respect to arc length s gives

$$\frac{d^2v}{dx^2}\frac{dx}{ds} = \sec^2 \phi \frac{d\phi}{ds}$$

or, that the curvature is

$$\frac{d\phi}{ds} = \frac{d^2v}{dx^2}\frac{dx}{ds}\cos^2 \phi$$

Now, from Fig. 8.2b we have

$$\cos \phi = \frac{dx}{ds} = \frac{1}{[1 + (dv/dx)^2]^{1/2}}$$

so that in terms of derivatives of the deflection, the curvature is

$$\frac{d\phi}{ds} = \frac{d^2v/dx^2}{[1 + (dv/dx)^2]^{3/2}} \tag{8.2}$$

If we substitute (8.2) into (8.1), we obtain a *nonlinear* differential equation for the determination of v once M_b is known as a function of x[1]

$$\frac{d^2v/dx^2}{[1 + (dv/dx)^2]^{3/2}} = \frac{M_b}{EI}$$

When the slope angle ϕ shown in Fig. 8.2 is small, then dv/dx is small compared to unity. If we neglect $(dv/dx)^2$ in the denominator of the right-hand term of (8.2), we obtain a simple approximation for the curvature

$$\frac{d\phi}{ds} \approx \frac{d^2v}{dx^2} \tag{8.3}$$

There is less than a 1 percent error involved in the approximation (8.3) to the exact curvature expression (8.2) when ϕ is less than 4.7°. In most engineering applications where relatively stiff beams are used, the slope angle *is* small, and we *can* use the approximation (8.3) for the curvature.

Substituting the approximate curvature (8.3) into (8.1) we obtain the linear differential equation

$$\frac{d^2v}{dx^2} = \frac{M_b}{EI} \tag{8.4}$$

which relates the bending moment to the transverse displacement. This equation is fundamental to our subsequent work on elastic beam deflections. Although (8.4) involves an approximation to the curvature which is valid only for small bending angles, we shall henceforth call it the *moment-curvature relation*. It is essentially a

[1] See W. Flügge, "Handbook of Engineering Mechanics," p. 45–48, McGraw-Hill Book Company, New York, 1962, for a discussion of this nonlinear problem. See also Prob. 8.61.

"force-deformation" or "stress-strain" relation in which the bending moment is the "force" or "stress" and the approximate curvature is the resulting "deformation" or "strain." The relation is a *linear* one; the constant of proportionality EI is sometimes called the *flexural rigidity* or the *bending modulus*.

The sign convention associated with (8.4) should be noted. We shall take the orientation shown in Figs. 8.1 and 8.2 as standard throughout this chapter. According to the sign convention of Chap. 3, the bending moment shown in Fig. 8.1 is considered *positive*. The corresponding curvature in Fig. 8.1, concave up, is *positive* and corresponds to a *positive* value of the approximate curvature (8.3).

In solving beam-deflection problems we will use the three steps of (2.1). We study the forces and use the *equilibrium* requirements to obtain M_b as a function of x. The *force-deformation* relation (8.4) then gives us the approximate curvature as a function of x. Finally, we study the geometry of deformation as we integrate (8.4); the constants of integration are evaluated by imposing requirements of *geometric compatibility*.

8.3 INTEGRATION OF THE MOMENT-CURVATURE RELATION

As outlined in the previous section, when an expression for the bending moment as a function of position along the beam has been obtained from force and moment equilibrium considerations, then direct integration of the moment-curvature relation leads to the correct deflection curve, provided that the integration constants are determined so as to make the deflection curve compatible with the external restraints. This procedure is illustrated in the following examples.

Example 8.1 The simply supported beam of uniform cross section shown in Fig. 8.3 is subjected to a concentrated load W. It is desired to obtain the deflection curve of the deformed neutral axis.

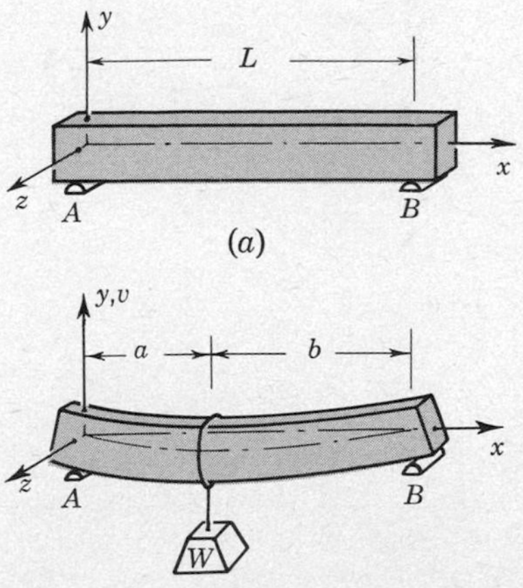

Fig. 8.3 Example 8.1. Simply supported beam (*a*) before and (*b*) after application of a concentrated load W.

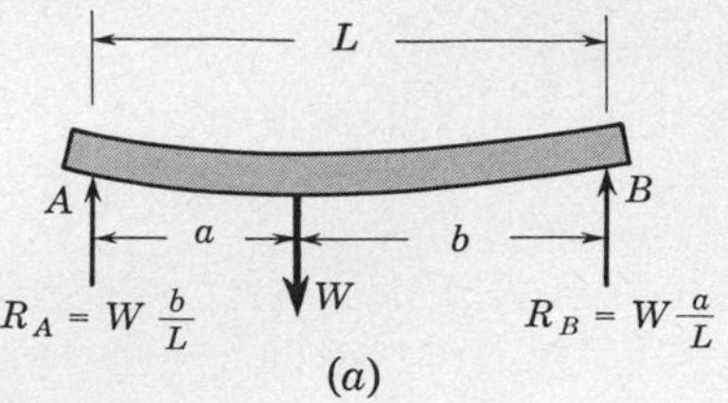

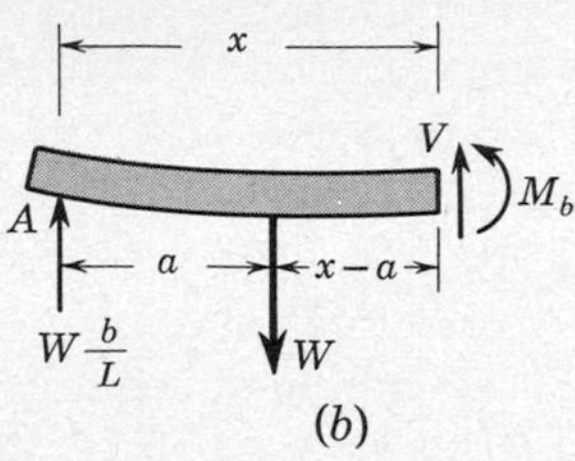

Fig. 8.4 Example 8.1. Free-body diagram of beam and segment of beam.

We begin the analysis by drawing a free-body diagram of the beam, as shown in Fig. 8.4*a*. The reactions R_A and R_B are obtained from the overall force and moment balance conditions. Using the singularity functions and bracket notation introduced in Sec. 3.6, we can write a single expression for the bending moment M_b directly from the free body of Fig. 8.4*b*.

$$M_b = \frac{Wb}{L}x - W\langle x - a\rangle^1 \tag{a}$$

which is valid for $0 \leqq x \leqq L$. The moment-curvature relation (8.4) combined with (*a*) leads to

$$EI\frac{d^2v}{dx^2} = M_b = \frac{Wb}{L}x - W\langle x - a\rangle^1 \tag{b}$$

Since the bending modulus EI is constant along the beam, integration of (*b*) yields

$$EI\frac{dv}{dx} = \frac{Wb}{L}\frac{x^2}{2} - W\frac{\langle x - a\rangle^2}{2} + c_1 \tag{c}$$

$$EIv = \frac{Wb}{L}\frac{x^3}{6} - W\frac{\langle x - a\rangle^3}{6} + c_1x + c_2 \tag{d}$$

where c_1 and c_2 are constants of integration.

The geometric boundary conditions for this problem are that there should be no transverse displacement over the supports; i.e.,

$$v = 0 \qquad \text{at } x = 0 \text{ and at } x = L \tag{e}$$

These conditions together with (d) give us the following relations for the determination of c_1 and c_2:

$$0 = c_2$$
$$0 = \frac{Wb}{6L}L^3 - \frac{Wb^3}{6} + c_1L \qquad (f)$$

If we insert the values for c_1 and c_2 determined from (f) into (d), we obtain the following deflection curve for the neutral axis of the beam:

$$v = -\frac{W}{6EI}\left[\frac{bx}{L}(L^2 - b^2 - x^2) + \langle x - a\rangle^3\right] \qquad (g)$$

To give some idea of order of magnitudes, let us consider the following particular case:

$$L = 12 \text{ ft}$$
$$a = b = 6 \text{ ft}$$
$$W = 400 \text{ lb} \qquad (h)$$
$$E = 1.6 \times 10^6 \text{ psi}$$
$$I = 57.1 \text{ in.}^4$$

These values correspond to a very common case in small-house construction. The beam is a nominal 2 × 8 in. (actually 1⅝ × 7½ in.) floor joist spanning 12 ft with a central load close to the maximum which would be considered for a single joist of this span in small-house design. The value of E listed is an intermediate value for the reciprocal of S_{11} given in Table 5.5 for Douglas-fir. If we insert the particular values (h) into (g), the greatest deflection occurs at the center and has the value

$$(v)_{x=L/2} = -\frac{400\,\dfrac{72 \times 72}{144}[(144)^2 - (72)^2 - (72)^2]}{6 \times 1.6 \times 10^6 \times 57.1}$$
$$= -0.27 \text{ in.} \qquad (i)$$

Another magnitude of interest is the greatest slope of the deformed neutral surface. For the particular case (h) the greatest slope magnitude occurs simultaneously at the two ends. To evaluate the slope, we can either substitute (f) back into (c) or differentiate (g). Inserting the values (h) and setting $x = 0$ yields

$$\left(\frac{dv}{dx}\right)_{x=0} = -0.0057 \qquad (j)$$

which may be taken as the value of the slope angle ϕ, in radians. Converted to degrees, this is 0.33°.

These numerical results provide a justification for the use of the approximate curvature relation (8.3) in connection with wooden structural members. Since wooden beams are about the most flexible structural members employed, we can conclude that (8.3) provides an adequate representation of the curvature of any structural element.

If in this example the bracket notation had not been used, it would have been necessary to represent the bending moment by separate analytical expressions valid on each side of the load and to integrate the two expressions separately, with the result that there would be *four* constants of integration to be evaluated. Two of these constants would be determined, as before, from the geometric boundary conditions at the ends of the beam, while the other two would be determined from the condition of geometric fit that the deflection and slope should be *continuous* at the load. We discussed this procedure in Example 3.6. These continuity conditions are automatically taken care of by using the bracket notation and integrating the singularity functions according to (3.16) and (3.17).

Example 8.2 A uniform cantilever beam has bending modulus EI and length L. It is built in at A and subjected to a concentrated force P and moment M applied at B, as shown in Fig. 8.5*a*. We shall find the deflection δ and the slope angle ϕ at B due to these loads.

In order to obtain the bending moment in the interior of the beam, we isolate the segment of length $L - x$ shown in Fig. 8.5*b*. From this free body we obtain the bending moment

$$M_b = -P(L - x) - M \tag{a}$$

which is valid for $0 \leq x \leq L$. Inserting (*a*) into the moment-curvature relation (8.4), we find the differential equation for the beam displacement $v(x)$,

$$EI\frac{d^2v}{dx^2} = -PL + Px - M \tag{b}$$

The geometric boundary conditions for this beam are that, at $x = 0$, the beam is built in with zero slope and zero deflection; that is,

$$\left(\frac{dv}{dx}\right)_{x=0} = 0 \qquad (v)_{x=0} = 0 \tag{c}$$

Integration of (*b*) subject to the boundary conditions (*c*) proceeds as follows. A first integration yields

$$EI\frac{dv}{dx} = -PLx + P\frac{x^2}{2} - Mx + c_1 \tag{d}$$

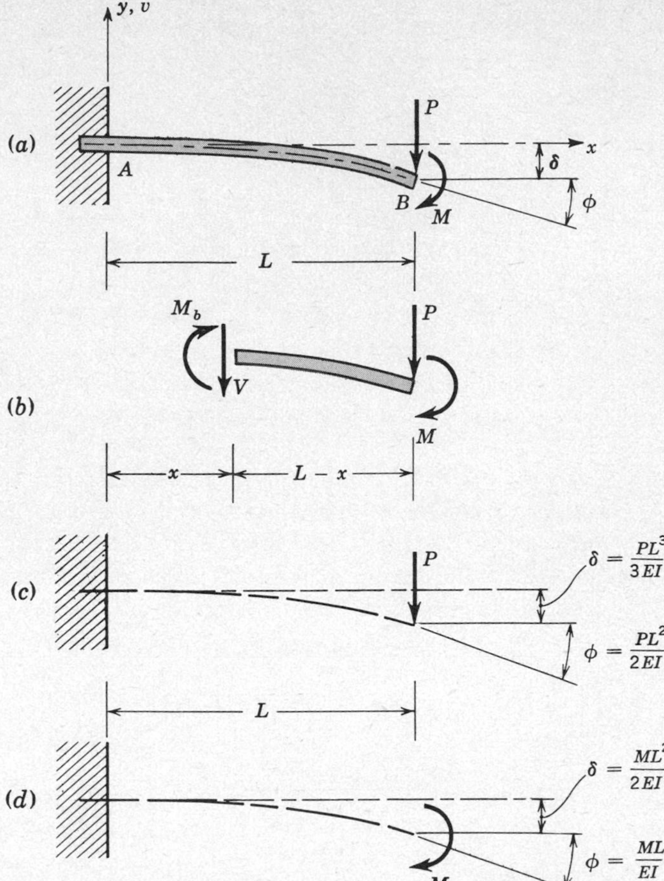

Fig. 8.5 Example 8.2. Cantilever beam with force and moment load.

where c_1 is a constant of integration. By inserting (*d*) into the slope boundary condition in (*c*), we find that c_1 must vanish. Integration of (*d*) then yields

$$EIv = -PL\frac{x^2}{2} + P\frac{x^3}{6} - M\frac{x^2}{2} + c_2 \tag{e}$$

where c_2 is a second constant of integration. In order for (*e*) to satisfy the displacement boundary condition in (*c*), it is necessary for c_2 to vanish. Thus the displacement $v(x)$ of the neutral axis of the cantilever is

$$v = -\frac{1}{EI}\left[P\frac{x^2}{6}(3L - x) + M\frac{x^2}{2}\right] \tag{f}$$

The terminal deflection δ in Fig. 8.5a is given by

$$\delta = -(v)_{x=L} = \frac{PL^3}{3EI} + \frac{ML^2}{2EI} \tag{g}$$

and the terminal slope ϕ is

$$\phi = -\left(\frac{dv}{dx}\right)_{x=L} = \frac{PL^2}{2EI} + \frac{ML}{EI} \tag{h}$$

These results are displayed in Fig. 8.5c and d for two special limiting cases. Figure 8.5c shows the case where the moment $M = 0$, and thus the only load on the cantilever is the force P. Figure 8.5d shows the case where $P = 0$, and the only load on the beam is the moment M.

Example 8.3 Figure 8.6a shows a cantilever beam built-in at A and subjected to a uniformly distributed load of intensity w per unit length acting on the segment BC. It is desired to obtain the deflection δ of the neutral axis at C due to the distributed load in terms of the constant bending modulus EI and the dimensions shown.

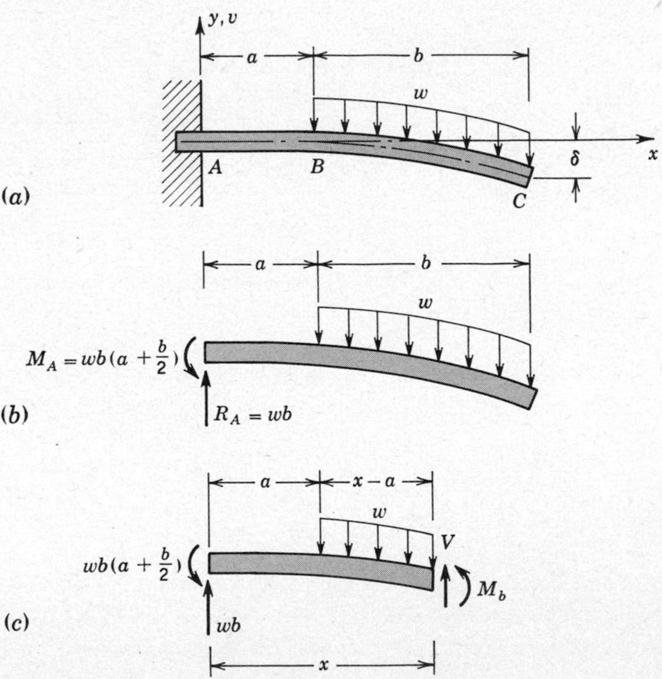

Fig. 8.6 Example 8.3.

The analysis begins with a study of the forces and equilibrium requirements in Fig. 8.6*b* and *c*. A free-body diagram of the entire beam, from which we compute the reactions, is shown in Fig. 8.6*b*. In Fig. 8.6*c* a free-body diagram of a segment of length x is shown, from which we obtain the bending moment to be inserted in (8.4).

$$EI\frac{d^2v}{dx^2} = M_b = wbx - wb\left(a + \frac{b}{2}\right) - \frac{w\langle x - a\rangle^2}{2} \tag{a}$$

One integration of (*a*) leads to

$$EI\frac{dv}{dx} = wb\frac{x^2}{2} - wb\left(a + \frac{b}{2}\right)x - \frac{w\langle x - a\rangle^3}{6} + c_1 \tag{b}$$

where c_1 is a constant of integration. We can evaluate c_1 at this time because one of the conditions of geometric constraint is that at the built-in end A the slope of the neutral axis should remain zero.

$$\left(\frac{dv}{dx}\right)_{x=0} = 0 \tag{c}$$

In order for (*b*) to satisfy (*c*) we must have $c_1 = 0$. One more integration of (*b*) then yields

$$EIv = wb\frac{x^3}{6} - wb\left(a + \frac{b}{2}\right)\frac{x^2}{2} - \frac{w\langle x - a\rangle^4}{24} + c_2 \tag{d}$$

The constant of integration c_2 is evaluated by applying the geometric requirement that the displacement of the neutral axis at the built-in end A should remain zero.

$$(v)_{x=0} = 0 \tag{e}$$

In order for (*d*) to satisfy (*e*) we must have $c_2 = 0$. We thus obtain from (*d*) an equation for the locus of the deformed neutral axis. The displacement labeled δ in Fig. 8.6*a* is the negative of the value of v at $x = a + b$. Substituting $x = a + b$ in (*d*) and simplifying, we obtain

$$\delta = -(v)_{x=a+b} = \frac{wb}{EI}\left(\frac{a^3}{3} + \frac{3a^2b}{4} + \frac{ab^2}{2} + \frac{b^3}{8}\right) \tag{f}$$

Two special cases of (*f*) are of interest. When $b = 0$, there is no loaded portion of the beam and according to (*f*) there is no deflection. When $a = 0$, the entire beam is loaded uniformly, and the deflection at the end is

$$\delta = \frac{wb^4}{8EI} \tag{g}$$

where now b is the entire length of the beam.

The preceding examples have been statically determinate; i.e., the bending moments could be explicitly determined from the equilibrium requirements. In a statically indeterminate problem the conditions of equilibrium are insufficient to determine the bending moment. We must take into account the geometrical restrictions and the moment-curvature relation as well as the equilibrium conditions before we can evaluate the bending moment. In other words, we must pursue all three of the steps (2.1) simultaneously. The following examples give an illustration of this procedure.

Example 8.4 Figure 8.7*a* shows a beam whose neutral axis coincided with the x axis before the load P was applied. The beam has a simple support at A and a clamped or built-in support at C. The bending modulus EI is constant along the length of the beam. It is desired to sketch the bending-moment diagram for the bending moments due to the load P.

Figure 8.7*b* shows a free-body diagram of the entire beam. Since P is given as vertical and R_A can only be vertical, the reaction at C can only consist of a vertical force R_C and a clamping moment M_C. There are no horizontal forces, and hence there are only *two* independent equilibrium requirements, but there are three unknowns: R_A, R_C, and M_C. The equilibrium conditions furnish only two relations between three quantities. The best we can do by considering only equilibrium is to take one of the reactions as an unknown and express the other two in terms of this unknown. For example,

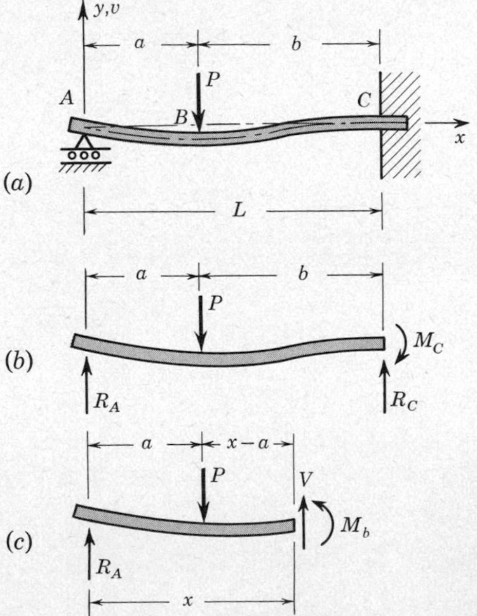

Fig. 8.7 Example 8.4.

taking R_A as unknown, the conditions of equilibrium applied to Fig. 8.7*b* yield

$$\begin{aligned} R_C &= P - R_A \\ M_C &= Pb - R_A L \end{aligned} \tag{a}$$

Similarly, applying the conditions of equilibrium to the segment of length x in Fig. 8.7*c* gives the following expression for the bending moment:

$$M_b = R_A x - P\langle x - a\rangle^1 \tag{b}$$

which is valid for $0 < x < L$.

Turning to the geometrical requirements for the deformed beam, we see that now we have *three* compatibility conditions

$$\begin{aligned} v &= 0 \qquad \text{at } x = 0 \\ v &= 0 \qquad \text{at } x = L \\ \frac{dv}{dx} &= 0 \qquad \text{at } x = L \end{aligned} \tag{c}$$

Thus, if we integrate the moment-curvature relation, we have enough conditions not only to evaluate the two constants of integration but also to evaluate the unknown reaction R_A which appears in (*b*). Setting up the moment-curvature relation (8.4) and carrying out one integration yields

$$\begin{aligned} EI\frac{d^2v}{dx^2} &= M_b = R_A x - P\langle x - a\rangle^1 \\ EI\frac{dv}{dx} &= R_A\frac{x^2}{2} - P\frac{\langle x - a\rangle^2}{2} + c_1 \end{aligned} \tag{d}$$

In order for the third of (*c*) to be satisfied, we must have

$$c_1 = \frac{Pb^2}{2} - \frac{R_A L^2}{2} \tag{e}$$

Inserting (*e*) in (*d*) and carrying out one more integration gives

$$EIv = R_A\frac{x^3}{6} - P\frac{\langle x - a\rangle^3}{6} + \frac{Pb^2x}{2} - \frac{R_A L^2 x}{2} + c_2 \tag{f}$$

In order for the first of (*c*) to be satisfied we must have $c_2 = 0$. Finally, to satisfy the second of (*c*) we must have

$$0 = R_A\frac{L^3}{6} - P\frac{b^3}{6} + \frac{Pb^2L}{2} - \frac{R_A L^3}{2} \tag{g}$$

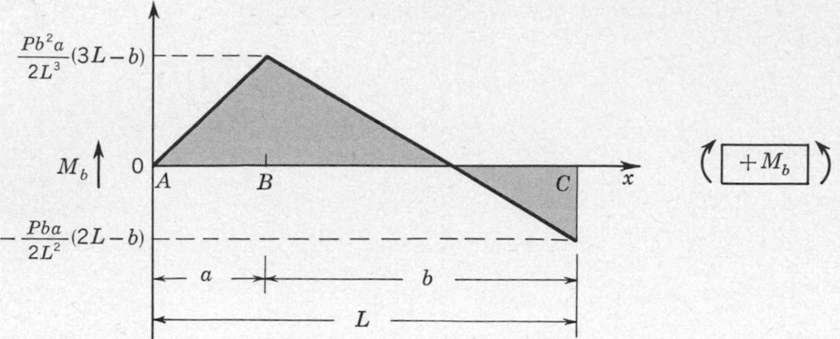

Fig. 8.8 Example 8.4. Bending-moment diagram for the beam of Fig. 8.7.

from which we find

$$R_A = \frac{Pb^2}{2L^3}(3L - b) \tag{h}$$

Thus, to complete the force analysis, we had to bring in the geometric restrictions and the moment-curvature relation. Now we can return to (*a*) with the value (*h*) to obtain explicit results for the reactions. With these values it is an easy matter to sketch the bending-moment diagram shown in Fig. 8.8. An interesting question is: Where is the location of the greatest bending moment? From Fig. 8.8 it is clear that the greatest bending moment occurs either at B or at C depending on the relative position of P. By equating the magnitudes given in Fig. 8.8, we find that when $a = (\sqrt{2} - 1)L = 0.414L$ the bending moments at B and at C have equal magnitude. When a is smaller than this, the greatest bending moment is at B under the load, and, when a is larger than this, the greatest bending moment is at C at the built-in support.

The final example in this section is also statically indeterminate. The situation in this example is somewhat unusual in that the *nature* of the reactions cannot be determined without consideration of all three of the steps (2.1).

Example 8.5 A long uniform rod of length L, weight w per unit length, and bending modulus EI is placed on a rigid horizontal table such that a short segment CD of length a overhangs the table, as shown in Fig. 8.9*a*. It is required to find the length b of the segment BC which lifts up from the table.

The difficult part of this example is the determination of the reactions with the table. There is clearly a concentrated vertical reaction at the edge of the table at C, but the nature of the reaction between the table and the segment AB is not at all clear, as is indicated in Fig. 8.9*b* by the arbitrary

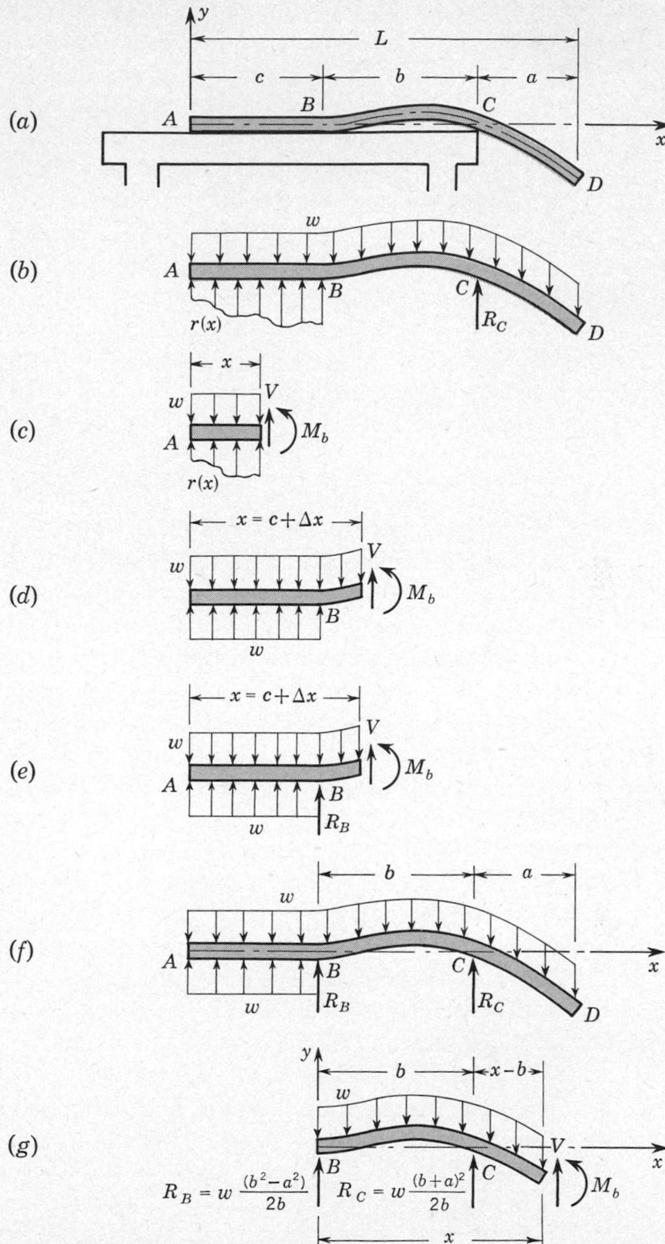

Fig. 8.9 Example 8.5. Beam overhanging edge of the table causes segment BC to lift up from the table.

shape shown for the reaction distribution $r(x)$. It is possible, however, to deduce the nature of this reaction distribution by considering the requirements of all three of the steps (2.1); i.e., by considering the equilibrium, geometric compatibility, and moment-curvature requirements for the beam segment AB.

We begin by observing that, since the table is flat, the beam must have zero curvature in the region AB. Then from the moment-curvature relation (8.4) we conclude that the bending moment must be zero throughout the segment of the beam from A to B. Also, since the bending moment is constant (zero) along the beam, we reason from (3.12) that the shear force must also be zero in this region. Pursuing our reasoning one step further, we conclude from (3.11) that the net intensity of loading must be zero in the region AB because of the constant (zero) shear force. Thus for the free body of Fig. 8.9*c* we must have

$$\begin{aligned} M_b &= V = 0 \\ r(x) &= w \end{aligned} \tag{a}$$

In the free body of Fig. 8.9*d* we have included the results (*a*); that is, we show the reaction in the region between A and B to be of magnitude w per unit length. If we now satisfy the requirement of moment equilibrium for this free body, we shall obtain a *negative* value for the bending moment M_b. A negative bending moment at this point (a distance Δx to the right of B) is not compatible with the requirement that the beam must have a positive curvature in order to leave the surface, since a positive curvature implies a *positive* bending moment. A positive bending moment a distance Δx to the right of B requires the existence of a preponderantly upward external force in this interval, and therefore we conclude that there must be a *concentrated* upward reaction force at point B, as indicated in Fig. 8.9*e*. It is to be noted that the presence of R_B is not in conflict with any of our previous arguments which led to the uniformly distributed reaction in the region between A and B. Thus, after a rather lengthy series of arguments, we have determined that the reaction with the table must be as shown in Fig. 8.9*f*. By applying the equilibrium requirements to this free body, we obtain the magnitudes of the reactions shown in Fig. 8.9*g*.

To proceed further, it is convenient to deal only with the segment of the beam between B and D. Relocating our coordinate system to measure x from point B, we obtain M_b from the free body of Fig. 8.9*g* and insert in (8.4).

$$EI\frac{d^2v}{dx^2} = M_b = \frac{w(b^2 - a^2)}{2b}x + \frac{w(b+a)^2}{2b}\langle x - b\rangle^1 - \frac{wx^2}{2} \tag{b}$$

Integrating (*b*), we find

$$EI\frac{dv}{dx} = \frac{w(b^2 - a^2)}{4b}x^2 + \frac{w(b+a)^2}{4b}\langle x - b\rangle^2 - \frac{wx^3}{6} + c_1 \tag{c}$$

Since the beam is tangent to the table at $x = 0$, we conclude that $c_1 = 0$. Integrating once more, we then obtain

$$EIv = \frac{w(b^2 - a^2)}{12b} x^3 + \frac{w(b+a)^2}{12b} \langle x - b \rangle^3 - \frac{wx^4}{24} + c_2 \tag{d}$$

The constant of integration c_2 is zero since $v = 0$ at $x = 0$.

Finally, we can evaluate b from Eq. (d) by requiring the condition that $v = 0$ at $x = b$.

$$0 = \frac{w(b^2 - a^2)}{12b} b^3 + 0 - \frac{wb^4}{24} \tag{e}$$

Solving (e) for b, we find

$$b = \sqrt{2}\, a \tag{f}$$

Thus when a long, uniform, flexible rod overhangs a rigid table by a distance a, the rod is not in continuous contact with the table until a distance $\sqrt{2}\, a$ back from the edge. It is instructive to review again the conditions at point B in Fig. 8.9 where the rod separates from the table. The deflection and slope are both zero since the curved segment BC must join smoothly with the uncurved segment AB. The bending moment is zero since there is no bending moment in the segment AB and there is no mechanism for introducing a sudden change in bending moment at B. There is, however, a sudden appearance of shear force at B since the table can exert a concentrated upward reaction force.

8.4 SUPERPOSITION

If solutions to a number of deflection problems involving simple conditions of load and support are available, then a convenient method for the solution of beam-deflection problems consists of using the principle of superposition. This method depends upon the linearity of the governing relations between the load and deflection, and it involves the reduction of complex conditions of load and support into a combination of simple loading conditions for which solutions are available. The solution of the original problem then takes the form of a superposition of these solutions.

As may be noted by examining the results of the foregoing examples, the deflection of a beam is linearly proportional to the applied load. This linearity depends upon two factors: (1) the linearity between bending moment and curvature expressed in Eq. (8.1), and (2) the linearity between curvature and deflection expressed in Eq. (8.3).

The linearity of the moment-curvature relation depends upon the fact that we are considering linearly elastic materials. The effect of a nonlinear moment-curvature relation is illustrated in Fig. 8.10. In the linear case in Fig. 8.10*a* it may be seen that a given increment ΔM_b in the bending moment results in the same increment $\Delta(1/\rho)$ in the curvature regardless of the value of M_b to which ΔM_b is added. Thus if we put a load on a beam with a linear moment-curvature relation, the increment of curvature added is independent of whether or not there are other loads acting on the beam. In the nonlinear case in Fig. 8.10*b*, however, we see that the increment of curvature added depends on the magnitude of the other loads acting on the beam. Thus, in the case of nonlinear materials we cannot "superpose" curvatures; i.e., we cannot say that the curvature of two loads acting together is the sum of the curvatures of the two loads acting separately.

The linearity between the curvature and the deflection depends upon the assumption that the deflections are so small that the approximate curvature (8.3) can be used in place of the true curvature $1/\rho$.

When we combine the linear relation between moment and curvature and the linear relation between curvature and deflection, we obtain Eq. (8.4).

$$EI\frac{d^2v}{dx^2} = M_b \tag{8.4}$$

which is a linear differential equation for the deflection v. From the foregoing discussion it may be seen that the linear nature of (8.4) allows superposition of deflections; i.e., the total deflection due to a number of loads is equal to the sum of the deflections due to each load acting separately.

The validity of superposition for beams which satisfy (8.4) can also be demonstrated analytically. Let the total bending moment be the sum of a number of contributions due to separate loadings.

$$M_b = M_{b1} + M_{b2} + \cdots$$

and let

$$v_1, v_2, \ldots$$

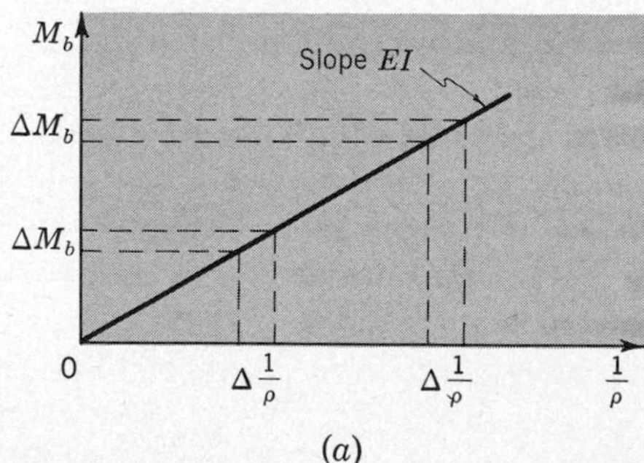

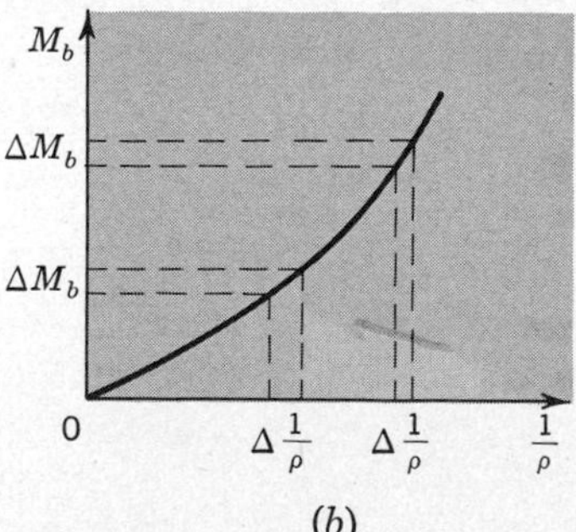

Fig. 8.10 Moment-curvature relations. (*a*) Linear; (*b*) nonlinear.

be the separate solutions of

$$EI\frac{d^2v_1}{dx^2} = M_{b1}$$

$$EI\frac{d^2v_2}{dx^2} = M_{b2}$$

$$\cdots\cdots\cdots\cdots$$

Then the sum

$$v = v_1 + v_2 + \cdots$$

is also a solution of (8.4) since

$$\begin{aligned}EI\frac{d^2v}{dx^2} &= EI\frac{d^2}{dx^2}(v_1 + v_2 + \cdots)\\ &= EI\frac{d^2v_1}{dx^2} + EI\frac{d^2v_2}{dx^2} + \cdots\\ &= M_{b1} + M_{b2} + \cdots\\ &= M_b\end{aligned}$$

In order to make use of the superposition principle to solve beam-deflection problems, it is convenient to have a catalog of solutions for certain standard cases such as those displayed in Fig. 8.5*c* and *d*. In Table 8.1, solutions of several such simple beam-deflection problems are given[1] for reference. These will be employed in the following illustrations of the use of superposition.

Example 8.6 The cantilever beam shown in Fig. 8.11*a* carries a concentrated load P and an end moment M_o. It is desired to predict the deflection δ at the free end C in terms of the constant bending modulus EI and the dimensions shown.

Using the superposition principle, we break up the combined loading of Fig. 8.11*a* into the two separate cases shown in Fig. 8.11*b* and *c*. It is clear that the bending moments for these two cases do in fact add up to give the bending moment for the combined loading in Fig. 8.11*a*. For simplicity we have sketched only the trace of the neutral axis in each of these cases.

[1] For a larger collection, see R. J. Roark, "Formulas for Stress and Strain," 3d ed., p. 100, McGraw-Hill Book Company, New York, 1954; see also W. Flügge, *op. cit.*, chap. 32.

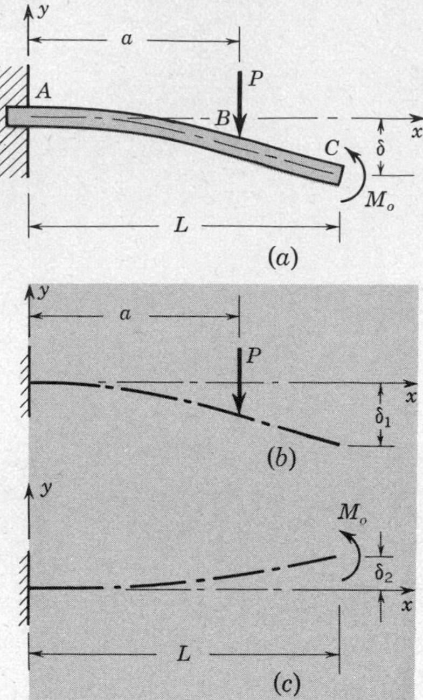

Fig. 8.11 Example 8.6. Illustrating superposition.

Now referring to Table 8.1, cases 1 and 3, we obtain directly the following individual deflections:

$$\delta_1 = \frac{Pa^2}{6EI}(3L - a)$$
$$\delta_2 = \frac{M_o L^2}{2EI} \tag{a}$$

The resultant deflection is obtained by superposing these.

$$\delta = \delta_1 - \delta_2 = \frac{Pa^2(3L - a) - 3M_o L^2}{6EI} \tag{b}$$

If the complete deflection curves had been required, we would simply have combined, in the same way, the entries in Table 8.1 for the deflections as functions of x.

In applying superposition, it is necessary not only to have the separate loads combine to give the original load but also to have the separate deflection curves combine to give a resulting deflection curve which meets the geometric-compatibility requirements for the original problem. In this example the built-in boundary conditions used in Fig. 8.11*b* and *c* can be superposed to give the built-in boundary condition at A in Fig. 8.11*a*.

Table 8.1 Deflection formulas for uniform beams

δ is positive downward

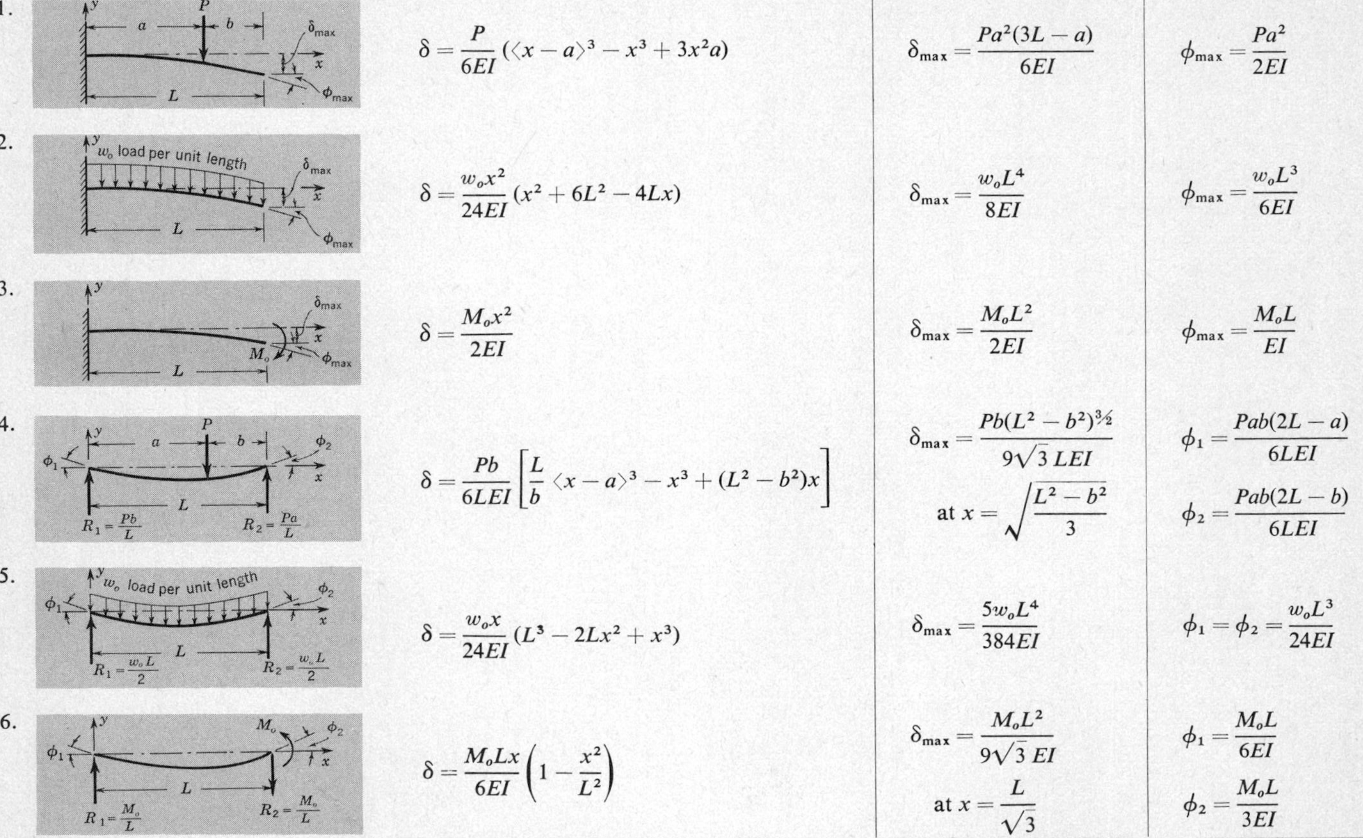

	δ	δ_{max}	ϕ
1.	$\delta = \dfrac{P}{6EI}(\langle x-a\rangle^3 - x^3 + 3x^2a)$	$\delta_{max} = \dfrac{Pa^2(3L-a)}{6EI}$	$\phi_{max} = \dfrac{Pa^2}{2EI}$
2.	$\delta = \dfrac{w_o x^2}{24EI}(x^2 + 6L^2 - 4Lx)$	$\delta_{max} = \dfrac{w_o L^4}{8EI}$	$\phi_{max} = \dfrac{w_o L^3}{6EI}$
3.	$\delta = \dfrac{M_o x^2}{2EI}$	$\delta_{max} = \dfrac{M_o L^2}{2EI}$	$\phi_{max} = \dfrac{M_o L}{EI}$
4.	$\delta = \dfrac{Pb}{6LEI}\left[\dfrac{L}{b}\langle x-a\rangle^3 - x^3 + (L^2-b^2)x\right]$	$\delta_{max} = \dfrac{Pb(L^2-b^2)^{3/2}}{9\sqrt{3}\,LEI}$ at $x = \sqrt{\dfrac{L^2-b^2}{3}}$	$\phi_1 = \dfrac{Pab(2L-a)}{6LEI}$ $\phi_2 = \dfrac{Pab(2L-b)}{6LEI}$
5.	$\delta = \dfrac{w_o x}{24EI}(L^3 - 2Lx^2 + x^3)$	$\delta_{max} = \dfrac{5w_o L^4}{384EI}$	$\phi_1 = \phi_2 = \dfrac{w_o L^3}{24EI}$
6.	$\delta = \dfrac{M_o Lx}{6EI}\left(1 - \dfrac{x^2}{L^2}\right)$	$\delta_{max} = \dfrac{M_o L^2}{9\sqrt{3}\,EI}$ at $x = \dfrac{L}{\sqrt{3}}$	$\phi_1 = \dfrac{M_o L}{6EI}$ $\phi_2 = \dfrac{M_o L}{3EI}$

It should be noted that in using superposition to obtain the solution to a beam-deflection problem we again are using the three steps in (2.1) and nothing additional. Equilibrium is satisfied by requiring that the separate loadings add up to the original load condition. Geometric compatibility is assured by requiring that the sum of the separate deflection conditions add up to the original geometric constraints on the beam. Finally, the force-deformation relations used are those given in Table 8.1 or an equivalent table.

Superposition can also be used in situations where the beam is statically indeterminate. A common procedure is to temporarily remove enough unknown reactions to make the beam statically determinate and then to calculate the deflection due to the loads and the remaining reactions. The actual loads are then removed, and each previously removed reaction is then treated as a load and the beam deflection calculated. The sum of all calculated deflections is then made to fit the conditions of geometric constraint imposed on the original beam; it will be found that these conditions will be just sufficient to evaluate the indeterminate reactions. The following example illustrates this procedure.

Example 8.7 A uniform beam which is built-in at the ends carries a concentrated load P, as shown in Fig. 8.12*a*. It is desired to obtain the bending-moment diagram.

If the walls at A and C are not free to change their separation distance, then it is possible that the built-in supports could sustain horizontal reactions. We shall make the assumption that when $P = 0$ there is no tension or compression in the beam. We further assume that the deflection under P is sufficiently small that any longitudinal tension has a negligible effect on the bending. (See Example 8.8 for an order-of-magnitude verification of this kind of assumption.) We thus consider that the supports exert only vertical reactions R_A and R_C and clamping moments M_A and M_C. Since only two conditions of equilibrium are available to obtain four reaction components, we say that the beam has two degrees of statical indeterminacy.

We can make the system statically determinate by removing the unknown reaction components R_C and M_C at C. The resulting statically determinate system is shown in Fig. 8.12*b*. In Fig. 8.12*c* and *d* we have removed the load P and introduced the unknown reaction components R_C and M_C as loads on the statically determinate cantilever of Fig. 8.12*b*. Superposition of the three loadings of Fig. 8.12*b*, *c*, and *d* gives the loading of Fig. 8.12*a*. Superposition of the geometric conditions at C will lead to the compatibility requirement of zero slope and zero displacement at C in Fig. 8.12*a*, provided that

$$\begin{aligned} \delta_1 - \delta_2 + \delta_3 &= 0 \\ \phi_1 - \phi_2 + \phi_3 &= 0 \end{aligned} \qquad (a)$$

Now consulting Table 8.1, cases 1, 1, and 3, respectively, we find

$$\delta_1 = \frac{Pa^2(3L-a)}{6EI} \qquad \phi_1 = \frac{Pa^2}{2EI}$$

$$\delta_2 = \frac{R_C L^3}{3EI} \qquad \phi_2 = \frac{R_C L^2}{2EI} \tag{b}$$

$$\delta_3 = \frac{M_C L^2}{2EI} \qquad \phi_3 = \frac{M_C L}{EI}$$

Substitution of (*b*) into (*a*) gives a pair of simultaneous algebraic equations for R_C and M_C. Their solution is

$$R_C = \frac{Pa^2(3L-2a)}{L^3}$$

$$M_C = \frac{Pa^2(L-a)}{L^2} \tag{c}$$

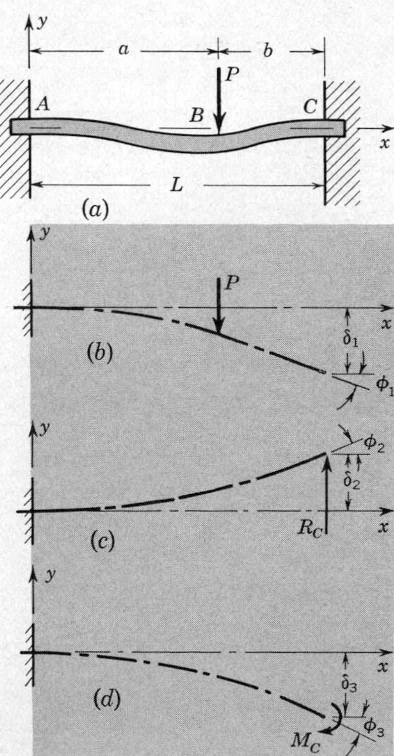

Fig. 8.12 Example 8.7.

Now that we have these statically indeterminate reactions, we can use the equilibrium requirements to obtain the remaining reactions and to calculate the bending moment in the usual manner. An interesting alternate way to obtain the bending moment is to extend the superposition argument, as shown in Fig. 8.13. In Fig. 8.13*b*, *c*, and *d* we sketch the bending-moment diagrams corresponding to the separate loadings of Fig. 8.12*b*, *c*, and *d*. These are then superposed to obtain the resulting bending-moment diagram of Fig. 8.13*a* which corresponds to the original problem of Fig. 8.12*a*.

Example 8.8 We return to the problem pictured in Fig. 8.14, which we have already discussed in Example 1.3 and Example 2.4. In these previous examples we obtained the support reactions and the displacement of point *D* after making the idealization that the bolted joint at *C* could be treated as a frictionless pinned joint. We now have developed our subject to the point where it is profitable to reexamine this question. A complete theory of the behavior of bolted joints is still beyond us, but we can analyze a new limiting case based on the assumption that the bolted joint is completely effective in *clamping* the beam at *C*. This is just about as unrealistic as our earlier

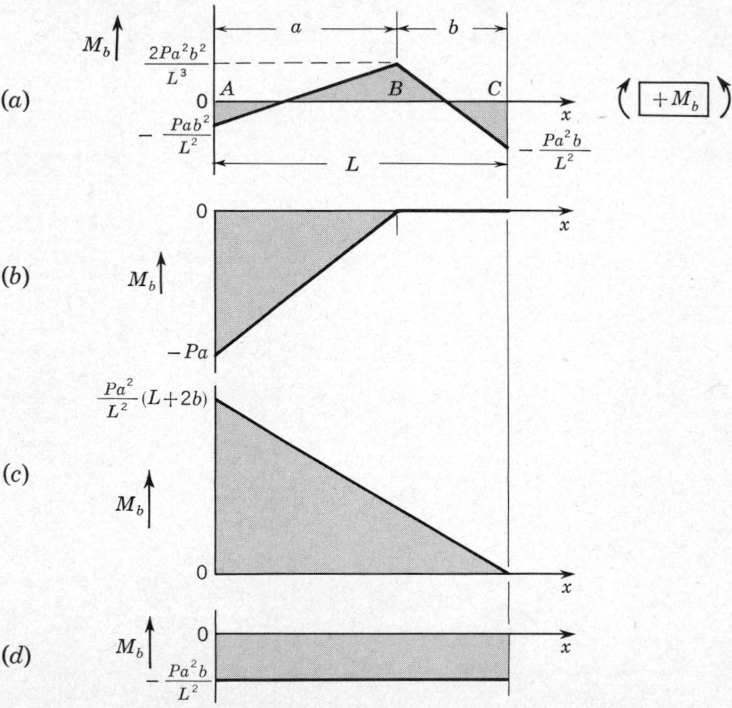

Fig. 8.13 Example 8.7. Superposition of bending-moment diagrams.

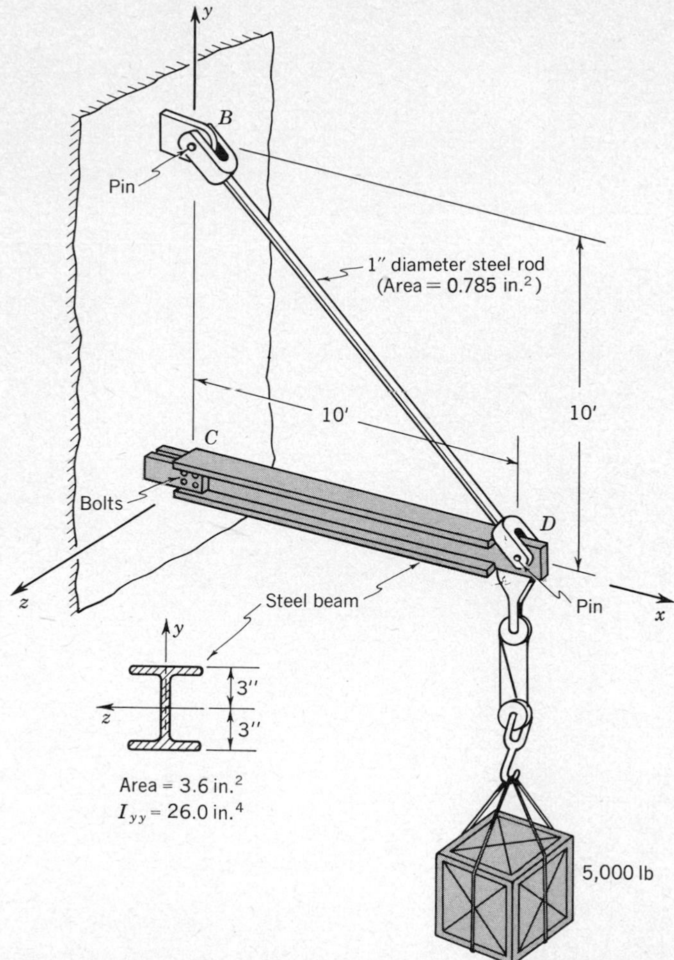

Fig. 8.14 Example 8.8.

assumption of a *pinned* joint, but the two solutions together provide two extremes between which the actual case must lie.

Figure 8.15*a* shows the idealized model with which we shall work. The beam CD is taken as if built in or clamped at C. In Fig. 8.15*b* are isolated free-body sketches of BD and CD showing the forces acting. The system is statically indeterminate. The compression in the horizontal beam CD has been called X. All other forces and moments can be expressed in terms of X, as shown, by using equilibrium conditions.

The geometry of the deformation is sketched in Fig. 8.15*c*, and an enlarged view of the neighborhood of D is shown in Fig. 8.15*d*. This very

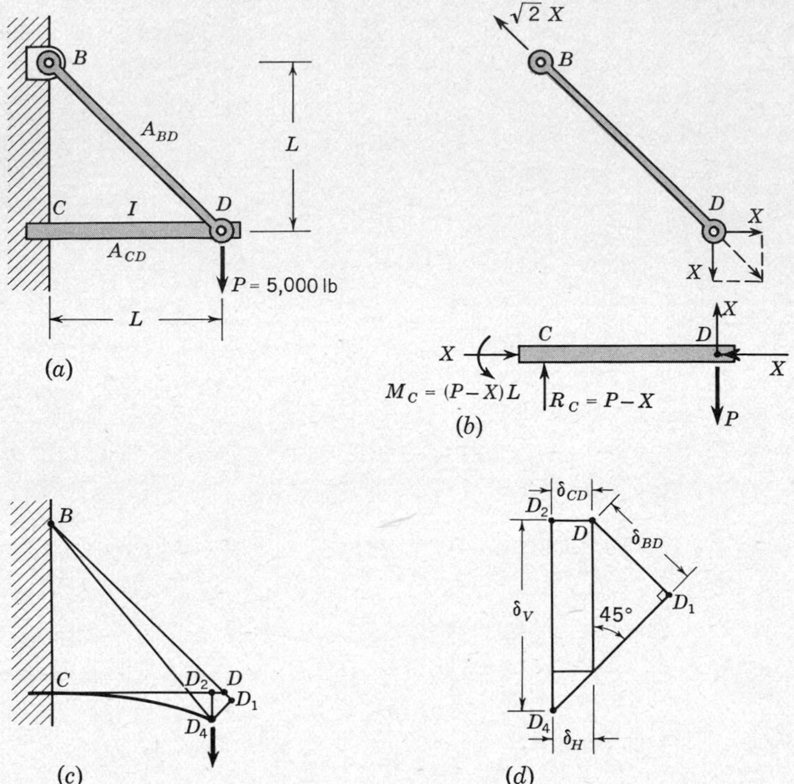

Fig. 8.15 Example 8.8. Force analysis and geometric analysis for a model based on clamping assumption at C.

closely resembles Fig. 2.7*d*. The essential difference is that δ_V does not now result from the rigid-body rotation of CD about the end C but rather from the vertical deflection of the cantilever beam CD. The compatibility relationship

$$\delta_V = \delta_{CD} + \sqrt{2}\,\delta_{BD} \tag{a}$$

still remains valid.

Now having studied the forces and the deformations, we turn to the last of the three steps of (2.1). We relate the forces to the deformations by using the appropriate force-deformation law. For the bar BD in tension we have

$$\delta_{BD} = \frac{\sqrt{2}\,X}{EA_{BD}}\sqrt{2}\,L = \frac{2XL}{EA_{BD}} \tag{b}$$

as a statement of Hooke's law for uniaxial loading. For the beam CD in compression we have, similarly,

$$\delta_{CD} = \frac{XL}{EA_{CD}} \tag{c}$$

Finally, for the vertical deflection of the cantilever beam CD, we can use case 1 of Table 8.1 to obtain

$$\delta_V = \frac{(P - X)L^3}{3EI} \tag{d}$$

Here we have made the assumption that the bending of the beam is unaffected by the compressive load. This is not completely true, as we shall see in the next chapter. We shall, however, proceed on the basis of this assumption, and then after obtaining a solution, we can reconsider this question and estimate the order of magnitude of error involved.

We thus take (*b*), (*c*), and (*d*) to represent adequately the deflections in terms of the forces. Inserting these into the compatibility relation (*a*) yields an equation for determining X, from which we obtain

$$X = \frac{P}{1 + 3I/A_{CD}L^2 + 6\sqrt{2}\, I/A_{BD}L^2} \tag{e}$$

Particularizing this for the case shown in Fig. 8.14, we find

$$X = \frac{5{,}000}{1 + 0.0015 + 0.0195} = 4{,}897 \text{ lb} \tag{f}$$

Insertion of this value into (*c*) and (*d*) yields the following displacements for point D:

$$\begin{aligned} \delta_H &= 0.0054 \text{ in.} \\ \delta_V &= 0.0761 \text{ in.} \end{aligned} \tag{g}$$

Before comparing these with the results of Examples 1.3 and 2.4, we shall estimate the error involved in our assumption that the compression in CD did not affect the bending. In Fig. 8.16 a free-body diagram of CD is shown with the loads X and $P - X$ and the deflection δ_V which were obtained in the above analysis. We assumed that the bending in CD was

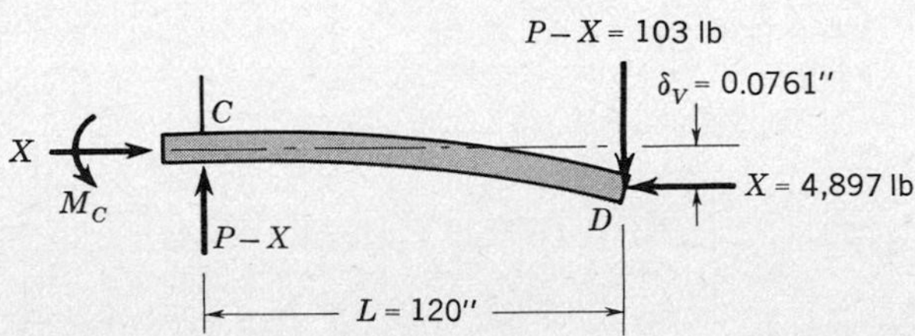

Fig. 8.16 Example 8.8. Estimation of interaction between compression and bending.

due entirely to the transverse force $P - X$. To obtain a rough order-of-magnitude check on this, let us compare the contributions of the compressive load and the transverse load to the bending moment at C.

$$\frac{M_C \text{ (due to compressive load)}}{M_C \text{ (due to transverse load)}} = \frac{(4{,}897)(0.0761)}{(103)(120)} = 0.03 \qquad (h)$$

This indicates that any additional bending in CD because of the compressive load would only be of the order of 3 percent of that due to the transverse force. Since we assumed that all the bending was due to the transverse force, our solution for the transverse force $P - X$ may be in error by about 3 percent. This would, however, make very little difference in the value of X or in the deflections (g). We conclude that the effect of the compressive load on the bending can be safely neglected in this case.

Now to compare our results based on the assumption of clamping at C with the earlier results of Examples 1.3 and 2.4, which were based on the assumption of a pinned joint at C, we first compare the deflections (g) directly with the results of Example 2.4. We note that the deflections in the clamped case are only about 2 percent smaller than in the pinned case. If we next compare the wall reactions, we find that at B there has been a 2-percent decrease in the tensile force. At C the situation is more complicated. Where there was simply a horizontal compressive force of 5,000 lb in the pinned case, there is, in the clamped case, a vertical component and a clamping moment in addition to the horizontal component. Figure 8.17*a* shows these separate components, and their resultant in the form of a single force is pictured in Fig. 8.17*b*. The offset distance CC' is obtained from the requirement that the moment about C should be the same in Fig. 8.17*a* and *b*. If we compare the single force in Fig. 8.17*b* with a 5,000-lb force acting horizontally at C,

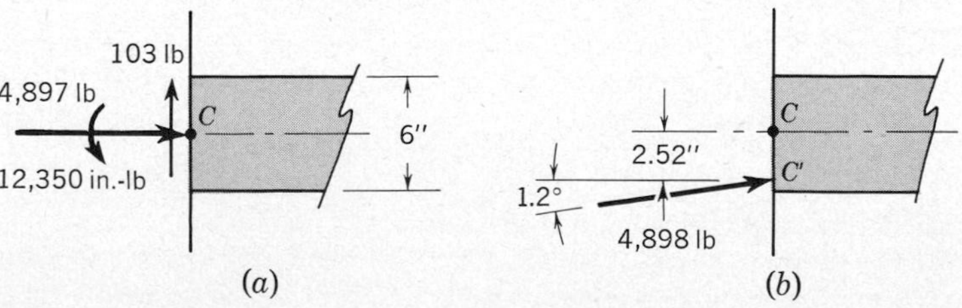

Fig. 8.17 Example 8.8. Reactions at C shown as a statically equivalent, single force resultant.

we can see the differences between the two cases. The magnitude of the force has changed by about 2 percent, its line of action has been tipped through a small angle, and the point of application has been shifted through a distance which is small compared with the distance BC between wall reactions.

As a result of the comparisons made above, we would be led to the conclusion that the precise nature of the connecting joints in a structure made up of slender elements loaded only at the joints has very little influence on the deflections of the structure or on the external reactions. This is, in fact, the case. As a result, the designer is justified in basing his calculations of these quantities on the simplest model of joint behavior. The assumption of frictionless pinned joints is usually the easiest to deal with.

Although the choice of joint idealization has led to little change in the displacements and reactions, it would be *wrong* to say that there has been *no* significant change. We have not yet considered the *stresses* in the individual members. The member BD which is in uniaxial tension in both cases has had only a 2-percent change in stress. The member CD, however, is in uniaxial compression in the first case and in combined compression and bending in the second case. The combined stress would be greatest in the bottom flange of the beam, where the direct compressive stress and the compressive stress due to bending are additive, and would have its maximum value at C. This peak stress in CD in the clamped case would be

$$\sigma = \frac{4{,}897}{3.6} + \frac{(12{,}350)(3)}{26} = 2{,}780 \text{ psi} \qquad (i)$$

which is about *double* the compressive stress in CD in the pinned case.

$$\sigma = \frac{5{,}000}{3.6} = 1{,}388 \text{ psi} \qquad (j)$$

Thus the type of joint restraint can cause significant *local* changes in stress distribution even though the overall effects are small.

A wide variety of structures (e.g., buildings, bridges, transmission towers) are assembled from slender members. If the joints are *pinned*, the structure is called a *truss*. If the joints are *rigid*, the structure is called a *frame*. The members of a truss carry axial loads, while the members of a frame generally carry shear forces and bending moments as well as axial loads. The structure just analyzed (see Fig. 8.15) is a simple example of a mixed structure. The joints B and D are pinned as in a truss, but the joint C is rigid as in a frame.

The procedure followed in analyzing this simple structure can be extended to structures with many members and any combination of joint restraints. Although the process is straightforward, the amount of calculation required for a complex structure can be enormous. During the past century, structural engineers developed an array of ingenious techniques to ease the burden of calculation. A major breakthrough has occurred in the last decade with the systematic application of digital computers to structural analysis problems. A large number of general-purpose and special-purpose structural analysis programs are now available to the engineer.

To illustrate the use of such a program, we show how the STRESS program can be used to solve the preceding example. The input to the computer shown in Table 8.2 is obtained directly from Fig. 8.14. The statements under joint releases and member releases are necessary to describe the pinned joints at B and D. If these joints were rigid the structure would be a true frame. For a complete explanation of how joint release and member release statements must be written, one should consult the STRESS manual.[1]

The printout of the STRESS program is displayed in Table 8.3. The forces and moments acting on each member are given with respect to a local coordinate system for that member, as described in Example 2.4 and shown in Fig. 8.18. Note that the forces agree (to within 1 lb) with the results displayed in Fig. 8.17*a*. Also note that the displacements agree with (*g*) above.

Table 8.2 Input to STRESS for Example 8.8

```
        STRUCTURE EXAMPLE 8.8
        TYPE PLANE FRAME
        NUMBER OF JOINTS      3
        NUMBER OF MEMBERS     2
        NUMBER OF SUPPORTS    2
        NUMBER OF LOADINGS    1
        JOINT COORDINATES
        1   0.0    0.0    S
        2 120.0    0.0
        3   0.0  120.0    S
        JOINT RELEASES
        3 MOMENT Z
        MEMBER INCIDENCES
        1 1 2
        2 2 3
        MEMBER PROPERTIES    PRISMATIC
        1 AX 3.6    IZ 26.0
        2 AX 0.785 IZ  0.049
        MEMBER RELEASES
        2 START MOMENT Z
        CONSTANTS   E 30000. ALL
        LOADING 1 VERTICAL CONCENTRATED LOAD
        JOINT LOADS
        2 FORCE Y -5.0
        TABULATE ALL
        SOLVE
   PROBLEM CORRECTLY SPECIFIED, EXECUTION TO PROCEED.
```

[1] See footnote on page 103.

Table 8.3 Output from STRESS for Example 8.8

MEMBER FORCES

MEMBER	JOINT	AXIAL FORCE	SHEAR FORCE	MOMENT
1	1	4.897	0.102	12.35
1	2	-4.897	-0.102	0.00
2	2	-6.925	-0.000	-0.00
2	3	6.925	0.000	-0.00

APPLIED JOINT LOADS, FREE JOINTS

JOINT	FORCE X	FORCE Y	MOMENT Z
2	0.000	-5.000	0.00

REACTIONS,APPLIED LOADS SUPPORT JOINTS

JOINT	FORCE X	FORCE Y	MOMENT Z
1	4.897	0.102	12.35
3	-4.897	4.897	-0.00

FREE JOINT DISPLACEMENTS

JOINT	X-DISPLACEMENT	Y-DISPLACEMENT	ROTATION
2	-0.0054	-0.0760	-0.0009

SUPPORT JOINT DISPLACEMENTS

JOINT	X-DISPLACEMENT	Y-DISPLACEMENT	ROTATION
1	0.0000	0.0000	0.0000
3	0.0000	0.0000	-0.0003

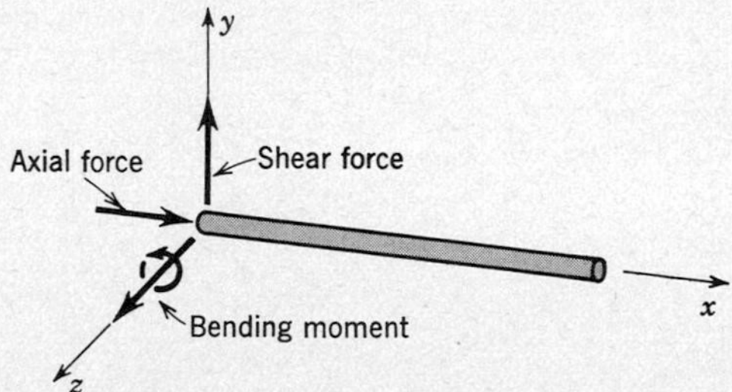

Fig. 8.18 Local axes used to define member forces and bending moment.

8.5 THE LOAD-DEFLECTION DIFFERENTIAL EQUATION

As an alternative to using the moment-curvature equation (8.4) to solve beam-deflection problems, we can make use of an equation which directly relates the external loading to the beam deflection. This equation incorporates the force and moment equilibrium conditions *and* the moment-curvature relation in a single differential equation.

The load-deflection differential equation is derived by starting with the differential equations of force and moment equilibrium which were derived in Chap. 3.

$$\frac{dV}{dx} + q = 0 \tag{3.11}$$

$$\frac{dM_b}{dx} + V = 0 \tag{3.12}$$

We can express the bending moment in terms of the loading by eliminating V in (3.11) and (3.12).

$$\frac{d^2 M_b}{dx^2} = q \tag{8.5}$$

If we now combine (8.4) and (8.5), we obtain a single differential equation relating the transverse load-intensity function q to the transverse deflection v.

$$\frac{d^2}{dx^2}\left(EI\,\frac{d^2 v}{dx^2}\right) = q \tag{8.6}$$

Note that in (8.6) the flexural rigidity EI has been left within the parentheses; in this general form the equation applies to beams with variable EI.

It should be emphasized that we have used both *equilibrium* relations and *deformation* relations in deriving (8.6). The boundary conditions for (8.6) will, in general, include both equilibrium conditions and geometric-compatibility conditions. The geometric conditions will involve restrictions on the deflection and slope at certain points. The equilibrium conditions will involve restrictions on the shear and bending moment at certain points.

Conditions on M_b can be written in terms of v by means of the moment-curvature relation, which gives

$$M_b = EI\,\frac{d^2 v}{dx^2} \tag{8.4}$$

while the shear force V can be expressed in terms of v by using (3.12) and (8.4) to get

$$V = -\frac{d}{dx}\left(EI\,\frac{d^2 v}{dx^2}\right) \tag{8.7}$$

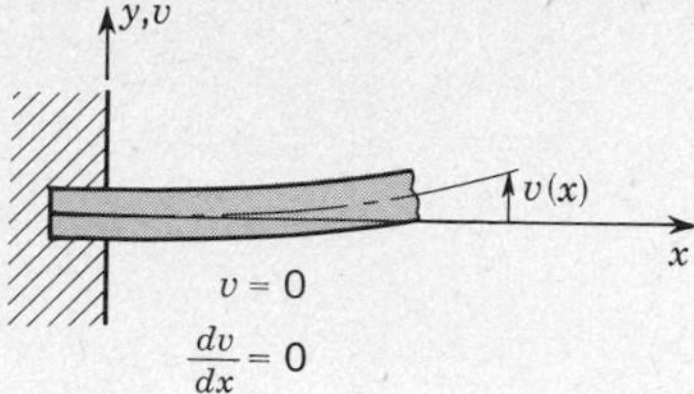

Fig. 8.19 Built-in or clamped end.

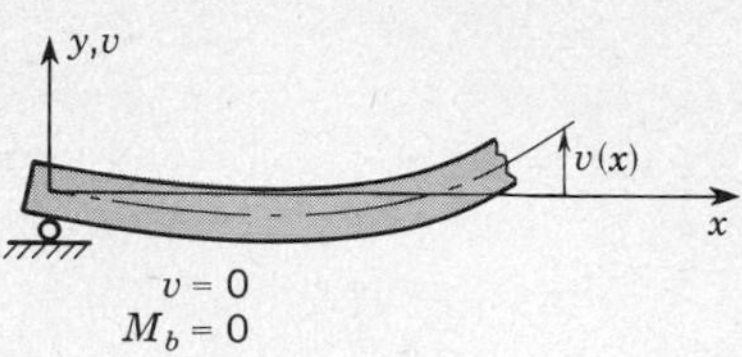

Fig. 8.20 Simply supported end.

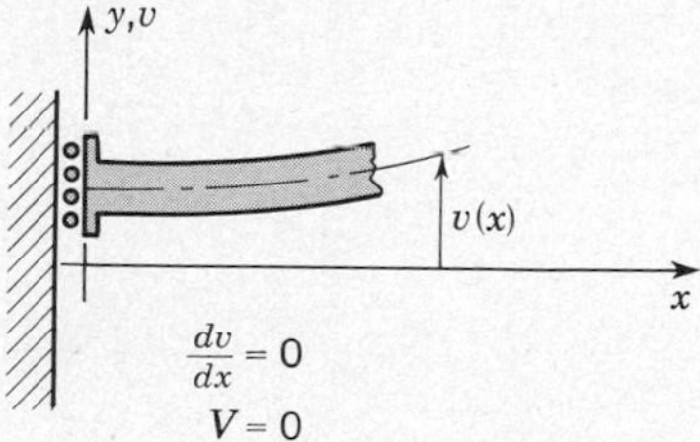

Fig. 8.21 End restrained against rotation but free to displace.

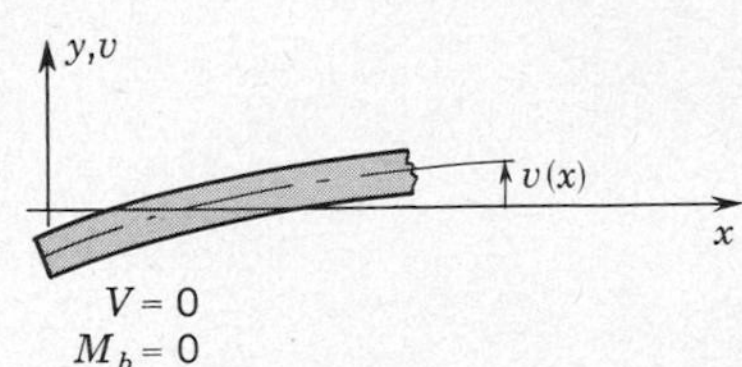

Fig 8.22 Free end.

In Figs. 8.19 to 8.22 we show the appropriate boundary conditions which correspond to four types of support conditions which are frequently used in the analysis of beams to represent actual physical supports.

In order to solve a beam-deflection problem using (8.6), it is first necessary to obtain an expression for the loading intensity $q(x)$ which is valid over the length of the beam (the singularity functions may be useful for this purpose). Then integration of (8.6) introduces *four* constants of integration which must be evaluated by applying the appropriate combination of boundary conditions. The method is quite general in that it applies equally well to statically indeterminate as well as statically determinate beams. This approach reduces beam-deflection problems to a routine procedure in which physical considerations are concentrated in the selection of boundary conditions and the establishment of $q(x)$, and the algebraic manipulations are concentrated in the process of evaluating the constants of integration.

Example 8.9 The beam shown in Fig. 8.23*a* is built-in at A and D and has an offset arm welded to the beam at the point B with a load W attached to the arm at C. It is required to find the deflection of the beam at the point B.

The effect of the arm on the beam is to supply a vertical force W and a couple $WL/3$ at B, as shown in Fig. 8.23*b*. With this replacement, the load-intensity function q for $0 < x < L$ is

$$q = \frac{WL}{3}\langle x - L/3\rangle_{-2} - W\langle x - L/3\rangle_{-1} \tag{a}$$

Because of the built-in supports the boundary conditions are

$$v = 0 \text{ and } \frac{dv}{dx} = 0 \qquad \text{at } x = 0 \text{ and } L \tag{b}$$

Insertion of (a) into the load-deflection differential equation (8.6) yields

$$EI\frac{d^4v}{dx^4} = W\left[\frac{L}{3}\langle x - L/3\rangle_{-2} - \langle x - L/3\rangle_{-1}\right] \tag{c}$$

Expressions for dv/dx and v are obtained by integrating (c).

$$\frac{dv}{dx} = \frac{W}{EI}\left[\frac{L}{3}\langle x - L/3\rangle^1 - \frac{\langle x - L/3\rangle^2}{2} + c_1\frac{x^2}{2} + c_2x + c_3\right] \tag{d}$$

$$v = \frac{W}{EI}\left[\frac{L}{6}\langle x - L/3\rangle^2 - \frac{\langle x - L/3\rangle^3}{6} + c_1\frac{x^3}{6} + c_2\frac{x^2}{2} + c_3x + c_4\right] \tag{e}$$

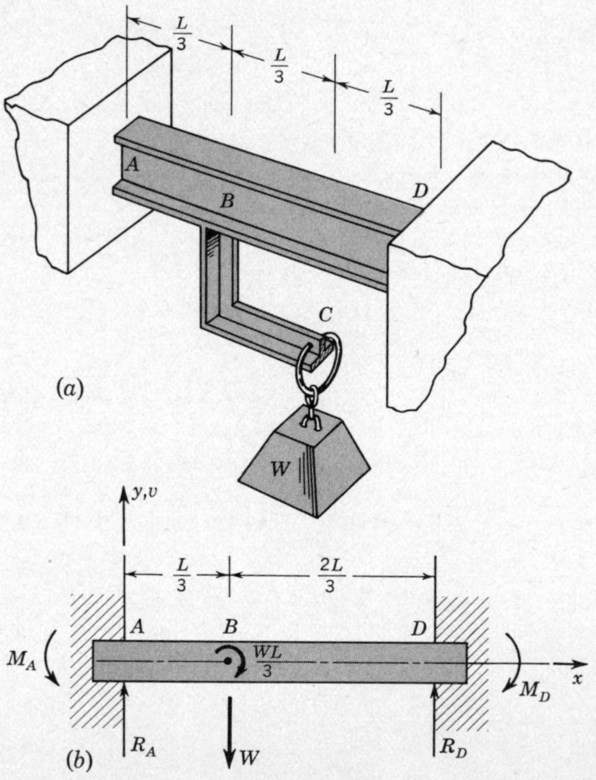

Fig. 8.23 Example 8.9. Offset loading is equivalent to a force and a couple at B.

Substitution of (d) and (c) into the boundary conditions (b) gives four simultaneous equations for the constants of integration. Their solution is

$$\begin{aligned} c_1 &= 8/27 \\ c_2 &= -4/27\,L \\ c_3 &= 0 \\ c_4 &= 0 \end{aligned} \tag{f}$$

Inserting these in (e) we find

$$v = \frac{W}{27EI}\left[\frac{9}{2}L\langle x - L/3\rangle^2 - \frac{9}{2}\langle x - L/3\rangle^3 + \frac{4}{3}x^3 - 2Lx^2\right] \tag{g}$$

We obtain the desired deflection by setting $x = L/3$.

$$\delta_B = -(v)_{x=L/3} = \frac{14WL^3}{2{,}187EI} \tag{h}$$

8.6 ENERGY METHODS

In Sec. 2.6, Castigliano's theorem is used to evaluate deflections in simple elastic systems and to obtain equations for determining statically indeterminate reactions. In this section the formulas for strain energy in torsion and bending, developed in Secs. 6.8 and 7.8, are used to illustrate the application of Castigliano's theorem to more complicated elastic systems.

If a slender elastic shaft oriented along the axis of x carries a tensile force $F(x)$, a twisting moment $M_t(x)$, and a bending moment $M_b(x)$, then according to (2.11), (6.13), and (7.31) the total strain energy in the member is

$$U = \int_L \frac{F^2}{2AE}\,dx + \int_L \frac{M_t^{\,2}}{2GI_x}\,dx + \int_L \frac{M_b^{\,2}}{2EI}\,dx \tag{8.8}$$

where the integrations are along the length L of the shaft, and where A, I_x, and I are, respectively, the area, the polar moment of inertia, and the diametral moment of inertia of the shaft cross section, and where E and G are the tension and shear moduli of the shaft material. Such a member may also be subjected to transverse shear, but as indicated in Sec. 7.8, the corresponding contribution to the total strain energy of a slender member can usually be neglected in comparison with the bending or twisting contribution.[1]

We recall that Castigliano's theorem states that if the total elastic energy in a system is expressed in terms of external loads P_i, the corresponding in-line deflections δ_i are given by the partial derivatives

$$\delta_i = \frac{\partial U}{\partial P_i} \tag{8.9}$$

[1] See also Probs. 7.27 and 8.42.

If a deflection δ is desired at a point where there is no load (or in a direction which is not in line with a load), it is only necessary to introduce a fictitious load Q in the desired direction at the desired point. Then if the elastic energy is expressed in terms of the P_i and Q, the desired deflection δ is given by differentiating with respect to Q and *then* setting Q equal to zero.

$$\delta = \left(\frac{\partial U}{\partial Q}\right)_{Q=0} \tag{8.10}$$

Castigliano's theorem may also be used to obtain statically indeterminate reactions. A statically indeterminate system can always be reduced to a statically determinate one if enough statically indeterminate reactions X_i are temporarily considered to be known external loads. Then, if the elastic energy is expressed in terms of the P_i *and* the X_i, a set of equations for determining the X_i can be obtained from the condition that there be no in-line deflection at each of the statically indeterminate reaction points

$$\frac{\partial U}{\partial X_i} = 0 \tag{8.11}$$

Example 8.10 To illustrate the application of Castigliano's theorem to beam-deflection problems, we reconsider Example 8.2, in which a cantilever beam is loaded by a force P and a moment M, as indicated in Fig. 8.5a. In this case the strain energy (8.8) consists only of the bending contribution with the bending moment

$$M_b = -P(L-x) - M \tag{a}$$

obtained from Fig. 8.5b; that is,

$$U = \int_0^L \frac{{M_b}^2}{2EI}\,dx = \int_0^L \frac{[P(L-x)+M]^2}{2EI}\,dx \tag{b}$$

The terminal deflection δ is in line with the load P, and so according to (8.9),

$$\begin{aligned} \delta = \frac{\partial U}{\partial P} &= \int_0^L \frac{P(L-x)^2 + M(L-x)}{EI}\,dx \\ &= \frac{PL^3}{3EI} + \frac{ML^2}{2EI} \end{aligned} \tag{c}$$

The terminal slope ϕ may be considered to be the in-line displacement corresponding to the moment load M [see Eq. (2.8)]. Thus a second application of (8.9) yields

$$\begin{aligned} \phi = \frac{\partial U}{\partial M} &= \int_0^L \frac{P(L-x)+M}{EI}\,dx \\ &= \frac{PL^2}{2EI} + \frac{ML}{EI} \end{aligned} \tag{d}$$

The results (c) and (d) just obtained are identical with (g) and (h) of Example 8.2. Note the relative simplicity of the calculation using Castigliano's theorem as compared with direct integration of the moment-curvature relation.

Example 8.11 To illustrate the application of energy methods to statically indeterminate systems in which slender members carry both longitudinal and bending loads, we reconsider Example 8.8, shown in Figs. 8.14 and 8.15. In order to obtain the horizontal deflection of point D by Castigliano's theorem, we insert the fictitious force Q in Fig. 8.24a. If the horizontal reaction X at point C is temporarily considered as an external load, the system becomes statically determinate. Using the equilibrium requirements it is possible to express all external forces and moments on the members CD and BD in terms of P, Q, and X, as indicated in the free-body diagrams of Fig. 8.24b and c. The internal forces and moments acting on the section at the location x in CD are displayed in Fig. 8.24d, and the internal force acting in BD is indicated in Fig. 8.24e. The total strain energy in this case is due to longitudinal and bending loading in CD and to longitudinal loading in BD,

$$U = \int_0^L \frac{X^2}{2A_{CD}E}\,dx + \int_0^L \frac{(P - X + Q)^2(L - x)^2}{2EI}\,dx + \int_0^{\sqrt{2}L} \frac{[\sqrt{2}(X - Q)]^2}{2A_{BD}E}\,dx \tag{a}$$

The vertical and horizontal deflections at point D are

$$\begin{aligned}\delta_V = \left(\frac{\partial U}{\partial P}\right)_{Q=0} &= \int_0^L \frac{(P - X)(L - x)^2}{EI}\,dx \\ &= \frac{(P - X)L^3}{3EI}\end{aligned} \tag{b}$$

$$\begin{aligned}\delta_H = \left(\frac{\partial U}{\partial Q}\right)_{Q=0} &= \int_0^L \frac{(P - X)(L - x)^2}{EI}\,dx - \int_0^{\sqrt{2}L} \frac{2X}{A_{BD}E}\,dx \\ &= \frac{(P - X)L^3}{3EI} - \frac{2\sqrt{2}\,XL}{A_{BD}E}\end{aligned} \tag{c}$$

in terms of the statically indeterminate reaction X. To determine X we use the fact that the deflection at point C in Fig. 8.24a is zero; i.e., according to (8.11) we have

$$\begin{aligned}0 = \left(\frac{\partial U}{\partial X}\right)_{Q=0} &= \int_0^L \frac{X}{A_{CD}E}\,dx - \int_0^L \frac{(P - X)(L - x)^2}{EI}\,dx + \int_0^{\sqrt{2}L} \frac{2X}{A_{BD}E}\,dx \\ &= \frac{XL}{A_{CD}E} - \frac{(P - X)L^3}{3EI} + \frac{2\sqrt{2}\,XL}{A_{BD}E}\end{aligned} \tag{d}$$

from which we solve for X,

$$X = \frac{P}{1 + 3I/A_{CD}L^2 + 6\sqrt{2}\,I/A_{BD}L^2} \tag{e}$$

Note that this result agrees with (*e*) in Example 8.8 and also that (*b*) above

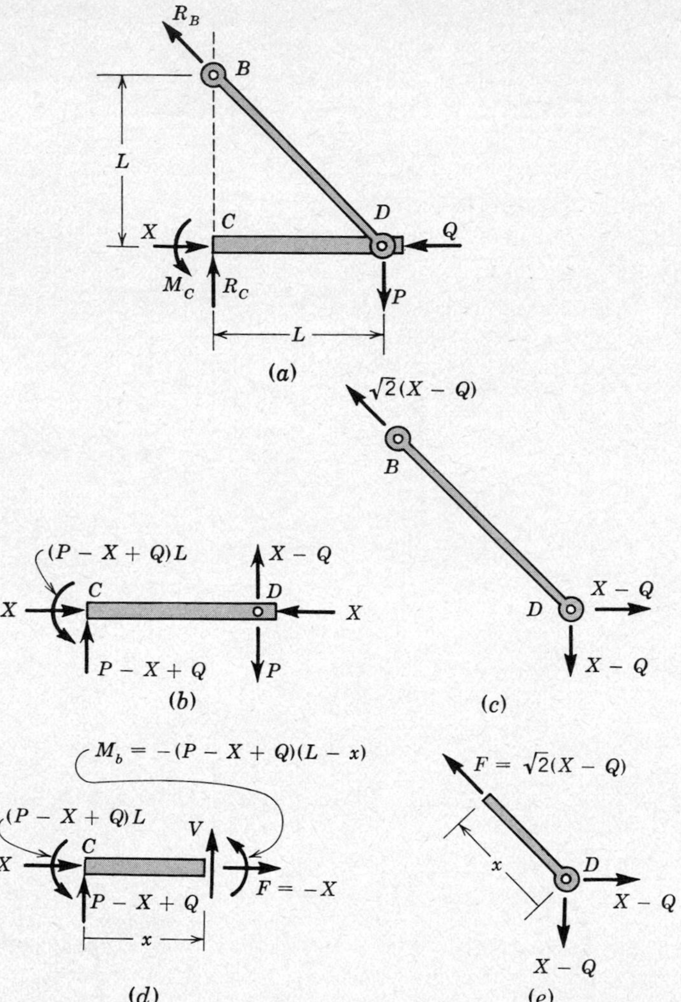

Fig. 8.24 Example 8.11. Free-body diagrams of: (*a*) complete structure, showing statically indeterminate reaction X and fictitious load Q; (*b*) and (*c*) separate members; (*d*) and (*e*) segments of members to obtain internal forces and moments.

agrees with (*d*) in Example 8.8. The horizontal deflection (*c*) above can be rewritten in the following form by using (*d*),

$$\delta_H = \frac{XL}{A_{CD}E} \tag{f}$$

This form is equivalent to the result (*c*) in Example 8.8.

The application of energy methods has thus led to the same results as the direct procedure described in Example 8.8. The reader should compare the relative convenience of the two approaches.

Example 8.12 To illustrate the application of Castigliano's theorem to a member subjected to simultaneous bending and twisting, we calculate the deflection δ of the tightly coiled spring in Fig. 8.25*a*. The wire has radius *r* and is formed into *n* complete turns of radius *R*. The spring is identical to that considered in Example 6.4 (see Fig. 6.18), except that here the ends of the spring are not brought into the center of the coil but extend directly from the rim of the coil. We shall see that this "small" difference has an important effect on the stiffness of the spring.

At each section of the spring, the wire carries a transverse shear force **P**, a twisting moment $\mathbf{M}_t$, and a bending moment $\mathbf{M}_b$, as indicated in Fig. 8.25*b*. By applying the equilibrium requirements to this free body, we find

$$M_t = PR(1 - \cos\theta) \qquad M_b = PR\sin\theta \tag{a}$$

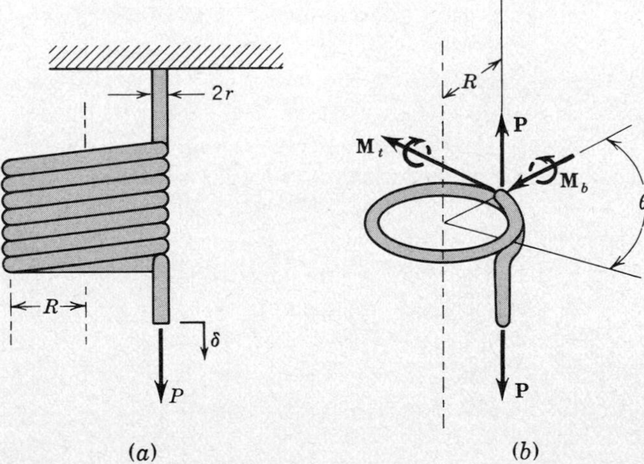

Fig. 8.25 Example 8.12.

The total strain energy (8.8) in the wire (of uncoiled length $2n\pi R$) due to the twisting and bending contributions is

$$U = \int_0^{2\pi n} \frac{P^2R^2(1 - \cos\theta)^2}{2GI_x} R\,d\theta + \int_0^{2\pi n} \frac{P^2R^2\sin^2\theta}{2EI} R\,d\theta$$

$$= \frac{P^2R^3}{2GI_x} 3\pi n + \frac{P^2R^3}{2EI} \pi n \tag{b}$$

where $I_x = \pi r^4/2$ and $I = \pi r^4/4$. The deflection δ in Fig. 8.25*a* is, according to (8.9),

$$\delta = \frac{\partial U}{\partial P} = PR^3\pi n\left(\frac{3}{GI_x} + \frac{1}{EI}\right)$$

$$= \frac{4PR^3n}{Gr^4}\left(\frac{3}{2} + \frac{G}{E}\right)$$

$$= \frac{4PR^3n}{Gr^4}\left(\frac{4 + 3\nu}{2 + 2\nu}\right) \tag{c}$$

where we have used (5.3) to introduce Poisson's ratio. Note that under the same load this spring deflects nearly twice as much as the spring with centered ends in Example 6.4.

8.7 LIMIT ANALYSIS

In Sec. 6.12 limit analysis was introduced in connection with torsion problems. We now reconsider the technique as applied to bending problems. There are two essential steps in limit analysis. The first concerns the geometry of the structure. It is necessary to determine what part or parts of the structure must undergo large deformation in order for the structure as a whole to suffer large deformations. The second step involves an equilibrium study to determine what external loads are needed to balance the limit loadings on the individual parts.

In a structure which is a beam (or several interconnected beams), the individual part which undergoes large deformation is the *plastic hinge*. In Sec. 7.10 we saw that as the curvature at a section of a beam of elastic–perfectly plastic material was increased, the bending moment grew linearly until the value M_Y was reached, which marked the onset of yielding. As the curvature was further increased, the bending moment asymptotically approached the *limit* or *fully plastic* bending moment M_L. The ratio between M_L and M_Y was denoted by the factor K and was tabulated for a few typical beam cross sections in Table 7.1. Whenever a section of a beam undergoes a curvature which is large compared with the curvature which first causes yielding, we call that section of the beam a plastic hinge, and we make very little error if we assume that the bending moment of this section is M_L.

A geometrically compatible large deformation of a structure is usually called a *collapse mechanism*. Two simple examples of collapse mechanisms for beams are shown in Fig. 8.26. The simply supported beam in Fig. 8.26*a* can undergo large deflections as soon as a plastic hinge develops at H. The statically indeterminate beam in Fig. 8.26*b* cannot undergo large deflections with just a single hinge at H_1 because the cantilever section AH_1 will not undergo large deflections until a hinge develops somewhere along its length. With two hinges H_1 and H_2, as shown in Fig. 8.26*c*, large deflections of the beam are geometrically compatible. Thus Fig. 8.26*c* represents a possible collapse mechanism while Fig. 8.26*b* does not.

In very general structures the location of possible hinges can be a difficult question. In the case of beams subjected to concentrated force loads and concentrated reactions, however, the situation is fairly simple. In Sec. 3.8 it was pointed out that for such cases the greatest bending moments always occur at a loading point or a reaction point. Plastic hinges can, therefore, develop only at these points. In a complicated beam of this type there may be more than one collapse mechanism. It is then necessary to compare the external limit loads for each possible collapse mechanism. The mechanism with the smallest limit load is the actual collapse mechanism.

Example 8.13 Figure 8.27 shows a beam built-in at C, simply supported at A, and subjected to a concentrated load P at B. It is desired to find the magnitude of the limit load P_L which corresponds to the condition of plastic collapse. Let the bending moment corresponding to the onset of yielding

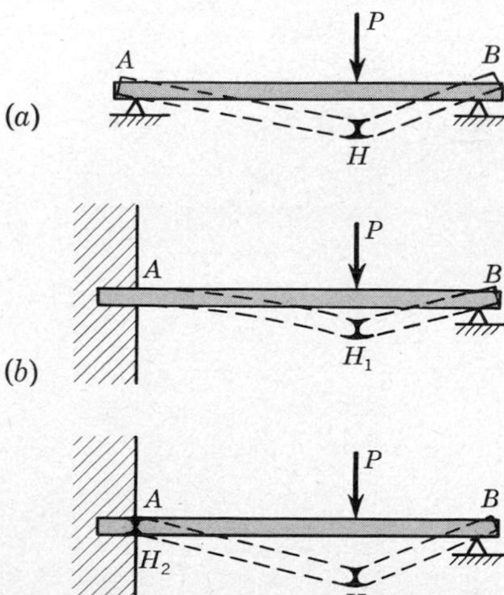

Fig. 8.26 One plastic hinge causes collapse in (*a*). Two plastic hinges are required for collapse of the beam shown in (*b*) and (*c*).

for the beam section be M_Y, and let the limiting or fully plastic bending moment be M_L.

In Fig. 8.27*a* the reactions are calculated according to the equilibrium requirements in terms of the statically indeterminate quantity M_C. The equilibrium analysis is extended in Fig. 8.27*b* where free bodies of the segments *AB* and *BC* are shown. Note that all shear forces and bending moments depend on the statically indeterminate quantity M_C. It should be emphasized again that this equilibrium analysis is valid independently of the stress-strain law. The stress-strain law and the conditions of geometric compatibility enter only in fixing the magnitude of M_C. In Example 8.4 we have already studied this problem for the purely elastic case and determined M_C from a condition of geometric compatibility with the support configuration. Here our only condition for the determination of M_C is the requirement that sufficient hinges have formed so that *P* causes collapse.

In Fig. 8.27*c* we show the collapse geometry with plastic hinges at *B* and *C*. We obtain our quantitative result by observing that the magnitudes of the bending moments at *B* and *C* must both be M_L. Referring to Fig. 8.27*b*, we have

$$\begin{aligned} \frac{2Pa}{3} - \frac{2M_C}{3} &= M_L \\ M_C &= M_L \end{aligned} \qquad (a)$$

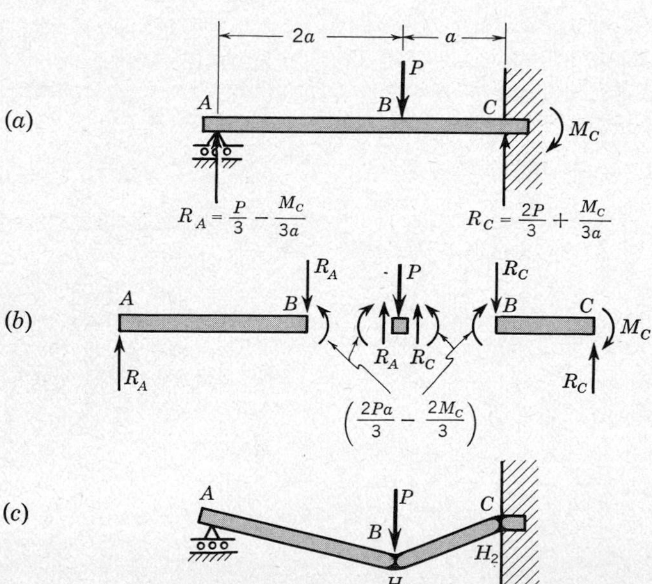

Fig. 8.27 Example 8.13. Equilibrium analysis of statically indeterminate beam, (*a*) and (*b*). Geometry of collapse, (*c*).

Eliminating M_C gives

$$P_L = 2.5 \frac{M_L}{a} \tag{b}$$

as the plastic limit load.

To compare this result with that of the purely elastic case, we can go back to Fig. 8.7 and use the formula for the maximum bending moment (which occurs at C) to obtain the load P_Y which corresponds to the initiation of yielding.

$$P_Y = 1.8 \frac{M_Y}{a} \tag{c}$$

Thus in terms of $K = M_L/M_Y$ we can write

$$P_L = 2.5 \frac{KM_Y}{a} = \frac{2.5}{1.8} KP_Y$$

or

$$P_L = 1.39KP_Y \tag{d}$$

There are thus two factors which make P_L larger than P_Y. One factor, K, is due to the redistribution of stresses in the cross section as the bending moment increases from M_Y to M_L. The other factor (1.39 in this case) is due to the redistribution of bending moments along the length of the beam. When the beam acts elastically, the bending moment at B is smaller than at C. As plastic flow progresses, these two bending moments tend to equalize as they both approach the same limit M_L in the condition of plastic collapse.

Example 8.14 The structure shown in Fig. 8.28 consists of two equal cantilever beams AC and CD with roller contact at C. Given the limiting bending moment M_L for the beams, it is desired to find the limiting value of the load P which corresponds to plastic collapse of the structure.

The forces and equilibrium requirements are analyzed in Fig. 8.28*b* which shows free-body diagrams of the various beam segments. The structure is statically indeterminate, but all forces and moments can be expressed in terms of the single unknown F, which is the magnitude of the interaction at C.

Turning next to the geometry of collapse, we find that there are two different geometrically admissible modes of collapse, as shown in Fig. 8.28*c* and *d*. In Fig. 8.28*c* plastic hinges form at A and B, causing collapse under the load but without incurring large deformations in the cantilever CD. In Fig. 8.28*d* plastic hinges form at A and D, causing large deformations of both beams. To decide which collapse mechanism would actually occur, we

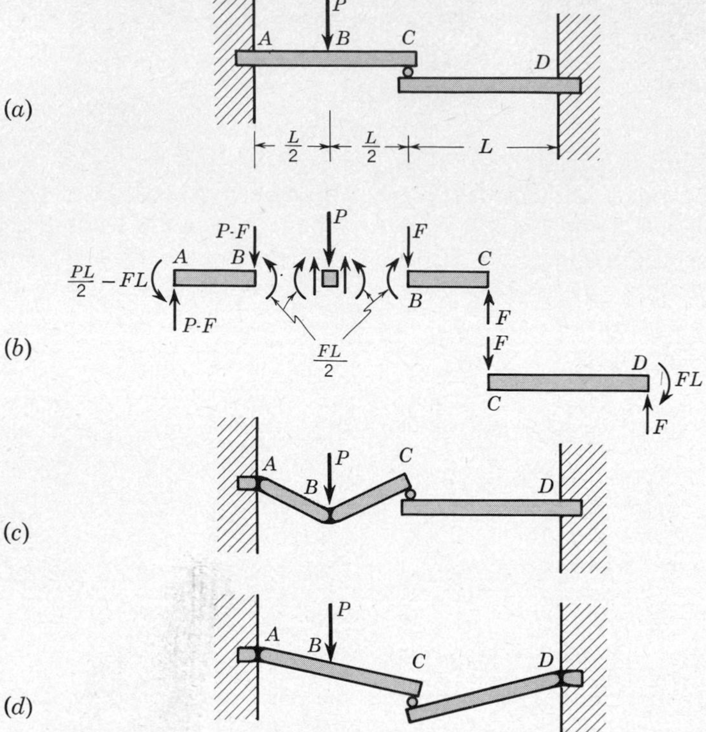

Fig. 8.28 Example 8.14. Structure with two possible modes of collapse.

obtain the value of P corresponding to each mode of collapse. The mode with the *smaller* value of P is the actual collapse mechanism.

For the mechanism of Fig. 8.28*c* we set the bending moments at A and B equal to the limiting bending moment M_L.

$$\frac{PL}{2} - FL = M_L$$
$$\frac{FL}{2} = M_L \tag{a}$$

Eliminating F, we obtain

$$P = 6\frac{M_L}{L} \tag{b}$$

as the load corresponding to Fig. 8.28*c*.

Similarly, for the mechanism of Fig. 8.28*d*, we set the bending moments at A and D equal to the limiting bending moment M_L.

$$\frac{PL}{2} - FL = M_L \qquad (c)$$

$$FL = M_L$$

Eliminating F, we obtain

$$P = 4\frac{M_L}{L} \qquad (d)$$

as the load corresponding to Fig. 8.28*d*. Since (*d*) is smaller than (*b*), the structure collapses in the mechanism of Fig. 8.24*d* under the limit load

$$P_L = 4\frac{M_L}{L} \qquad (e)$$

An alternative procedure for deciding against the result (*b*) is to continue the force analysis in Fig. 8.28*c*, obtaining the bending moment at D which corresponds to (*b*). If we do this we find that the magnitude of the bending moment at D must be $2M_L$, which is incompatible with the fact that the maximum bending moment can be developed in these beams is M_L. This indicates that a hinge will form at D before the mode of Fig. 8.28*c* can ever develop.

PROBLEMS

8.1. (*a*)–(*h*) Find the deflection of the neutral axis of each of the beams shown in Probs. 3.1 to 3.8. Take the bending modulus EI to be constant in each beam.

8.2. Find the central deflection of the uniform, simply supported beam due to the uniformly distributed load over the right half of the beam.

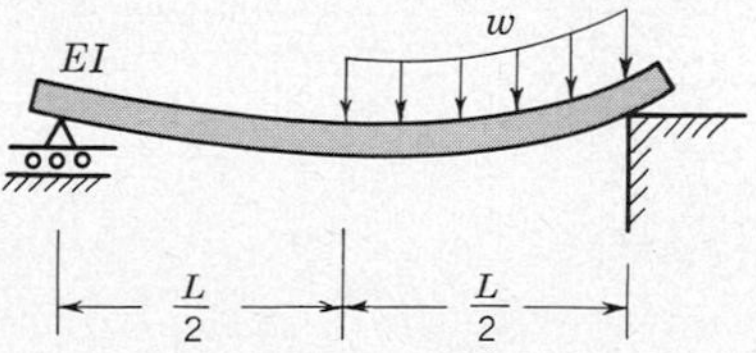

Prob. 8.2

8.3. Find the deflection at C due to the load P in terms of the length L and bending modulus EI of the uniform beam AC.

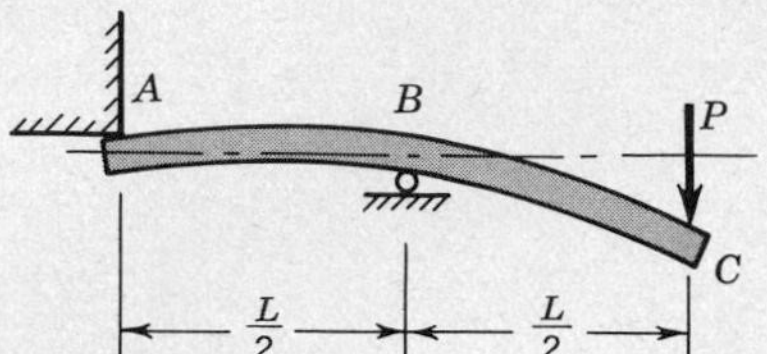

Prob. 8.3

8.4. Find the central deflection of the uniform, simply supported beam due to the pair of equal, symmetrically placed loads.

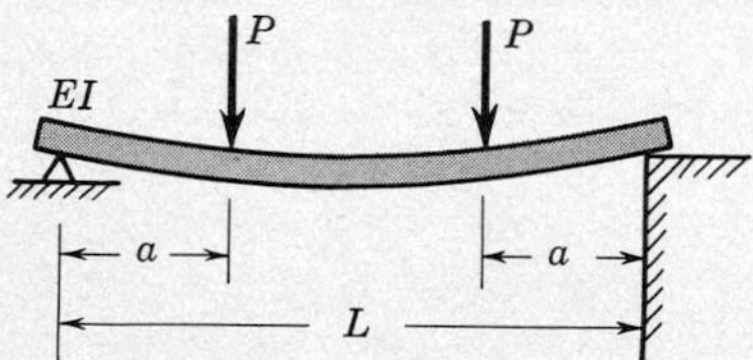

Prob. 8.4

8.5. Find the reactions at the clamped ends of the pair of equal, uniform, elastic cantilevers due to the load P, assuming that the beams just held the roller lightly at B before the load was applied.

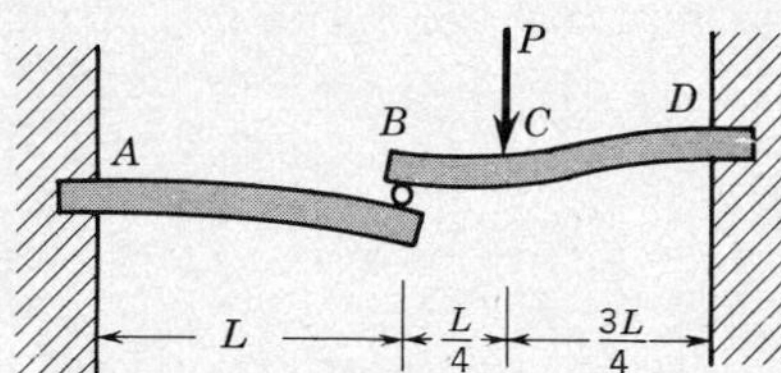

Prob. 8.5

8.6. The member ABC consists of two straight legs AB and BC of round bar stock of radius r welded together at B. It is built-in at A and lies in a horizontal plane when it is unloaded. Find the deflection at C under the vertical load P in terms of the dimensions given, and the elastic constants E and ν of the bar stock.

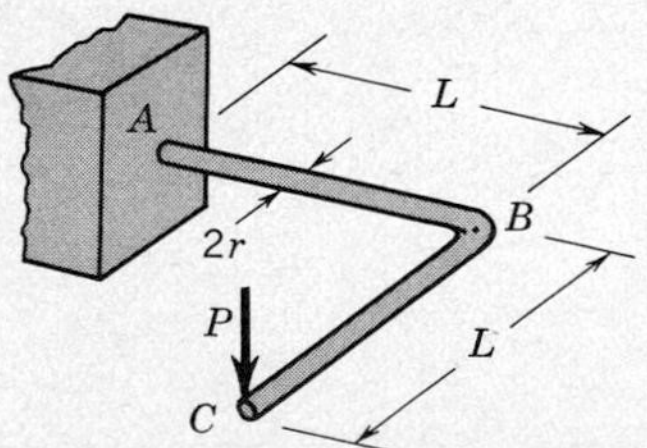

Prob. 8.6

8.7. Find the slope angle ϕ at the point A due to the applied couple M_o in terms of the dimensions shown, and the bending modulus EI of the uniform beam.

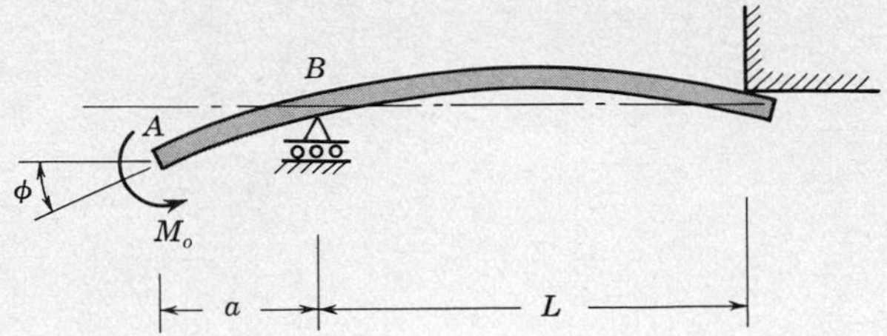

Prob. 8.7

8.8. In the five-storied structure shown, the vertical columns, into which the flexible horizontal beams are built, are very rigid. The diagram shows the original configuration of the structure as built. Estimate the maximum bending moment that would be induced in the horizontal beams if the foundation B were to settle a distance δ.

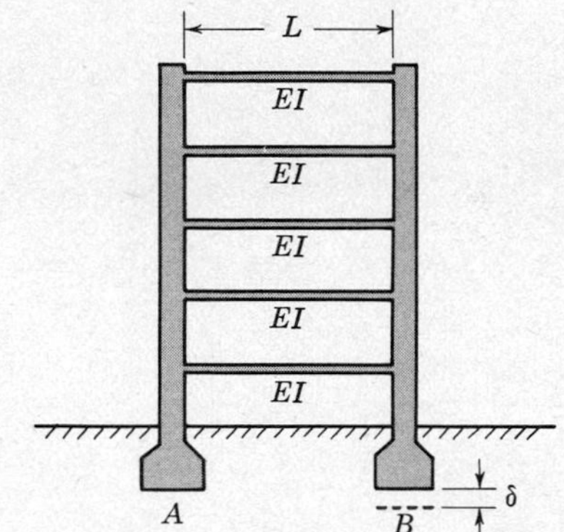

Prob. 8.8

8.9. The uniform beam shown has pinned supports at A, B, and C. Find the slope angle ϕ at C due to the applied couple M_o.

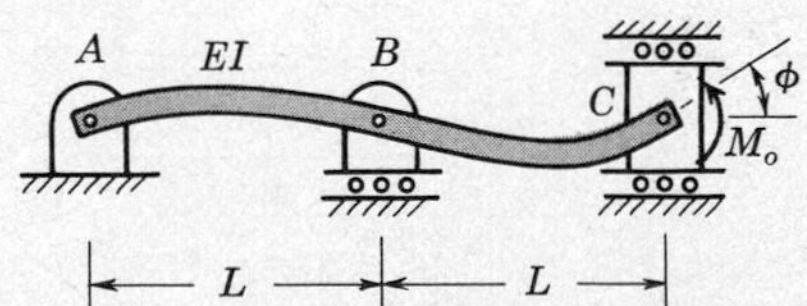

Prob. 8.9

8.10. Before the load P was applied to the beam shown, the beam was straight and the spring was unstretched. Find the deflection under the load P in terms of EI, L, and the spring constant k.

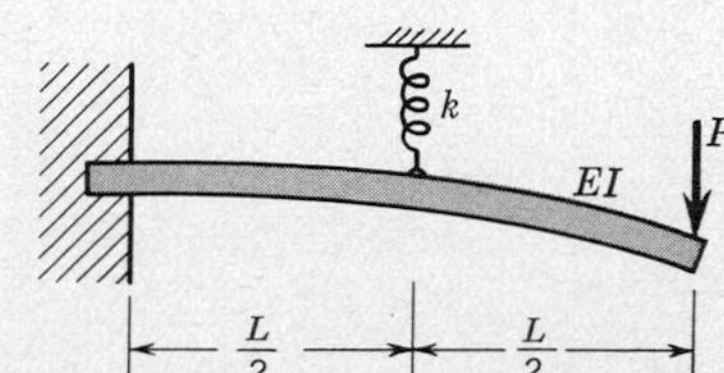

Prob. 8.10

8.11. A uniform beam is built-in at both ends. Find the maximum bending moment and the maximum deflection due to a uniformly distributed load of intensity w per unit length.

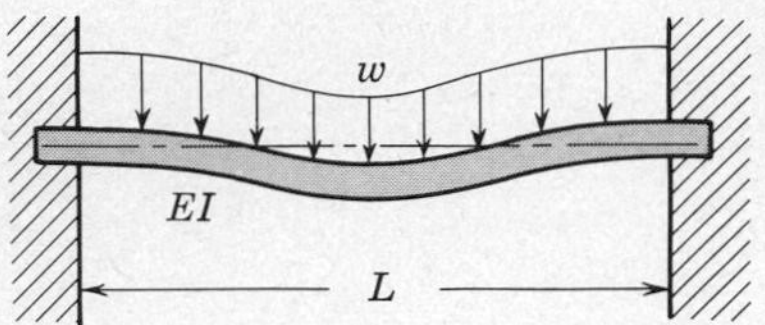

Prob. 8.11

8.12. A uniform cantilever beam is loaded by a total force W which is distributed uniformly over the middle half of the beam as shown. Show that the deflection at the right end is

$$\delta = \frac{7WL^3}{64EI}$$

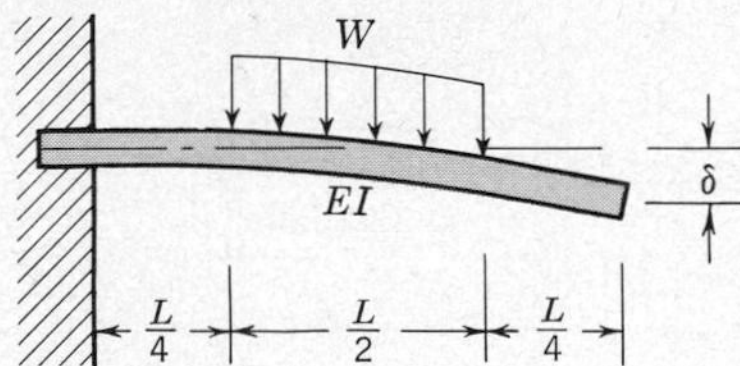

Prob. 8.12

8.13. The uniform beam shown has pinned supports at A, B, and C. Show that the deflection under the load is

$$\delta = \frac{23PL^3}{1{,}536EI}$$

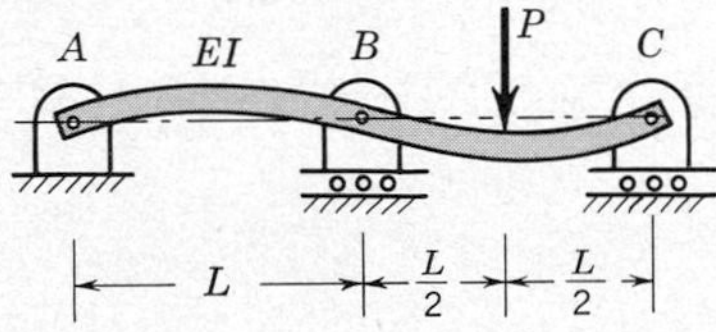

Prob. 8.13

8.14. A uniform cantilever beam carries a total load of W, which is distributed in the linearly varying fashion shown. Find the deflection at the right end.

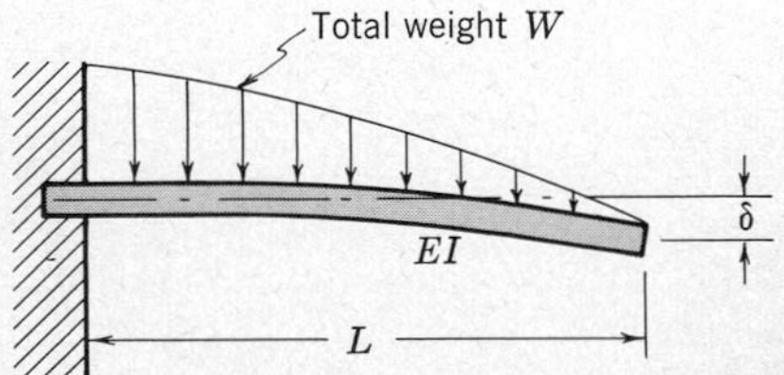

Prob. 8.14

8.15. The two cantilevers shown are identical in every respect except that in (*b*) there is a simple support at the right end. They are both loaded with a uniformly distributed load of intensity w. Compare

(1) The maximum bending moments

(2) The maximum deflections

in the two cases.

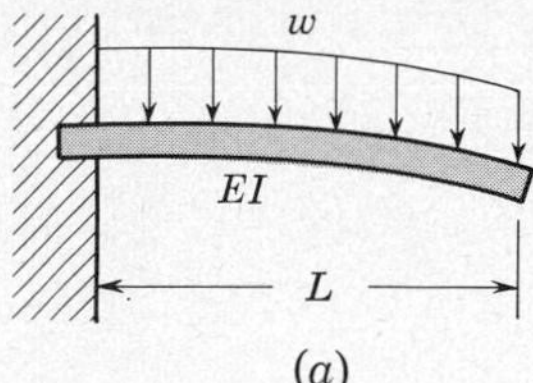

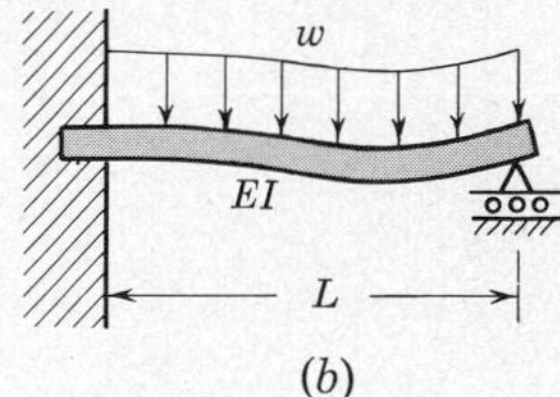

Prob. 8.15

8.16. The two systems shown are identical in every respect except in (*a*) there are two beams of length L while in (*b*) there is a single beam of length $2L$. The uniformly distributed load of intensity w is the same in both cases. Compare

(1) The reaction at the central support

(2) The maximum bending moments

(3) The maximum deflections

in the two cases.

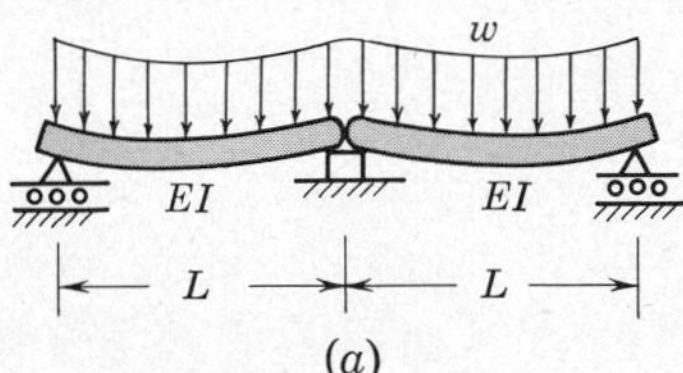

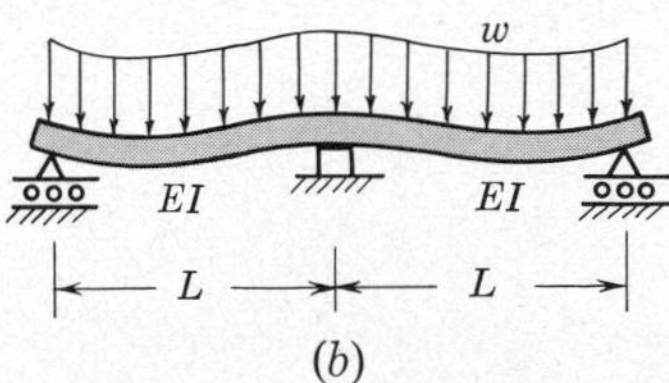

Prob. 8.16

8.17. Find the deflection at A for the two cases shown. Compare the results as the ratio $a/L \rightarrow 0$.

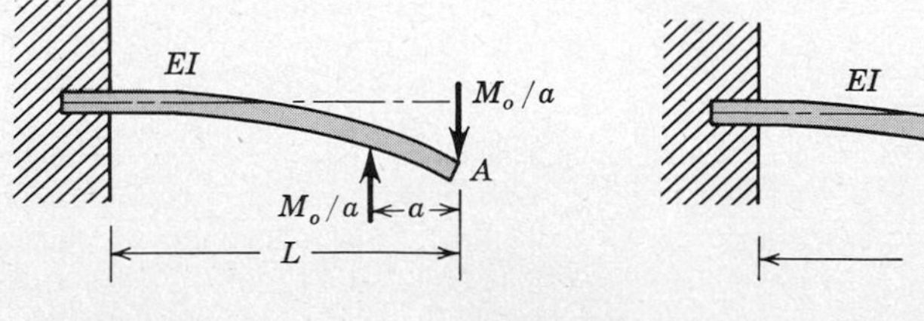

Prob. 8.17

8.18. The singularity functions introduced in Sec. 3.6 are useful for writing the bending moment at a section x based on a free body which extends to the *left* end of the beam. Investigate the family of reversed singularity functions

$$g_n(x) = \langle a - x \rangle^n$$

and show that they are useful for writing the bending moment at a section x based on a free body which extends to the *right* end of the beam. Show that when $n \geqq 0$

$$\int_x^{\infty} \langle a - x \rangle^n \, dx = \frac{\langle a - x \rangle^{n+1}}{n+1}$$

and

$$\int \langle a - x \rangle^n \, dx = -\frac{\langle a - x \rangle^{n+1}}{n+1} + c$$

8.19. Verify that the bending moment in the system shown (which is the mirror image of Fig. 8.7) can be written as

$$M_b(x) = R_A(L - x) - P\langle b - x \rangle^1$$

in terms of the reversed singularity functions of Prob. 8.18. Continue the solution to obtain the statically indeterminate reaction R_A. Compare corresponding steps with Example 8.4.

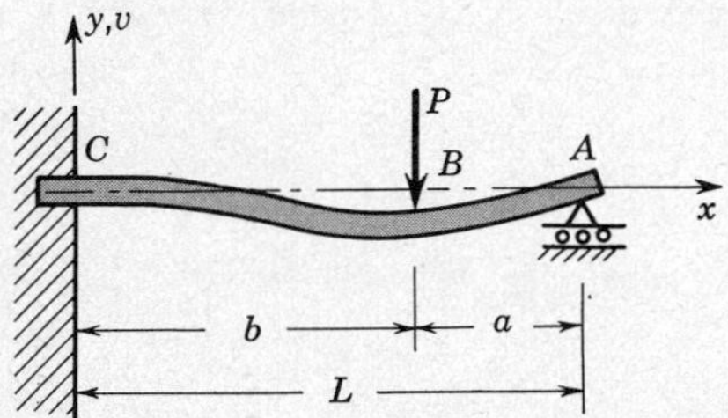

Prob. 8.19

8.20. The uniform beam is supported on three equal, equidistant springs so that when there is no load the beam is straight and horizontal. Find the forces in the three springs due to the load P.

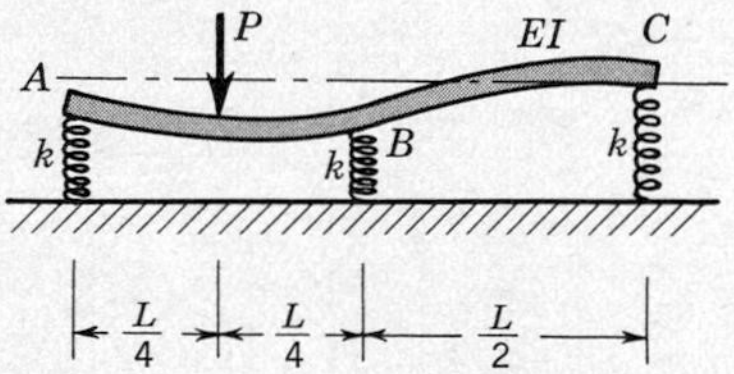

Prob. 8.20

8.21. Precision measurements of molecular-beam momentum are made by using a tiny aluminum blade which is built-in at one end and simply supported at the other. The blade's length is $L = 30$ cm and its cross section is rectangular, 0.5×0.01 cm. Young's modulus for aluminum is $E = 6.9 \times 10^{11}$ dynes/cm^2. The molecular beam impinges on one side of the blade at a distance $0.5255L$ (for greatest sensitivity) and is equivalent to a force P. A light beam is reflected from the other side at a distance βL where the beam's angular deflection ϕ is maximum.

(*a*) Find the location βL for the maximum angular deflection.
(*b*) Find the value of the maximum angular deflection (ϕ_{max}).
(*c*) Find the sensitivity ratio $k = S/P$.

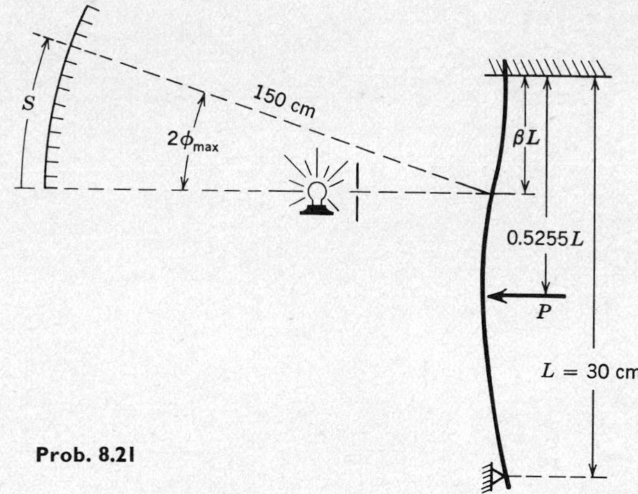

Prob. 8.21

8.22. By using Castigliano's theorem find the deflection at the end of the cantilever beam shown.

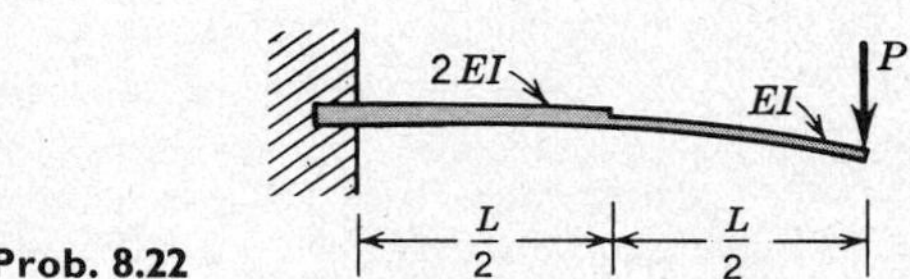

Prob. 8.22

8.23. An elastic wire of radius r has the form of a quarter circle of radius R. Obtain the deflection δ in the direction of the load P, taking into account bending *and* axial loading. Show that the ratio of the axial contribution to the bending contribution is

$$\frac{\delta_a}{\delta_b} = \frac{1}{4}\frac{r^2}{R^2}$$

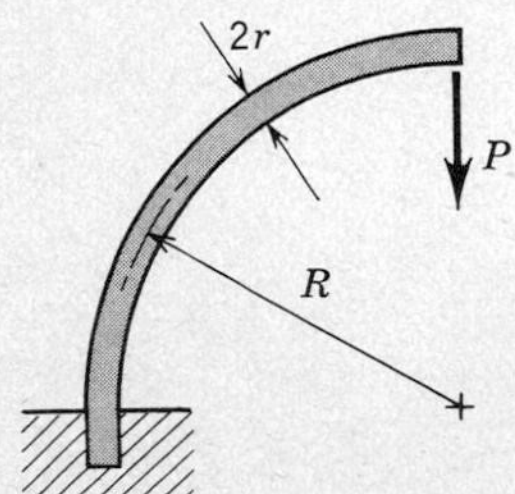

Prob. 8.23

8.24. Derive a formula for the deflection δ of the structural member shown.

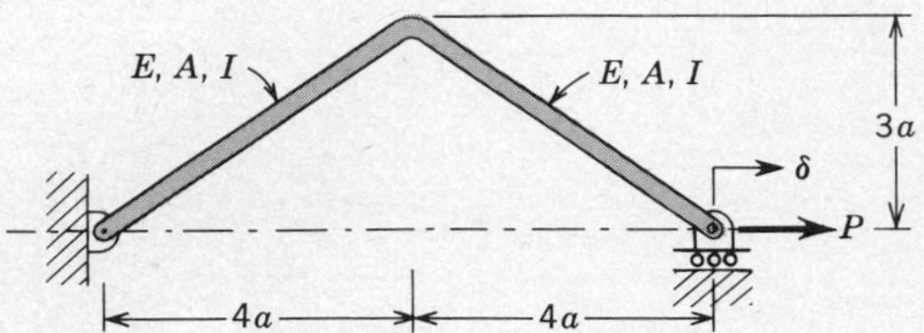

Prob. 8.24

8.25. An elastic wire of radius r has the form of a sector of a circle of radius R as shown. Determine the vertical and horizontal deflection of the point C as a function of the position θ of the load P along the wire.

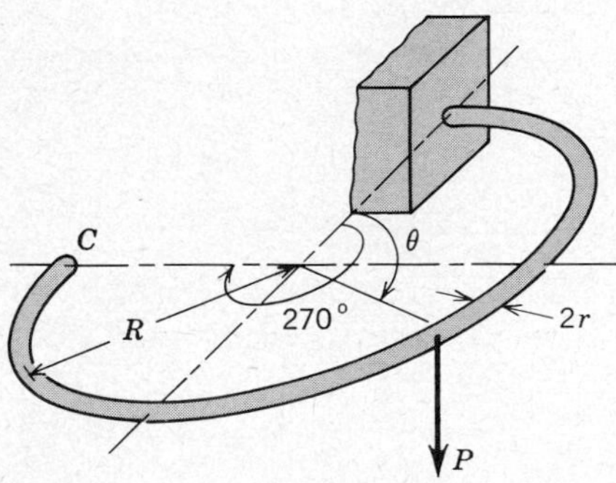

Prob. 8.25

8.26. A pair of round steel rods ½ in. in diameter are welded together at right angles at B and built into the wall at A. The other end is pinned to an object that moves horizontally to the right by an amount δ. What is the maximum stress set up in the rods if $\delta = 1$ in. and $L = 4$ ft?

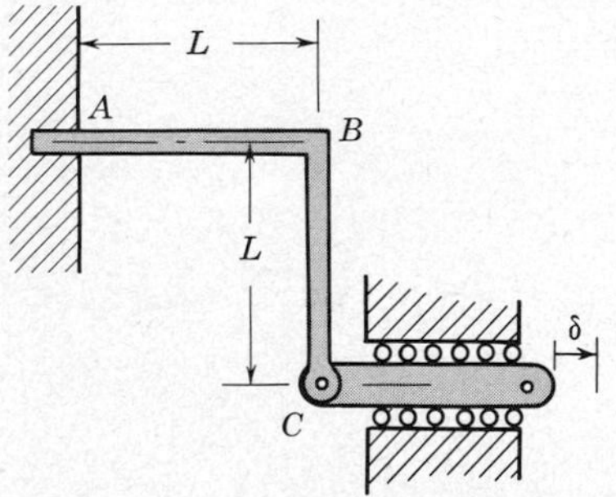

Prob. 8.26

8.27. In the course of a laser experiment, a beam of light travels in a 6-in.-diameter steel tube of ⅛-in. wall thickness supported at its ends. Because the tube is long, the sag in the middle is such that only half the light fed into the tube gets out without hitting the sides of the tube. What is the length L of the tube? In order to reduce this sag, it is decided to use a thicker tube. If a steel tube with the same outside diameter but with a wall thickness of ½ in. were used, how much would the sag be reduced and how much more light would go through?

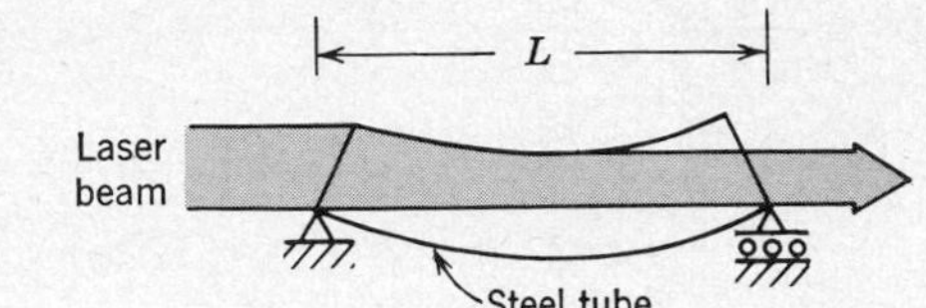

Prob. 8.27

8.28. A simply supported floor beam carries a uniformly distributed loading of 1,000 lb/ft. In order to avoid possible cracking of the plaster on the ceiling beneath the beam, it is desired that the deflection should not exceed 1/360 of the span length L. If $L = 12$ ft and $E = 10^6$ psi, what is the minimum allowable value of the section moment of inertia I?

8.29. A steel shaft 0.50 in. in diameter passes concentrically through a rigid housing whose inside diameter is 0.90 in. The shaft may be considered as simply supported at the bearings. What is the maximum bending moment M_o which can be applied to the end of the shaft if the clearance between shaft and housing is to be not less than 0.10 in.?

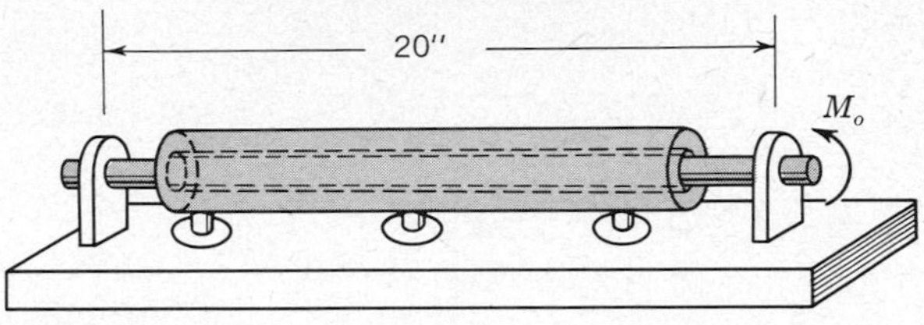

Prob. 8.29

8.30. The beams AB and CD are each of length L and both have the same bending stiffness EI. The beam CD is simply supported, while AB is built-in at A. Before the load P is applied the two beams make light contact at their midpoints. Find the deflection under the load P.

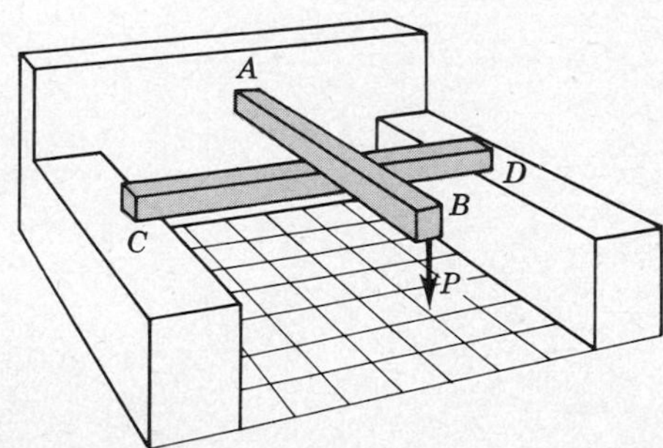

Prob. 8.30

8.31. A torque wrench is a wrench which uses the deflection of a flat steel bar to measure the torque being exerted on the bolt or nut. A typical wrench is shown in the accompanying sketch.

Calculate the scale which should be used to represent the torque; i.e., calculate how much deflection occurs on the scale for a given applied torque.

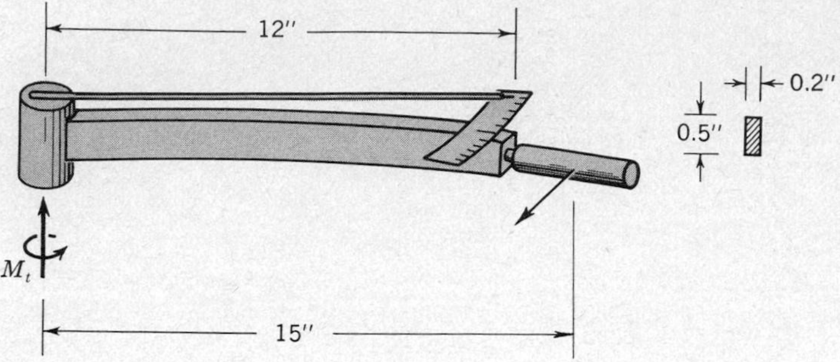

Prob. 8.31

8.32. The signpost is made of steel pipe with 4-in. outside diameter and 3½-in. inside diameter and is set in a concrete foundation. Estimate the lateral deflection at the top of the post when a 50-lb sign is hung on the arm as shown.

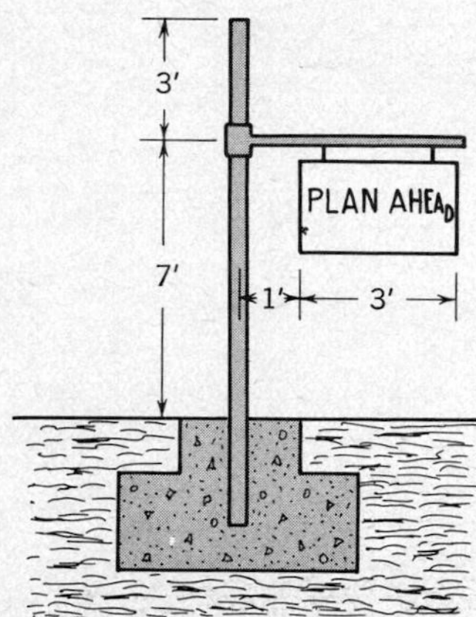

Prob. 8.32

8.33. A girder bridge is simply supported at both ends and is supported by a pontoon in the center of the bridge. How much will the pontoon sink into the water if the bridge is loaded by a uniformly distributed load w per unit of length? Let the length of the bridge be L, the weight per unit volume of water be γ, and the water-line area of the pontoon be A.

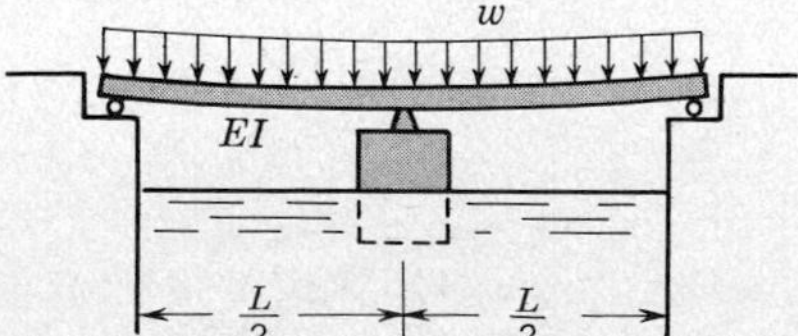

Prob. 8.33

8.34. A simply supported beam is constructed by welding a very stiff beam to a beam which is relatively much less stiff in bending. What is the deflection under a load P applied in the middle of the stiff part if we assume that this part carries a bending moment without any resulting curvature and the flexural rigidity of the other part is EI?

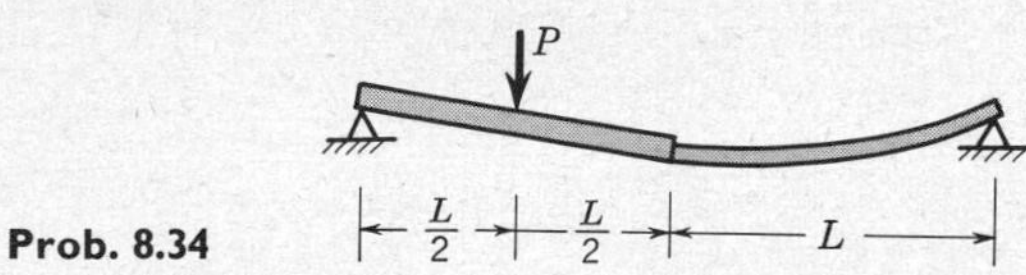

Prob. 8.34

8.35. A more-or-less flexible uniform box of weight W and length $L/3$ rests in the middle of a simply supported beam of length L. Estimate the deflection at mid-span and compare it with an approximate solution obtained by replacing the box by a concentrated load W at the center.

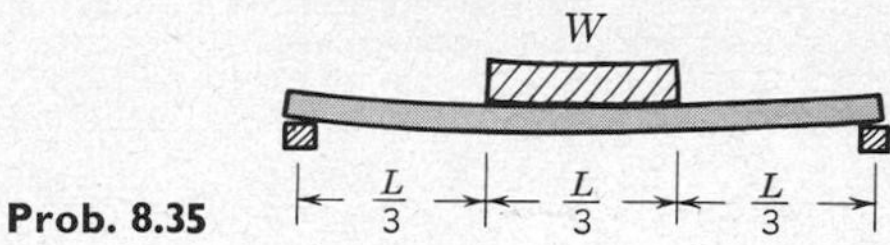

Prob. 8.35

8.36. The uniform beam shown is supported above a rigid plane by two rollers of small diameter d. Find the magnitude P of the pair of equal forces which are just sufficient to force the middle quarter of the beam into contact with the surface.

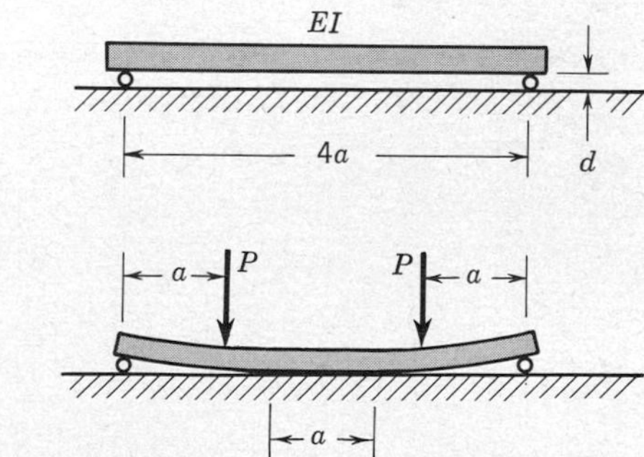

Prob. 8.36

8.37. Two separate cantilever beams of equal bending modulus EI are built-in, as shown in (*a*), and then loaded with an end load P, as shown in (*b*). Find the deflection δ under P. Assume that the two beams remain in contact at the points $x = 0$ and $x = L$. Then obtain the deflection curves for both beams in the region $0 < x < L$ and prove the correctness of the assumption.

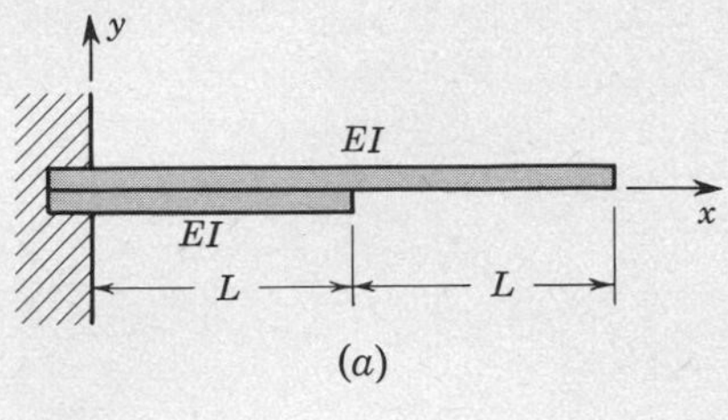

(a)

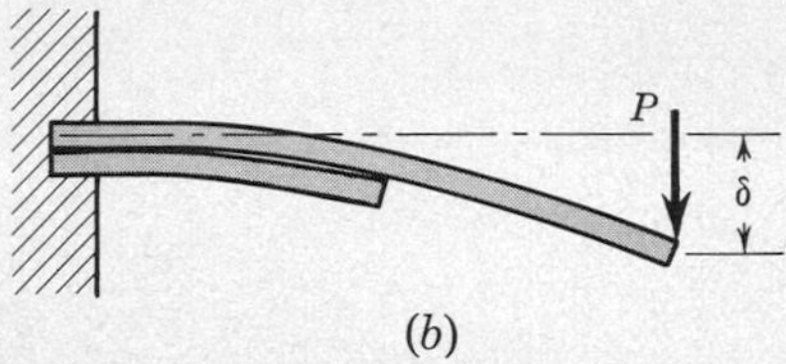

(b)

Prob. 8.37

8.38. Calculate the end deflection of the thin steel strip in Prob. 7.19 when the force P is such as to bring the strip into contact with the block over a distance c, as indicated in the sketch for Prob. 7.19.

8.39. Find the deflection at the tip of the tapered cantilever beam of Prob. 7.24.

8.40. The elastic flexure system shown is designed to allow the lower rigid rod to move horizontally without changing its angular orientation. The flexures are made from phosphor bronze with dimensions $2 \times \frac{1}{4} \times 0.060$ in. The flexures are rigidly attached to the support and to the rod. $E = 15 \times 10^6$ psi and $\sigma_Y = 50{,}000$ psi.

(*a*) What is the force-deflection relation for the system?

(*b*) What is the maximum deflection δ of the rod which can occur before the stress in the flexures reaches the yield stress σ_Y?

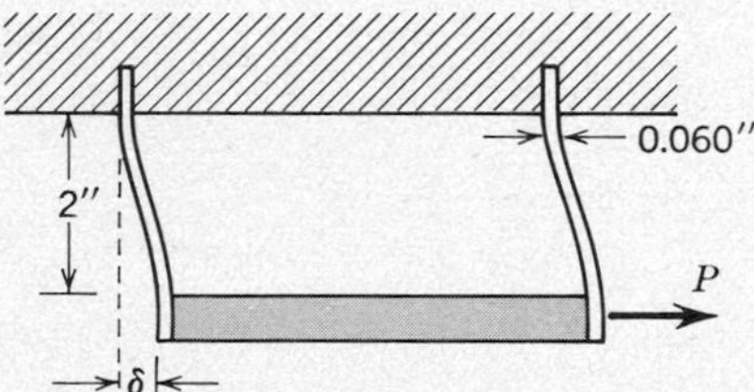

Prob. 8.40

8.41. In the text the deflection of a beam was considered to be due only to the bending, i.e., the effect of shear was neglected. An estimate of the order of magnitude of the shear deflection can be obtained as follows. Let the total deflection $v(x)$ be considered as the sum of $v_b(x)$ due to bending and $v_s(x)$ due to shear. The bending deflection is obtained, as in the text, by neglecting the presence of the shear forces. The shear deflection, as indicated in the figure, may be estimated by neglecting the presence of the bending moments. The deflection model shown in (*b*) is based on the simplifying assumption that the originally vertical plane cross sections remain plane and vertical, and that the element is subjected to a *uniform* shear strain γ_{xy}. This is an oversimplification because, as indicated in Fig. 7.16, the shear strain varies across the section, and plane sections do not remain plane. This model will, however, *overestimate* the shear deflection if we take the uniform shear strain γ_{xy} to be equal to the maximum shear strain in the actual distribution. If $\tau_{xy}(x)$ represents the maximum shear stress in the cross section as obtained by the equilibrium analysis of Chap. 7, the shear deflection of the model pictured here then satisfies the following differential equation:

$$\frac{dv_s}{dx} = \frac{\tau_{xy}}{G}$$

For a rectangular cantilever beam of the dimensions given in (*a*), show that the shear and bending deflections at $x = L$ have the following ratio:

$$\left(\frac{v_s}{v_b}\right)_{x=L} = \frac{3(1+\nu)}{4}\frac{h^2}{L}$$

and thus that the shear deflection is less than 1 percent of the bending deflection if the beam length is more than 10 times the beam cross-sectional height.

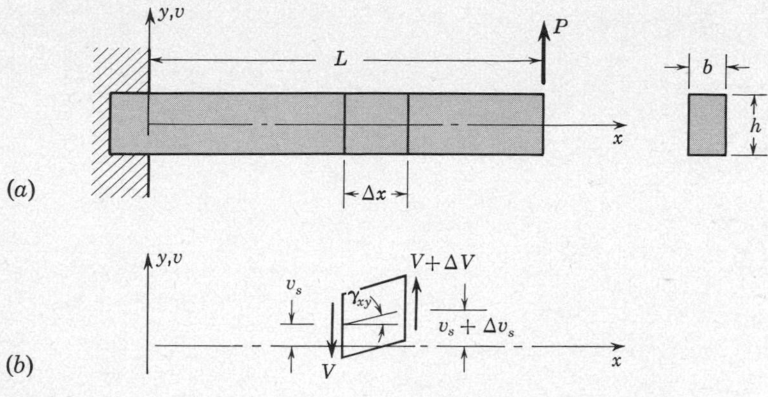

Prob. 8.41

8.42. The relative importance of transverse shear deflection in the deflection of beams is estimated in Prob. 8.41. A somewhat better approximation can be obtained by applying Castigliano's theorem to the energy expressions developed in Sec. 7.8. Rework Prob. 8.41, using the energy approach and show that this leads to the estimate

$$\left(\frac{v_s}{v_b}\right)_{x=L} = \frac{3(1+\nu)}{5}\frac{h^2}{L^2}$$

8.43. In (*a*) the trace of the neutral axis of a beam which has been bent by a bending moment $M_b(x)$ is shown. In (*b*) the corresponding values of M_b/EI are shown. We consider that the bending is small enough so that the approximation (8.3) for the curvature is valid. Use these diagrams to derive the following *area-moment* theorems: (1) The angle ϕ_{12} between tangents to the neutral axis at x_1 and x_2 is equal to the cross-hatched area under the M_b/EI curve between x_1 and x_2. (2) The distance ϵ_{12} between the tangent emanating from x_1 and the neutral axis at x_2 is equal to the *first moment* of the cross-hatched area under the M_b/EI curve between x_1 and x_2 *about the line* $x = x_2$.

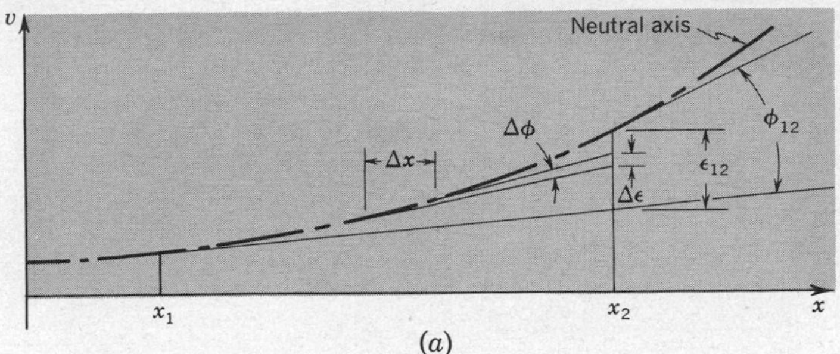

(a)

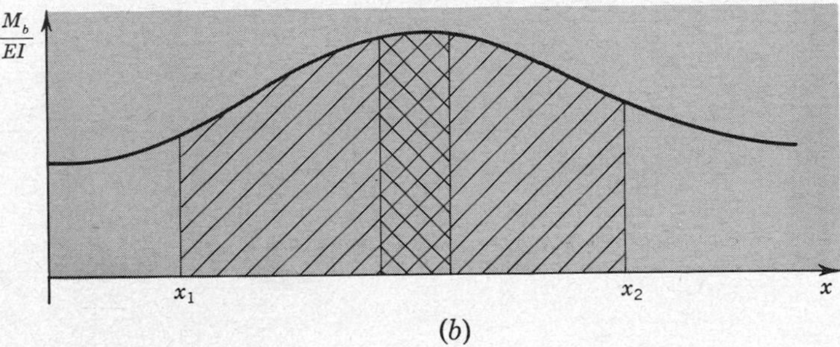

(b)

Prob. 8.43

8.44. In a precision telescope, a long, uniform tube is to be simply supported by a pair of symmetrically placed supports, as shown. The tube is made as stiff as possible to minimize deflections of the lenses at the ends of the tube; however, as the angle θ is varied from 0 to 90°, gravity will cause different amounts of bending in the tube. This bending will have a minimum effect on the optical alignment if during any such bending the two lenses *remain exactly parallel*. Find the location of the supports which achieves this condition. Neglect the weight of the lenses in comparison with the tube.

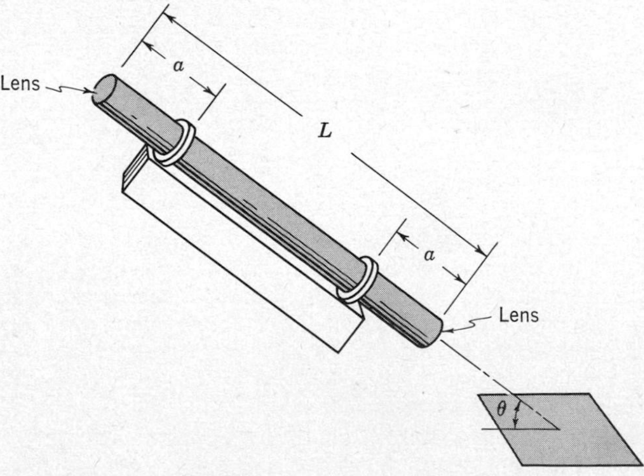

Prob. 8.44

8.45. A strip of mirror glass in a precision optical instrument is to be simply supported by a pair of symmetrically placed supports, as shown. During operation the angle θ will vary from 0 to 90° so that the bending due to gravity will vary. For what position of the supports will the deviations from *flatness* be minimized; i.e., at any fixed θ for what value of a will the greatest bending slope angle in the mirror be minimized?

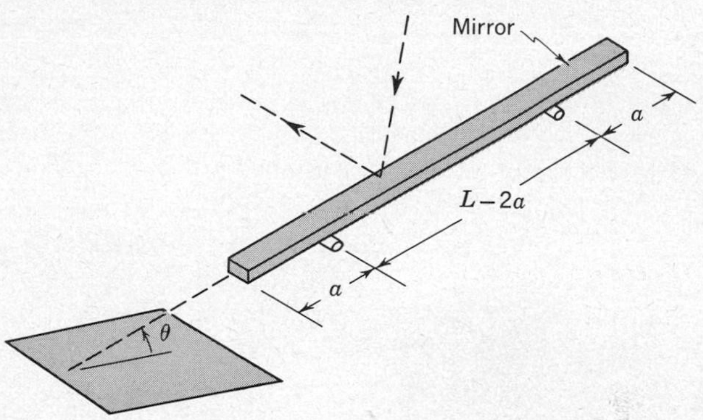

Prob. 8.45

8.46. A simple, flexible French curve consists of a steel strip ½ in. wide × 1/32 in. thick (similar in cross section to a hacksaw blade) which has the shape illustrated in (*a*). It may be used to draw a curve joining the points *A*, *B*, and *C*, which are nearly in a straight line. To accomplish this, forces are applied by fingers placed as shown in (*b*) until the strip bends into a curve passing through the three points. What is the curve taken up by the strip? Derive an equation for the deflection of the center of the strip as a function of the applied force *P*, making plausible estimates for any dimensions not given in the problem. Which of the types of steel shown in Fig. 5.5*a* would be most suitable? Estimate the largest force *P* which can be applied before the strip deforms plastically.

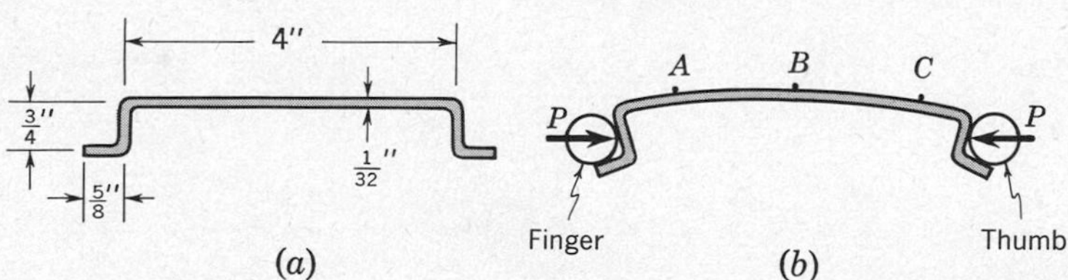

Prob. 8.46

8.47. A bimetallic strip consists of two equal-sized strips of different materials which have been bonded together. If the elastic moduli and thermal expansion coefficients are E_1, E_2, α_1, and α_2, estimate the tip deflection of the strip due to a temperature rise T.

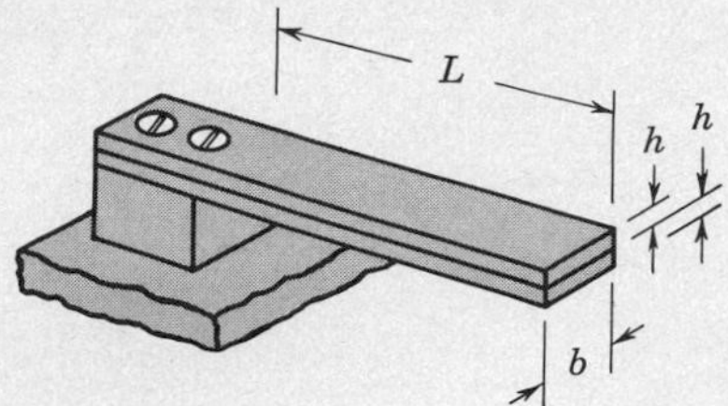

Prob. 8.47

8.48. A rigid disk of radius R is supported by two flexible strips, built-in as shown. When a couple M_o is applied to the disk, it rotates through a small angle ϕ_o. Find the torsional spring constant $k = M_o/\phi_o$ in terms of R, L, and EI.

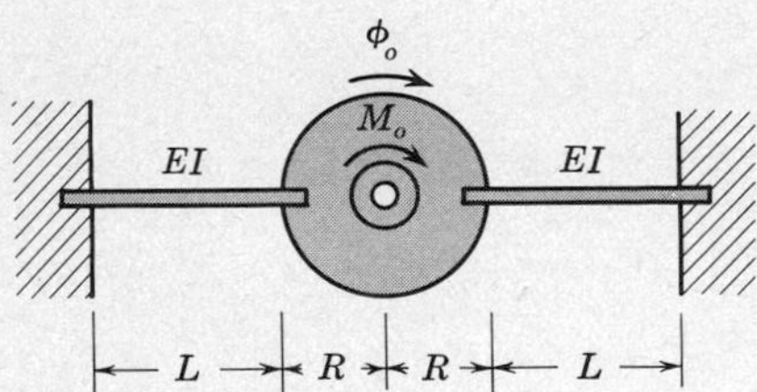

Prob. 8.48

8.49. The rectangular cantilever beam shown is made of material with elastic constants E and ν. It is loaded by an off-center force at the end. Estimate the vertical deflection directly under the load.

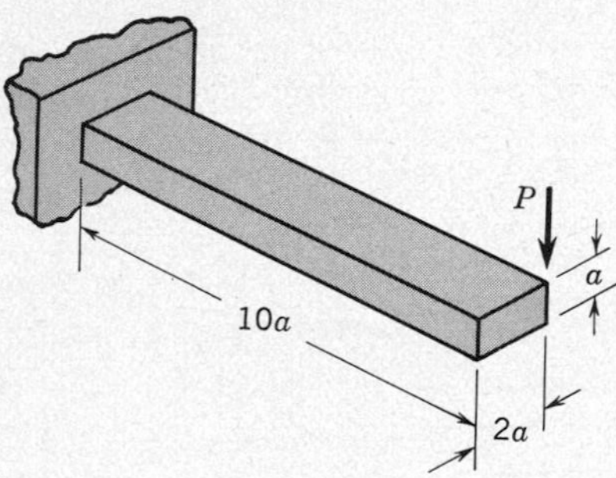

Prob. 8.49

8.50. Suppose that both beams in the system of Prob. 8.5 have the same fully plastic or limit bending moment M_L. Find the limiting value of P which corresponds to plastic collapse.

8.51. If the beam in Prob. 8.11 has a plastic-limit moment M_L, what value of load intensity w corresponds to plastic collapse?

8.52. The beam AD is built-in at both ends. The magnitude of its fully plastic bending moment is M_L. Find the limiting value of the load P on the swing seat which corresponds to plastic collapse, as a function of the dimension a, where $0 < a < L$. Assume that large deformations do not occur in the chains or in the seat.

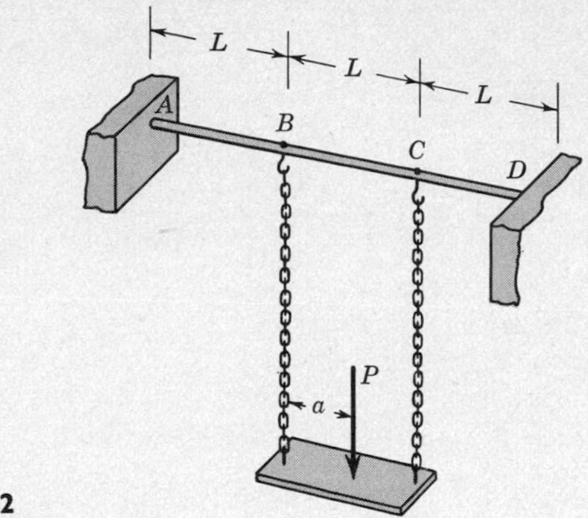

Prob. 8.52

8.53. The beam shown is constructed by welding together two square bars of the same size but of different materials. The yield stress of material 1 is Y, and the yield stress of material 2 is $0.5Y$. Estimate the limit load P which corresponds to plastic collapse.

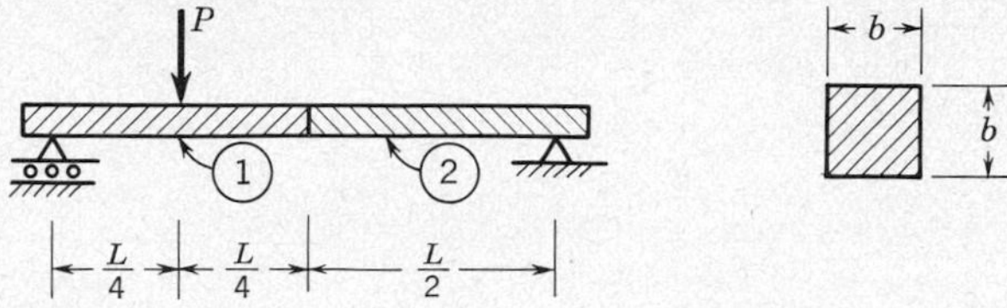

Prob. 8.53

8.54. The beam shown is built-in at A and pinned at B and C. The magnitude of the fully plastic bending moment is M_L. Find the magnitude of P which corresponds to plastic collapse and sketch the mode of collapse.

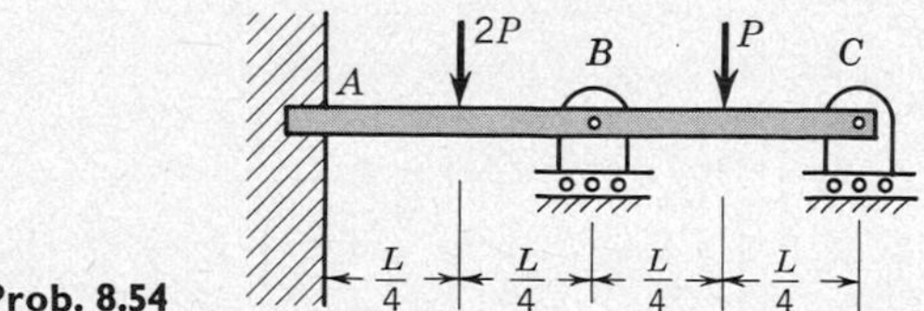

Prob. 8.54

8.55. The limit bending moment in the beam shown is $M_L = KM_Y$, where M_Y is the bending moment corresponding to the first appearance of yielding. Show that

$$\frac{P_L}{P_Y} = \frac{4}{3} K$$

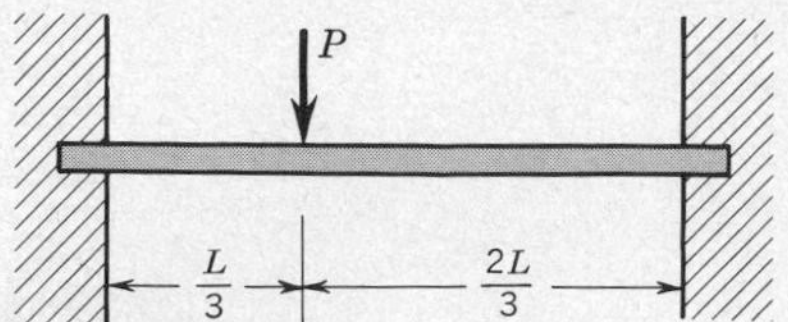

Prob. 8.55

8.56. Consider the system of Prob. 8.55, except now let the load P act at the center of the span. Show that in this case

$$\frac{P_L}{P_Y} = K$$

8.57. The limit bending moment in the beam shown is $M_L = KM_Y$, where M_Y is the bending moment corresponding to the first appearance of yielding. Show that

$$\frac{P_L}{P_Y} = \frac{9}{8}K$$

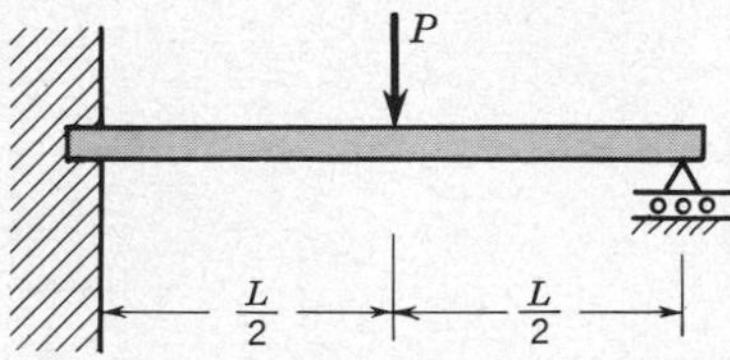

Prob. 8.57

8.58. In Example 8.5 it was shown that when a flexible rod overhangs a table by a length a, it does not come in continuous contact with the table until a distance $\sqrt{2}a$ back from the edge. The length of the rod, L, must clearly be greater than $(1 + \sqrt{2})a$ for this solution to be valid. If the length of the rod is less than $2a$, equilibrium is impossible: the rod falls off the table. Analyze the intermediate situation where $2a < L < (1 + \sqrt{2})a$.

8.59. In Chap. 7 it was shown that for unsymmetrical beams the resultant curvature of the neutral axis is not, in general, in the plane perpendicular to the bending-moment vector. It was shown, however, that the resultant curvature could be defined by giving the curvatures of the projections of the neutral axis on the xy and xz planes (see Fig. 7.35). For small deflections these curvatures can be expressed as follows in terms of the transverse displacement components v and w of the neutral axis in the y and z directions:

$$\lim_{\Delta s_1 \to 0} \frac{\Delta\alpha}{\Delta s_1} \approx \frac{d^2v}{dx^2}$$

$$\lim_{\Delta s_2 \to 0} \frac{\Delta\beta}{\Delta s_2} \approx \frac{d^2w}{dx^2}$$

Using the moment-curvature relations (7.63) in Chap. 7, the following differential equations can be obtained to relate the transverse displacement of the neutral axis to the bending-moment components.

$$\frac{d^2v}{dx^2} = \frac{1}{E}\frac{I_{yz}M_{by} + I_{zz}M_{bz}}{I_{yy}I_{zz} - I_{yz}^2}$$

$$\frac{d^2w}{dx^2} = -\frac{1}{E}\frac{I_{yy}M_{by} + I_{yz}M_{bz}}{I_{yy}I_{zz} - I_{yz}^2}$$

Using these results, calculate the displacement of the end of the Z-shaped cantilever beam of Prob. 7.77.

8.60. Determine the horizontal and vertical components of the deflection of the end of the cantilever beam of Prob. 7.78.

8.61. An initially straight cantilever beam of length L is built-in at A and loaded by a moment M_o at B. Find the slope and deflection at B by noting that the beam bends into an arc of a circle of radius ρ. For small M_o show that the results for the slope and deflection approach those given in Table 8.1, case 3.

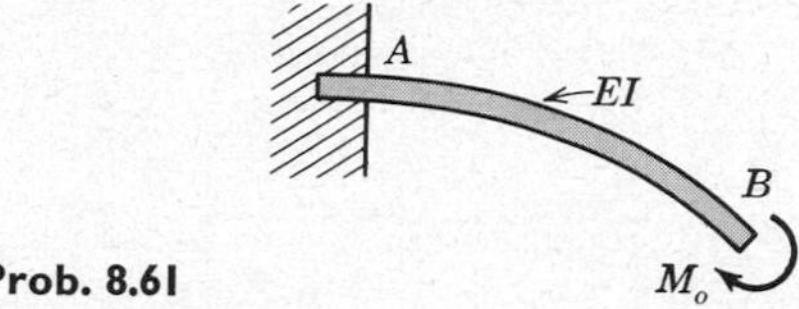

Prob. 8.61

8.62. The steering mechanism shown consists of a relatively rigid rim of radius R and a relatively rigid column joined by a flexible crossarm built-in to the rim and the column, as shown. The crossarm is of circular cross section with radius r and is made of material with elastic constants E and ν. (*a*) If a couple M_x is applied to the rim with the column fixed, the rim will rotate through a small angle ϕ_x. Estimate the stiffness $k_x = M_x/\phi_x$ for rotation about the x axis. (*b*) If a couple M_y is applied to the rim with the column fixed, the rim will rotate through a small angle ϕ_y. Estimate the stiffness $k_y = M_y/\phi_y$ for rotation about the y axis.

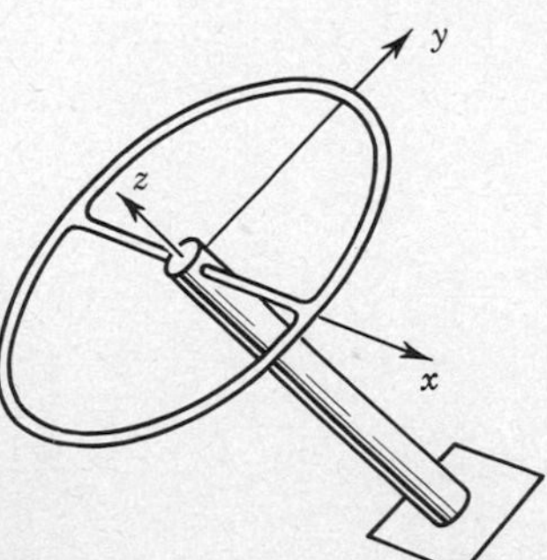

Prob. 8.62

8.63. Consider the simply supported, elastic–perfectly plastic rectangular beam of Fig. 7.30 at the stage where the plastic zones have just joined. The moment-curvature relation for the middle third of the beam is (7.45) while the moment-curvature relation for the remainder of the beam is (8.1). Using the approximation (8.3) for the curvature, show that if the origin of coordinates is taken at the center of the span, the locus of the deformed neutral axis is

$$v = -\frac{M_Y a^2}{EI}\left[\frac{20}{27} - \frac{4}{3\sqrt{3}}\left(\frac{x}{a}\right)^{3/2}\right] \qquad \text{when } 0 < x < \frac{a}{3}$$

$$v = -\frac{M_Y a^2}{4EI}\left[1 - \left(\frac{x}{a}\right)^2\right]\left(3 - \frac{x}{a}\right) \qquad \text{when } \frac{a}{3} < x < a$$

Verify that the central deflection under this condition is $^{20}/_{9}$ times the central deflection at the onset of yielding.

8.64. Consider a very long, uniform beam of bending modulus EI which rests on a uniformly distributed *elastic foundation*. Assume that the foundation exerts a distributed reaction of intensity $-kv$ per unit length, where the displacement of the beam is $v(x)$. Verify that when an external loading $q(x)$ is applied, the displacement satisfies the following differential equation:

$$EI\frac{d^4v}{dx^4} + kv = q$$

Show that when $q = 0$ the general solution of this equation is

$$v = e^{-\beta x}(C_1 \cos \beta x + C_2 \sin \beta x) + e^{\beta x}(C_3 \cos \beta x + C_4 \sin \beta x)$$

where $\beta^4 = k/4EI$. Verify that the deflection under a single concentrated load P at the middle of the very long beam is

$$\delta = \frac{P}{(64EIk^3)^{1/4}}$$

8.65. A linear-elastic beam has a load P applied first at $x = x_1$ and then at $x = x_2$. Maxwell's *reciprocal principle* asserts that the deflection at $x = x_2$ when the load is at x_1 is equal to the deflection at $x = x_1$ when the load is at x_2; that is, $\delta_1(x_2) = \delta_2(x_1)$.

(*a*) Verify the reciprocal principle for the simply supported beam by using the solution in case 4 of Table 8.1.

(*b*) Prove the reciprocal principle by using Castigliano's theorem.

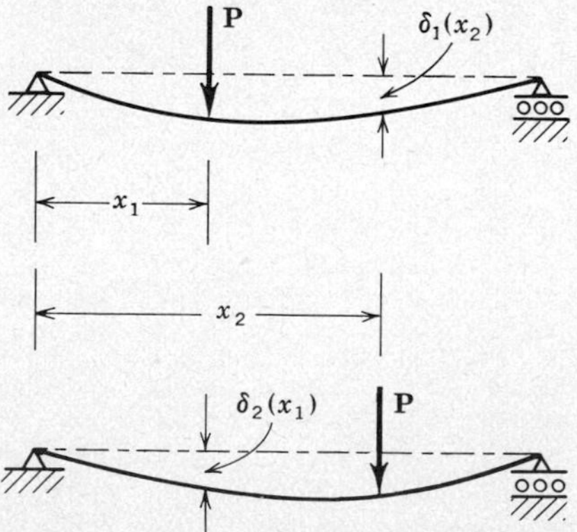

Prob. 8.65

8.66. Leonardo da Vinci suggested that concave bronze mirrors of very large focal lengths could be produced by rotating a thin horizontal bronze disk on a potter's wheel against a copper rod which was simply supported at its ends while a slurry of a polishing compound was fed into the interface between the rod and the disk. If the copper rod has a diameter of 1 in. and a length of 8 ft, calculate the radius of curvature of the mirror.

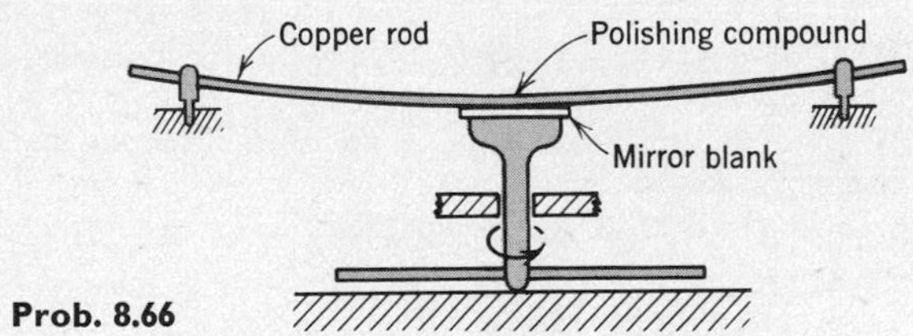

Prob. 8.66

8.67. In one of his discourses, Galileo tells a story about some Roman engineers who had to transport a large stone column to the construction site of a new temple. The standard method at that time was to place the column on two large logs and draw it forward with oxen. However, previous experience with columns of this size showed that they would break over one of the rollers. In order to avoid this problem, the Romans added a third support. Galileo reports that the column " broke upon the middle support." Can you explain why?

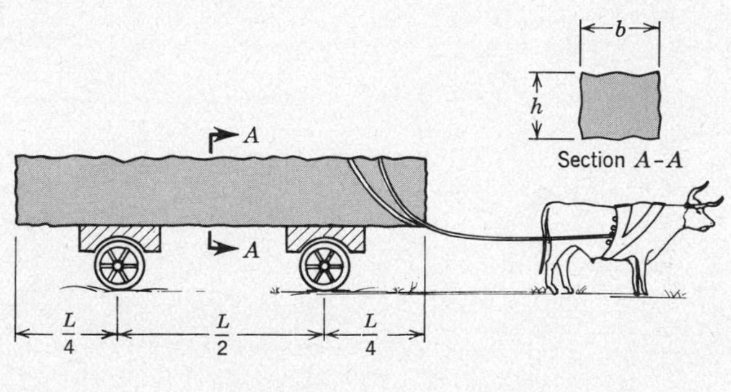

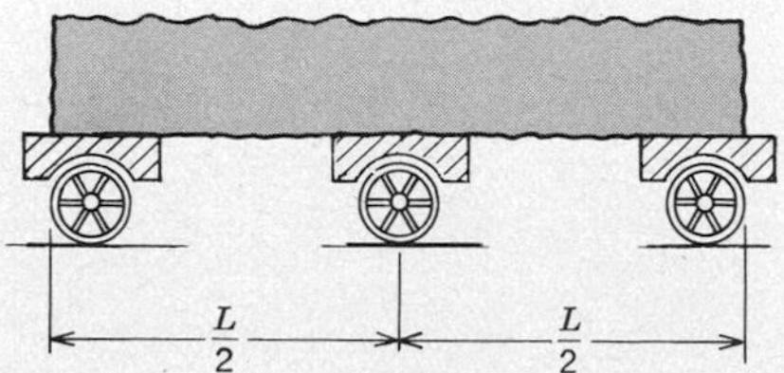

Prob. 8.67

8.68. A rectangular beam of minimum weight is to be designed to satisfy the following requirements:

1. The span L and the depth h of the beam are fixed.
2. The beam is to carry a given total load W, uniformly distributed along the length of the beam. The weight of the beam itself W_B is negligible in comparison with W.
3. Several materials having various values of maximum allowable stress σ_{max}, modulus E, and weight density γ are to be considered.

(*a*) In the first design, deflection is not considered to be important and the beam is designed strictly on the basis of strength. Show that the maximum value of W/W_B is obtained by using the material with the largest value of σ_{max}/γ.

(*b*) In the second design the beam is deflection-limited; i.e., the maximum beam deflection cannot be greater than a specified value δ_{max}, where for every material to be considered the deflection limit is reached before the stress limit. In this case show that W/W_B is maximized by the material with the largest value of E/γ.

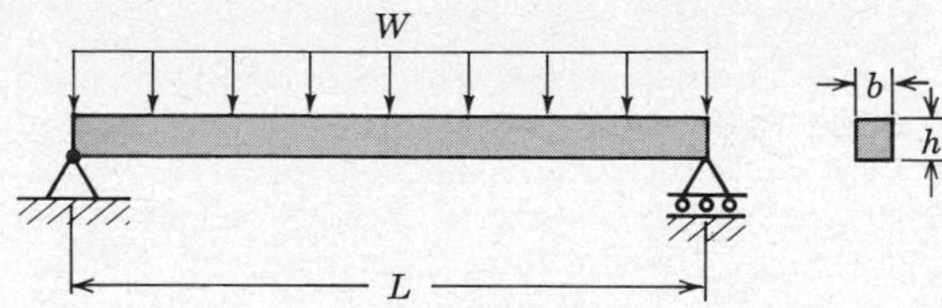

Prob. 8.68

9
Stability of Equilibrium: Buckling

9.1 INTRODUCTION

In the previous chapters we have investigated the configurations and stress distributions in systems which were in equilibrium. We said that a system was in equilibrium if the forces on every subdivision were balanced. In this chapter we shall widen the scope of our investigation by considering the behavior of systems when slightly disturbed from their equilibrium configurations. When the forces no longer balance within a system, there will be accelerations and, in general, a complicated resulting motion. We shall not attempt to obtain a complete picture of this motion but shall restrict our attention to the following question: *When slightly disturbed from an equilibrium configuration, does a system tend to return to its equilibrium position or does it tend to depart even further?* It is often possible to answer this without going into an elaborate study of the accelerated motions.

For example, consider the small weight on the frictionless surfaces of Fig. 9.1. The forces on the particle (gravity and the normal reaction of the surface) are clearly in balance wherever the surface is horizontal. These balance positions are indicated by the letter O. In Fig. 9.1a the weight is shown displaced slightly from

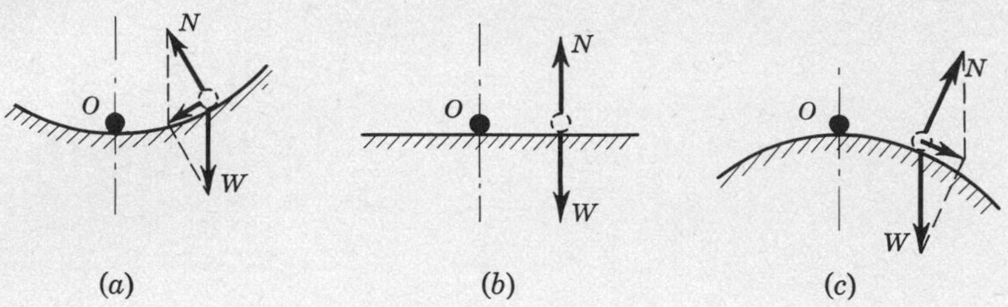

Fig. 9.1 Examples of (*a*) stable, (*b*) neutral, and (*c*) unstable equilibrium.

its equilibrium position. The forces on it no longer balance, but the resultant unbalance is a *restoring force*; i.e., the particle is accelerated back toward the equilibrium position. Such an equilibrium is called *stable*. In Fig. 9.1*c* we have the opposite situation. The resultant unbalance is an *upsetting force*; i.e., it accelerates the particle away from the equilibrium position. Such an equilibrium is called *unstable*. In Fig. 9.1*b* we have the border line between the two previous cases; when the particle is displaced slightly from an equilibrium position, it is again in equilibrium, and there is no tendency either to return or to go further. Such an equilibrium is said to possess *neutral stability*.

By generalizing from the example above, we arrive at the following definition of *stability of equilibrium*. A system is said to be in a *state of stable equilibrium* if, for all possible geometrically admissible small displacements from the equilibrium configuration, restoring forces arise which tend to accelerate the system back toward the equilibrium position.

A load-carrying structure which is in a state of unstable equilibrium is unreliable and hazardous. A small disturbance can cause a cataclysmic change in configuration. This represents a new and different mode of structural failure. We shall give several examples in the following sections.

9.2 ELASTIC STABILITY

In Fig. 9.2 AB is a rigid, weightless bar with a frictionless pin joint at B. When the bar is vertical, a vertical force P can be applied at A, and there will be equilibrium whether P is upward, as in Fig. 9.2*a*, or downward, as in Fig. 9.2*b*. If P remains vertical, however, it is easy to see that a small rotation of AB will give rise to a restoring torque in (*a*) and an upsetting torque in (*b*). The strut under the load P in Fig. 9.2*b* is thus *unstable*.

This unstable structure can be stabilized by adding guy wires or transverse springs, as shown in Fig. 9.3*a*. The vertical force P is carried entirely by AB. The function of the springs is simply to maintain the alignment of AB. In Fig. 9.3*b* we analyze the effect of a *small* transverse displacement x. The pair of springs

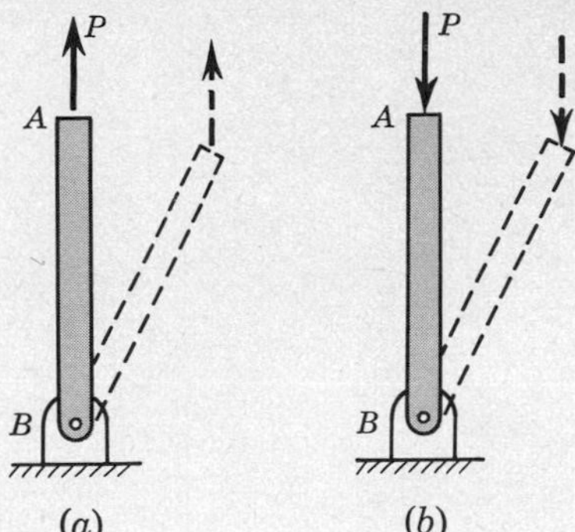

Fig. 9.2 Hinged bar is (*a*) stable for tensile load, and (*b*) unstable for compressive load.

exert a total force $2kx$ which has a *restoring* torque of $2kxL$ about B. The upsetting torque of P about B is Px. The stability of the structure for small deflections depends on which of these is the larger, as shown below.

$$\begin{aligned} Px &< 2kxL \qquad \text{(stable)} \\ Px &> 2kxL \qquad \text{(unstable)} \end{aligned} \tag{9.1}$$

Thus the springs have stabilized the strut for *small* loads P, but it is still unstable for *large* loads. The border line between stability and instability occurs when $P = 2kL$. This value is designated as the *critical* load or the *buckling* load.

An alternative method of studying this structure, which gives further insight into the stability problem, is indicated in Fig. 9.4. Here we take the force P to be applied slightly off-center. The small distance ϵ is called the *eccentricity* of the load. The assumption of some eccentricity provides a useful model for many types of *imperfections* which occur in actual structures; e.g., the bar might not be exactly

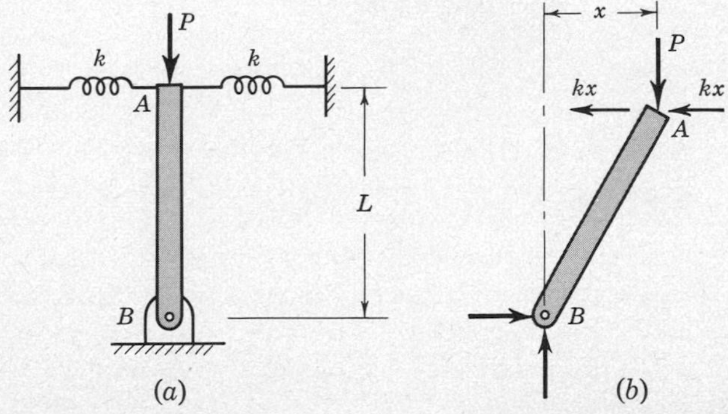

Fig. 9.3 Analysis of hinged bar in compression stabilized by springs.

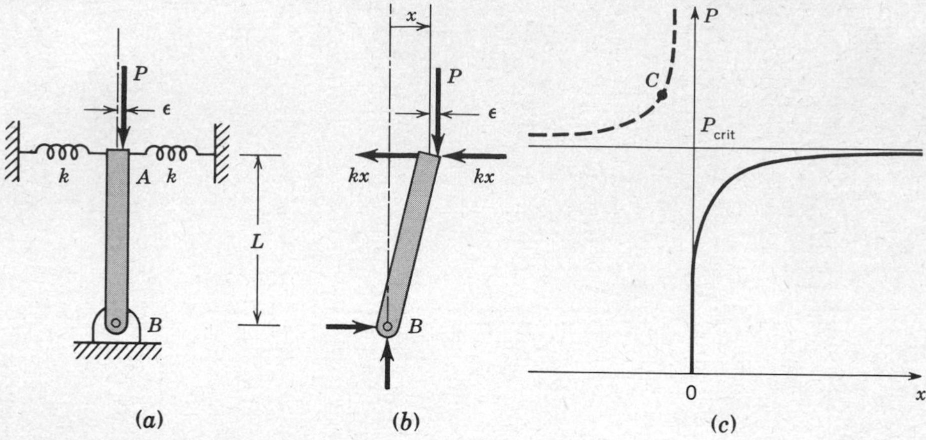

Fig. 9.4 Transverse displacement x due to load eccentricity ϵ.

straight, the springs might not be perfectly balanced, etc. We can solve for the transverse displacement x in the equilibrium position by balancing torques about B in Fig. 9.4*b*.

$$P(x + \epsilon) = 2kxL$$

$$x = \epsilon \frac{P}{2kL - P} \tag{9.2}$$

The values of load P and transverse displacement x which satisfy (9.2) are plotted in Fig. 9.4*c*. For small values of P, the displacement x is very small, but when P is close to the critical value $P_{crit} = 2kL$, the displacement is large.[1]

The dashed curve in Fig. 9.4*c* satisfies (9.2) and thus represents equilibrium positions, but these are *unstable* equilibrium positions. This can be seen by considering Fig. 9.5, which shows the configuration corresponding to the point C in Fig. 9.4*c*. Imagine a small disturbance of position in which the strut moves to the left; that is, x becomes more negative. According to Fig. 9.4*c* a *smaller* value of P is required for balance in the disturbed position. Since the load P is actually unchanged during the disturbance, the new position is one of unbalance, and the direction of the unbalance is such as to move the strut further from the original equilibrium configuration. Thus when P is greater than the critical load $P_{crit} = 2kL$, the equilibrium positions for small x are unstable.

[1] Strictly speaking, Eqs. (9.2) are valid only for small x. See Sec. 9.5 for a discussion of large deflections. See also Prob. 9.15.

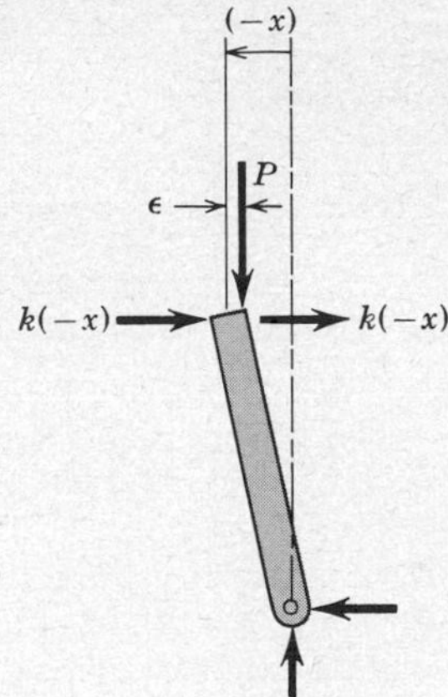

Fig. 9.5 The equilibrium position corresponding to point C in Fig. 9.4c.

When P is less than the critical load, the equilibrium positions are stable. Furthermore, the equilibrium displacements are *small* if P is not too close to the critical load (for example, $x < \epsilon$ if $P < \frac{1}{2}P_{\text{crit}}$). The simple model of Fig. 9.4 is a good prototype for understanding the nature of elastic stability in more complicated cases. In the *ideal* case in which the strut is exactly straight and the load is exactly centered, no transverse displacement (buckling) is possible until the critical load is reached. When a small imperfection is present, there is always a transverse displacement, but it remains small until the load approaches the critical load. In one respect the foregoing analysis differs from that of the earlier chapters. The equilibrium requirement (9.2) results from balancing torques *in the deformed configuration* of Fig. 9.4b. In the preceding chapters we have always assumed that, because the deformation was small, the equilibrium requirements could be applied in the undeformed configuration. Generally this is acceptable if the system is elastically stable and not in the neighborhood of a critical load condition. In order to investigate stability, however, it is essential to apply the equilibrium requirements in the deformed configuration even though the deformations are small.

9.3 EXAMPLES OF INSTABILITY

Several instances of instability are described briefly in this section to provide the reader with a qualitative appreciation of the extent of the problem. One of these cases will be studied in detail in Sec. 9.4.

In Fig. 9.6 a thin, deep, cantilever beam is shown subjected to a vertical end load. As long as the beam sections remain vertical, they resist the bending action of P very effectively, the deformation being of the type considered in Chap. 8. This is what happens for small values of P. If P becomes larger than a certain critical value, the vertical configuration becomes unstable, and a small disturbance

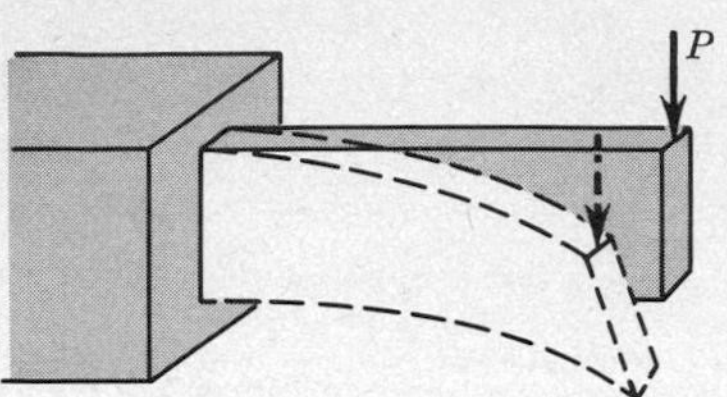

Fig. 9.6 Twist-bend buckling of a deep, narrow beam.

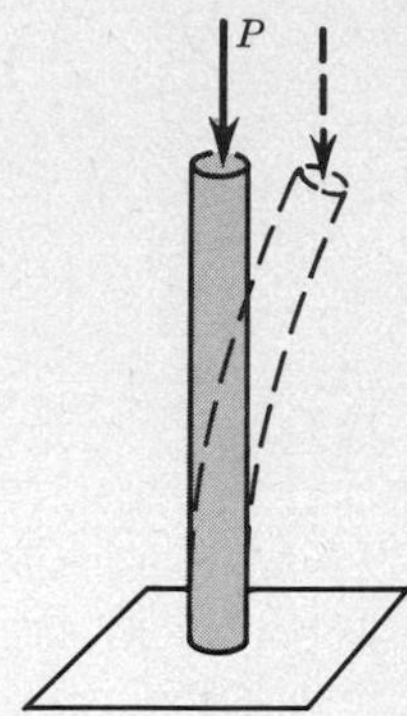

Fig. 9.7 Buckling of a column under a compressive load.

such as an accidental side load or tremor causes the beam to twist and bend sideward.

In Fig. 9.7 a flexible column is shown subjected to a vertical compressive load. For small values of P the column remains straight and is compressed uniformly. If P becomes larger than a certain critical value, the vertical position of the column becomes unstable. A small disturbance causes the column to bend out or *buckle*.

A somewhat similar situation is shown in Fig. 9.8*a*, where a thin-walled cylinder (e.g., a beer can) is subjected to a compressive load. For small values of P the vertical walls remain cylindrical and are compressed uniformly in the vertical direction. If P becomes too large, this position becomes unstable. A small disturbance causes the vertical walls to bend in and out and to eventually crumple in a rather complicated pattern, as indicated in Fig. 9.8*b*.

Another instance of buckling of a thin-walled cylinder was mentioned in Sec. 6.6. When such a cylinder is subjected to *torsion*, it will remain cylindrical,

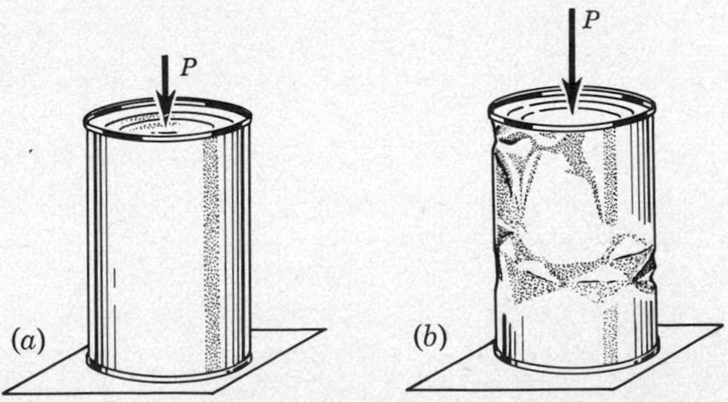

Fig. 9.8 Buckling and crumpling of the cylindrical walls of a can subjected to compressive force.

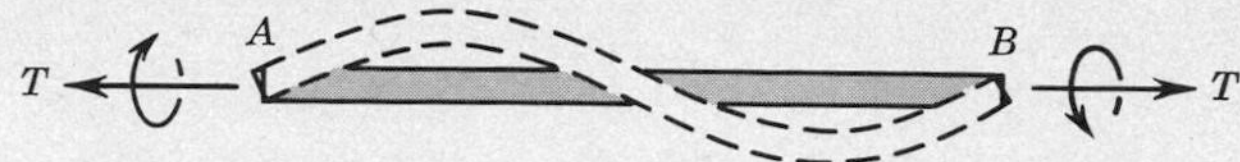

Fig. 9.9 Twist-bend buckling of a shaft in torsion.

and the deformation will be of the type considered in Chap. 6 if the applied twisting moment is small. When the twisting moment surpasses a certain critical value, the symmetrical deformation becomes unstable, and a small disturbance will cause the walls to buckle into wavy shapes, where the crests of the waves lie along helices which are at 45° with the axis of the cylinder.

It was noted in Sec. 7.6 that if the compression flange or the web of an I beam is made too thin, there is a strong likelihood that there will be local buckling when a bending load is placed on the beam. A similar type of buckling can sometimes be observed by a passenger sitting over the wing of an airplane when the air is rough. During momentary peak loads the top surface of the wing will distort into a rippled shape as the thin skin buckles locally under compression.

In Fig. 9.9 a flexible shaft is being twisted by equal and opposite moments directed along the line AB. If the twisting moment is small, the axis of the shaft remains collinear with AB, and the shaft is twisted uniformly. When T becomes larger than a certain critical value, the straight configuration becomes unstable, and a small disturbance will cause the shaft to bend out into a spiral-shaped space curve. An example of this is the twisting of long cables.

A somewhat different sort of instability is shown in Fig. 9.10. A shallow, curved member is subjected to a load which tends to straighten out the member. For small values of P, only small deflections result, as indicated in Fig. 9.10*a*, but at a critical value of P the member suddenly "snaps through" to a position of opposite curvature, as shown in Fig. 9.10*b*. This phenomenon is commonly called "oil-canning" because the bottom of an oil can behaves in this fashion. A similar type of "snap-through" buckling is utilized in electric light switches. See Prob. 9.7.

9.4 ELASTIC STABILITY OF FLEXIBLE COLUMNS

Out of all the examples described in Sec. 9.3 we shall investigate only one in full detail; namely, that shown in Fig. 9.7, the buckling of a column which is built-in at its base. Our reasoning and general approach can be extended to many other

Fig. 9.10 "Snap-through" instability of a shallow curved member.

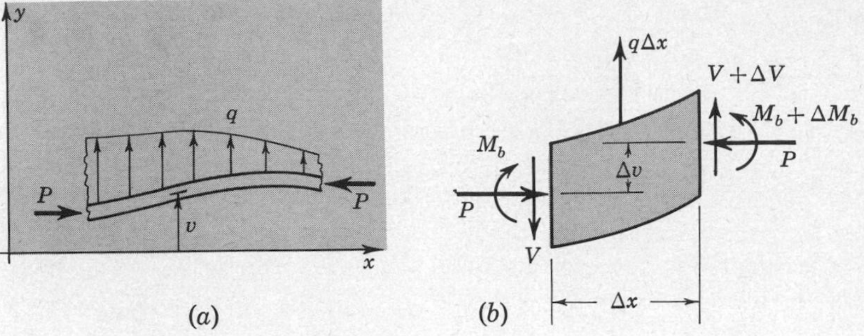

Fig. 9.11 (*a*) Beam subjected to longitudinal and transverse loads; (*b*) free-body sketch of element of beam.

cases, but the mathematical complications increase rapidly when we consider systems with more complex geometry.

Before considering the stability problem, we first derive the equations governing the bending of a beam subject to *longitudinal* as well as transverse loads. In Fig. 9.11 we show a portion of such a beam and an enlarged element isolated from the rest of the beam. For the element to be in equilibrium, we must have

$$\begin{aligned} &V + \Delta V - V + q\,\Delta x = 0 \\ &M_b + \Delta M_b - M_b + V\frac{\Delta x}{2} + (V + \Delta V)\frac{\Delta x}{2} + P\,\Delta v = 0 \end{aligned} \tag{9.3}$$

or, on passing to the limit and discarding the infinitesimal of higher order,

$$\begin{aligned} &\frac{dV}{dx} + q = 0 \\ &\frac{dM_b}{dx} + V + P\frac{dv}{dx} = 0 \end{aligned} \tag{9.4}$$

These should be compared with (3.11) and (3.12), which are the corresponding equilibrium equations when $P = 0$. Note that the deformation of the beam enters into the moment-equilibrium equation of (9.4) by way of the slope dv/dx. We still assume that the bending moment is responsible for the deformation of the beam; i.e., we neglect the effect of shear on the deformation. According to (8.4),

$$EI\frac{d^2v}{dx^2} = M_b \tag{9.5}$$

where EI is the bending stiffness of the section. Eliminating M_b and V from (9.4) and (9.5), we arrive at the governing equation

$$\frac{d^2}{dx^2}\left(EI\frac{d^2v}{dx^2}\right) + \frac{d}{dx}\left(P\frac{dv}{dx}\right) = q \tag{9.6}$$

for the *small* deflections of a beam subject to transverse load $q(x)$ per unit length and axial compressive force $P(x)$.

We now return to the elastic-stability problem for the cantilever beam (column) of Fig. 9.7, shown here again in Fig. 9.12. We take the beam to have uniform bending stiffness EI and assume that the buckling occurs in the plane of the sketch. If the beam should be accidentally displaced from a straight position, the force P produces a bending moment along the beam which causes the beam to bend even further, while the elastic forces in the beam tend to restore the original position. For small loads the straight position is stable and the beam is subjected to uniform compression. For large loads the straight position is unstable and the beam buckles. The most important single result to be obtained is the value of the critical load which marks the border between stability and instability. We can get this result by arguing that, when the critical load is acting, the *restoring* tendencies just balance the *upsetting* tendencies, and the system is in a state of *neutral* equilibrium. In Fig. 9.12 we assume that the critical load is acting and that it is, in fact, holding the beam in equilibrium in a displaced position. The magnitude of the critical load and the shape of the bent beam are initially unknown. We shall find them by requiring that the governing equation (9.6) and the following boundary conditions be satisfied:

$$\left.\begin{aligned} v &= 0 \\ \frac{dv}{dx} &= 0 \end{aligned}\right\} \text{ at } x = 0 \qquad \left.\begin{aligned} M_b &= 0 \\ V &= 0 \end{aligned}\right\} \text{ at } x = L \tag{9.7}$$

The boundary conditions at $x = L$ may be expressed in terms of v by substituting in the second of (9.4) and in (9.5).

$$\left.\begin{aligned} M_b &= EI\frac{d^2v}{dx^2} = 0 \\ -V &= \frac{d}{dx}\left(EI\frac{d^2v}{dx^2}\right) + P\frac{dv}{dx} = 0 \end{aligned}\right\} \text{ at } x = L \tag{9.8}$$

The governing equation (9.6) takes the following form when EI and P are constants and there is no transverse load:

$$EI\frac{d^4v}{dx^4} + P\frac{d^2v}{dx^2} = 0 \tag{9.9}$$

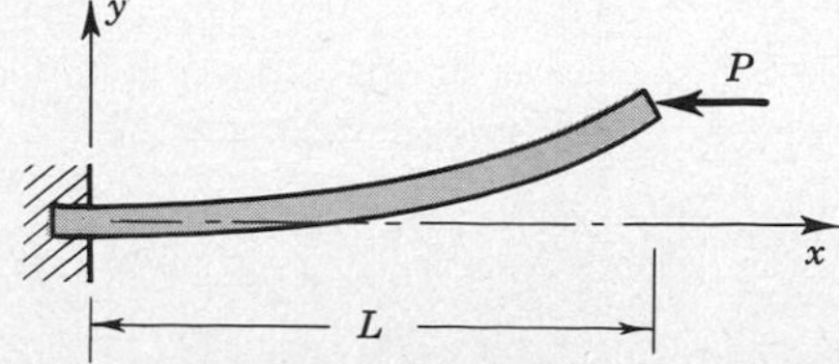

Fig. 9.12 Column in a state of neutral equilibrium in the bent position.

Our problem is to find the equilibrium configuration $v(x)$ and the critical load P which simultaneously satisfy (9.9) and the boundary conditions at $x = 0$ and $x = L$. It is interesting to note that $v(x) = 0$ satisfies these for any value of P. This means simply that the *straight position* of the column is always a possible equilibrium position. As far as the buckling problem goes, this must be considered a *trivial* solution. We are seeking the *nonstraight* neutral equilibrium position with its critical load which marks the border line between stability and instability.

The theory of differential equations teaches us that the most general solution to an equation having the form of (9.9) contains *four* independent constants of integrations. For certain equations (including this one) there exist routine methods for obtaining general solutions. We shall not here actually follow through the details of such a method but shall simply write down an expression which does contain four independent constants and which does satisfy (9.9).

$$v = c_1 + c_2 x + c_3 \sin \sqrt{\frac{P}{EI}}\, x + c_4 \cos \sqrt{\frac{P}{EI}}\, x \tag{9.10}$$

The reader should verify that (9.10) is, in fact, a solution to (9.9) for arbitrary values of the four constants. Substituting (9.10) into the four boundary conditions of (9.7) and (9.8), we obtain the following four simultaneous equations for the constants of integration.

$$\begin{aligned}
c_1 \qquad\qquad\qquad\qquad &+ c_4 = 0 \\
c_2 + c_3 \sqrt{\frac{P}{EI}} \qquad\qquad &= 0 \\
- c_3 \frac{P}{EI} \sin \sqrt{\frac{P}{EI}}\, L &- c_4 \frac{P}{EI} \cos \sqrt{\frac{P}{EI}}\, L = 0 \\
c_2 P \qquad\qquad\qquad\qquad &= 0
\end{aligned} \tag{9.11}$$

This algebraic problem is of a rather unusual sort (it is known as an *eigenvalue* problem). Because all the right-hand members are zero, an obvious solution is $c_1 = c_2 = c_3 = c_4 = 0$. This is a true equilibrium solution (it says the beam does not bend), but it is the trivial solution again. Our objective is to find another solution. One way to do this is to note that the fourth and second equations imply $c_2 = c_3 = 0$, that the first equation implies $c_4 = -c_1$, and that then the third equation becomes simply

$$c_1 \frac{P}{EI} \cos \sqrt{\frac{P}{EI}}\, L = 0 \tag{9.12}$$

This can be satisfied by setting $c_1 = 0$, which is the trivial solution again, *or* by having a value of P such that

$$\cos \sqrt{\frac{P}{EI}} L = 0 \tag{9.13}$$

The smallest value of P meeting this condition[1] is

$$P = \frac{\pi^2}{4} \frac{EI}{L^2} \tag{9.14}$$

Substituting back into (9.10), the corresponding deflection curve is

$$v = c_1 \left(1 - \cos \frac{\pi}{2} \frac{x}{L}\right) \tag{9.15}$$

We have thus found the *critical load* (9.14) and the shape (9.15) which the beam bends into when balancing the critical load.

Note that in (9.15) the constant c_1 has not been fixed. The deflection v is an equilibrium position for any arbitrary value of c_1 so long as the deflection remains within the validity of the small deflection theory. It can be shown that these are all *neutral* equilibrium positions; i.e., the critical load will hold any of these deflections in equilibrium. For smaller values of P the straight column is stable; i.e., if any accidental bending occurs, the restoring tendencies overcome the upsetting tendencies. For larger values of P the straight position is no longer stable; any small disturbance will result in buckling of the column.

Additional insight into column buckling can be obtained by considering a case where there is a small *imperfection* in either the column or the loading; e.g., the column is slightly bent in its unloaded state or the load is not precisely centered. As an illustration we examine the case shown in Fig. 9.13*a*, where the compressive load P is applied with a small *eccentricity* ϵ. This is statically equivalent to a central force P and a moment $M_o = P\epsilon$, as indicated in Fig. 9.13*b*. A graphical demonstration of the equivalence is displayed in Fig. 9.13*c*. We shall now investigate the relation between the transverse deflection δ and the compressive force P in the presence of the bending moment M_o due to the eccentricity ϵ.

[1] The other solutions to (9.13), namely, $P = EI\pi^2(2n-1)^2/4L^2$, correspond to loads which will hold the bent shapes $v = c_1\left(1 - \cos \frac{2n-1}{2} \frac{x}{L}\right)$ for $n = 2, 3, 4, \ldots$, in equilibrium. These configurations (called *higher modes*) are not of much practical significance because they are *violently* unstable.

In the analysis of *vibrating* systems a similar eigenvalue problem occurs, but there the higher modes do have practical significance. See, for example, S. H. Crandall, "Engineering Analysis," chap. 5, McGraw-Hill Book Company, New York, 1956.

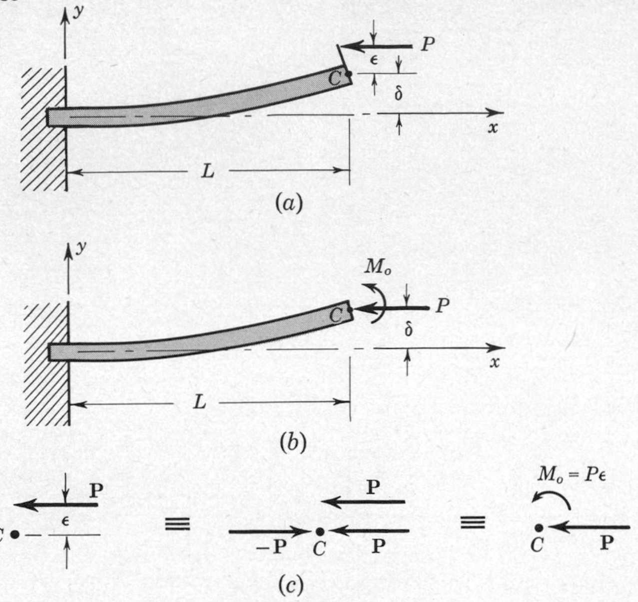

Fig. 9.13 Flexible column held in equilibrium by (*a*) a longitudinal compressive force P with eccentricity ϵ and (*b*) the same compressive force P plus an end moment M_o. The equivalence of the two loadings is shown in (*c*).

The governing equation is still (9.9), and all the boundary conditions are the same, except that at $x = L$ the bending moment is now M_o instead of zero. The general solution (9.10) still applies. The boundary conditions provide the following equations for the constants of integration:

$$\begin{aligned}
c_1 \qquad\qquad\qquad + c_4 &= 0 \\
c_2 + c_3\sqrt{\frac{P}{EI}} \qquad\qquad &= 0 \\
- c_3 \frac{P}{EI}\sin\sqrt{\frac{P}{EI}}\,L - c_4 \frac{P}{EI}\cos\sqrt{\frac{P}{EI}}\,L &= \frac{M_o}{EI} \\
c_2 P \qquad\qquad\qquad &= 0
\end{aligned} \tag{9.16}$$

Solving these (which now have a unique solution) and inserting in (9.10) yields

$$v = \frac{M_o}{P}\frac{1 - \cos\sqrt{P/EI}\,x}{\cos\sqrt{P/EI}\,L}$$

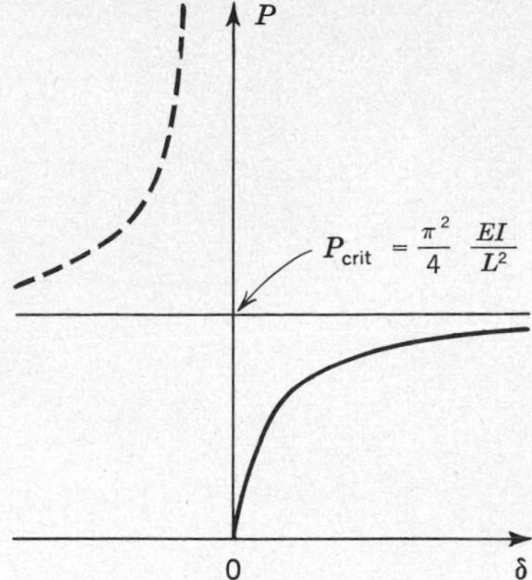

Fig. 9.14 Relation between compressive force P and transverse deflection δ due to eccentricity ϵ.

Putting $x = L$ gives

$$\delta = \frac{M_o}{P}\left(\sec\sqrt{\frac{P}{EI}}L - 1\right) \tag{9.17}$$

which reduces to

$$\delta = \epsilon\left(\sec\sqrt{\frac{P}{EI}}L - 1\right) \tag{9.18}$$

in the case of Fig. 9.13*a*, where $M_o = P\epsilon$. This relationship is sketched in Fig. 9.14. For small values of P the transverse deflection is very nearly zero (for example, $\delta < \epsilon$ for $P < \frac{4}{9}P_{crit}$). As P approaches the critical load, the deflection δ becomes large. Furthermore, it can be shown that when P is greater than the critical load, the equilibrium positions represented by (9.18) are *unstable*. Thus a flexible column can be used as a reliable structural element only when the axial compressive force which it carries remains somewhat below the critical load.

The critical load of a column is quite sensitive to the nature of the supports at the ends of the column. In Fig. 9.15 a number of different methods of column support are shown, together with the critical loads which were obtained from analyses similar to the one just given (see Probs. 9.8 and 9.9). Note that the critical load of a column which is clamped at both ends is 16 times larger than that of the column just discussed, which has one end free. In practical structures the idealized types of supports shown in Fig. 9.15 almost never occur. Most column supports provide more restraint than a hinged joint but not as much restraint as a clamped joint. In many cases a designer can use his judgment and experience to interpolate between tabulated results such as those given in Fig. 9.15.

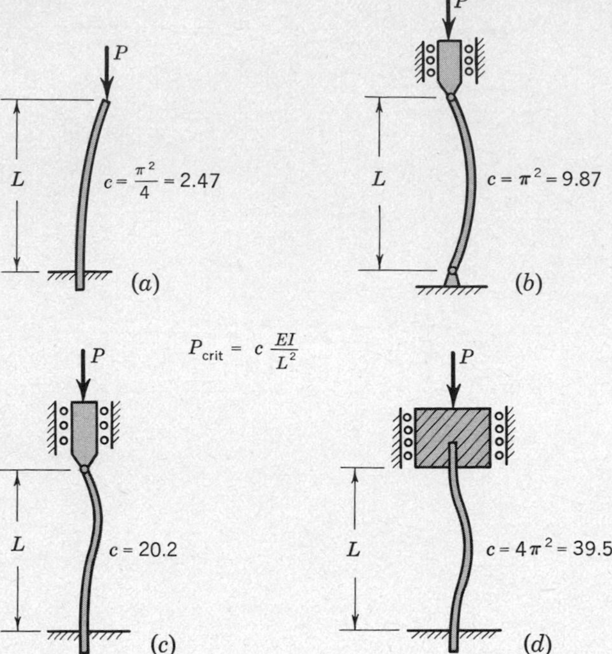

Fig. 9.15 Critical loads for (*a*) clamped-free, (*b*) hinged-hinged, (*c*) clamped-hinged, and (*d*) clamped-clamped columns. In each case the constant c shown is to be inserted in the formula $P_{crit} = cEI/L^2$.

9.5 ELASTIC POSTBUCKLING BEHAVIOR

In the preceding analyses of the initiation of buckling, the equilibrium requirements have been applied in the deformed configuration, but the assumption of small deformations has been retained. In order to explain the behavior of a structure once buckling has begun, it is necessary to include the effects of larger deformations. This generally requires the introduction of *geometrical nonlinearities*; e.g., in the case of a flexible column the nonlinear curvature-deflection relation (8.2) must be used in place of the linear approximation (8.3). The resulting analysis is quite complicated and beyond the scope of this text.

We can, however, study a highly simplified model whose postbuckling behavior is qualitatively similar to that of many actual structures, such as flexible columns, plates, and shells. The model is shown in Fig. 9.16. It is the same as the spring-stabilized strut of Fig. 9.3, except that here the springs are taken to be *nonlinear*.

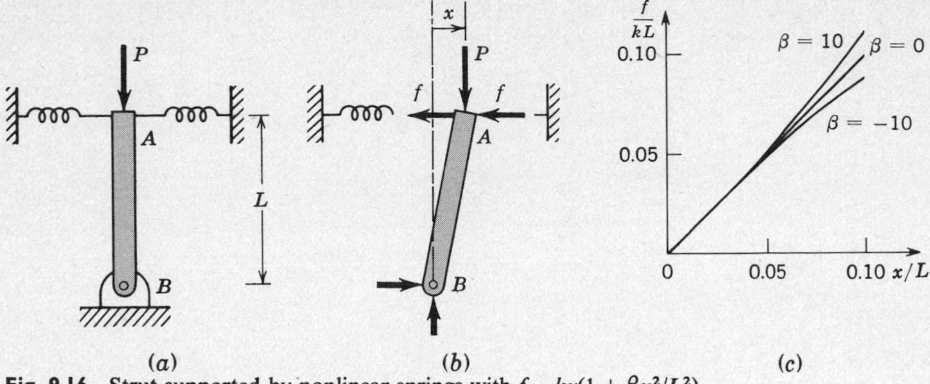

Fig. 9.16 Strut supported by nonlinear springs with $f = kx(1 + \beta x^2/L^2)$.

The geometrical nonlinearity of an actual structure is thus modeled by the analytically simpler mechanism of a nonlinear force-deformation relation. We take the force in each spring to be given by

$$f = kx\,(1 + \beta x^2/L^2) \tag{9.19}$$

where β is a parameter which fixes the nature of the nonlinearity. When $\beta > 0$ the spring is said to be a *stiffening* spring. When $\beta < 0$ the spring is said to be a *softening* spring. The relation (9.19) is plotted in Fig. 9.16*c* for $\beta = 10$, 0, and -10 over the small deflection range $0 < x < L/10$.

We first explore the possible equilibrium positions of the strut for the *ideal* case in which the strut is initially exactly vertical and the load P is exactly centered. Balance of moments about point B in Fig. 9.16*b* yields

$$Px - 2kLx\,(1 + \beta x^2/L^2) = 0$$

from which we conclude that either $x = 0$ or $P = 2kL(1 + \beta x^2/L^2)$. These loci are plotted in Fig. 9.17 for $\beta = 10$, 0, and -10. For small values of P the only stable equilibrium position is along AB, where $x = 0$. When the load reaches the value $P_{crit} = 2kL$, transverse deflection becomes possible. The point B in the diagrams of Fig. 9.17 is called a *bifurcation point* because the load-deflection path AB splits into two branches, BC and BD, at this point. In every case the branch BD represents *unstable* equilibrium positions. There is a fundamental difference in the stability of the branch BC, depending on the nature of the nonlinearity. For stiffening nonlinearity, $\beta > 0$, the branch BC represents *stable* equilibrium positions. This can be demonstrated by considering the strut in Fig. 9.16*b* to be in equilibrium, corresponding to a point on BC in Fig. 9.17*a*, and then imagining that the strut is moved slightly so as to increase x without changing P. Since a larger value of P is required for equilibrium, the unbalanced forces in the new position will tend to move the strut back to its original equilibrium position. The solid line BC in Fig. 9.17*a* thus represents stable equilibrium positions. For softening nonlinearity, $\beta < 0$, the branch BC represents *unstable* equilibrium

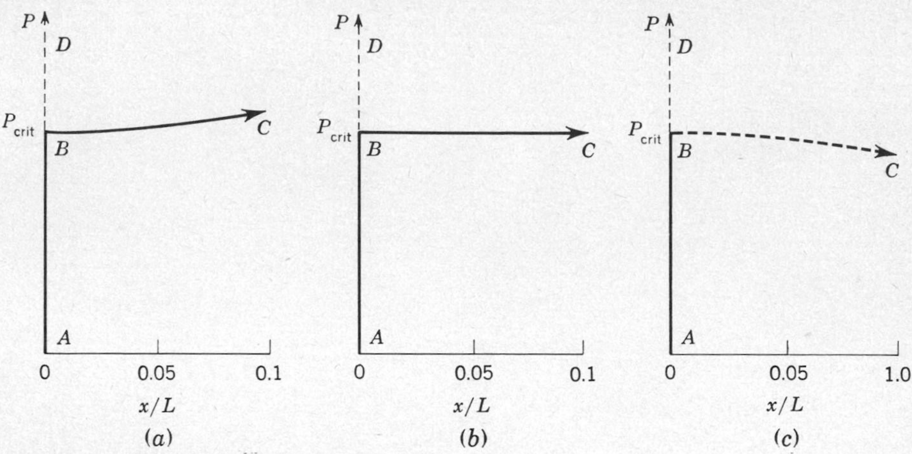

Fig. 9.17 Ideal postbuckling curves for (*a*) $\beta = 10$, (*b*) $\beta = 0$, (*c*) $\beta = -10$.

positions, as indicated by the dashed line in Fig. 9.17*c*. For the linear case, $\beta = 0$, the branch BC in Fig. 9.17*b* represents *neutral* equilibrium positions. The nature of the stability of the ideal postbuckling curve BC strongly influences the behavior of the corresponding system with small imperfections.

As a particular example of an imperfection, we next consider the behavior of the system of Fig. 9.16*a* when the load is positioned slightly off-center. The eccentricity ϵ is indicated in Fig. 9.18*a*. The equilibrium relation between the displacement x and the load P is obtained by balancing moments about point B in Fig. 9.18*b*,

$$P(x + \epsilon) = 2kLx(1 + \beta x^2/L^2) \tag{9.20}$$

This relation between P and x is plotted for three values of the imperfection parameter ϵ/L in each of the cases $\beta = 10$, 0, and -10 in Fig. 9.19. It is seen that as $\epsilon/L \to 0$, the curves approach the postbuckling curves BC of Fig. 9.17 for the ideal system. In the case of stiffening nonlinearity, $\beta > 0$, the equilibrium positions represented by (9.20) are stable (see Fig. 9.19*a*). Furthermore, the transverse displacement x remains small until the load P gets fairly close to P_{crit}.

The situation is quite different for the case of softening nonlinearity, $\beta < 0$. As indicated in Fig. 9.19*c*, the curves representing (9.20) rise to a maximum and then decrease. This is shown again in Fig. 9.20*a*, where OMN represents one of these curves. The corresponding equilibrium positions are *stable* along OM and *unstable* along MN. In this case there is a maximum load P_{max} which the strut can support with small transverse displacement x. When $P = P_{max}$, the slightest disturbance will cause the system to snap-through to some large displacement equilibrium position represented by point S. We shall not enter into a

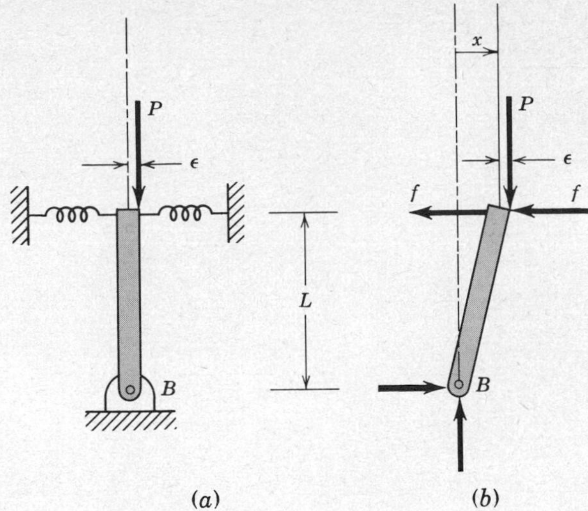

Fig. 9.18 Eccentric load on strut supported by nonlinear springs.

large displacement analysis here[1] to derive the equilibrium positions represented by the branch *RS*. In most structures to which this model applies, however, the equilibrium positions corresponding to *RS* represent collapse failures. To avoid

[1] For such an analysis, see Prob. 9.15.

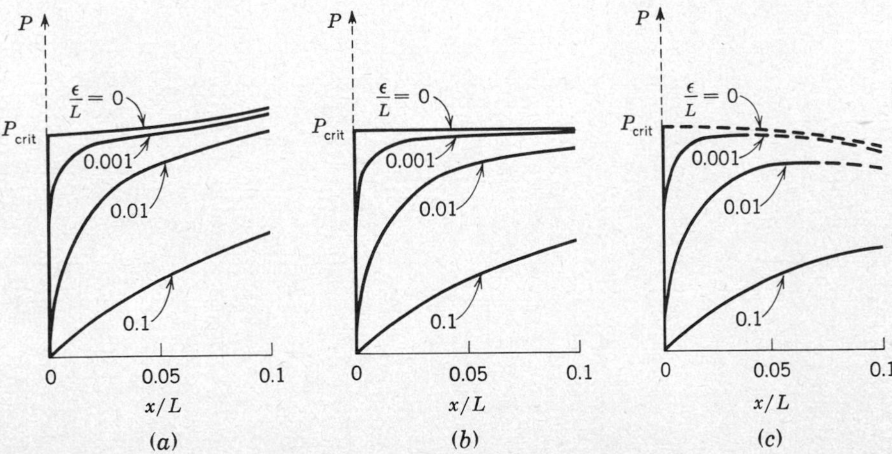

Fig. 9.19 Effect of imperfection parameter ϵ/L on postbuckling behavior for (*a*) $\beta = 10$, (*b*) $\beta = 0$, (*c*) $\beta = -10$.

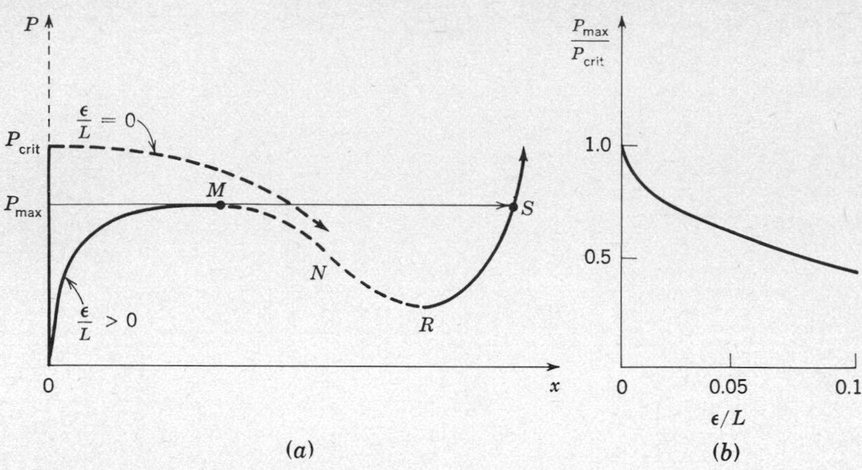

Fig. 9.20 Maximum load for softening nonlinearity ($\beta = -10$) depends on magnitude of imperfection.

such failures, it is necessary to keep the load P smaller than P_{max}. Thus for the case of softening nonlinearity, the significant buckling load is not P_{crit}, determined by the bifurcation phenomenon, but is P_{max}, which is always less than P_{crit}.

From Fig. 9.19*c* it is seen that P_{max} depends on the magnitude of the imperfection parameter ϵ/L. This relationship[1] is plotted in Fig. 9.20*b*. Note that $P_{max} \to P_{crit}$ as the imperfection parameter approaches zero, but that P_{max} is extremely sensitive to small imperfections. Structures whose postbuckling behavior can be modeled by a softening nonlinearity are said to be *imperfection-sensitive*.

The model just studied provides a good qualitative description of postbuckling behavior of elastic beams, plates, and shells under compressive loads. In each case the actual nonlinearity is introduced by the geometry of large deflections. For beams and plates the nonlinearity is weak but of a stiffening nature. As indicated in Fig. 9.19*a*, the transverse displacements for such structures with small imperfections grow slowly with load until the load approaches the critical load. There is no sharp indication as the load passes through the critical value, just a gradual increase in the rate of increase in displacement with increase in load.

For many shell structures the nonlinearity is quite strong and is of a softening nature. Such shells buckle with a sudden snap-through at an imperfection-sensitive maximum load which can be considerably smaller than the critical bifurcation load. For example, the circularly cylindrical shell under axial load, shown in Fig. 9.8, buckles in most tests at loads which are $\frac{1}{3}$ to $\frac{1}{2}$ of the critical load.

[1] See Probs. 9.20 and 9.21.

Throughout this section it has been assumed that the structure remains elastic in the postbuckling regime. In many structures the buckling problem is further complicated by the fact that the large deformations involved can cause yielding and plastic flow. This in turn can cause permanent deformations which may be undesirable, and it can initiate collapse due to plastic buckling at lower loads than predicted for elastic buckling. An introduction to plastic buckling appears in Sec. 9.8.

9.6 INSTABILITY AS A MODE OF FAILURE

The designer of a structure must always be alert to the possibility of buckling. The various elements must be *strong* enough to carry their share of any possible load, and the structure must be *rigid* enough to remain in stable equilibrium under any contemplated loading.

In recent years the availability of high-strength alloys plus careful design for minimum weight have led to structures which are inherently less rigid than those of the past. For example, compare an airplane wing with a Gothic tower. This decreased rigidity has greatly increased the possibilities for instability and buckling. To give a brief introduction to the problem of designing for both strength and stability, we consider in highly simplified form the problem of selecting a flexible column to support a compressive load P. We take the column to be exactly straight and assume that the load is exactly centered. As long as the column remains straight, the direct compressive stress is P/A, where A is the area of the cross section. This stress cannot be too large or the material will fail. For example, if the material is ductile and we did not wish plastic flow to occur, we would have to keep

$$P < YA \tag{9.21}$$

where Y is the yield stress (in compression).

On the other hand, the column will not remain straight unless it is stable. Assuming that the column is supported as shown in Figs. 9.7 and 9.12, the critical load is given by (9.14). To maintain stability, we must keep

$$P < \frac{\pi^2}{4}\frac{EI}{L^2} \tag{9.22}$$

The designer must check to see that both (9.21) and (9.22) are satisfied. These two requirements are shown in the diagram of Fig. 9.21, where P is plotted against the length of the column. The combinations of P and L which simultaneously satisfy (9.21) and (9.22) lie in the region below the line BCD. For short columns the yield condition limits the allowable load, while for long columns the stability condition governs. The contour BCD is the failure border line for the *ideal* case. Deviations from the assumed boundary conditions lead to wide

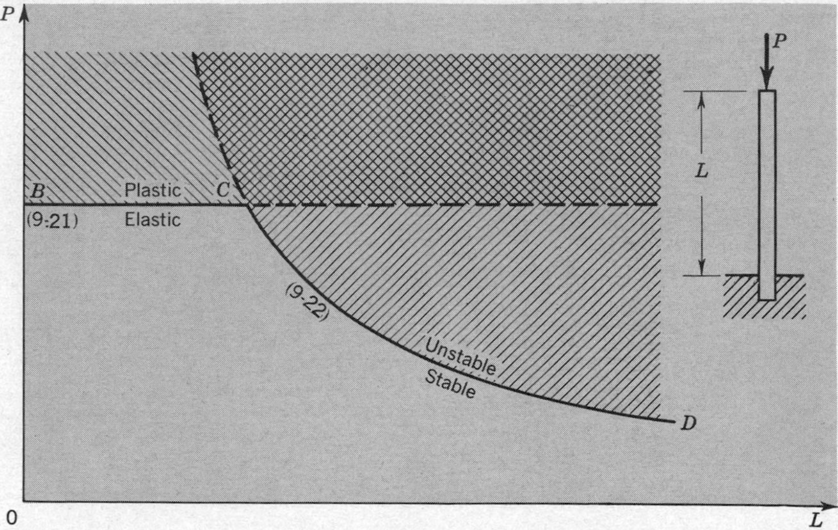

Fig. 9.21 Column may fail either by yielding or by buckling.

variations in the position of CD, while small eccentricities in the loading or in the shape of the column will alter the position of the entire failure border line. Because of these uncertainties, the column designer has been forced to rely on fairly conservative modifications[1] of idealized results such as those shown in Fig. 9.21.

When the column material has strain-hardening characteristics, a short column will not necessarily fail at the onset of yielding, as assumed by Eq. (9.21). In some applications it is possible to take advantage of the reserve strength obtained by permitting some plastic flow. This cannot be pushed too far, however, because of the possibility of *plastic buckling*. This is discussed in Sec. 9.8.

9.7 NECKING OF TENSION MEMBERS

In the above examples the *initiation* of instability occurs while the material is acting elastically, although the large deflections which occur during the subsequent buckling may involve inelastic action. We consider next a case of purely plastic instability. This is the phenomenon of necking which occurs during the final stages of a tension test of a ductile specimen (see Sec. 5.12).

When a uniform rod is subjected to axial tension, the stress may be assumed to be uniformly distributed across the cross-sectional area of the rod. During plastic flow, the cross-sectional area gets smaller as the rod elongates (plastic flow is essentially a constant-volume process). To investigate whether uniform elonga-

[1] See, for example, E. P. Popov, "Mechanics of Solids," p. 535, Prentice-Hall, Inc., Englewood Cliffs, N.J., 1968.

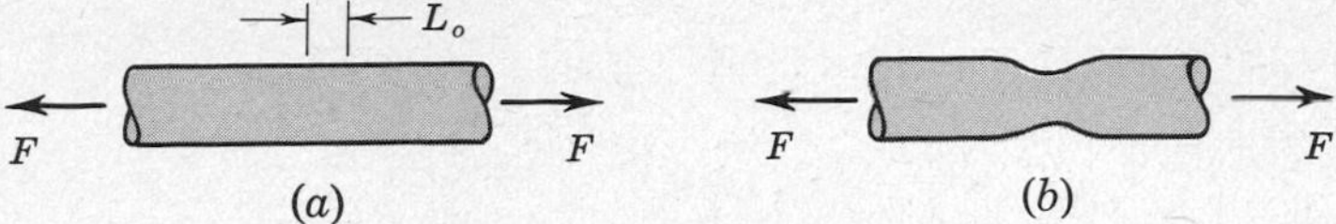

Fig. 9.22 Necking of a plastic rod under tension.

tion is stable, we analyze what would happen if, by some accident, a particular cross section became infinitesimally smaller than the remainder of the rod. Since, as the rod is stretched, the same axial force is transmitted through all sections, the smaller cross section would be subjected to a slightly higher stress than its neighbors. Whether this would intensify the deviation from uniform elongation depends on the strain-hardening properties of the material. If, along with the original decrease in area, there went a sufficient local increase in strain-hardening to compensate for the increase in stress, there would be no additional flow at this section until the other sections had deformed and hardened to the same extent. Under these circumstances uniform elongation is stable.

If, on the contrary, the material at the smaller section does not strain-harden sufficiently to compensate for the increase in stress, then there will be local axial strain with additional decrease in area in response to the increase in stress. Under these circumstances uniform elongation is unstable. All subsequent deformation becomes concentrated at the section in question, and the specimen is said to undergo *necking*, as indicated in Fig. 9.22*b*.

Let us make a quantitative study of this for a material whose *true-stress–engineering-strain* curve is shown in Fig. 9.23. Suppose that a *small* element of the rod in Fig. 9.22*a* has length L_o and area A_o when the rod is unstrained. If, under plastic strain (the elastic strain is considered negligible in comparison), the length is $L_o + \delta$ and the area is A, the local engineering strain ϵ and the true stress σ under an axial force F are

$$\epsilon = \frac{\delta}{L_o} \qquad \sigma = \frac{F}{A} \tag{9.23}$$

Assuming that the plastic flow involves no volume change, we have

$$A(L_o + \delta) = A_o L_o = \text{constant} \tag{9.24}$$

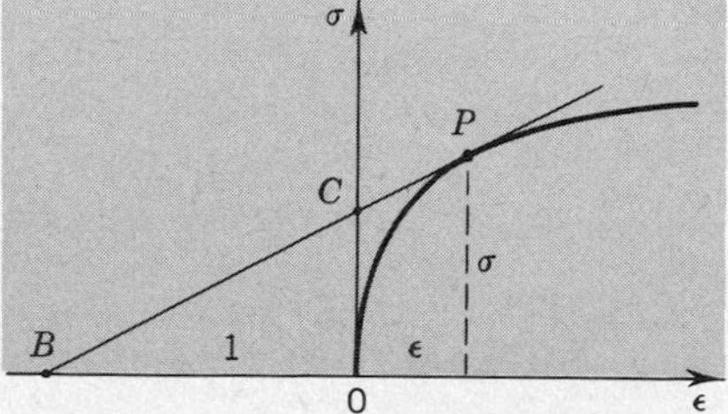

Fig. 9.23 True-stress–engineering-strain curve of a strain-hardening plastic material. The tangent *BP* determines the point at which local necking begins.

We now consider the consequences of an infinitesimal increase in strain ϵ. The rate of change of area is given by differentiating (9.24) and substituting for $d\delta/d\epsilon$ from (9.23).

$$\frac{dA}{d\epsilon}(L_o + \delta) + A\frac{d\delta}{d\epsilon} = 0$$

$$\frac{dA}{d\epsilon} = \frac{-A}{L_o + \delta}\frac{d\delta}{d\epsilon} = \frac{-AL_o}{L_o + \delta}$$

$$= \frac{-A}{1 + \epsilon} \tag{9.25}$$

The axial force F required to maintain the stress-strain relation of Fig. 9.23 is

$$F = A\sigma \tag{9.26}$$

from (9.23). The rate of change of F is

$$\frac{dF}{d\epsilon} = A\frac{d\sigma}{d\epsilon} + \sigma\frac{dA}{d\epsilon}$$

$$= A\left(\frac{d\sigma}{d\epsilon} - \frac{\sigma}{1 + \epsilon}\right) \tag{9.27}$$

where we have substituted from (9.25). Now, if $dF/d\epsilon$ is positive, we know that the element under question will not undergo further plastic strain without an increase in the axial load. Furthermore, if $dF/d\epsilon$ remains positive as ϵ is increased from zero, the axial force F will be an increasing function of the strain ϵ. The greater the strain of the element, the larger will be the force required to cause further strain.

Now, if we consider the rod to be made up of a large number of such elements in series, we can see that the strain will tend to be uniform along the entire rod so long as $dF/d\epsilon$ is positive. As F is increased, the element which begins to strain first is always the one which has had the smallest prior strain. If by accident one element is strained more than its neighbors, it will not participate in further elongation until the rest of the rod has "caught up with it." Thus when $dF/d\epsilon$ is positive, uniform strain is *stable* in the sense that any departures from uniformity in strain are decreased by further elongation of the rod.

On the contrary, if $dF/d\epsilon$ is negative, uniform strain is unstable. In this case, if by some accident the strain in a particular element becomes greater than that in the rest, the force required to cause further flow of this element would be less than in any other. All subsequent elongation would then be concentrated in this element and necking would occur. Whether the necking proceeds in a controlled manner or in a sudden catastrophic manner depends on how the tension test is conducted. If the testing machine applies a controlled total elongation to the specimen, then nothing spectacular happens. As the total elongation is increased beyond the point where necking begins, the axial force F decreases, thus unloading the rest of the specimen, while in the necked portion the local strain increases and the cross-sectional area is further reduced. On the other hand, if the testing machine applies a controlled force (e.g., a hanging weight), then as soon as necking begins it proceeds catastrophically to fracture. As long as the resistance to flow of the material in the necked portion remains less than the fixed force applied, the flow will continue to accelerate.

The border line between stability and instability for the uniform strain process occurs when $dF/d\epsilon = 0$, or when

$$\frac{d\sigma}{d\epsilon} = \frac{\sigma}{1 + \epsilon} \tag{9.28}$$

Under this condition a local increase in strain can occur without any change in the axial force F. This is the largest force, F_{max}, which the rod can withstand. The ratio F_{max}/A_o is called the *tensile strength* of the material.

Given the true-stress–engineering-strain curve of a material, the point at which (9.28) is satisfied may be determined by the graphical construction shown in Fig. 9.23. The point B is located a unit distance to the left of the origin. The tangent BP to the stress-strain curve then will have the slope $\sigma/(1 + \epsilon)$ as required by (9.28). To the left of P, $dF/d\epsilon$ is positive and uniform strain is stable. To the right of P, $dF/d\epsilon$ is negative and uniform strain is unstable. It is left to the reader to show that in this same construction the *tensile strength* is given by the intercept OC (see Prob. 9.17).

9.8 PLASTIC BUCKLING

Consider the column shown in Fig. 9.24*a* which is made of a *strain-hardening* material. The stress-strain curve (in compression) is drawn in Fig. 9.24*b*. We shall study the behavior of the column as the load P is increased. If the column is not too long, the compressive stress in the material will reach the yield stress without elastic instability. If P is increased further, the column can remain straight while the material flows plastically. Since the material strain-hardens, it is necessary at any stage to increase P in order to cause further plastic flow.

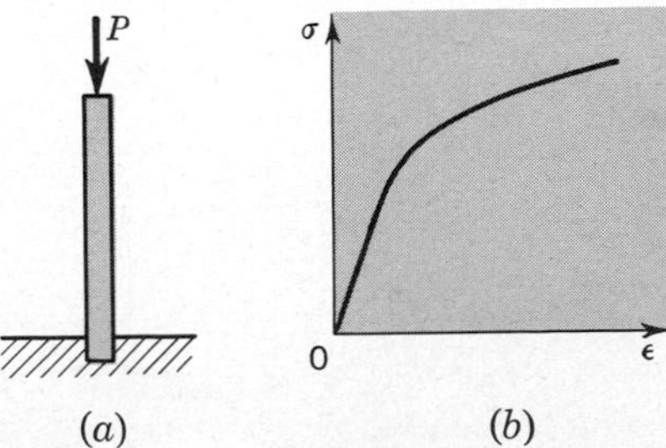

Fig. 9.24 Column of strain-hardening material subjected to a compressive force P.

Fig. 9.25 Strain-hardening column in slightly bent equilibrium configuration.

We next consider the possibility of the column bending out of its straight position. Following the approach used in describing elastic buckling, we should look for a critical value of P at which bifurcation can occur for the ideal case of a centered load on a perfect column. We should also examine the behavior of the column when there is a small imperfection.

To begin the analysis, we should study the possibility of a plastically deformed column being in equilibrium in the slightly bent shape shown in Fig. 9.25. To do this completely, taking into account the stress-strain curve of Fig. 9.24*b*, the variation of strain in each cross section, and the variations from section to section along the length of the column, is beyond our present scope. In order to gain some insight into this phenomenon, we can attempt to find a simpler situation which retains, at least qualitatively, the basic features of our problem. In Fig. 9.26*a* we show a rigid member ABC supported by a strain-hardening spring at A and at B. In this model the deformations can occur only in the springs. This simplifies both the problem of strain distribution along the column (it is all concentrated at the bottom) and the problem of strain distribution throughout the cross section (it is all concentrated in the two extreme fibers). We do, however, retain the strain-hardening aspects of the original problem by taking the force-deformation relation for the springs to have the same form as the stress-strain curve for the column

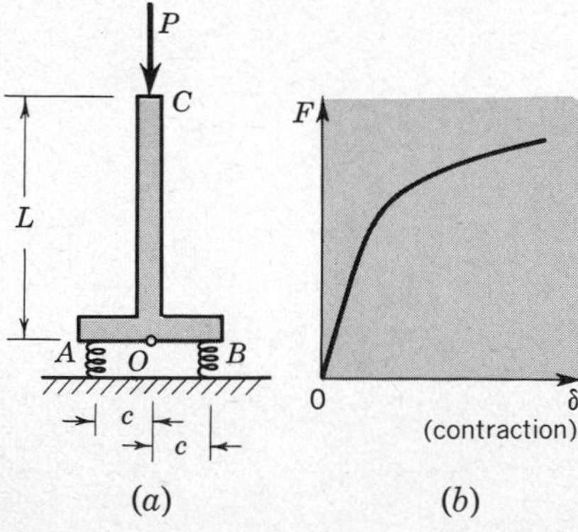

Fig. 9.26 (*a*) Simplified model of plastic buckling using springs at A and B which have the force-deformation relation shown in (*b*).

material. We now analyze the model of Fig. 9.26 to investigate the possibility of equilibrium in a *slightly* tipped position in the ideal case where there are no imperfections.

Suppose that when the load P has reached the value $P = P_o$, the system is in the position shown in Fig. 9.27*b*. Both springs have been compressed an amount δ_o and the column is still straight. The force in each spring is $F_o = P_o/2$, as indicated in Fig. 9.27*c*. We now examine the possibility of bifurcation at the load P_o; i.e., we investigate the possibility of a tipped equilibrium position, as shown in Fig. 9.27*d*, which involves only *small* changes in the spring forces and deflections. Following the first two steps of (2.1), we apply the equilibrium and geometric compatibility requirements to the slightly tipped position with $\theta > 0$. Figure 9.28 shows a free-body diagram of ABC. For small θ, balance of forces and moments require

$$\begin{aligned} P &= F_B + F_A \\ PL\theta &= (F_B - F_A)c \end{aligned} \tag{9.29}$$

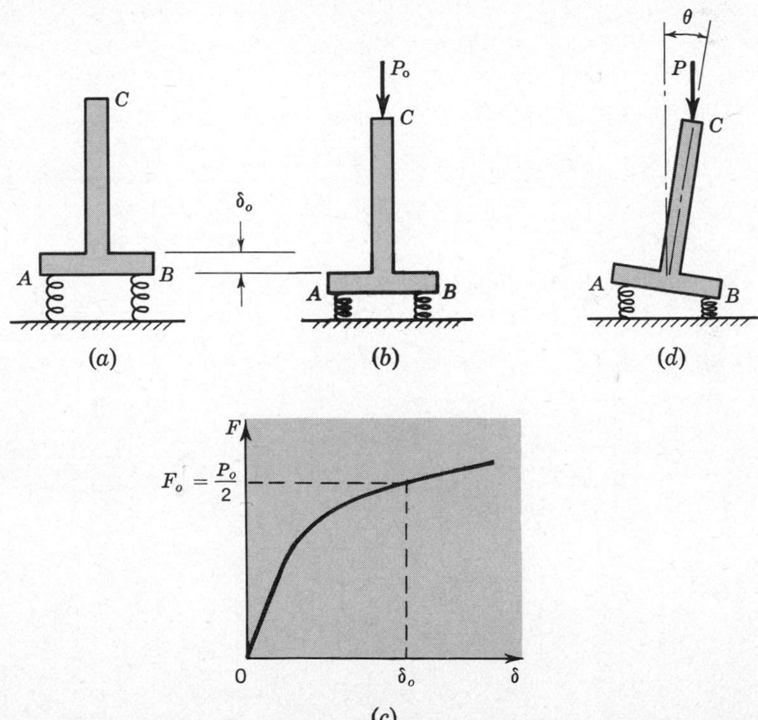

Fig. 9.27 When column model (*a*) deflects δ_o under load P_o as shown in (*b*), each spring is in state (δ_o,F_o), as indicated in (*c*) prior to tipping through angle θ in (*d*).

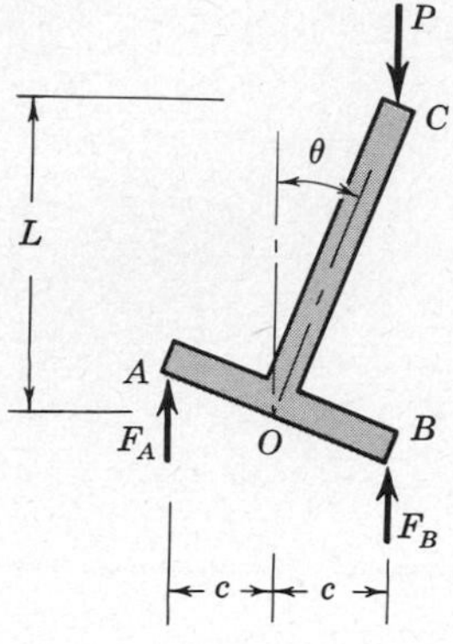

Fig. 9.28 Free body of rigid member in column of Fig. 9.26.

Figure 9.29 shows the geometry of deformation. It is assumed that AB tips about a neutral point N, whose location (defined by the distance x to the left of the center O) is unknown. For small θ, the increments in spring deflection are

$$\begin{aligned} \delta_B - \delta_o &= (c + x)\theta \\ \delta_A - \delta_o &= -(c - x)\theta \end{aligned} \tag{9.30}$$

The relations (9.29) and (9.30) apply independently of the force-deformation relation.

Figure 9.30 shows the force-deformation relation for compression of the springs. To simplify the discussion of the behavior of the springs in the neighborhood of the state (δ_o, F_o), we have introduced the straight line OL, defined by

$$F = F_o + k_t(\delta - \delta_o) \tag{9.31}$$

where k_t is the slope of the tangent at δ_o, as an approximation to the curve for

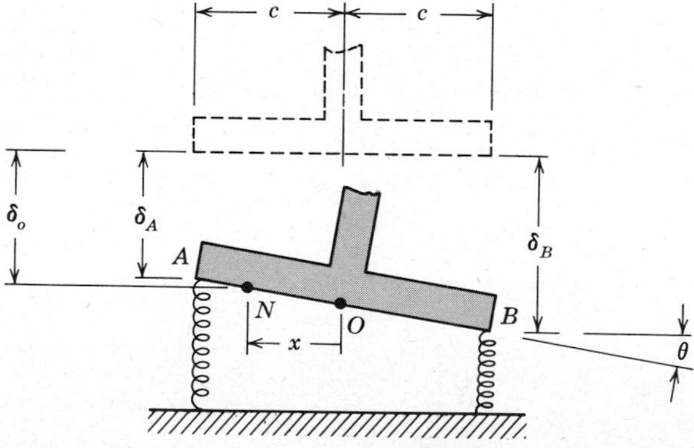

Fig. 9.29 Geometry of spring deformation.

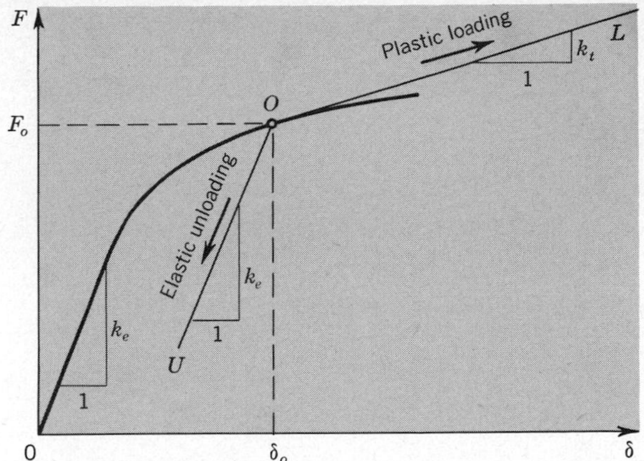

Fig. 9.30 Approximations of the force-deformation curve by straight lines in the neighborhood of δ_o.

continued compressive loading. We also show the *elastic unloading* relation as the straight line OU, defined by

$$F = F_o + k_e(\delta - \delta_o) \tag{9.32}$$

where k_e is the elastic spring constant.

When the column model begins to tip, there are three possible mechanisms: (1) both springs can continue to compress but at different rates, (2) one spring can continue to compress while the other begins to decompress, and (3) both springs can begin to decompress but at different rates. From Fig. 9.29 we see that, to have mechanism (1), it is necessary for the neutral point N to be to the left of A; that is, $c < x < \infty$. For mechanism (2) the neutral point must lie between A and B $(-c < x < c)$, and for mechanism (3) the neutral point must lie to the right of B $(-\infty < x < -c)$. For a complete analysis it is necessary to examine all three mechanisms. It turns out that the analysis for the first and for the third is simpler than that for the second, but the most important results depend on the second mechanism. We therefore proceed first to the case where, during tipping, spring B continues to compress according to (9.31), while spring A starts to decompress according to (9.32). Setting $F_o = P_o/2$ and using (9.30), we find the following force-deformation relations for this case

$$\begin{aligned} F_B &= \frac{P_o}{2} + k_t\theta(c + x) \\ F_A &= \frac{P_o}{2} - k_e\theta(c - x) \end{aligned} \tag{9.33}$$

When Eqs. (9.33) are substituted into the first of (9.29), we obtain

$$P - P_o = \theta[(k_e + k_t)x - (k_e - k_t)c] \tag{9.34}$$

which states that, for fixed x, the increment in column load is proportional to the tip angle θ. Next we substitute (9.33) into the second of (9.29) to get

$$P = \frac{c}{L}[(k_e + k_t)c - (k_e - k_t)x] \tag{9.35}$$

which can be considered as a relation between the column load P and the position x of the neutral point. The two relations (9.34) and (9.35) provide a complete description of the bifurcations which are possible for mechanism (2).

We see from (9.34) that when $\theta \to 0$, $P \to P_o$. If we consider (9.35) to apply in this same limit, it defines a range of possible bifurcation loads P_o as x varies in the range $-c < x < c$. This range of bifurcation loads extends from point $B_t(x = c)$ to point $B_e(x = -c)$ in Fig. 9.31. The corresponding loads from (9.35) are

$$P_t = \frac{2c^2}{L} k_t \tag{9.36}$$

$$P_e = \frac{2c^2}{L} k_e \tag{9.37}$$

These are called, respectively, the *tangent-modulus load* and the *critical elastic load.* The latter would have been the critical load if the springs had remained linearly

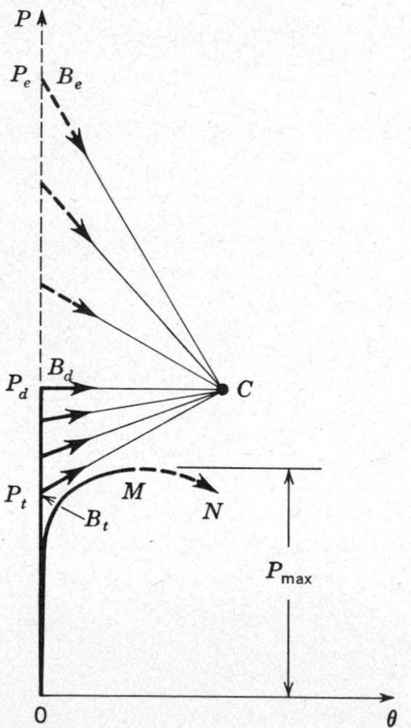

Fig. 9.31 Bifurcation points for ideal strain-hardening-column model and maximum load in presence of imperfection.

elastic up to the point of buckling (see Prob. 9.18). For each bifurcation point B, the relation (9.34) defines a straight line BC if x retains the same value it has at bifurcation. The initial portions of a number of such lines are indicated in Fig. 9.31. Between B_t and B_d the initial increment of P is positive; between B_d and B_e the initial increment of P is negative. To obtain the value of P_d corresponding to B_d, we set $P = P_o$ in (9.34), with $\theta \neq 0$, to obtain

$$x_d = \frac{k_e - k_t}{k_e + k_t} c \tag{9.38}$$

and then insert this value of x in (9.35) to get

$$P_d = \frac{2c^2}{L} \frac{2k_e k_t}{k_e + k_t} \tag{9.39}$$

which is called the *double-modulus load.* In Fig. 9.31 the equilibrium positions for $\theta = 0$ are stable for $P < P_d$ in the sense that it is necessary to *increase* the load P in order to initiate tipping. Conversely, the equilibrium positions for $\theta = 0$ with $P > P_d$ are unstable. This accounts for the dashed lines in Fig. 9.31.

We thus find the unusual result that there is a range of loads between the tangent-modulus load P_t and the double-modulus load P_d for which the position $\theta = 0$ is stable but for which tipping of the perfect column may begin whenever the load P is increased.

To complete the analysis we must examine the possibilities of mechanism (1), where (9.31) is used for both springs, and mechanism (3), where (9.32) is used for both springs. It can be shown that for the first mechanism, there is only a single bifurcation point at B_t and that for the third mechanism there is only a single bifurcation point at B_e. These are simply the end points of the range already discussed for mechanism (2).

We shall not enter into a detailed analysis of the behavior of this system when a small imperfection (such as an eccentricity ϵ in the position of the load P) is introduced.[1] The load-deflection curve in this case has the form indicated by OMN in Fig. 9.31. For a small imperfection the column does not tip appreciably until the load approaches the value P_t. The deflections then increase rapidly with load until the point M is reached where a maximum load P_{max} is attained. The equilibrium positions corresponding to points on OM are *stable* under fixed load, but those corresponding to points on MN are *unstable.* The magnitude of P_{max} is always less than the double-modulus load P_d, but it can be larger or smaller than the tangent-modulus load P_t, depending on the size of the imperfection.

The above analysis may be taken as a satisfactory explanation of the behavior of the simplified model of Fig. 9.26 and as a qualitative guide to the behavior of

[1] See N. J. Hoff, Inelastic Buckling of Columns in the Conventional Testing Machine, pp. 383–402, in "Proceedings—Symposium on the Theory of Shells to Honor Lloyd Hamilton Donnell," University of Houston, 1967.

the original column of Fig. 9.24. There are, however, many questions about buckling of structures which remain unanswered. Buckling problems constitute one of the most active areas of current research in applied mechanics.[1]

PROBLEMS

9.1. The small weight can slide without friction on the surface AOB. Is the equilibrium at O stable?

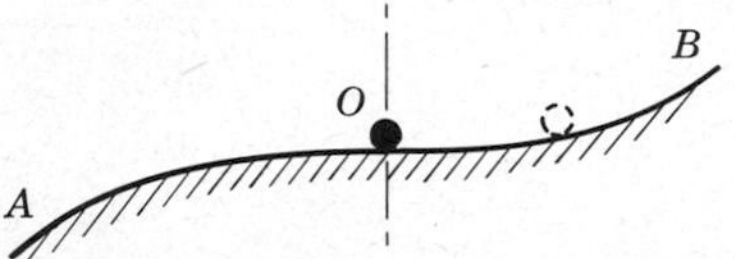

Prob. 9.1

9.2. The rigid strut AB is pivoted without friction at B and is stabilized by a spring of stiffness k which is unstretched when $x = 0$. Find the critical value for a load P which is delivered by a rope passing through a guide at O.

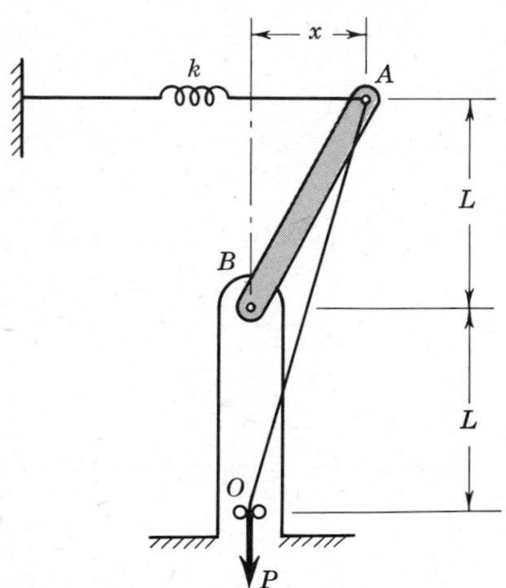

Prob. 9.2

9.3. The system consists of three identical rigid bars pinned together and stabilized by the spring of constant k. A moment M is applied to the central bar, as shown. Find the critical value of M which marks the border line of elastic stability. What is the stability limit when M is reversed?

[1] See, for example, B. Budiansky and J. Hutchinson, A Survey of Some Buckling Problems, *AIAA J.*, vol. 4, pp. 1505–1515, 1966; J. Hutchinson and W. Koiter, Postbuckling Theory, *Appl. Mech. Rev.*, vol. 23, no. 12, pp. 1353–1356, 1970.

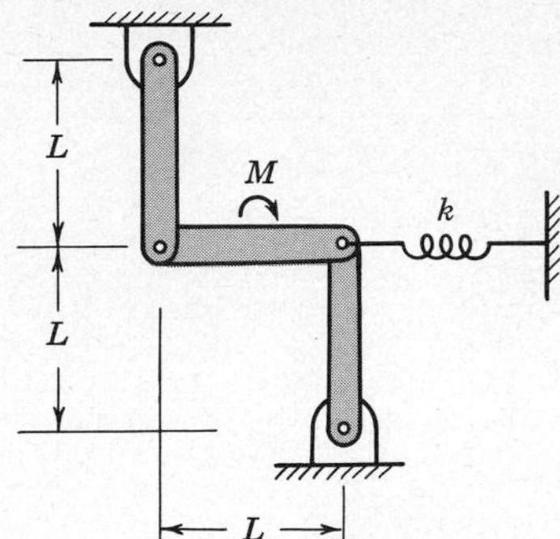

Prob. 9.3

9.4. The rigid bar of length L is stabilized by a spiral spring which exerts a torque $k\theta$ when the bar is turned through the angle θ. Find the equilibrium deflection angle as a function of the magnitude of P. For what values of P is the equilibrium stable? Assume that ϵ is small and that the analysis can be limited to small angles θ.

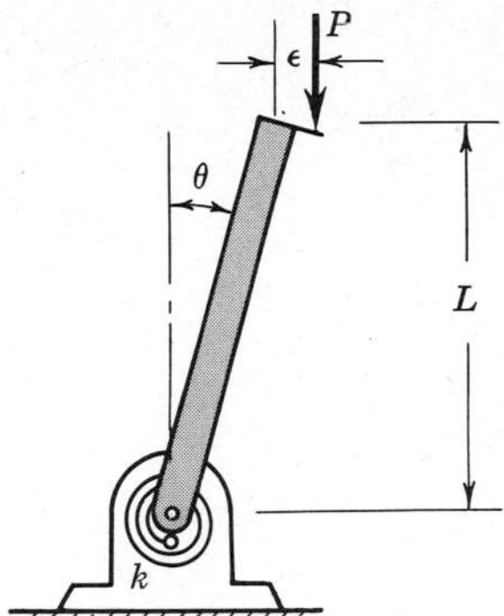

Prob. 9.4

9.5. The pivoted rigid strut is stabilized by the pair of linear springs oriented at the angle θ shown. Find the critical value of the load P.

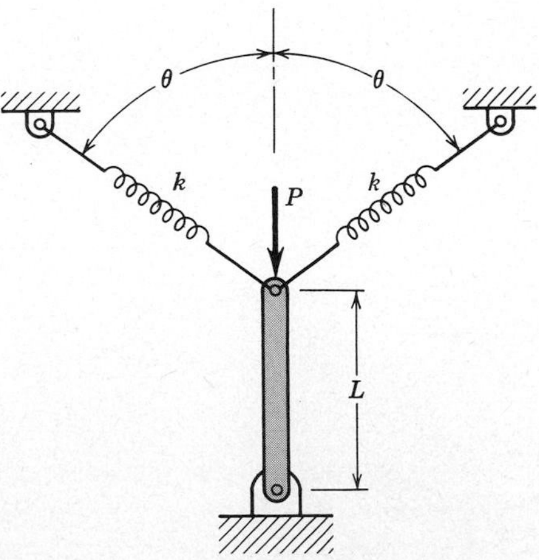

Prob. 9.5

9.6. Show that the equations for small-displacement equilibrium of the system shown may be written as follows:

$$\left(2 - \frac{kL}{P}\right)x_1 - \qquad x_2 = 0$$

$$-x_1 + \left(1 - \frac{2}{3}\frac{kL}{P}\right)x_2 = 0$$

Find the critical value of P which just holds the system in equilibrium in a displaced position.

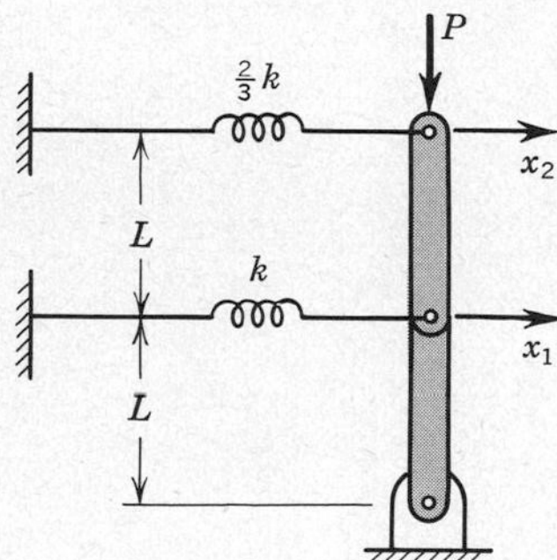

Prob. 9.6

9.7. The springs AB and BC are weightless and make the *small* angle ϕ when $P = 0$. Obtain the relation between the equilibrium angle θ and the load P. Verify that it has the shape shown in the diagram. Discuss the stability on the various branches of this curve. What value of P causes "snap-through"?

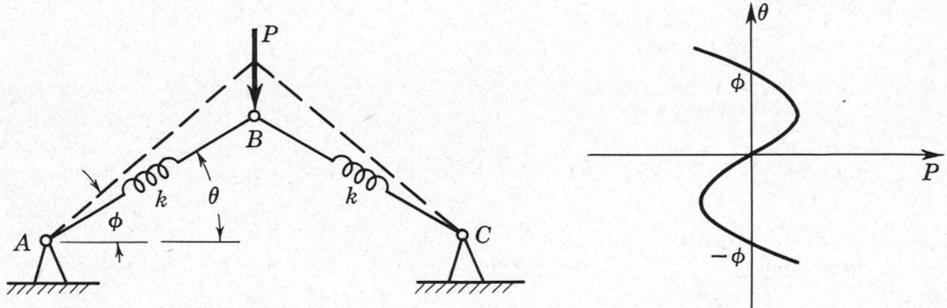

Prob. 9.7

9.8. Find the critical elastic compressive load for a uniform flexible beam which is *hinged* at both ends.

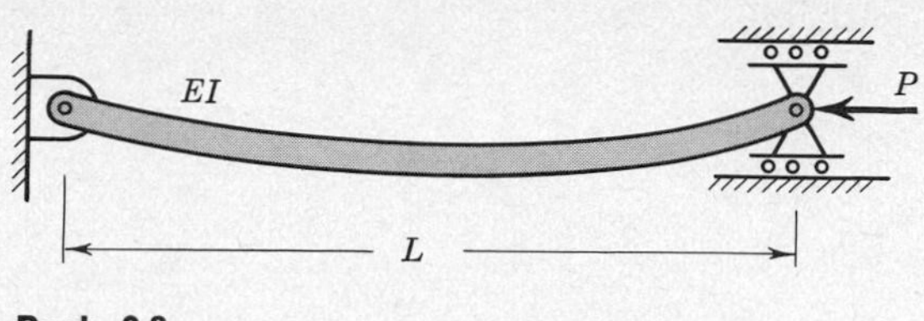

Prob. 9.8

9.9. Find the critical elastic compressive load for a uniform flexible beam which is *clamped* at both ends.

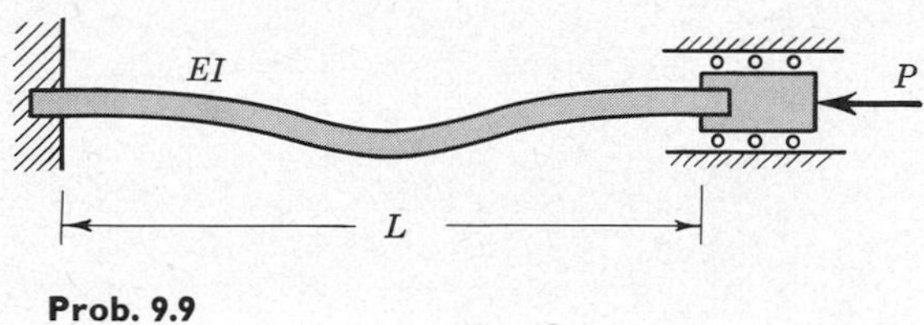

Prob. 9.9

9.10. Obtain an expression for the equilibrium deflection δ at the end of a cantilever column subjected to a compressive load P and a force F. Sketch the graph of δ against P for constant F.

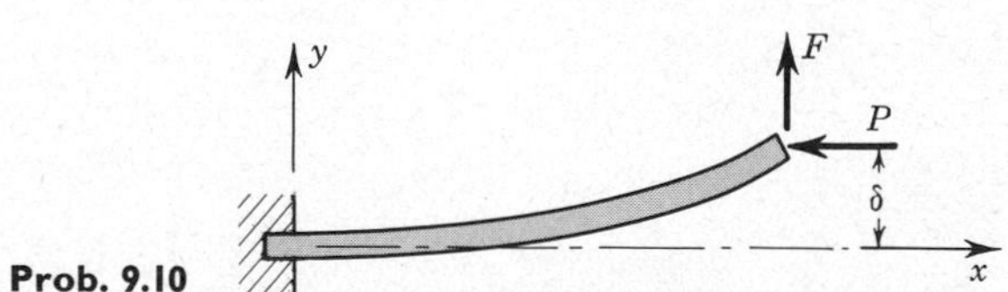

Prob. 9.10

9.11. The diagram of Fig. 9.21 can be made much more useful by plotting it in terms of *dimensionless* variables. Show that the instability locus is the same for all materials and all geometrical combinations if we plot P/AE against L/r, where r is the *radius of gyration* defined by $I = Ar^2$. If we used such a diagram, what would be the significance of the ordinate corresponding to BC in Fig. 9.21?

9.12. A 1020 HR steel beam of square cross section is to be used as a cantilever column to support a weight of 3,000 lb. The length of the column is to be 5 ft. What should the cross-sectional dimension be so that the load which would cause yielding or buckling is 12,000 lb?

9.13. The system shown is similar to that considered in Sec. 9.2 except that the strut is not rigid and therefore may act like the hinged column of Prob. 9.8. Ascertain the mode of failure and the corresponding critical load P when the strut is a 1-in.-diameter 2024-T4 aluminum rod.

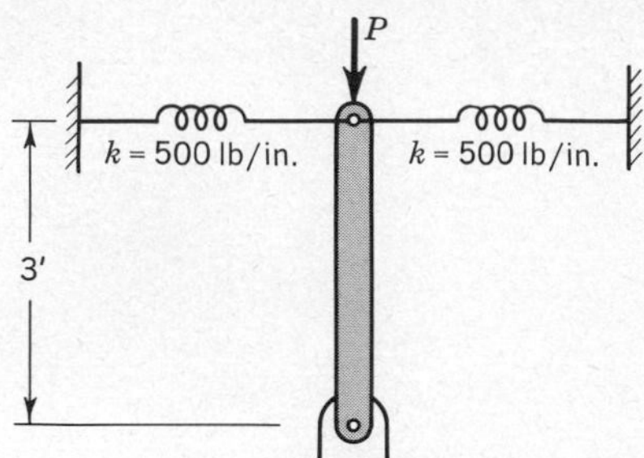

Prob. 9.13

9.14. Use the first two terms of the series expansion for cosine to verify that when P is small the deflection given by (9.17) approaches $M_oL^2/2EI$.

9.15. Let the position of the pivoted strut be denoted by the angle θ. Assume that the linear spring is effective in tension and in compression and that it always remains horizontal. Consider first that the spring is unstretched when $\theta = 0$.

(*a*) What is the critical value for the vertical load P when $\theta = 0$?

(*b*) Determine the equilibrium value of P for large θ; i.e., for any θ in $0 < \theta < 2\pi$.

(*c*) For what range, or ranges, of θ are the equilibrium positions in (*b*) stable?

Next consider that the system has a slight imperfection so that the spring is unstretched when $\theta = \epsilon$, where ϵ is a very small angle.

(*d*) Determine the equilibrium value of P as a function of θ.

(*e*) For what range, or ranges, of θ are the equilibrium positions in (*d*) stable?

(*f*) If P is slowly increased from zero, what is the value P_{max} at which "snap-through" occurs?

(*g*) Show that θ jumps from θ_1 to $\theta_2 = \pi - \theta_1$ during snap-through, where $\sin^3 \theta_1 = \sin \epsilon$.

(*h*) Show that for small θ the geometrical nonlinearity in this problem can be modeled by the system of Fig. 9.16 with a nonlinearity parameter of $\beta = -\frac{1}{2}$.

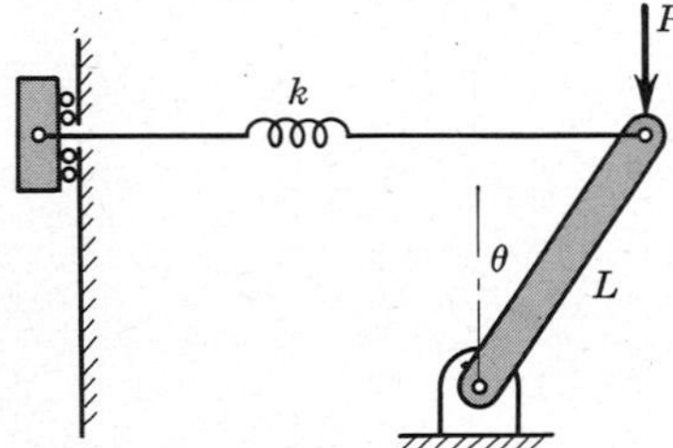

Prob. 9.15

9.16. A spherical shell of radius r_o has a thin wall of thickness t_o. The stress-strain curve of the (strain-hardening) material is known in the form of a true-stress $\bar{\sigma}$ versus true-strain $\bar{\epsilon}$ curve for a uniaxial test specimen. When internal pressure p is introduced into the shell, it develops a state of biaxial tension and expands. Using (5.29) and (5.30) to define $\bar{\sigma}$ and $\bar{\epsilon}$ and assuming that the metal in the wall of the sphere deforms with no volume change, show that initially the pressure must increase to cause further expansion but that a maximum pressure occurs when

$$\frac{d\bar{\sigma}}{d\bar{\epsilon}} = \frac{3}{2}\bar{\sigma}$$

For further expansion smaller pressures are required. What is the maximum pressure that the sphere can hold?

9.17. In connection with necking of a tension member show that the tensile force which the member carries is given by

$$F = \frac{A_o \sigma}{1+\epsilon}$$

and hence that the tensile strength F_{max}/A_o is given by the intercept OC in Fig. 9.23.

9.18. Find the elastic buckling load for the system of Fig. 9.26 on the assumption that the springs are completely linear; that is, $F = k_e \delta$ for all deflections δ.

9.19. The sketch shows a free body of the end portion of the cantilever column of Fig. 9.13*b*. Show that substitution of the bending moment M_b into the moment-curvature relation (8.4) leads to the following differential equation:

$$EI \frac{d^2 v}{dx^2} + Pv = P(\epsilon + \delta)$$

Verify that

$$v = c_1 \sin \sqrt{\frac{P}{EI}}\, x + c_2 \cos \sqrt{\frac{P}{EI}}\, x + \epsilon + \delta$$

where c_1 and c_2 are arbitrary constants, is a solution to this differential equation. Show also that when the geometric boundary conditions at $x = 0$ and $x = L$ are satisfied, the result (9.18) is obtained for the tip deflection δ.

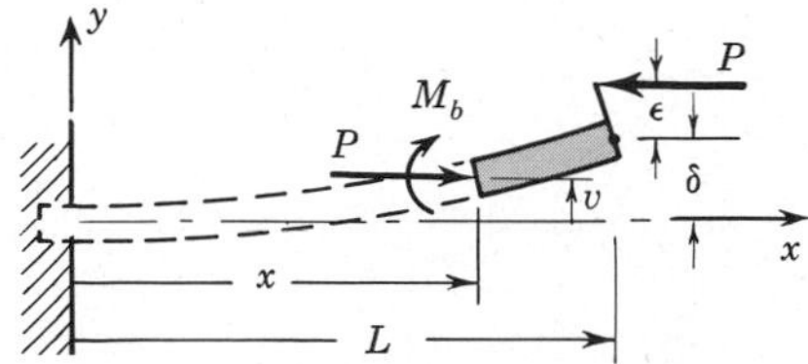

Prob. 9.19

9.20. Consider the nonlinear buckling model of Fig. 9.16 in the case $\beta < 0$. Show that the maximum point M in Fig. 9.20*a* for any value of the imperfection parameter ϵ/L lies on the curve

$$1 - \frac{P_{max}}{P_{crit}} = -3\beta \frac{x^2}{L^2}$$

Verify that the equation for the curve displayed in Fig. 9.20*b* is

$$\left(1 - \frac{P_{max}}{P_{crit}}\right)^{3/2} = \frac{3\sqrt{-3\beta}}{2} \frac{\epsilon}{L} \frac{P_{max}}{P_{crit}}$$

9.21. Reconsider the nonlinear buckling model of Fig. 9.16 for the case where the spring relation is

$$f = kx(1 + \alpha x/L)$$

in place of (9.19). Sketch the ideal postbuckling curves for $\alpha > 0$ and $\alpha < 0$. Rework Prob. 9.20

for the case $\alpha < 0$. Show that for $x > 0$ the maximum point corresponding to M in Fig. 9.20*a* lies on the curve

$$1 - \frac{P_{\text{max}}}{P_{\text{crit}}} = -2\alpha \frac{x}{L}$$

and the equation for the curve corresponding to that in Fig. 9.20*b* is

$$\left(1 - \frac{P_{\text{max}}}{P_{\text{crit}}}\right)^2 = -4\alpha \frac{\epsilon}{L} \frac{P_{\text{max}}}{P_{\text{crit}}}$$

9.22. Consider Fig. 9.31 for the strain-hardening-column model. Show that the abscissa of the point C is given by

$$\theta = \frac{c}{L} \frac{k_e - k_t}{k_e + k_t}$$

9.23. When the load P in Fig. 9.26 is applied with an eccentricity ϵ, the resulting force-deflection curve has the form OMN in Fig. 9.31. For small P both springs compress, but at different rates, when P is increased. For values of P approaching P_{max}, spring B compresses and spring A decompresses when P is increased. Show that the load P_r at which the force in spring A first begins to decrease is smaller than the tangent-modulus load; that is, $P_r < P_t$.

Answers to Selected Problems

CHAPTER 1

1.9. $F_A = 200$ lb, $F_B = 400$ lb, $F_C = 200$ lb

1.11. $F_{AB} = F_{BC} = 57.8$-lb compression
$F_{AC} = 71.1$-lb compression

1.13. (*a*) 414 lb (*b*) 135 lb

1.18. $\mathbf{F} = 100\mathbf{i} - 1{,}000\mathbf{k}$ lb
$\mathbf{M} = -2{,}000\mathbf{i} + 2{,}500\mathbf{j} + 2{,}700\mathbf{k}$ ft-lb

1.23. $|\mathbf{F}| = 140.6$ lb
$|\mathbf{M}| = 145.5$ ft-lb

1.26. (*a*) 4.5 lb (*b*) 1.0 lb

1.27. (*b*) $f = 0.2$

1.29. $\mathbf{F}_A = 10\mathbf{j} - 48.5\mathbf{k}$ lb
$\mathbf{F}_B = 10\mathbf{j} + 63.5\mathbf{k}$ lb
$\mathbf{F}_C = 41.0\mathbf{k}$ lb

CHAPTER 2

2.2. 370 psi

2.6. (*b*) 0.024 in. to left, 0.11 in. down

2.7. 0.14 in.
2.16. 5,700 lb
2.19. $\frac{3}{2}a$
2.22. 1.7 in.
2.24. 29,400 lb
2.27. $P = \frac{W}{f} e^{-f\pi}$
2.34. 59.11 ft
2.36. 0.077 in. if case slides down the ramp and comes to rest against the upper end of the wood; 0.135 in. if the bottom end of the wood is placed in hole in ramp and the upper end is wedged against the case until it is in the position shown in the sketch
2.39. 0.0048 in.; 65 lb

CHAPTER 3

3.6. $V_{max} = w_o L$, $(M_b)_{ax} = \frac{1}{2} w_o L^2$, at wall
3.9. $F = -P \cos \theta$, $V = P \sin \theta$, $M_b = -PR \cos \theta$
3.14. At 1, $F_x = 100$ lb, $M_b = 600$ in.-lb
At 2, $F_x = 100$ lb, $F_a = 37.5$ lb, $M_b = 375$ in.-lb
3.25. $a = 0.586L$
3.26. $x = 0.7L$
3.28. $T = 0.432D$
3.29. $M_t = 0.988PR$, $M_b = 0.157PR$
3.30. $M_b = 6{,}700$ in.-lb, $M_t = 2{,}120$ in.-lb in the 18-in. bar; $M_b = 15{,}000$ in.-lb, $M_t = 6{,}370$ in.-lb in the 24-in. bar
3.37. $V_{max} = 339$ lb (A to B)
$(M_b)_{max} = 4{,}060$ in.-lb (at B)
$(M_t)_{max} = 900$ in.-lb (B to C)
3.39. $V = 1{,}530$ lb, $(M_b)_{max} = 7{,}650$ in.-lb,
$M_t = 3{,}600$ in.-lb
3.40. $V = 1{,}000$ lb, $(M_b)_{max} = 2{,}000$ in.-lb, $M_t = 2{,}850$ in.-lb (in right section)

CHAPTER 4

4.7. (*a*) $\sigma_1 = 10{,}250$ psi
$\sigma_2 = -6{,}250$ psi
$\theta_1 = 36.0°$
(*b*) $\sigma_1 = 16{,}500$ psi
$\sigma_2 = -500$ psi
$\theta_1 = -22.5°$
(*c*) $\sigma_1 = 9{,}600$ psi
$\sigma_2 = -16{,}600$ psi
$\theta_1 = 65.2°$
4.8. $\sigma_x = 15{,}000$ psi
$\theta_1 = -26.6°$
4.11. $F = -2\pi r^2 p$
$F \geqq 0$
4.15. $\sigma_a = 1{,}000$ psi
$\sigma_b = 7{,}000$ psi
$\tau_{ab} = -5{,}200$ psi

4.20. $\epsilon_{I} = 981 \times 10^{-6}$
$\epsilon_{II} = -81 \times 10^{-6}$
$\theta_{I} = -24.4°$
4.22. (*a*) $\epsilon_{I} = 1{,}000 \times 10^{-6}$
$\epsilon_{II} = 0$
4.25. $w = 4.9r$
4.26. (*b*) $\theta_1 = 28.2°$
4.27. (*a*) $p = 0$, $F = 126{,}000$ lb
4.32. $\alpha = 54.8°$

CHAPTER 5

5.9. 15,000 psi
5.10. $\sigma_x = -3{,}585$ psi, $\sigma_a = -25$ psi, $\tau_{xy} = 3{,}070$ psi
5.12. $p = 2tE\epsilon_o/(1-2\nu)r$
5.14. 340 psi
5.16. Approximately 55°F for $\nu = 1/3$
5.20. $F = (1-2\nu)\pi r^2 p$
5.23. $(2-\nu)/(1-\nu)$
5.24. $u = \dfrac{P}{bhE}x$, $v = -\nu\dfrac{P}{bhE}y$, $w = -\nu\dfrac{P}{bhE}z$
5.27. (*a*) No (*b*) Yes
5.29. Zero
5.30. (*a*) −5,000 psi in cladding, zero in core
(*b*) −5,000 psi in cladding, 1,220 psi in core
5.31. (*a*) $2AY\cos\theta$
(*b*) AY increase
(*c*) $\sigma = \dfrac{PL}{AE(1+2\cos^3\theta)}$
(*d*) $\sigma_{res} = -Y\left(\dfrac{1+2\cos\theta}{1+2\cos^3\theta} - 1\right)$

CHAPTER 6

6.3. $\phi = \dfrac{M_t L}{G_1 I_1 + G_2 I_2}$

$\tau_{\theta z} = \dfrac{M_t G_1 r}{G_1 I_1 + G_2 I_2}$ for $0 < r < r_i$

$\tau_{\theta z} = \dfrac{M_t G_2 r}{G_1 I_1 + G_2 I_2}$ for $r_i < r < r_o$

where $I_1 = \dfrac{\pi r_i^4}{2}$ $I_2 = \dfrac{\pi}{2}(r_o^4 - r_i^4)$

6.4. $\tau = 3{,}140$ psi, $\phi = 0.60°$
6.8. $\tau = 6{,}700$ psi in top shaft, $\tau = 4{,}700$ psi in bottom shaft
6.13. (*a*) $M_t = 3{,}300$ in.-lb (*b*) $M_t = 11{,}000$ in.-lb; $\phi_Y = 12.7°$
6.21. (*a*) Impossible
6.23. 466 lb
6.24. $L = 123$ in., $2\phi = 468°$

6.25. 2.07°
6.26. 7,500 in.-lb
6.28. 1,570 psi
6.40. $\tau = \dfrac{7M_t}{4\pi r_o^3}$, $\phi = \dfrac{49}{16\pi^2}\dfrac{kLM_t^2}{r_o^7}$

CHAPTER 7

7.8. $I_{yy} = 56.4$ in.^{4}
7.15. $t = 0.016$ in.; 20,000 psi
7.18. $\alpha = 3/2$; $w_o = \dfrac{t\sigma_T}{10}$
7.19. $P = \dfrac{Ewt^3}{36Rc}$
7.24. $x = L$; $(\sigma_x)_{max} = \dfrac{3}{4}\dfrac{PL}{bd^2}$
7.29. 4; 0.5
7.35. 3.0 lb/in.
7.40. 1.40; 1.02
7.41. 2.94 in.
7.45. $\dfrac{2}{7}\dfrac{L}{h}$
7.48. 2.5; 0.34
7.52. 5.80×10^5 in.-lb
7.53. 0.84 in.; 7.25×10^5 in.-lb
7.54. 1.09×10^6 in.-lb
7.56. 2.2 in.
7.57. 13,200-psi compression
7.59. 45,000 psi
7.63. $P = 0.770\,\dfrac{r^3 Y}{a}$
7.65. $K = 1.71$
7.66. $M_L = 24{,}000$ in.-lb

CHAPTER 8

8.2. $\dfrac{5wL^4}{768EI}$
8.3. $\dfrac{PL^3}{12EI}$
8.4. $\dfrac{Pa}{24EI}(3L^2 - 4a^2)$
8.7. $\dfrac{M_o}{3EI}(L + 3a)$
8.8. $\dfrac{6EI\delta}{L^2}$

8.9. $\dfrac{7M_oL}{24EI}$

8.10. $\dfrac{PL^3}{3EI}\left[1-\dfrac{25}{32(24EI/kL^3+1)}\right]$

8.14. $\dfrac{WL^3}{15EI}$

8.26. 5,600 psi

8.28. 1,170 in.^{4}

8.34. $\dfrac{PL^3}{48EI}$

8.35. $\dfrac{17}{18}\dfrac{WL^3}{48EI}$

8.39. $\dfrac{3}{2}\left(\log_e 3-\dfrac{8}{9}\right)\dfrac{PL^3}{Ebd^3}$

8.44. $a=0.211L$

8.45. $a=0.223L$

8.48. $\dfrac{8EI}{L^3}(L^2+3RL+3R^2)$

8.50. $\dfrac{8M_L}{3L}$

8.51. $\dfrac{16M_L}{L^2}$

CHAPTER 9

9.2. $2kL$

9.3. $\frac{1}{2}kL^2$

9.4. $P<k/L$

9.6. $\frac{1}{3}kL$

9.7. "Snap-through" occurs when $P=\dfrac{2}{3\sqrt{3}}kL\phi^3$

9.13. 4,120 lb

9.18. $P_{crit}=\dfrac{2kc^2}{L}$

Index

Boldface numbers refer to the problems at the end of each chapter and are not page references.